# Theory and Applications of Computability

In cooperation with the association Computability in Europe

Books published in this series will be of interest to the research community and graduate students, with a unique focus on issues of computability. The perspective of the series is multidisciplinary, recapturing the spirit of Turing by linking theoretical and real-world concerns from computer science, mathematics, biology, physics, and the philosophy of science. The series includes research monographs, advanced and graduate texts, and books that offer an original and informative view of computability and computational paradigms.

Damir D. Dzhafarov • Carl Mummert

# Reverse Mathematics

## Problems, Reductions, and Proofs

Damir D. Dzhafarov
Department of Mathematics
University of Connecticut
Storrs, CT, USA

Carl Mummert
Department of Computer
and Information Technology
Marshall University
Huntington, WV, USA

ISSN 2190-619X ISSN 2190-6203 (electronic)
Theory and Applications of Computability
ISBN 978-3-031-11369-7 ISBN 978-3-031-11367-3 (eBook)
https://doi.org/10.1007/978-3-031-11367-3



Cover illustration: The "Wright Flying Machine". Illustration from the Wright Brothers' French patent application, submitted 18 November 1907. Image from the George Grantham Bain Collection at the United States Library of Congress.

This Springer imprint is published by the registered company Springer Nature Switzerland AG
The registered company address is: Gewerbestrasse 11, 6330 Cham, Switzerland

*Damir dedicates this book to Veronica.*

*Carl dedicates this book to Anna.*

# Preface

When we set out to write this book in the spring of 2011, the only introduction to reverse mathematics was Simpson's *Subsystems of Second Order Arithmetic* [288]. Our motivation was to write a complementary text, rather than a replacement. We planned on a more introductory treatment, necessarily less encyclopedic, that would offer a computability theoretic approach to the subject along with recent examples from the literature. But, almost as soon as we began writing, reverse mathematics started changing in fairly dramatic ways.

Ideally, books in active research subjects should be written during lulls. A subject will grow rapidly, slow down, expand again, and then after a while, perhaps, reach a period of stability. New ideas and results continue to emerge, but not ones that upend the subject by suddenly changing its scope or direction. This is when it makes sense to write a book. If the authors are very lucky, they may even finish it before the next major expansion begins.

We were not so lucky. The new expansion was major indeed and did not stop for many years. There was a sudden infusion of ideas from computable analysis centered around Weihrauch reducibility. There was the introduction of powerful new tools, such as preservation techniques and the probabilistic method. And there was a steady buildup of important new results, including the resolution of many longstanding open problems. Collectively, these developments reshaped and redefined reverse mathematics. In this way, our original conception began to seem less relevant.

And so we gradually aimed the project in a different direction, and reworked it from the ground up. In particular, we are happy to present many new developments in the subject for the first time in book form, and we hope that the overall outcome better reflects the state of the subject today. Ultimately, our main goal for this book is to convey just that, and thereby to provide a springboard for reading papers in reverse math and, for those interested, for *doing* some, too.

As mentioned, our treatment is based in computability theory. This has the disadvantage of making the contents less accessible as compared to the more syntactic treatments of Simpson and others, which only rely on a basic background in logic. But computability and reverse mathematics are naturally complementary—as Shore [282] put it, there is "a rich and fruitful interplay" between the two—and this interaction has only become stronger over time. Indeed, concepts and results from all parts of computability are increasingly finding applications to problems in reverse mathematics. This makes a contemporary account of the subject that omits or obscures the computability much more difficult, and much less useful. The combined perspective we have adopted treats reverse mathematics as overlapping with a large chunk of computable mathematics, Weihrauch style analysis, and other parts of computability that have become truly integral to the work of most researchers in the area.

We assume a basic background in logic, as covered in standard first year graduate courses following texts like Enderton's [93] or Mileti's [210]. For the reasons discussed above, a previous introduction to computability theory (e.g., Soare [295] or Downey and Hirschfeldt [83]) will undoubtedly be helpful. Chapter 2 provides an overview of the subject that can serve as a standalone introduction or refresher. Our notation will for the most part be standard, and is summarized in Section 1.5.

This book is still by no means meant as a replacement for Simpson's text, which includes many examples and results that we have chosen to omit. More generally, we have tried as much as possible to avoid duplicating material that is already well covered elsewhere. There are now a number of other texts focusing, in whole or in part, on reverse mathematics. These include *Slicing the Truth* by Hirschfeldt [147]; *Calculabilité* by Monin and Patey [219]; and *Mathematical Logic and Computation* by Avigad [8]. We recommend all of these as excellent companion texts. Stillwell's *Reverse mathematics: proofs from the inside out* [304] also provides a nice introduction to the subject aimed at a more general audience. We will mention a number of other references throughout the text.

The content of this book is organized into four parts. Part I includes the aforementioned background chapter on computability theory; a chapter on instance–solution problems, which are a main object of study in computable mathematics; and a chapter developing various reducibilities between such problems, including computable reducibility and Weihrauch reducibility. Part II introduces second order arithmetic and the major subsystems used in reverse mathematics. This is followed by a chapter on induction and other first order considerations, which are covered in other texts in more generality but deserve a concise treatment that makes them more accessible to the reverse mathematics community. We then move to a chapter on forcing in computability theory and arithmetic, with applications to conservation results. These first two parts lay out the bulk of the general theory of reverse mathematics, with the remainder of the book dedicated to specific case studies.

Part III focuses on the reverse mathematics of combinatorics, with one chapter dedicated exclusively to Ramsey's theorem, and another to many other combinatorial principles that have been studied in the literature. Part IV contains chapters on the reverse math of analysis, algebra, and set theory, descriptive set theory and more advanced topics, including a short introduction to higher order reverse mathematics. Exercises at the end of each chapter cover supplementary topics and fill in details omitted for brevity in the main text.

In terms of how to read this book, Parts I and II stand largely apart from Parts III and IV. The reader looking to learn the subject for the first time should therefore start in the first two parts, choosing specific chapters according to their background and interest. A reader already acquainted with computable mathematics and the reverse mathematics framework, on the other hand, can advance directly to the latter parts for an overview of research results, or to learn advanced techniques used in the reverse math of different areas. Dependencies between chapters are shown in the diagram below, where Chapter X → Chapter Y means that the material in Chapter Y is necessary (in substantial part) for Chapter X.

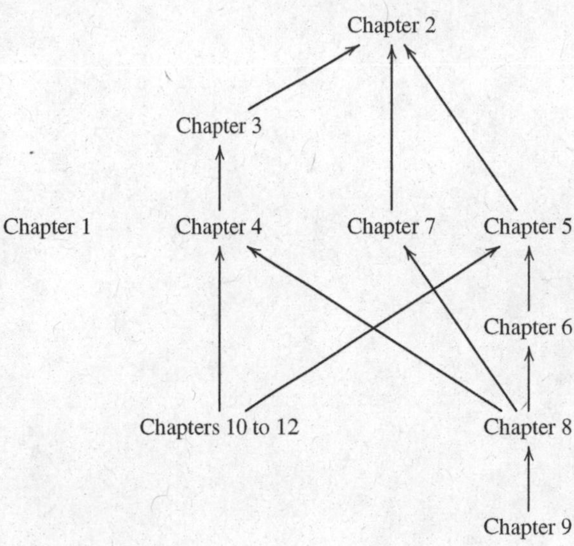


The way that ideas evolve and connect over time is messy. As a result, our presentation is not always chronological. This goes for results—e.g., a newer theorem sometimes being presented ahead of an older one—as well as broader themes. For example, we give an account of Weihrauch reducibility before second order arithmetic, even though the latter was developed earlier, and certainly was considered a part of reverse mathematics long before the former. In a similar spirit, outside of a section in Chapter 1 and some brief remarks here and there whose content is already well known, our account will be largely ahistoric. That said, there are we few places we could not resist the dramatic buildup offered by giving an account of how a theorem was arrived at, including prior partial results, failed attempts, etc. A good book should have some drama.

We do not, however, wish to *cause* any drama with this book. We wholly admit that the presentation of the subject here, and the choice of topics included and omitted, reflects only a particular point of view, which is our own. There is certainly much more we wish we could have included. The fact is that reverse mathematics, as a scientific field, is now quite broad and multifaceted, and it is too tall an order to try to squeeze all of it into one book. We are after the core of the subject here, from our perspective. Simpson [288] has said of reverse mathematics, and of the foundations of mathematics more generally, that it is the "study of the most basic concepts and logical structure of mathematics, with an eye to the unity of human knowledge". Shore [282] has called it, somewhat less grandly, the "playground of logic". It is both of these things: a foundationally important endeavor, and also one that is just plain fun. We hope this book conveys a bit of each. So with that, let's go play.

# Acknowledgments

We are immensely grateful to Peter Cholak, Denis Hirschfeldt, Jeff Hirst, Richard Shore, and Stephen Simpson for supporting our initial plan to write this book, and encouraging our continued work on it over the years. These were our earliest reverse mathematics teachers and colleagues, and having their confidence meant the world to us—and still does. We wish to express a special thanks to Barry Cooper, the driving force behind Computability in Europe, first the network and conference series, later the professional association, and of course this book series. His excitement for this book rivaled our own, and talking about it with him was always heartening and energizing. Sadly, Barry died in 2015. Though he only ever saw a very early draft of this book, we hope he would be pleased with the final result.

We acknowledge with gratitude the support of various funding agencies and organizations that have helped us complete this project. The first author was supported by several grants from the U.S. National Science Foundation, most recently the Focused Research Grant (FRG) "Computability-Theoretic Aspects of Combinatorics", which includes the second author as a key collaborator. The aim of this grant, which comes at an incredibly active time in reverse mathematics, is to tackle some of the major open problems in the subject. Being able to document those problems, and the progress made on them during the course of the grant, was a major motivation for us in the last months. (More information about the FRG is available at `ccfrg.computability.org`.)

We also acknowledge various forms of support during the preparation of this book from the American Association of University Professors, the Simons Foundation, and our home institutions, the University of Connecticut and Marshall University.

We thank the Banff International Research Station, the Schloss Dagstuhl–Leibniz-Zentrum für Informatik, and the Mathematisches Forschungsinstitut Oberwolfach. Though we did much of the writing for this book while physically apart, these organizations made it possible for us to periodically meet up and work side by side. Being able to do so in beautiful locations, like the Casa Matemática Oaxaca in Mexico, or the middle of the Black Forest in Germany, gave us all the more reason to look forward to these opportunities.

The first author also thanks the Czech–U.S. Fulbright Commission, and the Department of Theoretical Computer Science and Mathematical Logic at Charles University in Prague, for supporting and hosting him in the fall of 2022, in large part to work on this book. Thanks in particular to Antonín Kučera for making that visit possible and very memorable.

We thank the staff at Springer for their guidance and patience, and chiefly our editors, Ronan Nugent and Wayne Wheeler, for seemingly never losing faith that this book would someday get finished (in spite of our best efforts, at times).

The material in Section 12.3 of this book is based in part on lectures given by Antonio Montalbán at the University of Chicago in 2010, from notes taken by the first author and David Diamondstone. We are grateful to Antonio and David for agreeing for us to include this material.

A large number of colleagues over the years have read and commented on early drafts, answered our questions, and generally had interesting conversations with us about reverse mathematics and all things related. One of our worries at the start of this project was that there were too many things in the subject we did not know. Because of these folks we now know a lot more; enough to write this book. We thank Paul-Elliot Anglés d'Auriac, Eric Astor, Jeremy Avigad, David Belanger, Heidi Benham, Laurent Bienvenu, Vasco Brattka, Katie Brodhead, Mingzhong Cai, Lorenzo Carlucci, Doug Cenzer, Peter Cholak, Chi Tat Chong, Jennifer Chubb, Chris Conidis, Keith Conrad, Barry Cooper, Barbara Csima, Adam Day, Martin Davis, David Diamondstone, Natasha Dobrinen, François Dorais, Rod Downey, Benedict Eastaugh, Stephen Flood, Marta Fiori Carones, Ekaterina Fokina, Johanna Franklin, Harvey Friedman, Emanuele Frittaion, David Galvin, Jun Le Goh, Noam Greenberg, Marcia Groszek, Kirill Gura, Joel David Hamkins, Valentina Harizanov, Matthew Harrison-Trainor, Denis Hirschfeldt, Jeff Hirst, Rupert Hölzl, Noah Hughes, Greg Igusa, Corrie Ingall, Carl Jockusch, Bjørn Kjos-Hanssen, Julia Knight, Ulrich Kohlenbach, Leszek Kołodziejczyk, Roman Kossak, Antonín Kučera, Karen Lange, Steffen Lempp, Manny Lerman, Wei Li, Benedikt Löwe, Robert Lubarsky, Alberto Marcone, Andrew Marks, Marianna Antonutti Marfori, Adrian Mathias, Joe Mileti, Joe Miller, Russell Miller, Benoît Monin, Antonio Montalbán, Keng Meng Ng, David Nichols, Marie Nicholson, Ludovic Patey, Arno Pauly, Chris Porter, Pavel Pudlák, Michael Rathjen, Jan Reimann, Sarah Reitzes, Dave Ripley, Marcus Rossberg, Dino Rossegger, Sam Sanders, Jim Schmerl, Noah Schweber, Paul Shafer, Stephen Simpson, Richard Shore, Ted Slaman, Robert Soare, Reed Solomon, Mariya Soskova, Sean Sovine, Rachel Stahl, Frank Stephan, Jacob Suggs, Andrew Tedder, Neil Thapen, Morgan Thomas, Henry Towsner, Dan Turetsky, Manlio Valenti, Java Darleen Villano, Peter Vojtáš, Sean Walsh, Wei Wang, Lucas Wansner, Bartosz Wcisło, Andreas Weiermann, Rose Weisshaar, Linda Brown Westrick, Yue Yang, and Keita Yokoyama.

It is a cliché to end with the words "last but not least". It is also a cliché to thank one's family and friends for their love and emotional support. But we cannot help ourselves on either count. So, last but not least, we thank them.

# Contents

**Preface** .................................................... vii

**Acknowledgments** .................................................... xi

**1 Introduction** .................................................... 1
- 1.1 What is reverse mathematics? .................................................... 1
- 1.2 Historical remarks .................................................... 3
- 1.3 Considerations about coding .................................................... 6
- 1.4 Philosophical implications .................................................... 8
- 1.5 Conventions and notation .................................................... 10

**Part I Computable mathematics**

**2 Computability theory** .................................................... 15
- 2.1 The informal idea of computability .................................................... 15
- 2.2 Primitive recursive functions .................................................... 17
  - 2.2.1 Some primitive recursive functions .................................................... 19
  - 2.2.2 Bounded quantification .................................................... 20
  - 2.2.3 Coding sequences with primitive recursion .................................................... 22
- 2.3 Turing computability .................................................... 23
- 2.4 Three key theorems .................................................... 26
- 2.5 Computably enumerable sets and the halting problem .................................................... 31
- 2.6 The arithmetical hierarchy and Post's theorem .................................................... 33
- 2.7 Relativization and oracles .................................................... 36
- 2.8 Trees and PA degrees .................................................... 37
  - 2.8.1 $\Pi^0_1$ classes .................................................... 39
  - 2.8.2 Basis theorems .................................................... 43
  - 2.8.3 PA degrees .................................................... 46
- 2.9 Exercises .................................................... 48

**3 Instance–solution problems** ..... 51
3.1 Problems ..... 51
3.2 ∀∃ theorems ..... 53
3.3 Multiple problem forms ..... 56
3.4 Represented spaces ..... 59
3.5 Representing $\mathbb{R}$ ..... 61
3.6 Complexity ..... 63
3.7 Uniformity ..... 68
3.8 Further examples ..... 70
3.9 Exercises ..... 74

**4 Problem reducibilities** ..... 77
4.1 Subproblems and identity reducibility ..... 78
4.2 Computable reducibility ..... 81
4.3 Weihrauch reducibility ..... 84
4.4 Strong forms ..... 88
4.5 Multiple applications ..... 91
4.6 $\omega$-model reducibility ..... 95
4.7 Hirschfeldt–Jockusch games ..... 100
4.8 Exercises ..... 103

**Part II Formalization and syntax**

**5 Second order arithmetic** ..... 107
5.1 Syntax and semantics ..... 107
5.2 Hierarchies of formulas ..... 111
5.2.1 Arithmetical formulas ..... 111
5.2.2 Analytical formulas ..... 113
5.3 Arithmetic ..... 113
5.3.1 First order arithmetic ..... 113
5.3.2 Second order arithmetic ..... 115
5.4 Formalization, and on $\omega$ and $\mathbb{N}$ ..... 117
5.5 The subsystem $\mathsf{RCA}_0$ ..... 118
5.5.1 $\Delta^0_1$ comprehension ..... 118
5.5.2 Coding finite sets ..... 119
5.5.3 Formalizing computability theory ..... 122
5.6 The subsystems $\mathsf{ACA}_0$ and $\mathsf{WKL}_0$ ..... 123
5.6.1 The subsystem $\mathsf{ACA}_0$ ..... 124
5.6.2 The subsystem $\mathsf{WKL}_0$ ..... 127
5.7 Equivalences between mathematical principles ..... 129
5.8 The subsystems $\Pi^1_1$-$\mathsf{CA}_0$ and $\mathsf{ATR}_0$ ..... 132
5.8.1 The subsystem $\Pi^1_1$-$\mathsf{CA}_0$ ..... 132
5.8.2 The subsystem $\mathsf{ATR}_0$ ..... 136
5.9 Conservation results ..... 141
5.10 First order parts of theories ..... 143

5.11 Comparing reducibility notions ... 144
5.12 Full second order semantics ... 146
5.13 Exercises ... 148

**6 Induction and bounding** ... 153
6.1 Induction, bounding, and least number principles ... 153
6.2 Finiteness, cuts, and all that ... 158
6.3 The Kirby–Paris hierarchy ... 162
6.4 Reverse recursion theory ... 166
6.5 Hirst's theorem and $\mathsf{B}\Sigma^0_2$ ... 168
6.6 So, why $\Sigma^0_1$ induction? ... 172
6.7 Exercises ... 174

**7 Forcing** ... 175
7.1 A motivating example ... 175
7.2 Notions of forcing ... 177
7.3 Density and genericity ... 182
7.4 The forcing relation ... 186
7.5 Effective forcing ... 189
7.6 Forcing in models ... 194
7.7 Harrington's theorem and conservation ... 196
7.8 Exercises ... 199

**Part III Combinatorics**

**8 Ramsey's theorem** ... 203
8.1 Upper bounds ... 204
8.2 Lower bounds ... 209
8.3 Seetapun's theorem ... 212
8.4 Stability and cohesiveness ... 216
8.4.1 Stability ... 217
8.4.2 Cohesiveness ... 221
8.5 The Cholak–Jockusch–Slaman decomposition ... 223
8.5.1 A different proof of Seetapun's theorem ... 225
8.5.2 Other applications ... 228
8.6 Liu's theorem ... 231
8.6.1 Preliminaries ... 232
8.6.2 Proof of Lemma 8.6.6 ... 234
8.6.3 Proof of Lemma 8.6.7 ... 234
8.7 The first order part of $\mathsf{RT}$ ... 238
8.7.1 Two versus arbitrarily many colors ... 238
8.7.2 Proof of Proposition 8.7.4 ... 241
8.7.3 Proof of Proposition 8.7.5 ... 245
8.7.4 What else is known? ... 249
8.8 The $\mathsf{SRT}^2_2$ vs. $\mathsf{COH}$ problem ... 251

8.9 Summary: Ramsey's theorem and the "big five" ... 256
8.10 Exercises ... 257

**9 Other combinatorial principles** ... 259
9.1 Finer results about RT ... 259
9.1.1 Ramsey's theorem for singletons ... 259
9.1.2 Ramsey's theorem for higher exponents ... 265
9.1.3 Homogeneity vs. limit homogeneity ... 270
9.2 Partial and linear orders ... 271
9.2.1 Equivalences and bounds ... 272
9.2.2 Stable partial and linear orders ... 275
9.2.3 Separations over $\mathsf{RCA}_0$ ... 279
9.2.4 Variants under finer reducibilities ... 286
9.3 Polarized Ramsey's theorem ... 291
9.4 Rainbow Ramsey's theorem ... 294
9.5 Erdős–Moser theorem ... 303
9.6 The Chubb–Hirst–McNicholl tree theorem ... 305
9.7 Milliken's tree theorem ... 309
9.8 Thin set and free set theorems ... 312
9.9 Hindman's theorem ... 317
9.9.1 Apartness, gaps, and finite unions ... 319
9.9.2 Towsner's simple proof ... 323
9.9.3 Variants with bounded sums ... 329
9.9.4 Applications of the Lovász local lemma ... 336
9.10 Model theoretic and set theoretic principles ... 338
9.10.1 Languages, theories, and models ... 338
9.10.2 The atomic model theorem ... 340
9.10.3 The finite intersection principle ... 345
9.11 Weak weak König's lemma ... 349
9.12 The reverse mathematics zoo ... 354
9.13 Exercises ... 355

**Part IV Other areas**

**10 Analysis and topology** ... 363
10.1 Formalizations of the real line ... 364
10.2 Sequences and convergence ... 366
10.3 Sets and continuous functions ... 368
10.3.1 Sets of points ... 368
10.3.2 Continuous functions ... 369
10.4 The intermediate value theorem ... 371
10.5 Closed sets and compactness ... 373
10.5.1 Separably closed sets ... 376
10.5.2 Uniform continuity and boundedness ... 379
10.6 Topological dynamics and ergodic theory ... 380

10.6.1 Birkhoff's recurrence theorem ... 381
10.6.2 The Auslander–Ellis theorem and iterated Hindman's theorem ... 385
10.6.3 Measure theory and the mean ergodic theorem ... 388
10.7 Additional results in real analysis ... 388
10.8 Topology, MF spaces, CSC spaces ... 390
10.8.1 Countable, second countable spaces ... 390
10.8.2 MF spaces ... 393
10.8.3 Reverse mathematics of MF spaces ... 395
10.9 Exercises ... 397

**11 Algebra** ... 401
11.1 Groups, rings, and other structures ... 401
11.2 Vector spaces and bases ... 403
11.3 The complexity of ideals ... 406
11.4 Orderability ... 410
11.5 The Nielsen–Schreier theorem ... 415
11.6 Other topics ... 420
11.7 Exercises ... 423

**12 Set theory and beyond** ... 427
12.1 Well orderings and ordinals ... 427
12.1.1 The $\Sigma^1_1$ separation principle ... 428
12.1.2 Comparability of well orderings ... 431
12.1.3 Proof of Proposition 12.1.12 ... 435
12.2 Descriptive set theory ... 439
12.3 Determinacy ... 443
12.3.1 Gale–Stewart games ... 443
12.3.2 Clopen and open determinacy ... 445
12.3.3 Gödel's constructible universe ... 448
12.3.4 Friedman's theorem ... 449
12.4 Higher order reverse mathematics ... 452
12.5 Exercises ... 457

**References** ... 459

**Index** ... 473

# List of Figures


2.1 The arithmetical hierarchy. . . . . . . . . . . . . . . . . . . . . . . . . . . . . . . . . . . 34

3.1 An illustration to Proposition 3.7.4 . . . . . . . . . . . . . . . . . . . . . . . . . . . . . . 70
3.2 Construction from the proof of Lemma 3.8.2. . . . . . . . . . . . . . . . . . . . . 72

4.1 Diagram of a general reduction of problems. . . . . . . . . . . . . . . . . . . . . . 77
4.2 Diagrams of identity and uniform identity reductions. . . . . . . . . . . . . 79
4.3 Diagram of a computable reduction. . . . . . . . . . . . . . . . . . . . . . . . . . . . 82
4.4 Diagram of a Weihrauch reduction. . . . . . . . . . . . . . . . . . . . . . . . . . . . 84
4.5 Diagrams of strong computable and strong Weihrauch reductions. . . . 89
4.6 Relationships between various problem reducibilities. . . . . . . . . . . . . 101

5.1 Relationships between $\leqslant_c$, $\leqslant_\omega$, and $\leqslant_{\mathsf{RCA}_0}$. . . . . . . . . . . . . . . . . . . . . . 145
5.2 Subsystems of second order arithmetic. . . . . . . . . . . . . . . . . . . . . . . . 152

6.1 The Kirby–Paris hierarchy. . . . . . . . . . . . . . . . . . . . . . . . . . . . . . . . . . 162

7.1 An illustration of a non-generic set. . . . . . . . . . . . . . . . . . . . . . . . . . . 177

8.1 An illustration of a Seetapun tree. . . . . . . . . . . . . . . . . . . . . . . . . . . . . 217
8.2 The location of versions of Ramsey's theorem below $\mathsf{ACA}_0'$. . . . . . . . . 258

9.1 The location of CAC, ADS, and their variants below $\mathsf{ACA}_0$. . . . . . . . . . 287
9.2 Relationships between variants of CAC and ADS under $\leqslant_{\mathrm{W}}$. . . . . . . . . 289
9.3 An illustration of a strong subtree. . . . . . . . . . . . . . . . . . . . . . . . . . . . 310
9.4 Relationships between versions of HT for bounded sums. . . . . . . . . . 335
9.5 A snapshot of the reverse mathematics zoo. . . . . . . . . . . . . . . . . . . . . 357

# Chapter 1
# Introduction

## 1.1 What is reverse mathematics?

For most of its existence as a subject, reverse mathematics had a clear and unambiguous definition as a program in the foundations of mathematics concerned with the question of which axioms are necessary (as opposed to sufficient) for proving various mathematical theorems. *Which* theorems this includes is a question in its own right. Simpson [288, p. 1] refers to theorems of "ordinary mathematics", by which he means mathematics "prior to or independent of the introduction of abstract set theoretic concepts". This means mathematics that does not principally involve sets of arbitrary set theoretic complexity—modern algebra, geometry, logic, number theory, real and complex analysis, and some limited topology (but certainly not all of any of these fields).

The remarkable observation, due to Friedman, is that "When the theorem is proved from the right axioms, the axioms can be proved from the theorem". This phenomenon appears with many theorems: once the theorem is proved from an appropriate axiomatic system, it is also possible to show that the axioms of that system are derivable from the theorem itself, over a weak base theory. In this sense, it is said that the theorem "reverses" to the axioms, giving us a possible etymology for the name "reverse mathematics". The other possibility, of course, is that it is a play on "reverse engineering", which is quite apt as well.

The proof that the theorem implies particular axioms over a base theory is known as a *reversal* to those axioms. A reversal provides a measurement of a theorem's *axiomatic strength*: a theorem requiring a stronger system to prove is *stronger* than a theorem that can be proved in a weaker system. In particular, if a theorem $T$ implies a set of axioms $S$ over a base system, then any other proof of $T$ in an extension of the base system must use axioms that imply $S$. This gives a methodological conclusion about the minimum axioms needed to prove the theorem.

Reverse mathematics traditionally, but not always, uses subsystems of second order arithmetic for these axiom systems. Second order arithmetic is useful for this purpose because of its intermediate strength: it is strong enough to formalize much of

 
D. D. Dzhafarov, C. Mummert, *Reverse Mathematics*, Theory and Applications of Computability, https://doi.org/10.1007/978-3-031-11367-3_1

" ordinary mathematics", but weak enough that it does not overshadow the theorems being studied in the way that set theory does.

Over time, reverse mathematics has become more difficult to distinguish from *computable mathematics*, sometimes called *applied computability theory*. This should be distinguished from *constructive mathematics*, which aims to study mathematical theorems using wholly constructive means, which do not include the law of the excluded middle or other nonconstructive axioms. Computable mathematics, by contrast, seeks to *measure* the constructive content of mathematical theorems (or lack thereof) using the rich assortment of tools developed in classical computability theory. A prominent point of focus is on theorems of the form

$$(\forall x)[\varphi(x) \rightarrow (\exists y)\psi(x, y)], \tag{1.1}$$

where $x$ and $y$ are understood to range over the elements of some ambient set, and $\varphi$ and $\psi$ are properties of $x$ and $y$ of some kind. For obvious reasons, we will refer to statements of this form as $\forall\exists$ *theorems*.

When $x$ and $y$ range over numbers or sets of numbers, the theorem lends itself naturally to computability theoretic (or *effective*) analysis. Two commonly asked questions include the following.

- Given a computable $x$ such that $\varphi(x)$ holds, is there always a computable $y$ such that $\psi(x, y)$ holds?
- Does there exist a computable $x$ such that $\varphi(x)$ holds, and such that every $y$ for which $\psi(x, y)$ holds computes the halting set?

As it turns out, more often than not, the answers to these questions (and others like them) are directly reflected in the axiomatic strength of the theorem, as measured using the frameworks of reverse mathematics. This is not an accident, but rather a consequence of the setup of reverse mathematics, particularly of how the formal systems involved are defined. For example, the base system $\mathsf{RCA}_0$ typically used in reverse mathematics is often viewed as a formalization of computable mathematics. This makes it possible to directly translate many results from computability theory into reverse mathematics, and *vice versa*. The tie between computability theory and reverse mathematics is a key source of interest for many reverse mathematicians.

To accommodate both the axiomatic and computability theoretic viewpoints, we will understand reverse mathematics in this book as denoting a program concerned, quite generally, with the complexity of *solving problems*.

Intuitively, to *solve* a problem means to have a method to produce a solution to each given instance of that problem. For example: knowing how to factor composite integers means being able to produce a nontrivial factor of a given composite integer; knowing how to construct maximal ideals of nonzero unital rings means being able to produce such an ideal from a given such ring; and so forth. Depending on the problem and the instance, there may be many possible solutions, or a unique one. The task of "solving" a problem means be able to produce at least one solution for each instance, not necessarily to produce all of the solutions.

Conceptually, perhaps the first evolution of the idea of solving a problem is the idea of solving it *well*. This can mean different things for different problems and in different contexts. We all know how to factor integers, for example, but it would be quite a revolution if someone figured out how to do it quickly, say in polynomial time relative to the integer to be factored. As of this writing, the best algorithm for integer factorization is the *general number field sieve*, which is far slower. But this raises an equally interesting question: could this actually be the best possible algorithm? It is often the case that a mathematical problem is first solved somewhat crudely, and then over time the solution is refined in various ways, perhaps many times over. Is there a sense in which we could say that we have reached an optimal refinement, that we have found the "most efficient" solution possible?

This is a much more encompassing approach than may appear at first glance. Most noticeably, perhaps, it covers $\forall\exists$ theorems, and with them a great swath of "ordinary mathematics". In particular, for each a theorem having form (1.1) we may associate the problem "given $x$ such that $\varphi(x)$ holds, find a $y$ such that $\psi(x, y)$ holds". Moving further, we can also consider problems *about* problems, and in that way accommodate an even broader discussion. It is entirely common, after all, to reduce one problem to another, or to show that solving two problems are equivalent in some sense. Even finding a proof of a theorem in a given formal system is an example of such a reduction, too: "reduce this theorem to these axioms, via a proof". And if we think of the task of finding such reductions as problems in their own right, we can again ask about efficiency and optimality (e.g., is there a shorter proof? is there a more constructive one? etc.).

We do not mean to suggest a single abstract framework for reverse mathematics, computable mathematics, and everything surrounding them. Far from it. But the view that we are interested in problems, and how difficult they are to solve, conveys a theme that will serve us well as motivation and intuition in the pages ahead.

## 1.2 Historical remarks

A full account of the history of reverse mathematics would require its own book. Here, we give a brief sketch of the origins and development of reverse mathematics from its prehistory, through its origins in the 1970s, and to the present. We mention a few contributions that we view as important waypoints in the development of the field, with no intention to minimize the many contributions not mentioned. The results presented in the remainder of this book help to fill out this historical sketch.

Before reverse mathematics became a separate field, Weyl, Feferman, Kreisel, and others investigated the possibility of formalizing mathematics in systems of second order or higher order arithmetic, in the spirit of *arithmetization of analysis*. In the mid-20th century, early results in computable and constructive mathematics explored the ability to carry out mathematics effectively, as in the work of Specker [299] to construct a bounded, increasing, computable sequence of rational numbers whose limit is not a computable real number. In 1967, Bishop's seminal *Foundations of Constructive Analysis* demonstrated the breadth of mathematics that could be studied constructively. For a much deeper account of this "prehistory" of reverse mathematics, see Dean and Walsh [64].

Reverse mathematics itself was at least partially inspired by an ancient problem: given a mathematical theorem, can we specify precisely which axioms are needed to prove it? Some two millennia ago, Greek logicians asked this question about theorems in Euclid's geometry. A hundred years ago, similar questions were being raised about the axiom of choice, and about which fragments of set theory could be retained without it. Historically, this study of *logical strength* (of axioms, of theorems, of parts of logic or mathematics) has had profound foundational consequences, enhancing our understanding of our most basic assumptions and the most complex theorems that depend on them.

Combining these antecedents, Friedman proposed a program to measure the axiomatic strength of mathematical theorems using various subsystems second order arithmetic as benchmarks. This program was originally introduced in a series of talks at the 1975 International Congress of Mathematicians and the 1976 Annual Meeting of the Association for Symbolic Logic. Friedman [108, 109] identified a particular collection of subsystems of second order arithmetic—the so-called *"big five"*—and a number of mathematical theorems equivalent to each over the weak base theory $\mathsf{RCA}_0$. In his 1976 PhD thesis, Steel [302] (working under the supervision of Simpson) presented another early example. Descriptions of the emergence of reverse mathematics are given by Friedman and Simpson [116] and Friedman [113].

In the 1980s, significant progress was made in formalizing theorems of "ordinary" mathematics into second order arithmetic and showing these theorems equivalent to one of the "big five" systems. Simpson, along with several PhD students, was key in developing reverse mathematics into an active research field. A principal focus in this period was expanding the field, which now included algebra, real and complex analysis, differential equations, logic, and descriptive set theory. A limited number of combinatorial results were also studied, including Hindman's theorem and Ramsey's theorem. This work, and much more, is well documented in Simpson [288].

A key motivation emphasized in the literature from this time is the philosophical and foundational information gained in reverse mathematics. In particular, when a seemingly powerful theorem is provable in a weak system, it shows that a person willing to accept the weak system (in some philosophical sense) should be led to accept the theorem in the same sense. The weakest three systems of the "big five", in particular, have small proof theoretic ordinals and are generally viewed as predicative.

By the late 1990s, reverse mathematics had drawn the attention of computability theorists. They realized that second order arithmetic provided a "playground", in the words of Shore [282], where established computability techniques could be applied and new techniques could be developed. In particular, it was commonly noted at the time that the reverse mathematics results being obtained rarely required the complicated priority arguments that had become the norm in degree theory. In many cases, reverse mathematics proofs began to be written in a more informal, computability oriented manner. At the same time, researchers also began to look at implications that hold over $\omega$-models, rather than only at implications provable in $\mathsf{RCA}_0$. Implications over $\omega$-models do not have the same foundational implications, but they have stronger ties to computability.

A significant breakthrough related to combinatorics occurred near the end of the 1990s. Work of Seetapun [275] and Cholak, Jockush, and Slaman [33] showed that much more could be done with Ramsey's theorem than had been realized. Seetapun's work, in particular, showed the utility of effective forcing as a method to build models of second order arithmetic. This led to an explosion of results in the reverse mathematics of combinatorial principles. These techniques and results form a significant portion of this text. They also show that, although priority arguments never made a significant appearance in reverse mathematics (we will see a few in Chapter 9), the method of forcing turns out to be extremely applicable.

Another development was the discovery of a close connection between reverse mathematics and computable analysis. The possibility of this was first suggested by Gherardi and Marcone [122], and later independently by Dorais, Dzhafarov, Hirst, Mileti, and Shafer [72]. The latter group rediscovered a reducibility notion originally described 20 years earlier by Weihrauch [324]. The programs of Weihrauch-style computable analysis and reverse mathematics had developed separately, with little to no overlap, until the deep relationships between them were suddenly realized. This unanticipated relationship opened many research problems for both reverse mathematics and Weihrauch-style computability. It also led to an adoption of several reducibility notions as tools within reverse mathematics.

We view reverse mathematics broadly, as a program including all these influences and viewpoints. One common thread is the goal, when possible, to find the *minimum* resources needed to prove a theorem or solve a problem. This distinguishes reverse mathematics from many other programs of computable or constructive analysis, where the goal is to find a way to make theorems provable with a given framework, often by modifying the theorems or adding hypotheses.

There are several branches of reverse mathematics and logic that pull in other directions, and which we will not discuss in detail. The program of *constructive reverse mathematics* studies the axioms needed to prove theorems within a constructive base theory (which may be informal); Diener and Ishihara [67] provide an introduction. The program of *higher order reverse mathematics* [185] uses third order or higher order arithmetic. This allows for theorems to be represented differently, and can help to avoid some of the coding needed merely to state theorems in the context of second order arithmetic. Recent work of Normann and Sanders illustrates the range of possible applications of higher order techniques. We survey some of these results in Section 12.4. Higher order reverse mathematics is also linked to the *proof mining* program of Kohlenbach [183]. *Strict reverse mathematics*, also developed by Friedman [112], uses a multisorted free logic instead of second order arithmetic, and attempts to minimize or eliminate coding and logical formalisms in order to remain "strictly mathematical". In proof theory, there has been substantial work on subsystems of second order arithmetic, specifically including characterizations of the proof theoretic ordinals of many subsystems [250, 255].

## 1.3 Considerations about coding

We have explained that reverse mathematics results typically use methods from computability theory and second order arithmetic to study theorems and their associated problem forms. However, neither the fundamental notions of computability nor second order arithmetic refer directly to the mathematical objects we wish to study, such as groups, fields, vector spaces, metric spaces, continuous functions, ordinals, Borel sets, etc. Therefore, it is necessary to *code* these objects in ways that are more amenable to our analysis. We need to represent the objects of interest with more basic objects such as natural numbers or sets of naturals.

The first consequence of this is that classical reverse mathematics can only work with *miniaturizations* of many theorems of "ordinary mathematics", meaning versions formulated in terms of countable or separable objects. Thus we can only consider theorems about countable groups, complete separable metric spaces, and so forth. The second consequence is that, formally, we only ever work with *representations* or *codings* of mathematical objects, rather than those objects directly.

This leads to a key challenge in reverse mathematics. Some coding systems, such as the method for representing continuous functions, are highly nontrivial. But the use of nontrivial coding systems leads to a question of how much our results reflect the strength of the original theorems (or at least, the strength of their miniaturizations), and how much the results are influenced by the specific choices of coding system. (Eastaugh and Sanders [92] have termed this *coding overhead*.) Just as the observer effect is a fundamental limitation for physics, the need for coding is a fundamental limitation of reverse mathematics.While higher order reverse mathematics and strict reverse mathematics attempt to reduce the coding, some amount of coding in the broadest sense is inevitable.

Because coding cannot be avoided when formalizing theorems, the use of nontrivial coding systems leads naturally to questions about the optimality of the codings being employed. It is difficult to find mathematical criteria to determine which coding system is the best. This problem—to find a rigorous way to compare coding systems—is not itself a topic of reverse mathematics, but it is sometimes considered an open problem in the broad spirit of foundations of mathematics. Progress on this problem through mathematical methods seems to require a breakthrough in new ways to formally compare coding systems.

In practice, the choice of coding systems comes down to a combination of factors we describe as utility, directness, fullness, and professional judgment. These are not formal, mathematical criteria, but they nevertheless illustrate the ways that researchers in the literature evaluate and choose coding systems.

The *utility* of a coding system can be seen in the results it allows us to formalize and analyze. This is an aesthetic judgment, of course. But, in the end, the purpose of a coding system is to allow us to obtain results, and a system that allows few results to be obtained will be of correspondingly little interest. Utility can also present itself as uniformity. For example, the coding of real numbers with quickly converging Cauchy sequences leads to the operations being uniformly computable. The coding of continuous functions allows us to evaluate and compose functions in $\mathsf{RCA}_0$.

By *directness*, we mean the close relationship between the coding used and what is being coded. For example, there is significant directness in representing a countably infinite group by numbering the elements and representing the group operation as a function from $\omega \times \omega$ to $\omega$. On the other hand, our coding for continuous functions is arguably not as direct as simply coding the function's value for each element in a dense subset. But that alternate system makes it much more challenging to compose functions, lacking utility.

Directness also relates to the inclusion (or not) of additional information in the coding. For example, we can ask whether continuous real-valued functions on $[0, 1]$ are coded in a way that includes a modulus of uniform continuity. It is common in reverse mathematics to attempt to minimize the amount of additional information carried directly in the coding. This allows us to analyze the difficulty of obtaining that information from the coding, and also allows us to include the additional information in the hypothesis of theorems when we choose.

The other side of directness is whether the coding somehow removes information that would have been available with the original objects. In constructive mathematics, it is common to include a modulus of continuity with each continuous function. In that setting, it is assumed that a constructive proof that a function is continuous would demonstrate the modulus.

*Fullness* is the property that every object of the desired type has a code of the desired type in the standard model. This is a vital property for a coding system, as it shows that our formalization does not exclude any actual objects of interest. For example, every real number is the limit of some quickly converging Cauchy sequence of rationals, and every open set of reals is the union of a sequence of open rational intervals. Thus, when interpreted in the standard model, a theorem referring to coded objects like these retains its original scope. Of course, nonstandard models may have additional codes for nonstandard objects, and submodels of the standard model may not include codes for all objects.

One example of a system that lacks fullness comes from certain schools of constructive analysis, where a real number is coded as an algorithm or Turing machine that produces arbitrarily close approximations to the number. This coding limits the set of coded reals to the set of computable real numbers. Some theorems from ordinary mathematics are thus false in this framework, such as the theorem that every bounded increasing sequences of rationals converges.

The final test is professional judgment, which includes an aspect of opinion but cannot be disregarded. As with any field, the choice of which problems to study and which frameworks to use is not purely mathematical, but includes a significant aesthetic element.

When multiple coding systems are possible, there is an opportunity to compare the coding systems in various ways. This can lead to interesting mathematics on its own. An example in reverse mathematics can be seen in Hirst [158]. Weihrauch reducibility, with its focus on representations, gives particularly powerful methods to compare coding systems. For example, we can consider the problem of computing the identity function $f(x) = x$ with the input and output in different representations.

In Section 12.4, we discuss work in higher order arithmetic that provides a different approach to examining coding systems.

We will see numerous coding systems throughout the remainder of this text. The preceding comments provide a framework for the reader to consider whenever a new coding system is presented. In particular, although the term is not often used in the literature, fullness is a key consideration that all or nearly all of our coding systems will possess.

## 1.4 Philosophical implications

In the wake of the foundational crisis of mathematics in the early part of the 20th century, the mathematical community saw the emergence of a controversy between several schools of thought about the philosophy of mathematics and mathematical practice. Among these were formalism, championed by Hilbert, and various forms of constructivism, including the version of intuitionism championed by Brouwer. Attempts to reconcile these opposing views ultimately led Hilbert in the 1920s to formulate what is now called *Hilbert's program* for the foundations of mathematics. This included as a key tenet the existence of a formalization wherein mathematics could be proved to be consistent using wholly constructive "finitistic" means. Any hope of completely realizing Hilbert's program was famously dashed a decade later by Gödel with his incompleteness theorems. However, Simpson [286] and others have argued that formalizations of theorems in weak subsystems of second order arithmetic provide a *partial* realization of Hilbert's program.

The program of *predicativism*, advocated by Poincaré and Weyl, was also influential. This program seeks to formalize mathematics in ways that avoid impredicative definitions. An example of an impredicative definition is the definition of the least upper bound of a nonempty bounded set $A$ of reals as the smallest element of the set of all upper bounds of $A$. In contrast, one predicative construction of the least upper bound begins by forming a sequence of rationals $(r_i)$, each of which is an upper bound for $A$, so that there is an element of $A$ within distance $2^{-i}$ of $r_i$, and then forming the least upper bound as the limit of the sequence $(r_i)$. This construction produces the least upper bound in a way that only quantifies over the rationals and the set $A$ (which we are given from the start) rather than the set of all real numbers.

Each of the "big five" subsystems of reverse mathematics has a philosophical interpretation, based on the proof theoretic ordinal of the system, and broader experience with the system.

- $\mathsf{RCA}_0$ has an interpretation as effective (computable) mathematics, including the law of the excluded middle.
- $\mathsf{WKL}_0$ has an interpretation described by Simpson as *finitistic reductionism.* While $\mathsf{WKL}_0$ is not finitistic, it is conservative over the finitistic system $\mathsf{PRA}$ for $\Pi^0_2$ formulas. Thus, even though a person embracing finitism might not accept $\mathsf{WKL}_0$ as meaningful, they should so accept a proof of a $\Pi^0_2$ sentence in $\mathsf{WKL}_0$.

- $\mathsf{ACA}_0$ has an interpretation as (a portion of) predicative mathematics. This system is conservative for arithmetical formulas over first order Peano arithmetic.
- $\mathsf{ATR}_0$ has an interpretation via *predicative reductionism*. Although $\mathsf{ATR}_0$ is generally viewed as impredicative, it is conservative over a predicative system $\mathsf{IR}$ of Feferman (see [115]).
- Theorems requiring $\Pi^1_1$-$\mathsf{CA}_0$ or stronger systems are more fully impredicative.

Overall, these systems form a small portion of a large hierarchy of formal systems, ranging from extremely weak theories of bonded arithmetic through large cardinals in set theory. Simpson [289] call this the *Gödel hierarchy*. A more recent account can be found in Sanders [267].

Beyond the foundational aspects of each of the "big five" subsystems lie considerations about the "big five" phenomenon itself. While its historical prominence is indisputable, the emergence of more and more counterexamples to this classification over the past twenty years—resulting in what is now called the *reverse mathematics zoo* (see Section 9.12)—raises questions about the nature and significance of this phenomenon. There is no *a priori* reason one should expect a small number of axiom systems to be as ubiquitous and (very nearly) all encompassing as the "big five". Nor is there any reason the systems that appear should necessarily be linearly ordered in strength, as the "big five" subsystems are.

Slaman (personal communication) has suggested the phenomenon is an artifact of the choice of theorems and available techniques; as the subject expands and techniques develop, the phenomenon will grow less pronounced. Eastaugh [91] provides a different pushback against the standard view of the "big five" phenomenon, arguing that it is better understood in terms of closure conditions on the power set of the natural numbers. In the other direction, Montalbán [220] proposed that the "big five" subsystems owe their importance to being *robust*, meaning invariant under "small perturbations". This makes a plausible case for why, when a theorem is equivalent to one of the "big five" subsystems, so are its natural variations and elaborations. As for linearity of the subsystems, and the seemingly hierarchal nature of reverse mathematics, Normann and Sanders [234] and Sanders [272] have argued that this is a consequence of the choice of working in second order arithmetic; see Section 12.4 for a further discussion.

Another interesting question concerns the growing collection of exceptions to the "big five" phenomenon, and the curious fact that they come predominantly from combinatorics. As noted in [31], computability theoretically natural notions tend to be combinatorially natural, and *vice versa*. This makes combinatorial theorems interesting from the point of view of computability, and hence also reverse mathematics. One possibility for why so many theorems from combinatorics (particularly, extremal combinatorics) fall outside the "big five" is that they require little to no coding to formalize in second order arithmetic. Indeed, codings often directly feature in the reversals of theorems from other areas—even when, with reference to Section 1.3, the coding overhead is minimal. In effect: with simple mathematical objects, there is not much to code, but also not much to code *with*.

A different take is that combinatorics itself may play a special role within reverse mathematics. There is the notion in mathematics of the "combinatorial core" of a theorem, which is usually understood to be an amalgam of the combinatorial properties needed to work with and manipulate the theorem. As pointed out by Hirschfeldt [147], a common perspective is that it is the "combinatorial core" that the analysis of a theorem in reverse mathematics actually reveals. For this reason, equivalent theorems (in the sense of reverse mathematics) are sometimes said to have the same "underlying combinatorics". Of course, on this view, it would stand to reason that combinatorics offers the widest assortment of theorems with "distinct" combinatorial cores.

Which of these questions are most compelling, and which proposed solution, if any, holds the most explanatory value, are of course primarily for philosophers to discuss. We are not philosophers, and will generally focus on the mathematical side of reverse mathematics, without drawing philosophical consequences in the remainder of the text. But we consider these types of reflections important, and encourage the reader to keep them in mind (at least in a general sense) while working through this book. There is room for much philosophical analysis of reverse mathematics—far beyond the smattering of topics presented above—and this analysis would benefit from open and thoughtful participation by mathematicians. Conversely, being aware of some of these philosophical considerations can only help better inform and ground the mathematician working in the subject.

## 1.5 Conventions and notation

We let $\omega$ denote the set of natural numbers, $\{0, 1, 2, \ldots\}$, as well as the usual ordinal number. Thus, we may write $n \in \omega$ and $n < \omega$ interchangeably. For a set $X \subseteq \omega$ and $n \in \omega$, we write $X \upharpoonright n$ for the set $\{x \in X : x < n\}$. (Thus, $X \upharpoonright 0 = \emptyset$.) For convenience, we implicitly assume a mild typing wherein the elements of $\omega$ are distinguished from sets and other set theoretic constructs. So, for example, $\emptyset$ should never be confused with $0 \in \omega$, etc.

For subsets $X$ and $Y$ of $\omega$ we write $X < Y$ as shorthand for $\max X < \min Y$. For $x \in \omega$, we write $x < X$ or $X < x$ as shorthand for $x < \min X$ and $\max X < x$, respectively. We write $X \subseteq^* Y$ if $X \setminus Y$ is finite, and say $X$ is *almost contained* in $Y$. We write $X =^* Y$ if $X \subseteq^* Y$ and $Y \subseteq^* X$; in this case $X$ and $Y$ are *almost equal*. For finite sets $X$, we write $|X|$ for the *cardinality* of $X$.

The *principal function* of an infinite set $X = \{x_0 < x_1 < \cdots\} \subseteq \omega$ is the function $p_X : \omega \to X$ such that $p_X(n) = x_n$ for all $n$; thus $p_X$ enumerates $X$ in increasing order. Given two functions $f, g : X \to \omega$, we say $f$ *dominates* $g$ if $f(n) \geqslant g(n)$ for all sufficiently large $n$.

**Definition 1.5.1 (Finite sequences).** Fix $X \subseteq \omega$.

1. A *finite sequence* or *string* or *tuple* in $X$ is a function $\alpha : \omega \upharpoonright n \to X$ for some $n \in \omega$. The *length* of $\alpha$ is $n$, denoted by $\mathrm{lh}(\alpha)$, or more commonly, $|\alpha|$. We write $\alpha = \langle x_0, \ldots, x_{n-1} \rangle$ if $\alpha(i) = x_i$ for all $i < n$.
2. The unique string of length 0 is denoted $\langle\rangle$.

3. If $\alpha, \beta$ are finite sequences with $|\alpha| \leqslant |\beta|$ and $\alpha(i) = \beta(i)$ for all $i < k$, then $\alpha$ is an *initial segment* of $\beta$, and $\beta$ is an *extension* of $\alpha$, written $\alpha \preceq \beta$. If $\alpha \neq \beta$, then $\alpha$ is a *proper* initial segment of $\beta$, written $\alpha \prec \beta$.
4. If $\alpha, \beta$ are finite sequences then the *concatenation of $\alpha$ followed by $\beta$*, written $\alpha\beta$ or $\alpha \frown \beta$, is the string $\gamma$ of length $|\alpha| + |\beta|$ with $\gamma(i) = \alpha(i)$ for all $i < |\alpha|$ and $\gamma(|\alpha| + i) = \beta(i)$ for all $i < |\beta|$. If $|\beta| = n$ and $\beta(i) = n$ for all $i < n$, we write simply $\alpha x^n$ or $\alpha \frown x^n$.
5. If $|\alpha| \geqslant 1$ then $\alpha^\#$ denotes $\alpha \upharpoonright |\alpha| - 1$, i.e., $\alpha$ with its last element (as a sequence) removed. In particular, for every $x \in \omega$, $(\alpha x)^\# = \alpha$.
6. The set of all finite sequences in $X$ is denoted by $X^{<\omega}$, the set of all finite sequences of length less than $n$ by $X^{<n}$, and the set of all finite sequences of length exactly $n$ by $X^n$. If $X = \{0, \ldots, k-1\}$ for some $k \in \omega$, we write $k^{<\omega}$, $k^{<n}$, and $k^n$ instead.

We typically use lowercase Greek letters to denote elements of $X^{<\omega}$. In general, these are letters near the beginning of the alphabet ($\alpha, \beta, \gamma, \ldots$), but when $X = \{0, \ldots, k-1\}$ for some $k \in \omega$ we usually use letters near the middle and end of the alphabet ($\rho, \sigma, \tau, \ldots$). For elements of $X^n$, we sometimes also use the notation $\vec{x}, \vec{y}$, etc., and when $n$ is fixed but unspecified we may abuse notation slightly and write, e.g., $\vec{x} \in X$ instead of $\vec{x} \in X^n$. Elements of $2^{<\omega}$ are called *binary* strings.

**Definition 1.5.2 (Infinite sequences).** Fix $X \subseteq \omega$.

1. An *infinite sequence* in $X$ is a function $f\colon \omega \to X$. The set of all infinite sequences in $X$ is denoted $X^\omega$.
2. If $\alpha \in X^{<\omega}$ and $f \in X^\omega$ then $\alpha$ is an *initial segment* of $f$, and $f$ is an *extension* of $\alpha$, written either as $\alpha \preceq f$ or $\alpha \prec f$, if $\alpha = f \upharpoonright |\alpha|$. Though we allow the notation $\alpha \preceq f$, of course we can never have $\alpha = f$ for $\alpha \in X^{<\omega}$ and $f \in X^\omega$.

**Definition 1.5.3 (Cantor space and Baire space).** Fix $X \subseteq \omega$.

1. For $\alpha \in X^{<\omega}$, the *cylinder set* of $\alpha$ is the set $[[\alpha]] = \{f \in X^\omega : \alpha \prec f\}$.
2. The *(clopen) topology on $X^\omega$* is generated by the cylinder sets $[[\alpha]]$ for $\alpha \in X^{<\omega}$.

When $X = \{0, \ldots, k-1\}$ for some $k$ (most often $k = 2$), the resulting space is called *Cantor space*. When $X = \omega$, the resulting space is called *Baire space*. A standard fact is that $X^\omega$ is compact if and only if $X$ is finite.

**Definition 1.5.4 (Finitary functions).** Fix $X \subseteq \omega$.

1. A *finitary function on a set* $X$ is a function $f\colon X^n \to X$ for some $n \in \omega$, in which case it is also called an *$n$-ary function on $X$*.
2. Fix $k \in \omega$ and $X \subseteq \omega$. An $n$-ary function $f$ on $X$ is *$k$-valued* if $f(\vec{x}) < k$ for all $\vec{x} \in X^n$. We sometimes write $f\colon X^n \to k$ in place of $f\colon X^n \to \{0, \ldots, k-1\}$. Thus, when $X = \omega$ and $n = 1$, a function $f$ is $k$-valued precisely if $f \in k^\omega$.
3. A *finitary relation on a set* $X$ is a subset $R$ of $X^n$ for some $k \in \omega$, in which case it is also called an *$n$-ary relation on $X$*.
4. The *characteristic function* of an $n$-ary relation $R$ on $X$ is the function $\chi_R\colon X^n \to 2$ such that for all $\vec{x} \in X^n$, $\chi_R(\vec{x}) = 1$ if and only if $\vec{x} \in R$.

A 1-ary relation on a set $X$ is thus just a subset of $X$. A 0-ary function is technically a function $f: \{\langle\rangle\} \to X$, but we identify it with the unique element of $X$ given by $f(\langle\rangle)$. Since $X \subseteq \omega$ and $\langle\rangle \notin \omega$, the only 0-ary relation is $\emptyset$. Notice that if $\{0, 1\} \subseteq X$ then the characteristic function of an $n$-ary relation on $X$ is an $n$-ary function on $X$.

*Convention 1.5.5 (Identification of sets and characteristic functions).* We identify all finitary relations on $\omega$ with their characteristic functions. In particular, we use the term "set" (or, more precisely, "subset of $\omega$") interchangeably with "element of $2^\omega$". So for example, if $A$ is a set we can write either $x \in A$ or $A(x) = 1$, as convenient; given $f \in \omega^\omega$ we write $f(x) \neq A(x)$ to mean that if $f(x) = 0$ then $x \in A$ and if $f(x) \neq 0$ then $x \notin A$, etc. In this way, too, any definition or result formulated for finitary functions on $\omega$ automatically extends to finitary relations.

We fix an effective *pairing function*, which is a bijection $\omega^2 \to \omega$ coding pairs of numbers by numbers. For a concrete example, see, e.g., Soare [295, p. xxxii]. For an example formalized in second order arithmetic, see Theorem 5.3.4 and Exercise 5.13.25 below. For our discussion outside of formal systems, the specific choice does not matter. We denote the code for the pair $(x, y)$ by $\langle x, y\rangle$.

**Definition 1.5.6 (Effective joins).** Fix $X_0, X_1, \ldots \subseteq \omega$.

1. For each $n \in \omega$, the *(finite) join of* $X_0, \ldots, X_{n-1}$ is

$$\bigoplus_{i<n} X_i = X_0 \oplus \cdots \oplus X_{n-1} = \{\langle i, x\rangle : i < n \wedge x \in X_i\}.$$

2. The *(infinite) join of* $X_0, X_1, \ldots$ is $\bigoplus_{i<\omega} X_i = \{\langle i, x\rangle : i < \omega \wedge x \in X_i\}$.

The join is also called the *Turing join*. We may also use the notation $\langle X_0, \ldots, X_{n-1}\rangle$ in place of $X_0 \oplus \cdots \oplus X_{n-1}$, and $\langle X_i : i \in \omega\rangle$ in place of $\bigoplus_{i<n} X_i$.

The join $X_0 \oplus X_1$ of two sets is sometimes defined as $\{2x : x \in X_0\} \cup \{2x+1 : y \in X_1\}$. This has the advantage of being notationally lighter since it avoids the use of the pairing function. The definition in (1) above, by contrast, has the advantage of being consistent with the definition of the infinite join. But the choice will not matter for any part of our discussion below, as we can move between the two without affecting any of our arguments. So we will use them interchangeably, as convenient.

**Definition 1.5.7 (Projections).** For $X \subseteq \omega \times \omega$ and $i \in \omega$, $X^{[i]} = \{x \in \omega : \langle i, x\rangle \in X\}$. Thus, if $X = \bigoplus_{i<n} X_i$ or $X = \bigoplus_{i<\omega} X_i$ then $X^{[i]} = X_i$ for all $i$.

**Definition 1.5.8 ($\lambda$ notation).** If $f$ is a function of multiple parameters $f(\vec{x}, \vec{y})$, the notation

$$\lambda\vec{x}.f(\vec{x}, \vec{y})$$

refers to the function of tuples $\vec{x}$ induced by fixing particular values for $\vec{y}$. For example, $f(x) = \lambda x.x^y$ is a function which computes the $y$th power of its input, assuming $y$ is fixed by context.

We use $\equiv$ to denote literal equality of formulas in formal languages. For example, $\Phi \equiv X = Y$ means that $\Phi$ is the formula "$X = Y$".

# Part I
# Computable mathematics

# Chapter 2
# Computability theory

One of our primary tools for studying the difficulty of producing a solution to a problem will be *computability*. The theory of computability is a field in its own right, of which we will need only certain pieces. For a complete introduction to classical computability, we refer to Soare [295] or Downey and Hirschfeldt [74]. A reader who is familiar with the basics of computability theory, or who is willing to take them for granted temporarily, may wish to skip to Chapter 3.

## 2.1 The informal idea of computability

Classical computability theory begins with the question of which functions from $\omega$ to $\omega$ are "effectively calculable" or "algorithmically computable by a human". Of course, these terms appear in quotes because they are informal, intuitive concepts, and as such can never be completely answered by mathematics. But the early part of the 20th century saw a flurry of attempts at giving formal models of computation to capture all those functions anyone could reasonably regard as "effectively calculable" in the informal sense. This culminated, in 1936, with Turing's now-famous model of *Turing machines* and, perhaps equally importantly, his highly compelling philosophical argument for why his model succeeds in the above regard (see Turing [314]). Today, the premise that this is the case—that the functions that can be computed via Turing's model are exactly the "effectively calculable" functions—is called the *Church–Turing thesis*. For an account of Turing's argument, and more about the history of its development, see Soare [294].

Functions from $\omega$ to $\omega$ are a natural place to begin because there are many clearly reasonable ways to represent natural numbers (e.g. binary notation) that allow them to be manipulated concretely. This allows us to focus more directly on the algorithmic natural of the computation. In more general settings, the individual objects of study (e.g. real numbers or continuous functions from $\mathbb{R}$ to $\mathbb{R}$) will need to be represented in a more concrete way to allow us to compute with them.

There are a few key facets of algorithmic computation that bear mentioning.



D. D. Dzhafarov, C. Mummert, *Reverse Mathematics*, Theory and Applications of Computability, https://doi.org/10.1007/978-3-031-11367-3_2

- The computation is deterministic: if the same input is provided multiple times, precisely the same computational process will be followed, and same result (if any) will be produced.
- Resource limitations are ignored. A successful computation may take an arbitrary long (though finite) time to perform, and may require an arbitrary amount of temporary storage space (scratch paper, in the case of a human).
- The computation must be algorithmic: no creativity or ingenuity is required of the person performing the computation. In principle, the entire procedure can be written in a finite amount of space and conveyed to another person.
- The algorithm must include a way for the person performing the computation to determine when it is finished, and determine the specific value of the function that has been obtained.

Turing proposed a particular type of mechanical computer, the *Turing machine*, that models computations with these properties. He gave a compelling argument that any function computable by a human in an algorithmic fashion is computable by a Turing machine.

For the purposes of formalized arithmetic, it will be more convenient for us to consider a different formalization of computability. We will begin with a class of functions, the *primitive recursive* functions, which is a proper subset of the class of computable functions, but which has the advantage that each primitive recursive function is computable using particularly restricted means. In particular, it will be clear that the computation of a primitive recursive function on any input value will produce a result. In this sense, all primitive recursive functions are *total*.

We will then define the *partial computable functions* as the closure of the primitive recursive functions under operations including an *unbounded search operator*, the $\mu$ operator. This definition is particularly useful for our overall goal of relating computability with formal theories of arithmetic, because the $\mu$ operator is closely tied to numerical quantification. This comes at the cost of having to consider *partial* functions, but these are important and useful in their own right. In any case, we can always consider just the total functions when desired.

It will be clear from the definition that each such total function is "effectively calculable": computable, in principle, by a human with pencil and paper. The converse is of course more difficult, as it is not a mathematical argument but a philosophical one. In this regard, the definition using the $\mu$ operator is not very persuasive, in spite of its technical advantages. Instead, the standard argument here is to show that the total computable functions are co-extensive with the total functions that can be computed by Turing machines, and then appeal to Turing's original argument about the latter class. The more skeptical reader may, perhaps, be better convinced by an empirical fact: in the 90 years since Turing gave this argument, no counterexamples to the Church–Turing thesis have emerged.

## 2.2 Primitive recursive functions

The class of primitive recursive functions is the smallest collection of functions containing certain basic functions and closed under certain operations. We begin with the former.

**Definition 2.2.1 (Basic functions).** The *basic functions* are a particular class of finitary functions on $\omega$. They are:

1. For each $n > 0$ and $i < n$, a *projection function* $P_i^n$ defined by
$$P_i^n(x_0, \ldots, x_{n-1}) = x_i$$
for all $\langle x_0, \ldots, x_{n-1}\rangle \in \omega^n$.
2. The successor function defined by $S(x) = x + 1$ for all $x \in \omega$.
3. The constant unary zero function $Z(0) = 0$ and the constant 0-ary function $Z_0 = 0$.

The two basic operations for primitive recursive functions are generalized composition and primitive recursion.

**Definition 2.2.2 (Generalized composition).** Fix $n, m \in \omega$. If $f$ is an $n$-ary function on $\omega$ and, for each $i < n$, $g_i$ is an $m$-ary function on $\omega$, the *generalized composition* function $C(f, g_0, \ldots, g_{n-1})$ is the $m$-ary function satisfying

$$C(f, g_0, \ldots, g_{n-1})(x_0, \ldots, x_{m-1}) = f(g_0(x_0, \ldots, x_{m-1}), \ldots, g_{n-1}(x_0, \ldots, x_{m-1}))$$

for all $\langle x_0, \ldots, x_{m-1}\rangle \in \omega^m$. A class of functions $\mathcal{F}$ is closed under generalized composition if it is closed under the operator $C$.

**Definition 2.2.3 (Primitive recursion).** Fix $n \in \omega$ along with $f\colon \omega^n \to \omega$ and $g\colon \omega^{n+1} \to \omega$. Then $R_n(f, g)$ is the unique function satisfying

$$R_n(f, g)(0, x_0, \ldots, x_{n-1}) = f(x_0, \ldots, x_{n-1}),$$
$$R_n(f, g)(y + 1, x_0, \ldots, x_{n-1}) = g(R_n(f, g)(y, x_0, \ldots, x_{n-1}), x_0, \ldots, x_{n-1}),$$

for all $\langle x_0, \ldots, x_{n-1}\rangle \in \omega^n$ and $y \in \omega$. A class of functions $\mathcal{F}$ is closed under primitive recursion if it is closed under the operator $R$. This operator is a special case of the general concept of a *recursor* from proof theory.

It is immediate that the function $R_n(f, g)$ is well-defined and unique for all appropriate $f$ and $g$. Moreover, there is a natural algorithm for computing it. Given inputs $y, x_0, \ldots, x_{n-1}$, we first compute $s_0 = f(x_0, \ldots, x_{n-1})$, then we compute $s_1 = g(s_0, x_0, \ldots, x_{n-1})$, then $s_2 = g(s_1, x_0, \ldots, x_{n-1})$, and we continue this way until we have computed $s_y$, which is the value of $R_n(f, g)(y, x_0, \ldots, x_{n-1})$.

**Definition 2.2.4.** The class of *primitive recursive functions* is the smallest class of finitary functions on $\omega$ that contains the basic functions and is closed under generalized composition and under primitive recursion.

We are often interested in the ability to "decide" a set, that is, to calculate the characteristic function of the set. What we are doing, then, is finding an algorithm to correctly determine which elements are members of the set. Of course, using classical logic, each $n \in \omega$ is, or is not, an element of a set $A$. The question of interest is: how hard is it to determine which option holds? For our purposes, the easiest sets to decide have primitive recursive characteristic functions. This method of converting questions about sets into corresponding questions about functions appears very often in computability.

A key insight, which arrived early in the study of computability theory, is that little changes if we allow extra functions as basic functions. This concept, known as *relativization*, now appears throughout computability theory. Rather than only studying functions that are computable, the field is interested in which functions are computable "relative to" other functions which may not themselves be computable. (See the discussion in Section 2.7.) For this reason, it is often said, only partially in jest, that computability theory is primarily about noncomputable functions.

**Definition 2.2.5.** Let $g$ be a finitary function on $\omega$. The class of functions *primitive recursive relative to* $g$ (or *primitive recursive in* $g$) is the smallest class of functions that contains $g$ and all the basic functions and is closed under generalized composition and primitive recursion.

When we take an arbitrary function $g\colon \omega \to \omega$ and use it as if it was a basic function, we describe $g$ as an "oracle". The idea is that we may need to occasionally "consult" $g$ for information that we cannot obtain by primitive recursive means alone. And indeed, if $g$ is not itself primitive recursive, then the overall function we compute might not be primitive recursive.

We often want to view oracle functions like $g$ above as higher-level parameters to the function we are computing. Suppose $f(\vec{n})$ is a primitive recursive function relative to a function $g$. Because the class of functions primitive recursive in $g$ is defined with a closure condition, $f$ can be associated with a particular finite construction tree. Each node of the tree is labeled with a basic function, one of the construction rules $C$ or $R$, or with $g$. We can thus view $g$ is a kind of parameter to the tree: for every $h$ the same tree would define a function primitive recursive in $h$, if we simply replaced $g$ with $h$.

We can thus view $f$ as a function $f(\vec{n}, g)$ of both $\vec{n}$ and $g$. In particular, if $\vec{n}$ is a single variable $n$, we have an induced map $F(g) = \lambda n.f(n, g)$ from functions to functions. We call $F$ a *primitive recursive functional*. In general, we will use the notation $h(\vec{n}, g)$ to refer to a primitive recursive function defined by some (unnamed) construction tree in which some leaves are labeled with the oracle function $g$.

### 2.2.1 Some primitive recursive functions

In this section, we show that certain functions are primitive recursive. These functions will be required for the deeper results in the following sections.

The direct way to show that a function is primitive recursive is to write a definition of $f$ in terms of the basic functions and the operators $C$ and $R$. Such definitions are difficult to read, however. Instead, we typically write the definition in ordinary mathematical notation, via a set of recursion equations, relying on the reader to translate the informal definition into a formal one.

To provide an example of this method, we now show that the identity function $I(x) = x$ is primitive recursive. We typically do this by giving a set of recursion equations. For $I$, the recursion equations are

$$I(0) = 0,$$
$$I(x+1) = I(x) + 1.$$

To translate these equations into a formal definition, we first identify the functions on the right hand sides of the equations as $Z_0$ and $S(I(x))$. Thus

$$I(x) = R(Z_0, S)(x).$$

Of course, we also have $I(x) = P_1^1(x)$. In general, there can be many ways to construct a primitive recursive function.

The trick in the next theorem is to use the two cases of the primitive recursion operator as a substitute for the "if" statement of a programming language.

**Lemma 2.2.6.** *The modified subtraction function* $\dot{-}$ *defined for all* $x, y \in \omega$ *by*

$$x \dot{-} y = \begin{cases} 0 & \textit{if } y > x, \\ x - y & \textit{otherwise,} \end{cases}$$

*is primitive recursive.*

*Proof.* We first define an auxiliary function $P(k)$ with the recursion equations

$$P(0) = 0,$$
$$P(k+1) = k.$$

Thus $P(k)$ is $k - 1$ if $k > 0$ and 0 if $k = 0$. Now we define $f(m, n) = m \dot{-} n$ as

$$f(m, 0) = m,$$
$$f(m, k+1) = P(f(m, k)).$$

Formally, the $m$ in the first case should be $I(m)$, where $I$ is the identity function defined above. We will not comment again on trivialities such as this. □

**Lemma 2.2.7.** *The function $E(n, m)$ defined so that*

$$E(n, m) = \begin{cases} 0 & \text{if } m = n, \\ 1 & \text{otherwise}, \end{cases}$$

*is primitive recursive.*

*Proof.* We first define an auxiliary function $E_0(n)$ by primitive recursion, as follows:

$$E_0(0) = 0,$$
$$E_0(k + 1) = 1.$$

Then $E(n, m)$ is defined as

$$E(n, m) = E_0((m \dot{-} n) + (n \dot{-} m)).$$

Now $E_0(k) = 0$ if and only if $k = 0$, and $(n \dot{-} m) + (m \dot{-} n)$ equals 0 if and only if $m = n$, so the definition is correct. □

### 2.2.2 Bounded quantification

Per Convention 1.5.5, the definition of being primitive recursive extends to sets and other finitary relations by identifying these with their characteristic functions. Thus, for example, the relation $x = y$ is primitive recursive, since its characteristic function is the function $1 \dot{-} E(x, y)$ as defined in the previous section. Finitary relations on $\omega$ are closely related to syntax, because the standard connectives from propositional logic all have direct set theoretic interpretations. Indeed, if $R$ and $S$ are both $n$-ary relations on $\omega$, we can define $R \vee S = R \cup S$, $R \wedge S = R \cap S$, and $\neg R = \omega^n \setminus R$. The following is an easy but important observation.

**Proposition 2.2.8.** *Fix $n \in \omega$ and let $R$ and $S$ be $n$-ary relations on $\omega$. If $R$ and $S$ are both primitive recursive then so are $R \vee S$, $R \wedge S$, and $\neg R$. Thus the class of primitive recursive relations is closed under logical connectives.*

*Proof.* We have

$$R \vee S = \min\{1, R + S\}$$
$$R \wedge S = R \cdot S$$
$$\neg R = 1 \dot{-} R,$$

where on the right hand side we are treating $R$ and $S$ as functions. (Note that min, +, and $\cdot$ are all primitive recursive functions.) □

If $R(\vec{x})$ is a relation and $m \in \omega$ is fixed, we may ask whether every $\vec{x} < m$ satisfies $R$, or ask whether there is at least one $\vec{x} < m$ satisfying $R(k)$. (Recall that $\vec{x} < m$

means that every element of the tuple $\vec{x}$ is bounded by $m$.) These *bounded quantifiers* will play an important role in the sequel because, if $R$ is effectively calculable and $m$ is known, then the relations

$$(\exists \vec{x} < m)R(\vec{x}) = \{m : (\exists \vec{x})[\vec{x} < m \wedge R(\vec{x})]\},$$
$$(\forall \vec{x} < m)R(\vec{x}) = \{m : (\forall \vec{x})[\vec{x} < m \rightarrow R(\vec{x})]\}$$

are also calculable. Intuitively, this is because deciding the formula for a particular $m$ only requires testing the finite number of tuples less than $m$. This stands in contrast to the usual "unbounded" quantifiers: even if we have an algorithm to determine whether $R(m, \vec{x})$ holds for each $m$ and $\vec{x}$, there may be no algorithm for the relation $E(m) = (\exists \vec{x})R(m, \vec{x})$ or the relation $A(m) = (\forall \vec{x})R(m, \vec{x})$. We will see many concrete examples of this in Section 2.5.

**Theorem 2.2.9.** *If $R(\vec{x})$ is a primitive recursive relation, then so are $(\forall \vec{x} < m)R(\vec{x})$ and $(\exists \vec{x} < m)R(\vec{x})$, as functions of $m$ and any other inputs of $R$.*

*Proof.* First, by duality, we have

$$(\exists \vec{x} < m)R(\vec{x}) \leftrightarrow \neg(\forall \vec{x} < m)[\neg R(\vec{x})].$$

Thus, because the class of primitive recursive relations is closed under negation, it is enough to consider only the bounded universal quantifier. By definition, we have

$$(\forall \vec{x} < m)R(\vec{x}) = \prod_{\vec{x} < m} R(\vec{x}).$$

Therefore it is enough to show that if $f : \omega \rightarrow \omega$ is primitive recursive then so is $g(m) = \prod_{k<m} f(k)$. This is straightforward. First define a function $h$ by the recursion equations

$$h(0) = f(0),$$
$$h(k + 1) = f(k + 1) \cdot h(k).$$

and then define

$$g(0) = 1,$$
$$g(k + 1) = h(k).$$

*Remark 2.2.10.* Although the use of recurrence relations in the previous proof is already a simplification from using the generalized composition and primitive recursion operators directly, the proofs may still appear formalistic. As in other areas of logic, some formalism is unavoidable, especially when setting up the basic theorems. We will soon have enough tools, however, to avoid needed to write complex recurrence relations. There was a period in the mid 20th century when papers relied on derivations using recurrence relations (for example, Kleene [181]), but contemporary work in computability theory tends to avoid them unless needed.

The closure of primitive recursive relations under bounded quantification makes it much easier to prove that relations of interest are primitive recursive. The following result, which will be important in the next section, demonstrates how this method is used.

**Theorem 2.2.11.** *The relation "m is a prime number" is primitive recursive.*

*Proof.* First, note that the relation $A(p, q, m) \equiv (p \cdot q = m)$ is primitive recursive. Thus we may define a primitive recursive relation

$$R(m) \equiv m \geqslant 2 \wedge \neg(\exists p < m)(\exists q < m)[p \cdot q = m].$$

It is immediate that $R(m)$ represents the relation "$m$ is prime", because a number $m \geqslant 2$ is the product of two strictly smaller factors if and only if it is composite. □

### 2.2.3 Coding sequences with primitive recursion

*Coding* is a key technique in reverse mathematics: a mathematical object of one kind is represented as an object of a second kind, so that manipulating objects of the second kind indirectly manipulates objects of the first kind. The most basic kind of coding happens outside our formalization: we treat natural numbers as atomic entities with no internal structure, but a human computing with natural numbers will code them using binary or decimal representation (or some other method) to write them down.

The next most basic kind of coding allows us to represent a tuple of natural numbers using a single natural number. To achieve this, we will need a *coding function* that produces a number ("code") for the tuple; a *length function* that tells the length of the original tuple given its code; and a *decoding function* that gives back the elements of the tuple from the code. The definition below is designed in a way that each function has codomain $\omega$, matching the way the class of primitive recursive functions was defined.

**Definition 2.2.12.** A *coding system for finite sequences* consists of the following.

- *(Coding):* A sequence $\langle f_n : n > 0\rangle$ of injective functions $f_n \colon \omega^n \to \omega \setminus \{0\}$ such that every $s \in \omega^+$ is in the range of at most one $f_n$.
- *(Length):* A function $\mathrm{lh}(s)$ which, given $s \in \omega^+$, returns the unique $n$ such that $s$ is in the range of $f_n$, or 0 if $s$ is not in the range of any $f_n$.
- *(Decoding):* A function $\pi(s, i)$ such that if $i < \mathrm{lh}(s)$ and

$$f_n(x_0, \ldots, x_{k-1}) = s,$$

then $\pi(s, i) = x_i$.

We will be interested in coding systems for sequences in two settings: the primitive recursive functions and arithmetic. The role of these coding systems is, in a sense, to allow for a function or formula with a fixed number of parameters to act as if it has a variable number of parameters, by taking as input a code for a sequence and then pretending that the values of that sequence were all passed as parameters. In this way, we can limit our considerations to functions that take at most one input.

**Theorem 2.2.13.** *There is a coding system for sequences in which all the functions $f_n$, the function lh, and the function $\pi$ are all primitive recursive.*

One option is to use our pairing function (defined in Chapter 1) to code pairs of numbers, and then define codes for longer sequences recursively. (See also Theorem 5.3.4.) Another option is to use functions of the form

$$f_n(x_0, \ldots, x_{n-1}) = \prod_{i=1}^{n} p_i^{1+x_i},$$

where $p_i$ is the $i$th prime. In this case, the function $\mathrm{lh}(s)$ will determine the number of distinct prime factors of a number $s$, and the function $\pi(s, i)$ will return the exponent of the $i$th prime in the prime factorization of $s$. The proof that these functions are primitive recursive is Exercise 2.9.2. Although this provides a simple coding scheme, the functions here require more work in the context of formal arithmetic, where it is not as straightforward to work with an enumeration of the prime numbers. In that context, we use a coding system based on Gödel's $\beta$ function, as we will see in Definition 5.5.5.

With Theorem 2.2.13 in hand, we can talk about finite sequences informally, as in Definition 1.5.1. In actuality, we will always implicitly be referring to codes. For example, if we consider the function $g : 2^{<\omega} \to \omega$ that assigns each finite binary string its length, then formally $g$ is defined on the set of codes, i.e., the set of $s \in \omega$ with $\mathrm{lh}(s) \neq 0$.

## 2.3 Turing computability

As we continue our study of computability, it will be necessary to consider *partial* functions on $\omega$. These are functions defined on a subset of $\omega$ but possibly not on all on $\omega$. The fundamental reason we need these functions is that an algorithm may only work correctly for certain input values. For example, considering natural numbers, we can only find square roots for numbers that are perfect squares. Thinking of the algorithm as giving a function, we will call the set of legitimate inputs the *domain*. In the case of the square root function, we can effectively determine if a natural number has a natural number square root. However, we will soon see algorithms for which there is no way to predetermine effectively whether a given input is in the domain or not. If we want to think of these algorithms as giving functions of some kind, they must give partial functions.

There is standard notation in computability to work with partial functions. We write $f(x)\downarrow$ (which is read as "$f(x)$ converges") to indicate that $f$ is defined on input $x$; we write $f(x)\downarrow= y$ to indicate that $f(x)$ is defined and equals $y$. We write $f(x)\uparrow$ (read "$f(x)$ diverges") to indicate that $f$ is not defined on input $x$. We write $f \simeq g$ for functions of the same arity $n$ to indicate that for all $\vec{x} \in \omega^n$, if $f(\vec{x})$ and $g(\vec{x})$ both converge then they take the same value. This is different from writing $f = g$, by which we mean that for all $\vec{x} \in \omega^n$, either both $f(\vec{x})$ and $g(\vec{x})$ both diverge, or they both converge and take the same value.

In computer programming, the "while" loop construct makes loops that may run forever. The $\mu$ operator, which we define now, is most naturally programmed using this kind of loop. The $\mu$ operator says, informally, to perform a certain calculation for $x = 0$, $x = 1$, $x = 2$, ..., and halt once we find a value where the calculation succeeds.

**Definition 2.3.1.** Let $f$ be a partial $(n+1)$-ary function on $\omega$. The function $M(f)$ is a partial $k$-ary function defined as follows: for $x_0, \dots, x_{n-1} \in \omega$, $M(f)(x_0, \dots, x_{n-1})$ is the smallest $y$ such that $f(y, x_0, \dots, x_{n-1})\downarrow= 0$ and $f(y^*, x_0, \dots, x_{n-1})\downarrow\neq 0$ for all $y^* < y$, if such a $y$ exists, and $M(f)(x_0, \dots, x_{n-1})\uparrow$ if no such $y$ exists. We often write $M(f)$ using the $\mu$ operator, so that

$$M(f)(x_0, \dots, x_{n-1}) = (\mu y)[f(y, x_0, \dots, x_{n-1})\downarrow= 0].$$

**Definition 2.3.2.** The class of *partial computable functions* is the smallest class of finitary functions on $\omega$ which includes all primitive recursive functions and which is closed under generalized composition, primitive recursion, and applying the $M$ operator. A function is *computable* if it is partial computable and total.

To avoid a common confusion with the terminology, we emphasize that the term "partial" in the previous definition refers to "partial function", not "partially computable"! For a *set*, it only makes sense to be (total) computable, not partial computable, because characteristic functions are always total.

Let us now turn to *relative computability*, i.e., computability of one set *from* another. To get a picture of what this refers to, consider an arbitrary $A \subseteq \omega$. Although $A$ may not be computable itself, there is a natural sense in which "if $A$ was computable, then $\omega \setminus A$ would also be computable". We again describe this in terms of oracles. Imagine we have an oracle that can tell us whether each given number is in $A$—essentially, we assume the oracle can somehow calculate the characteristic function of $A$. To tell whether a number $n$ is in $\omega \setminus A$, we first ask the oracle whether the number is in $A$, and then we return the opposite answer. In a similar fashion, many other sets can be computed "from $A$", e.g., $\{2n : n \in A\}$, the set of prime numbers in $A$, etc.

Of course, we might use a much more complicated procedure to compute one set or function from another. Our definition of computable functions can be extended to make the notion of relative computability precise.

**Definition 2.3.3.** Suppose that $g$ is a finitary function on $\omega$. The class of functions *partial computable from* $g$ (the class of *partial* $g$*-computable functions*) is the smallest class which includes every function that is primitive recursive in $g$, and is closed under generalized composition, primitive recursion, and applying the $\mu$ operator.

A function $f$ is *computable from g* (or *g-computable*, or *Turing reducible to g*), written $f \leqslant_T g$, if it is partial computable from $g$ and total.

As with primitive recursion, if $g$ is not itself computable then there will be functions computable from $g$ that are not computable in the ordinary sense, including, of course, $g$ itself. The next theorem follows immediately from our definitions.

**Theorem 2.3.4.** *The relation $\leqslant_T$ satisfies the following.*

1. (Reflexivity): *For every $g$, we have $g \leqslant_T g$.*
2. (Transitivity): *For all $f, g, h$, if $f \leqslant_T g$ and $g \leqslant_T h$, then $f \leqslant_T h$.*

*Remark 2.3.5 (Other computable objects).* We overload the term "computable", or more generally, "$g$-computable", by using it for anything that can be coded by a finitary function on $\omega$. For example, we say a *sequence of numbers* $\langle x_i : i \in \omega\rangle$ is computable if there is a computable function $f : \omega \to \omega$ with $f(i) = x_i$ for all $i$. We say a *sequence of sets* $\langle X_i : i \in \omega\rangle$ is computable if there is a computable set $X \subseteq \omega$ such that $X^{[i]} = X_i$ for all $i$. (Recall that $X^{[i]} = \{x : \langle i, x\rangle \in X\}$, where $\langle i, x\rangle$ denotes a coded pair as in Section 2.2.3. So, $\langle X_i : i \in \omega\rangle$ is computable if $\bigoplus_{i\in\omega} X_i$ is computable.) We will see more complicated ways of coding mathematical structures by sets in Section 3.4.

In general, a *reducibility notion* on a particular class of problems is a reflexive, transitive relation between problems. Relations of this kind are sometimes called *preorders* (also *quasiorders*). The following proposition explains the way that so-called *degree structures* are formed from preorders. We will see this phenomenon many times as we study additional reducibility notions. The result is so well known in the field that the corresponding conclusions for a new reducibility notion are often taken for granted with no explicit statement.

**Proposition 2.3.6 (Degree structures).** *Let $\preceq$ be a preorder on a set A.*

1. *Define $a, b \in A$ to be* $\preceq$-equivalent *if $a \preceq b$ and $b \preceq a$. Then $\preceq$-equivalence is an equivalence relation on A. We will write $[a]$ for the equivalence class of A.*
2. *Let $A^*$ be the set of equivalence classes of $\preceq$-equivalence. Then $\preceq$ induces a relation $\preceq^*$ on $A^*$ with the rule that $[a] \preceq^* [b]$ if $a \preceq b$. The relation $\preceq^*$ is a partial order on $A^*$, that is, a reflexive, antisymmetric, and transitive relation. We almost always use the same symbol $\preceq$ to refer to $\preceq^*$ as well.*

*We call the equivalence classes of A under $\preceq$ the* $\preceq$-degrees.

Applying the proposition to the relation $\leqslant_T$, with $A = \mathcal{P}(\omega)$, gives the definition of the Turing degrees.

- Two functions (or sets) are *Turing equivalent* if each is computable from the other. We denote Turing equivalence by $\equiv_T$.
- The *Turing degree* of a function $g$ is the collection $\deg(g)$ (also denoted $\deg_T(g)$) of all functions Turing equivalent to $g$.
- The Turing degrees are themselves partially ordered by $\leqslant_T$.

The structure of the partial order of Turing degrees is one of the central questions in "classical" computability theory. It has been the focus of enormous amounts of research (see Simpson [284] for a partial survey).

At the same time, the Turing degrees give us a precise way to measure the computational difficulty of mathematical problems. If one particular problem $P$ entails computing a function $f$, and another problem $Q$ entails producing a function $g$, with $f \leqslant_T g$, then the second problem is computationally stronger in a formally defined sense. Most mathematical problems of interest do not simply ask us to produce a single function, however. In the next chapter, we will discuss *instance–solution problems*, which are more complex. Nonetheless, the methods of computability theory will still be vital to studying them.

## 2.4 Three key theorems

In this section, we will state three key theorems about the class of partial computable functions. We will provide brief sketches of the key ideas of each proof. Complete details are given by Soare [295, 293] and many other texts on computability.

The first of our three theorems shows that there is an effective indexing of the partial computable functions: i.e., a way to assign a natural number *index* $e$ to each partial computable function $f$. This number $e$ is essentially a program for computing $f$, encoded as a natural number. The primitive recursive functions $T$ and $U$ from the theorem carry out the computation, given $e$ and the inputs and oracles of the function. The main complication in the proof is that we are working with numbers, rather than with binary strings that electronic computers typically manipulate. Using numbers will facilitate formalizing computability into arithmetic later.

**Theorem 2.4.1 (Kleene's normal form theorem).** *Let $k > 0$ be fixed. There are primitive recursive functions $T = T^k$ and $U = U^k$ such that, for every partial computable $k$-ary function $f$ with oracle function $g$ there is an $e$ such that for all $\vec{n}$ of length $k$:*

$$f(\vec{n}) \simeq U(\mu s\, [T(s, e, \vec{n}, g \upharpoonright s) = 0]).$$

*Proof (sketch).* The proof is based on the fact that computation is carried out in a step-by-step manner. Essentially, $T(s, e, \vec{n}, \sigma)$ returns 0 if $s$ is a code for a sequence of computational states for program $e$ with input $\vec{n}$ and an oracle extending $\sigma$, so that the computation enters (and thus remains in) a halting state within the sequence of steps coded by $s$. If any step attempts to use oracle information beyond the finite sequence $\sigma$, $T$ views the computation as nonhalting. The function $U(s)$ takes a sequence of these computational steps, verifies that the computation entered a halting state, and then returns the value produced when the computation halted.

The value of $s$ plays two roles. The number $s$ itself encodes the sequence of steps that the computation performed, so the number of possible steps increases without bound as $s \longrightarrow \infty$. In addition, the value of $s$ such that $T(s, e, \vec{n}, g[s]) = 0$ serves as a bound on how much information was used from the oracle $g$ during the computation.

Again, this limit increases without bound as $s \to \infty$. In order for $T(s, e, \vec{n}, g[s])$ to return 0, $s$ must be large enough to code the entire sequence of steps and also large enough that no oracle information was used beyond $g \upharpoonright s$. $\square$

Although the $s$ in our statement of the theorem is a code for a sequence of states, not a simple count, we often refer to $s$ as if it counts the "steps" of the computation. Formally speaking, for each $e$, $\vec{n}$, and $g$ we can effectively produce a sequence $s_0$, $s_1$, ... that corresponds to the sequence of steps taken in the computation. Other variations of the theorem use $s$ to count the steps, and include additional parameters to $U$.

Because the parameter $k$ is always clear from context, we typically do not mention it explicitly. Thus, for example, the following definition should really be read as a family of definitions, one for each $k > 0$.

**Definition 2.4.2 (Universal computable function).** We define the partial function $\Xi(e, \vec{n}, g)$ as

$$\Xi(e, \vec{n}, g) = U(\mu s\, [T(s, e, \vec{n}, g \upharpoonright s) = 0]).$$

The function $\Xi$ is a "universal" function for all $k$-ary partial computable functions: by simply changing the value of the parameter $e$, every partial computable $k$-ary function with oracle $g$ can be obtained. On the other hand, $\Xi$ itself is immediately seen to be a partial computable $(k+1)$-ary function relative to $g$.

As with primitive recursive functions, we can view $g$ as a parameter to $\Xi$. This gives us a collection of Turing functionals, each of which takes a function from $\omega$ to $\omega$ as input and produces a partial function from $\omega$ to $\omega$ as output.

**Definition 2.4.3.** A *(Turing) functional* is a function $\Phi$ defined on $\omega^\omega$ such that for some $e \in \omega$, $\Phi(g) = \lambda\vec{n}.\Xi(e, \vec{n}, g)$ for all $g \in \omega^\omega$. We call $e$ an *index* for $\Phi$, and if $\Phi(g) = f$ then we call $e$ a $\Delta^{0,g}_1$ *index for* $f$ (or $\Delta^0_1$ *index for* $f$, if $g = \emptyset$). (We will better understand the reason for this terminology in Section 2.6.) We say $\Phi$ is $k$-ary, where $k$ is the length of $\vec{n}$.

*Convention 2.4.4 (Notation for functionals).* We reserve $\Xi$ for the function in Definition 2.4.2. We let $\Phi_e$ denote the Turing functional with index $e \in \omega$. We also use this to denote the partial computable *function* $\Phi_e(\emptyset)$, i.e., the partial computable function whose $\Delta^0_1$ index relative to $\emptyset$ is $e$. Which of these senses the notation is being used in will always be clear from context. When we wish to talk about Turing functionals without needing to specify their indices, we will use other capital Greek letters ($\Delta, \Gamma, \Phi, \Psi$, etc.). In general, if $\Phi$ is a Turing functional and $g$ is an oracle function, we use $\Phi^g$ and $\Phi(g)$ interchangeably. Thus, we write things like $\Phi_e(g)(0)$, $\Delta^g(17)$, etc., according to what is most convenient.

It is a standard fact, known as "padding", that every Turing functional has infinitely many different indices. A related result is the following.

**Proposition 2.4.5 (Indices and uniformity).** *Let $g$ be a finitary function on $\omega$ and $\langle f_i : i \in \omega\rangle$ a sequence of functions. Then $\langle f_i : i \in \omega\rangle$ is $g$-computable if and only if there is a computable function $h \colon \omega \to \omega$ such that for every $i \in \omega$, $\Phi^g_{h(i)} = f_i$.*

The contrast here is between $g$ knowing only how to compute each $f_i$ individually (meaning that, for each $i$, there is some $e$ which is a $\Delta_1^{0,g}$ index for $f_i$), as opposed to knowing how to compute all of the $f_i$ in sequence (so that for each $i$, a $\Delta_1^{0,g}$ index for $f_i$ can be found computably). The latter property is known as *uniformity*: if $\langle f_i : i \in \omega \rangle$ is $g$-computable, then we say that $f_i$ is *uniformly $g$-computable in $i$*. As a concrete example, fix a noncomputable set $A$, and consider the sequence of initial segments, $\langle A \upharpoonright i : i \in \omega \rangle$. For each $i$, $A \upharpoonright i$ is finite, and hence computable, but it is not uniformly computable (in $i$). Indeed, $\langle A \upharpoonright i : i \in \omega \rangle$ computes $A$, and so cannot itself be computable.

An important consequence of our definitions is that a halting computation, being a finite process, only "uses" a finite amount of information from the oracle, because the computation has only a finite number of steps when it could query specific values of the oracle. This gives the name to the proposition below. First, we add the following definition.

**Definition 2.4.6 (Use of a computation).** Fix $e \in \omega$, let $\Phi$ be the Turing functional with index $e$.

1. For $n \in \omega$ and $\alpha \in \omega^{<\omega}$, we write $\Phi^\alpha(n) \downarrow$ if there is a sequence $s$ with $\text{lh}(s) \leqslant |\alpha|$ such that $T(s, e, \vec{n}, \alpha) = 0$ and $\Phi^\alpha(n) \uparrow$ otherwise. In the former case, we also write $\Phi^\alpha(n) = y$ for $y = U(\mu s\,[\text{lh}(s) \leqslant |\alpha| \wedge T(s, e, \vec{n}, \alpha \upharpoonright s) = 0])$.
2. We write $\Phi^g(n)[t]$ for $\Phi^{g \upharpoonright t}(n)$. The least $t$, if it exists, such that $\Phi^g(n)[t] \downarrow$ is called the *use* of the computation $\Phi^g(n)$.

**Proposition 2.4.7 (Use principle).** *Let $\Phi$ be a Turing functional.*

1. *Monotonicity of computations: For $\alpha \in \omega^{<\omega}$ and $\vec{n} \in \omega$, if $\Phi^\alpha(\vec{n}) \downarrow = y$ then $\Phi^\beta(\vec{n}) \downarrow = y$ for all $\beta \in \omega^{<\omega}$ such that $\alpha \preceq \beta$, and also $\Phi^g(\vec{n}) \downarrow = y$ for all $g \in \omega^\omega$ such that $\alpha \preceq g$.*
2. *Every convergent computation has a use: For all $g \in \omega^\omega$, $\Phi^g(\vec{n}) \downarrow = y$ if and only if $\Phi^g(n)[s] \downarrow = y$ for some $s \in \omega$.*

Part (2) above expresses that Turing functionals are continuous as functions $\omega^\omega \to \omega^\omega$ in the Baire space topology discussed in Section 1.5. We add one more convention, largely for convenience.

*Convention 2.4.8.* If $\Phi_e$ is a Turing functional, we follow the convention that for all $g \in \omega^\omega$ and $\vec{n} = \langle n_0, \dots, n_{k-1} \rangle \in \omega$, if $\Phi_e^g(\vec{n})[s] \downarrow$ then $n_i < s$ for all $i < k$. If $\Phi_e$ is unary and $\Phi_e^g(n) \downarrow$, we also assume that for every $n^* \leqslant n$ there is an $s^* \leqslant s$ such that $\Phi^g(n^*)[s^*] \downarrow$.

The next two theorems establish additional properties of the effective indexing from the normal form theorem. Their proofs use additional properties of the indexing not evident in the statement of the theorem. However, the ability to prove these theorems is a key property of the indexing that is constructed in the proof of the normal form theorem.

As with primitive recursive functions, the key tool for proving these theorems is the definition of the partial computable functions as the closure of certain basic functions under certain operations. This gives a finite construction tree for each partial computable function $f$ in which some nodes are labeled with oracle functions. Given a way to compute the oracle functions, we can then use the tree to compute arbitrary values of $f$. The index $e$ constructed in Kleene's normal form theorem is essentially a concrete representation of the construction tree.

In this way, the index is a natural number that does not itself contain an oracle function. Instead, the index—which is a kind of representation of a computer program—includes instructions to query the oracle functions at specific points during a computation. The person computing the function is then responsible for producing the correct values of the oracle function as required during the computation process.

The following proposition shows that the indexing from the normal form theorem is "effective on indices": if we have an operation of generalized composition, primitive recursion, or minimization, we can compute the index for the resulting function from the indices of the functions the operation is applied to. Moreover, we can effectively produce an index for a function that chooses between two given functions based on whether two inputs are equal. The proof follows from our ability to manipulate the finite construction trees in an effective way.

**Proposition 2.4.9.** *The indexing from the normal form theorem has the following properties. Here g is an arbitrary oracle.*

- *Effective branching: There is a primitive recursive function $d(i, j)$ so that*

$$\Phi^g_{d(i,j)}(x, y, \vec{z}) = \begin{cases} \Phi^g_i(x, \vec{z}) & \text{if } x = y, \\ \Phi^g_j(x, \vec{z}) & \text{if } x \neq y. \end{cases}$$

- *Effective composition: There is a primitive recursive function c so that*

$$\Phi^g_{c(i,j_1,\ldots,j_k)}(\vec{z}) = \Phi^g_i(\Phi_{j_1}(\vec{t}), \ldots, \Phi_{j_n}(\vec{z})).$$

- *Effective primitive recursion: There is a primitive recursive function r that, given an index i for a function $\Phi_i(n, x, \vec{t})$ and an index j for a function $\Phi_j(\vec{t})$, produces an index $r(i, j)$ that applies the primitive recursion scheme to $\Phi_i$ and $\Phi_j$:*

$$\Phi^g_{r(i,j)}(n, \vec{z}) = \begin{cases} \Phi^g_j(\vec{z}) & \text{if } n = 0, \\ \Phi^g_i(n, \Phi_{r(i,j)}(n-1, \vec{z})) & \text{if } n > 0. \end{cases}$$

  *Here $\Phi^g_{r(i,j)}(n, \vec{t}) \uparrow$ if any of the computations along the way diverges.*
- *Effective minimization: There is a primitive recursive function m so that, for all i and all $\vec{z}$,*

$$\Phi^g_{m(i)}(\vec{z}) = (\mu x)[\Phi^g_i(x, \vec{z}) = 0].$$

The second of our three key theorems is the *parameterization theorem*, more commonly known as the $S^m_n$ *theorem* after the functions that Kleene originally used to state it. The theorem says that if we have a computable function $f(s, t)$, we can hard-code a value $k$ for the input $s$ to produce a computable function $g(t) = \lambda t.f(k, t)$.

Moreover, the index for $g$ can be computed by a primitive recursive function given the index of $f$ and the value for $k$. The proof follows, again, from our ability to effectively manipulate the construction tree encoded by the index.

**Theorem 2.4.10 ($S_n^m$ theorem, Kleene).** *For all $m, n \in \omega$, there is an $(m+1)$-ary primitive recursive function $S_n^m$ such that for all $e \in \omega$ and $\vec{x} \in \omega^m$, $S_n^m(e, \vec{x})$ is an index of an $n$-ary partial computable function satisfying*

$$\Phi_e^g(\vec{x}, \vec{y}) = \Phi^g_{S_n^m(e,\vec{x})}(\vec{y}).$$

*for all finitary functions $g$ and all $\vec{y} \in \omega^n$.*

The final key theorem allows us to create programs that, in a way, have access to their own index. This is the *recursion theorem*, also known as *Kleene's fixed point theorem.*

**Theorem 2.4.11 (Recursion theorem, Kleene [180]).** *For every computable function $f : \omega \to \omega$ there is an index $e$ such that $\Phi_e^g = \Phi_{f(e)}^g$ for every oracle $g$.*

The fixed point theorem is extremely useful for defining functions by recursion. The next proposition gives a concrete example, and its proof illustrates the utility of Proposition 2.4.9.

**Proposition 2.4.12.** *The function $A(n, x)$ defined by the following recurrence is computable:*

$$\begin{aligned} A(0,x) &= 2x, \\ A(n+1,x) &= \underbrace{A_n(A_n(\cdots A_n(1)\cdots))}_{x \text{ times}}, \end{aligned}$$

*where $A_n(x)$ denotes $A(n, x)$.*

*Proof.* There is a primitive recursive function $f(i)$ that creates an index for a function $\Phi_{f(i)}(s, x)$ that does the following.

- If $s = 0$, return $2x$.
- If $s = n + 1$, use primitive recursion to compute a sequence $r_0, \ldots, r_x$ where $r_0 = 1$ and $r_{m+1} = \Phi_i(n, r_m)$ for $1 \leqslant m \leqslant x$. Then return $r_x$.

To show that $f$ is computable, we first note that we can compute an index to perform the operation in each bullet. The first bullet is a primitive recursive function of $x$, and thus has an index by the statement of the normal form theorem. To show there is a primitive recursive function to find an index of a function to perform the primitive recursion in the second bullet from the index $i$, we use the effective composition and effective primitive recursion properties from Proposition 2.4.9. We also use the fact the each of the basic primitive recursive functions has an index.

Once we have a primitive recursive function to compute an index for each bullet, it follows from the effective branching property that $f$ is primitive recursive.

By the recursion theorem, there is an index $p$ so that $\Phi_p = \Phi_{f(p)}$. A straightforward induction on $n$ shows that $\Phi_p(n,x) = A(n,x)$ for all $n$. Essentially, given a value for $n$, the index $p$ can "use itself" in place of index $i$ to compute any needed results for smaller values of $n$. □

A less useful but perhaps equally interesting consequence of the fixed point theorem is the ability to produce functions the return their own index. Programs corresponding to indices of this kind are often called "quines" in honor of Willard Van Orman Quine.

**Proposition 2.4.13.** *There is an index e so that* $\Phi_e(0) = e$.

*Proof.* Let $i$ be an index of the computable function $(s,t) \mapsto s$. Let $f(s)$ be the function $\lambda s.S^1_1(i,s)$, where $S^1_1$ is from the $S^m_n$ theorem. Then $f$ is computable (in fact, primitive recursive). Apply the recursion theorem to $f$ to obtain an index $e$ with $\Phi_e = \Phi_{f(e)}$. Then

$$\Phi_e(0) = \Phi_{f(e)}(0) = \Phi_{S^1_1(i,e)}(0) = \Phi_i(e,0) = e.$$

## 2.5 Computably enumerable sets and the halting problem

For a computable set, we have an effective procedure for deciding which numbers are and are not in the set. But sometimes, what we have instead is an effective procedure for listing, or *enumerating*, the elements of the set. This notion turns out to capture a broader collection of sets.

**Definition 2.5.1.** A set $X \subseteq \omega$ is *computably enumerable* (or *c.e.* for short) if there is a partial computable function with $X$ as its domain. More generally, if $g$ is a finitary function on $\omega$ and $P$ is a finitary relation, then $P$ is *$g$-computably enumerable* ($g$-*c.e.*) if there is a partial $g$-computable function with $\{\vec{n} : P(\vec{n})\}$ as its domain.

The previous definition may seem confusing, since it does not seem to agree with the intuitive description of "enumerating" a set we started with. The next lemma shows that it agrees with what we expect.

**Lemma 2.5.2.** *A nonempty set $X \subseteq \omega$ is c.e. if and only if there is a computable function whose range is X. Moreover, if X is infinite then we may assume this function is injective.*

Crucially, we *cannot* assume that the function above is increasing. Indeed, if a set could be effectively enumerated in increasing order, it would be computable: to effectively decide whether or not a given a number $x$ is in the set we simply wait either for $x$ to be enumerated, or for some $y > x$ to be enumerated without $x$ having been enumerated earlier.

But there are c.e. sets that are not computable. The next definition gives perhaps the most famous example. Its name comes from an interpretation in which we are asking for a way to determine whether a particular program will eventually halt (stop running), or whether it will continue forever without stopping, as in an infinite loop.

**Definition 2.5.3 (Halting problem).** The *halting problem* is the set $\varnothing' = \{e \in \omega : \Phi_e(e)\downarrow\}$. More generally, the *halting problem relative to a finitary function* $g$, also called the *Turing jump of* $g$, is the set $g' = \{e \in \omega : \Phi^g_e(e)\downarrow\}$.

It is common to call the map $' \colon \omega^\omega \to 2^\omega, g \mapsto g'$, the *jump operator*. We collect some of the most well-known establishing properties concerning this operator in the next theorem. Part (1), implicit in the original groundbreaking paper of Alan Turing [314], was the first clue that the Turing degrees are a nontrivial structure.

**Theorem 2.5.4.** *Fix a finitary function* $g$.

1. $g \leqslant_T g'$ *but* $g' \nleqslant_T g$.
2. *If* $h$ *is a finitary function and* $g \equiv_T h$ *then* $g' \equiv_T h'$.
3. $g'$ *is* $g$*-c.e.*
4. *If* $A \subseteq \omega$ *is* $g$*-c.e. then* $A \leqslant_T g'$.
5. *A set* $A \subseteq \omega$ *is* $g$*-computable if and only if* $A$ *and* $\overline{A}$ *are both* $g$*-c.e.*

Combining (3), and (4), we see that $g'$ is a kind of "universal $g$-c.e." set. We will see a further elucidation of the properties in the above theorem in Post's theorem (Theorem 2.6.2) below. That theorem will make key use of iterated Turing jumps, as in the following definition.

**Definition 2.5.5.** Let $g$ be a finitary function. We define $g^{(n)}$, for each $n \in \omega$, as follows: $g^{(0)} = g$, and $g^{(n+1)} = (g^{(n)})'$.

By Theorem 2.5.4, for every $g$ we have a strictly increasing hierarchy

$$g <_T g' <_T \cdots g^{(n-1)} <_T g^{(n)} <_T \cdots .$$

The jump operator also gives rise to the following hierarchy of functions which are "close to computable" in a certain sense.

**Definition 2.5.6 (Low$_n$).** Fix $n \in \omega$ and a finitary function $g$. A set $X$ is *low*$_n$ *relative to* $g$ if $(g \oplus X)^{(n)} \leqslant_T g^{(n)}$. If $g = \varnothing$, then $X$ is simply *low*$_n$.

Thus, the computable functions are the same as the low$_0$ functions (and so the latter term is not used). The low$_1$ functions are typically just called *low*. Thus a function $g$ is low if $g' \equiv_T \varnothing'$. If a function is low$_n$ it is of course low$_{n+1}$, but the converse need not be true.

## 2.6 The arithmetical hierarchy and Post's theorem

Post's theorem establishes a tight link between iterations of the Turing jump and numerical quantifiers. The connection with quantifiers is of particular interest in reverse mathematics because, in the context of formal theories of arithmetic, we can count quantifiers in specific formulas of interest. The basic conclusion is that the number of iterated Turing jumps required to compute a set is tied to the number of alternations of universal and existential quantifiers required to define the set.

There are many versions of the "arithmetical hierarchy". Some are syntactical hierarchies that assign classifications to formulas; others are semantical hierarchies that assign classifications to relations. Different versions also look at different families of formulas or different families of relations. In this section, we define a semantical hierarchy that assigns classifications to certain relations on $\omega$, beginning with the classification for computable relations. In Chapter 5, we will define a closely related syntactical hierarchy for formulas in second order arithmetic.

This arithmetical hierarchy assigns one or more classifications, relative to an oracle $g$, to certain relations on $\omega$. The possible classifications are $\Sigma^{0,g}_n$, $\Pi^{0,g}_n$, and $\Delta^{0,g}_n$, for $n \in \omega$, which may be alternatively written as $\Sigma^0_n$, $\Pi^0_n$, and $\Delta^0_n$ *relative to* $g$, or (less commonly) as $\Sigma^0_n(g)$, $\Pi^0_n(g)$, and $\Delta^0_n(g)$. When $g$ is computable (or when no oracle is intended), it is not written, giving simply $\Sigma^0_n$, $\Pi^0_n$, and $\Delta^0_n$.

**Definition 2.6.1 (Arithmetical hierarchy for relations on $\omega$).**
Fix $n \in \omega$, let $g$ be a finitary function on $\omega$, and let $P$ be a finitary relation.

1. $P$ is $\Sigma^{0,g}_0$ and also $\Pi^{0,g}_0$ if it is $g$-computable.
2. $P$ is $\Sigma^{0,g}_{n+1}$ if $P(\vec{x}) = (\exists \vec{y})Q(\vec{y},\vec{x})$ for some $\Pi^{0,g}_n$ relation $Q(\vec{y},\vec{x})$.
3. $P$ is $\Pi^{0,g}_{n+1}$ if $P(\vec{x}) = (\forall \vec{y})Q(\vec{y},\vec{x})$ for some $\Sigma^{0,g}_n$ relation $Q(\vec{y},\vec{x})$.
4. $P$ is $\Delta^{0,g}_n$ if it is $\Sigma^{0,g}_n$ and $\Pi^{0,g}_n$.

Any relation that receives a classification in this hierarchy is an *arithmetical relation* (relative to $g$). A relation is *properly* $\Sigma^{0,g}_n$ or *properly* $\Pi^{0,g}_n$ if it is $\Sigma^{0,g}_n$ or $\Pi^{0,g}_n$, respectively, but not $\Delta^{0,g}_n$. For $n \geqslant 1$, a relation is *properly* $\Delta^{0,g}_n$ if it is $\Delta^{0,g}_n$ but not $\Sigma^{0,g}_{n-1}$ or $\Pi^{0,g}_{n-1}$.

Because we can introduce and quantify over dummy variables, every $\Sigma^{0,g}_n$ or $\Pi^{0,g}_n$ relation is $\Delta^{0,g}_{n+1}$. We often identify $\Sigma^{0,g}_n$, $\Pi^{0,g}_n$, and $\Delta^{0,g}_n$ with the collections of relations they describe, so that, e.g., $\Delta^{0,g}_n = \Sigma^{0,g}_n \cap \Pi^{0,g}_n$.

Note that the complexity of a relation in the arithmetical hierarchy can be measured by the number of blocks of alternating quantifiers that appear in its definition. For example, if $P$ is $\Sigma^0_1$ then it has the form $(\exists \vec{x})R(\vec{x})$ for some computable predicate $R$, and if $P$ is $\Sigma^0_2$ it has the form $(\exists \vec{x})(\forall \vec{y})R(\vec{y},\vec{x})$. By induction, if $P$ is $\Sigma^0_n$ for some $n > 2$ then it has the form

$$(\exists \vec{x}_{n-1})(\forall \vec{x}_{n-2})(\exists \vec{x}_{n-3})\cdots(\mathsf{Q}\vec{x}_0)R(\vec{x}_0,\ldots,\vec{x}_{n-2},\vec{x}_{n-1}),$$

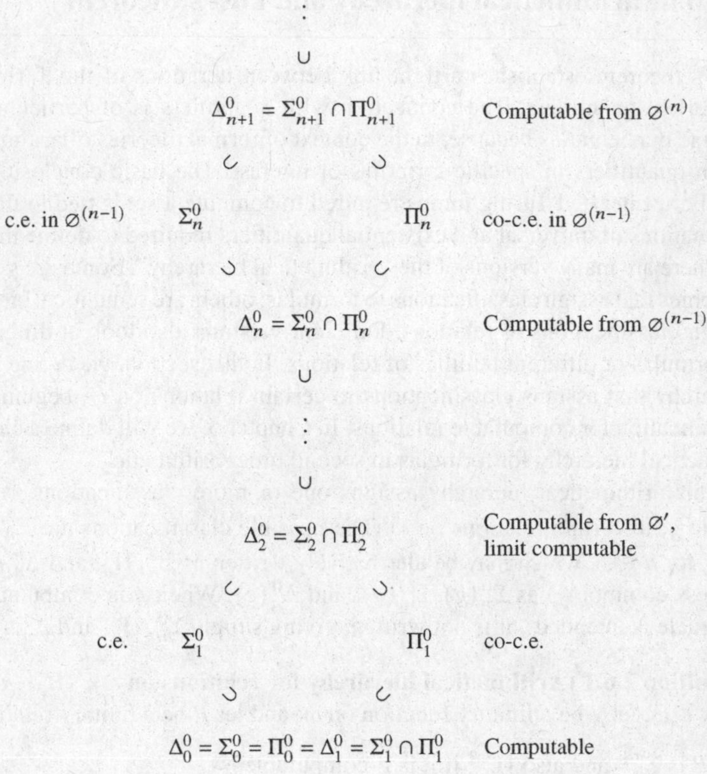


**Figure 2.1.** The arithmetical hierarchy, showing classes of finitary relations on $\omega$ under inclusion. Here, $n \geqslant 2$ is arbitrary. All containments are strict. The labels restate the characterizations of each class from Post's theorem and the limit lemma.

where Q is $\forall$ or $\exists$ depending as $n$ is even or odd. A similar observation holds for $\Pi^0_n$ formulas, with the quantifiers interchanged. Geometrically, $\Sigma^{0,g}_{n+1}$ relations are the projections of higher-dimensional $\Delta^{0,g}_n$ relations, and $\Pi^{0,g}_n$ relations are the complements of $\Sigma^{0,g}_n$ relations.

The next theorem characterizes the overall structure of the arithmetical hierarchy. Figure 2.1 provides an illustration.

**Theorem 2.6.2 (Post's theorem).** *Fix $n \in \omega$, let $g$ be a finitary function on $\omega$, and let $P$ be a finitary relation.*

1. *$P$ is $\Delta^{0,g}_{n+1}$ if and only if it is computable from $g^{(n)}$. In particular, a relation is $\Delta^{0,g}_1$ if and only if it is computable from $g$.*
2. *$P$ is $\Sigma^{0,g}_{n+1}$ if and only if it is c.e. in $g^{(n)}$.*
3. *$P$ is $\Pi^{0,g}_{n+1}$ if and only if its complement is c.e. in $g^{(n)}$.*

*In addition, for $n > 0$, the class of $\Delta_n^{0,g}$ relations is a proper subclass of the classes of $\Sigma_n^{0,g}$ and $\Pi_n^{0,g}$ relations. In particular, neither of the latter two classes is contained in the other.*

Importantly, we see the identification of $\Delta_1^0$ relations with computable relations. This is part of Theorem 2.5.4 above, and is where we get the term "$\Delta_1^0$ index" in Definition 2.4.3. We also see the identifications of $\Sigma_1^0$ relations with c.e. relations, and $\Pi_1^0$ relations the complements of c.e. relations. In addition, every $\Sigma_n^0$ relation, being also $\Delta_{n+1}^0$, is computable in $\varnothing^{(n)}$, which generalizes Theorem 2.5.4. Thus, each jump can be thought of as helping to "answer" questions with one additional alternating quantifier in the definition of the relation.

Post's theorem has the following immediate corollary, which establishes a firm connection between computational strength and logical syntax.

**Corollary 2.6.3.** *Fix $n \in \omega$, let $g$ be a finitary function on $\omega$, and let $P$ be a finitary relation. Each of the classes of $\Sigma_n^{0,g}$, $\Pi_n^{0,g}$, and $\Delta_n^{0,g}$ relations is closed under the standard connectives of propositional logic and under bounded quantification.*

Formally, this allows us to represent various natural statements *about* arithmetical objects by other arithmetical objects. For example, given a computable set $A$, the statement that $A$ is infinite, $(\forall x)(\exists y)[y > x \wedge y \in A]$, defines a $\Pi_2^0$ set, $R_A(z)$. Since $z$ does not appear in the definition of $R_A$, it follows that $R_A$ is either all of $\omega$ or the empty set depending as $A$ is or is not infinite. By Post's theorem, there is an $e$ such that $R_A = \Phi_e^{\varnothing''}$, so $\varnothing''$ can check whether or not, say, $0 \in R$, whereby it can "answer" whether or not $A$ is infinite. (A point of caution: each of $\varnothing$ and $\omega$ is computable, but there is no single index $e$ such that $R_A = \Phi_e$! In the parlance of the remark following Proposition 2.4.5 above, $R_A$ is uniformly $\varnothing''$-computable from a $\Delta_1^0$ index for $A$; it is not uniformly computable from this index. So, $\varnothing$ cannot provide the "answer".) In practice, we usually forego such details, and simply say things like, "The question whether a $\Sigma_n^0$ property is true can be answered by $\varnothing^{(n)}$", etc.

We wrap up this section with one further, equally important classification specifically for $\Delta_2^0$ relations. This uses *limit approximations*.

**Definition 2.6.4.** Let $f$ be a 2-ary function. For each $x \in \omega$, we write $\lim_y f(x,y)\downarrow$ if there is a $z$ such that $(\exists w)(\forall y \geqslant w)[f(x,y) = z]$. In this case, we also write $\lim_y f(x,y) = z$.

**Theorem 2.6.5 (Limit lemma, Shoenfield).** *Let $g$ be a finitary function on $\omega$. A finitary relation $P$ is $\Delta_2^{0,g}$ if and only if there is a $g$-computable function $f: \omega^2 \to 2$ such that* $\lim_y f(x,y)\downarrow$ *for all $x$, and* $\lim_y f(x,y) = 1$ *if and only if $P(x)$.*

The next corollary, which follows immediately from the limit lemma and Post's theorem, is a commonly used application.

**Corollary 2.6.6.** *Let $g$ be a finitary function on $\omega$. For every $g'$-computable function $h: \omega \to \omega$ there exists a $g$-computable function $f: \omega^2 \to \omega$ such that $h(x) =$* $\lim_y f(x,y)$ *for all $x$.*

We can extend the definition of the arithmetical hierarchy to describe relations on functions (and so, sets), which will be useful for our work in Section 2.8.

**Definition 2.6.7.** Fix $n \in \omega$ and let $g$ be a finitary function on $\omega$.

1. A $\Sigma^{0,g}_0$ (also $\Pi^{0,g}_0$) *relation on* $\omega^\omega$ is an $e \in \omega$ such that $\Phi_e^{f \oplus g}(0) \downarrow$ for all $f \in \omega^\omega$. We identified $e$ with the relation $R$ that holds of $f \in \omega^\omega$ if and only if $\Phi_e^{f \oplus g}(0) \downarrow = 1$.
2. A $\Sigma^{0,g}_{n+1}$ *relation on* $\omega^\omega$ is a $g$-computable sequence $\langle 0, \langle e_i : i \in \omega \rangle \rangle$ such that for each $i$, $e_i$ is a $\Pi^{0,g}_n$ relation on $\omega^\omega$. If $e_i$ is identified with the relation $R_i$, then we identify $\langle 0, \langle e_i : i \in \omega \rangle \rangle$ with the relation $R$ that holds of $f$ if and only if $R_i$ holds of $f$ for some $i$.
3. A $\Pi^{0,g}_{n+1}$ *relation on* $\omega^\omega$ is a $g$-computable sequence $\langle 1, \langle e_i : i \in \omega \rangle \rangle$ such that for each $i$, $e_i$ is a $\Sigma^{0,g}_{n+1}$ relation on $\omega^\omega$. If $e_i$ is identified with the relation $R_i$, then we identify $\langle 0, \langle e_i : i \in \omega \rangle \rangle$ with the relation $R$ that holds of $f$ if and only if $R_i$ holds of $f$ for all $i$.

$\Sigma^{0,g}_n$ and $\Pi^{0,g}_n$ relations on $2^\omega$ (or some other subset of $\omega^\omega$) are defined in the obvious way. For instance, for each function $g$, the relation "$f(x) \leqslant g(x)$ for all $x$" is a $\Pi^{0,g}_1$ relation on $\omega^\omega$. For an infinite set $X$, the relation "$Y$ is an infinite subset of $X$" is a $\Pi^{0,X}_2$ relation on $2^\omega$.

## 2.7 Relativization and oracles

As we have already remarked, the central focus of modern computability theory is really *relative* computability. This is especially the case for the aspects of computability theory that pertain to reverse mathematics, as we will discover in Chapter 5. In a certain sense, everything *can* be relativized. For example, if we wish to relativize to some finitary function $g$, we would change all instances of "computable" to "$g$-computable", "$\varnothing'$" to "$g'$", "$\Sigma^0_n$" to "$\Sigma^{0,g}_n$", etc. This much we have seen throughout this chapter.

It is thus common in the subject to formulate definitions and results in unrelativized form, meaning without the use of oracles. Typically, it is easy and straightforward for these to then be relativized to an arbitrary oracle $g$, simply by "inserting the oracle" appropriately, as above. Some care must be taken of course: for example, relativizing "$X$ is low" to $g$ does not mean "$X' \leqslant_T g'$" but "$(g \oplus X)' \leqslant_T g'$" (see Definition 2.5.6).

We will generally follow this practice moving forward, beginning already in the next section. So, though a result may be stated and proved in unrelativized form, we may later appeal to the relativized version by saying, e.g., "Relativizing such and such theorem to $g$, ...". Even when a theorem is formulated explicitly in terms of an arbitrary oracle, we will often prove only the unrelativized version, and then simply remark that "the full version follows by relativization".

To be sure, there are also senses in which *not* everything relativizes. Most famously, it is known that the different "upper cones" in the Turing degrees (meaning, sets of the form $\{\mathbf{d} : \mathbf{a} \leqslant \mathbf{d}\}$ for different degrees $\mathbf{a}$) need not be isomorphic or even elementarily equivalent as partial orders, as shown by Feiner [105] and Shore[280, 281]. However, virtually everything we consider in this book will act in accordance with the following maxim:

*Maxim 2.7.1.* All natural mathematical statements about $\omega$ relativize.

This is unavoidably (intentionally) vague and subjective, but it serves as a useful starting point to frame our discussions. Certainly, the few examples we will see of statements that do not relativize (e.g., Example 4.6.8) will not be "natural" or "naturally occurring" in any way.

## 2.8 Trees and PA degrees

We now come to one of the first and most important points of intersection between computability and combinatorics: *trees*. Of course, trees and tree-like structures permeate many branches of mathematics. In computability theory and reverse mathematics, they arise as frameworks for building approximations. For example, to produce a function $f\colon \omega \to \omega$ with certain properties, we might first define $f(0)$, then define $f(1)$ using $f(0)$, then define $f(2)$ using $f(1)$ and $f(0)$, etc. If we have only one choice for $f(s+1)$ for each $s$, we will build a simple sequence of approximations. But suppose we have multiple choices (perhaps even infinitely many) for $f(s+1)$ given the sequence $f(0), \ldots, f(s)$. In that case, we will end up with is a *tree of approximations* to $f$, and there are multiple possibilities for what $f$ can be depending on which "path" we take through this tree.

The following definitions will help make these comments more precise.

**Definition 2.8.1.** Fix $X \in 2^{\omega}$ and $T \subseteq X^{<\omega}$.

1. $T$ is a *tree* if it is closed under initial segments, i.e., if $\alpha \in X$ and $\beta \preceq \alpha$ then $\beta \in X$. If $U \subseteq T$ is also a tree then $U$ is a *subtree* of $T$.
2. $T$ is *finitely branching* if for each $\alpha \in T$ there are at most finitely many $x \in X$ such that $\alpha x \in T$.
3. $T$ is *bounded* if there is a function $b\colon \omega \to \omega$ such that for every $\alpha \in T$, $\alpha(i) < b(i)$ for all $i < |\alpha|$, in which case we also say $T$ is *b-bounded*.
4. An *infinite path* (or just *path*) through $T$ is a function $f\colon \omega \to X$ such that $f \upharpoonright k \in T$ for every $k \in \omega$. The set of all paths is denoted $[T]$.
5. $T$ is *well founded* if $T$ has no path, i.e., if $[T] = \emptyset$.
6. $\alpha \in T$ is *extendible* if $\alpha \preceq f$ for some $f \in [T]$. The set of all extendible $\alpha \in T$ is denoted $\operatorname{Ext}(T)$.

Naturally, $X^{<\omega}$ itself is a tree. Also, every bounded tree is finitely branching. The elements of a tree $T$ are also sometimes called *nodes*. If $\alpha \preceq \beta$ belong to $T$ and

$|\beta| = |\alpha| + 1$, then $\beta$ is an *immediate successor of* $\alpha$ *in* $T$. If $\alpha \in T$ and there is no $x \in X$ such that $\alpha x \in T$ (or equivalently, there is no $\beta \in T$ such that $\beta \succ \alpha$) then $\alpha$ is called a *leaf* or *end node* of $T$.

The following shows that the sets of paths through trees have a natural connection to the clopen topology on $X^\omega$ (Definition 1.5.3).

**Proposition 2.8.2.** *Fix* $X \in 2^\omega$ *and* $C \subseteq X^\omega$. *Then* $C$ *is closed in the clopen topology if and only if* $C = [T]$ *for some tree* $T \subseteq X^{<\omega}$.

*Proof.* First suppose $C$ is closed. Write $X^\omega \smallsetminus C$ as $\bigcup_{\alpha \in U} [[\alpha]]$ for some $U \subseteq X^{<\omega}$. Then $T = \{\alpha \in X^{<\omega} : (\forall k \leqslant |\alpha|)[\alpha \upharpoonright k \notin S]\}$ is a tree with $C = [T]$. Now suppose $C = [T]$. We may assume $\langle\rangle \in T$, since otherwise $T = \emptyset$ and $C = \emptyset$. Let $U = \{\alpha \in X^{<\omega} : \alpha \notin T \wedge \alpha \upharpoonright |\alpha| - 1 \in T\}$. Then $X^\omega \smallsetminus C = \bigcup_{\alpha \in U} [[\alpha]]$. □

As is evident from the discussion above, our main interest in trees is in the objects we can construct using them, i.e., the paths. In general, there is no simple combinatorial criterion for a tree to have a path. But for finitely branching trees, there is, as given by the following hallmark result.

**Theorem 2.8.3 (König's lemma).** *Let* $T \subseteq \omega^{<\omega}$ *be a finitely branching tree. The following are equivalent.*

1. *$T$ is not well founded.*
2. *$T$ is infinite.*
3. *For every $k \in \omega$, $T$ contains a string of length $k$.*

*Proof.* (1) $\rightarrow$ (2) : If $f \in [T]$ then $f \upharpoonright k \in T$ for each $k \in \omega$, and these form infinitely many nodes in $T$.

(2) $\rightarrow$ (3) : We prove the contrapositive. Suppose $k \in \omega$ is such that $T$ has no node of length $k$. Being a tree, $T$ is closed under $\preceq$, and therefore every node of $T$ must have length some $j < k$. But as $T$ is finitely branching, and contains at most one node of length 0 (the empty string), it follows by induction that $T$ has finitely many nodes of length any such $j$. Ergo, $T$ is finite.

(3) $\rightarrow$ (1) : We define a sequence $\alpha_0 \preceq \alpha_1 \preceq \cdots$ of elements of $T$ such that, for all $s \in \omega$, $|\alpha_s| = s$ and there are infinitely many $\beta \in T$ with $\alpha_s \preceq \beta$.

This suffices, since the first property ensures $\bigcup_{s \in \omega} \alpha_s$ is a path through $T$. By assumption, the empty string, $\langle\rangle$, belongs to $T$. Let this be $\alpha_0$, noting that this has the properties. Assume next that $\alpha_s$ has been defined for some $s$. By hypothesis, there are infinitely many $\beta \in T$ such that $\alpha_s \preceq \beta$. Since $T$ is a tree, this means in particular that $\alpha_s$ has at least one extension $\alpha \in T$ of length $s + 1$. But since $T$ is finitely branching, there must be at least one such $\alpha$ such that $\alpha \preceq \beta$ for infinitely many $\beta \in T$. Let this be $\alpha_{s+1}$. Clearly, $\alpha_{s+1}$ has the desired properties. □

It is worth emphasizing a restatement of the equivalence of (1) and (2) above, which is that $[T]$ is nonempty if and only if $T$ is infinite. We often invoke König's lemma in this form.

More commonly, "König's lemma" is used to refer just to the implication from (2) to (1) above, under the assumption that $T$ is finitely branching. Importantly, this implication fails if $T$ is *not* finitely branching. For example, consider the set containing the empty node, $\langle\rangle$, and for each $x \in \omega$ the singleton node $\langle x \rangle$. This is an infinite subtree of $\omega^{<\omega}$, but clearly it has no path. A different example can be used to show that if $T$ is not finitely branching then the implication from (3) to (1) fails as well.

**Corollary 2.8.4.** *If $T \subseteq \omega^{<\omega}$ is an infinite, finitely branching tree, then there is an $f \in [T]$ such that $f \leqslant_T \mathrm{Ext}(T)$.*

*Proof.* By the proof of the implication (3) $\to$ (1) in the theorem, noting that for every $s$, we can take $\alpha_{s+1} = (\mu\alpha \in \mathrm{Ext}(T))[\alpha_s \preceq \alpha]$. □

Per our discussion in Section 2.2.3, if $X$ is a computable set we say that a tree $T \subseteq X^{<\omega}$ is *computable* if (the set of codes of elements of) $T$ is a computable subset of $\omega$. We say $T$ is *computably bounded* if it is $b$-bounded for some computable function $b\colon \omega \to \omega$. We will see later, in Exercise 4.8.5, that for the purposes of understanding the complexity of the paths through such trees it suffices to look at computable subtrees of $2^{<\omega}$. (Of course, we also care about relativizing the above effectivity notions, both for definitions and results, but these are straightforward to formulate and prove from the unrelativized versions, as discussed in the previous section.)

Computable, computably bounded trees arise naturally throughout computability theory, with the interest usually being in the paths rather than the trees themselves. To understand the complexity of such paths, we begin with the following.

**Proposition 2.8.5.** *Suppose $T \subseteq \omega^{<\omega}$ is a tree that is infinite, computable, and computably bounded. Then $\mathrm{Ext}(T) \leqslant_T \varnothing'$, and so $[T]$ contains a member $f \leqslant_T \varnothing'$.*

*Proof.* For each $k \in \omega$, there are only finitely many strings of length $k$, so the set of codes of such strings is bounded. Moreover, this bound is primitive recursive in $k$. It follows that if $T$ is a computable subtree of $2^{<\omega}$, there is a computable relation $P$ such that for all $k \in \omega$ and all (codes for) $\alpha \in 2^{<\omega}$, $P(\alpha, k)$ holds if and only if $(\exists \beta \in T)[\beta \succeq \alpha \wedge |\beta| = k]$. Thus, $\alpha \in \mathrm{Ext}(T) \leftrightarrow \alpha \in T \wedge (\forall k \geqslant |\alpha|)P(\alpha, k)$, meaning $\mathrm{Ext}(T)$ is a $\Pi^0_1$ set, and hence is computable in $\varnothing'$ by Post's theorem (Theorem 2.6.2). The rest follows by Corollary 2.8.4. □

### 2.8.1 $\Pi^0_1$ classes

Recall the definition of $\Sigma^0_n$ and $\Pi^0_n$ relations on $\omega^\omega$ or $2^\omega$ from Definition 2.6.7. We use this to define the following central concept.

**Definition 2.8.6.** Fix $n \in \omega$ and let $\Gamma \in \{\Sigma^0_n, \Pi^0_n\}$. A subset $C$ of $\omega^\omega$ (or $2^\omega$) is a $\Gamma$ *class* if there is a $\Gamma$ relation $R$ on $\omega^\omega$ (respectively, $2^\omega$) such that $f \in C \leftrightarrow R(f)$, for all $f$ in $\omega^\omega$ (respectively, $2^\omega$).

Thus, $\Sigma^0_n$ and $\Pi^0_n$ relations are associated to $\Sigma^0_n$ and $\Pi^0_n$ classes, respectively, in a way similar to how formulas are associated with the sets they define. In Chapter 5, we will see how to make this analogy formal. Primarily, we will be interested in $\Pi^0_1$ classes, which are related to trees through the following result.

**Proposition 2.8.7.** *Let $C$ be a subset of $\omega^\omega$. The following are equivalent.*

1. *$C$ is a $\Pi^0_1$ class.*
2. *There is a computable tree $T \subseteq \omega^{<\omega}$ such that $C = [T]$.*
3. *There is a computable set $A \subseteq \omega^{<\omega}$ such that $C = \{f : (\forall k)[f \upharpoonright k \in A]\}$.*

*The same result holds if $\omega^\omega$ is replaced by $2^\omega$ and $\omega^{<\omega}$ by $2^{<\omega}$.*

*Proof.* We prove (1) → (2). That (2) → (3) is obvious, since $T$ itself is a set $A$ of the desired form. And that (3) → (1) follows from the fact that the relation $R(f) \leftrightarrow (\forall k)[f \upharpoonright k \in A]$ is $\Pi^0_1$ (in the sense of Definition 2.6.7).

So fix a $\Pi^0_1$ class $C \subseteq \omega^\omega$. By definition, there is a computable sequence $\langle e_0, e_1, \ldots \rangle$ of indices such that $\Phi^f_{e_i}(0)\downarrow$ for every $f$ and $f \in C$ if and only if $(\forall i)[\Phi^f_{e_i}(0) \downarrow= 1]$. Define $T$ to be the set of all $\alpha \in \omega^{<\omega}$ such that for all $i$, if $\Phi^\alpha_{e_i}(0)\downarrow$ then $\Phi^\alpha_{e_i}(0) \downarrow= 1$. By the use principle (Proposition 2.4.7), $T$ is computable, and clearly if $\alpha$ is in $T$ then so is every $\beta \preceq \alpha$. Hence, $T$ is a tree. Now by assumption on the $e_i$,

$$\begin{aligned} f \in C &\leftrightarrow (\forall i)(\exists s)[\Phi^f_{e_i}(0)[s] \downarrow= 1] \\ &\leftrightarrow (\forall i)(\forall s)[\Phi^f_{e_i}(0)[s] \downarrow\rightarrow \Phi^f_{e_i}(0)[s] = 1]. \end{aligned}$$

Hence by monotonicity of computations,

$$\begin{aligned} f \in C &\leftrightarrow (\forall k)(\forall i)[\Phi^{f \upharpoonright k}_{e_i}(0) \downarrow\rightarrow \Phi^{f \upharpoonright k}_{e_i}(0) = 1] \\ &\leftrightarrow (\forall k)[f \upharpoonright k \in T] \\ &\leftrightarrow f \in [T]. \end{aligned}$$

The proof is complete. □

$\Pi^0_1$ classes allows us to refer to elements of $[T]$ (for a tree $T$) without first needing to define $T$ as a set of strings. For example, given a computable set $X$, we may wish to consider the class $C$ of all $Y \subseteq X$. Then $C$ is a $\Pi^0_1$ class (in $2^\omega$), since "$Y$ is a subset of $X$" is a $\Pi^0_1$ relation. Indeed, we could write

$$C = [\{\sigma \in 2^{<\omega} : (\forall i < |\sigma|)[\sigma(i) = 1 \rightarrow i \in X]\}].$$

Sometimes we will write things like this out explicitly, but usually it will be more convenient to describe directly the property we want the elements of $C$ to have.

We can assign indices to $\Pi^0_1$ classes, as follows.

**Definition 2.8.8.** An *index* of a $\Pi^0_1$ class $C$ is a $\Delta^0_1$ index for a computable tree $T$ as in Proposition 2.8.7 (2).

Alternatively, we could code a $\Pi^0_1$ class by the index of the computable sequence $\langle e_0, e_1, \dots \rangle$ specifying a $\Pi^0_1$ relation that defines $C$. But the proof of Proposition 2.8.7 is completely uniform, so we may move between these two types of indices uniformly computably, and therefore the specific choice does not matter. Under either choice, every $\Pi^0_1$ class has infinitely many indices.

As indicated at the end of the last section, our main interest in the sequel will be in bounded trees, typically subtrees of $2^{<\omega}$. In this setting, combining König's lemma (Theorem 2.8.3) with Proposition 2.8.7 yields the following simple fact, which is so commonly used that it deserves a separate mention.

**Corollary 2.8.9.** *If $C$ is a $\Pi^0_1$ class in $2^\omega$ and $T \subseteq 2^{<\omega}$ is a tree with $C = [T]$, then $C \neq \emptyset$ if and only if $T$ is infinite.*

Proposition 2.8.2 has the following corollary, using the fact that the Cantor space is compact.

**Corollary 2.8.10.**

1. *If $C$ is a $\Pi^0_1$ class in $2^\omega$ then $C$ is a compact set.*
2. *If $C_0 \supseteq C_1 \supseteq \cdots$ is a sequence of nonempty $\Pi^0_1$ classes in $2^\omega$ then $\bigcap_s C_s$ is nonempty.*

*Proof.* Both parts are general topological properties of compact spaces. Part (2) can also be seen as follows. For each $s$, fix a computable tree $T_s$ with $[T_s] = C_s$. We may assume $T_0 \supseteq T_1 \supseteq \cdots$. By hypothesis, each $T_s$ is infinite. Suppose $\bigcap_s T_s$ is finite. Since $T \subseteq 2^{<\omega}$, this means there is a $k$ such that $T$ contains no string $\sigma \in 2^k$ (i.e., no binary string of length $k$). For each such $\sigma$, let $s(\sigma)$ be the least $s$ such that $\sigma \notin T_s$. Let $s_0 = \max\{s(\sigma) : \sigma \in 2^k\}$. Then $T_{s_0}$ contains no string of length $k$ and so is finite, a contradiction. □

*Remark 2.8.11 (Compactness).* Suppose that $C \subseteq 2^\omega$ is an empty $\Pi^0_1$ class. By Proposition 2.8.7 (2), there is a computable $A$ (not necessarily a tree) such that $C = \{f \in 2^\omega : (\forall k)[f \upharpoonright k \in A]\}$. Thus for each $f \in 2^\omega$, since $f \notin C$, it follows that there is a $k$ such that $f \upharpoonright k \notin A$. But in fact, there is a *single* such $k$ that works for *all* $f \in 2^\omega$. If $A$ is not a tree, we can let $T = \{\sigma \in 2^{<\omega} : (\exists \tau \in A)[\sigma \preceq \tau]\}$. Then $T$ is not necessarily computable, but it is a tree. If $A$ is a tree, let $T = A$. Either way, $[T] = C$ and the existence of $k$ follows just as in the preceding proof. Typically, for brevity, we say $k$ exists "by compactness".

*Example 2.8.12.* Fix a Turing functional $\Phi$, and consider the $\Pi^0_1$ class $C$ of all $f \in 2^\omega$ such that $\Phi^f(0)\uparrow$. To be precise, $f \in C$ if and only if $(\forall s)[\Phi^f(0)[s]\uparrow]$. (Note that the matrix of this formula is a computable predicate in $f$.) Suppose $C = \emptyset$. Then by compactness, there is a $k$ such that for all $f \in 2^\omega$, $(\exists s \leqslant k)[\Phi^f(0)[s]\downarrow]$. (This illustrates an interesting computability theoretic fact: a functional which converges on *every* oracle, cannot converge arbitrarily late.)

To see this another way, consider the computable tree $T$ such that $C = [T]$. $T$ is the set of all $\sigma \in 2^{<\omega}$ such that $\Phi^\sigma(0)\uparrow$. If $C = \emptyset$, let $k$ be such that no $\sigma$ of length $k$ belongs to $T$, meaning $\Phi^\sigma(0)\downarrow$ and hence $(\exists s \leqslant k)[\Phi^{\sigma \upharpoonright s}(0)\downarrow]$. Then for all $f \in 2^\omega$, $(\exists s \leqslant k)[\Phi^f(0)[s]\downarrow]$, as before.

We will encounter many examples of $\Pi^0_1$ classes in this book. An important one is provided by the following definition.

**Definition 2.8.13.** A function $f\colon \omega \to \omega$ is *diagonally noncomputable* (or *DNC* for short) if, for every $e \in \omega$, if $\Phi_e(e)\downarrow$ then $f(e) \neq \Phi_e(e)$.

**Proposition 2.8.14.** *The set of all* 2*-valued DNC functions is a* $\Pi^0_1$ *class.*

*Proof.* We claim that the class $C$ of all $f \in 2^\omega$ such that

$$(\forall e)(\Phi_e(e)\uparrow \vee \Phi_e(e) \neq f(e)]$$

is a $\Pi^0_1$ class. Define $T$ to be the set of all $\sigma \in 2^{<\omega}$ such that

$$(\forall e < |\sigma|)[\Phi_e(e)[|\sigma|]\downarrow \rightarrow \sigma(e) \neq \Phi_e(e)]. \tag{2.1}$$

Then $T$ is a computable subtree of $2^{<\omega}$. If $f$ is 2-valued and DNC, then (2.1) obviously holds for $\sigma = f \upharpoonright k$, for every $k$. Hence, $f \in [T]$.

Conversely, suppose $f \in [T]$ and fix $e$ such that $\Phi_e(e)\downarrow$. Let $s$ be such that $\Phi_e(e)[s]\downarrow$ and consider any $k > s$. Then also $\Phi_e(e)[k]\downarrow$. Now, by Convention 2.4.8 we also have $k > e$. Since $f \upharpoonright k \in T$, taking $\sigma = f \upharpoonright k$ in (2.1) we obtain that $f(e) \neq \Phi_e(e)$. Thus, $f$ is DNC. Hence, $C = [T]$. □

It is instructive to reflect on this proof for a moment. Crucially, $T$ above is *not* the tree of all $\sigma \in 2^{<\omega}$ such that for all $e < |\sigma|$, $\sigma(e) \neq \Phi_e(e)$; that is $\mathrm{Ext}(T)$. $T$ itself contains many other nodes. To better understand this, let us present the proof in slightly different terms.

Imagine we are building $T$ by *stages*. At stage $s$, we must decide which strings $\sigma$ of length $s$ to put into $T$. Of course, we must do this consistently with making $T$ be a tree, so if $s > 0$ we cannot add any string that does not already have an initial segment in $T$ of length $s - 1$. But in addition, we must make sure that $[T]$ ends up being the set of 2-valued DNC functions.

To this end, we can check which $e < s$ satisfy that $\Phi_e(e)[s]\downarrow$, and then exclude any $\sigma$ of length $s$ such that $\sigma(e) = \Phi_e(e)$ for some such $e$. Any other $\sigma$ of length $s$ we must add to $T$, however. This is because, as far as we can tell at stage $s$, $\sigma$ looks like an initial segment of a DNC function. And we cannot add $\sigma$ at any later stage, since that is reserved for longer strings. This allows the possibility that $\sigma$ is put into $T$, and there is still an $e < |\sigma|$ with $\Phi_e(e)\downarrow= \sigma(e)$, only the least $s^*$ such that $\Phi_e(e)[s^*]\downarrow$ is larger than $|\sigma|$. Thus, $\sigma$ is in fact *not* an initial segment of any DNC function. Now, we cannot remove $\sigma$ from $T$. But we *can* ensure $\sigma \notin \mathrm{Ext}(T)$, by "pruning" $T$ above $\sigma$ to ensure no path through $T$ extends it. That is, at the stage $s^*$, we exclude all $\tau \succeq \sigma$ of length $s^*$ from $T$. Then for every function $f \succeq \sigma$, we have $f \upharpoonright s^* \notin T$ and hence $f \notin [T]$.

Note that, by definition, $\varnothing'$ is the set of those $e \in \omega$ such that $\Phi_e(e)\downarrow$, and so $\varnothing'$ can compute a 2-valued DNC function. In particular, the $\Pi^0_1$ class of Proposition 2.8.14 is nonempty. Moreover, it is clear that no DNC function can be computable.

**Corollary 2.8.15.** *There is a nonempty* $\Pi^0_1$ *class with no computable member.*

### 2.8.2 Basis theorems

Although $\Pi^0_1$ classes do not always have computable members, can anything be said about their elements in general? It is easy to see that for every $f \in 2^\omega$ there is a nonempty $\Pi^0_1$ class $C$ that does not contain $f$. (Let $C$ be the set of all $g \in 2^\omega$ with $g(0) \neq f(0)$.) Hence, positive results about members of $\Pi^0_1$ classes are usually presented in terms of *collections* of functions. nonempty $\Pi^0_1$ class contains some function from the collection.

**Definition 2.8.16.** A collection $\mathcal{B}$ of functions $f \in 2^\omega$ closed downward under $\leqslant_T$ is a *basis for* $\Pi^0_1$ *classes* if every nonempty $\Pi^0_1$ class $C$ contains an element in $\mathcal{B}$.

Of course, different $\Pi^0_1$ classes may contain different elements of a basis $\mathcal{B}$. Perhaps the easiest basis to recognize, in light of Proposition 2.8.5, is the set of functions computable from $\varnothing'$.

**Theorem 2.8.17 (Kleene).** *The collection of $\varnothing'$-computable functions is a basis for $\Pi^0_1$ classes.*

*Proof.* By Proposition 2.8.5, $\varnothing'$ can compute the sequence

$$\sigma_0 \preceq \sigma_1 \preceq \cdots$$

in the proof of the implication (3) $\to$ (1) of Theorem 2.8.3. Hence, the path $\bigcup_{s\in\omega} \sigma_s$ is computable in $\varnothing'$. □

This result can be improved in many ways. The literature is full of basis theorems of various kinds, with applications in many different areas of computability theory. For a partial survey, see Diamondstone, Dzhafarov, and Soare [66]. For our purposes, we now state and prove two especially prominent basis theorems which we will apply repeatedly in the sequel.

The method of proof here is known as *forcing with* $\Pi^0_1$ *classes* or *Jockusch–Soare forcing*. We will study forcing systematically in Chapter 7, and see further applications of Jockusch–Soare forcing there (particularly in Section 7.7). But we can use the basic method already without needing to know the theory of forcing as a whole. The idea is as follows. We are given a computable tree $T$, and wish to produce a $g \in [T]$ with certain computational and combinatorial properties. We build a nested sequence

$$T_0 \supseteq T_1 \supseteq T_2 \supseteq \cdots$$

of infinite computable subtrees of $T$, with each $T_e$ ensuring some more of the properties we wish $g$ to have in the end. Typically, these properties are organized into infinitely many requirements, $\mathcal{R}_0, \mathcal{R}_1, \ldots$, and each element of $[T_e]$ satisfies, e.g., $\mathcal{R}_i$ for all $i < e$. In this way, any element of $\bigcap_e [T_e]$ will satisfy all the $\mathcal{R}_e$, as desired. (Of course, the fact that there is at least one element in $\bigcap_e [T_e]$ follows by Corollary 2.8.10.)

We begin with arguably the most famous basis theorem of all, due to Jockusch and Soare [172].

**Theorem 2.8.18 (Low basis theorem; Jockusch and Soare [172]).** *The collection of low functions is a basis for $\Pi^0_1$ classes.*

*Proof.* Let $C$ be a nonempty $\Pi^0_1$ set, and fix an infinite computable tree $T \subseteq 2^{<\omega}$ such that $C = [T]$. As described above, we build a sequence

$$T_0 \supseteq T_1 \supseteq T_2 \supseteq \cdots$$

of infinite computable subtrees of $T$. In this proof, we also build a $\emptyset'$-computable function $j \colon \omega \to 2$, such that along with our sequence the following requirement is satisfied for each $e \in \omega$:

$$\mathcal{R}_e \colon (\forall f \in [T_{e+1}])[e \in f' \leftrightarrow j(e) = 1].$$

Thus, we will ensure that either $e \in f'$ for all $f \in [T_{e+1}]$, or $e \notin f'$ for all $f \in [T_{e+1}]$, and which of these is the case is recorded by the $\emptyset'$-computable function $j$. By taking $g \in \bigcap_e [T_e]$, it follows that for all $e$ we have $e \in g'$ if and only if $j(e) = 1$, so $g' \leqslant_T j \leqslant_T \emptyset'$. In fact, $j$ just *is* the jump of $g$.

We proceed to the details. Let $T_0 = T$, and suppose that for some $e$ we have defined $T_e$ and $j \upharpoonright e$. We define $T_{e+1}$ and $j(e)$. Define $U = \{\sigma \in T_e : \Phi^\sigma_e(e)\uparrow\}$. This is a computable subtree of $T_e$, so by Proposition 2.8.5, $\emptyset'$ can determine whether or not $U$ is infinite. If it is, let $T_{e+1} = U$. Now if $f$ is any element of $[T_{e+1}]$ then $\Phi^f_e(e)\uparrow$, so $e \notin f'$. We then define $j(e) = 0$, and now we have clearly satisfied $\mathcal{R}_e$. Suppose next that $U$ is finite. Then there is a $k$ such that no string $\sigma \in T_e$ of length $k$ belongs to $U$, meaning that every such string satisfies $\Phi^\sigma_e(e)\downarrow$. Choose any string $\sigma \in \mathrm{Ext}(T_e)$ of length $k$, and let $T_{e+1} = \{\tau \in T_e : \tau \preceq \sigma \vee \sigma \preceq \tau\}$. This time, if $f \in [T_{e+1}]$ then $e \in f'$. We let $j(e) = 1$, so again $\mathcal{R}_e$ is satisfied. □

Actually, the $g$ we chose above is the *unique* path through $\bigcap_e T_e$. (See Exercise 2.9.11.) Some proofs of the low basis theorem emphasize this by adding intermediate steps that determine "more and more" of $g$ along the way explicitly, but this is not needed.

As a side note, combining Theorem 2.8.18 and Corollary 2.8.15 yields a (somewhat roundabout) proof that there exist noncomputable low sets.

The next basis theorem we look at uses the following computability theoretic notion.

**Definition 2.8.19.** Fix $S \in 2^\omega$.

1. $S$ is *hyperimmune* if no computable function dominates its principal function, $p_S$.
2. $S$ is of, or has, *hyperimmune free degree* if no $S^* \leqslant_T S$ is hyperimmune. Otherwise, $S$ is of, or has, *hyperimmune degree*.

In general, to show that $S$ is hyperimmune we exhibit for each $e \in \omega$ an $x$ such that $\Phi_e(x)\uparrow$ or $\Phi_e(x) < p_S(x)$. The following shows that such sets can be found Turing below every noncomputable $\Delta^0_2$ set.

**Theorem 2.8.20 (Miller and Martin [213]).** *If $\emptyset <_T S \leqslant_T \emptyset'$ then $S$ has hyperimmune degree.*

On the flip side, we have the following famed characterization of having hyperimmune free degree.

**Theorem 2.8.21 (Miller and Martin [213]).** *A set $S$ has hyperimmune free degree if and only if every $S$-computable function is dominated by a computable function.*

For this reason, some authors use the term "computably dominated" instead of "hyperimmune free".

**Theorem 2.8.22 (Hyperimmune free basis theorem; Jockusch and Soare [172]).** *The collection of functions of hyperimmune free degree is a basis for $\Pi^0_1$ classes.*

*Proof.* As above, let $C$ be given and fix $T$ so that $C = [T]$. We again define a sequence

$$T_0 \supseteq T_1 \supseteq T_2 \supseteq \cdots,$$

of infinite computable subtrees of $T$. Our goal this time is to satisfy the following requirement for each $e \in \omega$:

$$\begin{aligned}\mathcal{R}_e\colon\ &(\forall f \in [T_{e+1}])[\Phi^f_e \text{ is not total}]\\ &\vee (\exists h \leqslant_T \emptyset)(\forall f \in [T_{e+1}])[\Phi^f_e \text{ is total} \wedge (\forall x)[\Phi^f_e(x) \leqslant h(x)]].\end{aligned}$$

If, at the end of the construction, we choose some $g \in \bigcap_e [T_e]$, then we will have that for all $e \in \omega$, if $\Phi^g_e$ is total then it is dominated by a computable function. Hence, by Theorem 2.8.21, $g$ will have hyperimmune free degree, as desired.

Let us proceed to the construction. Let $T_0 = T$ and suppose inductively that for some $e$ we have defined $T_e \subseteq T$. For each $x$, define $U_x = \{\sigma \in T_e : \Phi^\sigma_e(x)\uparrow\}$, which is a subtree of $T_e$ uniformly computable in $x$. If, for some $x$, $U_x$ is infinite, we let $T_{e+1} = U_x$. Now every $f \in [T_{e+1}]$ satisfies $\Phi^f_e(x)\uparrow$, hence $\mathcal{R}_e$ is satisfied trivially.

Suppose that $U_x$ is finite for every $x$. We let $T_{e+1} = T_e$. Then in particular, $\Phi^f_e$ is total for every $f \in [T_{e+1}]$. We now define a computable function $h$ as follows. On input $x$, $h$ searches for the least $\ell$ such that $\Phi^\sigma_e(x)\downarrow$ for all $\sigma \in T_{e+1}$ of length $\ell$, and then it outputs the supremum of the values of all these computations. (Notice that $\ell$ must exist, else $U_x$ would be infinite.) Since every $f \in [T_{e+1}]$ extends some $\sigma \in T_{e+1}$ of length $\ell$, we have $\Phi^f_e(x) \leqslant h(x)$. And since this holds for all $x$, $\mathcal{R}_e$ holds. □

The final basis theorem we prove will have important applications in Chapters 8 and 9. This theorem is also due to Jockusch and Soare [172].

**Theorem 2.8.23 (Cone avoidance basis theorem).** *Fix $C \nleqslant_T \emptyset$. The collection of functions $g \in 2^\omega$ such that $C \nleqslant_T g$ is a basis for $\Pi^0_1$ classes.*

*Proof.* Fix $C$ and $T$ with $C = [T]$. We define a sequence

$$T_0 \supseteq T_1 \supseteq T_2 \supseteq \cdots,$$

of infinite computable subtrees of $T$. We aim to satisfy the following requirement for each $e \in \omega$:

$$\mathcal{R}_e\colon (\forall f \in [T_{e+1}])[\Phi_e^f \neq C].$$

Choosing any $g \in \bigcap_e [T_e]$, it follows that $C \not\leqslant_T g$, as desired. Let $T = T_0$, and suppose that for some $e$ we have defined $T_e$. For each $x$, define $U_x$ as in the proof of the hyperimmune free basis theorem: that is, $U_x = \{\sigma \in T_e : \Phi_e^\sigma(x)\uparrow\}$, which is a computable subtree of $T_e$. As before, if $U_x$ is infinite for some $x$, we let $T_{e+1} = U_x$, so that every $f \in [T_{e+1}]$ satisfies $\Phi_e^f(x)\uparrow$. In this case, then, $\mathcal{R}_e$ is satisfied.

Suppose that $U_x$ is finite for each $x$. We claim there is an $x$ and a $\sigma \in \mathrm{Ext}(T_e)$ such that $\Phi_e^\sigma(x)\downarrow\neq C(x)$. Suppose not. Then for each $x$, there exist numbers $k_x$ and $y_x$ such that for every $\sigma \in T_e$ of length $k_x$, we have $\Phi_e^\sigma(x)\downarrow= y_x$. This is because $U_x$ is finite, so there is a $k$ such that $\Phi_e^\tau(x)\downarrow$ for every $\tau \in T_e$ of length $k$. By hypothesis, no such $\tau$ with $\Phi_e^\tau(x) \neq C(x)$ is extendible, so by König's lemma, there is a $k^* \geqslant k$ such that no $\sigma \in T_e$ of length $k^*$ extends any such $\tau$. Thus, every $\sigma \in T_e$ of length $k^*$ satisfies $\Phi_e^\sigma(x)\downarrow= C(x)$, and we can consequently take $k_x = k^*$ and $y_x = C(x)$. But $k_x$ and $y_x$ can be searched for, and found, computably. (Notice that finding them does *not* require knowing the value of $C(x)$.) Thus, $y_x = C(x)$ can be found computably for each $x$, which makes $C$ computable, a contradiction. This proves the claim, that there is an $x$ and a $\sigma \in \mathrm{Ext}(T_e)$ such that $\Phi_e^\sigma(x)\downarrow\neq C(x)$. Fix such an $\sigma$ and let $T_{e+1} = \{\tau \in T_e : \tau \preceq \sigma \vee \sigma \preceq \tau\}$. Now every $f \in [T_{e+1}]$ satisfies $\Phi_e^f(x)\downarrow= \Phi_e^\sigma(x) \neq C(x)$, so $\mathcal{R}_e$ is satisfied. □

### 2.8.3 PA degrees

In our discussion above, we saw one oracle, $\varnothing'$, that could compute a member of every nonempty $\Pi_1^0$ class. As it turns out, there are many such oracles.

**Definition 2.8.24.** A function $f \in 2^\omega$ is of, or has, *PA degree*, written $f \gg \varnothing$, if the collection of $f$-computable functions is a basis for $\Pi_1^0$ classes. Relativized to an oracle $g$, we say $f$ has *PA degree relative to* $g$, written $f \gg g$.

Thus, $f \gg \varnothing$ if and only if every nonempty $\Pi_1^0$ classes has an $f$-computable member, if and only if every infinite computable subtree of $2^{<\omega}$ has an $f$-computable path. Note that if $f \equiv_T f^*$, $g \equiv_T g^*$, and $f \gg g$ then $f^* \gg g^*$. So being "of PA degree" really is a degree property, justifying the terminology.

The abbreviation "PA" stands for "Peano arithmetic", as it follows by an old result of Scott [274] and Solovay (unpublished) that $f$ has PA degree if and only if $f$ computes a complete consistent extension of Peano arithmetic. For a modern proof of this equivalence, as well as several other characterizations of having PA degree, see Soare [295, Section 10.3]). Here, we restrict to the properties that will be most useful to us in the rest of the book.

**Theorem 2.8.25.** *Fix $f \in 2^\omega$. The following are equivalent.*

1. $f \gg \emptyset$.
2. *$f$ computes a 2-valued DNC function.*
3. *Every 2-valued partial computable function has a total $f$-computable extension.*
4. *There is an $f$-computable function $d\colon \omega^2 \to 2$ such that $d(e_0, e_1) \in \{e_0, e_1\}$ for all $e_0, e_1 \in \omega$, and if $(\exists i < 2)[\Phi_{e_i}(0)\uparrow]$ then $\Phi_{d(e_0,e_1)}(0)\uparrow$.*

*Proof.* (1) → (2) : Immediate from Proposition 2.8.14.

(2) → (3) : Suppose $\Phi_e$ is a 2-valued partial computable function. By the $S^m_n$ theorem (Theorem 2.4.10), we can fix a computable function $S\colon \omega \to \omega$ such that $\Phi_{S(x)}(z) = \Phi_e(x)$, for all $x$ and $z$. Then in particular $\Phi_e(x)\downarrow$ if and only if $\Phi_{S(x)}(S(x))\downarrow$, in which case the two computations agree. By (2), let $g$ be an $f$-computable 2-valued DNC function, and define $h = 1 - (g \circ S)$. Then $h$ is $f$-computable and 2-valued, and for all $x$ such that $\Phi_e(x)\downarrow$ we have $g(S(x)) = 1 - \Phi_{S(x)}(S(x)) = 1 - \Phi_e(x)$, so $h(x) = \Phi_e(x)$.

(3) → (4) : Consider the partial computable function defined by

$$p(e_0, e_1) = (\mu i < 2)(\exists s)[\Phi_{e_i}(0)[s]\downarrow \wedge (\forall t < s)\Phi_{e_{1-i}}(0)[t]\uparrow]$$

for all $e_0, e_1 \in \omega$. By (3), let $h$ be an $f$-computable 2-valued function extending $p$, and define $d(e_0, e_1) = e_{1-h(e_0,e_1)}$ for all $e_0, e_1$. Now if $\Phi_{e_i}(0)\uparrow$ and $\Phi_{e_{1-i}}(0)\downarrow$ the $p(e_0, e_1) = e_{1-i}$, so $d(e_0, e_1) = e_i$, as desired.

(4) → (1) : Fix an infinite computable tree $T$. By the $S^m_n$ theorem, we can fix a computable function $S\colon \omega^2 \to \omega$ such that for every $\sigma \in 2^{<\omega}$ and $i < 2$,

$$\Phi_{S(\sigma,i)}(x) = \begin{cases} (\mu k \in \omega)(\forall \tau)[(\tau \succeq \sigma i \wedge |\tau| = k) \to \tau \notin T] & \text{if } x = 0, \\ i & \text{otherwise,} \end{cases}$$

for all $x \in \omega$. Thus, $\Phi_{S(\sigma,i)}(0)\uparrow$ if and only if $\sigma i \in \mathrm{Ext}(T)$. Let $d \leqslant_T f$ be as given by (4). Then $d$ can computably construct a sequence $\sigma_0 \preceq \sigma_1 \preceq \cdots$ of elements of $T$ with $|\sigma_s| = s$ for all $s$. Namely, let $\sigma_0 = \langle\rangle$, and suppose inductively that we have defined $\sigma_s \in \mathrm{Ext}(T)$ for some $s \in \omega$. Then there must be an $i < 2$ such that $\sigma_s i \in \mathrm{Ext}(T)$, and hence such that $\Phi_{S(\sigma,i)}(0)\uparrow$. By assumption, $d(S(\sigma_s, 0), S(\sigma_s, 1)) = S(\sigma_s, i)$ for some such $i$, and $d$ can compute this $i$ since $\Phi_{d(S(\sigma_s,0),S(\sigma_s,1))}(1) = \Phi_{S(\sigma_s,i)}(1)\downarrow= i$. We then let $\sigma_{s+1} = \sigma_s i$. So $\bigcup_s \sigma_s$ is a $d$-computable (hence $f$-computable) path through $T$, and since $T$ was arbitrary, $f \gg \emptyset$. □

The equivalence of (1) and (2) means that there is a tree (namely, the one constructed in Proposition 2.8.14) whose paths are, up to Turing equivalence, precisely the functions of PA degree. It follows that if $\mathcal{B}$ is *any* basis for $\Pi^0_1$ classes then there is some $f \in \mathcal{B}$ of PA degree. So, for example, not only does every nonempty $\Pi^0_1$ class have a low member, but there is a *single* low $f$ that computes a member of every $\Pi^0_1$ class.

Part (4) merits a bit of elucidation. One application is as follows. Suppose we are given a computable sequence of pairs of computable sets $\langle\langle A_{0,s}, A_{1,s}\rangle : s \in \omega\rangle$. Then $\varnothing'$ can tell us, for each $s$, whether $A_{0,s}$ is empty, whether $A_{1,s}$ is empty, or whether both are empty. But if we simply wanted to know one of $A_{0,s}$ or $A_{1,s}$ that is empty, assuming at least one is, then (4) says exactly that we do not need $\varnothing'$, as any $f \gg \varnothing$ will do. Or suppose we wanted to know one of $A_{0,s}$ or $A_{1,s}$ that is *infinite*, assuming at least one is. Of course, $\varnothing''$ can answer this for each $s$, but by (4), relativized to $\varnothing'$, so can any $f \gg \varnothing'$. Note that being infinite is a $\Pi^0_2$ property of a computable set, hence by Post's theorem (Theorem 2.6.2), a $\Pi^{0,\varnothing'}_1$ property.

We conclude this section by looking at some properties of $\gg$ as an ordering of $2^\omega$. It is easy to see that $\gg$ is transitive. By relativizing Corollary 2.8.15, we see that $\gg$ is also anti-reflexive and hence anti-symmetric. The following establishes that $\gg$ is dense.

**Proposition 2.8.26 (Density of the PA degrees, Simpson [284]).** *Fix $f, g \in 2^\omega$. If $f \gg g$, there exists $h \in 2^\omega$ such that $f \gg h \gg g$.*

*Proof.* Relativize Proposition 2.8.14 and Theorem 2.8.25 to $g$ to obtain an infinite $g$-computable tree $T \subseteq 2^{<\omega}$ such that if $h$ is any path through $T$ then $h \gg g$. Define $U$ to be the set of all pairs of binary strings $(\sigma, \tau)$ such that $|\sigma| = |\tau|$, $\tau \in T$, and furthermore,

$$(\forall e < |\sigma|)[\Phi^\tau_e(e)\downarrow \to \sigma(e) \neq \Phi^\tau_e(e)]. \tag{2.2}$$

Clearly, $U$ is a $g$-computable set, and if $(\sigma, \tau) \in U$ then also $(\sigma \upharpoonright k, \tau \upharpoonright k) \in T$ for all $k \leqslant |\sigma| = |\tau|$. We may thus regard $U$ as a $g$-computable *tree* in the obvious way. Since $T$ is infinite, it also clear that so is $U$, i.e., that for every $k$ there is a $(\sigma, \tau) \in U$ with $|\sigma| = |\tau| = k$. Thus, as $f \gg g$, $f$ can compute a path through $U$, which is a pair of functions $(h_0, h)$ such that $(h_0 \upharpoonright k, h \upharpoonright k) \in U$ for all $k$. By definition, $h$ is also a path through $T$, so $h \gg g$. Moreover, $h_0$ satisfies that for all $e$, $h_0(e) \neq \Phi^h_e(e)$ if the latter converges, so $h_0$ is DNC relative to $h$ and hence by Theorem 2.8.25, $h_0 \gg h$. Since $h_0 \leqslant_T f$, it follows that $f \gg h \gg g$, as was to be shown. □

One interesting consequence of this is the following, seemingly stronger property of having PA degree. Namely, consider any $f \gg \varnothing$, and fix $g$ so that $f \gg g \gg \varnothing$. Then not only does every nonempty $\Pi^0_1$ class have a member $h \leqslant_T f$, but it has a member $h \leqslant_T g$ and therefore a member $h$ so that $f \gg h$. This is a useful tool in certain model constructions in reverse mathematics that we will encounter in Section 4.6.

## 2.9 Exercises

**Exercise 2.9.1.** Show that the following functions and relations are primitive recursive:

1. $f(s) = 2^s$.
2. $R(s) \equiv$ "$s$ is a power of 3".

3. $M(s,t) \equiv$ "$s$ is a multiple of $t$".
4. $g(s)$ returns the smallest prime factor of $s$.
5. If $R(n)$ is primitive recursive, and so are $f, g : \omega \to \omega$, then the function

$$h(s,t) = \begin{cases} f(s) & R(t) = 0, \\ g(s) & R(t) = 1. \end{cases}$$

is also primitive recursive.

**Exercise 2.9.2.** Show that the following functions and relations are primitive recursive, and then complete the proof of Theorem 2.2.13.

1. $p(i)$ is the $(i+1)$st prime: $p(0) = 2$, $p(1) = 3$, $p(3) = 5$, ....
2. $a(s)$ returns the number of distinct prime factors of $s$.
3. $\pi(s,i)$ as in Definition 2.2.12.

**Exercise 2.9.3.** Show that every infinite c.e. set has an infinite computable subset.

**Exercise 2.9.4.** Prove that a set $A \subseteq \omega$ is computable if and only if $A$ is finite or $A$ can be computably enumerated in strictly increasing order.

**Exercise 2.9.5.** Justify the second sentence of Theorem 2.5.4 by explicitly showing how to construct the function $f$ described, using the effective indexing from the normal form theorem.

**Exercise 2.9.6.** This is a warmup for Exercise 2.9.7. Show that the following sets are not computable:

1. $\{e \in \omega : 0 \in \text{range}(\Phi_e)\}$.
2. $\{e \in \omega : \text{range}(\Phi_e) \text{ is bounded}\}$.

**Exercise 2.9.7 (Rice's theorem, Rice [258]).** Suppose $C$ is a set of partial computable functions and let $I_C = \{n : \Phi_n \in C\}$. If $I_C$ is computable then $C = \varnothing$ or $C$ contains all the partial computable functions.

1. Prove this result using the recursion theorem.
2. Re-prove the result by showing that, if $C$ is not empty and does not contain every partial computable function, then $\varnothing' \leqslant_T I_C$.

**Exercise 2.9.8.** Suppose that $F(s, t_1, \ldots, t_n)$ is a partial computable function. There is an index $e$ for which

$$\Phi_e(t_1, \ldots, t_n) \simeq \lambda t_1, \ldots, t_n \,.\, F(e, t_1, \ldots, t_n).$$

This was the form of the fixed-point theorem originally proved by Kleene [180].

**Exercise 2.9.9.** Show there are indices $m$ and $n$ so that $\varphi_m(0) = n$ and $\varphi_n(0) = m$.

**Exercise 2.9.10.** Prove that there is a partial computable function $f\colon \omega \to \omega$ for which there is no total function $g\colon \omega \to \omega$ that agrees with $f$ whenever $f$ is defined. Colloquially: not every partial computable function can be extended to a (total) computable function. (Hint: Consider $f(e) = (\mu s)[T(s, e, e) = 0]$.)

**Exercise 2.9.11.** Show that if $X, Y \in 2^\omega$ and $X' = Y'$ then $X = Y$.

**Exercise 2.9.12.** In Definition 2.3.1, the operator $M$ looks for the least $y$ such that $f(y, \vec{x}) \downarrow= 0$ and $f(y^*, \vec{x}) \downarrow\neq 0$ for all $y^* < y$. We could define an alternate operator $M'$ such that $M'(f)(\vec{x})$ is simply the least $y$ such that $f(y, \vec{x}) \downarrow= 0$, with no consideration of whether $f(y^*, \vec{x}) \uparrow$ for $y^* < y$. Let $C$ be the class of partial computable functions.

- Let $C'$ be the smallest class of finitary functions which contains the primitive recursive functions and is closed under generalized composition, primitive recursion, and applying $M'$ to functions that are total.
- Let $C''$ be the smallest class of finitary partial functions which contains the primitive recursive functions and is closed generalized composition, primitive recursion, and applying $M'$ to arbitrary functions (partial or total).

Prove that $C = C'$ and $C$ is a proper subset of $C''$.

**Exercise 2.9.13.** Suppose that $T$ is a tree that is definable by $\Pi^0_1$ formula. That is, the set of nodes that are not in the tree is c.e. Show that there is a computable tree $T'$ that has the same set of paths as $T$.

**Exercise 2.9.14.** Let $T$ be the tree constructed in Proposition 2.8.14. Prove that $\mathrm{Ext}(T)$ is not computable.

**Exercise 2.9.15.** Show that for all $k \geqslant 2$ there exists a $\Delta^0_2$ sequence of sets $\langle P_0, \ldots, P_{k-1}\rangle$ whose members partition $\omega$ and for each $i < k$, $\omega \setminus P_i$ is hyperimmune. (A set $P$ such that it and its complement are hyperimmune is said to be *bi-hyperimmune*. With $k = 2$, this results gives the existence of $\Delta^0_2$ bi-hyperimmune sets.)

**Exercise 2.9.16.** A function $f\colon \omega \to \omega$ is *fixed point free* (or FPF) if $(\forall e)[W_{f(e)} \neq W_e]$. Show that every DNC function computes an FPF function, and conversely. (Hint: For the converse, apply the $S^m_n$ theorem to get a computable function $h$ such that $W_{h(e)} = W_{\Phi_e(e)}$ if $\Phi_e(e) \downarrow$, and $W_{h(e)} = \varnothing$ if $\Phi_e(e) \uparrow$. Let $g$ be FPF and consider $g \circ h$.)

# Chapter 3
# Instance–solution problems

As mentioned in the introduction, there is a natural way to translate mathematical theorems into *problems* and vice versa. In this chapter, we investigate this relationship and begin to collect some of its implications for the program of reverse mathematics. As we will see via numerous examples, the translation is not always straightforward, and not always unique. We also study *coding*, or *representations*, of problems in terms of numbers and sets of numbers, which in turn enables us to obtain our first assessments of a problem's computability theoretic strength. The measures of complexity introduced in this connection will be important throughout the sequel.

## 3.1 Problems

We start our discussion with the central definition.

**Definition 3.1.1 (Instance–solution problem).** An *instance–solution problem*, or just *problem*, is a partial function $\mathsf{P}\colon \mathcal{A} \to \mathcal{P}(\mathcal{B})$ for some sets $\mathcal{A}$ and $\mathcal{B}$. The elements of $\mathrm{dom}(\mathsf{P})$ are called the *instances of* $\mathsf{P}$, or $\mathsf{P}$-*instances* and, for each $x \in \mathcal{I}$, the elements of $\mathsf{P}(x)$ are called the *solutions to* $x$ *in* $\mathsf{P}$, or $\mathsf{P}$-*solutions to* $x$.

Following Blass [16, Section 4], it is instructive to think of a problem as a two player game between a "challenger" and a "responder", the former being tasked with playing instances of the problem (the "challenges") and the latter with playing solutions to these in turn (the "responses"). Heuristically, the "hardest" problems are then those that are the most difficult for the "responder" (e.g., a hypothetical problem having an instance with no solutions), and the "easiest" are those that are the most difficult for the "challenger" (e.g., a hypothetical problem having no instances).

Definition 3.1.1 is quite general, allowing for both of these extreme possibilities. That is, it could be that $\mathrm{dom}(\mathsf{P}) = \varnothing$, or that there is an $x \in \mathrm{dom}(\mathsf{P})$ with $\mathsf{P}(x) = \varnothing$. Clearly, these are somewhat unusual cases and, indeed, they only arise in specialized situations. For our purposes, all problems will be assumed to have at least one

 
D. D. Dzhafarov, C. Mummert, *Reverse Mathematics*, Theory and Applications of Computability, https://doi.org/10.1007/978-3-031-11367-3_3

instance, and for each instance, at least one solution. We will nonetheless develop a variety of measures to gauge the "difficulty" of various problems.

For completeness, we make some remarks about notation in the literature. The possible partiality of $\mathsf{P}$ is sometimes expressed by writing $\mathsf{P}\colon \subseteq \mathcal{A} \to \mathcal{P}(\mathcal{B})$. When, as in our case, it is assumed that $\mathsf{P}(x) \neq \emptyset$ for every $x \in \mathrm{dom}(\mathsf{P})$ then $\mathsf{P}$ is regarded by some authors instead as a multivalued function (*multifunction*), written $\mathsf{P}\colon \subseteq \mathcal{A} \rightrightarrows \mathcal{B}$, with the values of $\mathsf{P}(x)$ being the $\mathsf{P}$-solutions to $x$. (This is the practice in computable analysis in the style of Weihrauch [323, 324].) Other authors prefer to regard $\mathsf{P}$ as a binary relation, consisting of pairs $(x, y)$ with $x \in \mathrm{dom}(\mathsf{P})$ and $y \in \mathsf{P}(x)$. (In a different context, this conception of an instance–solution problem was independently proposed by Vojtáš [318]; see Blass [16], Section 4.)

In our case, we will largely suppress all the formalism. We will refer simply to problems, their instances and solutions, without naming the sets $\mathcal{A}$ and $\mathcal{B}$ or committing to how exactly $\mathsf{P}$ is built out of them. The only exception will be when we wish to underscore that the $\mathsf{P}$-instances are a subset of some larger set, so that thinking of $\mathsf{P}$ explicitly as a partial function becomes useful.

*Example 3.1.2.* Let $X$ be any nonempty set. Then $\mathsf{Id}_X$ is the problem whose instances are all $x \in X$, with each such $x$ having itself as a unique solution. As a function, we have $\mathsf{Id}_X : x \mapsto \{x\}$.

*Example 3.1.3.* The $\mathsf{GCD}$ problem is the problem whose instances are all pairs $(x, y)$ of nonzero integers, with each such pair having $\gcd(x, y)$, the greatest common divisor of $x$ and $y$, as its unique solution.

The preceding example illustrate an important aspect for our discussion, which is that problems can be obtained from theorems. (Here, $\mathsf{GCD}$ is a problem form of the result that every pair of nonzero integers has a greatest common divisor.) We discuss this further in Section 3.2. The following is a more interesting example which also underscores an important caveat: the way a problem is obtained from a theorem need not be unique. We will explore this issue in detail in Section 3.3.

*Example 3.1.4.* Consider a partial order $\leqslant_P$ on a set $X$. An *infinite descending sequence* in $\leqslant_P$ is a sequence $\langle x_n : n \in \omega \rangle$ of elements of $X$ such that $x_{n+1} <_P x_n$ for all $n$. The partial order is *well founded* if it has no infinite descending sequence. A well ordering is thus a well founded linear order. A *linear extension* of $\leqslant_P$ is a linear order $\leqslant_L$ on $X$ such that $x \leqslant_L y$ for all $x, y \in X$ with $x \leqslant_P y$. It is not difficult to see that every partial order admits a linear extension. But Bonnet [18] showed that every well founded partial order on $\omega$ has a well founded linear extension.

There are several ways to think of Bonnet's theorem as a problem. The most obvious is to take the instances to be all well founded partial orders on $\omega$, and the solutions to each such instance to be its well founded linear extensions. Alternatively, we may feel it more fitting to regard this is as a partial problem on the set of all partial orders on $\omega$, with domain the subset of partial orders that happen to be well founded. Of course, as noted above, this is really just a notational distinction.

Yet a third possible way to think of this as a problem, that differs from the previous two more significantly, is the following: the instances are all partial orders on $\omega$, and the solutions to a given instance are its infinite descending sequences (in the case that the instance is not well founded), or its well founded linear extensions (in the case that it is).

Problems can be combined with themselves or other problems to produce new problems. We list a few examples.

**Definition 3.1.5 (Parallelization).** Let P and Q be problems.

1. The *parallel product* of P with Q, denoted $\mathsf{P} \times \mathsf{Q}$, is the problem whose instances are all pairs $(X_0, X_1)$ where $X_0$ is a P-instance and $X_1$ is a Q-instance, with the solutions to any such $(X_0, X_1)$ being all pairs $(Y_0, Y_1)$ such that $Y_0$ is a P-solution to $X_0$ and $Y_1$ is a Q-solution to $X_1$.
2. The *parallelization* of P, denoted $\widehat{\mathsf{P}}$, is the problem whose instances are sequences $\langle X_i : i \in \omega \rangle$ where each $X_i$ is a P-instance, with the solutions to any such $\langle X_i : i \in \omega \rangle$ consisting of all sequences $\langle Y_i : i \in \omega \rangle$ such that $Y_i$ a P-solution to $X_i$ for each $i$.

The name "parallelization" comes from the interpretation of $\widehat{\mathsf{P}}$ in which we are trying to solve an infinite collection of instances of P simultaneously, "in parallel". We will see that this can be much more difficult than solving each instance on its own. In a sense, to solve the parallelization of a problem we must be able to solve individual instances "uniformly".

For this chapter and the next, we will use the notions of parallel product and parallelization largely in a technical way, to illustrate various concepts. More natural examples of these operations will appear later on.

We will typically only be interested in problems all of whose instances and solutions are subsets of $\omega$. As we will see, many mathematical objects can be represented by subsets of $\omega$, so this still leaves a very broad range of problems. For example, this is the case for all three of the problems mentioned in the preceding example. We will discuss various such representations in Section 3.4 below.

## 3.2 ∀∃ theorems

As hinted in Chapter 1, the vast majority of problems we will consider come from ∀∃ theorems. Recall that these are problems having the form

$$(\forall x)[\varphi(x) \to (\exists y)\psi(x,y)]. \tag{3.1}$$

There is a canonical way to translate (3.1) into a problem: we take as its instances all $x$ such that $\varphi(x)$ holds, and as the solutions to an instance $x$ are all $y$ such that $\psi(x,y)$ holds.

We can illustrate this by looking first at König's lemma (Theorem 2.8.3), discussed in Section 2.8. Recall that this asserts that every infinite, finitely branching tree has an infinite path. We restate this below to make it more clear that this has the same form as (3.1). We also give an important variant, called *weak König's lemma*, which we will encounter frequently in the sequel.

**Definition 3.2.1 (König's lemma and weak König's lemma).**

1. KL is the following statement: for every infinite, finitely branching tree $T \subseteq \omega^{<\omega}$, there exists an $f \in \omega^\omega$ which is a path through $T$.
2. WKL is the following statement: for every infinite tree $T \subseteq 2^{<\omega}$, there exists an $f \in 2^\omega$ which is a path through $T$.

We can then define the associated problem forms.

**Definition 3.2.2 (König's lemma and weak König's lemma, problem forms).**

1. KL is the problem whose instances are all infinite, finitely branching trees $T \subseteq \omega^{<\omega}$, with the solutions to any such $T$ being all its paths.
2. WKL is the problem whose instances are all infinite trees $T \subseteq 2^{<\omega}$, with the solutions to any such $T$ being all its paths.

We deliberately use the same abbreviation for the $\forall\exists$ theorem and its associated problem form. This cuts down on notation, but also reflects the practice of shifting freely between the two perspectives. As we will see, for the vast majority of examples we encounter it will be convenient to use the theorem and problem forms interchangeably.

For another example, consider the well-known *infinitary pigeonhole principle* asserting that a finite partition of the natural numbers must have an infinite part. Here, we think of a partition of $\omega$ into $k$ parts as a function $f : \omega \to k$, so that two numbers belong to the same part provided they are assigned the same value $i < k$ by $f$. A finite partition is thus more generally a function $f : \omega \to \omega$ with bounded range. So formally, we take the infinitary pigeonhole principle (IPHP) to be the statement: for every $f : \omega \to \omega$ with bounded range, there exists an $i \in \omega$ so that $f^{-1}\{i\} \subseteq \omega$ is infinite. The problem version associated to this principle is then the following.

**Definition 3.2.3 (Infinitary pigeonhole principle).** IPHP is the problem whose instances are functions $f : \omega \to \omega$ with bounded range, with the solutions to any such $f$ being all $i \in \omega$ such that $f^{-1}\{i\} \subseteq \omega$ is infinite.

One point of consideration here is whether an instance of IPHP should instead be a pair, $(k, f)$, where $k \in \omega$ and $f : \omega \to k$, i.e., whether the number of parts of the partition should be given explicitly as part of the instance. At first blush, we may not consider this to make much of a difference—a function $\omega \to \omega$ with bounded range is a function $\omega \to k$ for some $k$, after all. But the issue is that there is no clear way to determine such a $k$ from $f$ alone (see Exercise 3.9.2). Indeed, as we will see in Chapter 4, there is a sense in which the version of IPHP with the number of parts specified is strictly weaker than the one in the definition above. We will explore other ways in which a principle can potentially have multiple interpretations as a problem in the next section.

We conclude this section with one final problem, which will be a prominent example throughout this chapter and the next, as well as the central focus of all of Chapters 8 and 9. This is the *infinitary Ramsey's theorem* (RT), which states, for each fixed $n \geqslant 1$, that if the unordered $n$-tuples of natural numbers are each assigned one of finitely many colors, then there exists an infinite subset of the natural numbers all $n$-tuples of which are assigned the same color. To make this more precise, we make the following definitions.

**Definition 3.2.4.** Fix numbers $n, k \geqslant 1$ and a set $X \subseteq \omega$.

1. $[X]^n = \{F \subseteq X : |F| = n\}$.
2. A *k-coloring of* $[X]^n$ is a map $c\colon [X]^n \to k$.
3. A set $Y \subseteq X$ is *homogeneous* for $c\colon [X]^n \to k$ if $c$ is constant on $[Y]^n$.

When $k$ is fixed or emphasis on it is unnecessary, we can speak simply of *finite colorings* or *colorings*, instead of $k$-colorings. If $Y \subseteq X$ is homogeneous for $c$, and $c$ takes the value $i < k$ on all elements of $[Y]^n$, then we also say $Y$ is *homogeneous for c with color i*.

Given $\{x_0, \dots, x_{n-1}\} \in [\omega]^n$, we usually write $c(x_0, \dots, x_{n-1})$ in place of $c(\{x_0, \dots, x_{n-1}\})$ for brevity. We write $\langle x_0, \dots, x_{n-1}\rangle \in [X]^n$ as shorthand for $\{x_0, \dots, x_{n-1}\} \in [X]^n$ and $x_0 < \cdots < x_{n-1}$. In this way, we tacitly identify $[X]^n$ with the set of *increasing n-tuples* of elements of $X$, and so sometimes also use $\vec{x}, \vec{y}, \dots$ to denote elements of $[X]^n$. If $n > 1$ and we are given $\vec{x} = \langle x_0, \dots, x_{n-2}\rangle \in [X]^{n-1}$ and $y > x_{n-2}$, we may also write $c(\vec{x}, y)$ as shorthand for $c(x_0, \dots, x_{n-2}, y)$.

**Definition 3.2.5 (Infinitary Ramsey's theorem).**

1. For $n, k \geqslant 1$, $\mathsf{RT}^n_k$ is the problem whose instances are all colorings $c\colon [\omega]^n \to k$, with the solutions to any such $c$ being all its infinite homogeneous sets.
2. For $n \geqslant 1$, $\mathsf{RT}^n$ is the problem whose instances are all colorings $c\colon [\omega]^n \to k$ for some $k \geqslant 1$, with the solutions to any such $c$ being all its infinite homogeneous sets.
3. $\mathsf{RT}$ is the problem whose instances are all colorings $c\colon [\omega]^n \to k$ for some $n, k \geqslant 1$, with the solutions to any such $c$ being all its infinite homogeneous sets.

Thus, if we go back to thinking of problems as functions, then $\mathsf{RT}$ is just the union of $\mathsf{RT}^n$ for all $n \geqslant 1$, and $\mathsf{RT}^n$ for each fixed $n$ is the union of $\mathsf{RT}^n_k$ for all $k \geqslant 1$. Each instance of $\mathsf{RT}^n_k$ is an instance of $\mathsf{RT}^n$, which is in turn an instance of $\mathsf{RT}$.

As implied by the "infinitary" adjective, there is also a finitary analogue of RT, which we discuss in Definition 3.3.6 below. Moving forward, however, we will simply say "Ramsey's theorem" in place of "infinitary Ramsey's theorem", and use this to always refer to the problem RT defined above.

As a theorem, Ramsey's theorem is a powerful generalization of the infinitary pigeonhole principle, which is essentially Ramsey's theorem for singletons. However, as problems, $\mathsf{RT}^1$ and $\mathsf{IPHP}$ are quite different. For starters, there is the technical distinction between $[\omega]^1$ and $\omega$ itself, but we can ignore this and thereby regard the two problems as having exactly the same set of instances. The main difference is that a solution to an instance of $\mathsf{RT}^1$ is an infinite set of numbers, whereas a solution to an instance of IPHP is a single number. In particular, there are instances of IPHP having unique solutions, whereas there are no such instances of $\mathsf{RT}^1$. We will soon develop means to formally show that IPHP is a strictly "harder" problem than $\mathsf{RT}^1$, precisely because of this distinction. Yet another version of IPHP appears in Definition 3.3.2 below.

We add that the translation process, from $\forall\exists$ theorem to problem, also works in reverse. To each problem P we can associate the $\forall\exists$ theorem

$$(\forall x)[x \text{ is a P-instance} \to (\exists y)[y \text{ is a P-solution to } x]].$$

This motivates the following convention.

*Convention 3.2.6 (Defining theorems and problems).* In the sequel, when the translations are clear, we will usually define either a problem or the statement of an $\forall\exists$ theorem. Except in cases where more than one translation is possible, we will then use the same initialisms and abbreviations for both.

As in the "forward" translation, we will typically only look at problems whose instances and solutions are "represented" by subsets of $\omega$.

## 3.3 Multiple problem forms

The method of deriving a problem from an $\forall\exists$ theorem described above is very specific to the syntactic form of (3.1). Different but logically equivalent syntactic forms can thereby give rise to different problems. For example, if we rewrite (3.1) as

$$(\forall x)(\exists y)[\neg\varphi(x) \lor \psi(x,y)], \tag{3.2}$$

then the above method results in the problem having as instances all $x$ (in the ambient set), with an instance $x$ having as solutions either all $y$ if $\varphi(x)$ does not hold, or all $y$ such that $\psi(x,y)$ holds if $\varphi(x)$ does hold. We saw this concretely in Example 3.1.4.

Such distinctions are not always purely formal, however. While many theorems we encounter have a canonical presentation in the form of (3.1), for others, multiple forms will be equally natural, and these theorems will thus not admit any single "right" problem version. An interesting example comes from looking at contrapositives. For instance, consider the simple principle that every finite union of finite subsets of $\omega$ is finite. Its problem form is the following.

**Definition 3.3.1 (Finite unions principle).** FUF is the problem whose instances are families $\{F_i : i \in \omega\}$ of finite sets with $F_i = \emptyset$ for almost all $i$, with the solutions to any such collection being all numbers $b > \bigcup_{i\in\omega} F_i$.

Note that the instances above are thus all possible finite collections of finite subsets of $\omega$, again without having to explicitly specify how many sets we are dealing with.

On the other hand, we could easily restate the principle by saying that in any finite collection of subsets of $\omega$ whose union is infinite, at least one of the subsets must be infinite. The problem form is then the following generalized version of IPHP.

**Definition 3.3.2 (Infinitary pigeonhole principle over general sets).** General-IPHP is the problem whose instances are families $\{X_i : i \in \omega\}$ of subsets of $\omega$ whose union is infinite and $X_i = \emptyset$ for almost all $i$, with the solutions to any such collection being all $i$ such that $X_i$ is infinite.

Thus IPHP is the restriction of General-IPHP to the case where all the $X_i$ in the instance are disjoint and $\bigcup_{i\in\omega} X_i = \omega$.

The two principles FUF and General-IPHP are equivalent, of course, as mathematical statements. Not so for the associated problems, which can be discerned not just by their different sets of instances, but also—more tellingly—by the complexity of their solutions. Given a finite collection of finite subsets of $\omega$, determining that a number is an upper bound on the union is $\Pi^0_1$ relative to the collection. By contrast, given a finite collection of subsets of $\omega$ with infinite union, determining one of the sets that is infinite is in general $\Pi^0_2$ relative to the collection. Based on this, we should expect General-IPHP to be "harder" than FUF in a precise sense, and this is indeed the case (Proposition 4.3.5).

By varying a problem, we can also similarly gleam something about the complexity of a problem's instances (as opposed to just its solutions, or the relationship between the two). In Example 3.1.4, we saw two problem versions of Bonnet's theorem, one whose instances were well founded partial orderings and solutions were well founded linear extensions, and one whose instances additionally included ill founded partial orderings, with solutions to these being infinite descending sequences. We can consider a similar "disjunctive version" of WKL.

**Definition 3.3.3 (Weak König's lemma, disjunctive form).** Disj-WKL is the problem whose instances are all trees $T \subseteq 2^{<\omega}$, with the solutions to any such $T$ being either all its infinite paths (if $T$ is infinite) or all $\ell \in \omega$ such that $T$ contains no string of length $\ell$ (if $T$ is finite).

For another example, we can turn to the well-known Heine–Borel theorem.

**Definition 3.3.4 (Heine–Borel theorem for $[0, 1]$).**

1. $\mathsf{HBT}_{[0,1]}$ is the problem whose instances are pairs of sequences $\langle (x_k, y_k) : k \in \omega\rangle$ of real open intervals covering $[0, 1]$, with solutions being all $\ell \in \omega$ such that each $x \in [0, 1]$ belongs to $(x_k, y_k)$ for some $k < \ell$.
2. $\mathsf{Disj\text{-}HBT}_{[0,1]}$ is the problem whose instances are pairs of sequences $\langle (x_k, y_k) : k \in \omega\rangle$ of real open intervals, with solutions being all $\ell \in \omega$ such that each $x \in [0, 1]$ belongs to $(x_k, y_k)$ for some $k < \ell$, or all $x \in [0, 1]$ that do not belong to $(x_k, y_k)$ for any $k$.

In all these examples, we thus have a version that places some (or more) of the onus of describing a problem's instances on the "solver", isolating a particular aspect of the instances without which the problem is trivial or uninteresting.

One way to think of the above issues is in terms of constructive mathematics. Without the law of excluded middle, we cannot simply assert that a partial ordering either is or is not well founded, that a collection of intervals does or does not cover the real unit interval, or that a tree is or is not infinite. On this view, if we are unable to produce a well founded linear extension of a given partial order, or a finite subcover of a given set of intervals, or an infinite path through a given tree, then we should construct an explicit witness for why this is the case. The question of how difficult exhibiting such a witness is says something about how "hard" the problem is, and so is of interest in many situations we encounter.

Likewise, it is well-known that contrapositives are not necessarily constructively equivalent. We saw a reflection of this between FUF and General-IPHP, but a more famous example is that of WKL and Brouwer's fan theorem. To state it, we recall that a *bar* is a subset $B$ of $2^{<\omega}$ such that every $X \in 2^{\omega}$ extends some $\sigma \in B$.

**Definition 3.3.5 (Brouwer's fan theorem).** FAN is the problem whose instances are bars, with a solution to any bar being all its finite subsets that are also bars.

See, e.g., Berger, Ishihara, and Schuster [15] for a discussion of the relationship between FAN and WKL in the constructive setting. We will compare WKL and FAN more carefully in Proposition 4.3.8.

Finally, a theorem may admit multiple problem forms simply because its usual statement has multiple interpretations, even from an ordinary mathematical viewpoint. An example of this is the *finitary* Ramsey's theorem. (This should be compared with RT, the infinitary Ramsey's theorem, defined in Definition 3.2.5). The usual way this is stated is as follows.

**Definition 3.3.6 (Finitary Ramsey's theorem).** FRT is the following statement: for all $n, k, m \geqslant 1$ with $m \geqslant n$, there is an $N \in \omega$ such that for every finite set $X$ with $|X| \geqslant N$, every $c\colon [X]^n \to k$ has a homogeneous set $H \subseteq X$ with $|H| \geqslant m$.

The least $N$ as above is called the *Ramsey number* for $n$, $k$, and $m$ (or more properly, the *hypergraph Ramsey number*, if $n > 2$). We will use the notation $R^n_k(m)$ to denote this number, though this is nonstandard. (In the combinatorics literature, it is usually denoted by $R_n(m, \ldots, m)$, with $k$ many $m$ terms.) What problem should we associate to FRT? Here it seems we have (at least) two possibilities.

- The *problem of bounding Ramsey numbers* is the problem whose instances are all triples of integers $\langle n, k, m\rangle$ with $m \geqslant n$, with the solutions to any such triple being all $N \geqslant R^n_k(m)$.
- The *problem of finding Ramsey numbers* is the problem whose instances all triples of integers $\langle n, k, m\rangle$ with $m \geqslant n$, with the solution to any such triple being $R^n_k(m)$.

And indeed, both of these are problems associated to the finitary Ramsey's theorem that mathematicians work on. (As an aside, they are also vastly different in difficulty: computing Ramsey numbers explicitly is much harder than bounding them. A notorious case in point is the problem of computing $R^2_2(6)$. This is still open, and has been since the 1950s, though it is known that $R_2^2(6) \leqslant 165$. For more about Ramsey numbers, see, e.g., the textbook of Graham, Rothschild, and Spencer [128].) We could also consider a third problem form, closer to that of RT, involving actually *finding* homogeneous sets for colorings defined on sufficiently large finite sets.

## 3.4 Represented spaces

While the general notion of a problem does not require that the instances and solutions belong to any particular set, our interest in analyzing problems from the point of view of computability theory requires these to be objects to which computability theoretic notions can be applied, which is to say, ultimately, subsets of $\omega$. This is facilitated by the notion of representation.

**Definition 3.4.1.** A *representation* of a set $\mathcal{X}$ is a partial surjection $\delta\colon 2^\omega \to \mathcal{X}$. The pair $(\mathcal{X}, \delta)$ is called a *represented space*.

The idea is that the elements of $\mathcal{X}$ are coded by subsets of $\omega$: if $\delta(X) = x$ for some $X \subseteq \omega$ and $x \in \mathcal{X}$ then $X$ is a *code* for $x$. Here, we will identify $n \in \omega$ with $\{n\} \subseteq \omega$, and thereby allow for codes to also be natural numbers (instead of only sets of natural numbers). Note that $\delta$ above need not be injective, so an element of $\mathcal{X}$ may have multiple codes under a given representation.

For example, the elements of each of $\mathbb{Z}$ and $\mathbb{Q}$ can be coded by natural numbers using the pairing function. Similarly for $\omega^{<\omega}$, using codes for sequences as in Definition 2.2.12. A different but equally straightforward representation is the following.

*Example 3.4.2.* Consider the set $\omega^\omega$ of all functions $f\colon \omega \to \omega$, which is clearly not a set of subsets of $\omega$. Let $\delta$ be the following map. The domain of $\delta$ consists of all $X \subseteq \omega$ such that for each $n \in \omega$ there is exactly one $m \in \omega$ with $\langle n, m\rangle \in X$. Given such an $X$, let $\delta(X)$ be the set of all ordered pairs of natural numbers $(n, m)$ such that $\langle n, m\rangle \in X$, noting that $\delta(X)$ is a function $\omega \to \omega$. Then $\delta$ is a representation of $\omega^\omega$.

Representations can be easily combined to produce new represented spaces from old ones. Recall that if we have a finite collection $X_0, \ldots, X_{n-1}$ of subsets of $\omega$ then $\langle X_0, \ldots, X_{n-1}\rangle$ denotes the join of these sets, i.e., $\{\langle x, i\rangle : i < n\}$, which is again a subset of $\omega$. And if instead we have an infinite collection $X_0, X_1, \ldots$, then $\langle X_i : i \in \omega\rangle$ denotes $\{\langle x, i\rangle : i \in \omega\}$.

**Definition 3.4.3.** Let $(\mathcal{X}_0, \delta_0)$ and $(\mathcal{X}_1, \delta_1)$ be represented spaces, and let $\delta\colon 2^\omega \to \mathcal{X}_0 \times \mathcal{X}_1$ be the partial map whose domain consists of all $\langle X_0, X_1\rangle$ with $X_0 \in \operatorname{dom}(\delta_0)$ and $X_1 \in \operatorname{dom}(\delta_1)$, and such that

$$\delta(\langle X_0, X_1\rangle) = (\delta_0(X_0), \delta_1(X_1)).$$

It is evident that $\delta$ above is a representation of the set $\mathcal{X}_0 \times \mathcal{X}_1$. In particular, if $(\mathcal{X}, \delta)$ is a represented space then this yields a natural representation of $\mathcal{X}^n$ for each fixed $n \in \omega$. We can also obtain representations of other kinds of self-products, as follows.

**Definition 3.4.4.** Let $(\mathcal{X}, \delta)$ be a represented space.

1. Let $\delta_{<\omega}\colon 2^\omega \to \mathcal{X}^{<\omega}$ be the partial map whose domain consists of all $\langle X_0, \ldots, X_{n-1}\rangle$, where $n \in \omega$, $X_i \in \mathrm{dom}(\delta)$ for all $i < n$, and
$$\delta_{<\omega}(\langle X_0, \ldots, X_{n-1}\rangle) = (\delta(X_0), \ldots, \delta(X_{n-1})).$$
2. Let $\delta_\omega\colon 2^\omega \to \mathcal{X}^\omega$ be the partial map whose domain consists of all $\langle X_i : i \in \omega\rangle$, where $X_i \in \mathrm{dom}(\delta)$ for all $i \in \omega$ and
$$\delta_\omega(\langle X_i : i \in \omega\rangle) = (\delta(X_i) : i \in \omega).$$

Here, we clearly have that $\delta_{<\omega}$ and $\delta_\omega$ are representations of $\mathcal{X}^{<\omega}$ and $\mathcal{X}^\omega$, respectively.

One takeaway of the above definitions for us is that some basic sets that ought to have representations indeed, do. Thus we can, for example, speak of pairs of functions $\omega \to \omega$, or tuples or finite or infinite sequences of the same, and formally this will refer to an element of one of the represented spaces per the preceding definition.

Representations are not unique, and a space may even admit multiple equally natural representations. For example, while each rational number $q$ can be represented by a number $\langle n, m\rangle \in \omega$ with $m \neq 0$ and $n/m = q$ (which defines a noninjective representation of $\mathbb{Q}$), the same rational could also be coded by the set of all such numbers $\langle n, m\rangle$ (and this would define an injective representation instead). In this specific instance, the choice is inconsequential for our purposes: from a code of the first kind we can uniformly compute the code of the second, and given a code of the latter kind, any element of this code (which is a set) serves as a code of the first kind.

In other situations, we cannot so easily move between different representations, so an explicit choice is required. A case in point is representing the real numbers and functions defined on the real numbers, which we discuss in the next section and then in far more depth in Chapter 10.

Still, once a representation has been established, we usually largely suppress it for ease of terminology. More precisely, we abide by the following.

*Convention 3.4.5.* When no confusion can arise, we identify elements of a represented space with their codes and forego mentioning the representation explicitly. We thus formally work with the domain of a representation rather than the represented set itself, and pretend that the two are one and the same for the purposes of stating definitions, theorems, and problems.

So, moving forward, we will use familiar notations like $\mathbb{Q}^{<\omega}$ or $\mathbb{Z} \times \omega^\omega$ and move freely between thinking of these as the original sets and as the represented spaces. Similarly, when talking about tuples of sequences of elements (from some represented space), we may use $(\cdots)$ and $\langle\cdots\rangle$ interchangeably, always formally referring to the latter. And when we say, e.g., that a subset of $\mathbb{Q}^{<\omega}$ is computable, or that an element of $\mathbb{Z} \times \omega^\omega$ computes $\varnothing'$, it should always be understood that we are actually referring to these objects' codes.

## 3.5 Representing $\mathbb{R}$

There are many possible representations of $\mathbb{R}$: any surjective map from $2^\omega$ will do. However, an arbitrary such map will not necessarily preserve any of the analytic properties of the reals, which will make working with this representation difficult. We consider this issue in greater detail in Section 10.1. For now, we fix the following.

**Definition 3.5.1 (Representation of $\mathbb{R}$).**

1. A *code for a real number* is a sequence $\langle q_n : n \in \omega \rangle$ of rational numbers such that $|q_n - q_{n+1}| < 2^{-n}$ for all $n$.
2. The map $\delta_\mathbb{R}$ has domain the set of all codes of real numbers, and if $\langle q_n : n \in \omega \rangle$ is any such code then $\delta_\mathbb{R}(\langle q_n : n \in \omega \rangle) = \lim_{n\to\infty} q_n \in \mathbb{R}$, in which case we also say that $\langle q_n : n \in \omega \rangle$ is a *code for* the real number $\lim_{n\to\infty} q_n$.

Using Cauchy sequences here is perhaps expected, but the bound of $2^{-n}$ may be less so. We could demand instead that $|q_n - q_m| < f(n)$ for all $n$ and $m > n$, where $f : \omega \to \mathbb{Q}$ is any nonincreasing computable function with $\lim_{n\to\infty} f(n) = 0$, and get basically the same representation (see Exercise 10.9.1). The real advantage here is that this definition is both structurally and computationally well-behaved, as we can see from the following definitions.

**Definition 3.5.2.** Let $x = \langle q_n : n \in \omega \rangle$ and $y = \langle r_n : n \in \omega \rangle$ be codes for real numbers.

1. $-x$ is the code $\langle -q_n : n \in \omega \rangle$.
2. $x + y$ is the code $\langle q_{n+1} + r_{n+1} : n \in \omega \rangle$.
3. $x < y$ if there exist $m, n_0 \in \omega$ such that $q_n + 2^{-m} < r_n$ for all $n > n_0$.
4. $x = y$ if for every $m \in \omega$ there exists $n_0 \in \omega$ such that $|q_n - r_n| < 2^{-m}$ for all $n > n_0$, and $x \neq y$ otherwise.

Now, from codes $x$ and $y$ (as oracles), we can uniformly compute the codes $-x$ and $x + y$. If $x \neq y$, we can also uniformly computably determine whether $x < y$ or $y < x$ (see Exercise 3.9.6). However, as is well-known from constructive mathematics, there is no computable procedure to tell whether or not $x = y$ (Exercise 3.9.7).

Since Definition 3.5.3 permits us to speak of sequences of codes of real numbers, we can also define the following.

**Definition 3.5.3.** Let $\langle x_k : k \in \omega \rangle$ be a sequence of codes of real numbers, with $x_k = \langle q_{k,n} : n \in \omega \rangle$ for each $k$.

1. The sequence $\langle x_k : k \in \omega \rangle$ is *bounded* if there exist codes for real numbers $y$ and $z$ such that $y < x_k < z$ for all $k$.

2. The sequence $\langle x_k : k \in \omega\rangle$ *converges* to a code for a real number $x = \langle q_n : n \in \omega\rangle$ if for each $m \in \omega$ there is a $k_0 \in \omega$ such that for each $k > k_0$ there is an $n_0 \in \omega$ with $|q_{k,n} - q_n| < 2^{-m}$ for all $n > n_0$.
3. A *subsequence* of $\langle x_k : k \in \omega\rangle$ is a sequence of codes for real numbers $\langle y_j : j \in \omega\rangle$ such that there is an infinite set $I = \{k_0 < k_1 < \cdots\} \subseteq \omega$ with $y_j = x_{k_j}$ for all $j$.

In a similar fashion, we can easily transfer a host of other familiar definitions from the reals to codes for reals, such as other arithmetical operations, the distance between two real numbers, etc.

Notably, all of these notions behave as the original notions do when passed through $\delta_{\mathbb{R}}$. For example, $x = y$ precisely when $\delta_{\mathbb{R}}(x) = \delta_{\mathbb{R}}(y)$; $\delta_{\mathbb{R}}(x + y) = \delta_{\mathbb{R}}(x) + \delta_{\mathbb{R}}(y)$; $\langle x_k : k \in \omega\rangle$ converges to $x$ precisely when $(\delta_{\mathbb{R}}(x_k) : k \in \omega)$ converges to $\delta_{\mathbb{R}}(x)$; and so on. We therefore need not dwell on the formal distinctions between the actual notions for the reals and the corresponding notions for the codes, and following Convention 3.4.5, will usually speak simply of "the reals" instead of "codes for reals", when convenient. Consider the following example:

**Definition 3.5.4 (Bolzano–Weierstrass theorem).** BW is the problem whose instances are bounded sequences of real numbers, with the solutions to any such sequence being all its convergent subsequences.

As it is appears here, this is the direct translation of the Bolzano–Weierstrass theorem into the parlance of instance–solution problems, with no heed for representations or similar concerns. But having now fixed a representation for $\mathbb{R}$, we may regard it instead as a statement about codes for reals, and the various properties and operations from Definitions 3.5.2 and 3.5.3 plugged into their respective locations.

Notice that unlike with functions on countable sets, cardinality considerations mean we cannot hope to represent all functions $\mathbb{R} \to \mathbb{R}$. However, we can hope to represent continuum-sized subclasses, of which the continuous functions are an especially important example. We will develop such a representation in Chapter 10, along with a generalization to other separable metric spaces. For now, we define a representation of continuous real-valued functions on the real closed unit interval, which the more general definition will extend.

**Definition 3.5.5.** A *code for a continuous real-valued function on* $[0, 1]$ is a uniformly continuous function $f : [0, 1] \cap \mathbb{Q} \to \mathbb{R}$.

Here, we are availing ourselves of the fact that every continuous real-valued function on $[0, 1]$ is the unique continuation of some code as above. If $x = \langle q_n : n \in \omega\rangle$ is a code for a real number with $0 \leqslant x \leqslant 1$, then there is an $n_0$ such that $0 \leqslant q_n \leqslant 1$ for all $n > n_0$, and we write $f(x)$ for $\lim_{n > n_0, n \to \infty} f(q_n)$ (which exists by uniform continuity of $f$). Note that real-valued functions on $\mathbb{Q}$ are just particular countable subsets of $\mathbb{Q} \times \mathbb{R}$, so the codes in Definition 3.5.5 can themselves by represented using Definitions 3.5.1 and 3.5.3.

With this in hand, we can formulate, e.g., the intermediate value theorem as a problem whose instances and solutions are (coded by) subsets of $\omega$.

**Definition 3.5.6 (Intermediate value theorem).** IVT is the problem whose instances are continuous functions $f\colon [0,1] \to \mathbb{R}$ with $f(0) < 0$ and $f(1) > 0$, with the solutions to any such $f$ being all real numbers $x$ such that $0 < x < 1$ and $f(x) = 0$.

## 3.6 Complexity

We now restrict to problems whose instances and solutions are subsets of $\omega$ in an attempt to measure the "difficulty" of solving such problems from the point of view of our computability theoretic investigation. Since solving a problem is the task of taking a given instance and producing a solution, our aim is essentially to understand the computational resources necessary to carry out this process. We begin with the "easiest" problems in this sense.

**Definition 3.6.1.** A problem P *admits computable solutions* if every instance $X$ of P has a solution $Y \leqslant_T X$.

It is worth emphasizing that the instances and solutions of a problem that admits computable solutions need not themselves be computable. Indeed, as sets of numbers they may be arbitrarily complicated. Rather, it is the relationship between the two that is at issue: more precisely, the fact that it requires no additional computational power to go from any instance to at least one solution to it. It might thus be more honest to say that "P admits solutions computable in its instances", but that would be rather unwieldy.

Many problems admit computable solutions simply because each instance (computable or not) has a computable solution. In particular, problems whose solutions are natural numbers, like IPHP or FUF or $\mathsf{HBT}_{[0,1]}$, admit computable solutions. For an example not of this sort, consider $\mathsf{RT}^1_2$: given a partition $c\colon \omega \to 2$, of arbitrary complexity, each of the sets $\{x : c(x) = 0\}$ and $\{x : c(x) = 1\}$ is computable from $c$, and at least one of the two is infinite set and so is a solution to $c$. This is one situation where the parallelization can be useful: although $\mathsf{RT}^1_2$ admits computable solutions, $\widehat{\mathsf{RT}^1_2}$ does not (Exercise 3.9.4).

It is also worth stressing that even problems with numerical solutions may not be objectively "easy" in any practical sense. A good example of this is the problem of finding Ramsey numbers, discussed following the statement of FRT in Definition 3.3.5.

In essentially all problems we encounter, including all those naturally derived from $\forall\exists$ theorems, the relationship between instances and solutions follows Maxim 2.7.1. This means that such a problem satisfies a computability theoretic property concerning its computable instances and their solutions if and only if, for every $A \subseteq \omega$, the problem satisfies the relativized property concerning its $A$-computable instances and their solutions. Thus, to describe more complex types of problems, we can look only at computable instances and then generalize by relativization.

For example, we expect that a problem whose computable instances all have at least one computable solution will actually satisfy that for every $A \subseteq \omega$, every

$A$-computable instance $X$ has a solution $Y \leqslant_T A$. This is equivalent to the problem admitting computable solutions. By the same token, a problem having at least one computable instance with no computable solution should satisfy that for every $A \subseteq \omega$, there is at least one $A$-computable instance $X$ having no solution $Y \leqslant_T A$. This is stronger than merely not admitting computable solutions. And since the instances of the problem may not be closed upward under $\leqslant_T$, it is also stronger than saying that for every $A$ there is an $A$-computable instance $X$ with no solution $Y \leqslant_T X$.

More generally, we make the following abstract definition.

**Definition 3.6.2.** Let $\mathcal{K} = \{\mathcal{K}(A) : A \in 2^\omega\}$ be a class of subsets of $2^\omega$. Let $\mathsf{P}$ be a problem.

1. $\mathsf{P}$ *admits solutions in* $\mathcal{K}$ if for every $A \subseteq \omega$, every $A$-computable instance $X$ of $\mathsf{P}$ has a solution $Y \in \mathcal{K}(A)$.
2. $\mathsf{P}$ *omits solutions in* $\mathcal{K}$ if for every $A \subseteq \omega$, there is an $A$-computable instance $X$ of $\mathsf{P}$ having no solution $Y \in \mathcal{K}(A)$.

Our preceding example thus works with $\mathcal{K}(A) = \{Y : Y \leqslant_T A\}$ for every $A$. For brevity, we can say $\mathsf{P}$ admits or omits *low solutions* if (1) or (2) above hold with $\mathcal{K}(A) = \{X : (A \oplus Y)' \leqslant_T A'\}$ for every $A$. Analogously, we can define what it should mean for a problem to admit or omit *low*$_2$ *solutions*, or $\Delta^0_2$ *solutions*, or *arithmetical solutions*, or solutions in any other natural computability theoretic class. Again, we should really perhaps say that "$P$ admits solutions that are low over its instances", "$P$ admits solutions that are arithmetical in its instances", etc., but we prefer the shorter nomenclature for brevity. In all these examples, we are thinking of $\mathcal{K}$ as (the class of sets satisfying) some relativizable property, and $\mathcal{K}(A)$ as (the class of sets satisfying) the relativization to the specific set $A$.

To be sure, it is possible for a problem to neither admit nor omit solutions in a class $\mathcal{K}$. In practice, this never happens outside of certain specific, and usually deliberately constructed, situations (see Exercise 3.9.3). For this reason, it is common simply to say that a problem *does not admit* solutions of a particular kind as a synonym for what we are calling here omitting. But technically, these are different notions.

The following observation is obvious:

**Proposition 3.6.3.** *Let* $\mathcal{K} = \{\mathcal{K}(A) : A \subseteq \omega\}$ *and* $\mathcal{K}' = \{\mathcal{K}'(A) : A \subseteq \omega\}$ *be classes of subsets of* $2^\omega$ *with* $\mathcal{K}(A) \subseteq \mathcal{K}'(A)$ *for all A. Let* $\mathsf{P}$ *be a problem. If* $\mathsf{P}$ *admits solutions in* $\mathcal{K}$ *then it admits solutions in* $\mathcal{K}'$*. If* $\mathsf{P}$ *omits solutions in* $\mathcal{K}'$ *then it omits solutions in* $\mathcal{K}$.

Typically, the classes $\mathcal{K}$ we consider in this context have $\mathcal{K}(A)$ closed under $\leqslant_T$ for every $A$. This way, admitting a solution in the given class can be regarded as an upper bound on the "difficulty" of the problem, and omitting solutions in it as a lower bound. So for example, if we know that a problem admits low solutions but not computable solutions, we may view the computability theoretic relationship between its instances and solutions as lying somewhere between the two classes. And we regard such a problem as "harder" than one that admits computable solutions, but "easier" than one that does not admit $\Delta^0_2$ solutions, and so on. This is our first means of comparing the relative complexities of (certain) problems.

We move to some specific examples. Three commonly encountered and related measures of complexity are the following, which use the notion of a set $D$ having PA degree relative to a set $A$, denoted $D \gg A$, as defined in Definition 2.8.24.

**Definition 3.6.4.** Let $\mathsf{P}$ be a problem.

1. $\mathsf{P}$ *admits solutions in PA* if for every $A \subseteq \omega$ and every set $D \gg A$, every $A$-computable instance $X$ of $\mathsf{P}$ has a solution $Y$ such that $A \oplus Y \leqslant_T D$.
2. $\mathsf{P}$ *admits PA avoidance* if for every $A \subseteq \omega$ and every $C \not\gg A$, every $A$-computable instance $X$ of $\mathsf{P}$ has a solution $Y$ such that $C \not\gg A \oplus Y$.
3. $\mathsf{P}$ *codes PA* if for every $A \subseteq \omega$, there is an $A$-computable instance $X$ of $\mathsf{P}$ with $A \oplus Y \gg A$ for every solution $Y$ to $X$.

We note for completeness that these can all be formulated in terms of Definition 3.6.2. (Doing so does not really add any insight to these notions, but we mention it to highlight that these classes fit the general framework.) However, for (1) and (2) we actually need to use *families* of classes of subsets of $2^\omega$. For example, for (1), first define $\mathcal{K}_f(A) = \{Y : Y \leqslant_T f(A)\}$ for every function $f : 2^\omega \to 2^\omega$ and set $A$. Then $\mathsf{P}$ admits solutions in PA precisely if it admits solutions in $\mathcal{K}_f = \{\mathcal{K}_f(A) : A \in 2^\omega\}$ for every $f$ satisfying $f(X) \gg X$ for every $X$. For (3), a simpler definition suffices: for each $A$ let $\mathcal{K}(A) = \{Y : A \oplus Y \not\gg A\}$. Then $\mathsf{P}$ codes PA precisely if it omits solutions in $\mathcal{K} = \{\mathcal{K}(A) : A \in 2^\omega\}$.

The most obvious problem to look at in connection with these notions is $\mathsf{WKL}$.

**Proposition 3.6.5.** $\mathsf{WKL}$ *admits solution in PA and codes PA.*

*Proof.* That $\mathsf{WKL}$ admits solutions in PA follows directly from the definition. That $\mathsf{WKL}$ codes PA follows from relativizing Proposition 2.8.14, that there exists an infinite computable tree each of whose paths has PA degree. □

Admitting solutions in PA is again an "upper bound" style result, while coding PA is a "lower bound" style result. So the preceding proposition can be seen as saying that the PA degrees precisely capture the complexity of the problem $\mathsf{WKL}$. This is what we should expect. The following is an easy corollary.

**Proposition 3.6.6.** *Let* $\mathsf{P}$ *be a problem and let* $\mathcal{K}$ *be any relativizable class that forms a basis for* $\Pi^0_1$ *classes. If* $\mathsf{P}$ *admits solutions in PA then* $\mathsf{P}$ *admits solutions in* $\mathcal{K}$.

Thus, admitting solutions in PA entails admitting low solutions as well as hyperimmune free solutions, etc. (In particular, $\mathsf{WKL}$ admits low solutions as well as hyperimmune free solutions, so this terminology reflects what we should expect.) As we will see, proving that a given problem admits solutions in PA often involves "reducing" the problem to $\mathsf{WKL}$ in an appropriate sense. This will be made precise in Proposition 4.2.5.

The next definition gives two additional notions of complexity of a problem.

**Definition 3.6.7.** Let P be a problem.

1. P *admits cone avoidance* if for every $A \subseteq \omega$ and every $C \nleqslant_T A$, every $A$-computable instance $X$ of P has a solution $Y$ such that $C \nleqslant_T A \oplus Y$.
2. P *codes the jump* if for every $A \subseteq \omega$, there is an $A$-computable instance $X$ of P with $A' \leqslant_T A \oplus Y$ for every solution $Y$ to $X$.

In part (1), *cone* refers to the set $\{Z : C \leqslant_T Z\}$—"the cone above $C$"—and cone avoidance to being able to stay outside of this set. In applications, we often contrast this with not being able to avoid one specific cone, namely that of the jump, which is how we obtain (2).

The parallel between cone avoidance and PA avoidance, and between coding the jump and coding PA, should be clear. We add the following more explicit connection.

**Proposition 3.6.8.** *If* P *admits solutions in PA then it admits cone avoidance. If* P *codes the jump then it codes PA.*

*Proof.* By the cone avoidance basis theorem (Theorem 2.8.23), for every noncomputable set $C$ there is a set $D \gg \emptyset$ such that $C \nleqslant_T D$. Now relativize to arbitrary sets $A \subseteq \omega$. This gives the first part. For the second, recall that $A' \gg A$ for every $A$. □

The canonical example of a problem coding the jump here is the following.

**Definition 3.6.9 (Existence of the Turing jump).** TJ is the problem whose instances are all subsets of $\omega$, with the unique solution to any such set being its Turing jump.

Clearly, TJ codes the jump. This is a somewhat technical principle that we will use mostly as a means of showing other problems code or do not code the jump. A more intrinsically interesting example is König's lemma.

**Proposition 3.6.10.** KL *codes the jump.*

*Proof.* We define a computable finitely-branching tree $T \subseteq \omega^{<\omega}$ with a unique path computing $\emptyset'$. The full result then follows by relativization. For ease of presentation, we assume $\Phi_e(e)[0] \uparrow$ for all $e$. Let $T$ consist of all $\alpha \in \omega^{<\omega}$ satisfying the following conditions for all $e < |\alpha|$:

1. if $\alpha(e) = 0$ then $\Phi_e(e)[|\alpha|] \uparrow$,
2. if $\alpha(e) = s > 0$ then $\Phi_e(e)[s] \downarrow$ and $\Phi_e(e)[s-1] \uparrow$.

Checking whether a given $\alpha$ satisfies these conditions is uniformly computable in $\alpha$, so $T$ is computable. It is readily seen that each $\alpha \in T$ has at most two immediate successors. And it is clear that $T$ contains a single path $f \in \omega^\omega$, with $f(e) = 0$ if $\Phi_e(e) \uparrow$, and $f(e) \neq 0$ if $\Phi_e(e) \downarrow$ and $f(e)$ is the least $s$ such that $\Phi_e(e)[s] \downarrow$. Hence, $f$ clearly computes $\emptyset'$ (in fact, $f \equiv_T \emptyset'$). □

In light of Propositions 3.6.5 and 3.6.8, we thus have a computability theoretically detectable distinction between weak König's lemma and König's lemma (even for 2-branching trees). As seen in the above proof, this stems from the fact that instances of the latter are not constrained to $2^{<\omega}$, and hence a string $\sigma$ can have immediate successors that are arbitrarily far apart.

We conclude this section by mentioning two additions to the framework of Definition 3.6.2. The first is a specialization of the definition to the case where $\mathcal{K}$ represents a notion of computational weakness, meaning a property of sets closed downward under $\leqslant_T$. The following definition was originally articulated by Wang [322] and Patey [243].

**Definition 3.6.11 (Preservation of properties).** Let $C$ be a class of sets closed downward under $\leqslant_T$. A problem $\mathsf{P}$ *admits preservation of* $C$ if for every $A \in C$, every $A$-computable $\mathsf{P}$-instance $X$ has a solution $Y$ with $A \oplus Y \in C$.

We can relate this to our earlier framework as follows. Given $C$, define the class $\mathcal{K} = \{\mathcal{K}(A) : A \in 2^\omega\}$, where $\mathcal{K}(A) = \{Y : A \oplus Y \in C\}$ if $A \in C$ and $\mathcal{K}(A) = 2^\omega$ otherwise. Then $\mathsf{P}$ admits preservation of $C$ if and only if it admits solutions in $\mathcal{K}$. And if $\mathsf{P}$ omits solutions in $\mathcal{K}$ then it does not admit preservation of $C$. The converse of this may not hold in general, but it is what we see in natural cases (as we should expect by Maxim 2.7.1). We give one important example.

**Proposition 3.6.12.** *Fix $D \gg \emptyset$ and let $C = \{A \in 2^\omega : D \gg A\}$. Then* $\mathsf{WKL}$ *admits preservation of $C$.*

*Proof.* Clearly $C$ is closed downward under $\leqslant_T$. Fix $A \in C$ and let $T \leqslant_T A$ be an instance of $\mathsf{WKL}$. Thus $T$ is an infinite subtree of $2^\omega$. By Proposition 2.8.26, there is a $D^*$ such that $D \gg D^* \gg A$. By definition of $\gg$, there is an $f \in [T]$ with $f \leqslant_T D^*$. Hence also $A \oplus f \leqslant_T D^*$ and so $A \oplus f \ll D$. Thus $A \oplus f \in C$, and since $f$ is a $\mathsf{WKL}$-solution to $T$, the proof is complete. □

The second addition we make to Definition 3.6.2 is a technical variation, employed heavily in shading out subtle distinctions between mathematical theorems.

**Definition 3.6.13 (Strong solutions).** Let $\mathcal{K} = \{\mathcal{K}(A) : A \in 2^\omega\}$ be a class of subsets of $2^\omega$. A problem $\mathsf{P}$ *admits strong solutions in* $\mathcal{K}$ if for every $A \subseteq \omega$, every instance $X$ of $\mathsf{P}$ has a solution $Y \in \mathcal{K}(A)$.

The emphasis is thus on the fact that the instance $X$ need *not* be computable from the set $A$. So, for example, $\mathsf{P}$ *admits strong PA avoidance* if for every set $A$, every instance $X$ (computable from $A$ or not) has a solution $Y$ such that $A \oplus Y \not\gg A$; $\mathsf{P}$ *admits strong cone avoidance* if for every set $A$ and every $C \nleqslant_T A$, every instance $X$ has a solution $Y$ such that $C \nleqslant_T A \oplus Y$.

It may be hard to believe that, outside of artificial examples, a problem could admit strong solutions (at least, in any interesting class). In fact, this is the case. We will revisit this notion, and see a number of natural examples and their applications, in Section 4.4 and Chapters 8 and 9.

## 3.7 Uniformity

In this section, we consider the following stronger form of admitting computable solutions.

**Definition 3.7.1.** A problem P *uniformly admits computable solutions* if there is a Turing functional $\Phi$ so that $\Phi(X)$ is a solution to every instance $X$ of P.

We could also formulate uniform versions of many of the other complexity measures described in the previous section, though we will not do this in lieu of a more nuanced approach that we develop in Section 4.3.

As we have seen, issues of uniformity often crop up in our discussion. Often, when constructing a solution computably from a given instance of some problem, we nonetheless break into cases that cannot be distinguished effectively. A basic example is $\mathsf{IPHP}$.

**Proposition 3.7.2.** $\mathsf{IPHP}$ *does not uniformly admit computable solutions.*

*Proof.* Seeking a contradiction, suppose $\mathsf{IPHP}$ uniformly admits computable solutions, with witness $\Phi$ as in Definition 3.7.1. Let $c_0 \colon \omega \to 2$ be defined by $c_0(x) = 0$ for all $x$. Then $c_0$ is an instance of $\mathsf{IPHP}$, and we must have $\Phi(c_0) = \{0\}$. Let $\sigma \in 2^{\omega}$ be a long enough initial segment of $c_0$ so that $\Phi(\sigma)(0) \downarrow= 1$. Then $\Phi(c)(0) \downarrow= 1$ for every $c \colon \omega \to 2$ that extends $\sigma$. Define $c \colon \omega \to 2$ by $c(x) = \sigma(x) = 0$ for $x < |\sigma|$ and $c(x) = 1$ for $x \geqslant |\sigma|$. Then $\Phi(c)(0) \downarrow= 1$, even though we should have $\Phi(c) = \{1\}$. □

Intuitively, to solve an instance $c \colon \omega \to 2$ of $\mathsf{IPHP}$ we break into cases: there are infinitely many $x$ such that $c(x) = 0$, there are infinitely many $x$ such that $c(x) = 1$. If exactly one of these cases is true, we cannot computably determine which. The above proposition basically says that this is intrinsic to the problem and cannot be eliminated (e.g., by a more clever argument).

Nonuniform reasoning can be rather subtle as well. We recall $\mathsf{IVT}$, the problem form of the intermediate value theorem, from Section 3.5.

**Proposition 3.7.3.** *The problem* $\mathsf{IVT}$ *admits computable solutions.*

*Proof.* Let $f$ be an instance of $\mathsf{IVT}$. If there is a $q \in \mathbb{Q}$ such that $f(q) = 0$ then we are done, so assume otherwise. We define two sequences of rational numbers, $\langle q_n : n \in \omega \rangle$ and $\langle r_n : n \in \omega \rangle$, as follows. Let $q_0 = 0$ and $r_0 = 1$. Given $q_n$ and $r_n$ for some $n \in \omega$, form their mean, $m_n = (q_n + r_n)/2$. Since $m_n \in \mathbb{Q}$, it follows by our assumption that either $f(m_n) < 0$ or $f(m_n) > 0$. In the first case, let $q_{n+1} = m_n$ and $r_{n+1} = r_n$, and in the second, let $q_{n+1} = q_n$ and $r_{n+1} = m_n$. By induction, it is easily verified that for every $n \in \omega$, we have $q_n < r_n$, $f(q_n) < 0$ and $f(r_n) > 0$, and each of the quantities $|q_{n+1} - q_n|$, $|r_{n+1} - r_n|$, and $|q_{n+1} - r_{n+1}|$ is smaller than $2^{-n}$. It follows that $\langle q_n : n \in \omega \rangle$ and $\langle r_n : n \in \omega \rangle$ are codes for one and the same real number, $x$, and by continuity $f(x) = 0$. It remains only to verify that (the code) $x$ is computable from $f$. By the remark following Definition 3.5.2, whether $f(m_n) < 0$ or $f(m_n) > 0$ above is uniformly computable in $n$. Hence, the sequences $\langle q_n : n \in \omega \rangle$ and $\langle r_n : n \in \omega \rangle$ are both computable, and so is $x$. □

Of course, the nonuniformity above is where we assume $f$ has no rational zeroes. Note that past this moment, our construction is entirely uniform. And indeed, the restriction of IVT to functions with no rational zeroes therefore *does* uniformly admit computable solutions. We can exploit this to construct instances of IVT (necessarily with rational zeroes) witnessing that IVT itself is not.

**Proposition 3.7.4.** IVT *does not uniformly admit computable solutions.*

*Proof.* For clarity, we will write $\langle x, y\rangle$ for points in $\mathbb{R}^2$, to distinguish from the open interval $(a, b)$. Suppose to the contrary that there is a Turing functional $\Phi$ mapping instances of IVT to solutions. We shall deal only with continuous piecewise linear functions on $[0,1] \cap \mathbb{Q}$ in this proof, which are all uniformly continuous. Given points $\langle x, y\rangle$ and $\langle w, z\rangle$ in $\mathbb{Q}^2$, let $\ell_{\langle x,y\rangle\to\langle w,z\rangle}$ denote the equation of the line segment connecting these two points.

For each $n \in \omega$, define $f_n : [0,1] \cap \mathbb{Q} \to \mathbb{R}$ by

$$f_n(q) = \begin{cases} \ell_{\langle 0,-1\rangle\to\langle\frac{1}{3},-2^{-n-3}\rangle}(q) & \text{if } 0 \leqslant q \leqslant \frac{1}{3}, \\ \ell_{\langle\frac{1}{3},-2^{-n-3}\rangle\to\langle\frac{2}{3},2^{-n-3}\rangle}(q) & \text{if } \frac{1}{3} < q < \frac{2}{3}, \\ \ell_{\langle\frac{2}{3},2^{-n-1}\rangle\to\langle 1,1\rangle}(q) & \text{if } \frac{2}{3} \leqslant q \leqslant 1. \end{cases}$$

We have $f_n(0) < 0$, $f_n(1) > 0$, and it is easy to check that $|f_{n+1}(q) - f_n(q)| < 2^{-n}$ for each rational $q$. Thus, we can define an instance $f$ of IVT by $f(q) = \langle f_n(q) : n \in \omega\rangle$.

Say $\Phi(f) = \langle q_n : n \in \omega\rangle$, and fix a finite initial segment $\sigma \preceq f$ long enough so that $\Phi(\sigma)$ produces $\langle q_0, q_1, q_2, q_3, q_4\rangle$. Since $\sigma$ is finite it can only determine the value of $f(q)$ for finitely many $q$, and for each of these only up to a finite degree of approximation. Hence, we may fix an $n$ such that $\sigma$ is also an initial segment of any function $g : [0,1] \cap \mathbb{Q} \to \mathbb{R}$ such that for all $q$, if $g(q) = \langle r_m : m \in \omega\rangle$ then $r_m = f_m(q)$ for all $m \leqslant n$. Call any such $g$ *good*.

We can view $f_n$ itself as good. (More precisely, we can define $g(q) = \langle r_m : m \in \omega\rangle$ where $r_m = f_m(q)$ for all $m \leqslant n$ and $r_m = f_n(q)$ for all $m > n$. Then as reals, $g(q) = f_n(q)$ for all $q$.) Since $f_n$ is an instance of IVT with a unique zero at $q = \frac{1}{2}$, it follows that $q_4$ lies within $2^{-4}$ of $\frac{1}{2}$. Hence, any $g$ which is good and also an instance of IVT must have a zero within $2^{-3}$ of $\frac{1}{2}$, i.e., in the interval $(\frac{3}{8}, \frac{5}{8})$.

Now define $h : [0,1] \cap \mathbb{Q} \to \mathbb{R}$ as follows:

$$h(q) = \begin{cases} \ell_{\langle 0,-1\rangle\to\langle\frac{1}{3},-2^{-n-3}\rangle}(q) & \text{if } 0 \leqslant q \leqslant \frac{1}{3}, \\ \ell_{\langle\frac{1}{3},-2^{-n-3}\rangle\to\langle\frac{5}{8},-2^{-n-3}\rangle}(q) & \text{if } \frac{1}{3} < q < \frac{5}{8}, \\ \ell_{\langle\frac{5}{8},-2^{-n-3}\rangle\to\langle\frac{2}{3},2^{-n-3}\rangle}(q) & \text{if } \frac{5}{8} \leqslant q < \frac{2}{3}, \\ \ell_{\langle\frac{2}{3},2^{-n-1}\rangle\to\langle 1,1\rangle}(q) & \text{if } \frac{2}{3} \leqslant q \leqslant 1. \end{cases}$$

That is, $h$ is obtained from $f_n$ by replacing $f_n(q)$ for each $q$ in the interval $(\frac{1}{3}, \frac{5}{8})$ by $-2^{-n-3}$, and then connecting the points $\langle\frac{5}{8}, -2^{-n-3}\rangle$ and $\langle\frac{2}{3}, 2^{-n-3}\rangle$ so that $h$ is continuous. (See Figure 3.1.) Now for each $q \in [0,1] \cap \mathbb{Q}$, define $g(q) = \langle r_m : m \in \omega\rangle$ where $r_m = f_m(q)$ for all $m \leqslant n$ and $r_m = h(q)$ for all $m > n$. Then

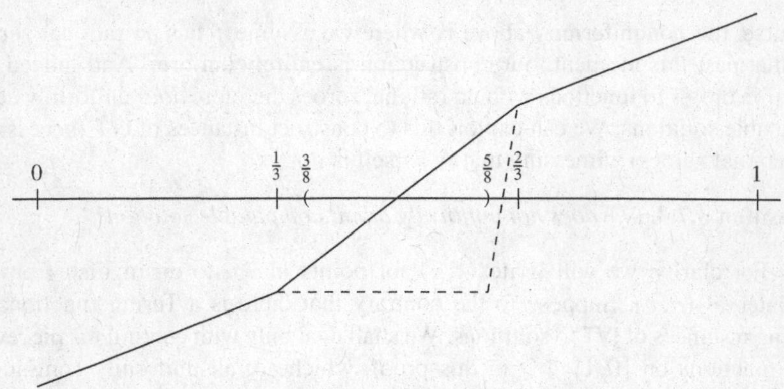


**Figure 3.1.** An illustration to Proposition 3.7.4. The functions $f_n$ (solid) and $h$ (dashed) overlaid.

$g(q)$ is a real number for each $q$. This is clear if $q$ lies outside the interval $(\frac{1}{3}, \frac{2}{3})$, since there $h = f_n$. If $q$ does lie in the interval, then since $|f_n(q)| < 2^{-n-1}$ we have $|h(q) - f_n(q)| < 2^{-n}$, as needed. Thus, $g$ is good and an instance of IVT, but it has no zero in $(\frac{3}{8}, \frac{5}{8})$. The proof is complete. □

## 3.8 Further examples

In this section, we consider two use cases of the framework for measuring complexity introduced in Section 3.6. Both are problems arising from compactness principles. The first is the "disjunctive form" of the Heine–Borel theorem, Disj-HBT$_{[0,1]}$, introduced earlier. We show Disj-HBT$_{[0,1]}$ has the same computability theoretic bounds as WKL, which is not surprising since these are both problems corresponding to basically the same mathematical statement (compactness of the closed unit interval or Cantor space).

The second principle is SeqCompact$_{2^\omega}$, which we define below. This corresponds to the statement of sequential compactness of Cantor space, and as we will see, it exhibits very different behavior. In this way, our investigation into the complexity of problems can also be reflective of underlying non-logical (in this case topological) properties.

**Proposition 3.8.1.** *The problem* Disj-HBT$_{[0,1]}$ *admits solutions in PA.*

*Proof.* We restrict to computable instances. The general case follows by relativization. So fix a computable instance of Disj-HBT$_{[0,1]}$, which is a sequence $\langle\langle x_k, y_k\rangle : k \in \omega\rangle$ of reals with $x_k < y_k$ for all $k$. Using the definition of codes of real numbers, we can fix a computable sequence of intervals with rational endpoints whose union covers $\bigcup_{k\in\omega}(x_k, y_k)$. Thus, without loss of generality, we may assume all the $x_k$ and $y_k$ are rational numbers.

First, we inductively define an interval $I_\sigma \subseteq [0,1]$ for each $\sigma \in 2^{<\omega}$. Let $I_{\langle\rangle} = [0,1]$, and having defined $I_\sigma = [a,b]$ for some $\sigma \in 2^{<\omega}$, let $I_{\sigma i}$ for $i < 2$ be $[a + \frac{a+b}{2}i, b - \frac{a+b}{2}(1-i)]$. Thus, $I_\sigma$ is the closed dyadic interval whose position within $[0,1]$ is determined by the bits of $\sigma$, with 0 denoting the "left" subinterval and 1 the "right".

Now construct a binary tree $T$ as follows. Let $\sigma \in 2^{<\omega}$ belong to $T$ if and only if $I_\sigma$ is not contained in $\bigcup_{k<|\sigma|}(x_k, y_k)$. All the $x_k$ and $y_k$ are rationals, so $T$ is computable in the sequence $\langle\langle x_k, y_k\rangle : k \in \omega\rangle$. If $T$ is finite, fix $\ell$ such that $T$ contains no strings $\sigma \in 2^{<\omega}$ of length $\ell$. Since all of $[0,1]$ is the union of $I_\sigma$ for $\sigma \in 2^\ell$, we have that every $x \in [0,1]$ belongs to the interval $(x_k, y_k)$ for some $k < \ell$. Then $\ell$ is a (computable) solution to our instance.

So suppose next that $T$ is infinite. Given any $D \gg \emptyset$, there is a path $f$ through $T$ computable from $D$. For each $n$, let $q_n$ be the left endpoint of the interval $I_{f \upharpoonright n}$. Then $|q_n - q_{n+1}| < 2^{-n}$ for all $n$, so $\langle q_n : n \in \omega\rangle$ defines a real number $x$. By construction, $x \in I_{f\upharpoonright n}$ for all $n$. Thus if $x$ belonged to $(x_k, y_k)$ for some $k$, then $I_{f\upharpoonright n}$ would be contained in $(x_k, y_k)$ for all sufficiently large $n$, which cannot be since $f \upharpoonright n \in T$. Thus, $x$ is an element of $[0,1]$ not covered by the intervals $(x_k, y_k)$, and hence a solution to our instance. It remains only to observe that $x \leqslant_T f \leqslant_T D$. □

We now describe a method of embedding subtrees of $2^{<\omega}$ into the interval $[0,1]$. This will be of independent interest to us in Chapter 10.

**Lemma 3.8.2.** *Given a tree $T \subseteq 2^{<\omega}$, there is a uniformly $T$-computable enumeration $\langle U_i : i \in \omega\rangle$ of open rational intervals in $[0,1]$ as follows. Let $U = \bigcup_{i\in\omega} U_i$. There is a computable injection from $[T]$ to $[0,1] \setminus U$ with a computable inverse from $[0,1] \setminus U$ to $[T]$. In particular, $[T] = \emptyset$ if and only if $U = [0,1]$.*

*Proof.* This construction, which is inspired by the construction of the middle-thirds Cantor set $C$, is illustrated in Section 3.8. Recall that $C \subseteq [0,1]$ is formed by creating a sequence $\langle C_n\rangle$ of closed sets and then letting $C = \bigcap_{n\in\mathbb{N}} C_n$. Each set $C_i$ is a finite collection of $2^i$ closed subintervals. In particular, we can identify a closed interval $C_\tau$ for each $\tau \in 2^{<\omega}$ so that $C_n = \bigcup_{|\tau|=n} C_\tau$ and each $C_\tau$ is $3^{-|\tau|}$ units in width.

We construct two sequences of open rational intervals, $\langle I_\sigma : \sigma \in 2^{<\omega}\rangle$ and $\langle J_\sigma : \sigma \in 2^{<\omega}\rangle$. For each $\sigma \in 2^{<\omega}$, let $I_\sigma$ be the open middle third interval of $C_\sigma$, so that $I_\sigma$ is removed from $C_\sigma$ in the next stage of the construction of $C$. In particular, $I_\sigma$ and $I_\tau$ are disjoint from one another and from $C$ for all distinct $\sigma, \tau \in 2^{<\omega}$. Let $I = \bigcup_\sigma I_\sigma$, so $I \cap C = \emptyset$ and $I \cup C = [0,1]$.

Let $J_\sigma$ be an open interval containing $C_\sigma$ and extending an additional $3^{-(|\sigma|+1)}$ units to either side. Thus $J_\sigma$ and $J_\tau$ are disjoint unless $\tau \preceq \sigma$ or $\sigma \preceq \tau$. The endpoints of $I_\sigma$ and $J_\sigma$ are rational and are uniformly computable from $\sigma$.

Now, given a tree $T \subseteq 2^{<\omega}$, we let the sequence $\langle U_i : i \in \omega\rangle$ be an effective enumeration of the set $\{I_\sigma : \sigma \in 2^{<\omega}\} \cup \{J_\sigma : \sigma \notin T\}$. It is immediate from the construction of the Cantor set that there is a computable function $F \colon 2^\omega \to C$ with a computable inverse. Let $U = \bigcup_i U_i$. Then, for each $f \in 2^\omega$, we have $f \in [T] \leftrightarrow F(f) \notin \bigcup_{\sigma\notin T} J_\sigma \leftrightarrow F(f) \in [0,1] \setminus U$. □

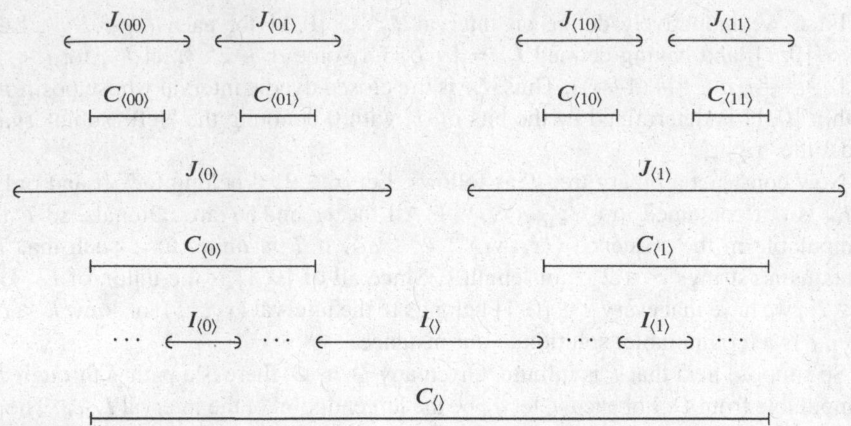


**Figure 3.2.** Construction from the proof of Lemma 3.8.2. The proof constructs an embedding of the Cantor space $2^\omega$ into $[0, 1]$. Each sequence $\tau \in 2^{<\mathbb{N}}$ corresponds to a closed interval $C_\tau$. The open intervals $I_\tau$ cover the complement of the Cantor set, and each open interval $J_\tau$ covers $C_\tau$. The interval $J_{\langle\rangle}$ is not shown.

**Proposition 3.8.3.** *The problem* $\mathsf{Disj\text{-}HBT}_{[0,1]}$ *codes PA.*

*Proof.* We construct a computable instance of $\mathsf{Disj\text{-}HBT}_{[0,1]}$ all of whose solutions have PA degree. The full result is again proved by relativizing the computable case.

We showed in Section 2.8.1 that DNC is a $\Pi^0_1$ classs, an in particular there is a computable tree $T \subseteq 2^{<\omega}$ so that every element of $[T]$ has PA degree. Apply Lemma 3.8.2 with this $T$ to obtain a computable instance $\langle U_i : i \in \omega\rangle$ of $\mathsf{Disj\text{-}HBT}_{[0,1]}$. Because $[T]$ is nonempty, $U = \bigcup_i U_i \neq [0, 1]$, and so each solution to this instance is an element of $[0, 1] \setminus U$. Hence, each solution computes a path through $T$ and therefore has PA degree. □

We now turn to the second of our examples in this section, which is a formulation of sequential compactness of $2^\omega$ introduced by Hamkins (personal communication) under the name $\mathsf{NonSplit}_\omega$. In the next chapter, we will give this another name, COH, which is a prominent problem in the reverse mathematics of Ramsey's theorem.

**Definition 3.8.4 (Sequential compactness of $2^\omega$).** $\mathsf{SeqCompact}_{2^\omega}$ is the principle whose instances are sequences $\vec{R} = \langle R_i : i \in \omega\rangle$ of elements of $2^\omega$, with solutions to any such sequence being all infinite $S \subseteq \omega$ so that for each $e$, the values $R_i(e)$ are the same for almost all $i \in S$.

**Proposition 3.8.5.** $\mathsf{SeqCompact}_{2^\omega}$ *admits PA avoidance.*

*Proof.* As usual, we work only with computable instances. Our proof will relativize. To show $\mathsf{SeqCompact}_{2^\omega}$ admits PA avoidance, fix a computable instance $\langle R_i : i \in \omega\rangle$ of $\mathsf{SeqCompact}_{2^\omega}$. For subsets $X$ and $Y$ of $\omega$, write $X \preceq Y$ if $X$ is finite, $X \subseteq Y$, and $Y - X > X$. In this proof, we follow our convention that any computation with a finite set as oracle is assumed to have use bounded by the maximum of the finite set.

We define a sequence $F_0 \preceq F_1 \preceq \cdots$ of finite sets, and a sequence $I_0 \supseteq I_1 \supseteq \cdots$ of infinite computable sets, satisfying the following properties.

1. $\lim_e |F_e| = \infty$.
2. $F_{e'} - F_e \subseteq I_e$ for all $e' > e$.
3. If $i, j \in I_e$ then $R_i(e) = R_j(e)$.
4. There is an $n$ such that either $\Phi_e^{F_e}(n) \downarrow = \Phi_n(n) \downarrow \in \{0, 1\}$, or there is no $X \succeq F_e$ with $X \subseteq I_e$ such that $\Phi_e^X(n) \downarrow \in \{0, 1\}$.

We let $S = \bigcup_{e \in \omega} F_e$, which is then clearly a solution to $\langle R_i : i \in \omega \rangle$. For each $e$, we have $F_e \preceq S \subseteq I_e$, and so condition (4) ensures that $S$ computes no 2-valued DNC function and hence that $S \not\gg \emptyset$.

For ease of notation, let $F_{-1} = \emptyset$ and $I_{-1} = \omega$, and assume that for some $e \in \omega$ we have defined $F_{e-1}$ and $I_{e-1}$. We consider two cases:

*Case 1: There exist $n \in \omega$ and a finite set $F \succeq F_{e-1}$ with $F \subseteq I_{e-1}$ such that $\Phi_e^F(n) \downarrow = \Phi_n(n) \downarrow \in \{0, 1\}$.* In this case, set $F_e = F$.

*Case 2: Otherwise.* In this case, set $F_e = F_{e-1}$.

In any case, we let $I_e$ consist of all $i > F_e$ in $I_{e-1}$ such that $R_i(e) = 0$ if there are infinitely many such $i$, and otherwise we let $I_e$ consist of all $i > F_e$ in $I_{e-1}$ such that $R_i(e) = 1$.

It is easy to check that we satisfy properties (1)–(3) above for every $e$, and if we are in Case 1, also (4). So say we are in Case 2. We claim that for some $n$, there is no finite $F \succeq F_{e-1}$ with $F \subseteq I_{e-1}$ such that $\Phi_e^F(n) \downarrow = \{0, 1\}$, which again gives (4). Suppose otherwise. Then for each $n$, we can computably search for the first $F \succeq F_e$ as above, and record the value of $\Phi_e^F(n)$ as $f(n)$. By assumption, this defines a 2-valued function $f$. Now if $\Phi_n(n) \downarrow \in \{0, 1\}$ for some $n$ then it must be that $f(n) \neq \Phi_n(n)$, else Case 1 would have applied. Thus, $f$ is DNC. But $f$ is computable, a contradiction. □

**Proposition 3.8.6.** SeqCompact$_{2^\omega}$ *omits PA solutions.*

*Proof.* We construct a computable instance $\langle R_i : i \in \omega \rangle$ of SeqCompact$_{2^\omega}$ and exhibit an $f \gg \emptyset$ that computes no solution to this instance. More specifically, let $T \subseteq 2^{<\omega}$ be the infinite computable tree whose paths are all the 2-valued DNC functions. Our $f$ will be a path through $T$, and will satisfy the following requirement for all $e \in \omega$:

$\mathcal{R}_e$: If $\Phi_e^f$ has unbounded range then for each $b < 2$ there are infinitely many $i \in \omega$ such that $\Phi_e^f(n) \downarrow = i$ for some $n$ and $R_i(e) = b$.

Clearly, this will have the desired effect.

We proceed in stages. For each $e$, we define numbers $n_e \in \omega$ and $b_e < 2$, which we may periodically redefine throughout the construction. To start, we set $n_e = b_e = 0$ for all $e$. Also, we initially declare all $\langle e, n \rangle \in \omega$ *active*.

At stage $s$, assume we have already defined $R_i \upharpoonright s$ for all $i < s$. We define $R_i(s)$ for all $i < s$, along with $R_s \upharpoonright s + 1$. For each $k < s$, let $U_{k,s}$ be the tree of all $\sigma \in T$ for which it is not the case that $\Phi_e^\sigma(n) \downarrow\geqslant n_e$ for any active $\langle e, n\rangle \leqslant k$. Thus if $j < k < s$ we have $U_{k,s} \subseteq U_{j,s}$. Say $\langle e, n\rangle$ *requires attention* at stage $s$ if it is active and $U_{\langle e,n\rangle,s}$ does not contain strings of length $s$ but each $U_{k,s}$ for $k < \langle e, n\rangle$ does. So if $k < \langle e, n\rangle$ and $\sigma \in U_{k,s}$ has length $s$, then $\Phi_e^\sigma(n) \downarrow= i$ for some $i \geqslant n_e$. By our use convention, we have $i < s$, so in particular $R_i(e)$ is already defined. In this case, we redefine $n_e = s$, redefine $b_e$ to be 0 or 1 depending as it was 1 or 0, respectively, and declare $\langle e, n\rangle$ *inactive*. We then set $R_{i^*}(e^*) = b_{e^*}$ for all $i^*, e^* \leqslant s$ where this is undefined.

The trees $U_{k,s}$ in the construction are all uniformly computable in $k$ and $s$. From this it is evident that $\langle R_i : i \in \omega\rangle$ is computable. Now, note that at most one $\langle e, n\rangle$ can require attention at any given stage. Thus, a pair $\langle e, n\rangle$ is either always active or it is inactive from some stage onward. Given $k$, we can thus take $s \geqslant k$ large enough so that every $\langle e, n\rangle \leqslant k$ active at stage $s$ is active forever. Then for all $t \geqslant s$, we have $U_{k,t} = U_{k,s}$, and this tree is infinite. Write $U_k$ in place of $U_{k,s}$ for some (any) such $s$. We thus have a nested sequence $U_0 \supseteq U_1 \supseteq \cdots$ of infinite subtrees of $T$. By Corollary 2.8.10, there is thus some $f \in 2^\omega$ which is a path through $U_k$ for all $k$.

Fix $e$. We claim $f$, which is also a path through $T$, satisfies $\mathcal{R}_e$. By construction, if some $\langle e, n\rangle$ requires attention at a stage $s$, then every $\sigma \in U_{\langle e,n\rangle,s}$ of length $s$ satisfies $\Phi_e^\sigma(n) \downarrow= i$ for some $i \geqslant n_e$ and $R_i(e) = b_e$ for the values of $n_e$ and $b_e$ current at stage $s$. At that point, $n_e$ is redefined to be a larger number, and $b_e$ is redefined to be the opposite bit, and neither of these numbers is redefined again until, if ever, $\langle e, n'\rangle$ requires attention for some other $n'$. But $f$ is itself an extension of some $\sigma \in U_{\langle e,n\rangle,s}$ of length $s$. It follows that for $f$ to fail to satisfy the conclusion of the statement of $\mathcal{R}_e$, it must be that there is an $n$ such that $\langle e, n\rangle$ never requires attention (and hence is active forever). Then $n_e$ reaches a limit value, and by definition, no $\sigma \in U_{\langle e,n\rangle}$ can satisfy $\Phi_e^\sigma(n) \downarrow\geqslant n_e$. Since $f$ is a path through $U_{\langle e,n\rangle}$, this implies $\Phi_e^f$ has bounded range. This completes the proof. □

## 3.9 Exercises

**Exercise 3.9.1.** Prove that there is a computable instance of KL having a single noncomputable set as a solution.

**Exercise 3.9.2.** Consider the following two problems.

- Instances are all functions $f\colon \omega \to \omega$ with bounded range, with a solution to $f$ being all $k \in \omega$ such that $f$ is $k$-bounded.
- Instances are all functions $f\colon \omega \to \omega$, with a solution to $f$ being all $k \in \omega$ such that $f$ is $k$-bounded, or arbitrary if $f$ has unbounded range.

Show that neither problem uniformly admits computable solutions.

**Exercise 3.9.3.** Find an example of a problem that neither admits nor omits computable solutions. (It need not be a problem from "ordinary mathematics".)

**Exercise 3.9.4.** Consider $\widehat{\mathsf{RT}^1_2}$, the parallelization of $\mathsf{RT}^1_2$ (Definition 3.1.5). Show that $\widehat{\mathsf{RT}^1_2}$ codes the jump.

**Exercise 3.9.5.** Find an example of two problems $\mathsf{P}$ and $\mathsf{Q}$ so that each admits cone avoidance but the parallel product, $\mathsf{P} \times \mathsf{Q}$, does not.

**Exercise 3.9.6.** Show that the problem whose instances are pairs $(x, y)$ of distinct real numbers, with solutions being 0 or 1 depending as $x < y$ or $y < x$, does not uniformly admit computable solutions.

**Exercise 3.9.7.** Show that the problem whose instances are real numbers $x$, with a solution being 0 or 1 depending as $x = 0$ or $x \neq 0$, does not uniformly admit computable solutions.

**Exercise 3.9.8 (Hirst [156]).** Given an infinite sequence $(x_i : i \in \omega)$ of reals, there is an infinite sequence $(y_i : i \in \omega)$ such that $y_n = \min\{x_i : i \leqslant n\}$ for all $n$. However, in general, there is no computable sequence $(j(n) : n \in \omega)$ such that $x_{j(n)} = \min\{x_i : i \leqslant n\}$ for all $n$.

**Exercise 3.9.9.** Let $\mathsf{C}_{\mathbb{N}}$ (the *choice problem on* $\mathbb{N}$) be the problem whose instances are all functions $e \colon \omega \times \omega \to 2$ as follows.

- $e(x, 0) = 0$ for all $x \in \omega$.
- For each $x \in \omega$ there is at most one $s$ such that $e(x, s) \neq e(x, s + 1)$.
- There is at least one $x \in \omega$ such that $e(x, s) = 0$ for all $s$.

For each $n \geqslant 2$, let $\mathsf{C}_n$ (the *choice problem on n*) be the restriction of $\mathsf{C}_{\mathbb{N}}$ to functions $e \colon n \times \omega \to 2$. Show that neither $\mathsf{C}_{\mathbb{N}}$ nor $\mathsf{C}_n$ uniformly admits computable solutions.

**Exercise 3.9.10.** Does there exist a problem $\mathsf{P}$ which is *not* uniformly computably true, but such that its parallelization, $\widehat{\mathsf{P}}$, is computably true?

# Chapter 4
# Problem reducibilities

The approach developed in Section 3.6 provides a means of measuring the complexity of an instance–solution problem, but it is too crude in general to adequately compare the complexities of two or more problems.As noted earlier, if one problem admits solutions in a particular class of sets and another does not, then we may view the latter as "harder" from a certain computational standpoint. But it is not obvious how to find such a class for a particular pair of problems, or whether such a class even exists. It is also unclear what relationship this kind of classification really expresses.

In this chapter, we develop direct methods of gauging the relative strengths of problems. We think of these as *reducibilities*, or *reductions*, in that we are defining various ways of reducing (the task of solving) one problem to (the task of solving) another. In schematic form, such reductions all roughly have the shape shown in Figure 4.1.

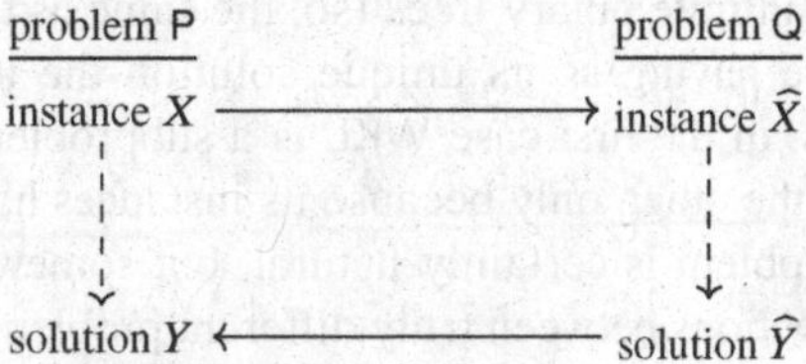


**Figure 4.1.** Generalized diagram of a reduction between problems.

Here, $X$ is a P-instance, $\widehat{X}$ is a Q-instance, $\widehat{Y}$ is a Q-solution to $\widehat{X}$, and $Y$ is a P-solution $Y$ to $X$. The dashed lines may be interpreted as going from an instance to a solution (within each problem). The solid arrows denote maps or transformations "forth and back" between the problems in such a way that we may (in a loose sense) regard the diagram as commuting. (At this level of generality, such maps have also been called *generalized Galois–Tukey connections* (Vojtáš [318]) or *morphisms* (Blass [16].)) There are many specific reducibilities fitting this shape. The ones of interest to us here

 
D. D. Dzhafarov, C. Mummert, *Reverse Mathematics*, Theory and Applications of Computability, https://doi.org/10.1007/978-3-031-11367-3_4

will of course be those arising in computable mathematics and reverse mathematics. We will also consider several variations and extensions.

## 4.1 Subproblems and identity reducibility

Arguably, the simplest nontrivial way in which one problem can reduce to another is by being a subproblem.

**Definition 4.1.1 (Subproblem).** A problem P is a *subproblem* of a problem Q if every P-instance $X$ is also a Q-instance, and every Q-solution to any such $X$ is also a P-solution to it.

For example, consider the following: first, the problem whose instances are all odd-degree polynomials with integer coefficients in one variable, with the unique solution to any such polynomial being its least zero; and on the other hand, the problem whose instances are all such polynomials of degree 3, with the solutions being all its zeroes. The second problem is then a subproblem of the first. This fits well with our abstract conception of problem reducibility because, intuitively, being able to find the least zero of any odd-degree polynomial entails being able to find some (any) zero of a third-degree polynomial.

Note that the second problem above has both fewer instances than the first, as well as more solutions to some of these. Among the problems we will encounter, this is actually not so common. Typically, a subproblem either restricts the domain (and so is a special case), or enlarges some or all of its values (and so is a less specific problem), but not both. For example, WKL is a subproblem of the version Disj-WKL discussed in Chapter 3. But WKL is also a subproblem of the problem whose instances are all infinite binary trees (so, the same as the instances of WKL), but with each such tree having as its unique solution the lexicographically least ("leftmost") path. Thus, in the first case WKL is a subproblem only because it has fewer instances, and in the other, only because its instances have more solutions.

The notion of subproblem is certainly natural, but somewhat limiting in that it cannot reveal any connections between truly different problems. A first step towards a more useful notion of reducibility is given by the following definition.

**Definition 4.1.2 (Identity reducibility).** Let P and Q be problems.

1. P is *identity reducible* to Q if every P-instance $X$ computes a Q-instance $\widehat{X}$ such that every Q-solution to $\widehat{X}$ is a P-solution to $X$.
2. P is *uniformly identity reducible* to Q if there is a Turing functional $\Phi$ such that $\Phi(X)$ is a Q-instance for every P-instance $X$, and every Q-solution to $\Phi(X)$ is a P-solution to $X$.

See Figure 4.2. Instead of asking for the instances of P to be included in those of Q, then, here we look at computably transforming the instances of P into instances of Q. The particular transformation may depend on the instance in the case of an identity

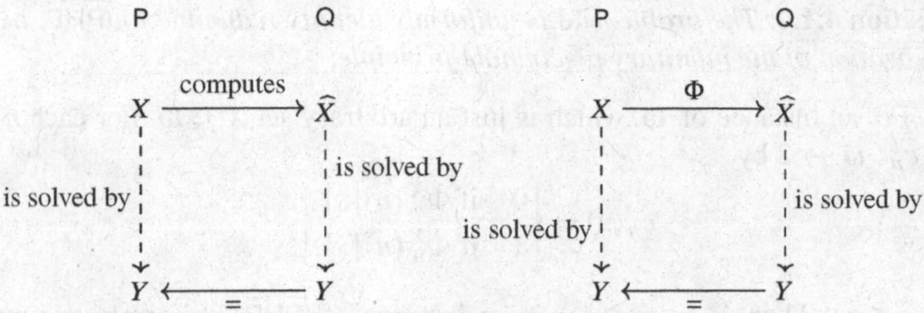


**Figure 4.2.** Diagrams of identity and uniform identity reductions. On the left, $X$ is a P-instance computing a Q-instance $\widehat{X}$ such that every Q-solution to $\widehat{X}$ is also a P-solution $Y$ to $X$. On the right, the same situation, but $\widehat{X} = \Phi(X)$ for a fixed Turing functional $\Phi$.

reduction, but not in the case of a uniform identity reduction. In particular, being a subproblem is a uniform identity reduction, witnessed by the identity functional.

**Proposition 4.1.3.** *The problem whose instances are all partial orders on $\omega$, with solutions to any such order being all its linear extensions, is uniformly identity reducible to* WKL.

*Proof.* Fix a partial order $\leqslant_P$ on $\omega$. (Recall that formally this is a subset of $\omega$ such that the set of $(x, y)$ with $\langle x, y\rangle \in \leqslant_P$ forms a partial order. We will write $x \leqslant_P y$ in place of $\langle x, y\rangle \in \leqslant_P$ for clarity.) Define $T$ to be the set of all $\sigma \in 2^{<\omega}$ that are consistent with being a linearization of $\leqslant_P$. That is, $\sigma \in T$ if and only if it satisfies the following conditions for all $x, y, z \in \omega$.

1. If $\langle x,x\rangle < |\sigma|$ then $\sigma(\langle x,x\rangle) = 1$.
2. If $\langle x,y\rangle, \langle y,x\rangle < |\sigma|$ and $\sigma(\langle x,y\rangle) = \sigma(\langle y,x\rangle) = 1$ then $x = y$.
3. If $\langle x,y\rangle, \langle y,z\rangle, \langle x,z\rangle < |\sigma|$ and $\sigma(\langle x,y\rangle) = \sigma(\langle y,z\rangle) = 1$ then $\sigma(\langle x,z\rangle) = 1$.
4. If $\langle x,y\rangle < |\sigma|$ and $x \leqslant_P y$ then $\sigma(\langle x,y\rangle) = 1$.
5. If $\langle x,y\rangle, \langle y,x\rangle < |\sigma|$ then either $\sigma(\langle x,y\rangle) = 1$ or $\sigma(\langle y,x\rangle) = 1$.

Then $T$ is uniformly computable from $\leqslant_P$, and it is infinite because every partial order can be linearized, and every such linearization is a path through $T$. Thus, $T$ is an instance of WKL. Let $L$ be any solution, i.e., any path through $T$. By construction, $L$ is a linearization of $\leqslant_P$. □

It may seem peculiar that, in the above proof, we used the existence of linearizations to conclude that $T$ is infinite. This is not actually a problem, since our goal is not to prove that linearizations exist, but rather, to show that the set of linearizations has a particular shape (namely, that it is the set of solutions to a particular instance of WKL). All the same, it is worth noting that we could prove $T$ to be infinite directly, without the need for any facts about partial orders. To see this, fix $n \in \omega$. We exhibit a $\sigma \in T$ of length $n^2$. With our usual coding of pairs, this means we need to define $\sigma$ on $\langle x, y\rangle$ for all $x, y < n$. Fix $\ell < n$, and suppose we have defined $\sigma$ on $\langle x, y\rangle$ for all $x, y < \ell$. We then set $\sigma(\langle \ell, \ell\rangle) = 1$, and for each $x < \ell$, set $\sigma(\langle \ell, x\rangle) = 1$ if $\ell <_P x$ and $\sigma(\langle x, \ell\rangle) = 1$ otherwise. It is easy to check that $\sigma \in T$. For another example along these lines, see the remarks following Theorem 11.3.1.

**Proposition 4.1.4.** *The problem* TJ *is uniformly identity reducible to* $\widehat{\mathsf{IPHP}}$*, i.e., the parallelization of the infinitary pigeonhole principle.*

*Proof.* Fix an instance of TJ, which is just an arbitrary set $X \subseteq \omega$. For each $n \in \omega$, define $c_n : \omega \to 2$ by

$$c_n(s) = \begin{cases} 0 & \text{if } \Phi_n^X(n)[s] \uparrow, \\ 1 & \text{if } \Phi_n^X(n)[s] \downarrow, \end{cases}$$

for all $s \in \omega$. Then $\langle c_n : n \in \omega \rangle$ is an instance of $\widehat{\mathsf{IPHP}}$, uniformly computable from $X$. Each $c_n$ is either constantly 0 if $\Phi_n^X(n) \uparrow$, or eventually 1 if $\Phi_n^X(n) \downarrow$. Hence, if $\langle i_n : n \in \omega \rangle$ is a solution to $\langle c_n : n \in \omega \rangle$ then it is unique and equal to the characteristic function of $X'$. As sets and their characteristic functions are identified, this completes the proof. □

To show that uniform identity reducibility is a strict refinement of identity reducibility in general, we turn again to WKL and Disj-WKL. As noted above, WKL is a subproblem of, and so is uniformly identity reducible to, Disj-WKL. Perhaps surprisingly, the identity reduction also holds in reverse.

**Proposition 4.1.5.** Disj WKL *is identity reducible to* WKL*, but not uniformly so.*

*Proof.* We begin by showing that Disj-WKL is identity reducible to WKL. Fix an instance of Disj-WKL. This is a binary tree $T$ that may or may not be infinite. If $T$ is infinite, we can simply view it also as an instance $U$ of WKL, and then any solution to it as such is also a Disj-WKL-solution. If $T$ is finite, let $\ell$ be such that $T$ contains no strings of length $\ell$. Then, let $U$ be a computable binary tree having as its only path the characteristic function of $\{\ell\}$. (Recall that for problem solutions, we identify numbers with the singletons containing them.) Clearly, the only WKL-solution to $U$ is a Disj-WKL-solution to $T$. In both cases we have $U \leqslant_{\mathrm{T}} T$, so we have described an identity reduction.

The lack of uniformity above stems from us using different procedures depending as $T$ is finite or infinite. To make this explicit, suppose to the contrary that Disj-WKL is identity reducible to WKL, say with witnessing Turing functional $\Phi$. Let $T_1 = \{1^n : n \in \omega\}$ be the tree consisting of all finite strings of 1s. Then by assumption, $\Phi(T_1)$ is an infinite binary tree whose only path is $1^\omega$. Let $\ell$ be large enough so that every string $\sigma \in \Phi(T_1)$ of height $\ell$ satisfies $\sigma(0) = \sigma(1) = 1$, and let $s$ be large enough so that $\Phi(T_1 \upharpoonright s)$ converges on all strings of length $\ell$. Take $T = T_1 \upharpoonright s$. Then $\Phi(T)$ is also an infinite binary tree, and every path $f$ through $\Phi(T)$ satisfies $f(0) = f(1) = 1$. But as $T$ is finite, any such $f$ must be (the characteristic function of) a singleton, which is a contradiction. □

We include one further example, meant mostly to introduce a problem that will play an important role in our work in Chapter 8, and to relate it to one we have already seen.

**Definition 4.1.6 (Cohesive principle).** COH is the problem whose instances are sequences $\vec{R} = \langle R_i : i \in \omega \rangle$ of elements of $2^\omega$, with solutions to any such sequence being all infinite sets $S \subseteq \omega$ such that for each $e$, either $S \subseteq^* R_i$ or $S \subseteq^* \overline{R_i}$.

A set $S$ as above is called *cohesive for the family $\vec{R}$*, or *$\vec{R}$-cohesive for short*. The terminology comes from the study of the c.e. sets in computability theory, where a set is called *p-cohesive* if it is cohesive for the family of all primitive recursive sets; *r-cohesive* if it is cohesive for the family of all computable sets; and simply *cohesive* if it is cohesive for the family of all c.e. sets.

**Proposition 4.1.7.** *Each of* $\mathsf{COH}$ *and* $\mathsf{SeqCompact}_{2^\omega}$ *is uniformly identity reducible to the other.*

*Proof.* Given a family of sets $\langle R_i : i \in \omega \rangle$, define a new family $\langle \widehat{R}_e : e \in \omega \rangle$ by $\widehat{R}_e(i) = R_i(e)$ for all $i, e \in \omega$. Now any $\mathsf{SeqCompact}_{2^\omega}$-solution to $\langle \widehat{R}_e : e \in \omega \rangle$ is a $\mathsf{COH}$-solution to $\langle R_i : i \in \omega \rangle$, and any $\mathsf{COH}$-solution to $\langle \widehat{R}_e : e \in \omega \rangle$ is a $\mathsf{SeqCompact}_{2^\omega}$-solution to $\langle R_i : i \in \omega \rangle$. □

As the proposition explicates, $\mathsf{COH}$ and $\mathsf{SeqCompact}_{2^\omega}$ are really the same principle, and each can be obtained from the other just by "turning the instances on their sides". $\mathsf{SeqCompact}_{2^\omega}$ is more obviously a compactness result, whereas $\mathsf{COH}$ is more convenient when working with combinatorial principles.

## 4.2 Computable reducibility

We now come to one of the main reducibilities between problems, *computable reducibility*. This vastly generalizes identity reducibility, and is certainly one of the most direct ways to compare the computational complexity of different problems. It does have certain limitations, however, which we will discuss in Section 4.5.

**Definition 4.2.1 (Computable reducibility).** Let $\mathsf{P}$ and $\mathsf{Q}$ be problems.

1. $\mathsf{P}$ is *computably reducible* to $\mathsf{Q}$, written $\mathsf{P} \leqslant_{\mathrm{c}} \mathsf{Q}$, if every $\mathsf{P}$-instance $X$ computes a $\mathsf{Q}$-instance $\widehat{X}$ such that if $\widehat{Y}$ is any $\mathsf{Q}$-solution to $\widehat{X}$ then $X \oplus \widehat{Y}$ computes a $\mathsf{P}$-solution to $X$.
2. $\mathsf{P}$ is *computable equivalent* to $\mathsf{Q}$, written $\mathsf{P} \equiv_{\mathrm{c}} \mathsf{Q}$, if $\mathsf{P} \leqslant_{\mathrm{c}} \mathsf{Q}$ and $\mathsf{Q} \leqslant_{\mathrm{c}} \mathsf{P}$.

We represent this visually in Figure 4.3. Basic facts are collected in Proposition 4.2.2.

**Proposition 4.2.2.** $\leqslant_{\mathrm{c}}$ *is a transitive relation, and* $\equiv_{\mathrm{c}}$ *is an equivalence relation. Moreover, if* $\mathsf{P}$ *and* $\mathsf{Q}$ *are problems and* $\mathsf{P}$ *is identity reducible to a problem* $\mathsf{Q}$ *then* $\mathsf{P} \leqslant_{\mathrm{c}} \mathsf{Q}$.

Intuitively, $\mathsf{P} \leqslant_{\mathrm{c}} \mathsf{Q}$ means that the problem $\mathsf{P}$ can be (effectively) *coded* into the problem $\mathsf{Q}$, and this is how the situation is often described. We have seen some examples of this already in Section 3.6, with problems that code the jump or code PA. Computable reducibility allows us to recast some of these complexity notions, and better understand some of the results employing them.

**Proposition 4.2.3.** *A problem* $\mathsf{P}$ *admits computable solutions if and only if* $\mathsf{P} \leqslant_{\mathrm{c}} \mathsf{Id}_{2^\omega}$.

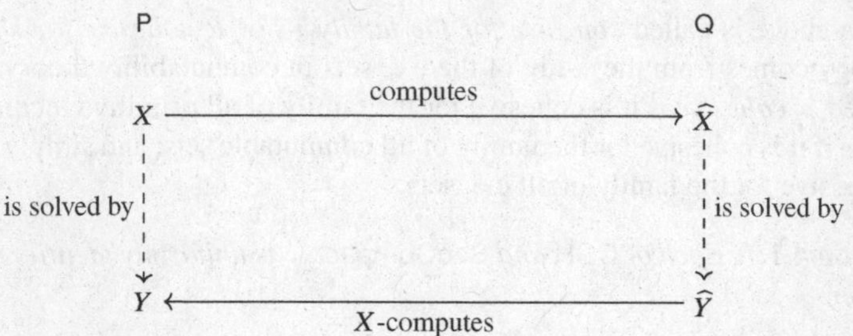


**Figure 4.3.** Diagram of a computable reduction: $X$ is a P-instance computing a Q-instance $\widehat{X}$; $\widehat{Y}$ is a Q-solution to $\widehat{X}$ so that $X \oplus \widehat{Y}$ computes a P-solution $Y$ to $X$.

**Proposition 4.2.4.** *A problem* P *codes the jump if and only if* $\mathsf{TJ} \leqslant_c \mathsf{P}$.

**Proposition 4.2.5.** *Let* P *be a problem.*

1. P *admits solutions in PA if and only if* $\mathsf{P} \leqslant_c \mathsf{WKL}$.
2. P *codes PA if and only if* $\mathsf{WKL} \leqslant_c \mathsf{P}$.

*Proof.* Part (1) is basically the definition. We prove (2). By definition, and using the existence of a subtree of $2^{<\omega}$ all of whose paths have PA degree, WKL is computably reducible to P if and only if for every $A \subseteq \omega$, there is an $A$-computable instance of P whose every solution $Y$ satisfies $A \oplus Y \gg A$. □

Recall the problems $\mathsf{Disj\text{-}HBT}_{[0,1]}$ and COH given by Definitions 3.3.4 and 4.1.6, respectively. Combining Proposition 4.2.5 with earlier results yields the following.

**Corollary 4.2.6.** $\mathsf{Disj\text{-}HBT}_{[0,1]} \equiv_c \mathsf{WKL}$.

*Proof.* By Propositions 3.8.1 and 3.8.3. □

**Corollary 4.2.7.** $\mathsf{COH} \nleqslant_c \mathsf{WKL}$ *and* $\mathsf{WKL} \nleqslant_c \mathsf{COH}$.

*Proof.* By Propositions 3.8.5, 3.8.6 and 4.1.7. □

**Corollary 4.2.8.** $\mathsf{KL} \nleqslant_c \mathsf{WKL}$.

*Proof.* By Proposition 3.6.10. □

And while we are discussing KL and Proposition 3.6.10, let us note that using computable reducibility, the latter can actually be improved to an equivalence. Thus, as far as their computability theoretic complexity, "solving König's lemma" and "solving the halting problem" are the same.

**Proposition 4.2.9.** $\mathsf{KL} \equiv_c \mathsf{TJ}$.

*Proof.* By Proposition 3.6.10, we only need to show $\mathsf{KL} \leqslant_c \mathsf{TJ}$. Let $T \subseteq \omega^{<\omega}$ be an infinite, finitely branching tree $T$. Define $p\colon \omega^{<\omega} \to \omega$ as follows:

$$p(\alpha) = \begin{cases} 0 & \text{if } \alpha \notin \operatorname{Ext}(T), \\ (\mu i)[\alpha i \in \operatorname{Ext}(T)] & \text{if } \alpha \in \operatorname{Ext}(T). \end{cases}$$

By König's lemma, $p$ is total, and by relativizing Proposition 2.8.5, $p \leqslant_T T'$. Now define $\alpha_0 = \langle\rangle$, and for $i \in \omega$, $\alpha_{i+1} = \alpha_i \frown p(\alpha_i)$. Then $\alpha_0 \prec \alpha_1 \prec \cdots \in T$, and $f = \bigcup_{i\in\omega} \alpha_i$ is an infinite path through $T$. We have $f \leqslant_T \langle \alpha_i : i \in \omega\rangle \leqslant_T p \leqslant_T T'$.□

At this point, we may believe that any two problems that admit computable solutions are computably equivalent. But consider, on the one hand, the problem whose instances are all sets, and on the other, the problem whose instances are all sets computing $\varnothing'$, with each instance of either problem having itself as its unique solution. Both problems admit computable solutions, but the former is not computably reducible to the latter since it has instances (e.g., the computable sets) that compute no instance of the latter. Once again this is a bit of an artificial situation, however. Natural problems do have $A$-computable instances for all $A \subseteq \omega$, and for these the above intuition is correct. In particular, we have the following.

**Proposition 4.2.10.** *The following are all computably equivalent.*

1. $\mathsf{RT}^1$.
2. $\mathsf{IPHP}$.
3. General-$\mathsf{IPHP}$.
4. $\mathsf{FUF}$.

We will look at these principles again with a finer lens in the next section. Specifically, this result should be compared with Proposition 4.3.4.

We next prove that the converse of Proposition 4.1.4 fails even under computable reducibility. Thus, from a computational standpoint, $\widehat{\mathsf{IPHP}}$ is a more complex problem than that of finding the Turing jump of a set.

**Proposition 4.2.11.** $\widehat{\mathsf{IPHP}} \nleqslant_c \mathsf{TJ}$.

*Proof.* We build a computable instance $\langle c_e : e \in \omega\rangle$ of $\widehat{\mathsf{IPHP}}$ with no $\varnothing'$-computable solution. More precisely, each $c_e$ will be a map $\omega \to 2$, and we will ensure that for each $e \in \omega$, if $\Phi_e^{\varnothing'}(e)\downarrow= i$ for some $i < 2$ then $c_e(s) = 1 - i$ for almost all $s$. An $\widehat{\mathsf{IPHP}}$-solution to $\langle c_e : e \in \omega\rangle$ is an $X \in 2^\omega$ with $X(e) = i$ only if $c_e(s) = i$ for infinitely many $s$. So this construction ensures that $\Phi_e^{\varnothing'}$ fails to be a solution to our $\widehat{\mathsf{IPHP}}$-instance for every $e$.

Each $c_e$ is defined uniformly computably in $e$, which ensures that $\langle c_e : e \in \omega\rangle$ is computable. To define $c_e(s)$ for some $s$, run the computation $\Phi_e^{\varnothing'}(e)[s]$, and if this converges to some $i < 2$ let $c_e(s) = 1 - i$. (Note that $\Phi_e^{\varnothing'}(e)[s]$ is run for at most $s$ many steps, so checking this convergence is computable.) Now if $\Phi_e^{\varnothing'}(e)\downarrow= i$ then there is an $s_0$ so that $\Phi_e^{\varnothing'}(e)[s] = i$ for all $s \geqslant s_0$, and therefore $c_e(s) = 1 - i$ for all such $s$, as desired. □

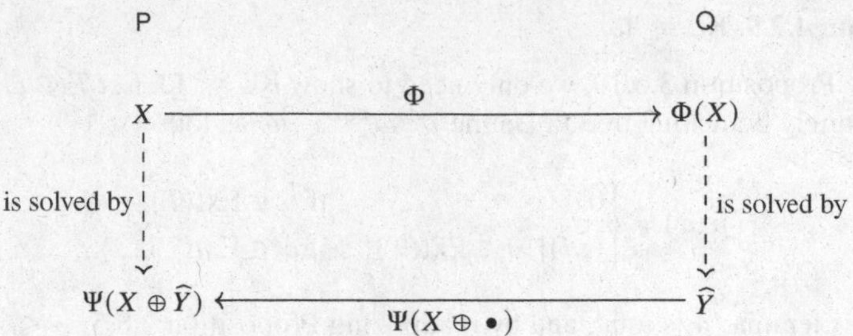


**Figure 4.4.** Diagram of a Weihrauch reduction: $X$ is a P-instance computing a Q-instance $\widehat{X}$, and $\widehat{Y}$ is a Q-solution to $\widehat{X}$ so that $X \oplus \widehat{Y}$ computes a P-solution $Y$ to $X$.

In the above proof, it is worth noting that if $\Phi_e^{\emptyset'}(e) \uparrow$ then $\Phi_e^{\emptyset'}(e)[s]$ may converge to 0 and 1 for infinitely many $s$ each, so $c_e$ will not be eventually constant. And indeed, we cannot modify the proof so that $c_e$ is eventually constant for *every* $e$ (see Exercise 4.8.3).

## 4.3 Weihrauch reducibility

The second main motion of problem reducibility we consider in this book is to computable reducibility as uniform identity reducibility is to identity reducible. That is, it is its uniform analogue. The origins of this reducibility are in computable analysis, going back to the work of Weihrauch [323]. Later, it was noticed by Gherardi and Marcone [122], and independently by Dorais, Dzhafarov, Hirst, Mileti, and Shafer [72], that this notion can also be a fruitful way of comparing problems that arise in the context of reverse mathematics.

**Definition 4.3.1 (Weihrauch reducibility).** Let P and Q be problems.

1. P is *Weihrauch reducible* to Q, written $\mathsf{P} \leqslant_{\mathrm{W}} \mathsf{Q}$, if there exist Turing functionals $\Phi$ and $\Psi$ satisfying the following: if $X$ is any P-instance then $\Phi(X)$ is a Q-instance, and if $\widehat{Y}$ is any Q-solution to $\Phi(X)$ then $\Psi(X \oplus \widehat{Y})$ is a P-solution to $X$.
2. P is *Weihrauch equivalent* to Q, written $\mathsf{P} \equiv_{\mathrm{W}} \mathsf{Q}$, if $\mathsf{P} \leqslant_{\mathrm{W}} \mathsf{Q}$ and $\mathsf{Q} \leqslant_{\mathrm{W}} \mathsf{P}$.

See Figure 4.4.

**Proposition 4.3.2.** $\leqslant_{\mathrm{W}}$ *is a transitive relations, and* $\equiv_{\mathrm{W}}$ *is an equivalence relation. Moreover, let* P *and* Q *be problems.*

1. *If* P *is uniformly identity reducible* Q *then* $\mathsf{P} \leqslant_{\mathrm{W}} \mathsf{Q}$.
2. *If* $\mathsf{P} \leqslant_{\mathrm{W}} \mathsf{Q}$ *then* $\mathsf{P} \leqslant_{\mathrm{c}} \mathsf{Q}$.

**Proposition 4.3.3.** *A problem* P *uniformly admits computable solutions if and only if* $\mathsf{P} \leqslant_{\mathrm{W}} \mathsf{Id}_{2^\omega}$.

If P $\leqslant_W$ Q then the instances of P should not split apart in some noncomputable way that affects how they compute instances of Q. Similarly, the solutions of Q should not split apart in such a fashion either. A Weihrauch reduction from one problem to another thus conveys a much closer relationship than a computable reduction alone. As a result, it is able to tease out subtler distinctions than computable reducibility.

We begin with the following analogue of Proposition 4.2.10. Due to the increased uniformity, this is no longer a completely obvious result.

**Proposition 4.3.4.** *The following are all Weihrauch equivalent.*

*1.* $\mathsf{RT}^1$.
2. IPHP.
*3.* General-IPHP.

*Proof.* We show $\mathsf{RT}^1 \leqslant_W$ IPHP $\leqslant_W$ General-IPHP $\leqslant_W \mathsf{RT}^1$.

First, fix an instance of $\mathsf{RT}^1$, which is a map $f: \omega \to \omega$ with bounded range. Then $f$ is also an instance of IPHP. If $i \in \omega$ is any IPHP-solution to $f$ then $\{x : f(x) = i\}$ is uniformly computable from $f$ and $i$, and is an $\mathsf{RT}^1$-solution to $f$.

That IPHP $\leqslant_W$ General-IPHP is immediate since IPHP is a subproblem of General-IPHP.

So now fix an instance of General-IPHP. This is a sequence $\langle X_i : i \in \omega\rangle$ of sets with infinite union and $X_i = \emptyset$ for almost all $i$. Define $f: \omega \to \omega$ as follows: given $x \in \omega$, search for the least $i \in \omega$ such that $X_i$ contains some $y \geqslant x$, and let $f(x) = i$. Clearly $f$ is uniformly computable from $\langle X_i : i \in \omega\rangle$, and since only finitely many of the $X_i$ are nonempty, $f$ has bounded range. Thus, $f$ is an $\mathsf{RT}^1$-instance, and if $H$ is any solution to it then $f(\min H)$ is a General-IPHP-solution to $\langle X_i : i \in \omega\rangle$. □

Notably absent above is FUF. Indeed, whereas FUF was equivalent to the three above under $\leqslant_c$, it is strictly weaker under $\leqslant_W$.

**Proposition 4.3.5.** FUF $\leqslant_W \mathsf{RT}^1$ *but even* $\mathsf{RT}^1_2 \nleqslant_W$ FUF.

*Proof.* Seeking a contradiction, suppose $\Phi$ and $\Psi$ witness that $\mathsf{RT}^1_2 \leqslant_W$ FUF as in Definition 4.3.1. We construct an instance $f: \omega \to 2$. By assumption, $\Phi(f)$ is an instance of FUF, meaning a family $\langle F_i : i \in \omega\rangle$ of finite sets with $F_i = \emptyset$ for almost all $i$. As $f \in 2^\omega$, initial segments of it are elements of $2^{<\omega}$. For $\sigma \in 2^{<\omega}$, say $\sigma$ *puts* $x \in F_i$ if $\Phi(\sigma)(\langle i, x\rangle)\downarrow= 1$. In this case, if $\sigma \prec f$ we will indeed have $x \in F_i$.

There must exist a $\sigma \in 2^{<\omega}$ and an $n \in \omega$ such that no $\tau \succeq \sigma$ puts any $x$ in any $F_i$ with $i > n$, and no $x > n$ in any $F_i$ with $i \leqslant n$. If not, we could define a sequence $\tau_0 \preceq \tau_1 \preceq \cdots$ such that for each $n$, $\tau_n$ puts some $x$ in some $F_i$ with $i > n$ or some $x > n$ in some $F_i$ with $i \leqslant n$. But then, if we let $f = \bigcup_{i \in \omega} \tau_i$, the family $\Phi(f) = \langle F_i : i \in \omega\rangle$ will either satisfy $F_i \neq \emptyset$ for infinitely many $i$, or else some $F_i$ will be infinite, which is impossible by our assumption on $\Phi$.

It follows that for every $f \succeq \sigma$, the FUF-instance $\Phi(f)$ has $n$ as a solution. By assumption on $\Psi$, then, $\Psi(f \oplus \{n\})$ is an $\mathsf{RT}^1_2$-solution to any such $f$. We can thus find $\tau \succeq \sigma$ so that $\Psi(\tau \oplus \{n\})\downarrow= \{i\}$ for some $i < 2$. But now let $f = \tau^\frown(1-i)^\omega$. That is, $f$ is the extension of $\tau$ with $f(x) = 1 - i$ for all $x \geqslant |\tau|$. Clearly, $i$ is not an $\mathsf{RT}^1_2$-solution to this $f$ even though $\Psi(f \oplus \{n\}) = i$, which is a contradiction. □

Weihrauch reducibility can elucidate further subtle differences between different formulations the same problem. Recall from Section 3.2 the version of $\mathsf{IPHP}$ whose instances, rather than being maps $\omega \to \omega$ with bounded range, are pairs $(k, f)$, where $k \in \omega$ and $f : \omega \to k$. Let us denote this variant by $\mathsf{IPHP}_+$. It is easy to see that $\mathsf{IPHP}_+ \equiv_c \mathsf{IPHP}$. Also, $\mathsf{IPHP}_+$ is trivially Weihrauch (in fact, uniformly identity) reducible to $\mathsf{IPHP}$.

**Proposition 4.3.6.** $\mathsf{IPHP} \nleqslant_W \mathsf{IPHP}_+$.

*Proof.* Suppose $\Phi$ and $\Psi$ are Turing functionals witnessing that $\mathsf{IPHP} \leqslant_W \mathsf{IPHP}_+$. We construct an instance $g : \omega \to \omega$ of $\mathsf{IPHP}$. Initial segments of $g$ are thus strings $\alpha \in \omega^{<\omega}$. By assumption on $\Phi$, we can fix some such $\alpha$ such that $\Phi(\alpha)(2k) \downarrow= 1$ for some $k$. (Note that the instances of $\mathsf{IPHP}_+$ are formally pairs $\langle k, f\rangle = \{k\} \oplus f$.) Thus, for any $\mathsf{IPHP}$-instance $g \succeq \alpha$, the $\mathsf{IPHP}_+$-instance $\Phi(g)$ will have some $i < k$ as a solution. By assumption on $\Psi$, we may thus fix $\alpha' \succeq \alpha$ such that for each $i < k$, either $\Psi(\alpha' \oplus \{i\}) \downarrow$ or $\Psi(\beta \oplus \{i\}) \uparrow$ for all $\beta \succeq \alpha'$. Let $b$ be the largest value of $\Psi(\alpha' \oplus \{i\})$ for $i < k$. Then, define $g = \alpha' {}^\frown (b+1)^\omega$. So $g(x) = b + 1$ for almost all $x$, hence $g$ has bounded range and is thus an instance of $\mathsf{IPHP}$. Since $g$ extends $\alpha'$ we have $\Phi(g) = \langle k, f\rangle$ for some $f$ and $\Psi(g \oplus \{i\}) \leqslant b$ for all $i < k$. Because there are not infinitely many $x$ with $g(x) \leqslant b$, this means $\Psi(g \oplus \{i\})$ is not an $\mathsf{IPHP}$-solution to $g$ for any $i < k$, a contradiction. □

A related example concerns Ramsey's theorem, $\mathsf{RT}^n_k$, for different values of $k$. Consider first the case $n = 1$. Clearly, if $j < k$ then $\mathsf{RT}^1_j$ is a subproblem of $\mathsf{RT}^1_k$. Since both principles also admit computable solutions, we also have that $\mathsf{RT}^1_k \leqslant_c \mathsf{RT}^1_j$, so the two are computably equivalent. As it turns out, this cannot be improved to a Weihrauch reduction.

**Proposition 4.3.7 (Dorais, Dzhafarov, Hirst, Mileti, and Shafer [72]; Hirschfeldt and Jockusch [148]; Brattka and Rakotoniaina [20]).** *For all* $k > j$, $\mathsf{RT}^1_k \nleqslant_W \mathsf{RT}^1_j$.

The proof is Exercise 4.8.4. In Section 4.5, we will define a notion of "applying a problem multiple times" and, among other results, establish that $\mathsf{RT}^1_k$ can be uniformly reduced to multiple applications of $\mathsf{RT}^1_j$. That argument helps elucidate the key obstacle to uniformly reducing $\mathsf{RT}^1_k$ to (a single application of) $\mathsf{RT}^1_j$, as above. In Theorem 9.1.11, we will also see a higher dimensional analogue of this result: that if $n \geqslant 2$ and $k > j$, then $\mathsf{RT}^n_k \nleqslant_c \mathsf{RT}^n_j$.

Interestingly, Weihrauch reducibility also captures many of the nuances of constructive/intuitionistic mathematics. For example, part of the relationship between $\mathsf{WKL}$ and the Brouwer fan theorem, $\mathsf{FAN}$, is reflected in the following result.

**Proposition 4.3.8.** $\mathsf{FAN} \leqslant_W \mathsf{WKL}$ *but* $\mathsf{WKL} \nleqslant_c \mathsf{FAN}$.

*Proof.* Fix an instance of $\mathsf{FAN}$, i.e., a bar $B \subseteq 2^{<\omega}$. Let $T$ be the set of all $\sigma \in 2^{<\omega}$ as follows: for each $n < |\sigma|$, if every $\sigma \in 2^n$ has an initial segment in $B$ then $\sigma(n) = 1$, and otherwise $\sigma(n) = 0$. Then $T$ is an infinite tree, uniformly computable from $B$. Clearly, $T$ has a single path, $f \in 2^\omega$, that is either equal to $0^\omega$ or to $0^n 1^\omega$ for some $n$.

We claim the former cannot happen. For otherwise, the set of all $\sigma \in 2^{<\omega}$ such that $\sigma$ has no initial segment in $B$ would be an infinite binary tree, and any path through it would thus contradict the fact that $B$ is a bar. So let $n$ be such that $f = 0^n 1^\omega$. Then the set $B'$ of all initial segments in $B$ of the strings $\sigma \in 2^n$ is a finite bar. Note that $n$ can be found uniformly computably from $f$, and so $B'$ is uniformly computable from $B \oplus f$.

To show that $\mathsf{WKL} \nleqslant_{\mathrm{c}} \mathsf{FAN}$ we just note that WKL has a computable instance with no computable solutions, whereas FAN admits computable solutions. □

We include one more example related to WKL, which further underscores the difference between bounded and unbounded strings and trees. As with $\mathsf{RCA}_0$, the "R" in DNR is due to tradition, apart from which the system could be called DNC.

**Definition 4.3.9 (Diagonally noncomputable problem).**

1. DNR is the problem whose instances are all sets $X \in 2^\omega$, with the solutions to any such $X$ being all functions $f : \omega \to \omega$ that are DNC relative to $X$.
2. For each $k \geqslant 2$, $\mathsf{DNR}_k$ is the problem whose instances are all sets $X \in 2^\omega$, with the solutions to any such $X$ being all $k$-valued functions $f : \omega \to \omega$ that are DNC relative to $X$.

The following basic facts are left to the reader.

**Proposition 4.3.10.**

*1.* $\mathsf{WKL} \equiv_{\mathrm{W}} \mathsf{DNR}_2$.
*2. For all* $j, k \geqslant 2$, $\mathsf{DNR}_j \equiv_{\mathrm{c}} \mathsf{DNR}_k$.

One half of part (2) follows simply from the fact that if $j < k$ then $\mathsf{DNR}_k$ is a subproblem of $\mathsf{DNR}_j$ (and so $\mathsf{DNR}_k \leqslant_{\mathrm{W}} \mathsf{DNR}_j$). Surprisingly, the other half of the equivalence cannot be similarly strengthened.

**Theorem 4.3.11 (Jockusch [170]).** *For all* $k > j$, $\mathsf{DNR}_j \nleqslant_{\mathrm{W}} \mathsf{DNR}_k$.

*Proof.* Suppose otherwise, as witnessed by functionals $\Phi$ and $\Psi$. Given $X \leqslant_{\mathrm{T}} Y$, any function which is DNC relative to $Y$ is also DNC relative to $X$. Thus, we may assume that $\Phi$ is the identity functional. Taking $\emptyset$ as a $\mathsf{DNR}_j$ instance, it thus suffices to exhibit a $k$-valued DNC function $f$ such that $\Psi(f)$ is not a $j$-valued DNC function. Let $T$ be the set of all $\sigma \in k^{<\omega}$ such that for all $e < |\sigma|$, if $\Phi_e(e)[|\sigma|] \downarrow$ then $\sigma(e) \neq \Phi_e(e)$. Then $T$ is a computable tree, and the paths through $T$ are precisely the $k$-valued DNC functions.

Fix $x \in \omega$. Since $T$ is bounded, it follows by assumption on $\Psi$ there is an $\ell \in \omega$ such that $\Psi(\sigma)(x) \downarrow < j$ for all $\sigma \in T$ with $|\sigma| = \ell$. Let $\ell_x$ be the least such $\ell$. For every $y < j$, let $T_{x,y} = \{\sigma \in T : |\sigma| = \ell_x \wedge \Psi(\sigma)(x) \downarrow = y\}$, so that

$$\{\sigma \in T : |\sigma| = \ell_x\} = \textstyle\bigcup_{y<j} T_{x,y}. \tag{4.1}$$

Let $D = \{e < \ell_x : \Phi_e(e)[|\ell_x|] \uparrow\}$, and call $T_{x,y}$ *large* if for every $\tau \in k^{<\omega}$ with $|\tau| = \ell_x$ there is a $\sigma \in T_{x,y}$ such that $\sigma(e) \neq \tau(e)$ for all $e \in D$.

We claim that there is a $y < j$ such that $T_{x,y}$ is large. Suppose not. Then for each $y$ we can fix $\tau_y \in k^{<\omega}$ such that $|\tau| = \ell_x$ and every $\sigma \in T_{x,y}$ satisfies $\sigma(e) = \tau_y(e)$ for some $e \in D$. Now, we define a string $\sigma \in k^{<\omega}$ of length $\ell_x$ as follows. Fix $e < \ell_x$. If $e \notin D$ then $\Phi_e(e) \downarrow$, and we let $\sigma(e)$ have any value different from this computation. If $e \in D$, then we let $\sigma(e)$ be any $i < k$ such that $\tau_y(e) \neq i$ for all $y < j$, which exists because $j < k$. But then $\sigma \in T$ and $\sigma \notin T_{x,y}$ for any $y < j$, which contradicts (4.1). So the claim holds.

Next, we claim that if $T_{x,y}$ is large then there exists $\sigma \in T_{x,y}$ with $\sigma(e) \neq \Phi_e(e)$ for all $e < \ell_x$. Indeed, let $\tau(e)$ for each $e < \ell_x$ be $\Phi_e(e)$ if the latter converges and 0 otherwise. By definition of $T$, every $\sigma \in T_{x,y}$ satisfies that $\sigma(e) \neq \tau(e)$ for all $e \notin D$. If $T_{x,y}$ is large, then there is a $\sigma \in T_{x,y}$ that additionally satisfies that $\sigma(e) \neq \tau(e)$ for all $e \in D$, which proves the claim.

To complete the proof, note that the level $\ell_x$ is uniformly computable from $x$, and that determining whether or not $T_{x,y}$ is large is uniformly computable from $x$ and $y < j$. Thus, by the first claim above, there is a computable function $g$ such that for every $x$, $g(x) = y$ for the least $y < j$ such that $T_{x,y}$ is large. Let $h$ be the computable function defined by setting $\Phi_{h(x)}(z) = g(x)$ for all $z$, and apply the recursion theorem (Theorem 2.4.11) to get an $e$ such that $\Phi_{h(e)} = \Phi_e$. By the second claim, there is a $k$-valued DNC function $f$ having an initial segment in $T_{e,g(e)}$. But now this function satisfies

$$\Psi(f)(e) \downarrow= g(e) = \Phi_{h(e)}(e) = \Phi_e(e),$$

so $\Psi(f)$ is not DNC. □

## 4.4 Strong forms

An important but somewhat subtle detail in both computable reducibility and Weihrauch reducibility concerns access to the original instances. Say P is being reduced to Q, and we are given a P-instance $X$. Then there is a Q-instance $\widehat{X}$ computable from $X$ (uniformly, in the case of Weihrauch reducibility) every solution to which, *together with* $X$, computes a P-solution to $X$. The access to $X$ as an oracle is used in many such reductions as part of a kind of "post-processing".

For example, consider the reduction $\mathsf{IPHP} \leqslant_{\mathrm{W}} \mathsf{RT}^1$. An IPHP-instance $c$ is viewed as an $\mathsf{RT}^1$-instance, and from an $\mathsf{RT}^1$-solution $H$ we uniformly compute an IPHP-solution by applying $c$ to an element of $H$. That is, $c$ tells us what color it takes on $H$, and this is the IPHP-solution. But without access to $c$, there is no apparent way to determine this information, and the reduction breaks down. Does a different reduction exist that avoids needing to consult $c$ at the end? We will see below that the answer is no.

Denying access to the original instance is thus an additional resource constraint, and determining when such access is essential helps further elucidate how much information is necessary to reduce one problem to another. We can measure when

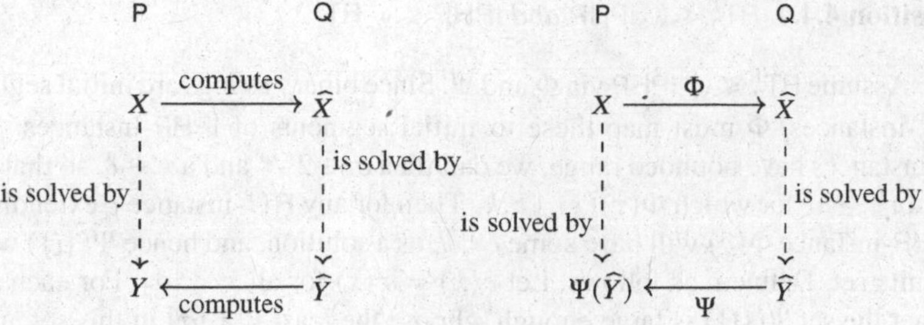


**Figure 4.5.** Diagrams of strong computable and strong Weihrauch reductions.

access to the original P-instance $X$ is indeed essential using the following *strong* forms of computable and Weihrauch reducibilities.

**Definition 4.4.1 (Strong reducibility forms).** Let P and Q be problems.

1. P is *strongly computably reducible* to Q, written P $\leqslant_{sc}$ Q, if every P-instance $X$ computes a Q-instance $\widehat{X}$ such that every Q-solution $\widehat{Y}$ to $\widehat{X}$ computes a solution $Y$ to $X$.
2. P is *strongly computably equivalent* to Q, written P $\equiv_{sc}$ Q, if P $\leqslant_{sc}$ Q and Q $\leqslant_{sc}$ P.
3. P is *strongly Weihrauch reducible* to Q, written P $\leqslant_{sW}$ Q, if there exist Turing functionals $\Phi$ and $\Psi$ satisfying the following: if $X$ is any P-instance then $\Phi(X)$ is a Q-instance, and if $\widehat{Y}$ is any Q-solution to $\Phi(X)$ then $\Psi(\widehat{Y})$ is a P-solution to $X$.
4. P is *strongly Weihrauch equivalent* to Q, written P $\equiv_{sW}$ Q, if both P $\leqslant_{sW}$ Q and Q $\leqslant_{sW}$ P.

The next proposition establishes the key properties of $\leqslant_{sc}$ and $\leqslant_{sW}$.

**Proposition 4.4.2.** *$\leqslant_{sc}$ and $\leqslant_{sW}$ are transitive relations, and $\equiv_{sc}$ and $\equiv_{sW}$ are equivalence relations. Moreover, let* P *and* Q *be problems.*

1. *If* P *is identity reducible to* Q *then* P $\leqslant_{sc}$ Q.
2. *If* P *is uniformly identity reducible to* Q *then* P $\leqslant_{sW}$ Q.
3. *If* P $\leqslant_{sW}$ Q *then* P $\leqslant_{sc}$ Q.
4. *If* P $\leqslant_{sc}$ Q *then* P $\leqslant_{c}$ Q.
5. *If* P $\leqslant_{sW}$ Q *then* P $\leqslant_{W}$ Q.

Identity and uniform identity reducibilities are of course the extreme examples of strong reductions, since they involve no "post-processing" at all.

We leave to Exercise 4.8.6 the following strong analogue of a part of Proposition 4.3.4. However, as the subsequent result shows, we cannot obtain a strong analogue of the full proposition.

**Proposition 4.4.3.** General-IPHP $\equiv_{sW}$ IPHP.

**Proposition 4.4.4.** $\mathsf{RT}^1_2 \nleqslant_{\mathrm{sW}} \mathsf{IPHP}$ *and* $\mathsf{IPHP} \nleqslant_{\mathrm{sW}} \mathsf{RT}^1$.

*Proof.* Assume $\mathsf{RT}^1_2 \leqslant_{\mathrm{sW}} \mathsf{IPHP}$ via $\Phi$ and $\Psi$. Since binary strings are initial segments of $\mathsf{RT}^1_2$-instances, $\Phi$ must map these to initial segments of $\mathsf{IPHP}$-instances. Since $\mathsf{IPHP}$-instances have bounded range, we can fix a $\sigma \in 2^{<\omega}$ and a $k \in \omega$ so that there is no $x$ or $\tau \succeq \sigma$ for which $\Phi(\tau)(x) \downarrow\geqslant k$. Then for any $\mathsf{RT}^1_2$-instance $c$ extending $\sigma$, the $\mathsf{IPHP}$-instance $\Phi(c)$ will have some $i < k$ as a solution, and hence $\Psi(\{i\})$ will be an infinite set. Define $c$ as follows. Let $c(x) = \sigma(x)$ for all $x < |\sigma|$. For each $i < k$ such that the set $\Psi(\{i\})$ is large enough, choose the least $x \geqslant |\sigma|$ in this set and set $c(x) = 0$. Then, let $c(x) = 1$ for all other $x$. Thus, all $\mathsf{RT}^1_2$-solutions to $c$ consist of elements colored 1 by $c$. But if $i < k$ is any solution to $\Phi(c)$, then $\Psi(\{i\})$ contains an element colored 0 by $c$, hence $\Psi(\{i\})$ is not a solution after all.

Next, assume $\mathsf{IPHP} \leqslant_{\mathrm{sW}} \mathsf{RT}^1$ via $\Phi$ and $\Psi$. We define an instance $g$ of $\mathsf{IPHP}$ as in the proof of Proposition 4.3.6. First, we can fix $\alpha \in \omega^{<\omega}$ and $k \in \omega$ such that for all $\beta \succeq \alpha$ and all $x$, if $\Phi(\beta)(x) \downarrow= i$ then $i < k$. Thus, for any $g \succeq \alpha$, $\Phi(g)$ will be an instance $c$ of $\mathsf{RT}^1_k$, and any infinite homogeneous set for $c$ with color $i < k$ will be mapped by $\Psi$ to an $\mathsf{IPHP}$-solution to $c$. We can now fix $\alpha' \succeq \alpha$ such that one of the following alternatives for each $i < k$.

1. There exists a finite set $F$ so that $\Phi(\alpha')(x) \downarrow= i$ for each $x \in F$ and such that $\Psi(F)(j) \downarrow= 1$ for some $j$.
2. There exists a $b \in \omega$ so that no $x > b$ satisfies $\Phi(\beta)(x) \downarrow= i$ for some $\beta \succeq \alpha'$.

Let $k$ be the maximum of all the $j$ obtained under Case 1. Define $g \colon \omega \to k+2$ by $g(x) = \alpha'(x)$ for all $x < |\alpha'|$, and $\alpha(x) = k+1$ for all other $x$. Thus, the only $\mathsf{IPHP}$-solution to $g$ is $k+1$. But if $H$ is any $\mathsf{RT}^1$-solution to $\Phi(g)$, then $\Psi(H) = j$ for some $j < k+1$, a contradiction. □

The preceding result illustrates an interesting subtlety concerning computable instances. If $X$ is computable and $Z$ is any set, then anything computable from $X \oplus Z$ is also computable from $Z$ alone. We might thus expect that if we restrict to computable instances, the strong versions of our reducibilities behave the same way as the originals. And indeed, this is the case for computable reducibility, in the following sense: if $\mathsf{P}$ and $\mathsf{Q}$ are problems and a computable $\mathsf{P}$-instance $X$ witnesses that $\mathsf{P} \nleqslant_{\mathrm{sc}} \mathsf{Q}$, then the same instance also witnesses that $\mathsf{P} \nleqslant_{\mathrm{c}} \mathsf{Q}$. But this fails for Weihrauch reducibility: the $\mathsf{RT}^1_2$-instance $c$ we built above to witness that $\mathsf{RT}^1_2 \nleqslant_{\mathrm{sW}} \mathsf{IPHP}$ was computable, and yet we still have $\mathsf{RT}^1_2 \leqslant_{\mathrm{W}} \mathsf{IPHP}$.

A closer look at the proof reveals the reason. By first finding the initial segment $\sigma$ of $c$ and determining the bound $k$, we were able to ask $\Psi$ to produce elements on which our coloring was not yet defined. We could then change the colors appropriately to diagonalize. Had $\Psi$ had access to $c$, it could have easily avoided this situation, for example by never outputting any number before $c$ colored it. In the proof, $c$ is computable, but not *uniformly* computable. The initial segment $\sigma$ requires a $\varnothing'$ oracle to find. As it turns out, this is in general the only obstacle to moving from $\nleqslant_{\mathrm{sW}}$ to $\nleqslant_{\mathrm{W}}$.

**Proposition 4.4.5.** *Let* P *and* Q *be problems. Suppose that for each pair of Turing functionals* $\Phi$ *and* $\Psi$ *there is a* P*-instance* $X_{\Phi,\Psi}$ *that is uniformly computable in (indices for)* $\Phi$ *and* $\Psi$ *and that witnesses that* P *is not strongly Weihrauch reducible to* Q *via these functionals. Then* $\mathsf{P} \nleqslant_{\mathrm{W}} \mathsf{Q}$.

*Proof.* Let $f$ be a computable function such that for all $e, i \in \omega$, $f(e,i)$ is an index for $X_{\Phi_e,\Phi_i}$. Now fix Turing functionals $\Phi_n$ and $\Phi_m$. Define a computable function $g$ such that for all $i \in \omega$ and all oracles $Z$,

$$\Phi_{g(i)}(Z) = \Phi_m(\Phi_{f(n,i)} \oplus Z).$$

By the recursion theorem, take a fixed point $i$ for $g$, so $\Phi_{g(i)}(Z) = \Phi_i(Z)$ for all oracles $Z$. Then we have

$$\Phi_m(X_{\Phi_n,\Phi_i} \oplus Z) = \Phi_m(\Phi_{f(n,i)} \oplus Z) = \Phi_{g(i)}(Z) = \Phi_i(Z).$$

for all $Z$. Now since $X_{\Phi_n,\Phi_i}$ witnesses that P is not strongly Weihrauch reducible to Q via $\Phi_n$ and $\Phi_i$, it also witnesses that P is not Weihrauch reducible to Q via $\Phi_n$ and $\Phi_m$. □

We conclude by mentioning, without proof for now, an example of a nonreduction under $\leqslant_{\mathrm{sc}}$. The following is analogous to Proposition 4.3.7, where we saw it for $\leqslant_{\mathrm{W}}$.

**Proposition 4.4.6.** *For all* $k > j$, $\mathsf{RT}^1_k \nleqslant_{\mathrm{sc}} \mathsf{RT}^1_j$.

The proposition can be proved by a direct stage-by-stage construction, but in that form it is somewhat cumbersome. (The enthusiastic reader is encouraged to try it!) A cleaner presentation is as a *forcing construction*. We introduce forcing in Chapter 7, and give a proof of the above proposition (along with generalizations) in Section 9.1.

## 4.5 Multiple applications

Computable reducibility seems to be a good measure of the computational complexity of a problem, but it has a particular limitation. A computable reduction measures the complexity of a problem's *one time* use. To explain, consider again Proposition 4.2.11, that $\widehat{\mathsf{IPHP}} \nleqslant_{\mathrm{c}} \mathsf{TJ}$. Knowing how to find the Turing jump of a given set is not enough to find solutions to arbitrary instances of $\widehat{\mathsf{IPHP}}$. But knowing how to find the *double jump* certainly is. And intuitively, if we have a method for finding $X'$ from $X$, then we ought to be able to obtain $X''$ by appealing to this method twice.

A different measure of a problem's overall computational strength is therefore one that allows us to "apply" problems multiple times in a reduction. The following generalization of computable reducibility is a step in this direction.

**Definition 4.5.1 (Computable reducibility to multiple applications).** Let P and Q be problems. For $m \geqslant 1$, P is *computably reducible to m applications of* Q if:

- every P-instance $X$ computes a Q-instance $\widehat{X}_0$,
- for each $i < m-1$, if $\widehat{Y}_i$ is any Q-solution to the Q-instance $\widehat{X}_i$ then $X \oplus \widehat{Y}_0 \oplus \cdots \oplus \widehat{Y}_i$ computes a Q-instance $\widehat{X}_{i+1}$,
- if $\widehat{Y}_{m-1}$ is any Q-solution to the Q-instance $\widehat{X}_{m-1}$, then $X \oplus \widehat{Y}_0 \oplus \cdots \oplus \widehat{Y}_{m-1}$ computes a P-solution $Y$ to $X$.

The following basic facts are left to Exercise 4.8.7.

**Proposition 4.5.2.** *Let* P, Q, *and* R *be problems.*

1. P $\leqslant_c$ Q *if and only if* P *is computably reducible to one application of* Q.
2. *For* $l, m \geqslant 1$, *if* P *is computably reducible to m applications of* Q, *and* Q *is computably reducible to l applications of* R, *then* P *is computably reducible to* $m \cdot l$ *applications of* R.

Now the objection raised above is alleviated.

**Proposition 4.5.3.** $\widehat{\mathsf{IPHP}}$ *is computably reducible to two applications of* TJ.

*Proof.* Fix an instance $X = \langle c_i : i \in \omega \rangle$ of $\widehat{\mathsf{IPHP}}$, so that each $c_i$ is a map $\omega \to \omega$ with bounded range. We pass to $X$ itself, viewed as an instance of TJ. The only TJ-solution to this is $X'$, and given it, we pass to it as another instance of TJ. Now the only TJ-solution is $X''$, from which we can compute a solution to $X$: for each $i$, $X''$ can uniformly compute the least $j = j_i$ such that $c_i(x) = j$ for infinitely many $x$. Then $\langle j_i : i \in \omega \rangle$ is an $X''$-computable $\widehat{\mathsf{IPHP}}$-solution to $X$. □

The preceding result could be easily expressed without any appeal to problem reducibilities. Indeed, as stated it is merely a basic computability theoretic observation in different, and arguably more convoluted, parlance. We will see shortly that this is not always the case. And in Chapter 9 we will encounter examples that are very difficult to express otherwise.

For completeness, we mention also the following basic fact, which basically says that Definition 4.5.1 can be "padded". In particular, we have that being computably reducible to $m$ applications of a problem is the same as being computable reducible to *at most m* applications.

**Proposition 4.5.4.** *Let* P *and* Q *be problems. If* P *is computably reducible to m applications of* Q, *then* P *is computably reducible to l applications of* Q *for every* $l \geqslant m$.

*Proof.* Suppose we have a P-instance $X$, along with Q-instances $\widehat{X}_0, \ldots, \widehat{X}_{m-1}$ with solutions $\widehat{Y}_0, \ldots, \widehat{Y}_{m-1}$, as in Definition 4.5.1. Thus, $X \oplus \widehat{Y}_0 \oplus \cdots \oplus \widehat{Y}_{m-1}$ computes a P-solution to $X$. Now for each $i$ with $m \leqslant i < l$, define $\widehat{X}_i = \widehat{X}_{m-1}$. For any choice of Q-solutions $\widehat{Y}_m, \ldots, \widehat{Y}_{l-1}$ to $\widehat{X}_m, \ldots, \widehat{X}_{l-1}$, we then have that $X \oplus \widehat{Y}_0 \oplus \cdots \oplus \widehat{Y}_{l-1}$ computes a P-solution. □

An interesting problem in this context is WKL. As it turns out, it always suffices just to apply it once. We will see this in a different guise also in Chapter 5.

**Proposition 4.5.5.** *If a problem* P *is computably reducible to m applications of* WKL *for any* $m \geqslant 1$, *then* $\mathsf{P} \leqslant_{\mathrm{c}} \mathsf{WKL}$.

*Proof.* Fix a P-instance $X$ and any $Y \gg X$. We claim $Y$ computes a P-solution to $X$, whence the conclusion of the proposition follows. Fix a WKL-instance $T_0 \leqslant_{\mathrm{T}} X$, as in Definition 4.5.1. Then in particular, $Y \gg T_0$. Now suppose that for some $i < m$, we have defined WKL-instances $T_0, \ldots, T_i$ and for each $j < i$, a path $f_j$ through $T_j$ such that $T_{j+1} \leqslant_{\mathrm{T}} X \oplus f_0 \oplus \cdots \oplus f_j$. If $i > 0$, we assume inductively that $Y \gg X \oplus f_0 \oplus \cdots \oplus f_{i-1}$. So $Y \gg T_i$, and by Proposition 2.8.26, there is a path $f_i$ through $T_i$ such that $Y \gg X \oplus f_0 \oplus \cdots \oplus f_i$. In this way, we find a P-solution to $X$ computable in $X \oplus f_0 \oplus \cdots \oplus f_{m-1}$ and hence, by construction, in $Y$. □

**Corollary 4.5.6.** *Neither* COH *nor* KL *is computably reducible to m applications of* WKL, *for any* $m \geqslant 1$.

Once again, we can also consider uniform versions.

**Definition 4.5.7 (Weihrauch reducibility to multiple applications).** Let P and Q be problems. For $m \geqslant 1$, P is *Weihrauch reducible to m applications of* Q if there is a Turing functional $\Delta$ satisfying the following.

- If $X$ is a P-instance then $\Delta(0, X)$ is a Q-instance $\widehat{X}_0$.
- For each $i < m - 1$, if $\widehat{Y}_i$ is any Q-solution to the Q-instance $\widehat{X}_i$, then
$$\Delta(i + 1, X, \widehat{Y}_0, \ldots, \widehat{Y}_i)$$
is a Q-instance $\widehat{X}_{i+1}$.
- If $\widehat{Y}_{m-1}$ is any Q-solution to the Q-instance $\widehat{X}_{m-1}$, then
$$\Delta(m, X, \widehat{Y}_0, \ldots, \widehat{Y}_{m-1})$$
is a P-solution $Y$ to $X$.

Thus, this is precisely the same as Definition 4.5.1, only the computations at each step are uniform. Naturally, we have the analogue of Proposition 4.5.2 (see Exercise 4.8.7).

**Proposition 4.5.8.** *Let* P, Q, *and* R *be problems.*

1. $\mathsf{P} \leqslant_{\mathrm{W}} \mathsf{Q}$ *if and only if* P *is Weihrauch reducible to one application of* Q.
2. *If* P *and* Q *are problems and* P *is Weihrauch reducible to m applications of* Q *then* P *is computably reducible to m applications of* Q.
3. *If* P *is Weihrauch reducible to m applications of* Q, *then* P *is Weihrauch reducible to l applications of* Q *for every* $l > m$.
4. *For* $l, m \geqslant 1$, *if* Q *is Weihrauch reducible to l applications of* R *and* P *is Weihrauch reducible to m applications of* Q, *then* P *is Weihrauch reducible to* $l \cdot m$ *applications of* R.

We illustrate this notion with Ramsey's theorem for different numbers of colors.

**Proposition 4.5.9 (Dorais, Dzhafarov, Hirst, Mileti, and Shafer [72]).** *For all $n \geqslant 1$ and $j \geqslant 2$ and $m \geqslant 1$, $\mathsf{RT}^n_{j^m}$ is Weihrauch reducible to $m$ applications of $\mathsf{RT}^n_j$.*

*Proof.* The proof is by induction on $m$. If $m = 1$, the result is clear. Assume the result is true for $m \geqslant 1$, and consider an instance $c$ of $\mathsf{RT}^n_{j^{m+1}}$. We define an instance $d_0$ of $\mathsf{RT}^n_j$ (and hence of $\mathsf{RT}^n_{j^m}$, since $j < j^m$) by letting

$$d_0(\vec{x}) = (\mu k < j)[j^m k \leqslant c(\vec{x}) < j^m(k+1)]$$

for all $\vec{x} \in [\omega]^n$. Then $d_0$ is uniformly computable from $c$. Given any infinite homogeneous set $H_0 = \{h_0 < h_1 < \cdots\}$ for $d_0$, say with color $k < j$, we define an instance $d_1$ of $\mathsf{RT}^n_{j^m}$ by letting

$$d_1(x_0, \ldots, x_{n-1}) = c(h_{x_0}, \ldots, h_{x_{n-1}}) - j^m k$$

for all $\langle x_0, \ldots, x_{n-1}\rangle \in [\omega]^n$. Then $d_1$ is uniformly computable from $c \oplus H_1$. We now apply the inductive hypothesis to $d_1$. So for $1 \leqslant i < m$, given an infinite homogeneous set $H_i$ for the $\mathsf{RT}^n_{j^m}$-instance $d_i$, we obtain another such instance $d_{i+1}$ uniformly computable in $c \oplus H_1 \oplus \cdots \oplus H_i$. And given an infinite homogeneous set $H_m$ for the $\mathsf{RT}^n_{j^m}$-instance $d_m$ we obtain an infinite homogeneous set $H$ for $d_1$ uniformly computable in $c \oplus H_1 \oplus \cdots \oplus H_m$. Then $\{h_x : x \in H\}$ is an infinite homogeneous set for $c$. □

**Corollary 4.5.10.** *For all $n \geqslant 1$ and all $k > j \geqslant 2$, there is an $m$ so that $\mathsf{RT}^n_k$ is Weihrauch reducible to $m$ applications of $\mathsf{RT}^n_j$.*

*Proof.* Fix $m$ so that $j^m \geqslant k$. □

We use this opportunity to introduce an important operator on problems from the computable analysis literature. Intuitively, the *compositional product* defined below represents the problem of applying one problem and then a second, in sequence.

**Definition 4.5.11 (Compositional product).** Let $\mathsf{P}$ and $\mathsf{Q}$ be problems. The *compositional product* of $\mathsf{Q}$ with $\mathsf{P}$, written $\mathsf{Q} * \mathsf{P}$, is the problem whose instances are pairs $\langle X, \Gamma\rangle$ satisfying the following.

- $X$ is a $\mathsf{P}$-instance.
- $\Gamma$ is a Turing functional.
- If $Y$ is any $\mathsf{P}$-solution to $X$ then $\Gamma(X, Y)$ is a $\mathsf{Q}$-instance.

The solutions to any such $\langle X, \Gamma\rangle$ are all pairs $\langle Y, \widehat{Y}\rangle$, where $Y$ is a $\mathsf{P}$-solution to $X$ and $\widehat{Y}$ is a $\mathsf{Q}$-solution to $\Gamma(X, Y)$.

It can be shown that $*$ is associative (Exercise 4.8.8). We can recast Definition 4.5.7 as follows: $\mathsf{P}$ is Weihrauch reducible to $m \geqslant 1$ applications of $\mathsf{Q}$ if $\mathsf{P}$ is Weihrauch reducible to the $m$-fold compositional product of $\mathsf{Q}$ with itself. But the compositional product is more expressive. For example, it is evident from our proof of Proposition 4.5.9 that if $k = ij$ then $\mathsf{RT}^n_k \leqslant_{\mathrm{W}} \mathsf{RT}^n_i * \mathsf{RT}^n_j$. Another observation is the following.

**Proposition 4.5.12.** *For all $n, k \geqslant 2$, $\mathsf{RT}^n_k \leqslant_{\mathrm{W}} \mathsf{RT}^{n-1}_k * \mathsf{TJ} * \mathsf{TJ}$.*

*Proof.* We first examine the classic inductive proof of Ramsey's theorem. Fix an instance $c$ of $\mathsf{RT}^n_k$. We define an infinite set $\{x_0, x_1, \ldots\}$ and a sequence of infinite sets $R_0 \supseteq R_1 \supseteq \cdots$ satisfying the following for all $s \in \omega$.

1. $x_s < x_{s+1}$.
2. $x_s < R_s$.
3. For all $\vec{x} \in [\{x_0, \ldots, x_s\}]^{n-1}$, the color $c(\vec{x}, y)$ is the same for all $y \in R_s$.

Of course, part (2) is vacuous if $s < n-2$. So for $s < n-2$, we simply let $x_s = s$ and $R_s = \{y \in \omega : y \geqslant n-2\}$. Now fix $s \geqslant n-2$, and suppose we have defined $x_t$ and $R_t$ for all $t < s$. Let $x_s = R_{s-1}$, and enumerate the elements of $[\{x_0, \ldots, x_s\}]^{n-1}$ as $\vec{x}_1, \ldots, \vec{x}_l$. Let $R^0 = R_{s-1}$, and given $R^i$ for some $i < l$, choose the least color $j$ so that $c(\vec{x}_{i+1}, y) = j$ for infinitely many $y \in R^i$ and set $R^{i+1} = \{y \in R^i_s : c(\vec{x}_{i+1}, y) = j\}$. Finally, let $R_s = R^l$. It is easily seen that this ensures the desired properties hold.

We next define an instance $d$ of $\mathsf{RT}^{n-2}_k$. For each $\langle s_0, \ldots, s_{n-2}\rangle \in [\omega]^{n-1}$, let

$$d(s_0, \ldots, s_{n-2}) = c(x_{s_0}, \ldots, x_{s_{n-2}}, x_s)$$

for some (any) $s > s_{n-2}$. By property (2) above, this uniquely defined. Now if $H$ is any infinite homogeneous set for $d$, it is clear that $\{x_s : s \in H\}$ is a homogeneous set for $c$.

Observe that $d$ and the set $\{x_0, x_1, \ldots\}$ are uniformly computable from $c''$, and the homogeneous set $\{x_s : s \in H\}$ is uniformly computable from $\{x_0, x_1, \ldots\}$ and $H$, hence from $c'' \oplus H$. Thus, $\mathsf{RT}^n_k$ can be Weihrauch reduced to $\mathsf{RT}^{n-1}_k * \mathsf{TJ} * \mathsf{TJ}$ as follows. First, the given $\mathsf{RT}^n_k$-instance $c$ is passed as a $\mathsf{TJ}$-instance, yielding $c'$ since this is the only solution. Next, $c'$ is passed as a $\mathsf{TJ}$-instance, yielding $c''$. Now, from $c''$ we uniformly compute the $\mathsf{RT}^{n-2}_k$-instance $d$, and from $c''$ and any $\mathsf{RT}^{n-2}_k$-solution $H$ to $d$ we compute an $\mathsf{RT}^n_k$-solution to $c$, as needed. □

## 4.6 $\omega$-model reducibility

The major limitation of computably reducing to $m$ applications is that the number $m$ is fixed. Consider problems $\mathsf{P}$ and $\mathsf{Q}$, with the instances of $\mathsf{P}$ partitioned into (finitely or infinitely many) sets $I_0, I_1, \ldots$. For each $i$, let $\mathsf{P}_i$ be the restriction of $\mathsf{P}$ to $I_i$, and suppose $\mathsf{P}_i$ is computably reducible to $m_i$ applications of $\mathsf{Q}$ for some $m_i$. If we are only looking at finitely many $i$, then $\mathsf{P}$ itself is computably reducible to $\max\{m_0, m_1, \ldots\}$ applications of $\mathsf{Q}$. But if there are infinitely many $i$, and it happens for example that $m_0 < m_1 < \cdots$, then there is no reason $\mathsf{P}$ should to be computably reducible to $m$ applications of $\mathsf{Q}$ for any $m$. And yet, each individual instance of $\mathsf{P}$ still requires only "finitely many applications of $\mathsf{Q}$" to solve.

There is thus a disconnect between the intuitive and formal notions here. In the example above, the latter seems only marginally better than just (ordinary) computable reducibility at describing the relationship between $\mathsf{P}$ and $\mathsf{Q}$.

Rather than further elaborating on our previous reducibilities, we find the right formalism by starting from scratch. We would like to say that $\mathsf{P}$ is reducible to $\mathsf{Q}$ if having the computational resources to solve every given instance of $\mathsf{Q}$ implies the same for every given instance of $\mathsf{P}$. What should having computational resources of a particular sort mean? We take a model theoretic view.

**Definition 4.6.1 ($\omega$-model).** An *$\omega$-model* is a nonempty collection $\mathcal{S} \subseteq 2^\omega$ closed downward under $\leqslant_T$ and $\oplus$.

We regard an $\omega$-model as representing a universe of sets that we, with our computational resources, know how to find. On this view, we can think of being able to "solve" a problem as a closure condition: our universe should be closed under finding solutions to instances of $\mathsf{Q}$.

**Definition 4.6.2.** An $\omega$-model $\mathcal{S}$ *satisfies* a problem $\mathsf{P}$, written $\mathcal{S} \vDash \mathsf{P}$, if for every $\mathsf{P}$-instance $X \in \mathcal{S}$ there is at least one solution $Y \in \mathcal{S}$.

Let us illustrate this concept with a familiar example.

**Proposition 4.6.3.** *Let $\mathcal{S}$ be an $\omega$-model. Then $\mathcal{S} \vDash \mathsf{WKL}$ if and only if for every $A \in \mathcal{S}$ there exists an $X \in \mathcal{S}$ such that $X \gg A$.*

*Proof.* First, suppose $\mathcal{S} \vDash \mathsf{WKL}$ and fix $A \in \mathcal{S}$. By Proposition 2.8.14, relativized to $A$, there exists an infinite $A$-computable tree each of whose paths has PA degree relative to $A$. Since $\omega$-models are closed under $\leqslant_T$, this tree belongs to $\mathcal{S}$. Since $\mathcal{S} \vDash \mathsf{WKL}$, some path through this tree belongs to $\mathcal{S}$. Hence, $\mathcal{S}$ contains a set of PA degree relative to $A$, as desired. Conversely, let $T \subseteq 2^{<\omega}$ be any infinite tree in $\mathcal{S}$, and fix $X \in \mathcal{S}$ which has PA degree relative to $T$. By definition, $X$ computes a path through $T$, which is then in $\mathcal{S}$. Since $T$ was arbitrary, $\mathcal{S} \vDash \mathsf{WKL}$. □

In computability theory, $\omega$-models are called *Turing ideals*. Those that satisfy $\mathsf{WKL}$ are called *Scott sets*, and those that satisfy the problem $\mathsf{TJ}$ are called *jump ideals*.

We can now define a reducibility based on the above notions.

**Definition 4.6.4 ($\omega$-model reducibility).** Let $\mathsf{P}$ and $\mathsf{Q}$ be problems.

1. $\mathsf{P}$ is *$\omega$-model reducible* to a problem $\mathsf{Q}$, written $\mathsf{P} \leqslant_\omega \mathsf{Q}$, if every $\omega$-model satisfying $\mathsf{Q}$ also satisfies $\mathsf{P}$.
2. $\mathsf{P}$ and $\mathsf{Q}$ are *equivalent over $\omega$-models*, written $\mathsf{P} \equiv_\omega \mathsf{Q}$, if $\mathsf{P} \leqslant_\omega \mathsf{Q}$ and $\mathsf{Q} \leqslant_\omega \mathsf{P}$.

In Chapter 5, we will find that $\omega$-models are indeed models in the usual sense, i.e., structures in a particular language satisfying a particular set of formulas. But for now we may think of them simply as mathematical objects facilitating the above definition.

**Proposition 4.6.5.**

1. *$\leqslant_\omega$ is a transitive relation and $\equiv_\omega$ is an equivalence relation.*
2. *If $A$ is any set, then $\{Y : Y \leqslant_T A\}$ is an $\omega$-model.*

3. *If* $A_0 \leqslant_T A_1 \leqslant_T \cdots$ *are sets, then* $\{Y : (\exists i)[Y \leqslant_T A_i]\}$ *is an* $\omega$*-model.*
4. *If* P *admits computable solutions then every* $\omega$*-model satisfies* P.

Note that every $\omega$-model contains the computable sets. In fact, from part (2) above, we see that the collection of all computable sets forms an $\omega$-model.

**Definition 4.6.6 (The model REC).**

1. The $\omega$-model consisting of precisely the computable sets is denoted REC.
2. A problem P is *computably true* if it is satisfied by REC.

The terminology here is potentially confusing, so it is important to be careful. If a problem P admits computable solutions then it is computably true. But it is possible for a problem to be computably true yet *not* admit computable solutions. Nonetheless, such cases are sufficiently unusual or unnatural that many authors use the terms interchangeably. The following examples illustrate this.

*Example 4.6.7.* Consider the assertion that every total function $\omega \to \omega$ is computable. In constructive mathematics, this principle is known as *Church's thesis* (not to be confused with the Church–Turing thesis from Chapter 2). From a classical viewpoint, this principle is false. But it is not constructively disprovable, and we can see this fact reflected in our framework as follows. We can view Church's thesis as a problem whose instances are all total functions $f : \omega \to \omega$, with the solutions to any such $f$ being all $e$ such that $f = \Phi_e$. This has instances with no solutions, but this is allowed for by the definition of instance–solution problems. In particular, these instances witness that the problem does not admit computable solutions. But every computable instance has a solution, and this is an element of $\omega$ and so is also computable. Hence, the problem form of Church's thesis is computably true.

*Example 4.6.8.* Let P be the problem whose instances are all sets, with the sole solution to any computable instance being itself, and the solution to any noncomputable instance $X$ being $X'$. Then P is computably true but it does not admit computable solutions. Note, too, that this problem witnesses a failure of relativization: the statement, "every computable instance of P has a computable solution" does *not* relativize. However, P is clearly designed solely for this purpose, and in that sense is not "natural" in any real sense. Ergo, it makes a very unconvincing counterexample to Maxim 2.7.1.

We move on to relate $\omega$-model reducibility with reducibility to multiple applications.

**Proposition 4.6.9.** *Let* P *and* Q *be problems. If* P *admits computable solutions or, for some* $m \geqslant 1$, P *is computably reducible to* $m$ *applications of* Q, *then* $\mathsf{P} \leqslant_\omega \mathsf{Q}$.

*Proof.* Fix any $\omega$-model $\mathcal{S} \vDash \mathsf{Q}$. If P admits computable solutions then it is satisfied by every $\omega$-model, so in particular $\mathcal{S} \vDash \mathsf{P}$. Assume therefore that P is computably reducible to $m$ applications of Q for some $m \geqslant 1$. Fix any instance $X \in \mathcal{S}$. By hypothesis, we obtain some Q-instance $\widehat{X}_0$ computable from $X$. Since $\omega$-models are closed under $\leqslant_T$, this $\widehat{X}_0$ belongs to $\mathcal{S}$, and so it has some solution $\widehat{Y}_0$ in $\mathcal{S}$. By hypothesis, we obtain some Q-instance $\widehat{X}_1$ computable from $X \oplus \widehat{Y}_0$ (assuming $m > 1$). Since $\omega$-models are closed under $\oplus$ and $\leqslant_T$, this $\widehat{X}_1$ again belongs to $\mathcal{S}$. Continuing, we find Q-instances $\widehat{X}_0, \ldots, \widehat{X}_{m-1}$ with solutions $\widehat{Y}_0, \ldots, \widehat{Y}_{m-1}$, all in $\mathcal{S}$. Now by hypothesis, $X \oplus \widehat{Y}_0 \oplus \cdots \oplus \widehat{Y}_{m-1}$ computes some P-solution $Y$ to $X$, necessarily in $\mathcal{S}$. Since $X$ was arbitrary, we conclude $\mathcal{S} \vDash \mathsf{P}$, as desired. □

It follows that if P is reducible to Q via *any* of the notions of reduction discussed above then $\mathsf{P} \leqslant_\omega \mathsf{Q}$. Thus, $\leqslant_\omega$ is the coarsest of our reducibilities.

We can show that $\omega$-model reducibility is a strict coarsening by looking at the following problem that is arguably somewhat artificial but illustrates the power of $\omega$-models.

**Proposition 4.6.10.** *Let* P *be the problem whose instances are all pairs* $\langle X, n\rangle$ *where* $X \in 2^\omega$ *and* $n \in \omega$, *with the unique solutions to any such* $\langle X, n\rangle$ *being* $X^{(n)}$. *Then* $\mathsf{P} \equiv_\omega \mathsf{TJ}$, *even though* P *neither admits computable solutions nor is it computably reducible to* $m$ *applications of* TJ *for any* $m$.

*Proof.* Note that TJ is identity reducible to P, so certainly $\omega$-model reducible. We next show that $\mathsf{P} \leqslant_\omega \mathsf{TJ}$. Fix any $\omega$-model $\mathcal{S} \vDash \mathsf{TJ}$ along with any P-instance $\langle X, n\rangle$ in $\mathcal{S}$. Then $X \in \mathcal{S}$, and since this is a TJ-instance, we must have $X' \in \mathcal{S}$. But $X'$ is also a TJ-instance, so we must have $X'' \in \mathcal{S}$, and so on. It follows that $X^{(m)} \in \mathcal{S}$ for all $m \geqslant 1$, and hence in particular, the P-solution $X^{(n)}$ to $\langle X, n\rangle$ belongs to $\mathcal{S}$. So $\mathcal{S} \vDash \mathsf{P}$. To conclude, fix $m \geqslant 1$. We show P is not computably reducible to $m$ applications of TJ. Consider the P-instance $\langle \varnothing, m+1\rangle$, which is computable and hence in $\mathcal{S}$. It is easy to see from Definition 4.5.1, that if P were reducible to $m$ applications of TJ then the instance $\langle \varnothing, m+1\rangle$ would necessarily have a solution $Y \leqslant_T \varnothing^{(m)}$. But this is false since its only solution is $\varnothing^{(m+1)}$. □

A more natural example is the following, which will be of further significance to us in Chapter 9, where we also give a proof.

**Proposition 4.6.11.** *For each* $n \geqslant 3$, *we have* $\mathsf{RT} \equiv_\omega \mathsf{RT}^n_2$, *but* RT *neither admits computable solutions nor is it computably reducible to* $m$ *applications of* $\mathsf{RT}^n_2$ *for any* $m$.

Speaking of Ramsey's theorem, notice that if we combine Proposition 4.5.9 with the fact that $\mathsf{RT}^n_j$ is a subproblem of $\mathsf{RT}^n_k$ for all $n$ and all $j < k$, we obtain the following result, which is of independent interest. It is instructive to look at another proof.

**Corollary 4.6.12.** *For all* $n \geqslant 1$ *and* $j \geqslant 2$, $\mathsf{RT}^n \equiv_\omega \mathsf{RT}^n_j$.

*Proof.* Fix an $\omega$-model $\mathcal{S}$ of $\mathsf{RT}^n_j$ along with $k \geqslant 2$. We claim that $\mathcal{S} \vDash \mathsf{RT}^n_k$. The proof is by induction on $k$. If $k \leqslant j$, the result is clear, and this includes the base case, $k = 2$. So, assume $k > j$ and the result is true for $k - 1$. Let $c\colon [\omega]^n \to k$ be an instance of $\mathsf{RT}^n_k$ in $\mathcal{S}$. Define $d\colon [\omega]^n \to k-1$ as follows:

$$d(\vec{x}) = \begin{cases} c(\vec{x}) & \text{if } c(\vec{x}) < k-1, \\ k-2 & \text{otherwise.} \end{cases}$$

Now, $d \leqslant_{\mathrm{T}} c$, so $d \in \mathcal{S}$. Thus, we may fix an infinite homogeneous set $H_d$ for $d$ in $\mathcal{S}$. If $H_d$ is homogeneous with color $i < k-2$, then it is also homogeneous for $c$, by construction. Otherwise, $c(\vec{x})$ is either $k-2$ or $k-1$, for all $\vec{x} \in [H_d]^n$. In this case, enumerate the elements of $H_d$ as $h_0 < h_1 < \cdots$, and define $e \colon [\omega]^n \to 2$ as follows:

$$e(x_0, \ldots, x_{n-1}) = \begin{cases} 0 & \text{if } c(h_{x_0}, \ldots, h_{x_{n-1}}) = k-2, \\ 1 & \text{otherwise,} \end{cases}$$

for all $\langle x_0, \ldots, x_{n-1} \rangle \in [\omega]^n$. We have $e \leqslant_{\mathrm{T}} c \oplus H$, hence $e \in \mathcal{S}$. Since $j \geqslant 2$, $\mathcal{S}$ satisfies $\mathsf{RT}^n_2$, and so we may fix an infinite homogeneous set $H_e$ for $e$ in $\mathcal{S}$. Let $H = \{h_x : x \in H_e\}$, which is computable from $H_e \oplus H_d$, and hence also belongs to $\mathcal{S}$. Now if $H_e$ is homogeneous for $e$ with color 0 then $H$ is homogeneous for $c$ with color $k-2$, and if $H_e$ is homogeneous for $e$ with color 1 then $H$ is homogeneous for $c$ with color $k-1$. □

We will reference this argument in Section 9.1.

In practice, it is common for a problem $\mathsf{P}$ which is $\omega$-model reducible to a problem $\mathsf{Q}$ to be in fact computably reducible, and often even Weihrauch reducible, to $\mathsf{Q}$. As illustrated in this chapter, the study of when one reduction holds and not another helps shade out differences (subtle and not subtle) between problems, and so it is also with $\omega$-model reducibility. But since this is the coarsest of our reducibilities, separating problems using it gives the strongest measure of difference.

To understand such separations better, we now introduce an important method, called *iterating and dovetailing*, for building $\omega$-models with prescribed computability theoretic properties. We recall the definition of *preservation* from Definition 3.6.11.

**Theorem 4.6.13.** *Let $\mathcal{C}$ be a class of sets closed downward under $\leqslant_{\mathrm{T}}$, and let $\mathsf{P}$ and $\mathsf{Q}$ be problems. If $\mathsf{P}$ admits preservation of $\mathcal{C}$ but $\mathsf{Q}$ does not then there is an $\omega$-model $\mathcal{S} \subseteq \mathcal{C}$ satisfying $\mathsf{P}$ but not $\mathsf{Q}$. In particular, $\mathsf{Q} \nleqslant_\omega \mathsf{P}$.*

*Proof.* We build an $\omega$-model satisfying $\mathsf{P}$ but not $\mathsf{Q}$. To this end, we first define sets $A_0 \leqslant_{\mathrm{T}} A_1 \leqslant_{\mathrm{T}} \cdots$, as follows. Let $A_0 \in \mathcal{C}$ be a witness to $\mathsf{Q}$ not admitting preservation of $\mathcal{C}$. Assume now that we have defined $A_i$ for some $i \in \omega$ and that $A_i \in \mathcal{C}$. Say $i = \langle e, s \rangle$, so that $s \leqslant i$. If $\Phi_e^{A_s}$ is not a $\mathsf{P}$-instance, let $A_{i+1} = A_i$. Otherwise, using the fact that $\mathsf{P}$ admits preservation of $\mathcal{C}$, let $Y$ be any solution to the $\mathsf{P}$-instance $\Phi_e^{A_s}$ (which is also an $A_i$-computable instance) with $A_i \oplus Y \in \mathcal{C}$. Let $A_{i+1} = A_i \oplus Y$.

Define $\mathcal{S} = \{Y \in 2^\omega : (\exists i)[Y \leqslant_{\mathrm{T}} A_i]\}$. By Proposition 4.6.5, $\mathcal{S}$ is an $\omega$-model, and we claim that it satisfies $\mathsf{P}$. If $X$ is any $\mathsf{P}$-instance in $\mathcal{S}$, then $X = \Phi_e^{A_i}$ for some $e$ and $i$. Then $A_{\langle e,i \rangle + 1}$ computes a solution to $X$, which is consequently in $\mathcal{S}$. This establishes the claim.

It remains only to show that $\mathcal{S}$ does not satisfy $\mathsf{Q}$. Let $\widehat{X}$ be the $A_0$-computable $\mathsf{Q}$-instance in $\mathcal{C}$ such that no solution $\widehat{Y}$ to $\widehat{X}$ satisfies $A_0 \oplus \widehat{Y} \in \mathcal{C}$. Then $\widehat{X} \in \mathcal{S}$, and we claim it has no solution in $\mathcal{S}$. Indeed, if some Q-solution $\widehat{Y}$ to $\widehat{X}$ were computable from some $A_i$, then so would $A_0 \oplus \widehat{Y}$. But as $A_i$ is in $C$, which is closed downward under $\leqslant_T$, this would imply that $A_0 \oplus \widehat{Y} \in C$. □

Here, we see the reason for the name: we iterate the fact that P admits preservation of $C$ to find the solutions $Y$ above, "dovetailing" them into the construction by joining with the $A_i$.

**Corollary 4.6.14.** *If* P *admits cone avoidance then for every* $C \nleqslant_T \emptyset$ *there is an* $\omega$*-model satisfying* P *that does not contain* $C$*. In particular,* TJ $\nleqslant_\omega$ P.

*Proof.* Let $C$ be the collection of sets $A$ so that $C \nleqslant_T A$. □

**Corollary 4.6.15.** *If* P *admits PA avoidance then there is an* $\omega$*-model satisfying* P *that does not contain any set of PA degree. In particular,* WKL $\nleqslant_\omega$ P.

*Proof.* Let $C$ be the collection of sets $A \not\gg \emptyset$. □

Another way to use Theorem 4.6.13 is to use complexity facts about a problem P to construct specific $\omega$-models satisfying that problem.

**Theorem 4.6.16.** *Fix any* $D \gg \emptyset$*. There exists an* $\omega$*-model* $\mathcal{S}$ *satisfying* WKL *such that* $D \gg Y$ *for every* $Y \in \mathcal{S}$.

*Proof.* Let $C$ be the class of all sets $Y$ such that $D \gg Y$ and apply Proposition 3.6.12 and Theorem 4.6.13. □

This result is often summarized by "WKL has $\omega$-models below every PA degree". By taking a low set of PA degree, say, we can obtain an $\omega$-model satisfying WKL contained entirely in the class of low sets.

In general, having many examples of $\omega$-models of a problem P (without explicitly involving another problem Q) is useful. It helps to produce separations when they are needed, and yields a more complete understanding of P.

## 4.7 Hirschfeldt–Jockusch games

We summarize all the reducibilities we have covered in Figure 4.6. There is an evolution of notions evident in the figure from left to right, with each reducibility being an elaboration on its predecessors, which are in turn special cases. We wrap up our discussion in this chapter with *Hirschfeldt–Jockusch games*, which encompass all our reducibilities in a single framework.

**Definition 4.7.1 (Hirschfeldt and Jockusch [148]).** Let P and Q be problems. The *Hirschfeldt–Jockusch game* $G(\mathsf{Q} \to \mathsf{P})$ is a two-player game in which Players I and II alternate playing subsets of $\omega$ as follows.

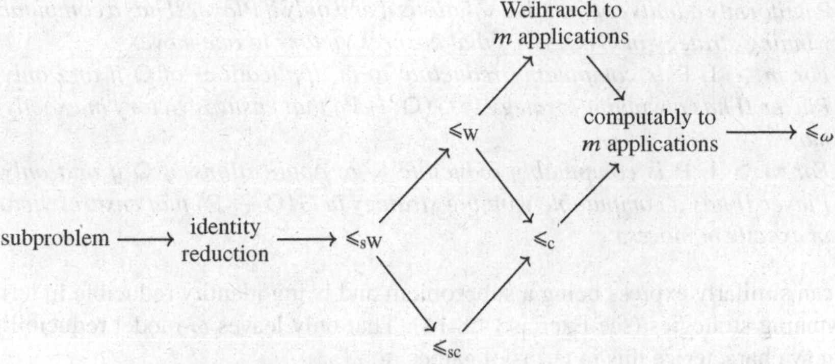


**Figure 4.6.** Relationships between various problem reducibilities. An arrow from one reducibility to another indicates that if P is reducible to Q in the sense of the first, then it is also reducible in the sense of the second. No additional arrows can be added.

1. On move 0, Player I plays a P-instance $X$.
2. On move 0, Player II plays either an $X$-computable P-solution $Y$ to $X$ and wins, or it plays an $X$-computable Q-instance $\widehat{X}_0$.
3. On move $i+1$, Player I plays a Q-solution $\widehat{Y}_i$ to Players II's Q-instance $\widehat{X}_i$.
4. On move $i+1$, Player II plays a either an $X \oplus \widehat{Y}_0 \oplus \cdots \oplus \widehat{Y}_i$-computable solution $Y$ to $X$, or an $X \oplus \widehat{Y}_0 \oplus \cdots \oplus \widehat{Y}_i$-computable Q-instance $\widehat{X}_{i+1}$.

The game ends if Player II ever plays a P-solution to $X$. If the game ends, Player II wins, and otherwise Player I wins.

There is a clear resemblance between these games and our reductions to multiple instances in Section 4.5. We make this explicit as follows.

**Definition 4.7.2.** Let P and Q be problems.

1. A *strategy for Player I* in $G(\mathsf{Q} \to \mathsf{P})$ is a function that, on input $i \in \omega$ and the sets played by Player II on moves $j < i$, outputs a set Player I can play on move $i$.
2. A *strategy for Player II* in $G(\mathsf{Q} \to \mathsf{P})$ is a function that, on input $i \in \omega$ and the sets played by Player I on moves $j \leqslant i$, outputs a set Player II can play on move $i$.
3. A strategy for Player I is*winning* if on each move $i$ when the game has not yet ended, playing the value of the strategy on $i$ and the sets played by Player II on moves $j < i$, ensures victory for Player I.
4. A strategy for Player II is if on each move $i$ when the game has not yet ended, playing the value of the strategy on $i$ and the sets played by Player I on moves $j \leqslant i$, ensures victory for Player II.
5. A strategy is *computable* if it is a Turing functional.

**Proposition 4.7.3 (Hirschfeldt and Jockusch [148]).** *Let* P *and* Q *be problems.*

1. P *admits computable solutions if and only if Player II has a winning strategy in* $G(\mathsf{Q} \to \mathsf{P})$ *that ensures victory in one move.*

2. P *uniformly admits computable solutions if and only if Player II has a computable winning strategy in* $G(\mathsf{Q} \to \mathsf{P})$ *that ensures victory in one move.*
3. *For* $m \geqslant 1$, P *is computably reducible to m applications of* Q *if and only if Player II has a winning strategy in* $G(\mathsf{Q} \to \mathsf{P})$ *that ensures victory in exactly m moves.*
4. *For* $m \geqslant 1$, P *is computably reducible to m applications of* Q *if and only if Player II has a computable winning strategy in* $G(\mathsf{Q} \to \mathsf{P})$ *that ensures victory in exactly m moves.*

We can similarly express being a subproblem and being identity reducible in terms of winning strategies (see Exercise 4.8.14). That only leaves $\omega$-model reducibility. We now characterize this in terms of games, too.

**Lemma 4.7.4.** *If* P *and* Q *are problems with* $\mathsf{P} \leqslant_\omega \mathsf{Q}$, *then either* P *admits computable solutions or every instance* $X$ *of* P *computes an instance of* Q.

*Proof.* Fix a P-instance $X$, and suppose $X$ computes no solution to itself. If $X$ computed no instance of Q, then $\{Y \in 2^\omega : Y \leqslant_T X\}$ would be an $\omega$-model that (trivially) satisfies Q but not P, which cannot be. □

**Theorem 4.7.5 (Hirschfeldt and Jockusch [148]).** *Let* P *and* Q *be problems. If* $\mathsf{P} \leqslant_\omega \mathsf{Q}$ *then Player II has a winning strategy in* $G(\mathsf{Q} \to \mathsf{P})$, *and otherwise Player I has a winning strategy.*

*Proof.* Suppose $\mathsf{P} \leqslant_\omega \mathsf{Q}$. We describe a strategy for Player II. On a move $i$ when the game has not yet ended, Player I will have played a P-instance $X$ and, for each $j < i$, a Q-solution $\widehat{Y}_j$. Now, if there is a P-solution $Y$ to $X$ computable from the join of $X$ and the $\widehat{Y}_j$ for $j < i$, then Player II can play it and win. If not, then fix an $X$-computable Q-instance $\widehat{X}$. Say $i = \langle e, s \rangle$, so that $s \leqslant i$. If $\Phi_e$, with oracle the join of $X$ and the $\widehat{Y}_j$ for $j < s$, is a Q-instance, then Player II plays it as $\widehat{X}_i$. Otherwise, Player II plays $\widehat{X}$ as $\widehat{X}_i$.

We claim Player II wins. Suppose not, so that the game goes on forever. Let $\mathcal{S}$ be the set of all sets computable from $X \oplus \widehat{Y}_0 \oplus \cdots \widehat{Y}_i$ for some $i$. Then $\mathcal{S}$ is an $\omega$-model, and by assumption, $\mathcal{S}$ contains $X$ but no P-solution to $X$. In particular, $\mathcal{S}$ does not satisfy P. Now suppose $\widehat{X}$ is any Q-instance in $\mathcal{S}$, say equal to $\Phi_e(X \oplus \widehat{Y}_0 \oplus \cdots \oplus \widehat{Y}_{s-1})$ for some $e$ and some $s > 0$. Then $\widehat{X} = \widehat{X}_i$ for $i = \langle e, s \rangle$, which means that $\widehat{Y}_i$ is a Q-solution to $\widehat{X}$. Since $\widehat{Y}_i \in \mathcal{S}$, and since $\widehat{X}$ was arbitrary, it follows that $\mathcal{S}$ satisfies Q. This contradicts P being $\omega$-model reducible to Q. We conclude that Player II wins, as claimed.

In the reverse direction, suppose that $\mathsf{P} \nleqslant_\omega \mathsf{Q}$. Fix an $\omega$-model $\mathcal{S}$ satisfying Q but not P. We describe a strategy for Player I. On move 0, Player I plays a P-instance $X$ in $\mathcal{S}$ with no solution in $\mathcal{S}$, which exists since $\mathcal{S} \nvDash \mathsf{P}$. Thus, Player II must play a Q-instance on move 0, and this Q-instance is necessarily in $\mathcal{S}$. Now suppose that the game has not ended through move $i$, and that all sets played on or before this move belong to $\mathcal{S}$. Then Player II must have played a Q-instance $\widehat{X}_i \in \mathcal{S}$ on move $i$, and since $\mathcal{S} \vDash \mathsf{Q}$, Player I can play a Q-solution $\widehat{Y}_i \in \mathcal{S}$ to $\widehat{X}_i$ on move $i+1$. Since Player II must play computably in the join of the finitely many sets played by Player I, it must play a set in $\mathcal{S}$. But then it cannot win on move $i+1$ either. We conclude that the game never ends, and hence Player I wins. □

## 4.8 Exercises

**Exercise 4.8.1.** A problem P *admits universal instances* if for every $A \in 2^\omega$, either P has no $A$-computable instance, or there exists an $A$-computable instance $X^u$ such that if $Y$ is any solution to $X^u$ then $A \oplus Y$ computes a solution to every other $A$-computable instance. Show that WKL admits universal instances. (We will see another example in Corollary 8.4.14.)

**Exercise 4.8.2.** Show that $\widehat{\mathsf{WKL}} \equiv_{\mathrm{sW}} \mathsf{WKL}$.

**Exercise 4.8.3.** Consider the restriction of $\widehat{\mathsf{IPHP}}$ to instances $\langle f_e : e \in \omega \rangle$ such that $\lim_x f_e(x)$ exists for every $e$ (i.e., each $f_e$ is eventually constant). Shows that this problem is computably reducible to TJ.

**Exercise 4.8.4.** Prove Proposition 4.3.7.

**Exercise 4.8.5.**

1. Show that if $X \gg \emptyset$, then every computable, computably bounded tree $T \subseteq \omega^{<\omega}$ has an $X$-computable path.
2. Let P be the problem whose instances are pairs $(T, b)$ such that $T$ is a $b$-bounded infinite tree, with the solutions to any such pair being all the infinite paths through $T$. Show that $\mathsf{P} \equiv_{\mathrm{c}} \mathsf{WKL}$.

**Exercise 4.8.6.** Prove Proposition 4.4.3.

**Exercise 4.8.7.** Prove Proposition 4.5.2 and formulate, and prove, a uniform analogue.

**Exercise 4.8.8.** Prove the compositional product, $*$, is associative.

**Exercise 4.8.9.** Construct a jump ideal that is not a Scott set.

**Exercise 4.8.10.** Recall the problems $\mathsf{C}_{\mathbb{N}}$ and $\mathsf{C}_n$ from Exercise 3.9.9. For all $n > m \geqslant 2$, prove the following.

1. $\mathsf{C}_m \leqslant_{\mathrm{sW}} \mathsf{C}_n \leqslant_{\mathrm{sW}} \mathsf{C}_{\mathbb{N}}$ for all $m < n$.
2. $\mathsf{C}_{\mathbb{N}} \nleqslant_{\mathrm{W}} \mathsf{C}_n \nleqslant_{\mathrm{W}} \mathsf{C}_m$ for all $m < n$.

**Exercise 4.8.11 (Pauly [247]; Brattka and Gherardi [19]).** Let P and Q be problems. Define the following problems.

1. $\mathsf{P} \sqcup \mathsf{Q}$ has domain all pairs $\langle 0, x \rangle$ for $x \in \mathrm{dom}(\mathsf{P})$ and $\langle 1, y \rangle$ for $y \in \mathrm{dom}(\mathsf{Q})$, with the solutions to $\langle 0, x \rangle$ being all elements of $\{0\} \times \mathsf{P}(x)$, and the solutions to $\langle 1, y \rangle$ being all elements of $\{1\} \times \mathsf{Q}(y)$.

2. $\mathsf{P} \sqcap \mathsf{Q}$ has domain $\mathrm{dom}(\mathsf{P}) \times \mathrm{dom}(\mathsf{Q})$, with the solutions to $\langle x, y \rangle$ being all elements of $(\{0\} \times \mathsf{P}(x)) \cup (\{1\} \times \mathsf{Q}(y))$.

Show that $\mathsf{P} \sqcup \mathsf{Q}$ is the join (i.e., least upper bound) of $\mathsf{P}$ and $\mathsf{Q}$ under $\leqslant_{\mathrm{W}}$, and that $\mathsf{P} \sqcap \mathsf{Q}$ is their meet (i.e., greatest lower bound). Show that the same is true under $\leqslant_{\mathrm{c}}$.

**Exercise 4.8.12.** Let $\mathcal{C}$ be a class of sets closed downward under $\leqslant_{\mathrm{T}}$. Let $\mathsf{P}_0$ and $\mathsf{P}_1$ be problems, each of which admits preservation of $\mathcal{C}$. Show that $\mathsf{P}_0 + \mathsf{P}_1$ admits preservation of $\mathcal{C}$. (So in particular, if $\mathsf{Q}$ is a problem that does not admit preservation of $\mathcal{C}$, then $\mathsf{Q} \nleqslant_{\omega} \mathsf{P}_0 + \mathsf{P}_1$.)

**Exercise 4.8.13.** Show that there is an $\omega$-model of $\mathsf{WKL}$ consisting entirely of sets of hyperimmune free degree.

**Exercise 4.8.14.** Formulate variants of the game $G(\mathsf{Q} \to \mathsf{P})$, or conditions on winning strategies, so as to obtain a characterize in the style of Proposition 4.7.3 of a problem $\mathsf{P}$ being a subproblem of $\mathsf{Q}$, or of being identity reducible to $\mathsf{Q}$.

# Part II
# Formalization and syntax

# Chapter 5
# Second order arithmetic

In the previous chapter, we discussed reducibility notions based on computability theory. Another method for comparing problems relies on provability: if we temporarily assume as an axiom that a problem P is solvable, how difficult is it to *prove* that a second problem Q is solvable? If we can prove that Q is solvable under the assumption that P is solvable, this gives us information that Q is "weaker" than P, at least modulo the other axioms used in our proof.

Viewing problems as representations of mathematical theorems, we can rephrase this second approach as: how hard is it to prove a particular theorem $T$ under the assumption that we are allowed to invoke a theorem $S$ as many times as we like? There is a clear analogy with $\omega$-model reducibility, but now we must also consider models with a nonstandard first-order part, and must *prove* $T$.

In this chapter we introduce the most common framework for this method: subsystems of second order arithmetic. The use of these subsystems has been part of the field since its inception, and the systems are sometimes viewed as a defining characteristic of reverse mathematics. In Section 5.11, we give a more detailed comparison of computability theoretic reducibilities with reducibilities based on formal systems.

## 5.1 Syntax and semantics

Second order arithmetic is a collection of theories in two sorted first order logic. Elements of the first sort will be called *numbers*, while elements of the second sort will be called *sets*. We assume a familiarity with basic definitions and results of first order logic. In the remainder of this section, we discuss how these definitions are specialized to the particular case of second order arithmetic.

**Definition 5.1.1 (Signature of $\mathcal{L}_2$).** The signature of second order arithmetic, $\mathcal{L}_2$, consists of the following.

- Constant number symbols 0 and 1.
- Binary function symbols + and $\cdot$ on numbers.



D. D. Dzhafarov, C. Mummert, *Reverse Mathematics*, Theory and Applications of Computability, https://doi.org/10.1007/978-3-031-11367-3_5

- Binary relation symbols $<$ and $=$ for numbers.
- A set membership relation $\in$ taking one number term and one set term.

The intended interpretation of the language, of course, is that number variables should range over the set $\omega$ of (standard) natural numbers, set variables should range over the powerset of $\omega$, the constants and arithmetic operations should be the standard ones, $<$ should be the order relation on $\omega$, and $\in$ should denote set membership.

The equality relation for sets is not included in the signature $\mathcal{L}_2$. Instead, we treat it as an abbreviation:

$$X = Y \equiv (\forall z)[z \in X \leftrightarrow z \in Y].$$

The reasons for this omission are discussed in Remark 5.1.8. We will silently assume the necessary definitional extension has been made whenever we use the set equality symbol in a formula.

The syntax of second order arithmetic, including the sets of terms and (well formed) formulas, is built up from the signature $\mathcal{L}_2$ in the usual manner. There are many numeric terms (for example, $1 + 1 + 1$ and $x + y + 0$) but, because there are no term forming operations for sets, the only terms for sets are the set variables themselves. Rather than using superscripts to indicate the sort of each variable, as might be common in type theory, we will usually use lowercase Roman letters for number variables and uppercase Roman letters for set variables. (We may deviate from this in specific instances, when the type is understood from context.)

Formulas in $\mathcal{L}_2$ may have several kinds of quantifiers. There are universal and existential quantifiers for number variables and for set variables. There are also the *bounded quantifiers* of the form $\forall x < t$ and $\exists x < t$, where $x$ is a number variable and $t$ is a number term. As usual, these are equivalent to slightly more complicated formulas with ordinary number quantifiers,

$$(\exists x < t)\varphi \equiv (\exists x)[x < t \wedge \varphi],$$
$$(\forall x < t)\varphi \equiv (\forall x)[x < t \rightarrow \varphi].$$

An $\mathcal{L}_2$ *theory* is simply a set of sentences in the signature $\mathcal{L}_2$. As usual, if $T$ is an $\mathcal{L}_2$ theory and $\varphi$ is an $\mathcal{L}_2$ sentence, we write $T \vdash \varphi$ if there is a formal proof of $\varphi$ from $T$ in one of the standard effective proof systems for two sorted first order logic.

The semantics of second order arithmetic are defined with a certain kind of first order structure.

**Definition 5.1.2 ($\mathcal{L}_2$ structures).** An $\mathcal{L}_2$ *structure* is a tuple

$$\mathcal{M} = \langle M, \mathcal{S}^{\mathcal{M}}, +^{\mathcal{M}}, \cdot^{\mathcal{M}}, 0^{\mathcal{M}}, 1^{\mathcal{M}}, <^{\mathcal{M}}\rangle,$$

where $M$ is a set that serves as the domain for quantifiers over number variables, $\mathcal{S}^{\mathcal{M}} \subseteq \mathcal{P}(M)$ serves as the domain for quantifiers over set variables, $0^{\mathcal{M}}$ and $1^{\mathcal{M}}$ are fixed elements of $M$, $+^{\mathcal{M}}$ and $\cdot^{\mathcal{M}}$ are fixed functions from $M \times M$ to $M$, and $<^{\mathcal{M}}$ is a binary relation on $M$.

The $\in$ relation is interpreted using the usual set membership relation. Exercise 5.13.7 explains why there is no loss of generality in this convention.

There are a number of standard effective deductive systems for two sorted first order logic, which are equivalent in terms of their provable sentences. We will assume one of these systems has been chosen, giving a notion of formal derivability and a notion of syntactic consistency. We will use the term *consistent* to refer to this notion of syntactic consistency.

**Definition 5.1.3 (Parameters).** Let $\mathcal{M}$ be an $\mathcal{L}_2$ structure and let $\mathcal{B}$ be a subset of $M \cup \mathcal{S}^{\mathcal{M}}$. Then $\mathcal{L}_2(\mathcal{B})$ denotes the language $\mathcal{L}_2$ expanded by adding a new constant symbol (of the appropriate sort) for each element of $\mathcal{B}$. The new constants are called (*first order* or *second order*) *parameters* from $\mathcal{M}$. If $\mathcal{B} = M \cup \mathcal{S}^{\mathcal{M}}$, we write $\mathcal{L}_2(\mathcal{M})$. If $\mathcal{B}$ is finite, say $\mathcal{B} = \{B_0, \ldots, B_{k-1}\}$, we write $\mathcal{L}_2(B_0, \ldots, B_{k-1})$.

**Definition 5.1.4 (Satisfaction).** Let $\mathcal{M}$ be an $\mathcal{L}_2$ structure and $\mathcal{B}$ a subset of $M \cup \mathcal{S}^{\mathcal{M}}$. Let $\varphi$ be a formula of $\mathcal{L}_2(\mathcal{B})$. Then the satisfaction relation $\mathcal{M} \vDash \varphi$ is defined in the usual way via the $T$-scheme, with each $B \in \mathcal{B}$ interpreted in $\mathcal{M}$ by $B$.

At the syntactic level, we always stay inside $\mathcal{L}_2$. But we say a formula $\varphi$ "has parameters", or words to this effect, to mean that it has free variables, which could be substituted by actual parameters in $\mathcal{L}_2(\mathcal{B})$ for some $\mathcal{M}$ and $\mathcal{B}$. If we *prove* $\varphi$ in some $\mathcal{L}_2$ theory $T$, then of course we are proving the universal closure of $\varphi$. So, if $\mathcal{M} \vDash T$, then $\mathcal{M}$ will satisfy every formula resulting from substituting actual parameters for the free variables of $\varphi$. In this sense, the "identity" of the parameters does not matter in our syntactic treatment, though of course their presence or absence does.

The main purpose for formalizing second order arithmetic in first order logic is to ensure the following fundamental metatheorems of first order logic are applicable to our system.

**Theorem 5.1.5.** *The following hold.*

- (Soundness theorem): *If an $\mathcal{L}_2$ theory is satisfied by any $\mathcal{L}_2$ structure then it is (syntactically) consistent.*
- (Completeness theorem): *If an $\mathcal{L}_2$ theory is consistent then it is satisfied by some $\mathcal{L}_2$ structure.*
- (Compactness theorem): *An $\mathcal{L}_2$ theory is consistent if and only if each of its finite subtheories is consistent.*
- (Downward Löwenheim–Skolem theorem): *If an $\mathcal{L}_2$ theory $T$ is consistent, then it is satisfied by some $\mathcal{L}_2$ structure $\mathcal{M}$ in which both $M$ and $\mathcal{S}^{\mathcal{M}}$ have cardinality $|T| + \aleph_0$.*

In Chapter 4, we discussed $\omega$-models in the guise of Turing ideals. In the context of second order arithmetic, there is a slightly broader notion: an $\omega$-model is typically defined to be an arbitrary submodel of the standard model, with the same first order part but possibly a smaller collection of sets. However, we will see in Theorem 5.5.3 that an $\omega$-model $\mathcal{M}$ in that sense satisfies the particular theory $\mathsf{RCA}_0$ if and only if $\mathcal{S}^{\mathcal{M}}$ is a Turing ideal. Because we will generally be concerned only with models of $\mathsf{RCA}_0$ or stronger theories, we will require the sets in an $\omega$-model to be a Turing ideal.

**Definition 5.1.6.**

1. An *ω-model* is an $\mathcal{L}_2$ structure $\mathcal{M}$ in which $M = \omega$, the arithmetical operations are the standard ones, and $\mathcal{S}^{\mathcal{M}}$ is a Turing ideal.
2. More generally, if $\mathcal{N}$ is a submodel of $\mathcal{M}$ in which $N = M$ and $\mathcal{S}^{\mathcal{N}} \subseteq \mathcal{S}^{\mathcal{M}}$, then $\mathcal{N}$ is an *ω-submodel* of $\mathcal{M}$, and $\mathcal{M}$ is an *ω-extension* of $\mathcal{N}$. In this case, it is not required that $M$ is actually $\omega$.

An important fact, which we will see generalized in Theorem 5.9.3, is that if $\mathcal{N}$ is an $\omega$-submodel of $\mathcal{M}$ then $\mathcal{N}$ and $\mathcal{M}$ satisfy the same arithmetical sentences (including with set parameters from $\mathcal{S}^{\mathcal{N}}$). See Exercise 5.13.1.

*Remark 5.1.7.* Because an $\omega$-model is completely determined by the collection of subsets of $\omega$ it contains, we usually identify the $\omega$-model with this collection of subsets. In this sense, our prior definition of $\omega$-model and the one given above are the same. We will follow this convention moving forward and not draw a distinction between the two notions of an $\omega$-model.

The *standard model of second order arithmetic* has $\omega$ as its first order part and the powerset of $\omega$ as its second order part. The class of nonstandard $\omega$-models is of particular interest because these share the same numerical properties as the standard model, differing only in the collection of sets. In particular, although any model can serve to show that a certain implication fails, a counterexample given by an $\omega$-model shows that the reason for the failure is second order rather than first order. We will see that there are implications which fail in general but which hold in every $\omega$-model, so knowing that a nonimplication is witnessed by an $\omega$-model provides additional information.

*Remark 5.1.8 (Set equality).* There is some variation in the literature about whether to include the set equality relation in the signature of second order arithmetic. From a purely foundational viewpoint, there is no harm in including it. There are two reasons we do not include it, which each involve the implicit quantifier in set equality.

The first reason relates to computability of $\Delta^0_1$ sentences. Provided set equality is not in the language, suppose that $\varphi(X, \vec{y}, \vec{Y})$ is a $\Sigma^0_1$ formula with a list of number parameters $\vec{y}$ and a list of set parameters $\vec{Y}$, and $\psi(X, \vec{y}, \vec{Y})$ is an equivalent formula that is $\Pi^0_1$. Then there is a computable functional $F$ with codomain $\{0, 1\}$ so that, for every $X$, $F(X, \vec{y}, \vec{Y}) = 1$ if and only of $\varphi(X, \vec{y}, \vec{Y})$ holds. On the other hand, given any set $Y$, there is no computable way to decide the formula $\varphi(X, Y) \equiv X = Y$. If set equality were included in the language, the formula $X = Y$ would be quantifier free, rather than $\Pi^0_1$ as it is under our conventions.

The second reason is the analogous issue with forcing. With our conventions, it is usually trivial to decide the relation $p \Vdash \varphi$ when $\varphi$ is quantifier free. If formulas such as $X = Y$ were quantifier free, the definition of forcing would be more complicated.

The set equality relation is closely related to axioms of extensionality. In second order arithmetic, because the only relation symbol involving sets is $\in$, we automatically have an axiom of extensionality (Exercise 5.13.8):

$$(\forall n)[n \in X \leftrightarrow n \in Y] \rightarrow [\varphi(X) \leftrightarrow \varphi(Y)].$$

In higher-order arithmetic or arithmetic based on function application rather than set membership, there are a number of extensionality axioms that arise in practice (see Kohlenbach [183] and Troelstra [313]).

## 5.2 Hierarchies of formulas

In Section 2.6, we classified sets and relations into an arithmetical hierarchy that measures the level of noncomputability of the relations included in that hierarchy, in a manner made precise by Post's theorem.

In this section, we define a parallel hierarchy that classifies formulas based on alternations of quantifiers. This leads to a second way to understand the complexity of sets and relations. The relationship between the two, which we discuss later in this chapter, is a key motivation for using second order arithmetic to study reverse mathematics.

### 5.2.1 Arithmetical formulas

An $\mathcal{L}_2$ formula is *arithmetical* if it has no set quantifiers. Arithmetical formulas may still have free set variables and, in the extended language for an $\mathcal{L}_2$ structure, they may have set constants. The next definition assigns finer classifications to some (but not all) arithmetical formulas. The possible classifications are "$\Sigma^0_n$" for $n \in \omega$ and "$\Pi^0_n$" for $n \in \omega$.

**Definition 5.2.1.** The *arithmetical hierarchy* is a classification of certain $\mathcal{L}_2$ formulas.

- A formula with only bounded quantifiers is given the classifications $\Sigma^0_0$ and $\Pi^0_0$.
- If $\varphi$ is $\Sigma^0_n$ then $(\forall \vec{x})\varphi$ is given the classification $\Pi^0_{n+1}$
- If $\varphi$ has classification $\Pi^0_n$ then $(\exists \vec{x})\varphi$ is given the classification $\Sigma^0_{n+1}$.

Here $(\forall \vec{x})$ and $(\exists \vec{x})$ represent finite sequences of universal existential number quantifiers, respectively. To emphasize that a formula has a set parameter $B$, we use the notations $\Sigma^{0,B}_n$, $\Pi^{0,B}_n$, $\Sigma^0_n(B)$, or $\Pi^0_n(B)$.

Not all arithmetical formulas are classified in the arithmetical hierarchy: the definition only assigns a classification to formulas in prenex normal form. However, a formula not in that form may still be logically equivalent to a formula with a classification. For example $(\exists x)(x = x) \rightarrow (\forall y)[y = y]$ is logically equivalent to $(\forall x)(\forall y)[x = x \rightarrow y = y]$, which is a $\Pi^0_1$ formula. By the usual result that every formula is equivalent to a formula in prenex form, every arithmetical formula is logically equivalent to some formula in the arithmetical hierarchy, possibly after rewriting bounded quantifiers in a equivalent way using unbounded quantifiers if necessary. A deeper analysis of the interaction between bounded quantifiers and unbounded quantifiers is given in Proposition 6.1.2.

Similarly, because we may prefix a formula with dummy quantifiers, every $\Pi^0_1$ formula is logically equivalent to a $\Sigma^0_2$ formula, a $\Pi^0_3$ formula, and so on. Thus, given a formula $\varphi$, we will usually be interested in the minimal $n$ such that $\varphi$ is equivalent to some $\Sigma^0_n$ or $\Pi^0_n$ formula.

We will sometimes abuse definitions by claiming that a formula $\varphi$ has a classification when in fact $\varphi$ is only equivalent to a formula with that classification. This is not as safe as it may seem: a key phenomenon is that a formula $\varphi$ may not be logically equivalent a formula $\psi$, but a nontrivial theory may prove that $\varphi \leftrightarrow \psi$, so that when we work within that theory we can treat $\varphi$ as if it has the classification of $\psi$. For example, the statement that a binary tree has an infinite path is not, naively, an arithmetical statement, but over the subsystem $\mathsf{WKL}_0$ this statement is equivalent to a $\Pi^0_2$ formula stating the tree is infinite. Similarly, we may have an equivalence $\varphi \leftrightarrow \psi$ under the combined assumptions at a particular place in a proof, although $\varphi$ and $\psi$ are not equivalent in general.

The following three results link the arithmetical hierarchy of relations from Definition 2.6.1 and the arithmetical hierarchy for formulas from Definition 5.2.1. They also give a sense of how computability theory can be formalized into second order arithmetic. The proofs are left to Exercise 5.13.17.

**Lemma 5.2.2.** *For each primitive recursive function $f(\vec{n}, h)$ there is a $\Sigma^0_1$ formula $\rho(\vec{n}, z, h)$ and a $\Pi^0_1$ formula $\psi(\vec{n}, z, h)$ such that, for all $h$, $\vec{n}$, and $z$,*

$$f(\vec{n}, h) = z \leftrightarrow \rho(\vec{n}, z, h) \leftrightarrow \psi(\vec{n}, z, h).$$

Applying the lemma to Kleene's $T$ and $U$ functions yields the following theorem on the formalization of computability theory.

**Theorem 5.2.3.** *For each $k$, there is a $\Sigma^0_1$ formula $\rho(e, \vec{n}, z, h)$ and a $\Pi^0_1$ formula $\psi(e, \vec{n}, z, h)$ such that for all $e$, $h$, $\vec{n}$ of length $k$, and $z$,*

$$\varphi^h_e(\vec{n}) = z \leftrightarrow \rho(e, \vec{n}, z, h) \leftrightarrow \psi(e, \vec{n}, z, h)$$

Combining the previous theorem with Post's theorem gives a tight connection between the two versions of the arithmetical hierarchy.

**Corollary 5.2.4.** *Let $B \subseteq \omega$ be arbitrary and $n > 0$. A set $X \subseteq \omega$ is $\Sigma^{0,B}_n$ (or $\Pi^{0,B}_n$) if and only if it is definable by a $\Sigma^{0,B}_n$ (or $\Pi^{0,B}_n$, respectively) formula of $\mathcal{L}_2$.*

This corollary is extremely useful for producing upper bounds on the classification of a set in the arithmetical hierarchy. Given a formula that defines a set, routine manipulations can put the formula into prenex normal form, at which point we can count the number of quantifier alternations to determine an upper bound. Rogers [259] called this method the *Tarski–Kuratowski algorithm*.

### 5.2.2 Analytical formulas

The arithmetical hierarchy is limited to formulas with no set quantifiers. The *analytical hierarchy* is an analogous classifications for formulas that may contain set quantifiers. While the arithmetical hierarchy begins with bounded quantifier formulas at the bottom level, the analytical hierarchy begins with arithmetical formulas at the bottom level. The possible classifications in the analytical hierarchy are $\Sigma^1_n$ for $n \in \omega$ and $\Pi^1_n$ for $n \in \omega$.

**Definition 5.2.5.** We inductively assign classifications as follows.

- Every arithmetical formula is given classifications $\Sigma^1_0$ and $\Pi^1_0$.
- If $\varphi$ is $\Sigma^1_n$ then $(\forall \vec{X})\varphi$ is given the classification $\Pi^1_{n+1}$.
- If $\varphi$ has classification $\Pi^1_n$ then $(\exists \vec{X})\varphi$ is given the classification $\Sigma^1_{n+1}$.

Here $(\forall \vec{X})$ and $(\exists \vec{X})$ represent finite sequences of universal and existential set quantifiers, respectively. To emphasize that a formula has a set parameter $B$, we write $\Sigma^{1,B}_n$, $\Pi^{1,B}_n$, $\Sigma^1_n(B)$, or $\Pi^1_n(B)$.

Each formula that has a classification in the analytical hierarchy is in a special kind of prenex form in which the set quantifiers appear first, followed by the number quantifiers, followed by a matrix which possesses only bounded quantifiers. Unlike the arithmetical case, we cannot simply use prenex normal form to place a formula into the analytical hierarchy, because we cannot easily move a number quantifier inside a set quantifier. We will soon discuss a system $\mathsf{ACA}_0$ which is strong enough to allow that kind of quantifier rearrangement.

## 5.3 Arithmetic

The goal of the next two sections is to describe axiom systems that are used to develop theories of second order arithmetic. The axioms can be divided into two kinds. The first order axioms ensure that basic number theoretic properties from $\omega$ will hold for the models we study, while the second order axioms describe the interaction between the numbers of a model and its sets.

### 5.3.1 First order arithmetic

We begin with the first order axioms, which are those that can be stated in a more restricted language.

**Definition 5.3.1.** The language $\mathcal{L}_1$ of first order arithmetic contains two binary function symbols, $+$ and $\cdot$; two constant symbols 0 and 1; an order relation symbol $<$; and an equality relation symbol $=$. The set of $\mathcal{L}_1$ formulas is defined using this signature and number quantifiers only. Thus an $\mathcal{L}_2$ formula is an $\mathcal{L}_1$ formula if and only if it contains no set variables whatsoever.

Every $\mathcal{L}_1$ formula is also an $\mathcal{L}_2$ formula. The definition of the arithmetical hierarchy (Definition 5.2.1) thus extends naturally also to $\mathcal{L}_1$ formulas. To be sure, there is a difference between $\mathcal{L}_1$ and $\mathcal{L}_2$ even for arithmetical formulas, since the latter can include parameters. But we tend to use the classes $\Sigma^0_n$ and $\Pi^0_n$ to refer to formulas both in $\mathcal{L}_1$ and $\mathcal{L}_2$, and rely on context to distinguish between the two when it matters. (See also the discussion at the beginning of Chapter 6.)

Peano arithmetic is often defined using only a successor operation and a few axioms including a second order induction axiom. In the first order setting, it is necessary to include the addition and multiplication relations from the start, because neither of these relations is first order definable over the structure $(\omega, S)$. To simplify working with restricted induction axioms, it is common to include a longer list of basic axioms (see also Kaye [176]).

**Definition 5.3.2.** The first order theory $\mathsf{PA}^-$ in the signature $\mathcal{L}_1$ with the following axioms, which describe a discrete ordered semiring.

1. $(\forall x, y, z)[(x+y)+z = x+(y+z)]$.
2. $(\forall x, y)[x+y = y+x]$.
3. $(\forall x, y, z)[(x \cdot y)\cdot z = x \cdot (y \cdot z)]$.
4. $(\forall x, y)[x \cdot y = y \cdot x]$.
5. $(\forall x, y, z)[x \cdot (y+z) = x \cdot y + x \cdot z]$.
6. $(\forall x)[x+0 = x \wedge x \cdot 0 = 0 \wedge x \cdot 1 = x]$.
7. $(\forall x, y, z)[x < y \wedge y < z \rightarrow x < z]$.
8. $(\forall x)[x \not< x]$.
9. $(\forall x, y)[x < y \vee x = y \vee y < x]$.
10. $(\forall x, y, z)[x < y \rightarrow x+z < y+z]$.
11. $(\forall x, y, z)[0 < z \wedge x < y \rightarrow x \cdot z < y \cdot z]$.
12. $(\forall x, y)(\exists z)[x < y \rightarrow x+z = y]$.
13. $(0 < 1) \wedge (\forall x)(x > 0 \rightarrow x \geqslant 1)$.
14. $(\forall x)[x = 0 \vee x > 0]$.

Here we have adopted the conventions that $a \not< b$ abbreviates $\neg(a < b)$ and $a \leqslant b$ abbreviates $a = b \vee a < b$.

The theory $\mathsf{PA}^-$ captures the basic properties of the natural numbers as a discrete ordered semiring. To verify stronger properties, we will use induction axioms.

**Definition 5.3.3.** *Peano arithmetic* is the first order theory in the signature $\mathcal{L}_1$ that is obtained by adjoining to $\mathsf{PA}^-$ every instance of the induction scheme

$$(\varphi(x) \wedge (\forall x)[\varphi(x) \rightarrow \varphi(S(x))]) \rightarrow (\forall x)\varphi(x) \tag{5.1}$$

in which $\varphi$ is a formula of $\mathcal{L}_1$.

The key role of first order induction axioms in reverse mathematics is to verify number theoretic properties of the model. We end the section with one of the most important of these, which is the existence of a *pairing function* that codes pairs of numbers by numbers. We already fixed such a function in Section 1.5, but we also need a pairing function in our formal theories of arithmetic. This is accommodated by the following theorem.

**Theorem 5.3.4.** $\mathsf{PA}^-$ *together with induction for all* $\Sigma^0_1$ *formulas of* $\mathcal{L}_1$ *proves that the function* $(n,m) \mapsto (n+m)^2+m$ *is an injection from pairs of numbers to numbers.*

Of course, $\mathsf{PA}^-$, as an $\mathcal{L}_1$ theory, cannot speak of functions directly. So what the above means is that $\mathsf{PA}^-$ can verify the properties of $(n+m)^2+m$ relative to $n$ and $m$ that make the map above an injective function. (For example, that for all $n, n^*, m, m^*$, if $(n+m)^2+m = (n^*+m^*)^2+m^*$, then $n = n^*$ and $m = m^*$.) The proof is standard, but of limited interest for our present purposes. We refer the reader Simpson [288, Theorem II.2.2] for complete details.

As in our informal discussion, we denote the code for the pair $(n,m)$ by $\langle n,m\rangle$. So, $\langle n,m\rangle = (n+m)^2+m$. This makes the definition of $\langle n,m\rangle$ simple, since it is just a term of $\mathcal{L}_1$, and therefore easy to work with. A downside is that, unlike the pairing function discussed in Chapter 1, this function is not surjective. This is not a problem for proving the basic results we need to get our discussion underway, however, most notably, the coding of finite sets in Section 5.5.2. In short order, we will have the tools to also define a bijection $\mathbb{N}\times\mathbb{N}\to\mathbb{N}$ (see Exercise 5.13.25), and at that point we may use $\langle n,m\rangle$ to refer to this bijection if convenient.

### 5.3.2 Second order arithmetic

In an arbitrary $\mathcal{L}_2$ structure $\mathcal{M}$, the collection of sets may be much smaller than the powerset of $M$. Thus, in defining theories of second order arithmetic, we must include explicit axioms stating that particular sets *exist*, to guarantee the sets we wish to use will be present in the models of our theories. Informally, axioms that imply the existence of sets will be called *set existence axioms*.

By carefully choosing which set existence axioms are included (along with other axioms), we will be able to construct theories of various strengths. In particular, we will consider axioms that state that if a set $X$ exists, and $Y$ is definable by a formula with $X$ as a parameter, then $Y$ also exists.

This manner of speaking deserves a small comment, because it is trivial that the set $Y$ will exist in the usual mathematical sense. When we speak of sets that "exist", we are thinking in the context of some fixed (possibly nameless) $\mathcal{L}_2$ structure: a set *exists* if it is a set in this structure.

Usually, we will try to be precise and only call something a *set* if it does exist in this sense. We will then refer to every other *collection* of numbers as just that (or, following set theoretic terminology, as a *class*). But occasionally, it may be convenient to use the word "set" informally, and here we will rely on the word "exists" to distinguish our meaning. For example, we will often encounter situations where not every $\Sigma^0_1$ definable collection of numbers exists. Here, we may refer to such a collection as a "$\Sigma^0_1$ definable set", so long as we are careful not to state that the collection exists, or to treat it as if it does.

**Definition 5.3.5.** Let $\varphi$ be a formula in the language $\mathcal{L}_2$.

1. The *comprehension axiom* for $\varphi$ is the universal closure of
$$(\exists X)(\forall x)[x \in X \Leftrightarrow \varphi(x)],$$
where $X$ is a set variable not mentioned in $\varphi$. The formula $\varphi$ may have free set variables, which serve as parameters relative to which $X$ is defined.
2. The *induction axiom* for $\varphi$ is the universal closure of
$$(\varphi(0) \wedge (\forall x)[\varphi(x) \rightarrow \varphi(x+1)]) \rightarrow (\forall x)\varphi(x). \tag{5.2}$$
3. For $\Gamma$ a collection of formulas of $\mathcal{L}_2$, $\Gamma$-$\mathsf{CA}$ is the axiom scheme consisting of the comprehension axiom for every $\varphi \in \Gamma$, while $\mathsf{I}\Gamma$ is the axiom scheme consisting of the induction axiom for every $\varphi \in \Gamma$.

Comprehension axioms are typical set existence axioms. There are other set existence axioms, however, as we will comment on below.

**Definition 5.3.6.** The theory $\mathsf{Z}_2$ of *full second order arithmetic* includes the axioms of $\mathsf{PA}^-$, the induction axiom scheme for all $\mathcal{L}_2$ formulas, and the comprehension axiom scheme for all $\mathcal{L}_2$ formulas.

$\mathsf{Z}_2$ is an amazingly strong theory. Indeed, it is somewhat challenging to find a mathematical theorem that can be naturally expressed as an $\mathcal{L}_2$ formula but cannot be proved in $\mathsf{Z}_2$. Of course, there are well known examples, such as $\mathrm{Con}(\mathsf{Z}_2)$, that cannot be proved in $\mathsf{Z}_2$ due to incompleteness. A somewhat more mathematical example, due to Friedman [107], is that determinacy for $\Sigma^0_5$ games is not provable in $\mathsf{Z}_2$ (see Section 12.3.4). Other results known to be provable in $\mathsf{ZFC}$ (Zermelo–Fraenkel set theory with the axiom of choice) but not $\mathsf{Z}_2$ are of similar character: they are familiar to set theorists but not to most undergraduate mathematics students.

Because $\mathsf{Z}_2$ is so strong, we consider fragments obtained by weakening the collection of comprehension and induction axioms that may be used. These *subsystems* of $\mathsf{Z}_2$ are a key subject of (classical) reverse mathematics. A typical such subsystem has the shape
$$\mathsf{PA}^- + \text{ a set existence axiom or scheme } + \mathsf{I}\Sigma^0_1,$$
with the set existence scheme often (but not always) being $\Gamma$-$\mathsf{CA}$ for some collection of formulas $\Gamma$. The motivations for restricting the induction axiom (5.2) to $\Sigma^0_1$ formulas are discussed in Section 6.6.

$\Gamma$-$\mathsf{CA}$ proves, for each $\varphi \in \Gamma$, that a set $X = \{n : \varphi(n)\}$ exists. The weak induction scheme $\mathsf{I}\Sigma^0_0$ already includes the *set induction* axiom
$$\mathsf{I}_{\mathsf{set}}: (0 \in X \wedge (\forall n)[n \in X \rightarrow n+1 \in X]) \rightarrow (\forall n)[n \in X].$$

Thus combining $\Gamma$-CA with $\mathsf{I}\Sigma^0_0$ gives all of $\mathsf{I}\Gamma$. In particular, if $\Gamma \supseteq \Sigma^0_0$, then all of the theories $\mathsf{PA}^- + \Gamma\text{-CA} + \mathsf{I}\Sigma^0_0$, $\mathsf{PA}^- + \Gamma\text{-CA} + \mathsf{I}_{\mathrm{set}}$, and $\mathsf{PA}^- + \Gamma\text{-CA} + \mathsf{I}\Gamma$ are the same.

*Remark 5.3.7 (The subscript 0).* The main subsystems of $\mathsf{Z}_2$ encountered in reverse mathematics are all named for their respective comprehension axiom or scheme and decorated with a subscript 0 to indicate the presence of $\mathsf{PA}^-$ and $\mathsf{I}\Sigma^0_1$. Thus, for example, $\Pi^1_1\text{-CA}_0$ denotes the system $\mathsf{PA}^- + \Pi^1_1\text{-CA} + \mathsf{I}\Sigma^0_1$, etc. Sometimes, however, dropping the subscript is used to indicate not the weaker system consisting of the comprehension axiom alone, but the *stronger* system obtained by the addition of induction for all formulas of $\mathcal{L}_2$. There is no good convention about this in the literature, which can lead to some confusion. We will largely avoid the latter practice in this book, but in any case will always be explicit when the subscript 0 is omitted.

## 5.4 Formalization, and on $\omega$ and $\mathbb{N}$

The language $\mathcal{L}_2$ lends itself naturally to *formalizing* various statements about $\omega$ and subsets of $\omega$. Once we develop a system for coding finite sequences by numbers in a weak subsystem of $\mathsf{Z}_2$ (see Section 5.5.2), we will be able to formalize more complicated objects defined in terms of $\omega$, like sequences of sets of numbers, the set $2^{<\omega}$, functions on $\omega$ of arbitrary arity, etc.

It is traditional, when arguing about some such formalized object in a subsystem of $\mathsf{Z}_2$, to use the symbol $\mathbb{N}$ in place of $\omega$. In this way, we reserve $\omega$ for the *bona fide* set of natural numbers. In $\mathsf{Z}_2$, we instead speak of $2^{<\mathbb{N}}$, about functions $\mathbb{N} \to \mathbb{N}$, etc., because the numbers in an arbitrary model may be nonstandard. Formally, these are abbreviations for certain definitions in $\mathcal{L}_2$, and hence care must taken that the objects so defined actually exist, in the sense discussed at the beginning of this section, in whatever formal theory we are employing.

Yet a third notation is used when dealing with structures. In an $\mathcal{L}_2$ structure $\mathcal{M}$ we will denote the "set of numbers" by $M$, and then speak of the set $2^{<M}$, of functions $M \to M$, etc. Here, we must again bear in mind that $M$ may be nonstandard, which can have implications for even some very basic notions (discussed in detail in Chapter 6). Of course, there is a tight connection between this and how much of $\mathsf{Z}_2$ holds in $\mathcal{M}$.

We move freely between these notations when discussing definitions, theorems, and proofs. In practice, this simply involves switching between $\omega$, $\mathbb{N}$, or $M$ (for a structure $\mathcal{M}$). Formally, of course, there is more going on. For example, consider the infinitary pigeonhole principle from Definition 3.2.3. As an $\forall\exists$ theorem, this states that for every function $f : \omega \to \omega$ with bounded range there is an $i \in \omega$ such that $f^{-1}\{i\}$ is infinite. In practice, we would write this in $\mathsf{Z}_2$ as "for every function $f : \mathbb{N} \to \mathbb{N}$ with bounded range there is an $i \in \mathbb{N}$ such that $f^{-1}\{i\}$ is infinite". But in actuality, we are using the fact that $\mathsf{Z}_2$ can code pairs of numbers, and interpreting the preceding as shorthand for the more opaque formula

$$
\begin{aligned}
(\forall X)[\ (\ &(\forall x)(\exists y)[\langle x,y\rangle \in X] \\
\wedge\ &(\forall x)(\forall y)(\forall z)[\langle x,y\rangle \in X \wedge \langle x,z\rangle \in X \to y = z] \\
\wedge\ &(\exists z)(\forall x)(\forall y)[\langle x,y\rangle \in X \to y < z]\ ) \\
\to\ &(\exists y)(\forall z)(\exists x)[x > z \wedge \langle x,y\rangle \in X]\ ].
\end{aligned}
$$

## 5.5 The subsystem $\mathsf{RCA}_0$

The weakest subsystem of $\mathsf{Z}_2$ we will consider is called $\mathsf{RCA}_0$. The initialism stands for "recursive comprehension axiom". As usual throughout the reverse mathematics and computability literature, "recursive" in this sense is a synonym for "computable". And, as we will see, $\mathsf{RCA}_0$ roughly corresponds to a formalization of computable mathematics.

### 5.5.1 $\Delta^0_1$ comprehension

**Definition 5.5.1.** The $\Delta^0_1$ comprehension scheme consists of the universal closure of each axiom of the form

$$(\forall n)[\varphi(n) \leftrightarrow \psi(n)] \to (\exists X)[n \in X \leftrightarrow \varphi(n)].$$

Intuitively, this axiom says that if a set $X$ is defined by a $\Sigma^0_1$ formula and also by a $\Pi^0_1$ formula, then we may assert that $X$ exists. We say that $X$ is $\Delta^0_1$ *definable* relative to the parameters of the formulas.

From the lens of computability theory, a set that is both $\Sigma^0_1$ and $\Pi^0_1$ is computable relative to the parameters of the formulas. Accordingly, the $\Delta^0_1$ comprehension scheme can be informally rephrased as: if a set $X$ is computable relative to other sets that exist, then $X$ also exists.

**Definition 5.5.2.** The formal system $\mathsf{RCA}_0$ consists of the basic numerical axioms $\mathsf{PA}^-$, the $\Delta^0_1$ comprehension scheme, and the induction axiom (5.2) restricted to $\Sigma^0_1$ formulas.

With reference to Remark 5.3.7, we note that $\mathsf{RCA}$ in the literature often refers to $\mathsf{RCA}_0$ together with full induction.

The analogy between $\mathsf{RCA}_0$ and computable mathematics is made more precise by the following theorem. It is common, when trying to understand the spirit of a subsystem, to consider the special case of $\omega$-models. This allows us to temporarily ignore the first order part of the theory and focus specifically on its set existence strength. The following theorem also justifies our restriction of $\omega$-models in Definition 5.1.6 to models in which $\mathcal{S}^{\mathcal{M}}$ is a Turing ideal.

**Theorem 5.5.3.** *Suppose that $\mathcal{M}$ is a submodel of the standard model of $\mathsf{Z}_2$. Then $\mathcal{M} \models \mathsf{RCA}_0$ if and only if $\mathcal{S}^{\mathcal{M}}$ is closed under Turing join and relative computability (i.e., if and only if $\mathcal{M}$ is a Turing ideal).*

*Proof.* First, assume $\mathcal{M}$ satisfies the $\Delta^0_1$ comprehension scheme. If $A, B \in \mathcal{S}^{\mathcal{M}}$, then $A \oplus B$ is in $\mathcal{S}^{\mathcal{M}}$, because $A \oplus B$ is $\Delta^0_1$ definable from $A$ and $B$. Similarly, if $A \in \mathcal{S}^{\mathcal{M}}$ and $B \leqslant_T A$ then $B$ is $\Delta^0_1$ definable from $A$ and hence $B \in \mathcal{S}^{\mathcal{M}}$.

Conversely, assume that $\mathcal{S}^{\mathcal{M}}$ is a Turing ideal. If a set $B$ is $\Delta^0_1$ definable relative to a sequence $A_1, \ldots, A_n$ of sets in $\mathcal{S}^{\mathcal{M}}$, then $B$ is $\Delta^0_1$ definable from $A = A_1 \oplus \cdots \oplus A_n \in \mathcal{S}^{\mathcal{M}}$. But then $B$ is computable from $A$, so $B$ is in $\mathcal{S}^{\mathcal{S}}$, as desired. □

In particular, we have the following basic but important observation. Recall the definition of REC from Definition 4.6.6.

**Corollary 5.5.4.**

1. REC $\models \mathsf{RCA}_0$.
2. *More generally, given any set $X \in 2^\omega$, the $\omega$-model (with second order part) $\mathcal{S} = \{Y \in 2^\omega : Y \leqslant_{\mathrm{T}} X\}$ is a model of* $\mathsf{RCA}_0$.

An $\omega$-model as in (2) is called a *topped model*, and is said to be *topped by* $X$. (Thus, REC is a topped model topped by any computable $X$.) We caution that this is *not* the same as an $\omega$-model $\mathcal{S}$ such that, for some $X$, we have $Y \leqslant_{\mathrm{T}} X$ for all $Y \in \mathcal{S}$. We will see many examples of models of the latter kind, but these will only be topped provided $X$ itself belongs to $\mathcal{S}$ (which it usually will not).

## 5.5.2 Coding finite sets

In contrast to $\omega$-models, nonstandard models of second order arithmetic, like non-standard models of first order arithmetic, exhibit unusual behaviors that one must be cautious about. Perhaps none is more striking than the distinction between "bounded" and "finite". For this reason, we need a more precise definition of *finite set*.

We previously described the pairing function $\langle n, m\rangle$, which will be a key tool. The method we will employ for coding finite sequences dates back to Gödel [125, 316], who showed that a version of the Chinese remainder theorem could be formalized in weak systems of arithmetic. To avoid repeating proofs that are present in numerous places in the literature, we will continue to refer to the development given by Simpson [288]. Another development of finite sets and sequences in weak systems of arithmetic is given by Hájek and Pudlák [134, Chapter I].

**Definition 5.5.5.** A number $c$ *represents* a set $X$ if there are $k$, $m$, and $n$ such that $c = \langle k, \langle m, n\rangle\rangle$ and, for all $i$, we have $i \in X$ if and only if

$$(i < k) \wedge (m \cdot (i + 1) + 1 \text{ divides } n). \tag{5.3}$$

The *code* for a set $X$ is the least $c$ that represents $X$, if such a number exists. A set $X$ is *coded* if it has a code.

The key property of the definition is that the property in (5.3) can be expressed with a formula of arithmetic that has only bounded quantifiers. Similarly, there is a primitive recursive function $\chi$ so that if a set $X$ is coded by a number $c$ then, for all $i$, we have $i \in X$ if and only if $\chi(i, c) = 1$. This allows us to handle coded sets within weak systems of arithmetic, which is the motivation for the specific coding method in (5.3).

When we work with nonstandard models of arithmetic, there is an important tension between sets that are finite, when viewed from outside the model, and sets that can be coded using numbers in the model.

**Definition 5.5.6.** Suppose $\mathcal{M}$ is a model of $\mathsf{RCA}_0$ and $X \subseteq M$. (We do not assume $X \in \mathcal{S}^{\mathcal{M}}$.)

1. $X$ is $\mathcal{M}$-*bounded* (or simply *bounded*) if there is a $n \in M$ such that $i \leqslant n$ for all $i \in X$.
2. $X$ is $\mathcal{M}$-*coded* (or simply *coded*) if there is a $c \in M$ that codes $X$ in the sense of Definition 5.5.5.

In the standard model, the two notions coincide. The following proposition is essentially the Chinese remainder theorem.

**Theorem 5.5.7.** *Every bounded subset of $\omega$ has a code, and every set that has a code is bounded.*

In arbitrary models, the relationship is more complicated. In a nonstandard model $\mathcal{M}$ of $\mathsf{RCA}_0$, there will always be subsets of $M$ that are bounded but not coded in the model, such as the set of standard natural numbers. We will discuss this phenomenon in detail in Section 6.2. The next result, which amounts to a formalization of the preceding one, shows that this cannot happen for sets in $\mathcal{S}^{\mathcal{M}}$.

**Theorem 5.5.8.** $\mathsf{RCA}_0$ *proves that, if $X$ is a bounded set, then $X$ has a code. That is, if $\mathcal{M} \vDash \mathsf{RCA}_0$, $X \in \mathcal{S}^{\mathcal{M}}$, and $X$ is $\mathcal{M}$-bounded, then $X$ is $\mathcal{M}$-coded.*

*Proof.* Arguing in $\mathsf{RCA}_0$, say $X$ is bounded by $k \in \mathbb{N}$. By primitive recursion, define $m = k!$. More precisely, define $f\colon k+1 \to \mathbb{N}$ as follows: let $f(0) = 1$, and for $i < k$ let $f(i+1) = f(i) \cdot (i+1)$; then let $m = f(k)$. By induction,

$$i + 1 \text{ divides } m \text{ for all } i < k. \tag{5.4}$$

We claim that for all $j < i < k$, $m(j+1)+1$ and $m(i+1)+1$ are relatively prime. Indeed, say $d$ divides both $m(j+1)+1$ and $m(i+1)+1$. Then $d$ also divides their difference, which is just $m(i-j)$. Setting $m(j+1)+1 = dq_j$ and $m(i-j) = dq$, we see that $m(j+1)+1$ divides $dq_jq = m(i-j)d_j$. But $m$ and $m(j+1)+1$ are relatively prime, so $m(j+1)+1$ must divide $(i-j)d_j$. (Check that $\mathsf{RCA}_0$ can prove this standard arithmetical fact!) On the other hand, $1 < i - j < k$ by hypothesis, so

$i - j$ divides $m$ by (5.4). Thus, $m(j+1)+1$ and $i-j$ must be relatively prime, and the former must consequently divide $d_j$. Since $d_j$ divides $m(j+1)+1$, it follows that $d = 1$, which proves the claim.

Now by primitive recursion, define $n = \prod_{i \in X}(m(i+1)+1)$. More precisely, define $g\colon k+1 \to \mathbb{N}$ as follows. Let $g(0) = 1$. For $i < k$, let

$$g(i+1) = \begin{cases} g(i) \cdot (m(i+1)+1) & \text{if } i \in X, \\ g(i) & \text{if } i \notin X. \end{cases}$$

And let $g(k) = n$.

Clearly, $m(i+1)+1$ divides $n$ if $i \in X$. An induction on $\ell \leqslant k$ shows that any prime factor of $g(\ell)$ must be a factor of $m(i+1)+1$ for some $i < \ell$ in $X$. By the claim above, it follows that $i \in X$ if $m(i+1)+1$ divides $n = g(k)$.

We conclude that $r = \langle k, \langle m, n\rangle\rangle$ represents $X$ in the sense of Definition 5.5.5. The set $R$ of all such representatives exists by $\Sigma^0_0$ comprehension, and hence is nonempty. Let $\varphi(x)$ be the formula $(\forall y \leqslant x)[y \notin R]$. Then we have $\neg\varphi(r)$, so by $\Sigma^0_0$ induction there is a $c \leqslant r$ such that $\neg\varphi(c)$ and either $c = 0$ or $c > 0$ and $\varphi(c-1)$. In any case, $c$ is a code for $X$. □

The statement of the previous theorem illustrates an important convention in the literature. When we consider an arbitrary model $\mathcal{M}$, there are two possible meanings for the word "set": arbitrary subsets of $M$, or elements of $\mathcal{S}^{\mathcal{M}}$. If we say that RCA$_0$ proves that all sets have a particular property $\Phi$, we mean that RCA$_0$ proves a sentence of the form $(\forall X)\Phi$, and thus proves the property for all sets in $\mathcal{S}^{\mathcal{M}}$. In contrast, the following theorem would often be stated in the literature as "RCA$_0$ proves that every coded set exists."

**Proposition 5.5.9.** *If $\mathcal{M}$ is a model of* RCA$_0$ *and $X \subseteq M$ is $\mathcal{M}$-coded, then $X \in \mathcal{S}^{\mathcal{M}}$.*

*Proof.* Suppose that $c$ is a code for $X$. The property in (5.3) can be expressed with only bounded quantifiers, and therefore $\Delta^0_1$ comprehension suffices to construct the set $X$. □

Combining the previous results gives a characterization of a particular class of sets in a model of RCA$_0$.

**Definition 5.5.10.** Suppose $\mathcal{M}$ satisfies RCA$_0$. A set $X \subseteq M$ is *$\mathcal{M}$-finite* if it satisfies the following equivalent conditions:

1. $X$ is $\mathcal{M}$-coded.
2. $X \in \mathcal{S}^{\mathcal{M}}$ and $X$ is $\mathcal{M}$-bounded.

When the model is clear from context, or if we are working within RCA$_0$, we may refer to these sets as simply *finite*. However, a set may be $\mathcal{M}$-finite but not finite in the standard sense. For example, if $\mathcal{M}$ is a nonstandard model and $n \in M$ is a nonstandard number, the set $L_n = \{i : i \leqslant n\}$ will be an $\mathcal{M}$-bounded set in $\mathcal{S}^{\mathcal{M}}$, and will thus be $\mathcal{M}$-coded with some nonstandard $c \in M$. But $L_n$ will contain every standard natural number, and will thus be infinite from an external viewpoint. This distinction between the sets that are finite from the perspective of a model, compared to sets that are finite from an external viewpoint, must be kept in mind when working with (possibly) nonstandard models, especially with regards to induction. We discuss this issue in more detail in Section 6.2.

The comprehension and induction axioms in $\mathsf{RCA}_0$ give us powerful tools to handle $\mathcal{M}$-finite sets. One example of this is the following theorem. We have seen that $\mathsf{RCA}_0$ does not include $\Sigma^0_1$ comprehension scheme, which is equivalent to $\mathsf{ACA}_0$. But $\mathsf{RCA}_0$ does include a version of $\Sigma^0_1$ comprehension applied to finite sets.

**Theorem 5.5.11 (Bounded $\Sigma^0_1$ comprehension).** $\mathsf{RCA}_0$ *proves the following. Suppose $\varphi(n)$ is a $\Sigma^0_1$ formula (which may have parameters), and $X$ is finite set. Then the set $\{n \in X : \varphi(n)\}$ exists. In particular, for all $z$ the set $\{n < z : \varphi(n)\}$ exists.*

We delay the proof to the next chapter, where Theorem 6.2.6 will provide a generalization. Just as bounded $\Sigma^0_1$ comprehension is linked to $\Sigma^0_1$ induction, we will see that bounded comprehension for formulas higher in the arithmetical hierarchy is linked to stronger induction principles.

### 5.5.3 Formalizing computability theory

In Theorem 5.2.3, we saw that every computable function is $\Delta^0_1$ definable. The proof relied on the fact that each primitive recursive function is $\Delta^0_1$ definable, and the way that the universal computable function $\Xi$ from Definition 2.4.2 is defined using the primitive recursive functions $U$ and $T$.

We can leverage these facts and their proofs to formalize computability theory within $\mathsf{RCA}_0$. As long as the natural $\Sigma^0_1$ and $\Pi^0_1$ formulas are used to represent a primitive recursive function in Lemma 5.2.2, $\mathsf{RCA}_0$ will be able to prove that the primitive recursive function satisfies the recursion equations used to construct that function. For example, if $f(x) = h(g_0(x), g_1(x))$ is formed by composition then, letting $\rho$ and $\psi$ be the formulas naturally used to prove Lemma 5.2.2, we will have

$$\mathsf{RCA}_0 \vdash (\forall n)(\forall x)([\rho(x,n) \leftrightarrow n = h(g_0(x), g_1(x))] \wedge [\rho(x,n) \leftrightarrow \psi(x,n)]).$$

The same holds for basic functions and for functions defined by primitive recursion: $\mathsf{RCA}_0$ will be able to prove that the corresponding properties hold for the functions defined by the formulas in Lemma 5.2.2. This is due to the way that the lemma is proved by structural induction.

This allows for a formalization of computability theory in $\mathsf{RCA}_0$. Once we have the natural representations of the primitive recursive functions $U$ and $T$, we have a $\Delta^0_1$ definition of $\Xi$ that allows us to verify many properties of computable functions in $\mathsf{RCA}_0$, including the use principle, the key properties listed in Proposition 2.4.9, the $S^m_n$ theorem, and the recursion theorem.

In this formalized computability theory, both the arguments to a computable function and the index for a computable function may be nonstandard numbers, in the context of a nonstandard model $\mathcal{M}$. For example, if $n$ is a nonstandard element of $M$, the constant function $f(t) = n$ is computable (and primitive recursive) from the perspective of $\mathcal{M}$, and hence has an index in $M$. That index is, necessarily, nonstandard. Similarly, if $e$ is the usual index for the function $\Phi_e(t) = \lambda t.2t$, then $\mathcal{M}$ will satisfy $\Phi_e(n) = 2n$.

In the previous paragraph, we referred to the "usual" index, that is, the one that would naturally be obtained from the equation shown. But there are infinitely many other indices for this function, some of which may behave differently in nonstandard models. In the context of formalized computability theory, it becomes even more important to distinguish between a computable function $f$ and a particular index $e$ such that $f = \Phi_e$. There are examples of $e, e' \in \omega$ as follows.

- $\Phi_e$ is total in the standard model of $\mathsf{RCA}_0$, but not total in some nonstandard models. For example, let $\Phi_e(n)$ converge if and only if $n$ does not code a proof of $0 = 1$ from the axioms of $\mathsf{RCA}_0$. This computable function is total in the standard model because there is no such coded proof, but is not total in a model of $\neg\mathrm{Con}(\mathsf{RCA}_0)$.
- $\Phi_e$ and $\Phi_{e'}$ are the same function in the standard model, total in all models of $\mathsf{RCA}_0$, but not equal in some nonstandard models. Examples again follow from the incompleteness theorem.

One property that does hold is $\Sigma^0_1$ *completeness*: if the standard model of $\mathsf{RCA}_0$ satisfies $\Phi_e(n) = m$ then so does each nonstandard model of $\mathsf{RCA}_0$. So a computable function on standard arguments cannot converge to different values in different nonstandard models, if the function converges to a value in the standard model.

The study of formalized computability is part of *reverse recursion theory*. For example, we can ask which induction axioms are needed to verify a particular constructions from classical recursion theory. We will present some results of this sort in Section 6.4. In most context of reverse mathematics, however, we do not need to refer directly to formalized computability. Instead, we view $\mathsf{RCA}_0$ itself as a different kind of formalization of computable mathematics.

## 5.6 The subsystems $\mathsf{ACA}_0$ and $\mathsf{WKL}_0$

The system $\mathsf{RCA}_0$ provides a convenient base system, because it is strong enough to formalize many of the routine coding methods from computable mathematics. At the same time, many well known mathematical theorems are not provable in $\mathsf{RCA}_0$. One goal of reverse mathematics is to identify the strength of these theorems. A remarkable phenomenon is that many theorems are provable in $\mathsf{RCA}_0$ or equivalent to one of four stronger subsystems over it. These subsystems, shown in Figure 5.2, are themselves linearly ordered by provability.

Not all theorems are equivalent to one of the "big five" subsystems, of course. For those principles, the "big five" serve as useful reference points. For example, a theorem that is stronger that $\mathsf{RCA}_0$ but weaker than $\mathsf{WKL}_0$ is different from a theorem that is weaker than $\mathsf{ACA}_0$ but neither implies or is implied by $\mathsf{WKL}_0$.

In this section we will describe the systems $\mathsf{ACA}_0$ and $\mathsf{WKL}_0$, which along with $\mathsf{RCA}_0$ make up the bottom half of the "big five". These systems are closely tied to Turing computability and subtrees of $2^{<\omega}$.

### 5.6.1 The subsystem $\mathsf{ACA}_0$

The initials ACA stand for "arithmetical comprehension axiom". The system we define now, $\mathsf{ACA}_0$, is a strengthening of $\mathsf{RCA}_0$ obtained by expanding the comprehension scheme to include all arithmetical formulas. As we will see, this is closely tied to the Turing jump operator.

**Definition 5.6.1.** $\mathsf{ACA}_0$ is the $\mathcal{L}_2$ theory that includes $\mathsf{RCA}_0$ and the comprehension axiom scheme for all arithmetical formulas.

That is, $\mathsf{ACA}_0$ consists of $\mathsf{PA}^-$, comprehension for arithmetical formulas, and $\mathsf{I}\Sigma^0_1$. As noted at the end of Section 5.3.2, we could replace $\mathsf{I}\Sigma^0_1$ by the set induction axiom or, alternatively, by the induction scheme for *all* arithmetical formulas (not just $\Sigma^0_1$). We will formally separate $\mathsf{ACA}_0$ from $\mathsf{RCA}_0$ in Proposition 5.6.14.

The next theorem gives a deeper result that comprehension for $\Sigma^0_1$ formulas is enough to give comprehension for all arithmetical formulas.

**Theorem 5.6.2.** $\mathsf{ACA}_0$ *is axiomatized by the theory* $\Sigma^0_1$-$\mathsf{CA}_0$ *consisting of* $\mathsf{PA}^-$*, the comprehension axiom scheme for* $\Sigma^0_1$ *formulas, and the set induction axiom.*

*Proof.* By the preceding discussion, it is enough to show that if $\varphi(n)$ is an arithmetical formula then $\Sigma^0_1$-$\mathsf{CA}_0$ proves that the set $\{n : \varphi(n)\}$ exists. Without loss of generality we may assume that $\varphi$ is in prenex normal form.

The $\Sigma^0_1$ comprehension scheme includes the comprehension axiom for taking the complement of a set:

$$(\forall X)(\exists Y)[n \in X \Leftrightarrow n \notin Y].$$

Thus, by taking complements, we may assume that $\varphi$ begins with a block of existential quantifiers, and is of the form $\varphi(n) \equiv (\exists \vec{x})\psi(n, \vec{x})$, possibly with parameters.

We proceed by induction the level of $\varphi$ in the arithmetical hierarchy. The base case, for $\Sigma^0_1$ formulas, follows by assumption. So, by induction, assume that we have comprehension for $\Sigma^0_k$ formulas, and assume $\varphi$ is $\Sigma^0_{k+1}$.

To proceed by induction, we make use of the pairing function to form the set $A = \{(n, x) : \neg\psi(n, x, \vec{y})\}$. This set is defined by the $\Sigma^0_k$ formula

$$(\exists t)(\exists n)(\exists x)[t = (n, x) \wedge \neg\psi],$$

which is $\Sigma^0_k$. By induction, we may form the set $A$ using repeated applications of $\Sigma^0_1$ comprehension. Then we may form our desired set $\{n : \varphi(n)\}$ using $\Sigma^0_1$ comprehension on the formula $(\exists x)[(n,x) \in A]$, which is $\Sigma^0_1$ with parameter $A$. □

Using the formalization of computability theory in $\mathsf{RCA}_0$, as discussed in Section 5.5.3, we get the following corollary to Theorem 5.6.2.

**Corollary 5.6.3.** $\mathsf{ACA}_0$ *is equivalent to* $\mathsf{RCA}_0 + (\forall X)[X'$ *exists*].

So, in the parlance of Section 4.6, we see that the $\omega$-models of $\mathsf{ACA}_0$ are precisely the jump ideals, i.e., the Turing ideals that are closed under the Turing jump. By Post's theorem (Theorem 2.6.2), we also have the following example.

**Corollary 5.6.4.** *Let $\mathcal{M}$ be the $\omega$-model with a second order part consisting of exactly the arithmetical sets. Then* $\mathcal{M} \vDash \mathsf{ACA}_0$.

There is a subtle point of caution here. Obviously, another characterization of the $\omega$-models of $\mathsf{ACA}_0$ is that they are those Turing ideals $\mathcal{S}$ such that for all $X \in \mathcal{S}$ and all $n \in \omega$, $X^{(n)} \in \mathcal{S}$. However, $\mathsf{ACA}_0$ is *not* equivalent to $\mathsf{RCA}_0 + (\forall n)[X^{(n)}$ exists]. Indeed, this characterizes another system.

**Definition 5.6.5.** $\mathsf{ACA}_0'$ is the $\mathcal{L}_2$ theory that includes $\mathsf{RCA}_0$ and the axiom

$$(\forall X)(\forall n)[X^{(n)} \text{ exists}].$$

The issue is that in a model, the quantifier on $n$ here may potentially range over nonstandard numbers. And indeed, it is not difficult to see that there is no way to get around this obstacle.

**Proposition 5.6.6.** $\mathsf{ACA}_0$ *does not prove* $(\forall X)(\forall n)[X^{(n)}$ *exists*]. *Therefore,* $\mathsf{ACA}_0'$ *is a strictly stronger system than* $\mathsf{ACA}_0$.

*Proof.* Start with a nonstandard model $M$ of $\mathsf{PA}^-$, and let $\mathcal{S}$ consist of all $X \subseteq M$ that are arithmetically definable in $M$. Let $\mathcal{M}$ be the $\mathcal{L}_2$ structure with first order part $M$ and $\mathcal{S}^{\mathcal{M}} = \mathcal{S}$. Then this is a model of $\mathsf{ACA}_0$ by Theorem 5.6.2.

Consider $\varnothing \in \mathcal{S}$ and let $n \in M$ be nonstandard. For a contradiction, assume there is $B \in \mathcal{S}^{\mathcal{M}}$ so that $\mathcal{M} \vDash B = \varnothing^{(n)}$. Choose $k \in \omega$ such that $B$ is $\Sigma^0_k$ definable over $\mathcal{M}$. By a formalization of Post's theorem, we have $\mathcal{M} \vDash B \leqslant_{\mathrm{T}} \varnothing^{(k)}$. Because $n$ is nonstandard, $k+1 <^{\mathcal{M}} n$. Hence, we have $\mathcal{M} \vDash \varnothing^{(k+1)} \leqslant_{\mathrm{T}} \varnothing^{(n)} = B \leqslant_{\mathrm{T}} \varnothing^{(k)}$. But $\mathsf{ACA}_0$ proves $\varnothing^{(k+1)} \nleqslant_{\mathrm{T}} \varnothing^{(k)}$, so we have a contradiction. □

By a clever compactness argument, even more can be said for $\Pi^1_2$ statements (which in particular include many sentences of $\mathcal{L}_2$ corresponding to $\forall\exists$ theorems of interest in reverse mathematics). We have the following somewhat surprising result.

**Theorem 5.6.7 (see Wang [320]).** *Suppose that $\varphi$ is an $\mathcal{L}_2$-sentence of the form $(\forall X)[\theta(X) \rightarrow (\exists Y)\psi(X,Y)]$, where $\theta$ and $\psi$ are arithmetical. If* $\mathsf{ACA}_0 \vdash \varphi$ *then there is a $k \in \omega$ such that* $\mathsf{ACA}_0 \vdash (\forall X)[\theta(X) \rightarrow (\exists Y \leqslant_{\mathrm{T}} X^{(k)})\psi(X,Y)]$.

So a problem (whose formalization into $\mathcal{L}_2$ is) provable in $\mathsf{ACA}_0$ must have the feature that not only does it admit arithmetical solutions, but for some (fixed!) $k \in \omega$, it admits $\Sigma^0_k$ solutions. An important example of a problem that enjoys the former property but not the latter is Ramsey's theorem (see Chapter 8).

The provenance of Theorem 5.6.7 is a bit hard to pin down. It seems to first show up in print in a 1981 book by Wang [320], who suggests it is folklore. (It appears to also be treated as such by Friedman in [110].) Wang gives two proofs, both unpublished and obtained by personal communication; one is due to Jockusch, and another, apparently earlier one, is due to Solovay. We give the proof by Jockusch.

*Proof (of Theorem 5.6.7; Jockusch, unpublished).* Suppose $\mathsf{ACA}_0 \vdash \varphi$. Let $\mathcal{L}_2'$ be the language obtained from $\mathcal{L}_2$ by the addition of a new constant symbol, $C$, of the second sort. For each $k \in \omega$, let $\zeta_k \equiv \neg(\exists Y \leqslant_{\mathrm{T}} C^{(k)})\psi(C,Y)$, and let $T$ be the $\mathcal{L}_2'$ theory $\mathsf{ACA}_0 + \theta(C) + \{\zeta_k : k \in \omega\}$. We claim that $T$ is inconsistent. From this it follows that there is a $k \in \omega$ such that $\mathsf{ACA}_0 \vdash \theta(C) \to (\exists Y \leqslant_{\mathrm{T}} C^{(k)})\psi(C,Y)$. Since $C$ is not in $\mathcal{L}_2$, this implies $\mathsf{ACA}_0 \vdash (\forall X)[\theta(X) \to (\exists Y \leqslant_{\mathrm{T}} X^{(k)})\psi(X,Y)]$, as desired.

Seeking a contradiction, suppose $T$ is consistent and fix $\mathcal{M} \vDash T$. Let $\mathcal{N}$ be the model with first order part $M$ and second order part all subsets of $M$ arithmetically definable in $\mathcal{M}$. By Theorem 5.6.2, $\mathcal{N} \vDash \mathsf{ACA}_0$, hence $\mathcal{N} \vDash \varphi$. Also, $\mathcal{N}$ is an $\omega$-submodel of $\mathcal{M}$, so by Exercise 5.13.1, the two models satisfy the same arithmetical sentences. Since $C \in \mathcal{S}^{\mathcal{N}}$, it follows in particular that $\mathcal{N} \vDash \theta(C)$. There must thus exist some $D \in \mathcal{S}^{\mathcal{N}}$ such that $\psi(C,D)$ holds in $\mathcal{N}$, and as $\mathcal{S}^{\mathcal{N}} \subseteq \mathcal{S}^{\mathcal{N}}$ and $\psi$ is arithmetical, this fact also holds in $\mathcal{M}$. But $D$ is arithmetically definable in $\mathcal{M}$, hence there must be some (standard) $k \in \omega$ such that $\mathcal{M} \vDash D \leqslant_{\mathrm{T}} C^{(k)}$. It follows that $\mathcal{M} \vDash \neg\zeta_k$, a contradiction. □

One more interesting strengthening of $\mathsf{ACA}_0$, which we will work with in Section 9.9, is the following.

**Definition 5.6.8.** $\mathsf{ACA}_0^+$ is the $\mathcal{L}_2$ theory that includes $\mathsf{RCA}_0$ and the axiom

$$(\forall X)(\exists Y)(\forall n)[Y^{[0]} = X \wedge Y^{[n+1]} = (Y^{[n]})'].$$

The $\omega$-models of $\mathsf{ACA}_0^+$ are precisely those classes $\mathcal{S}$ which, for every set $A$ they contain, contain also the *$\omega$ jump of $A$*, meaning $A^{(\omega)} = \langle A, A', A'', \ldots \rangle$.

**Proposition 5.6.9.** *There is an $\omega$-model of $\mathsf{ACA}_0$ (and hence of $\mathsf{ACA}_0'$) which is not closed under the map $A \mapsto A^{(\omega)}$. Hence, $\mathsf{ACA}_0^+$ is a strictly stronger system than $\mathsf{ACA}_0'$.*

*Proof.* Let $\mathcal{S}$ be the $\omega$-model consisting of all arithmetical subsets of $\omega$. Since $\varnothing^{(\omega)}$ is not arithmetical, we are done. □

### 5.6.2 The subsystem $\mathsf{WKL}_0$

We will leverage the formal definition of finite sets to work with finite sequences within subsystems of arithmetic. This will allow us to formalize trees and state the defining axiom for the subsystem $\mathsf{WKL}_0$.

We have been viewing finite sequences as (graphs of) functions on initial segments of $\omega$. To avoid using set quantifiers in formulas that quantify over finite sets (and, indirectly, finite sequences), we will define the collection of codes for finite sequences. As these codes are simply numbers, we can then quantify over codes with number quantifiers.

**Definition 5.6.10 (Coding strings in $\mathsf{RCA}_0$).** The following definition is made in $\mathsf{RCA}_0$. A (code for a) *finite sequence* or *string* is a (code for a) pair $\langle \ell, c \rangle$, where $c$ is a code for a finite set $X$ such that:

1. every element of $X$ is of the form $\langle i, n \rangle$ for some $i < \ell$ and $n \in \mathbb{N}$,
2. for every $i < \ell$ there is a unique $n \in \mathbb{N}$ such that $\langle i, n \rangle \in X$.

We call $\ell$ above the *length* of $\alpha$, denoted $|\alpha|$. If $(i, n) \in X$ we write $\alpha(i) = n$.

**Definition 5.6.11 (Coding functions in $\mathsf{RCA}_0$).** The following definition is made in $\mathsf{RCA}_0$. Fix sets $A, B \subseteq \mathbb{N}$. A set $X$ is a (code for a) *function from $A$ to $B$* if $X = \langle A, B, F \rangle$, where every element of $F$ is of the form $\langle n, m \rangle$ for some $n \in A$ and $m \in B$, and for every $n \in A$ there is a unique $m$ with $\langle n, m \rangle \in F$.

*Remark 5.6.12 (Notation for functions).* It is common to deviate from our notational conventions and use lowercase letters near the middle of the alphabet ($f$, $g$, $h$, etc.) when discussing coded functions in $\mathsf{RCA}_0$. This more closely aligns with standard usage, and as a result, makes the notation more evocative. We also write simply $f \colon A \to B$ or $f \in B^A$ instead of $f = \langle A, B, F \rangle$, and $f(n) = m$ instead of $\langle n, m \rangle \in F$, etc.

With these definitions and notational conventions, most other standard definitions formalize directly. This is the case, for instance, for the prefix relation $\preceq$, which enables us to define when a finite sequence is an initial segment of another, or of a function $f \colon \mathbb{N} \to \mathbb{N}$, etc. In $\mathsf{RCA}_0$, we can easily form the set of all finite sequences, as well as the set of all binary finite sequences, and following our discussion in Section 5.4, we denote these by $\mathbb{N}^{<\mathbb{N}}$ and $2^{<\mathbb{N}}$, respectively. A subset of one of these is a tree if it is closed downward under $\preceq$, as usual. An infinite path through a tree is defined in the standard way.

Using our formalism, we can naturally write König's lemma and weak König's lemma as $\Pi^1_2$ sentences in $\mathcal{L}_2$. We will refer to these using the abbreviations $\mathsf{KL}$ and $\mathsf{WKL}$, respectively, which we previously used in Chapter 3 to refer to the informal $\forall\exists$ theorem versions of these principles and their corresponding problem forms (Definitions 3.2.1 and 3.2.2).

**Definition 5.6.13.** $\mathsf{WKL}_0$ is the $\mathcal{L}_2$ theory that includes $\mathsf{RCA}_0$ and $\mathsf{WKL}$.

It follows immediately from definitions that an $\omega$-model $\mathcal{M}$ satisfies $\mathsf{WKL}_0$ if and only if $\mathcal{S}^{\mathcal{M}}$ satisfies the *problem* WKL discussed in Chapters 3 and 4. So by Proposition 4.6.3, we see that $\mathcal{M} \vDash \mathsf{WKL}_0$ if and only if (the second order part of) it is a Scott set (as defined in Section 4.6). Equivalently, by Theorem 4.6.16, $\mathcal{M} \vDash \mathsf{WKL}_0$ if and only if every $X \in \mathcal{S}^{\mathcal{M}}$ there is a $Y \in \mathcal{S}^{\mathcal{M}}$ with $X \ll Y$. In particular, $\mathsf{WKL}_0$ has no topped $\omega$-models. These considerations show that $\mathsf{WKL}_0$ lies strictly between $\mathsf{RCA}_0$ and $\mathsf{ACA}_0$ in strength.

**Proposition 5.6.14.**

1. $\mathsf{ACA}_0 \vdash \mathsf{WKL}_0$.
2. *There is an $\omega$-model of* $\mathsf{RCA}_0 + \neg\mathsf{WKL}_0$.
3. *There is an $\omega$-model of* $\mathsf{WKL}_0 + \neg\mathsf{ACA}_0$.

*Proof.* Part (1) follows by Proposition 2.8.5. For (2), let $\mathcal{M}$ be REC, which is a not a Scott set. For (3), apply the low basis theorem (Theorem 2.8.18, using the remark after the proof of Theorem 2.8.25) to fix a low $D \gg \emptyset$. By Theorem 4.6.16 there is an $\omega$-model of WKL consisting entirely of $D$-computable sets. In particular, every set in this model is low. It follows that $\emptyset'$ is not in this model, so by Corollary 5.6.3, $\mathsf{ACA}_0$ does not hold. □

It is interesting to note the $\omega$-models of $\mathsf{WKL}_0$ are structured differently from those of $\mathsf{RCA}_0$ and $\mathsf{ACA}_0$. For one, each of the latter two have *minimum $\omega$-models*, which are REC and the collection of arithmetical sets, respectively. By contrast, $\mathsf{WKL}_0$ does not have even *minimal* $\omega$-models. Informally, this suggests that there is no set $\Gamma$ of formulas so that $\mathsf{WKL}_0$ is equivalent to the set existence scheme restricted to $\Gamma$.

**Proposition 5.6.15 (Simpson [288]).** *Suppose $\mathcal{M}$ is an $\omega$-model of* $\mathsf{WKL}_0$*, with $X \in \mathcal{S}^{\mathcal{M}}$. There is an $\omega$-model $\mathcal{M}^*$ of* $\mathsf{WKL}_0$ *such that $X \in \mathcal{S}^{\mathcal{M}^*}$ but $\mathcal{S}^{\mathcal{M}^*}$ is a proper subset of $\mathcal{S}^{\mathcal{M}}$.*

*Proof.* Since $X \in \mathcal{S}^{\mathcal{M}}$, it follows by Proposition 4.6.3 that $\mathcal{S}^{\mathcal{M}}$ contains some $D \gg X$. By Proposition 2.8.26, there is a $D^*$ with $D \gg D^* \gg X$. Let $\mathcal{M}^*$ be the $\omega$-model consisting of all $Y \in 2^\omega$ such that $D^* \gg Y$. By Propositions 2.8.26 and 4.6.3, this model satisfies $\mathsf{WKL}_0$. We have $X \in \mathcal{S}^{\mathcal{M}^*}$ and $\mathcal{S}^{\mathcal{M}^*} \subsetneq \mathcal{S}^{\mathcal{M}}$ since $D^* \in \mathcal{S}^{\mathcal{M}}$, but $D \notin \mathcal{S}^{\mathcal{M}^*}$. □

A related structural difference is that the $\omega$-models of $\mathsf{WKL}_0$ are not closed under intersection. It is easy to see that this *is* the case for $\omega$-models of $\mathsf{RCA}_0$ and $\mathsf{ACA}_0$ (and indeed, their minimum $\omega$-models may be presented as the intersections of all their $\omega$-models).

**Proposition 5.6.16 (Simpson [288]).**

1. *There exist $\omega$-models $\mathcal{M}_0$ and $\mathcal{M}_1$ of* $\mathsf{WKL}_0$ *such that the $\omega$-model $\mathcal{M}$ with $\mathcal{S}^{\mathcal{M}} = \mathcal{S}^{\mathcal{M}_0} \cap \mathcal{S}^{\mathcal{M}_1}$ does not satisfy* $\mathsf{WKL}_0$.
2. *For every $X \nleq_T \emptyset$ there is an $\omega$-model of* $\mathsf{WKL}_0$ *not containing $X$.*

*Proof.* To prove part (1), fix $A_0, A_1 \in 2^\omega$ forming a minimal pair: that is, $A_0, A_1 \not\leqslant_T \emptyset$ and for all $Y$, if $Y \leqslant_T A_0$ and $Y \leqslant_T A_1$ then $Y \leqslant_T \emptyset$. Fix $i < 2$. By the cone avoidance basis theorem (Theorem 2.8.23) relativized to $A_i$ there exists $D_i \gg A_i$ such that $A_{i+1} \not\leqslant_T D_i$. Thus also $D_{i+1} \not\leqslant_T D_i$. Let $\mathcal{M}_i$ be the $\omega$-model consisting of all $Y \in 2^\omega$ such that $Y \leqslant_T A_i$ or $A_i \leqslant_T Y \ll D_i$. By Propositions 2.8.26 and 4.6.3, $\mathcal{M}_i \vDash \mathsf{WKL}_0$. Now consider an arbitrary $Y \in \mathcal{S}^{\mathcal{M}_0} \cap \mathcal{S}^{\mathcal{M}_1}$. By choice of $D_0, D_1$ and the fact that $Y \leqslant_T D_0, D_1$, it cannot be that $A_0, A_1 \leqslant_T Y$. We also cannot have $A_i \leqslant_T Y$ and $Y \leqslant_T A_{i+1}$ for any $i$, lest $A_i$ would be computable from $A_{i+1}$. Thus, it must be that $Y \leqslant_T A_0, A_1$, and hence that $Y \leqslant_T \emptyset$. In order words, $\mathcal{M} = \mathrm{REC}$, which is not a Scott set and hence not a model of $\mathsf{WKL}_0$.

To prove part (2), apply the cone avoidance basis theorem to find $D \gg \emptyset$ with $X \not\leqslant_T D$. Then, let $\mathcal{M}$ be the $\omega$-model consisting of all $Y \ll D$. This is a model satisfying $\mathsf{WKL}_0$ and $X \notin \mathcal{S}^{\mathcal{M}}$. □

## 5.7 Equivalences between mathematical principles

We are now ready to look at our first example of an equivalence between an arithmetical principle and a subsystem of second order arithmetic. Compared to the reducibility notions from Chapter 4, this gives a different way to compare the strengths of mathematical problems.

In particular, we represent the problems as subsystems $S$ and $T$ of second order arithmetic. We ask whether $\mathsf{RCA}_0 + S$ proves every axiom of $T$. If so, we see that $T$ is weaker than $S$ "relative to" $\mathsf{RCA}_0$. If it is also true that $\mathsf{RCA}_0 + T$ proves every axiom of $S$, then $S$ and $T$ are equivalent over $\mathsf{RCA}_0$, and otherwise $S$ is strictly stronger. In Section 5.11, we compare this new reducibility to those from Chapter 4 in more detail.

**Definition 5.7.1 (Ranges in $\mathsf{RCA}_0$).** The following definition is made in $\mathsf{RCA}_0$. Fix sets $A, B \subseteq \mathbb{N}$ and $f\colon A \to B$. A set $Y$ is the *range* of a function $X$ if $(\forall m)[m \in Y \leftrightarrow (\exists n)[\langle n, m\rangle \in X]$.

Note that as defined in Definition 5.6.11, each function comes together with its domain and codomain, and as such they both always exist. By contrast, ranges need not always exist, as the next theorem shows.

**Theorem 5.7.2.** *The following statements are equivalent over* $\mathsf{RCA}_0$.

1. $\mathsf{ACA}_0$.
2. *The range of every function from* $\mathbb{N}$ *to* $\mathbb{N}$ *exists.*

*Proof.* First, we assume $\mathsf{ACA}_0$ and prove that every function has a range. This follows immediately from the fact that the range of a function is defined by an arithmetical formula with the function as a parameter.

Second, we assume the axioms of $\mathsf{RCA}_0$ and also assume that every function has a range. We wish to prove that each axiom of $\mathsf{ACA}_0$ must hold. By Theorem 5.6.2

it is enough to show that each $\Sigma^0_1$ comprehension axiom holds. So let $\varphi(n)$ be a $\Sigma^0_1$ formula. We may assume $\varphi$ is of the form $(\exists m)\psi(m,n)$ where $\psi$ is $\Sigma^0_0$. We may also assume there is at least one $n_0$ such that $\varphi(n_0)$, because comprehension for $\varphi$ is trivial otherwise.

We first form the set $X$ consisting of all pairs $\langle\langle m,n\rangle, k\rangle$ such that

- $\psi(m,n)$ holds and $k = n$, or
- $\psi(m,n)$ does not hold and $k = n_0$.

This set can be formed by $\Delta^0_1$ comprehension. It follows from the definition that $X$ is a function, because each $z$ can be written in exactly one way in the form $\langle n,m\rangle$.

Now suppose that $Y$ is the range of $X$, and let $n \neq n_0$ be fixed. If $n$ is in the range of $X$ then, by construction, there is some $m$ such that $\psi(m,n)$ holds, so $\varphi(n)$ holds. Conversely, if $\varphi(n)$ holds then there is some $m$ such that $\psi(m,n)$ holds, which means $X(\langle m,n\rangle) = n$, so $n \in Y$. Thus $Y$ is the set $\{n : \varphi(n)\}$, which is what we wanted to form. □

The right way to think of this theorem is alongside Corollary 5.6.3. If we can prove that the range of any function exists, we can in particular prove it for the function that enumerates the halting set (relative to a given oracle), and conversely. In practice, this shows up as follows. To prove that $\mathsf{RCA}_0 \vdash \mathsf{P} \rightarrow \mathsf{ACA}_0$ for some $\Pi^1_2$ statement $\mathsf{P}$ (formalizing an $\forall\exists$ theorem), we begin by showing that $\mathsf{P}$ codes the jump. Then, formalizing the argument in $\mathsf{RCA}_0$ typically yields a proof that the range of every function from $\mathbb{N}$ to $\mathbb{N}$ exists. This formalized argument is made easier by referring to ranges rather than formalized Turing jumps.

The word "range" in connection with a function is also sometimes used to refer to the formula defining the range. For example, we may say "the range of $f$ is disjoint from the set $Z$" as shorthand for

$$(\forall y)[(\exists n)[f(n) = m \rightarrow m \notin Z].$$

Here again we will rely on usage of the word "exists", as discussed above, to make our meaning clear.

The proof of Theorem 5.7.2 has several aspects that are common to reverse mathematics. First, it shows an equivalence between a "mathematical" principle (that the range of a function exists) and a "logical" principle (the arithmetical comprehension scheme). The second part of the proof, which shows that if every function has a range then the arithmetical comprehension scheme is valid, is the *reversal*. The proof demonstrates two typical features of a reversal to $\mathsf{ACA}_0$. First, rather than directly establishing that every instance of arithmetical comprehension holds, the reversal makes use of a characterization of $\mathsf{ACA}_0$ given by Theorem 5.6.2. Even with this simplification, the proof must still show than an entire axiom scheme holds. The scheme is enumerated in the metatheory, rather than the object theory. Thus we do not attempt to prove in $\mathsf{RCA}_0$ that "if $\varphi$ is arithmetical then $\{n : \varphi(n)\}$ exists" under the assumption that every function has a range. Instead we verify that, for each $\varphi$, $\mathsf{RCA}_0$ is able to show that "$\{n : \varphi(n)\}$ exists" under the same assumption.

We will see that $\mathsf{WKL}_0$, too, admits a characterization in terms of ranges of functions, in the form of the $\Sigma^0_1$ *separation principle* (Exercise 5.13.18).

König's lemma gives another example of an equivalence between a mathematical theorem and a subsystem of second order arithmetic.

**Theorem 5.7.3.** *The following are equivalent over* $\mathsf{RCA}_0$.

*1.* $\mathsf{ACA}_0$.
*2.* $\mathsf{KL}$.
*3.* $\mathsf{KL}$ *restricted to trees in which each node has at most two immediate successors.*

*Proof.* We have essentially already proved this, in the guise of Propositions 3.6.10 and 4.2.9. We thus refer to these proofs, and indicate merely how to formalize them in $\mathsf{RCA}_0$. This gives us a first taste of a common practice in reverse mathematics: do the computability theoretic argument first, formalize second. A couple of technical steps in the formalization require facts we have not established yet; we point these out below with forward references the theorems that fill these in.

(1) $\to$ (2): We argue in $\mathsf{ACA}_0$. Let $T \subseteq \mathbb{N}^{<\mathbb{N}}$ be an infinite, finitely branching tree. Recall the definition of the function $p$ in the proof of Proposition 4.2.9. This function is arithmetically definable, and so exists. However, we must prove that it actually *is* a function, i.e., that it is total. (In the original proof, we relied on $\mathsf{KL}$ for this, but here of course that is not available to us since that is what we are proving.)

Fix $\alpha \in T$. If there is an $i$ such that $\{\beta \in T : \beta \succeq \alpha i\}$ is infinite then there is a least such $i$ and $p(\alpha) = i$. (The existence of this least $i$ requires justification. See Proposition 6.1.5.) So suppose $\{\beta \in T : \beta \succeq \alpha i\}$ is finite for all $i$. Since $T$ is finitely branching, we can form the finite set $F$ of all $i$ such that $\alpha i \in T$. For each $i \in F$, we can also form $S_i = \{\beta \in T : \beta \succeq \alpha i\}$, which is a finite set by assumption. Then $S = \bigcup_{i \in F} S_i$ is a finite union of finite sets, and hence is itself finite. (This is nontrivial because we are dealing with "finite" sets in potentially nonstandard models. It is generally false in $\mathsf{RCA}_0$, as discussed in detail in Section 6.5. A proof is given in Proposition 6.5.4 which shows, in particular, that this conclusion goes through in $\mathsf{ACA}_0$.) Hence, $\alpha$ has only finitely many successors in $T$, and therefore $p(\alpha) = 0$. With $p$ total, the rest of the proof of Proposition 4.2.9 formalizes easily.

The implication (2) $\to$ (3) is obvious. To prove (3) $\to$ (1), we now argue in $\mathsf{RCA}_0$ and formalize the proof of Proposition 3.6.10. By Theorem 5.7.2, it suffices to prove that the range of every injective function $f \colon \mathbb{N} \to \mathbb{N}$ exists. So fix such an $f$. Let $T$ consist of all $\alpha \in \mathbb{N}^{<\mathbb{N}}$ satisfying the following conditions for all $y < |\alpha|$.

1. If $\alpha(y) = 0$ then $f(x) \neq y$ for all $x \leqslant |\alpha|$.
2. If $\alpha(y) = x + 1$ then $f(x) = y$.

Then $T$ exists by $\Delta^0_1$ comprehension, and it is easily seen to be a tree. Moreover, using the fact that $f$ is injective, it follows that each $\alpha \in T$ has at most two immediate successors. Thus, by (3) we may fix an infinite path $g \in [T]$. Now it is easy to see that $y \in \operatorname{range}(f)$ if and only if $g(y) > 0$, so the range exists. $\square$

## 5.8 The subsystems $\Pi^1_1$-$\mathsf{CA}_0$ and $\mathsf{ATR}_0$

We now turn to the strongest two subsystems of the "big five": $\Pi^1_1$-$\mathsf{CA}_0$ and $\mathsf{ATR}_0$. While $\mathsf{ACA}_0$ and $\mathsf{WKL}_0$ are closely related to classical computability theory, and subtrees of $2^{<\omega}$, the stronger subsystems are closely related to hyperarithmetical computability and to subtrees of $\omega^{<\omega}$.

### 5.8.1 The subsystem $\Pi^1_1$-$\mathsf{CA}_0$

The subsystem $\Pi^1_1$-$\mathsf{CA}_0$ is simple to define via the ordinary comprehension scheme.

**Definition 5.8.1.** $\Pi^1_1$-$\mathsf{CA}_0$ is the subsystem consisting of $\mathsf{RCA}_0$ together with the comprehension scheme for $\Pi^1_1$ formulas.

Following the usual pattern discussed in Section 5.3.2, $\Pi^1_1$-$\mathsf{CA}_0$ proves the induction scheme for $\Pi^1_1$ formulas and for $\Sigma^1_1$ formulas as well. The starting point of our discussion of $\Pi^1_1$-$\mathsf{CA}_0$ is the following normal form theorem for $\Sigma^1_1$ formulas.

**Theorem 5.8.2 (Kleene's normal form theorem for $\Sigma^1_1$ formulas).** *For each $\Sigma^1_1$ formula $\varphi(X)$, there is a $\Sigma^0_0$ formula $\theta(x, y)$ such that*

$$\mathsf{ACA}_0 \vdash (\forall X)[\varphi(X) \leftrightarrow (\exists f \in \mathbb{N}^{\mathbb{N}})(\forall k)[\theta(X \upharpoonright k, f \upharpoonright k)]].$$

*Proof.* For ease of notation, we treat $\varphi$ as having no set variables, and show that there is a $\Sigma^0_0$ formula $\theta(x)$ such that

$$\mathsf{ACA}_0 \vdash \varphi \leftrightarrow (\exists f \in \mathbb{N}^{\mathbb{N}})(\forall k)[\theta(f \upharpoonright k)].$$

The proof is the same in the more general case. First, we exhibit a $\Pi^0_1$ formula $\psi(X, j)$ such that

$$\mathsf{ACA}_0 \vdash \varphi \leftrightarrow (\exists f \in \mathbb{N}^{\mathbb{N}})(\forall x)\psi(f, x).$$

Then $\psi(f, j)$ is equivalent to $(\forall \ell)\theta^*(f \upharpoonright \ell, j)$ for some $\Sigma^0_0$ formula $\theta^*$ by formalizing Proposition 2.8.7. Now let $\theta(\sigma)$ be the formula $(\exists i, j \leqslant |\sigma|)[|\sigma| = \langle i, j\rangle \wedge \theta^*(\sigma \upharpoonright i, j)]$. Then $\theta$ is $\Sigma^0_0$ and, as desired,

$$\begin{aligned}\mathsf{ACA}_0 \vdash \varphi &\leftrightarrow (\exists f \in \mathbb{N}^{\mathbb{N}})(\forall j)\psi(f, j)\\ &\leftrightarrow (\exists f \in \mathbb{N}^{\mathbb{N}})(\forall j)(\forall \ell)\theta^*(f \upharpoonright \ell, j)\\ &\leftrightarrow (\exists f \in \mathbb{N}^{\mathbb{N}})(\forall k)\theta(f \upharpoonright k).\end{aligned}$$

To define $\psi$, write $\varphi$ as

$$(\exists \vec{X})(\forall x_0)(\exists y_0)(\forall x_1)(\exists y_1)\cdots(\forall x_n)(\exists y_n) \quad \theta(\vec{X}, x_0, y_0, x_1, y_1, \ldots, x_n, y_n) \tag{5.5}$$

where $\theta$ is quantifier free. We claim that in $\mathsf{ACA}_0$, this is equivalent to

$$(\exists \vec{X})(\exists g_0, \ldots, g_n \in \mathbb{N}^{<\mathbb{N}})(\forall x_0, \ldots, x_n) \\ \theta(\vec{X}, x_0, g_0(x_0), x_1, g_1(x_0, x_1), \ldots, x_n, g_n(x_0, \ldots, x_n)) \tag{5.6}$$

where $g_i(x_0, \ldots, x_n)$ is shorthand for $g_i(\langle x_0, \ldots, x_n\rangle)$. That (5.6) implies (5.5) is clear. In the other direction, fix $\vec{X}$ witnessing (5.5). For each $i \leqslant n$, define $g_i \in \mathbb{N}^{\mathbb{N}}$ inductively as follows: for all $x_0, \ldots, x_i$, $g_i(x_0, \ldots, x_i)$ is the least $y_i$ such that

$$(\forall x_{i+1})(\exists y_{i+1}) \cdots (\forall x_{n-1})(\exists y_{n-1}) \\ \theta(\vec{X}, x_0, g_0(x_0), x_1, g_1(x_0, x_1), \ldots, x_{i-1}, g_{i-1}(x_i, \ldots, x_{i-1}), x_i, y_i, \ldots, x_n, y_n).$$

To be precise, the induction is external since $n$ is fixed (and standard), and arithmetical comprehension is used to prove that each $g_i$ exists. Now by induction on $i$, we obtain the matrix of (5.6).

Say the number of variables in $\vec{X}$ in $\theta$ is $m$. Define $\psi(f, x)$ to be the formula

$$f = \langle X_0, \ldots, X_{m-1}, g_0, \ldots, g_n\rangle \wedge x = \langle x_0, \ldots, x_n\rangle \\ \rightarrow (\forall y_0, \ldots, y_n)(\forall i \leqslant n)[\langle\langle x_0, \ldots, x_i\rangle, y_i\rangle \in g \\ \rightarrow \theta(X_0, \ldots, X_{m-1}, x_0, y_0, x_1, y_1, \ldots, x_n, y_n)].$$

Note that $\psi$ is $\Pi^0_1$. Here, we are thinking of $X_i(n)$ and $g_i(n)$ as $f(i, n)$, so if $f \in \mathbb{N}^{\mathbb{N}}$ then so are all the $X_i$ and $g_i$. Then it is not difficult to see that (5.6) is equivalent to $(\exists f \in \mathbb{N}^{\mathbb{N}})(\forall x)\psi(f, x)$, as wanted. □

Kleene's normal form theorem exposes an important connection between $\Sigma^1_1$ formulas and well founded relations, as we will see in Corollary 5.8.4. In $\mathcal{L}_2$, we formally think of a partial ordering as a (code for a) pair of sets $(X, \leqslant_X)$, such that $X \subseteq \mathbb{N}$ and $\leqslant_X$ is a set of (codes for) pairs of elements of $X$ satisfying the axioms of a partial order. As usual, for $n, m \in X$ we write $n <_X m$ if $n \leqslant_X m$ and $n \neq m$. We also typically abbreviate $(X, \leqslant_X)$ simply by $X$. It will be convenient to isolate the following definitions.

**Definition 5.8.3.**

1. If $X$ is a partial ordering, an *infinite descending sequence through* $X$ is a function $f \colon \mathbb{N} \rightarrow X$ such that $f(k+1) <_X f(k)$ for all $k \in \mathbb{N}$.
2. $\mathrm{WF}(X)$ is the formula asserting that $X$ is a partial ordering and there exists no infinite descending sequence through $X$.
3. $\mathrm{WO}(X)$ is the formula asserting that $X$ is a linear ordering and there exists no infinite descending sequence through $X$.

As usual, $\mathrm{WF}(X)$ we say $X$ is *well founded*. If $T \subseteq \mathbb{N}^{<\mathbb{N}}$ is a tree, we write $\mathrm{WF}(T)$ to mean $\mathrm{WF}(\langle T, \succeq\rangle)$, i.e., $T$ has no infinite path.

We now have the following consequence of Theorem 5.8.2. Essentially, it is a restatement, but it is very convenient in connection with $\Pi^1_1$-$\mathsf{CA}_0$.

**Corollary 5.8.4.** *For each $\Sigma^1_1$ formula $\varphi(X)$ of $\mathcal{L}_2$,* $\mathsf{ACA}_0$ *proves that there is a tree $T \subseteq \mathbb{N}^{<\mathbb{N}}$, uniformly $\Sigma^0_0$ definable from $\varphi$, as follows.*

1. $(\forall X)[\varphi(X) \leftrightarrow (\exists f \in \mathbb{N}^{\mathbb{N}})(X \oplus f \in [T])]$.
2. $(\exists X)\varphi(X) \leftrightarrow \neg\,\mathrm{WF}(T)$.

*Proof.* Part (2) is an immediate consequence of part (1), so we prove the latter. Let $\psi$ be the $\Sigma^0_0$ formula given by Theorem 5.8.2, and define $T$ to be the set of all sequences $\langle\langle \sigma \upharpoonright k, \alpha \upharpoonright k\rangle : k \leqslant t\rangle$ where $\sigma \in 2^t$, $\alpha \in \mathbb{N}^t$, and $(\forall k \leqslant t)\psi(\sigma \upharpoonright k, \alpha \upharpoonright k)$. Then $T$ is exists, and clearly it is a tree. We have

$$\begin{aligned}\varphi(X) &\leftrightarrow (\exists f \in \mathbb{N}^{\mathbb{N}})(\forall t)\psi(X \upharpoonright t, f \upharpoonright t)\\ &\leftrightarrow (\exists f \in \mathbb{N}^{\mathbb{N}})(\forall t)(\forall k \leqslant t)\psi(X \upharpoonright k, f \upharpoonright k)\\ &\leftrightarrow (\exists f \in \mathbb{N}^{\mathbb{N}})(\forall t)[\langle\langle X \upharpoonright k, f \upharpoonright k\rangle : k \leqslant t\rangle \in T]\\ &\leftrightarrow (\exists f \in \mathbb{N}^{\mathbb{N}})[X \oplus f \in [T]].\end{aligned}$$

This completes the proof. □

There is a point of subtlety in Definition 5.8.3 in the context of second order arithmetic because it is relatively easy for a model $\mathcal{M}$ to contain linear orderings that are not well founded when viewed externally, but which have no infinite descending sequences in $\mathcal{M}$. For example, a key fact in hyperarithmetical theory is the existence of computable *pseudowellorderings*, which are computable linear orderings of $\omega$ that are not well founded but have no hyperarithmetical (i.e., $\Delta^1_1$) infinite descending sequences. If $S$ is a pseudowellordering, the jump ideal below $S$ is an $\omega$-model of $\mathsf{ACA}_0$ that believes $S$ is a well ordering.

**Definition 5.8.5.** The *hyperjump* of a set $X$ is the set of numbers $e$ such that $\Phi^X_e$ is the characteristic function of a (total) well ordering of $\omega$.

The hyperjump operation plays an analogous role in hyperarithmetical theory to the Turing jump in classical computability. In particular, the hyperjump of a set $X$ is $\Pi^{1,X}_1$ complete, analogous to the fact that the Turing jump of $X$ is $\Sigma^{0,X}_1$ complete. This fact can be formalized to obtain the following equivalence.

**Theorem 5.8.6.** $\Pi^1_1$-$\mathsf{CA}_0$ *is equivalent over* $\mathsf{RCA}_0$ *to the principle that the hyperjump of every set exists.*

To prove the theorem, we recall the definition of the *Kleene–Brouwer order* on finite sequences.

**Definition 5.8.7.** The *Kleene–Brouwer ordering* $\leqslant_{\mathrm{KB}}$ is a linear ordering of $\omega^{<\omega}$ defined as follows: for $\alpha, \beta \in \omega^{<\omega}$, we have $\beta <_{\mathrm{KB}} \alpha$ if one of the following hold.

1. $\beta \succ \alpha$;
2. There is an $n$ such that $\tau \upharpoonright n = \alpha \upharpoonright n$ and $\beta(n) < \alpha(n)$.

Given a tree $T \subseteq \omega^{<\omega}$, $\mathrm{KB}(T)$ denotes the partial order $(T, \leqslant_{\mathrm{KB}})$.

It is easy to see that $\mathrm{KB}(T)$ is uniformly computable from $T$. The notion easily formalizes in $\mathsf{RCA}_0$, and we then have the following important observation, whose proof is left to Exercise 5.13.21.

**Proposition 5.8.8.** $\mathsf{ACA}_0$ *proves that for all trees* $T \subseteq \mathbb{N}^{<\mathbb{N}}$,

$$\mathrm{WF}(T) \leftrightarrow \mathrm{WO}(\mathrm{KB}(T)).$$

We turn to proving the theorem.

*Proof (of Theorem 5.8.6).* Using the formalization of computability theory described in Section 5.5.3, we can write down a $\Pi^1_1$ formula $\theta(X,e)$ stating that $\Phi_e(X)$ is the characteristic function of a well ordering of $\omega$.

First, assume $\Pi^1_1$-$\mathsf{CA}_0$ and fix any set $X$. Then $\{e \in \mathbb{N} : \theta(X,e)\}$ exists by $\Pi^1_1$ comprehension, and this is by definition the hyperjump of $X$.

In the other direction, we argue in $\mathsf{RCA}_0$. Assume that $\{e \in \mathbb{N} : \theta(X,e)\}$ exists for every $X$. First, we deduce $\mathsf{ACA}_0$. To this end, fix an injective function $f : \mathbb{N} \to \mathbb{N}$. We will show that the range of $f$ exists. By Theorem 5.7.2, this suffices. Using a formalization of the $S^m_n$ theorem (Theorem 2.4.10) in $\mathsf{RCA}_0$, we can fix a function $s : \mathbb{N} \to \mathbb{N}$ as follows: for all $y, n \in \mathbb{N}$, $\Phi^f_{s(y)}(\langle n,n\rangle) = 1$, and for all $m \neq n$,

$$\Phi^f_{s(y)}(\langle n,m\rangle) = \begin{cases} 1 - \chi_<(n,m) & \text{if } \neg(\exists x \leqslant \max\{n,m\})[f(x) = y], \\ \chi_<(n,m) & \text{otherwise.} \end{cases}$$

Here, $\chi_<$ refers to the characteristic function of $<$ (as a set of ordered pairs). Thus, $\Phi^f_{s(y)}$ orders $n$ and $m$ oppositely to their natural ordering if no number below $\max\{n,m\}$ witnesses that $y$ is in the range of $f$, and otherwise it orders them the same as their natural ordering. This means that if $y$ is not in the range of $f$ then $\Phi^f_{s(y)}$ is the characteristic function of $\omega^* = (\omega, \geqslant)$, which is of course not a well ordering. But if $y$ is in the range, say with $f(x) = y$, then $\Phi^f_{s(y)}$ is the linear ordering $\leqslant_X$ with

$$f(x-1) <_X f(x-1) - 1 <_X \cdots < 1 <_X 0 <_X f(x) <_X f(x) + 1 <_X \cdots,$$

which is just isomorphic to $(\omega, \leqslant)$. So $y \in \mathrm{range}(f) \leftrightarrow s(y) \in \{e \in \mathbb{N} : \theta(f,e)\}$, and thus $\mathrm{range}(f)$ exists.

We can now establish $\Pi^1_1$ comprehension. Let $\varphi(y)$ be any $\Pi^1_1$ formula of $\mathcal{L}_2$. Since $\mathsf{ACA}_0$ holds, we may appeal to Corollary 5.8.4 (and its uniformity) to find a family $T = \langle T_y : y \in \mathbb{N}\rangle$ of trees $T_y \subseteq \mathbb{N}^{<\mathbb{N}}$ such that $\varphi(y) \leftrightarrow \mathrm{WF}(T_y)$. Using again the $S^m_n$ theorem, let $s : \mathbb{N} \to \mathbb{N}$ be such that $\Phi^T_{s(y)}$ is the characteristic function of $\mathrm{KB}(T_y)$ for every $y$. Then by Proposition 5.8.8, we have

$$\varphi(y) \leftrightarrow \mathrm{WF}(T_y) \leftrightarrow \mathrm{WO}(\mathrm{KB}(T_y)) \leftrightarrow s(y) \in \{e \in \mathbb{N} : \theta(T,e)\}.$$

Hence, $\{y : \varphi(y)\}$ exists by $\Sigma^0_0$ comprehension. □

It follows that the $\omega$-models of $\Pi^1_1$-$\mathsf{CA}_0$ are exactly the Turing ideals that are closed under hyperjump.

The next characterization of $\Pi^1_1$-$\mathsf{CA}_0$ is useful in practice because it avoids referring directly to well orderings, while also giving a sentence equivalent to $\Pi^1_1$-$\mathsf{CA}_0$ over $\mathsf{RCA}_0$.

**Theorem 5.8.9.** *$\Pi^1_1$-$\mathsf{CA}_0$ is equivalent over $\mathsf{RCA}_0$ to the principle that, given a sequence $\langle T_i : i \in \mathbb{N}\rangle$ of subtrees of $\mathbb{N}^{<\mathbb{N}}$, there is a set $X$ so that, for all $i$, $i \in X$ if and only if $T_i$ has an infinite path.*

We leave the proof to Exercise 5.13.22. The key to the reversal is again first to deduce $\mathsf{ACA}_0$, which is a common and helpful step, and then to use Kleene's normal form theorem.

We will locate $\Pi^1_1$-$\mathsf{CA}_0$ alongside $\mathsf{RCA}_0$, $\mathsf{WKL}_0$, and $\mathsf{ACA}_0$ in the next section, after we have introduced the last major subsystem, $\mathsf{ATR}_0$.

### 5.8.2 The subsystem $\mathsf{ATR}_0$

We now turn to the subsystem $\mathsf{ATR}_0$. Like $\mathsf{WKL}_0$, this subsystem is not given by a restriction of the comprehension scheme, but by a separate kind of set existence principle. There are many more analogies between $\mathsf{WKL}_0$ and $\mathsf{ATR}_0$, in fact, although the mathematical principles equivalent to the subsystems are quite different.

The formal definition of $\mathsf{ATR}_0$ refers to well orderings of (nonempty) sets of natural numbers. In what follows, therefore, all well orderings are assumed to have field a nonempty subset of $\omega$. For ease of notation, if $L$ is a well ordering we use $L$ also for its field, and we use $\leqslant_L$ for the ordering relation. (Thus, $L$ as a well ordering refers to the pair $(L, \leqslant_L)$, coded as a subset of $\omega$.) We use $<_L$ for the strict version of $\leqslant_L$, and we write $0_L$ for the $\leqslant_L$-least element of $L$.

**Definition 5.8.10.** Let a well ordering $L$ and an operator $I\colon 2^\omega \to 2^\omega$ be given. For each $X \in 2^\omega$, we define $I^{(L)}(X)$ to be the unique $Y \in 2^\omega$ with the following inductively defined properties.

1. $Y^{[0_L]} = X$.
2. If $x \in L$ is the $\leqslant_L$-successor of $y \in L$ then $Y^{[x]} = I(Y^{[y]})$.
3. If $x \in L$ is a $\leqslant_L$-limit then
$$Y^{[x]} = \{\langle y, n\rangle : y <_L x \wedge n \in Y^{[y]}\}.$$

As an example, let $I$ be the Turing jump operator: $I(X) = X'$. (This is exactly the problem $\mathsf{TJ}$, when viewed as a function.) If we identify each $n \in \omega$ with the well ordering $(n, \leqslant)$ then we obviously have $I^{(n)}(X) = X^{(n)}$, the $n$th jump of $X$. And if we identify $\omega$ with $(\omega, \leqslant)$ then $I^{(\omega)}(X) = X^{(\omega)} = \bigoplus_{n\in\omega} X^{(n)}$. But now, of course, we can also define $X^{(L)} = I^{(L)}(X)$ for other well orderings, for example of order type $\omega + 1$ or $\omega + \omega$, etc.

By analogy with Post's theorem, the Turing jump is universal in the following sense for iterations of arithmetical functionals.

**Theorem 5.8.11.** *Suppose X and Y are sets. The following are equivalent.*

1. *X is computable from $I^{(L)}(Y)$ for some $I\colon 2^\omega \to 2^\omega$ arithmetical in Y and some well ordering L computable from Y.*
2. *X is computable from $Y^{(L)}$ for some well ordering L computable from Y.*

The field of hyperarithmetical theory begins with the study of which sets can be computed from these iterated Turing jumps of a given set, much as classical computability theory begins with the study of which sets are Turing reducible to a given set.

**Definition 5.8.12.** A set $X$ is *hyperarithmetical* in a set $Y$ if $X \leqslant_T Y^{(L)}$ for some well ordering $L$ computable from $Y$. If $X$ is hyperarithmetical in $Y$ we write $X \leqslant_{\mathrm{HYP}} Y$.

The relation $\leqslant_{\mathrm{HYP}}$ of *hyperarithmetical reducibility* is reflexive and transitive, so it induces an equivalence relation $\equiv_{\mathrm{HYP}}$ and a degree structure known as the *hyperdegrees*. There are many analogies between Turing reducibility and hyperarithmetical reducibility. In particular, just as Turing reducibility can be characterized by $\Delta^0_1$ definability, hyperarithmetical reducibility can be characterized by $\Delta^1_1$ definability.

**Theorem 5.8.13 (Kleene; see [264], Chapter II).** *A set X is hyperarithmetical in a set Y if and only if X is $\Delta^1_1$ definable relative to Y.*

The following subsystem of $\mathsf{Z}_2$ is a natural weakening of $\Pi^1_1$-$\mathsf{CA}_0$ that can thus be considered a formal analogue of hyperarithmetical theory, and serves as an intermediate step in our definition of $\mathsf{ATR}_0$.

**Definition 5.8.14.** $\Delta^1_1$-$\mathsf{CA}_0$ is the subsystem consisting of $\mathsf{RCA}_0$ together with the scheme of $\Delta^1_1$ comprehension: the universal closure of

$$(\forall n)[\varphi(n) \leftrightarrow \neg\psi(n)] \to (\exists X)[n \in X \leftrightarrow \varphi(n)],$$

where $\varphi$ and $\psi$ range over $\Sigma^1_1$ formulas.

The collection of all hyperarithmetical sets forms an $\omega$-model HYP that, by Theorem 5.8.13, is the $\subseteq$-minimum such model satisfying $\Delta^1_1$-$\mathsf{CA}_0$. Obviously, every arithmetical set is hyperarithmetical. But $\Delta^1_1$-$\mathsf{CA}_0$ is strictly stronger than $\mathsf{ACA}_0$, and in fact strictly stronger than $\mathsf{ACA}_0^+$.

**Proposition 5.8.15.**

1. $\Delta^1_1$-$\mathsf{CA}_0 \vdash \mathsf{ACA}_0^+$.
2. *There is an $\omega$-model of* $\mathsf{ACA}_0^+ + \neg\Delta^1_1$-$\mathsf{CA}_0$.

*Proof.* For (1), it is clear that $\Delta^1_1$-$\mathsf{CA}_0 \vdash \mathsf{ACA}_0$ since every arithmetical formula is trivially $\Delta^1_1$. Now fix $X$. We need to show that there is a $Y$ such that $(\forall n)[Y^{[0]} = X \wedge Y^{[n+1]} = (Y^{[n]})']$. Let $\varphi(\langle i,n\rangle)$ and $\psi(\langle i,n\rangle)$ be the formulas

$$(\exists Z)[Z^{[0]} = X \wedge (\forall j < i)[Z^{[j+1]} = Z^{[j]\prime}] \wedge n \in Z^{[i]}],$$

and

$$(\forall Z)[(Z^{[0]} = X \wedge (\forall j < i)[Z^{[j+1]} = Z^{[j]\prime}]) \rightarrow n \in Z^{[i]}],$$

respectively. By arithmetical induction on $i$, we have

$$(\forall n)[\varphi(\langle i, n\rangle) \leftrightarrow \psi(\langle i, n\rangle)].$$

Hence, by $\Delta^1_1$-$\mathsf{CA}_0$, the set $Y = \{\langle i, n\rangle : \varphi(\langle i, n\rangle)\}$ exists, and clearly this is the desired set.

For (2), let $\mathcal{M}$ be the $\omega$-model consisting of all $Y \leqslant_{\mathrm{T}} \varnothing^{(\omega n)}$ for some $n \in \omega$. Given any $X \in \mathcal{S}^{\mathcal{M}}$, say computable from $\varnothing^{(\omega n)}$, we have that

$$\begin{aligned} X^{(\omega)} &= \langle X, X', X'', \ldots\rangle \\ &\leqslant_{\mathrm{T}} \langle \varnothing^{(\omega n)}, \varnothing^{(\omega n+1)}, \varnothing^{(\omega n+2)}, \ldots\rangle \\ &\leqslant_{\mathrm{T}} \varnothing^{(\omega(n+1))}. \end{aligned}$$

Hence, $\mathcal{M} \vDash \mathsf{ACA}_0^+$. Now, let $L$ be any computable well ordering of order type $\omega^\omega$. Then $L$ belongs to $\mathcal{M}$ but $\varnothing^{(L)}$ does not because $\varnothing^{(L)} \geqslant_{\mathrm{T}} \varnothing^{(\omega n)}$ for all $n$. But $\varnothing^{(L)}$ is hyperarithmetical, so $\mathcal{M} \nvDash \Delta^1_1$-$\mathsf{CA}_0$. □

The subsystem $\mathsf{ATR}_0$ is also inspired by hyperarithmetical theory, but features closure properties that $\Delta^1_1$-$\mathsf{CA}_0$ lacks. This is because of the existence of hyperarithmetical pseudowellorderings, discussed in the preceding section. If $L$ is such an ordering then HYP $\vDash$ WO($L$) but HYP $\nvDash (\exists Y)[Y = \varnothing^{(L)}]$. $\mathsf{ATR}_0$ is basically designed to overcome this problem: informally, it states that we can iterate an arbitrary arithmetical functional along an arbitrary well ordering. In applications, this turns to be more flexible and more useful than being closed under $\Delta^1_1$ comprehension alone, and accounts for $\mathsf{ATR}_0$ being more ubiquitous system. (Hence the reason $\mathsf{ATR}_0$ makes the list of "big five" subsystems, while $\Delta^1_1$-$\mathsf{CA}_0$ does not.)

To begin, we can formalize Definition 5.8.10 in second order arithmetic for arithmetical operators. Formally, consider an arithmetical formula $\psi(i, X)$ of $\mathcal{L}_2$ (which may include parameters). We think of this as the operator $I_\psi$ such that $I_\psi(X) = \{i : \psi(i, X)\}$.

**Definition 5.8.16.** Let $\psi(i, X)$ be an arithmetical formula. The following definition is made in $\mathsf{RCA}_0$. Let $X, Y, L$ be sets. If WO($L$), then we say $Y = I_\psi^{(L)}(X)$ if the following hold.

1. $Y^{[0_L]} = X$.
2. If $x \in L$ is the $\leqslant_L$-successor of $y \in L$ then $i \in Y^{[x]}$ if and only If $\psi(i, Y^{[y]})$.
3. If $x \in L$ is a $\leqslant_L$-limit then

$$Y^{[x]} = \{\langle y, n\rangle : y <_L x \wedge n \in Y^{[y]}\}.$$

**Definition 5.8.17.** The subsystem $\mathsf{ATR}_0$ consists of $\mathsf{RCA}_0$ and the axiom scheme of *arithmetical transfinite recursion*, which states the following for each arithmetical formula $\psi(i, X)$: for all sets $X$ and $L$, if WO($L$) then there exists a $Y$ such that $Y = I_\psi^{(L)}(X)$.

It is not difficult to formalize Theorem 5.8.11 to obtain the following alternate characterization of $\mathsf{ATR}_0$, which is analogous to the jump characterization of $\mathsf{ACA}_0$ (Corollary 5.6.3).

**Theorem 5.8.18.** *The following statements are equivalent over* $\mathsf{RCA}_0$.

1. $\mathsf{ATR}_0$.
2. *For all sets X and L, if* $\mathrm{WO}(L)$ *then there is a set Y such that* $Y = X^{(L)}$.

*Here,* $X^{(L)}$ *abbreviates* $I^{(L)}_{\psi}(X)$ *for the formula* $\psi$ *defining* $X'$.

There is an important distinction between arithmetical transfinite recursion and arithmetical transfinite *induction*. The latter is provable in $\mathsf{ACA}_0$ (Exercise 5.13.19). In particular, $\mathsf{ACA}_0$ can prove that *if* there is a $Y$ such that $Y = I^{(L)}_{\psi}(X)$ then it is unique. However, $\mathsf{ATR}_0$ is strictly stronger than $\mathsf{ACA}_0$, and in fact, than $\Delta^1_1$-$\mathsf{CA}_0$, as we will see in Proposition 5.8.21.

The definition of $\mathsf{ATR}_0$ helps to explain its name, but it is not always convenient for reversals. We now state an equivalence that is much easier to work with in practice.

**Definition 5.8.19.** Let $\Gamma$ be a class of formulas of $\mathcal{L}_2$ in one free variable. The $\Gamma$ *separation principle* is the following scheme, for every pair $\varphi(x)$ and $\psi(x)$ in $\Gamma$: if $(\forall x)[\varphi(x) \rightarrow \neg\psi(x)]$, there exists a set $Z$ (called a *separating set*) such that $(\forall x)[\varphi(x) \rightarrow x \in Z \wedge \psi(x) \rightarrow x \notin Z]$.

Informally, this says that if $X$ and $Y$ are disjoint $\Gamma$ definable sets (which may not exist), then there exists a set $Z$ such that $X \subseteq Z$ and $Y \cap Z = \emptyset$.

Observe that if $Z$ satisfies $(\forall x)[\varphi(x) \rightarrow x \in Z \wedge \psi(x) \rightarrow x \notin Z]$ then $\omega \setminus Z$ satisfies $(\forall x)[\psi(x) \rightarrow x \in Z \wedge \varphi(x) \rightarrow x \notin Z]$. So, at least over $\mathsf{RCA}_0$, the order of $\varphi$ and $\psi$ does not matter. Also, note that if $\Gamma$ is a collection of formulas and $\neg\Gamma$ is the collection $\{\neg\varphi : \varphi \in \Gamma\}$ then the $\Gamma$ separation principle is equivalent over $\mathsf{RCA}_0$ to the $\neg\Gamma$ separation principle.

Our interest is in $\Gamma = \Sigma^1_1$. The Turing jump of a set $X$ is arithmetical in $X$, and so is trivially $\Sigma^1_1$ definable from $X$. This means that the $\Sigma^1_1$ separation principle implies $\mathsf{ACA}_0$ over $\mathsf{RCA}_0$. The next theorem is much less trivial. It provides a useful starting point for reversals to $\mathsf{ATR}_0$.

**Theorem 5.8.20 (Simpson [288]).** *The following are equivalent over* $\mathsf{RCA}_0$.

1. $\mathsf{ATR}$.
2. *The* $\Sigma^1_1$ *separation principle.*

We will prove this theorem in Chapter 12. Meanwhile, Exercise 5.13.18 shows that $\mathsf{WKL}_0$ is equivalent to the $\Sigma^0_1$ separation principle. This draws another analogy between $\mathsf{WKL}_0$ and $\mathsf{ATR}_0$. For now, we have the following consequence.

**Proposition 5.8.21.**

1. $\mathsf{ATR}_0 \vdash \Delta^1_1$-$\mathsf{CA}_0$.
2. *There is an* $\omega$*-model of* $\Delta^1_1$-$\mathsf{CA}_0 + \neg\mathsf{ATR}_0$.

*Proof.* For (1), we argue in $\mathsf{ATR}_0$. Suppose $\varphi$ and $\psi$ are $\Sigma^1_1$ and that $(\forall x)[\varphi(x) \leftrightarrow \neg\psi(x)]$. Then by Theorem 5.8.20 there is a separating set $Z$ for $\varphi$ and $\psi$, and clearly $(\forall x)[x \in Z \leftrightarrow \varphi(x)]$. For (2), consider the model HYP. As we remarked above, this satisfies $\Delta^1_1$-$\mathsf{CA}_0$ but not $\mathsf{ATR}_0$. □

Let us now look at $\mathsf{ATR}_0$ alongside $\Pi^1_1$-$\mathsf{CA}_0$. That $\Pi^1_1$-$\mathsf{CA}_0$ implies $\mathsf{ATR}_0$ is an easy consequence of the preceding theorem. The separation is analogous to Proposition 5.6.14 (2), which saw $\mathsf{WKL}_0$ separated from $\mathsf{ACA}_0$, only using higher computability theory analogues. In particular, instead of the low basis theorem for $\Pi^0_1$ classes, we will need the *Gandy basis theorem* for $\Sigma^1_1$ classes. This states the following: if $C$ is a nonempty $\Sigma^1_1$ definable subset of $2^\omega$ then there exists $X \in C$ that is *hyperarithmetically low*, meaning the hyperjump of $X$ is hyperarithmetically reducible to the hyperjump of $\emptyset$. (For a proof, see Sacks [264], Corollary III.1.5).

**Theorem 5.8.22.**

*1.* $\Pi^1_1$-$\mathsf{CA}_0 \vdash \mathsf{ATR}_0$.
*2. There is an* $\omega$*-model of* $\mathsf{ATR}_0 + \neg\Pi^1_1$-$\mathsf{CA}_0$.

*Proof.* For (1), we show that $\Pi^1_1$-$\mathsf{CA}_0$ implies the $\Sigma^1_1$ separation principle, which suffices by Theorem 5.8.20. $\Pi^1_1$-$\mathsf{CA}_0$ proves that every $\Sigma^1_1$ definable subset of $\mathbb{N}$ exists. Hence, it can serve as its own separating set (with respect to any other $\Sigma^1_1$ formula defining a disjoint set).

For (2), we build an $\omega$-model by dovetailing and iterating, as in Theorem 4.6.13. Let $\theta_0(x, X), \theta_1(x, X), \ldots$ be an enumeration of all $\Pi^1_1$ formulas in the displayed free variables and no other set parameters. Set $A_0 = \emptyset$, and assume that for some $s$ we have defined $A_s$. Say $s = \langle e, t, i, j\rangle$. If $\Phi^{A_t}_e$ is not an element of $2^\omega$, let $A_{s+1} = A_s$. Otherwise, let $C$ be the class of all $X \in 2^\omega$ such that for all $x$, if $\theta_i(x, \Phi^{A_t}_e)$ holds then $x \in X$, and if $\theta_j(x, \Phi^{A_t}_e)$ holds then $x \notin X$. Clearly, $C$ is nonempty since it contains $\{x : \theta_i(x, \Phi^{A_t}_e)\}$. By the Gandy basis theorem relative to $A_s$, there exists $Z \in C$ which is hyperarithmetically low relative to $A_s$. Let $A_{s+1} = A_s \oplus Z$.

We end up with $A_0 \leqslant_T A_1 \leqslant_T \cdots$. Let $\mathcal{M}$ be the $\omega$-model consisting of all $Y \in 2^\omega$ such that $Y \leqslant_T A_s$ for some $s$. By induction, each $A_{s+1}$ is hyperarithmetically low over $A_s$ and hence hyperarithmetically low. Thus, the hyperjump of $\emptyset$ does not belong to $\mathcal{S}^\mathcal{M}$, and therefore $\mathcal{M} \nvDash \Pi^1_1$-$\mathsf{CA}_0$ by Theorem 5.8.6. On the other hand, we claim that $\mathcal{M}$ satisfies the $\Pi^1_1$ separation principle, and therefore $\mathsf{ATR}_0$. Indeed, let $\varphi(x)$ and $\psi(x)$ be $\Pi^1_1$ formulas with parameters from $\mathcal{M}$. Let $Z$ be the join of all set parameters occurring in $\varphi$ and $\psi$. Then $Z \in \mathcal{S}^\mathcal{M}$, so we may fix a $t$ and $e$ such that $Z = \Phi^{A_t}_e$. We may also fix $i$ and $j$ so that $\varphi(x) = \theta_i(x, Z)$ and $\psi(x) = \theta_j(x, Z)$. Let $s = \langle e, t, i, j\rangle$. Then by construction, $A_{s+1} = A_s \oplus Z$ for some separating set $Z$ for $\varphi$ and $\psi$. Hence, $\mathcal{M}$ satisfies the $\Pi^1_1$ separation scheme, which is equivalent to $\mathsf{ATR}_0$ by Theorem 5.8.20. □

## 5.9 Conservation results

The previous sections gave examples of reverse mathematics results obtained from direct reversals. Although these are often the most interesting results, they can sometimes be difficult to obtain. A second class of results shows that two principles have the same consequences at some level of the arithmetical or analytical hierarchy, without showing that the two principles are equivalent.

**Definition 5.9.1 (Conservativity of theories).** Let $\Gamma$ be a set of $\mathcal{L}_2$ sentences. A $\mathcal{L}_2$ theory $T_1$ is $\Gamma$ *conservative* over a $\mathcal{L}_2$ theory $T_2$ if $T_2 \vdash \varphi$ whenever $T_1 \vdash \varphi$ for all $\varphi \in \Gamma$.

A *conservation result* shows that one principle is conservative over another for some class $\Gamma$ of formulas. Conservation results can be obtained using proof theoretic methods, which involve directly analyzing the structure of formal proofs. They can also be proved model theoretically, using the following definition.

**Definition 5.9.2 (Conservativity of structures).** Let $\Gamma$ be a set of $\mathcal{L}_2$ sentences, and let $\mathcal{N} < \mathcal{M}$ be structures. We say that $\mathcal{M}$ is $\Gamma$ *conservative* over $\mathcal{N}$ if, for every $\varphi \in \Gamma$, if $\mathcal{N} \vDash \varphi$ then $\mathcal{M} \vDash \varphi$.

The following theorem gives two initial results on conservation in submodels. Recall that a model $\mathcal{N}$ is an $\omega$-submodel of a structure $\mathcal{M}$ if $\mathcal{M}$ differs from $\mathcal{N}$ only in having additional sets.

**Theorem 5.9.3.** *Assume that $\mathcal{N}$ is an $\omega$-submodel of $\mathcal{M}$ and $\varphi$ is a sentence in $\mathcal{L}_2(\mathcal{N})$.*

1. *If $\varphi$ is $\Pi^1_1$ and $\mathcal{M} \vDash \varphi$ then $\mathcal{N} \vDash \varphi$.*
2. *If $\varphi$ is $\Sigma^1_1$ and $\mathcal{N} \vDash \varphi$ then $\mathcal{M} \vDash \varphi$.*

*Proof.* The key point is that for any arithmetical formula in $\mathcal{L}_2(\mathcal{N})$, $\mathcal{N} \vDash \psi$ if and only if $\mathcal{M} \vDash \psi$ (Exercise 5.13.1). For part (1), assume $\varphi = (\forall \vec{X})\psi(\vec{X})$ where $\psi$ is arithmetical. Let $\vec{C}$ be any sequence of sets from $\mathcal{S}^{\mathcal{N}}$ of the same length as $\vec{X}$. Because these sets are in $\mathcal{M}$, we have $\mathcal{M} \vDash \psi(\vec{C})$, and because $\psi$ is arithmetical this implies $\mathcal{N} \vDash \psi(\vec{C})$. Thus $\mathcal{N} \vDash (\forall \vec{X})\psi(\vec{X})$.

For part (2), assume $\varphi = (\exists \vec{X})\psi(\vec{X})$ where $\psi$ is arithmetical. Because $\mathcal{N} \vDash \varphi$, there is some $\vec{C}$ in $\mathcal{S}^{\mathcal{N}}$ with $\mathcal{N} \vDash \psi(\vec{C})$. Because $\mathcal{M}$ is an $\omega$-extension of $\mathcal{N}$, the $\vec{C}$ are in $\mathcal{S}^{\mathcal{M}}$ as well, and $\mathcal{M} \vDash \psi(\vec{C})$. Thus $\mathcal{M} \vDash (\exists \vec{X})\psi(\vec{X})$, as desired. □

Not all conservation theorems are this straightforward. We will see below that the system $\mathsf{WKL}_0$ is conservative over $\mathsf{RCA}_0$ for $\Pi^1_1$ sentences. It is also conservative over primitive recursive arithmetic (PRA) for $\Pi^0_2$ sentences. Both PRA and $\mathsf{RCA}_0$ are strictly weaker than $\mathsf{WKL}_0$. These conservation results require more than the formal manipulation of Theorem 5.9.3. These conservation results can be used to show that certain uses of $\mathsf{WKL}_0$ are dispensable, and the certain implications are impossible. For example, the conservation results just mentioned imply that if $\mathsf{WKL}_0$ is able to prove the consistency of a theory (which is expressed as a $\Pi^0_1$ sentence) then this is already provable in PRA. Hence the consistency strength of $\mathsf{WKL}_0$ is quite low, even though it is able to prove many sets exist that $\mathsf{RCA}_0$ cannot prove to exist. Conversely, we know that no $\Pi^1_1$ sentence can be equivalent to $\mathsf{WKL}_0$ over $\mathsf{RCA}_0$, because if the sentence is provable in $\mathsf{WKL}_0$ then it is already provable in $\mathsf{RCA}_0$.

A particularly striking application of this method is given by Kikuchi and Tanaka [179], who show that Gödel's second incompleteness theorem is provable in PRA by formalizing a model theoretic proof in the system $\mathsf{WKL}_0$, which is $\Pi^0_2$ conservative over PRA.

The next theorem establishes a useful method for proving $\Pi^1_1$ conservativity.

**Theorem 5.9.4.** *Suppose that $T_1$ and $T_2$ are $\mathcal{L}_2$ theories such that every countable $\mathcal{N} \vDash T_1$ is an $\omega$-submodel of some $\mathcal{M} \vDash T_2$. Then $T_2$ is $\Pi^1_1$ conservative over $T_1$.*

*Proof.* We proceed by contraposition, and assume that $T_1$ does not prove a $\Pi^1_1$ sentence $\varphi$. By the completeness theorem, there is a model $\mathcal{N}$ of $T_1$ such that $\mathcal{N} \vDash \neg\varphi$. Now $\neg\varphi$ is equivalent to a $\Sigma^1_1$ sentence $\rho$, so $\mathcal{N} \vDash \rho$ and thus, by the previous theorem $\mathcal{M} \vDash \rho$. This means that $\mathcal{M} \vDash \neg\varphi$. □

The theorem has an important corollary in the special but important case where $T_1$ is $\mathsf{RCA}_0$ and $T_2$ is $\mathsf{RCA}_0$ together with a $\Pi^1_2$ sentence. This serves as a template for many conservation results in reverse mathematics.

**Definition 5.9.5 (Model extension).** Fix an $\mathcal{L}_2$ structure $\mathcal{M}$ and a $G \subseteq M$ (which need not be in $\mathcal{S}^{\mathcal{M}}$). Then $\mathcal{M}[G]$ is the model with first order part $M$ and second order part consisting of all $A \subseteq M$ that are $\Delta^0_1$ definable with parameters from $M \cup \mathcal{S}^{\mathcal{M}} \cup \{G\}$.

Notice that for every $G \subseteq M$, $\mathcal{M}$ is a $\omega$-submodel of $\mathcal{M}[G]$. Furthermore, $\mathcal{M}[G]$ is automatically closed under $\Delta^0_1$ comprehension.

By Exercise 5.13.1, $\mathcal{M}$ and $\mathcal{M}[G]$ satisfy the same arithmetical formulas *with parameters from $\mathcal{M}$*. There are, however, many arithmetical formulas involving the set $G$, and we cannot conclude anything about these on general principles. For example, it may be that $\mathcal{M}$ satisfies $\mathsf{I}\Sigma^0_1$ but that $\mathcal{M}[G]$ does not. We would only know that induction in $\mathcal{M}[G]$ for those $\Sigma^0_1$ formulas that do not include $G$ as a parameter. In practice, this means that if we start with a model $\mathcal{M}$ of $\mathsf{RCA}_0$, and want to add a $G$ so that $\mathcal{M}[G]$ is also a model of $\mathsf{RCA}_0$, then we need to separately verify that $\mathsf{I}\Sigma^0_1$ holds in $\mathcal{M}[G]$.

In cases where we can do this verification, we get the following conservation result.

**Corollary 5.9.6.** *Let $\mathsf{P}$ be an $\mathcal{L}_2$ sentence of the form*

$$(\forall X)[\varphi(X) \to (\exists Y)\psi(X,Y)],$$

*where $\varphi$ and $\psi$ are arithmetical. Suppose that for every countable model $\mathcal{M}$ of $\mathsf{RCA}_0$ and every $A \in \mathcal{S}^{\mathcal{M}}$ such that $\mathcal{M} \vDash \varphi(A)$ there is a $G \subseteq M$ such that $\mathcal{M}[G] \vDash \mathsf{RCA}_0 + \psi(A,G)$. Then $T + \mathsf{P}$ is $\Pi^1_1$ conservative over $T$.*

*Proof.* Fix a countable model $\mathcal{M} \models \mathsf{RCA}_0$. We define a chain of models $\mathcal{M}_0 \preceq \mathcal{M}_1 \preceq \cdots$ of $\mathsf{RCA}_0$, each an $\omega$-submodel of the next. Let $\mathcal{M}_0 = \mathcal{M}$. Now fix $s \in \omega$, and assume by induction that have defined a countable model $\mathcal{M}_t$ of $T$ for each $t \leqslant s$, along with an enumeration $\langle A_{t,e} : e \in \omega \rangle$ of all $A \in \mathcal{S}^{\mathcal{M}_t}$ such that $\mathcal{M}_t \models \varphi(A)$. Say $s = \langle t, e \rangle$, so that $t, e \leqslant s$. Since $\mathcal{M}_t$ is an $\omega$-submodel of $\mathcal{M}_s$, it follows that $\mathcal{M}_s \models \varphi(A_{t,e})$. Hence, by assumption there is a $G \subseteq M$ such that $\mathcal{M}_s[Y] \models \psi(A, G)$. Let $\mathcal{M}_{s+1} = \mathcal{M}_s[G]$. This completes the construction. Now, let $\mathcal{M}^* = \bigcup_s \mathcal{M}_s$. (More precisely, $\mathcal{M}^*$ has first order part $M$, and $\mathcal{S}^{\mathcal{M}^*} = \bigcup_s \mathcal{S}^{\mathcal{M}_s}$.) Each $\mathcal{M}_s$, and in particular $\mathcal{M}$, is an $\omega$-submodel of $\mathcal{M}^*$. From here, it is easy to see that $\mathcal{M}^*$ is closed under $\Delta^0_1$ comprehension and satisfies $\mathsf{I}\Sigma^0_1$, so $\mathcal{M}^* \models \mathsf{RCA}_0$. Moreover, since $\varphi$ is arithmetical, each $A \in \mathcal{S}^{\mathcal{M}}$ such that $\mathcal{M}^* \models \varphi(A)$ must be equal to $A_{t,e}$ for some $t$ and $e$. Let $s = \langle t, e \rangle$. Then by construction, there is a $G \in \mathcal{S}^{\mathcal{M}_{s+1}}$ such that $\mathcal{M}_{s+1} \models \psi(A, G)$. Since $\mathcal{S}^{\mathcal{M}_{s+1}} \subseteq \mathcal{S}^{\mathcal{M}^*}$ and $\psi$ is arithmetical, it follows that $\mathcal{M}^* \models (\exists Y)\psi(A, Y)$. So $\mathcal{M}^* \models \mathsf{P}$. Now since $\mathcal{M}$ was arbitrary, the conclusion of the corollary follows by Theorem 5.9.4. □

## 5.10 First order parts of theories

This section will establish several conservation results of a certain form. If $T_1$ and $T_2$ are theories of second order arithmetic, $T_1$ has only arithmetical axioms, and $T_2$ is conservative over $T_1$ for $\Pi^1_1$ formulas, then we call $T_1$ the *first order part* of $T_2$. Similarly, the *first order part* of an $\mathcal{L}_2$ structure $\mathcal{M}$ is the $\mathcal{L}_1$ structure $\langle \omega^{\mathcal{M}}, +^{\mathcal{M}}, \cdot^{\mathcal{M}}, <^{\mathcal{M}} \rangle$.

Conservativity for $\Pi^1_1$ sentences is closely related to conservativity for arithmetical formulas with free set variables, and the following results could also be stated in terms of arithmetical formulas.

**Theorem 5.10.1.** *Suppose that $\mathcal{N}$ is an $\mathcal{L}_2$ structure that satisfies* $\mathsf{PA}^- + \mathsf{I}\Sigma^0_1$. *Then there is an $\mathcal{L}_2$ structure $\mathcal{M}$ that satisfies* $\mathsf{RCA}_0$ *and $\omega$-extends $\mathcal{N}$.*

*Proof.* Let $\mathcal{N}$ be as above. We let $\mathcal{M}$ have the same first order part as $\mathcal{N}$, and let the second order part of $\mathcal{M}$ consist of all sets that are $\Delta^0_1$ definable in $\mathcal{N}$ (with no set parameters). It is clear that $\mathcal{M}$ satisfies the $\mathsf{PA}^-$. Furthermore, $\mathcal{M}$ satisfies $\Delta^0_1$ comprehension scheme. Indeed, if $A$ is $\Delta^0_1$ definable from sets $B_0, \ldots, B_{n-1} \in \mathcal{S}^{\mathcal{M}}$ then each $B_i$ is $\Delta^0_1$ definable in $\mathcal{M}$ with no set parameters. Hence, by replacing references to the $B_i$ in the definition of $A$ by $\Sigma^0_1$ or $\Pi^0_1$ formulas as needed, we obtain a $\Delta^0_1$ definition of $A$ in $\mathcal{M}$ with no set parameters. Thus, $A \in \mathcal{S}^{\mathcal{M}}$.

We can similarly show that $\mathcal{M}$ satisfies the $\Sigma^0_1$ induction scheme. Indeed, suppose $\varphi$ is $\Sigma^0_1$ and $\mathcal{M} \models \varphi(0) \wedge (\forall x)[\varphi(x) \rightarrow \varphi(x+1)]$. Each set parameter in $\varphi$ is an element of $\mathcal{S}^{\mathcal{M}}$, so each reference to it in $\varphi$ may be replaced by a $\Sigma^0_1$ or $\Pi^0_1$ formula with no set parameters to obtain a $\Sigma^0_1$ formula $\widehat{\varphi}$ with no set parameters such that $\mathcal{M} \models (\forall x)[\varphi(x) \leftrightarrow \widehat{\varphi}(x)]$. In particular, $\mathcal{M}$ satisfies $\widehat{\varphi}(0) \wedge (\forall x)[\widehat{\varphi}(x) \rightarrow \widehat{\varphi}(x+1)]$. By identifying $\mathcal{N}$ with the $\mathcal{L}_2$ structure with first order part $N$ and second order part $\emptyset$, we can view $\mathcal{N}$ as an $\omega$-submodel of $\mathcal{M}$. Therefore, by Theorem 5.9.3, $\mathcal{N} \models \widehat{\varphi}(0) \wedge (\forall x)[\widehat{\varphi}(x) \to \widehat{\varphi}(x+1)]$. Thus $\mathcal{N} \models (\forall x)\widehat{\varphi}(x)$ by $\Sigma^0_1$ induction in $\mathcal{N}$. By Theorem 5.9.3 again, $\mathcal{M}$ satisfies $(\forall x)\widehat{\varphi}(x)$ and hence $(\forall x)\varphi(x)$. □

We will obtain several corollaries from this result. The first is a direct application of Theorem 5.10.1. The other two provide a link between $\mathsf{RCA}_0$ and first order arithmetic.

**Corollary 5.10.2.** $\mathsf{RCA}_0$ *is conservative over* $\mathsf{I}\Sigma^0_1$ *for* $\Pi^1_1$ *sentences.*

**Corollary 5.10.3.** *An* $\mathcal{L}_1$ *structure satisfies the* $\Sigma^0_1$ *induction scheme if and only if it is the first order part of some model of* $\mathsf{RCA}_0$.

*Proof.* Given a $\mathcal{L}_1$ structure $\mathcal{N}$ satisfying the $\Sigma^0_1$ induction scheme, consider the expansion of $\mathcal{N}$ to a $\mathcal{L}_2$ structure $\mathcal{N}'$ with no sets. Then $\mathcal{N}'$ satisfies the $\mathsf{I}\Sigma^0_1$ and has the same $\mathcal{L}_1$ theory as $\mathcal{N}$. Now, by the theorem, $\mathcal{N}'$ expands to a model $\mathcal{M}$ of $\mathsf{RCA}_0$, which has the same arithmetical theory. Thus, by the same □

**Corollary 5.10.4.** $\mathsf{RCA}_0$ *is conservative over* $\mathsf{PA}^- + \mathsf{I}\Sigma^0_1$ *(restricted to* $\mathcal{L}_1$ *sentences).*

We next obtain a parallel theorem showing that each model of arithmetical induction extends to a model of $\mathsf{ACA}_0$.

**Theorem 5.10.5.** *Suppose* $\mathcal{N}$ *is an* $\mathcal{L}_2$ *structure satisfying the arithmetical induction scheme. Then there is an* $\mathcal{L}_2$ *structure* $\mathcal{M}$ *that satisfies* $\mathsf{ACA}_0$ *and* $\omega$*-extends* $\mathcal{N}$.

**Corollary 5.10.6.** *An* $\mathcal{L}_1$ *structure satisfies the scheme for arithmetical induction if and only if it is the first order part of some model of* $\mathsf{ACA}_0$.

**Corollary 5.10.7.** $\mathsf{ACA}_0$ *is conservative over Peano arithmetic for sentences in* $\mathcal{L}_1$.

It should be noted that, although the results in this section might suggest a larger family of conservation results, the cases of $\mathsf{RCA}_0$ and $\mathsf{ACA}_0$ are somewhat special. For example, it is not true that every model of the full second order induction scheme extends to a model of $\mathsf{Z}_2$ satisfying full comprehension.

## 5.11 Comparing reducibility notions

At this point, we are ready to compare the reducibility notions of Chapter 4 with the reducibility notions based on provability in second order arithmetic.

In general, if we have a set $\mathcal{S}$ whose elements are viewed as problems in some sense, a *reducibility notion* (or just *reducibility*) on $\mathcal{S}$ is simply a reflexive, transitive relation on $\mathcal{S}$. Each of these requirements has a natural motivation. Reflexivity matches the informal concept that a problem should be reducible to itself. Transitivity matches the informal concept that, if we can solve problem P by appealing to problem

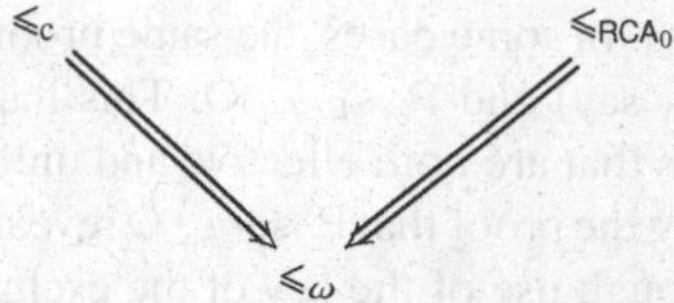


**Figure 5.1.** Relationships between $\leqslant_{c}$, $\leqslant_{\omega}$, and $\leqslant_{\mathsf{RCA}_0}$. An arrow from one reducibility to another indicates that if $\mathsf{P}$ and $\mathsf{Q}$ are $\Pi^1_2$ sentences of $\mathcal{L}_2$, regarded alternatively as problems, then if $\mathsf{P}$ is reducible to $\mathsf{Q}$ in the sense of the first, it is also reducible in the sense of the second. No additional arrows can be added.

$\mathsf{Q}$, and we can solve problem $\mathsf{Q}$ by appealing in the same manner to problem $\mathsf{R}$, then we should be able to solve $\mathsf{P}$ by appealing to $\mathsf{R}$ directly.

We may regard provability over $\mathsf{RCA}_0$ as a reducibility in the above sense: we can write $\mathsf{P} \leqslant_{\mathsf{RCA}_0} \mathsf{Q}$ if $\mathsf{RCA}_0 \vdash \mathsf{Q} \to \mathsf{P}$, that is, every model of $\mathsf{RCA}_0 + \mathsf{Q}$ satisfies $\mathsf{P}$. Here, $\mathsf{P}$ and $\mathsf{Q}$ range over sentences of $\mathcal{L}_2$. As discussed in Section 3.2, when these are formalizations of $\forall\exists$ theorems we may think of them naturally as instance–solution problems. Notably, this applies to $\Pi^1_2$ sentences of $\mathcal{L}_2$, of which we will see many examples throughout this book.

In this case, we may compare $\leqslant_{\mathsf{RCA}_0}$ directly with our prior reducibilities, of which the strongest to consider are $\leqslant_{c}$ and $\leqslant_{\omega}$. It is easy to see that if $\mathsf{P} \leqslant_{c} \mathsf{Q}$ or $\mathsf{P} \leqslant_{\mathsf{RCA}_0} \mathsf{Q}$ then $\mathsf{P} \leqslant_{\omega} \mathsf{Q}$, but that no additional relationships between the three reducibilities hold. (We have already seen examples of problems $\mathsf{P}$ and $\mathsf{Q}$ such that $\mathsf{P} \leqslant_{\omega} \mathsf{Q}$ but $\mathsf{P} \nleqslant_{c} \mathsf{Q}$. We will see examples of problems with $\mathsf{P} \leqslant_{\omega} \mathsf{Q}$ but $\mathsf{P} \nleqslant_{\mathsf{RCA}_0} \mathsf{Q}$ in the next chapter.) See Figure 5.1, and compare with Figure 4.6.

Reverse mathematics was first associated with reducibilities such as $\leqslant_{\mathsf{RCA}_0}$. Reducibilities such as $\leqslant_{\omega}$ were also of interest, especially for nonreducibility results: a result that $\mathsf{P} \leqslant_{\mathsf{RCA}_0} \mathsf{Q}$ gives more information than $\mathsf{P} \leqslant_{\omega} \mathsf{Q}$, while a result that $\mathsf{P} \nleqslant_{\omega} \mathsf{Q}$ gives more information than a result that $\mathsf{P} \nleqslant_{\mathsf{RCA}_0} \mathsf{Q}$. Interest in $\leqslant_{c}$ and related reducibilities ($\leqslant_{\mathrm{W}}$, etc.) came later. When we prove a result of the form $\mathsf{P} \leqslant_{c} \mathsf{Q}$, we are establishing the direct computability of solutions of $\mathsf{P}$ from solutions to $\mathsf{Q}$. If we instead establish $\mathsf{P} \leqslant_{\omega} \mathsf{Q}$, we are showing a more indirect form of computability of solutions, because $\mathsf{Q}$ may be applied many times to solve a single instance of $\mathsf{P}$. This form of reducibility requires the game theoretic approach of Theorem 4.7.5.

When we establish a result of the form $\mathsf{P} \leqslant_{\mathsf{RCA}_0} \mathsf{Q}$, we are certainly establishing an indirect form of computability, but perhaps more importantly we are establishing the *verifiability* of solutions to $\mathsf{P}$. This dichotomy between computability and verification is parallel to the dichotomy between computational complexity and verification in the analysis of algorithms. (For more on this dichotomy in proofs over $\mathsf{RCA}_0$, including some means of measuring "how much" of the use of a principle in a proof is for computability and "how much" for verification, see the recent work in [82, 151, 256].)

The lack of relationships between $\leqslant_{\mathsf{RCA}_0}$ and $\leqslant_{c}$ is a motivation for considering both of these reducibilities in practice. We can use precisely defined reducibility notions to describe the aspects of computability, verification, and uniformity that we see when studying theorems. In some cases, the same proof method establishes both $\mathsf{P} \leqslant_{\mathrm{c}} \mathsf{Q}$ (or even $\mathsf{P} \leqslant_{\mathrm{W}} \mathsf{Q}$, say) and $\mathsf{P} \leqslant_{\mathsf{RCA}_0} \mathsf{Q}$. This happens surprisingly often: many theorems have proofs that are both effective and uniform.

In other cases, examining the proof that $\mathsf{P} \leqslant_{\mathsf{RCA}_0} \mathsf{Q}$ reveals a nonuniformity. Often, nonuniformity appears through use of the law of the excluded middle in a situation where we cannot effectively decide which of a number of possible cases holds. We have seen examples of this in our dealings with $\leqslant_{\mathrm{c}}$ and $\leqslant_{\mathrm{W}}$. Here is another, using Ramsey's theorem for singletons (Definition 3.2.5). Consider the following proof in $\mathsf{RCA}_0$ that $\mathsf{RT}^1_2$ implies $\mathsf{RT}^1_3$. Given a coloring $f \colon \omega \to 3$, if $f(x) = 0$ for infinitely many $x$ then we are done. Otherwise, there is a number $m$ such that $f(x) > 0$ for all $x \geqslant m$. Then $f^*(x) = f(x+m) - 1$ induces a coloring $f^* \colon \omega \to 2$, and we can compute an infinite homogeneous set for $f$ from any such set for $f^*$. The nonuniformity here is in the question whether the first color is used infinitely often.

Of course, there are many alternate proofs of a theorem – especially if we consider all the possible formal proofs. We might ask whether there is an alternate proof that avoids this kind of nonuniformity. If we can show that $|\mathsf{P}| \nleqslant_{\mathrm{W}} |\mathsf{Q}|$, this shows in a precise way that nonuniformity cannot be avoided, in the sense that there is no proof uniform enough to give a uniform algorithm for reducing $\mathsf{P}$ to $\mathsf{Q}$.

Yet another possibility is that the proof that $\mathsf{P} \leqslant_{\mathrm{c}} \mathsf{Q}$ uses strong logical axioms, such as to verify its correctness. For example, the proof method of $\mathsf{RT}^1_2 \to \mathsf{RT}^1_3$ sketched above yields an inductive proof that $\mathsf{RT}^1_2$ implies $\mathsf{RT}^1$ as follows. Given $f \colon \omega \to k+1$ if $f(x) > 0$, if there is no infinite homogeneous set for $f$ with color 0, then by induction there is an infinite homogeneous set for $f$ with color $i > 0$, using the same technique as above with an alternate coloring $f^*$. In ordinary mathematics, this induction is hardly noticed. However, when formalized into second order arithmetic, this is induction on the existence of a coloring: a $\Sigma^1_1$ induction. We may ask if there is an alternate proof that stays within the resources of $\mathsf{RCA}_0$. The theorem that $\mathsf{RT}^1 \nleqslant_{\mathsf{RCA}_0} \mathsf{RT}^1_2$, in Chapter 6, shows that the answer is no. In fact, Theorem 6.5.1 characterizes precisely how much additional induction is required, beyond $\mathsf{RCA}_0$, to prove $\mathsf{RT}^1$.

## 5.12 Full second order semantics

By convention, second order arithmetic is formalized with first order semantics. This means simply that in any $\mathcal{L}_2$ structure $\mathcal{M}$ the collection of sets $\mathcal{S}^{\mathcal{M}}$ may be an arbitrary subset of the powerset $\mathcal{P}(M)$. There is an alternative semantics which can be employed, *full semantics*, in which $\mathcal{S}^{\mathcal{M}}$ is required to contain all subsets of $M$. Intuitively, this change will result in fewer possible structures, and thus fewer consistent theories. We will call an $\mathcal{L}_2$ structure $\mathcal{M}$ *full* if $\mathcal{S}^{\mathcal{M}} = \mathcal{P}(M)$.

If we only consider full models, rather than arbitrary models, the resulting semantics is known as *full second order semantics*. The next theorem will show that these semantics are more powerful than the usual semantics defined in the previous section (which are known as *Henkin semantics* in the literature). However, as we will also see, the cost of using full second order semantics is that Theorem 5.1.5 no longer holds.

The key example of full second order semantics is in Peano's original axiomatization of arithmetic. The following definition captures the essence of that axiomatization, but is not identical to Peano's original axiomatization.

**Definition 5.12.1.** The second order *Peano axioms*, $\mathsf{PA}^2$, are a set of axioms for arithmetic in a two sorted logic with one sort for numbers and a second sort for sets. The signature for $\mathsf{PA}^2$ has a constant symbol 0 of type 0 and a unary function symbol $S$ of type $0 \to 0$. There are only three axioms. The first two state that 0 is not in the range of $S$, and that $S$ is an injection.

1. $S(n) \neq 0$.
2. $S(n) = S(m) \to n = m$.

The third axiom is known as the *second order induction axiom*. It states that any set which contains 0 and is closed under $S$ must contain every element of the domain.

3. $(\forall X)[(0 \in X \land (\forall n)[n \in X \to S(n) \in X]) \to (\forall n)[n \in X]]$.

We define structures for the signature of $\mathsf{PA}^2$ as we did for the signature $\mathcal{L}_2$. The only difference is that $+$ and $\cdot$ have been replaced by $S$.

**Theorem 5.12.2.** *If $\mathcal{M}$ is a full structure satisfying the Peano axioms, then $M$ is isomorphic to $\omega$ with its standard successor operation. Hence $\mathcal{M}$ is isomorphic to an $\omega$-model.*

*Proof.* Suppose that $\mathcal{M} = \langle M, \mathcal{P}(M), S\rangle$ is a full model of the Peano axioms. Consider the set $D = \{0, S(0), S(S(0)), \ldots\} = \{S^k(0) : k \in \omega\}$, which is a subset of the domain of $\mathcal{M}$ and thus an element of $\mathcal{S}(M) = \mathcal{P}(M)$. Because this set contains 0 and is closed under $S$, and because $\mathcal{M}$ satisfies the induction axiom, $D = M$. Thus the map $f : k \mapsto S^k(0)$ is a surjection from $\omega$ to $M$.

We next prove that $f$ is injective. Suppose not. Then because $\omega$ is well founded there must be a minimal $l \in \omega$ such that there is some $k < l$ with $S^k(0) = S^l(0)$. It cannot be that $k = 0$, because of axiom 1. Thus we may write $k = p + 1$ and $l = q + 1$, so that $S(S^p(0)) = S(S^q(0))$. Then, by axiom 2, $S^p(0) = S^q(0)$. Because $k < l$, we have $p < q$, and thus we have a smaller counterexample to injectivity, which is impossible. Thus $f$ is a bijection from $\omega$ to $M$. Finally, by definition, we have $S(f(k)) = f(k+1)$, and thus $f$ is an isomorphism from $\omega$ with the standard successor operation to $\mathcal{M}$. □

Hence, in particular, if $\mathcal{M}$ is a full structure, then its first order part is standard. Hence, "full semantics" is sometimes also known as "standard semantics".

The key point is that any set that we can form during the proof is "already covered" by the quantifier in the third axiom, under full semantics. This mixture of the object theory (the Peano axioms) and the theory in which the proof is written is a unique feature of full semantics.

It is natural to ask what theory can be used to formalize the proof. One answer is that the proof can be formalized in ZFC if we also formalize second order semantics in ZFC. That is, if we write

$$\text{“}\mathcal{M} \text{ satisfies the Peano axioms under full semantics”} \tag{5.7}$$

as a sentence in the language of ZFC, then ZFC proves

$$\begin{array}{l}\text{“If } \mathcal{M} \text{ satisfies the Peano axioms under full semantics,} \\ \text{then } \mathcal{M} \text{ is isomorphic to } \omega \text{ with the usual successor operation”}\end{array} \tag{5.8}$$

as a theorem. Of course, there are nonstandard models of ZFC, each of which has its own $\omega$. Thus, when we read (5.8) translated into ZFC, it simply says that within a fixed model of ZFC, a set which appears to be a model of the second order Peano axioms is isomorphic to the $\omega$ of that model.

It is also possible to interpret (5.7) informally, in the usual style of mathematics. In this case, the universal set quantifier in the third axiom would range over all subsets of $M$ that exist, and the proof shows that $\mathcal{M}$ is isomorphic to the standard $\omega$. The reader may decide whether to interpret full second order semantics in the formal sense of the previous paragraph or in the informal sense of this one.

Theorem 5.12.2 also shows that the downward Löwenheim–Skolem theorem fails for second order theories with full semantics: although $\mathsf{PA}^2$ is a countable theory, it has no model $\mathcal{M}$ in which both $M$ and $\mathcal{S}^{\mathcal{M}}$ are countable. The exercises show that the completeness and compactness theorems also fail.

## 5.13 Exercises

**Exercise 5.13.1.** Let $\mathcal{N}$ and $\mathcal{M}$ be $\mathcal{L}_2$ structure with $\mathcal{N}$ an $\omega$-submodel of $\mathcal{M}$. Prove by induction on complexity that if $\varphi$ is an arithmetical formula, possibly with parameters from $N = M$ and $\mathcal{S}^{\mathcal{N}}$, then $\mathcal{N} \vDash \varphi$ if and only if $\mathcal{M} \vDash \varphi$.

**Exercise 5.13.2.** Show the following can be defined by bounded quantifier formulas.

1. The set of numbers that are codes for finite sets.
2. The set of numbers that are codes for finite sequences.
3. The set of $(y, m)$ such that $y$ is the code for a finite set $X$ and $m = \max X$.
4. The set of $(y, l)$ such that $y$ is a code for a finite sequence and $l$ is the length of the sequence.
5. The set of $(y, i, k)$ such that $y$ is a code for a finite sequence, $i < \operatorname{lh}(y)$, and $k$ is the $i$th element of the sequence.

**Exercise 5.13.3.** Prove that the following properties of a set $T \subseteq \mathbb{N}^{<\mathbb{N}}$ can be expressed as $\Pi^0_1$ formulas:

1. $T$ is a tree.
2. $T$ is a binary tree.

The following properties of a set $T$ can be expressed as $\Pi^0_2$ formulas:

1. $T$ is an infinite tree.
2. $T$ is a finitely branching tree.

**Exercise 5.13.4.** Show the following are provable in $\mathsf{RCA}_0$:

1. The graph of each primitive recursive function is coded by a set.
2. If $f$ is a function then, for each $k$ and each $n$ there is a finite sequence $\langle n, f(n), f(f(n)), \ldots, f^k(n)\rangle$.
3. If $\sigma$ and $\tau$ are finite sequences then so is the concatenation $\sigma\tau$.

**Exercise 5.13.5.** Show that if $\varphi(x)$ is a $\Sigma^0_1$ formula of $\mathcal{L}_2$ then $\mathsf{RCA}_0$ proves the following: if there are infinitely many $x$ such that $\varphi(x)$ then there exists an infinite set $X$ such that $\varphi(x)$ for all $x \in X$.

**Exercise 5.13.6.** Show that for every $n \geqslant 1$, every model $\mathcal{M}$ of $\mathsf{RCA}_0$, and every function $f\colon M \to M$ (which may or may not belong to $\mathcal{S}^{\mathcal{M}}$), the following are equivalent.

1. $f$ is $\Sigma^0_n$ definable in $\mathcal{M}$.
2. $f$ is $\Pi^0_n$ definable in $\mathcal{M}$.
3. $f$ is $\Delta^0_n$ definable in $\mathcal{M}$.

**Exercise 5.13.7.** In principle, we could consider an alternate definition of an $\mathcal{L}_2$ structure as a tuple

$$\mathcal{M} = \langle M, \mathcal{S}^{\mathcal{M}}, +^{\mathcal{M}}, \cdot^{\mathcal{M}}, 0^{\mathcal{M}}, 1^{\mathcal{M}}, <^{\mathcal{M}}, \in^{\mathcal{M}}\rangle,$$

where $\mathcal{S}^{\mathcal{M}}$ is now an arbitrary set of objects and $\in^{\mathcal{M}}$ is a fixed subset of $M \times \mathcal{S}^{\mathcal{M}}$.

Suppose that $\mathcal{M}$ is a structure in this alternate sense that satisfies the extensionality axiom

$$(\forall X, Y)[X = Y \Leftrightarrow (\forall n)[n \in X \leftrightarrow n \in Y]].$$

Prove that $\mathcal{M}$ is isomorphic to a structure as in Definition 5.1.2. Thus there is no loss of generality in assuming that $\mathcal{S}^{\mathcal{M}} \subseteq \mathcal{P}(M)$ and that $\in^{\mathcal{M}}$ is the set membership relation.

**Exercise 5.13.8.** Suppose that $\varphi(X)$ is an $\mathcal{L}_2$ formula with one free set variable, which may have parameters. Show by induction on the structure of $\varphi$ that

$$(\forall n)[n \in X \leftrightarrow n \in Y] \to [\varphi(X) \leftrightarrow \varphi(Y)].$$

**Exercise 5.13.9.** Show that there is an effective second order $T$ such that every finite subtheory of $T$ has a full model but $T$ itself does not. This shows that the compactness theorem will fail for second order logic with full semantics.

**Exercise 5.13.10.** Show that there is an effective second order theory $T$ that is consistent (that is, there is no formula $\varphi$ such that both $\varphi$ and $\neg\varphi$ are derivable from $T$) but $T$ has no full model. This shows that the completeness theorem fails for second order logic with full semantics.

**Exercise 5.13.11.** Prove that there is a definitional extension $T$ of the Peano axioms so that any full model of $T$ is isomorphic to the standard model of second order arithmetic $\langle \omega, \mathcal{P}(\omega), S, +, \times, <, = \rangle$.

**Exercise 5.13.12.** The results stated in Theorem 5.1.5 are typically stated and proved only for single sorted first order logic. We sketch how to reduce the two sorted logic to the single sorted one. First, extend the signature with an additional unary predicate $N$. Given an $\mathcal{L}_2$ formula $\varphi$, translate it recursively to a single sorted formula $\varphi'$ with the rules

$$\begin{aligned}((\forall n)\psi(n))' &= (\forall y)(Ny \rightarrow \psi'(y)),\\ ((\exists n)\psi(n))' &= (\exists y)(Ny \wedge \psi'(y)),\\ ((\forall X)\psi(X))' &= (\forall y)((\neg Ny) \rightarrow \psi'(y))),\\ ((\exists X)\psi(X))' &= (\exists y)((\neg Ny) \wedge \psi'(y)),\\ \psi' &= \psi \quad \text{otherwise.}\end{aligned}$$

Given an $\mathcal{L}_2$ theory $T$, let $T'$ be the single sorted theory $\{\varphi' : \varphi \in T\}$. Show that $T$ is consistent if and only if $T'$ is consistent in first order logic. Use this method to produce a proof of Theorem 5.1.5 using the corresponding results for single sorted logic as lemmas.

**Exercise 5.13.13.** Prove that there is a $\Sigma^0_1$ formula $\pi(e, n, X)$ such that for each $\Sigma^0_1$ formula $\psi(n, X)$ there is an $e^* \in \omega$ such that $\mathsf{RCA}_0$ proves

$$(\forall n)[\pi(e^*, n, X) \leftrightarrow \psi(n, X)].$$

This formula $\pi$ is one kind of *universal $\Sigma^0_1$ formula.*

More generally, for each $n > 0$ there is a $\Sigma^0_n$ formula $\varphi(e, n, X)$ such that for each $\Sigma^0_n$ formula $\psi(n, X)$ there is an $e^* \in \omega$ such that $\mathsf{RCA}_0$ proves

$$(\forall n)[\varphi(e^*, n, X) \leftrightarrow \psi(n, X)].$$

**Exercise 5.13.14.** Show that the systems $\mathsf{RCA}_0$, $\mathsf{WKL}_0$, and $\mathsf{ACA}_0$ are all finitely axiomatizable. (Use the universal formula from Exercise 5.13.13).

**Exercise 5.13.15.** Let $\varphi(X)$ be any $\Sigma^0_0$ formula of $\mathcal{L}_2$ with no free variables other than $X$. Prove that there exist formulas $\varphi_\Sigma(x)$ and $\varphi_\Pi(x)$ with the same parameters as $\varphi$ and no free variables other than $x$ such that $\mathsf{RCA}_0 \vdash (\forall X)[\varphi(X) \leftrightarrow (\exists j)[\varphi_\Sigma(X \upharpoonright j)] \leftrightarrow (\forall j)[\varphi_\Pi(X \upharpoonright j)]$.

**Exercise 5.13.16.** Let $\varphi(X)$ be a $\Sigma^0_1$ formula of $\mathcal{L}_2$. Prove by induction on the complexity of $\varphi$ that there is a $\Sigma^0_0$ formula $\theta(x)$ such that $\mathsf{RCA}_0 \vdash \varphi(X) \leftrightarrow (\exists k)\theta(X \upharpoonright k)$.

**Exercise 5.13.17.** Prove Lemma 5.2.2 by structural induction. Then prove Theorem 5.2.3. The key case is for $\Sigma^0_1$ formulas. Given a $\Sigma^0_1$ formula, it can be shown that the set being defined is enumerable. Conversely, given a $\Sigma^0_1$ set, use Kleene's $T$ and $U$ predicates and Lemma 5.2.2.

**Exercise 5.13.18.** Show the following statements are equivalent to $\mathsf{WKL}_0$ over $\mathsf{RCA}_0$.

1. The $\Sigma^0_1$ separation principle, i.e., for all pairs $\varphi$ and $\psi$ of $\Sigma^0_1$ formulas of $\mathcal{L}_2$, if $(\forall x)[\varphi(x) \to \neg\psi(x)]$ then there is a set $Z$ such that
$$(\forall x)[\varphi(x) \to x \in Z \land \psi(x) \to x \notin Z].$$
2. If $f : \mathbb{N} \to \mathbb{N}$ and $g : \mathbb{N} \to \mathbb{N}$ are functions with disjoint ranges, then there is a set $Z$ such that
$$(\forall y)[(\exists x)[f(x) = y] \to y \in Z \land (\exists x)[g(x) = y]) \to y \notin Z].$$

**Exercise 5.13.19 (Arithmetical transfinite induction in $\mathsf{ACA}_0$).** Prove the following transfinite induction scheme in $\mathsf{ACA}_0$. Suppose $\mathrm{WO}(X)$, and let $\psi$ be an arithmetical formula so that $\psi(0_z)$ holds and, for all $x \in X$, if $\psi(y)$ holds for all $y <_Z x$, then $\psi(x)$ holds. Then $\psi(x)$ holds for all $x$.

**Exercise 5.13.20.** The following exercise shows that although $\mathsf{ACA}_0$ is weaker than $\mathsf{ATR}_0$, it can prove the following restricted form of arithmetical transfinite recursion. Arguing in $\mathsf{ACA}_0$, suppose $\mathrm{WO}(X)$, and that $f : \mathbb{N} \to \mathbb{N}$ is a function such that for every $x \in X$ and every $e$, if $\Phi^X_e = \langle S_y : y <_X x\rangle$ then $\Phi^X_{f(e)} = \langle S_y : y \leqslant_X x\rangle$. Then $\langle S_x : x \in X\rangle$ exists.

**Exercise 5.13.21.**

1. Prove in $\mathsf{RCA}_0$ that $\leqslant_{\mathrm{KB}}$ is a dense linear order on $\mathbb{N}^{<\mathbb{N}}$.
2. Prove in $\mathsf{ACA}_0$ that for all trees $T \subseteq \mathbb{N}^{<\mathbb{N}}$, $\mathrm{WF}(T) \leftrightarrow \mathrm{WO}(\mathrm{KB}(T))$.

**Exercise 5.13.22.** Prove Theorem 5.8.9. There are two stages for the reversal. First, create a reversal to show that the principle implies $\mathsf{ACA}_0$. Then, working over $\mathsf{ACA}_0$, use Kleene's normal form theorem to convert an instance of $\Sigma^1_1$ comprehension into its associated sequence of trees.

**Exercise 5.13.23.** If $T \subseteq [\omega]^{<\omega}$ is a tree, a path $f \in [T]$ is the *leftmost path* if, for all $g \in [T]$ and all $n$, $f(n) \leqslant g(n)$. Prove the following.

1. Over $\mathsf{RCA}_0$, $\mathsf{ACA}_0$ is equivalent to the principle that every infinite tree on $\{0, 1\}$ has a leftmost path.
2. Over $\mathsf{RCA}_0$, $\Pi^1_1$-$\mathsf{CA}_0$ is equivalent to the principle that every infinite tree on $\omega$ has an infinite path.

**Exercise 5.13.24.** Prove in $\mathsf{RCA}_0$ that for every infinite set $S$, the principal function of $S$ exists: i.e., there exists a function $p_S : \mathbb{N} \to \mathbb{N}$ with $\mathrm{range}(p_S) = S$ and $p_S(n) < p_X(m)$ for all $n < m$.

**Exercise 5.13.25.** Prove in $\mathsf{RCA}_0$ that there is a bijection $\mathbb{N} \times \mathbb{N} \to \mathbb{N}$. (Hint: Start with the function $f(n, m) = \langle n, m\rangle$ defined in Theorem 5.3.4. Let $S$ be range of $f$ and prove that it exists. Then, consider $p_S^{-1} \circ f$.)

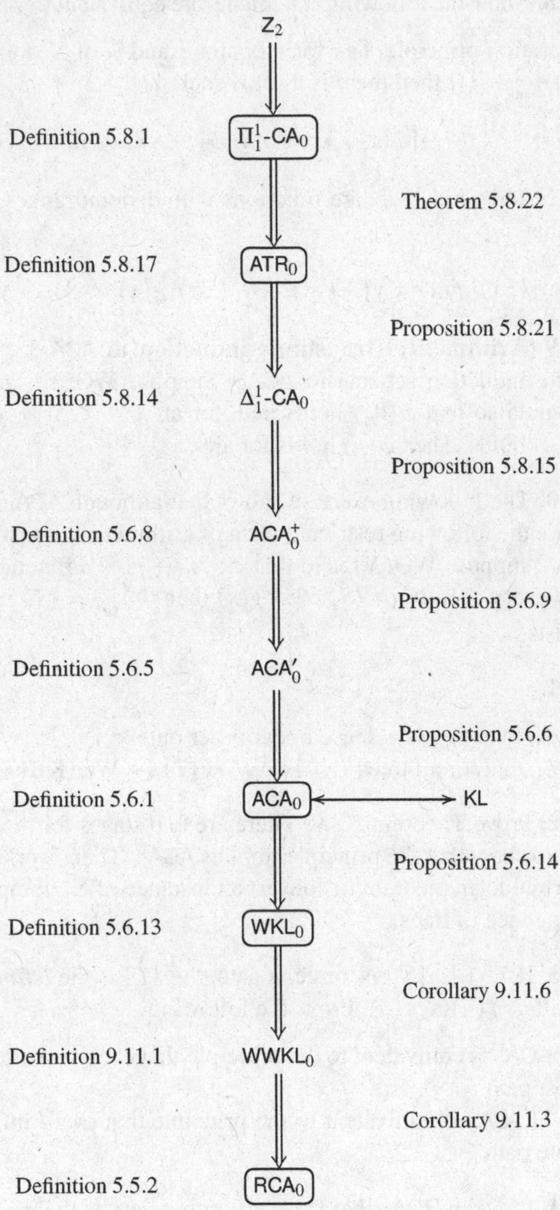


**Figure 5.2.** Various subsystems of $Z_2$, in order of strength. No double arrow can be reversed. All separations are witnessed by $\omega$-models except between $ACA_0$ and $ACA_0'$, whose $\omega$-models coincide; the references to separations are indicated. The so-called "big five" are circled. The system $WWKL_0$ will be defined in Chapter 9.

# Chapter 6
# Induction and bounding

The most common kind of reverse mathematics result shows that a given problem requires a particular set existence axiom to solve. Accordingly, we talk of the *second order strength* of the principle. A number of combinatorial principles also require additional induction axioms, beyond $\Sigma^0_1$ induction, to prove in second order arithmetic. A landmark example is $\mathsf{RT}^1$, which is equivalent to the bounding principle $\mathsf{B}\Sigma^0_2$ over $\mathsf{RCA}_0$. Being computably true, $\mathsf{RT}^1$ has no second order strength beyond $\mathsf{RCA}_0$. But it does consequently have additional *first order strength*. In this chapter, we describe the Kirby–Paris hierarchy of induction and bounding principles, which can be used to measure the first order parts of principles. We discuss reverse recursion theory, and prove the result mentioned above, due to Hirst, about $\mathsf{RT}^1$.

## 6.1 Induction, bounding, and least number principles

To begin our discussion of induction strength, we first define several important first order principles and establish the basic relationships between them. We open by recalling the definition of induction from Definition 5.3.5. Recall that for a collection $\Gamma$ of formulas of $\mathcal{L}_2$, the $\Gamma$ *induction scheme* ($\mathsf{I}\Gamma$) is the scheme over all $\varphi(x) \in \Gamma$ of sentences of the form

$$(\varphi(0) \wedge (\forall x)[\varphi(x) \rightarrow \varphi(x+1)]) \rightarrow (\forall x)\varphi(x).$$

Our interest here will be almost exclusively in cases where $\Gamma$ is a class of arithmetical formulas.

The notation $\mathsf{I}\Gamma$ originates in the study of models of first order arithmetic, where the class $\Gamma$ above would be a class of formulas of $\mathcal{L}_1$ rather than $\mathcal{L}_2$. For this reason, some authors prefer to reserve the notation $\mathsf{I}\Gamma$ for $\mathcal{L}_1$ formulas exclusively, and use instead the separate notion $\Gamma$-$\mathsf{IND}$ when dealing with $\mathcal{L}_2$ formulas. We will not do so, largely because the potential for confusion is minimal. In proofs over a theory (like $\mathsf{PA}$) that can be viewed as a theory in either language, the presence or absence of set

 
D. D. Dzhafarov, C. Mummert, *Reverse Mathematics*, Theory and Applications of Computability, https://doi.org/10.1007/978-3-031-11367-3_6

variables tends to make no difference. Since our interest is second order arithmetic, we can think of everything as happening in $\mathcal{L}_2$ for simplicity. But most results below were originally formulated in $\mathcal{L}_1$, and they hold equally well in either case.

To next principle $\mathsf{B}\Gamma$ may not immediately look like an induction statement.

**Definition 6.1.1.** Let $\Gamma$ be a collection of formulas of $\mathcal{L}_2$. The $\Gamma$ *bounding scheme* (or $\Gamma$ *collection scheme*) ($\mathsf{B}\Gamma$) is the scheme consisting over all $\varphi(x,y) \in \Gamma$ of sentences of the form

$$(\forall z)[(\forall x < z)(\exists y)\varphi(x,y) \to (\exists w)(\forall x < z)(\exists y < w)\varphi(x,y)].$$

One way of thinking about bounding is that, in the expression $(\forall x < z)(\exists y)\varphi(x,y)$, it allows us to shift the existential quantifier outside the scope of the bounded universal.

**Theorem 6.1.2 (Parsons [239]).** *Fix $n \geqslant 0$ and let $t$ be a term.*

1. *If $\varphi(x,\vec{z})$ is a $\Sigma^0_n$ formula, then $(\forall x < t)\,\varphi(x,\vec{z})$ is equivalent over $\mathsf{PA}^- + \mathsf{B}\Sigma^0_n$ to a $\Sigma^0_n$ formula.*
2. *If $\varphi(x,\vec{z})$ is a $\Pi^0_n$ formula, then $(\exists x < t)\,\varphi(x,\vec{z})$ is equivalent over $\mathsf{PA}^- + \mathsf{B}\Sigma^0_n$ to a $\Pi^0_n$ formula.*

*Proof.* The proof is by induction on $n$. The result is obvious for $n = 0$ since $\Sigma^0_0 = \Pi^0_0$ formulas are closed under bounded quantification. So fix $n > 0$, and assume the result is true for $n - 1$. We prove (1) for $n$, (2) being analogous. Say $\varphi(x,\vec{z})$ is $(\exists y)\,\psi(x,y,\vec{z})$, where $\psi$ is $\Pi^0_{n-1}$. Then in $\mathsf{PA}^- + \mathsf{B}\Sigma^0_n$, we have

$$(\forall x < t)\,\varphi(x,\vec{z}) \leftrightarrow (\exists w)(\forall x < t)(\exists y < w)\psi(x,y,\vec{z}).$$

Applying the inductive hypothesis to $(\exists y < w)\psi(x,y,\vec{z})$, we obtain an equivalent $\Pi^0_{n-1}$ formula $\theta(w,x,\vec{z})$. Thus, $(\forall x < t)\,\varphi(x,\vec{z})$ is equivalent to $(\exists w)(\forall x < t)\theta(w,x,\vec{z})$, which is $\Sigma^0_n$. □

Bounding is more than just a technical tool for making quantifiers behave, however. In Section 6.5, we will see several examples of how it shows up (almost insidiously) in the most common types of arguments.

With Proposition 6.1.2 in hand, we can now state and prove the following famous facts from the study of models of arithmetic.

**Theorem 6.1.3 (Parsons [239]; Kirby and Paris [237]).** *Fix $n \geqslant 0$. The following are provable in $\mathsf{PA}^-$.*

1. *$\mathsf{B}\Sigma^0_{n+1}$ is equivalent to $\mathsf{B}\Pi^0_n$.*
2. *$\mathsf{I}\Sigma^0_n + \mathsf{B}\Sigma^0_0$+ implies $\mathsf{B}\Sigma^0_n$.*
3. *$\mathsf{I}\Sigma^0_0 + \mathsf{B}\Sigma^0_{n+1}$ implies $\mathsf{I}\Sigma^0_n$.*

*Proof.* For (1), the implication $\mathsf{B}\Sigma^0_{n+1} \to \mathsf{B}\Pi^0_n$ is immediate. For the converse, suppose we are given a $\Sigma^0_{n+1}$ formula $\varphi(x,y) \equiv (\exists u)\psi(x,y,u)$, where $\psi$ is $\Pi^0_n$. Fix any $z$, and suppose that $(\forall x < z)(\exists y)\varphi(x,y)$ holds. Define $\theta(x,v)$ to be the formula

$$(\forall y, u \leqslant v)[v = \langle y, u\rangle \to \psi(x, y, u)],$$

which is $\Pi^0_n$. Then we have $(\forall x < z)(\exists v)\theta(x, v)$, and so by $\mathsf{B}\Pi^0_n$ we can fix $w$ such that $(\forall x < z)(\exists v < w)\theta(x, v)$. This means $(\forall x < z)(\exists y < w)\varphi(x, y)$.

Part (2) is proved by (external) induction on $n$. For $n = 0$, there is nothing to show since we are arguing over $\mathsf{PA}^- + \mathsf{B}\Sigma^0_0$. So fix $n > 0$, and assume the result is true for $n-1$: that is, $\mathsf{PA}^- + \mathsf{I}\Sigma^0_{n-1} + \mathsf{B}\Sigma^0_0$ implies $\mathsf{B}\Sigma^0_{n-1}$. Arguing in $\mathsf{PA}^- + \mathsf{I}\Sigma^0_n + \mathsf{B}\Sigma^0_0$, it suffices to prove $\mathsf{B}\Pi^0_{n-1}$, by part (1). So suppose $\varphi(x, y)$ is a $\Pi^0_{n-1}$ formula such that for some $z$,

$$(\forall x < z)(\exists y)\varphi(x, y).$$

Let $\psi(u)$ be the formula $u > z \vee (\exists w)(\forall x < u)(\exists y < w)\varphi(x, y)$. By inductive hypothesis, we have $\mathsf{B}\Sigma^0_{n-1}$, so we may apply Proposition 6.1.2 to conclude that $\psi$ is equivalent to a $\Sigma^0_n$ formula. Now obviously, $\psi(0)$ holds, and it is easy to see that so does $(\forall u)[\psi(u) \to \psi(u+1)]$. By $\Sigma^0_n$ induction, we conclude $(\forall u)\psi(u)$. Then $\psi(z)$ gives the conclusion of $\mathsf{B}\Sigma^0_n$ applied to $\varphi$.

For (3), we again proceed by induction on $n$. For $n = 0$, the result is trivial since we are working in $\mathsf{PA}^- + \mathsf{I}\Sigma^0_0$. So fix $n > 0$, and assume the result is true for $n-1$. We now argue in $\mathsf{PA}^- + \mathsf{B}\Sigma^0_{n+1}$. Let $\varphi(x)$ be a $\Sigma^0_n$ formula assume $\varphi(0) \wedge (\forall x)[\varphi(x) \to \varphi(x+1)]$. Write $\varphi(x)$ as $(\exists y)\psi(x, y)$ for some $\Pi^0_{n-1}$ formula $\psi$, and define

$$\theta(x, y) \equiv \psi(x, y) \vee (\forall v)\neg\psi(x, v),$$

which is a disjunction of a $\Pi^0_{n-1}$ formula and a $\Pi^0_n$ formula, and so is $\Pi^0_n$. Fix any $z$. Then we have

$$(\forall x < z)(\exists y)\theta(x, y),$$

and so by $\mathsf{B}\Pi^0_n$ (which is equivalent to $\mathsf{B}\Sigma^0_{n+1}$ by part (1)) there exists a $w$ such that

$$(\forall x < z)(\exists y < w)\theta(x, y).$$

In particular, for all $x < z$, if $\psi(x, y)$ holds for some $y$ then it holds for some $y < w$. Now, by Proposition 6.1.2, $(\exists y < w)\psi(x, y)$ is equivalent to a $\Pi^0_{n-1}$ formula. And since $\mathsf{B}\Sigma^0_{n+1} \to \mathsf{B}\Sigma^0_n$, it follows by inductive hypothesis that we may avail ourselves of $\mathsf{I}\Sigma^0_{n-1}$. So, by induction on the $\Sigma^0_{n-1}$ formula

$$\neg(\exists y < w)\psi(x, y) \to x \geqslant z,$$

we conclude that $(\forall x < z)(\exists y < w)\psi(x, y)$. In particular, $\varphi(x)$ holds for all $x < z$. Since $z$ was arbitrary, this yields $(\forall x)\varphi(x)$, as needed. □

We will see in Section 6.3 that the implication in parts (2) and (3) above are strict. That is, $\mathsf{B}\Sigma^0_{n+1}$ is intermediate between $\mathsf{I}\Sigma^0_n$ and $\mathsf{I}\Sigma^0_{n+1}$. This is complemented by a result of Slaman that we discuss at the end of this section.

We add a further principle to our discussion, this time one that is very clearly just a restatement of induction. We formally prove the equivalence below. Nonetheless, having this formulation explicitly is very convenient.

**Definition 6.1.4.** Let $\Gamma$ be a collection of formulas of $\mathcal{L}_2$. The $\Gamma$ *least number principle* ($\mathsf{L}\Gamma$) is the scheme over all $\varphi(x) \in \Gamma$ of sentences of the form

$$(\exists x)\,\varphi(x) \rightarrow (\exists x)[\varphi(x) \wedge \neg(\exists y < x)\,\varphi(y)].$$

**Proposition 6.1.5.** *For each $n \geqslant 1$, $\mathsf{PA}^- \vdash \mathsf{I}\Sigma^0_n \leftrightarrow \mathsf{I}\Pi^0_n \leftrightarrow \mathsf{L}\Sigma^0_n \leftrightarrow \mathsf{L}\Pi^0_n$.*

*Proof.* We argue in $\mathsf{PA}^-$.

($\mathsf{I}\Sigma^0_n \rightarrow \mathsf{I}\Pi^0_n$): Assume $\mathsf{I}\Sigma^0_n$ and suppose $\varphi(x)$ is a $\Pi^0_n$ formula such that $\varphi(0) \wedge (\forall x)[\varphi(x) \rightarrow \varphi(x+1)]$. If $\neg\varphi(x_0)$ for some $x_0$, define

$$\theta(x) \equiv x > x_0 \vee (\exists y)[x_0 = x + y \wedge \neg\varphi(y)]. \tag{6.1}$$

Thus we have $\theta(0)$. Now, suppose $\theta(x)$ holds for some $x$. If $x \geqslant x_0$ then obviously $\theta(x+1)$ holds as well. If $x < x_0$ then $\neg\varphi(x_0 - x)$ holds by definition, and hence we have $\neg\varphi((x_0 - x) - 1)$ by assumption on $\varphi$. So, again $\theta(x+1)$ holds. By $\Sigma^0_1$ induction, we conclude $(\forall x)\theta(x)$. In particular, we have $\theta(x_0)$ and hence $\neg\varphi(0)$, a contradiction.

($\mathsf{I}\Pi^0_n \rightarrow \mathsf{I}\Sigma^0_n$.) Assume $\mathsf{I}\Pi^0_n$ and suppose $\varphi(x)$ is a $\Sigma^0_n$ formula such that $\varphi(0) \wedge (\forall x)[\varphi(x) \rightarrow \varphi(x+1)]$. The proof is the same as that of the previous implication, except that (6.1) is replaced by

$$\theta(x) \equiv x > x_0 \vee (\forall y)[x_0 = x + y \rightarrow \neg\varphi(y)].$$

($\mathsf{I}\Sigma^0_n \rightarrow \mathsf{L}\Pi^0_n$): Assume $\mathsf{I}\Sigma^0_n$. Let $\varphi(x)$ be a $\Pi^0_n$ formula, and define $\psi(x)$ to be the formula $(\forall y \leqslant x)\neg\varphi(x)$. By Proposition 6.1.2, $\psi$ is equivalent to a $\Sigma^0_n$ formula. Now, if there is no least element satisfying $\varphi$ then we must have $\psi(0)$, and for every $x$, $\psi(x)$ must imply $\psi(x+1)$. By $\Sigma^0_n$ induction, this yields $(\forall x)\psi(x)$, hence $(\forall x)\neg\varphi(x)$.

($\mathsf{L}\Pi^0_n \rightarrow \mathsf{I}\Sigma^0_n$): Assume $\mathsf{L}\Pi^0_n$ and suppose $\varphi(x)$ is a $\Sigma^0_n$ formula. If $(\forall x)\varphi(x)$ does not hold, we may let $x_0$ be the least $x$ such that $\neg\varphi(x)$. Now either $x_0 = 0$, in which case $\neg\varphi(0)$ holds, or $x_0 > 0$, in which case $(\exists x)[\varphi(x) \wedge \neg\varphi(x+1)]$.

($\mathsf{I}\Pi^0_n \rightarrow \mathsf{L}\Sigma^0_n$): Same as the proof that $\mathsf{I}\Sigma^0_n \rightarrow \mathsf{L}\Pi^0_n$.

($\mathsf{L}\Sigma^0_n \rightarrow \mathsf{I}\Pi^0_n$.) Same as the proof that $\mathsf{L}\Pi^0_n \rightarrow \mathsf{I}\Sigma^0_n$. □

The following useful corollary is a formalization of the computability theoretic fact that every $\Sigma^0_1$ definable subset of $\omega$ is computably enumerable.

**Theorem 6.1.6.** *Let $\varphi(y)$ be a $\Sigma^0_1$ formula of $\mathcal{L}_2$. Then $\mathsf{RCA}_0$ proves there is a function $f\colon \mathbb{N} \rightarrow \mathbb{N}$ such that for every $y$, $\varphi(y)$ holds if and only if $y \in \mathrm{range}(f)$. Moreover, if $\varphi(y)$ holds for infinitely many $y$ then $f$ may be chosen to be injective.*

*Proof.* We reason in $\mathsf{RCA}_0$. First, suppose there is a finite set $F$ such that for all $y$, $y \in F$ if and only if $\varphi(y)$ holds. In this case, simply define $f$ by $f(y) = y$ for all $y \in F$ and $f(y) = \min F$ for all $y \notin F$. Clearly, $f$ exists and has the desired property.

So suppose next that no such $F$ exists. Write $\varphi(y)$ as $(\exists x)\psi(x,y)$, where $\psi$ is $\Pi^0_0$. Let $S$ be the set of all pairs $\langle x,y\rangle$ such that

$$\psi(x,y) \wedge (\forall x^* < x)\neg\psi(x^*,y).$$

By $\mathsf{L}\Pi^0_0$, we have for all $y$ that $\varphi(y) \leftrightarrow (\exists x)[\langle x,y\rangle \in S]$. Thus if $S$ were bounded by some $b$, we would have that $\varphi(y) \leftrightarrow (\exists x < b)[\langle x,y\rangle \in S]$. Hence, the set of $y$ such that $\varphi(y)$ holds would exist by $\Sigma^0_0$ comprehension, and being bounded, it would be finite (Definition 5.5.10). Since this contradicts our assumption, $S$ must be unbounded.

Let $p_S$ be the principal function of $S$ (Exercise 5.13.24). In particular, $p_S$ is injective and its range is $S$. Define $f$ by

$$\langle z,y\rangle \in f \leftrightarrow (\exists x)[\langle z,x,y\rangle \in p_S].$$

Note that $f$ is a function because $p_S$ is, and it is injective because $p_S$ is and because if $\langle x,y\rangle, \langle x^*,y\rangle \in S$ then necessarily $x = x^*$ by definition of $\psi$. Since this definition is $\Sigma^0_1$, it follows by Exercise 5.13.6 that $f$ is $\Delta^0_1$ definable. Hence, $f$ exists. We have $y \in \mathrm{range}(f) \leftrightarrow (\exists x)[\langle x,y\rangle \in \mathrm{range}(p_S)] \leftrightarrow (\exists x)[\langle x,y\rangle \in S] \leftrightarrow \varphi(y)$, as desired. □

We conclude this section by briefly turning to more advanced results, without any proofs. All of the principles we have discussed above have immediate $\Delta^0_n$ formulations. Namely, we define $\mathsf{I}\Delta^0_n$ to be the scheme consisting of all sentences of the form

$$(\forall x)[\varphi(x) \leftrightarrow \neg\psi(x)] \rightarrow [(\varphi(0) \wedge (\forall x)[\varphi(x) \rightarrow \varphi(x+1)]) \rightarrow (\forall x)\varphi(x)],$$

for $\Sigma^0_n$ formulas $\varphi(x)$ and $\psi(x)$. Similarly, we define $\mathsf{L}\Delta^0_n$ to be the scheme consisting of all sentences of the form

$$(\forall x)[\varphi(x) \leftrightarrow \neg\psi(x)] \rightarrow [(\exists x)\,\varphi(x) \rightarrow (\exists x)[\varphi(x) \wedge \neg(\exists y < x)\,\varphi(y)]],$$

again for $\Sigma^0_n$ formulas $\varphi$ and $\psi$. It turns out that considering this class nicely ties the induction and bounding hierarchy together.

**Theorem 6.1.7 (Gandy, unpublished).** *For each $n \geqslant 0$,*

$$\mathsf{PA}^- + \mathsf{I}\Sigma^0_0 \vdash \mathsf{B}\Sigma^0_n \leftrightarrow \mathsf{L}\Delta^0_n.$$

A full proof can be found in Chapter I of Hájek and Pudlák [134]. Surprisingly, the easy equivalence between induction and the least number principle we saw earlier does not go through in the $\Delta^0_n$ case. And indeed, the equivalence is only known to hold over a stronger base theory. In $\mathsf{PA}^- + \mathsf{I}\Sigma^0_0$ we can give a $\Sigma^0_0$ definition of the relation $z = y^x$, although we cannot prove the formula $\mathsf{Exp} \equiv (\forall x)(\forall y)(\exists z)[z = y^x]$ which expresses the totality of the exponential function.

**Theorem 6.1.8 (Slaman [291]).** *For each $n \geqslant 0$,*

$$\mathsf{PA}^- + \mathsf{I}\Sigma^0_0 + \mathsf{Exp} \vdash \mathsf{B}\Sigma^0_n \leftrightarrow \mathsf{L}\Delta^0_n \leftrightarrow \mathsf{I}\Delta^0_n.$$

For $n = 1$, a different proof, over a slightly weaker base theory, has been found by Thapen [309]. For our purposes, the moral is the same: the bounding scheme really is an induction principle after all, just for a restricted class of formulas.

## 6.2 Finiteness, cuts, and all that

We now turn to a discussion of what a failure of induction can look like in a nonstandard model of arithmetic.

**Definition 6.2.1.** Let $\mathcal{M}$ be an $\mathcal{L}_2$ (or $\mathcal{L}_1$) structure. Then $I \subseteq M$ is a *cut* of $\mathcal{M}$ if:

- for all $x, y \in M$, if $x \in I$ and $y <^{\mathcal{M}} x$, then $y \in I$,
- for all $x \in M$, if $x \in I$ then so is the successor of $x$ in $\mathcal{M}$.

$I$ is a *proper cut* of $\mathcal{M}$ if $I \neq M$.

Thus, one way $\Sigma^0_n$ induction can fail in a model $\mathcal{M}$ is if $\mathcal{M}$ has a proper $\Sigma^0_n$ definable cut. We will see in Theorem 6.2.3 that the converse is true as well. We begin this section with the following well-known result. The gist is that if $\mathcal{M}$ satisfies enough induction, then any arithmetical property holding inside or outside of a cut must *spill* across the cut.

**Proposition 6.2.2.** *Fix $n \geqslant 0$ and $\mathcal{M} \vDash \mathsf{PA}^- + \mathsf{I}\Sigma^0_n$. Suppose $I$ is a proper cut of $\mathcal{M}$, and let $\varphi(x)$ be a $\Sigma^0_n$ formula.*

1. (Overspill): *If $\mathcal{M} \vDash \varphi(a)$ for every $a \in I$, then $\mathcal{M} \vDash \varphi(b)$ for some $b \in M \setminus I$.*
2. (Underspill): *If $\mathcal{M} \vDash \varphi(b)$ for every $b \in M \setminus I$ then $\mathcal{M} \vDash \varphi(a)$ for some $a \in I$.*

*Proof.* For (1), suppose otherwise. So for all $a \in M$, we have $\mathcal{M} \vDash \varphi(a)$ if and only if $a \in I$. As $I$ is a cut, this means

$$\mathcal{M} \vDash \varphi(0) \wedge (\forall x)[\varphi(x) \to \varphi(x+1)].$$

But then by $\mathsf{I}\Sigma^0_n$ we have $I = M$, a contradiction. Part (2) is proved the same way, using the equivalence of $\mathsf{I}\Sigma^0_n$ with $\mathsf{I}\Pi^0_n$. □

The following fundamental result relates cuts, induction, and boundedness.

**Theorem 6.2.3 (Friedman, unpublished).** *Fix $n \geqslant 1$ and suppose $\mathcal{M} \vDash \mathsf{PA}^- + \mathsf{I}\Sigma^0_0$. The following are equivalent.*

1. *$\mathcal{M} \vDash \mathsf{I}\Sigma^0_n$.*
2. *$\mathcal{M}$ has no $\Sigma^0_n$ definable proper cut.*
3. *Every $\Sigma^0_n$ definable function on a proper cut of $\mathcal{M}$ has bounded range.*

*Proof.* The implication from (1) to (2) is clear: if $\mathcal{M}$ has a $\Sigma^0_n$ definable proper cut, the formula defining this cut witnesses a failure of $\Sigma^0_n$ induction in $\mathcal{M}$. Likewise for the implication from (2) to (3), since the domain of every $\Sigma^0_n$ definable function is $\Sigma^0_n$ definable. It therefore remains to prove that (3) implies (1).

We proceed by induction on $n$. If $n = 0$, there is nothing to prove since $\mathcal{M} \models \mathsf{I}\Sigma^0_0$. So fix $n > 0$, assume (3) for $n$, and assume that the implication holds for $n-1$. By inductive hypothesis, we have $\mathsf{I}\Sigma^0_{n-1}$ and hence also $\mathsf{B}\Sigma^0_{n-1}$. However, suppose $\mathsf{I}\Sigma^0_n$ fails. Let $\varphi(z)$ be a $\Sigma^0_n$ formula such that

$$\mathcal{M} \models \varphi(0) \wedge (\forall z)[\varphi(z) \rightarrow \varphi(z+1)] \wedge \neg\varphi(c)$$

for some $c \in M$. Say $\varphi(z)$ is $(\exists w)\psi(z, w)$, where $\psi$ is $\Pi^0_{n-1}$. Now define $\theta(x, y)$ to be the formula

$$y = \langle z, w\rangle \wedge x \leqslant z < c \wedge \psi(z, w) \wedge (\forall \langle z^*, y^*\rangle < \langle z, w\rangle)\neg\psi(z^*, w^*).$$

The rightmost conjunct here $\Sigma^0_{n-1}$ by Proposition 6.1.2, so the entire formula is $\Sigma^0_n$.

First, note that $\theta$ defines a function on a proper cut of $\mathcal{M}$. Indeed, the domain of $\theta$ is the set of $a \in M$ such that $\mathcal{M} \models (\exists z)[a \leqslant z < c \wedge \varphi(z)]$, which is a cut bounded by $c$. For any such $a$, it is clear that $\mathcal{M}$ satisfies $(\exists y)[y = \langle z, w\rangle \wedge a \leqslant z < c \wedge \psi(z, w)]$. Then $\mathsf{L}\Pi^0_{n-1}$ shows that $\mathcal{M}$ satisfies $(\exists y)\theta(a, y)$.

Next, we claim that $\theta$ has unbounded range. If not, fix $d \in M$ so that

$$\mathcal{M} \models (\forall x)(\forall y)[\theta(x, y) \rightarrow y < d].$$

In this case, let $\zeta(z)$ be the formula $(\exists w < d)\psi(z, w)$. Then, by Proposition 6.1.2 again, $\zeta$ is equivalent to a $\Pi^0_{n-1}$ formula, and it follows that

$$\mathcal{M} \models \zeta(0) \wedge (\forall z)[\zeta(z) \rightarrow \zeta(z+1)] \wedge \neg\zeta(c),$$

which contradicts $\mathsf{I}\Pi^0_{n-1}$. This proves the claim.

But now we have exhibited a function witnessing that (3) is false for $n$, contrary to assumption. Thus, $\mathsf{I}\Sigma^0_n$ must hold and the proof is complete. □

An important point of caution is that nonstandard models are *not* characterized exclusively by failures of induction. Indeed, even PA and RCA, which include induction for all $\mathcal{L}_2$ formulas, have nonstandard models. In particular, although it can be intuitively tempting to conclude that the models of RCA are precisely the $\omega$-models of $\mathsf{RCA}_0$, this is not the case.

It is interesting and instructive to compare the equivalence of (1) and (3) above with the equivalence in the following.

**Proposition 6.2.4.** *Fix $n \geqslant 1$ and $\mathcal{M} \models \mathsf{PA}^- + \mathsf{I}\Sigma^0_{n-1} + \mathsf{B}\Sigma^0_0$. The following are equivalent.*

1. *$\mathcal{M} \models \mathsf{B}\Sigma^0_n$.*
2. *For every $b \in M$, every $\Sigma^0_n$ definable function on $[0, b)$ has bounded range.*

*Proof.* That (1) implies (2) follows immediately from the definition of $\mathsf{B}\Sigma^0_n$. Conversely, assume (2) for $n$. We show $\mathsf{B}\Pi^0_{n-1}$ holds in $M$. Let $\varphi(x, y)$ be a $\Pi^0_{n-1}$ formula such that for some $b \in M$,

$$\mathcal{M} \vDash (\forall x < b)(\exists y)\varphi(x, y).$$

Define $\theta(x, y)$ to be the formula

$$x < b \wedge \varphi(x, y) \wedge (\forall y^* < y)\neg\varphi(x, y).$$

Since $\mathcal{M}$ satisfies $\mathsf{I}\Sigma^0_{n-1}$, it satisfies $\mathsf{B}\Sigma^0_{n-1}$. By Proposition 6.1.2, the rightmost conjunct above is equivalent in $\mathcal{M}$ to a $\Sigma^0_{n-1}$ formula, and therefore $\theta$ is $\Sigma^0_n$. Moreover, $\theta$ defines a function on $[0, b)$. Indeed, $\mathcal{M}$ satisfies that for every $x < b$ there is a $y$ such that $\varphi(x, y)$ holds, hence there is a least such $y$ by $\mathsf{L}\Sigma^0_{n-1}$. By (2) we can fix a $c \in M$ bounding the range of $\theta$. This means $\mathcal{M} \vDash (\forall x < b)(\exists y < c)\varphi(x, y)$. □

We saw in Section 5.5.2 that, in a model $\mathcal{M}$ of $\mathsf{RCA}_0$, we cannot treat every bounded subset of $M$ as we would a bounded (and hence finite) subset of $\omega$. Theorem 5.5.8 shows that we can do so if $X \in \mathcal{S}^{\mathcal{M}}$. The following result extends the theorem to all $\Sigma^0_n$-definable subsets of $M$ under the additional hypothesis that $\mathcal{M}$ satisfies $\Sigma^0_n$ induction.

**Proposition 6.2.5.** *Fix $n \geqslant 1$ and $\mathcal{M} \vDash \mathsf{RCA}_0$. The following are equivalent.*

*1. $\mathcal{M} \vDash \mathsf{I}\Sigma^0_n$.*
*2. Every $\Sigma^0_n$ definable bounded subset of $M$ is $\mathcal{M}$-finite.*

*Proof.* To prove (1) implies (2), assume $\mathcal{M} \vDash \mathsf{I}\Sigma^0_n$. Let $\varphi(x)$ be a $\Sigma^0_n$ formula and fix $k \in M$. Let $m = k!$, which can be defined in $M$ by primitive recursion as in the proof of Theorem 5.5.8, and consider the following formula in which $n$ is free:

$$(\forall i < k)[\varphi(i) \rightarrow m(i + 1) + 1 \text{ divides } n]. \tag{6.2}$$

Obviously, (6.2) holds in $\mathcal{M}$ of $\prod_{i<k} m(i + 1) + 1$. Moreover, by $\mathsf{B}\Sigma^0_n$ in $\mathcal{M}$, (6.2) is $\Pi^0_n$. Hence by $\mathsf{L}\Pi^0_n$, there is a least $n \in M$ satisfying (6.2). Suppose there is an $i <^{\mathcal{M}} k$ such that $\mathcal{M} \vDash m(i+1)+1$ divides $n$ but $\neg\varphi(i)$, and write $n = (m(i+1)+1)n^*$ for some $n^* <^{\mathcal{M}} n$. By the proof of Theorem 5.5.8, $\mathcal{M}$ satisfies

$$(\forall j < k)[j \neq i \rightarrow m(j + 1) + 1 \text{ and } m(i + 1) + 1 \text{ are relatively prime}].$$

Hence, $n^*$ still satisfies (6.2) in $\mathcal{M}$, contradicting the choice of $n$. We conclude that in $\mathcal{M}$,

$$(\forall i < k)[\varphi(i) \leftrightarrow m(i + 1) + 1 \text{ divides } n].$$

In other words, $\mathcal{M}$ satisfies that $\langle k, \langle m, n\rangle\rangle$ represents $\{i < k : \varphi(i)\}$ in the sense of Definition 5.5.5, and hence the latter set is $\mathcal{M}$-finite.

Conversely, let $\varphi(x)$ be a $\Sigma^0_n$ formula such that $\mathcal{M} \vDash \varphi(0) \wedge (\forall x)[\varphi(x) \rightarrow \varphi(x+1)]$ and fix an arbitrary $a \in M$. By (2), the set $F_a$ of $b \in M$ such that $\mathcal{M} \vDash b < a+1 \wedge \varphi(b)$

is $\mathcal{M}$-finite. Using a code for $F_a$ allows us to express $x \in F_a$ as a $\Sigma^0_0$ formula. Let $\psi(x)$ be the $\Sigma^0_0$ formula $x > a \vee x \in F_a$. Then $\mathcal{M} \models \psi(0) \wedge (\forall x)[\psi(x) \rightarrow \psi(x+1)]$. By $\mathsf{I}\Sigma^0_0$ in $\mathcal{M}$, we conclude that $\mathcal{M} \models \varphi(a)$. □

We can use this result to generalize bounded $\Sigma^0_1$ comprehension (Theorem 5.5.11) to higher levels of the arithmetical hierarchy. Of course, $\mathsf{ACA}_0$ includes regular (unbounded) $\Sigma^0_n$ comprehension for each $n \in \omega$, while $\mathsf{WKL}_0$ only includes $\Sigma^0_1$ induction. The generalization here is useful if we work in systems such as $\mathsf{RCA}_0 + \mathsf{I}\Sigma^0_2$.

**Theorem 6.2.6 (Bounded $\Sigma^0_n$ comprehension).** *For every $n \geqslant 1$, and every $\Sigma^0_n$ formula $\varphi(x)$,*

$$\mathsf{RCA}_0 + \mathsf{I}\Sigma^0_n \vdash (\forall z)(\exists X)(\forall x)[x \in X \leftrightarrow x < z \wedge \varphi(x)].$$

*Proof.* Fix $\mathcal{M} \models \mathsf{RCA}_0 + \mathsf{I}\Sigma^0_n$ along with $a \in M$. By Proposition 6.2.5, there is a code $\sigma \in M$ for the set $F = \{x < a : \varphi(x)\}$. As before, the code allows us to express $x \in F$ as a $\Sigma^0_0$ formula. Thus the set $F$ can be formed using $\Sigma^0_0$ comprehension. □

Another nice corollary of Proposition 6.2.5 is the following variation of the finitary pigeonhole principle.

**Proposition 6.2.7.** *Fox each $n \geqslant 1$, the following is provable in $\mathsf{RCA}_0 + \mathsf{I}\Sigma^0_n$: for each $x \geqslant 1$, there is no $\Sigma^0_n$ definable injection $x \longrightarrow y$ for any $y < x$.*

*Proof.* We argue in $\mathsf{RCA}_0 + \mathsf{I}\Sigma^0_n$. By Proposition 6.2.5, every $\Sigma^0_n$ definable function $x \longrightarrow y$ for $x, y \in \mathbb{N}$ is coded by a number. Hence, the statement to be proved is a $\Pi^0_1$ formula in $x$, and so we can prove it using $\mathsf{I}\Pi^0_1$. Clearly, the result holds for $x = 1$. So fix $x > 1$ and assume the result holds for $x - 1$. If the results fails for $x$, we can fix a (code for a) witnessing injection $f : x \longrightarrow y$ for some $y < x$. Without loss of generality, we can take $y - 1$ to be in the range of $f$. (More precisely, by $\mathsf{L}\Pi^0_n$ we can fix the least $y^* < y$ such that $(\forall z > y^*)(\forall w < x)[f(w) \neq z]$, and then replace $y$ by $y^*$ if necessary.) Now, we must have $y = x - 1$, as otherwise $f \upharpoonright x - 1$ would be an injection from $x - 1$ into a number smaller than $x - 1$, which cannot be. Likewise, if $f(x-1) = x - 2$ then $f \upharpoonright x - 1$ would be an injection from $x - 1$ to $x - 2$, again an impossibility. Thus, there exists $w^* < x - 1$ so that $f(w^*) = x - 2$ and $f(w) < x - 2$ for all $w < x$ different from $w^*$. We can then define $g : x - 1 \rightarrow x - 2$ by

$$g(w) = \begin{cases} f(w) & \text{if } w \neq w^*, \\ f(x-1) & \text{if } w = w^*, \end{cases}$$

for all $w < x - 1$. Now $g$ is obviously injective on all $w < x - 1$ different from $w^*$ since it agrees with $f$. And we also cannot have $g(w) = g(w^*)$ for any $w \neq w^*$, since otherwise we would have $f(w) = f(x-1)$ for some $w \neq x - 1$. Thus, $g$ is injective, which contradicts the choice of $a$. □

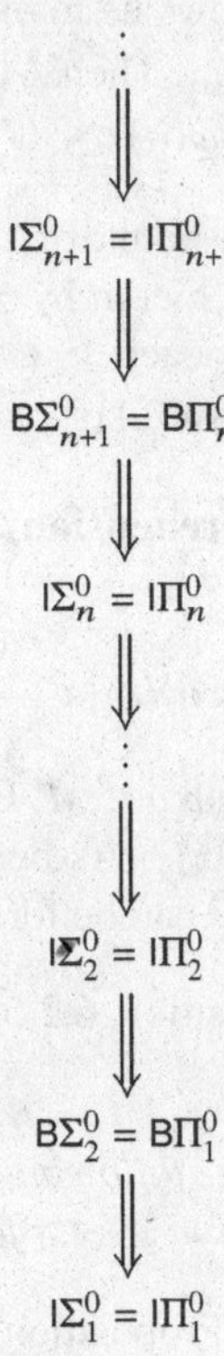


**Figure 6.1.** The Kirby–Paris hierarchy; $n \geqslant 1$ is arbitrary. An arrow from one principle to another indicates an implication provable in $\mathsf{RCA}_0$. No arrows can be reversed.

## 6.3 The Kirby–Paris hierarchy

The aim of this section is to show that the implications between induction and bounding principles at corresponding levels of the arithmetical hierarchy are strict. Thus, we have an increasing sequence under provability, as illustrated in Figure 6.1. In the literature, this is often called the *Kirby–Paris hierarchy* after the seminal paper of Paris and Kirby [237] in which it was first established. We follow the standard outline for separating the principles in the hierarchy, which is roughly the same as that of the original proofs in [237], with minor deviations. Similar treatments can be found in Kaye [176] and Hájek and Pudlák [134].

We first prove separations of the Kirby–Paris hierarchy in $\mathcal{L}_1$, and then apply conservativity to extend the results to $\mathcal{L}_2$ and $\mathsf{RCA}_0$. Thus, in the following definition, lemmas, and theorems, all structures should be understood to be $\mathcal{L}_1$ structures. We write $M$ rather than $\mathcal{M}$ for these structures, to emphasize that they can be identified with their first order parts. Exercise 6.7.7 shows the following definition is nontrivial.

**Definition 6.3.1.** Fix $n \geqslant 1$ and $M \vDash \mathsf{PA}^- + \mathsf{I}\Sigma^0_n$.

1. $K_n(M)$ is the set of all $\Sigma^0_n$ definable elements of $M$ (without parameters).
2. For $B \subseteq M$, $K_n(M, B)$ is the set of all $\Sigma^0_n(B)$ definable elements of $M$.

We may regard $K_n(M, B)$ as a substructure of $M$.

**Lemma 6.3.2.** *For each $n \geqslant 1$, there exists $M \vDash \mathsf{PA}^- + \mathsf{I}\Sigma^0_0$ such that $K_n(M, \varnothing)$ contains nonstandard elements.*

*Proof.* Consider any $\Pi^0_1$ sentence not provable in, but consistent with, PA, e.g., Con(PA). Write this as $\neg(\exists x)\varphi(x)$ where $\varphi$ is $\Sigma^0_0$, and fix $M \vDash \mathsf{PA} + (\exists x)\varphi(x)$. Let $\psi(x)$ be the $\Sigma^0_0$ formula $\varphi(x) \wedge \neg(\exists y < x)\varphi(x)$. By Theorem 6.1.3, $M \vDash \mathsf{L}\Sigma^0_0$, and so $M \vDash \psi(a)$ for some $a$. Clearly, $\psi$ defines $a$ in $M$, so $a \in K_0(M, \varnothing)$. But $a$ cannot be standard, else $\varphi(a)$ could be turned into an actual proof (in PA) of Con(PA). □

**Lemma 6.3.3.** *Fix $n \geqslant 1$ and $M \vDash \mathsf{PA}^- + \mathsf{I}\Sigma^0_n$. For every $B \subseteq M$, $K_n(M, B)$ is a $\Sigma^0_n$ elementary substructure of $M$ and $K_n(M, B) \vDash \mathsf{PA}^-$.*

*Proof.* To show $K_n(M, B)$ is a $\Sigma^0_n$ elementary substructure of $M$, we use the Tarski–Vaught test. Let $\varphi(x, z_0, \ldots, z_{k-1})$ be a $\Pi^0_{k-1}$ formula and suppose $M \vDash (\exists x)\varphi(x, b_0, \ldots, b_{k-1})$ for some $b_0, \ldots, b_{k-1} \in K_n(M, B)$. We exhibit an $a \in K_n(M, B)$ such that $M \vDash \varphi(a, b_0, \ldots, b_{k-1})$. By Theorem 6.1.3, $M \vDash \mathsf{L}\Sigma^0_{n-1}$ and so $M$ satisfies

$$(\exists x)[\varphi(x, b_0, \ldots, b_{k-1}) \wedge \neg(\exists y < x)\varphi(y, b_0, \ldots, b_{k-1})]. \tag{6.3}$$

We claim that $(\exists y < x)\varphi(y, z_0, \ldots, z_{k-1})$ is equivalent in $M$ to a $\Pi^0_{n-1}$ formula $\psi(x, y_0, \ldots, y_{k-1})$. If $n = 1$, this is obvious because $\varphi$ is then $\Pi^0_0$ and this class of formulas is closed under bounded quantification. If $n > 1$, we instead obtain it from Proposition 6.1.2 using the fact that, by Theorem 6.1.3, $M \vDash \mathsf{B}\Sigma^0_{n-1}$. Now, we can fix a tuple of elements $\vec{c}$ from $K_n(M, B)$ and, for each $i < k$, a $\Sigma^0_n$ formula $\theta_i(x, \vec{z})$ such that $\theta_i(x, \vec{c})$ defines $b_i$ in $M$. Let $\chi(x)$ be the formula

$$(\exists z_0, \ldots, z_{k-1})\left[\bigwedge_{i<k} \theta_i(x, \vec{c}) \wedge \varphi(x, z_0, \ldots, z_{k-1}) \wedge \neg\psi(x, z_0, \ldots, z_{k-1})\right].$$

Then (6.3) is equivalent to $(\exists x)\chi(x)$, so $\chi$ defines the $\leqslant^M$-least element $a \in M$ such that $M \vDash \varphi(a, b_0, \ldots, b_{k-1})$. But since $\chi$ is $\Sigma^0_n$, it follows that $a \in K_n(M, B)$, as was required.

To prove that $K_n(M, B) \vDash \mathsf{PA}^-$, we refer to Definition 5.3.2. All axioms except one are $\Pi^0_1$, and so since $n \geqslant 1$, they hold in $K_n(M, B)$ because they hold in $M$. The remaining axiom is

$$(\forall x, y)(\exists z)[x < y \rightarrow x + z = y].$$

To verify this holds in $K_n(M, B)$, fix $a, b, c \in K_n(M, B)$ such that $a <^M b$ and $M \vDash a + c = b$. Since $M$ is a model of PA, it satisfies $(\forall x, z, z^*)[x + z = x + z^* \rightarrow z = z^*]$.

Thus, $c$ is $\Sigma_0^0$ definable in $M$ using $a$ and $b$ as parameters. Since $a$ and $b$ are $\Sigma_n^0(B)$ definable in $M$, it follows that so is $c$. Hence, $c \in K_n(M, B)$ and $K_n(M, B) \vDash a+c = b$ by elementarity. □

We now proceed to the main results. In Theorem 6.3.4 we show that induction for $\Sigma_{n-1}^0$ formulas is weaker than bounding for $\Sigma_n^0$ formulas, and in Theorem 6.3.5 we show that bounding for $\Sigma_n^0$ formulas is weaker than induction for $\Sigma_n^0$ formulas.

**Theorem 6.3.4 (Parsons [239]).** *For each $n \geqslant 1$, there exists a model of*

$$\mathsf{PA}^- + \mathsf{I}\Sigma_{n-1}^0 + \neg\mathsf{B}\Sigma_n^0.$$

*Proof (Paris and Kirby [237]).* Fix a model $M \vDash \mathsf{PA}$ containing nonstandard $\Sigma_0^0$ (and hence $\Sigma_n^0$) definable elements, which exists by Lemma 6.3.2. Let $K_n = K_n(M, \varnothing)$. We claim this is the desired model.

By Lemma 6.3.3, $K_n \vDash \mathsf{PA}^-$ and $K_n$ is a $\Sigma_n^0$ elementary substructure of $M$. Also, if $\varphi(x)$ is a $\Sigma_{n-1}^0$ formula such that $K_n \vDash (\exists x)\, \varphi(x)$ then $M \vDash (\exists x)\varphi(x)$. By $\mathsf{L}\Sigma_{n-1}^0$ in $M$ we have that $M \vDash (\exists x)[\varphi(x) \wedge (\forall y < x)\, \neg\varphi(x)]$. But the latter is a $\Sigma_n^0$ sentence, and so holds in $K_n$ by elementarity. By Theorem 6.1.3, we conclude that $K_n \vDash \mathsf{I}\Sigma_{n-1}^0$.

Next, we claim that $K_n \vDash \neg\mathsf{B}\Sigma_n^0$. Fix a universal $\Sigma_n^0$ formula of the form $(\exists z)\, \psi(w, x, z)$, where $\psi$ is $\Pi_{n-1}^0$. So for each $a \in K_n$ there is an $e \in \omega$ such that $(\exists z)\, \psi(e, x, z)$ defines $a$ in $M$. Let $\theta(w, x)$ be the formula

$$\theta(w, x) = (\exists z)[\psi(w, x, z) \wedge \neg(\exists\langle x^*, z^*\rangle < \langle x, z\rangle)\psi(w, x^*, z^*)].$$

Proposition 6.1.2 implies that $\theta(w, x)$ is equivalent in $M$ to a $\Sigma_n^0$ formula $\upsilon(w, x)$. By $\mathsf{L}\Sigma_n^0$, for each $a \in K_n$ there is an $e \in \omega$ such that $\theta(e, a)$ is true in $M$, and by elementarity, also in $K_n$. In particular, for any nonstandard $b \in K_n$ we have

$$K_n \vDash (\forall x < b + 1)(\exists w < b)\theta(x, w).$$

Now if $\mathsf{B}\Sigma_n^0$ held in $K_n$, then by Proposition 6.1.2, the left hand side above would be equivalent in $K_n$ to a $\Sigma_n^0$ formula. Hence, it would also be true in $M$, and so

$$M \vDash (\forall x < b + 1)(\exists w < b)\theta(x, w).$$

But clearly $M \vDash \neg(\exists x, x^*, w)[x \neq x^* \wedge \theta(x, w) \wedge \theta(x^*, w)]$, so $\theta$ would be an arithmetical definition of an injection from $b + 1$ into $b$ in $M$. This is impossible by Proposition 6.2.7 since $M \vDash \mathsf{PA}$. □

**Theorem 6.3.5 (Lessan [196], Paris and Kirby [237]).** *For each $n \geqslant 1$, there exists a model of* $\mathsf{PA}^- + \mathsf{B}\Sigma_n^0 + \neg\mathsf{I}\Sigma_n^0$.

*Proof.* By Lemma 6.3.2, fix $M \vDash \mathsf{PA}$ with nonstandard $\Sigma_{n-1}^0$ definable elements. Now, set $B_0 = K_{n-1}(M, \varnothing)$, and for each $i \in \omega$, inductively define

$$B_{2i+1} = \{x \in M : (\exists y \in M_{2i})[x \leqslant y]\}$$

and

$$B_{2i+2} = K_{n-1}(M, B_{2i+1}).$$

Clearly, $B_i \subseteq B_{i+1}$ for all $i$. Let $B = \bigcup_{i\in\omega} B_i$. Every tuple of parameters from $B$ is then contained in $B_{2i}$ for some $i$, and so since $B_{2i}$ is a $\Sigma^0_{n-1}$ elementary substructure of $M$ by Lemma 6.3.3, it follows that so is $B$.

To show $B \models \mathsf{B}\Sigma^0_n$, let $\varphi(x, 0)$ be a $\Pi^0_{n-1}$ formula and suppose that for some $d \in B$ we have that $B \models (\forall x < d)(\exists y)\varphi(x, y)$. Since $B$ is closed downward under $\leqslant^M$, it follows by elementarity that $M \models (\forall x < d)(\exists y)\varphi(x, y)$. Now $M \models \mathsf{PA}$, so there is a $b \in M$ so that $M \models (\forall x < d)(\exists y \leqslant b)\varphi(x, y)$. We can also fix the $\leqslant^M$-least such $b$; denote this by $b^*$. Then there must be an $a^* <^M d$ so that $M \models \varphi(a^*, b^*) \wedge (\forall y < b^*)\neg\varphi(a^*, y)$. Since $B \models (\exists y)\varphi(a^*, y)$ there is a $b \in B$ such that, by elementarity, $M \models \varphi(a^*, b)$, and so necessarily $b^* \leqslant^M b$. This implies that $b^* \in B$, and hence $B \models (\forall x < d)(\exists y \leqslant b^*)\varphi(x, y)$. Thus, $B \models \mathsf{B}\Pi^0_{n-1}$, which is equivalent to $\mathsf{B}\Sigma^0_n$ by Theorem 6.1.3 (1).

To show that $B \models \neg\mathsf{I}\Sigma^0_n$, we prove that the standard part of $B$ is $\Sigma^0_n$ definable. Fix a universal $\Sigma^0_{n-1}$ formula of the form $(\exists z)\psi(w, x, y, z)$, where $\psi$ is $\Pi^0_{n-2}$. Define $\theta(u, x)$ to be the following:

$$\begin{aligned}(\exists i < u)(\exists\langle w_0, \ldots, w_i\rangle < u)(\forall j < i)(\exists x_j)(\exists\vec{y}_j < x_j)\\ [(j = 0 \rightarrow (\exists z)\psi(w_0, x_0, 0, z))\\ \wedge (j > 0 \rightarrow (\exists z)\psi(w_j, x_j, \vec{y}_{j-1}, z)) \wedge (\exists z)\psi(w_i, x, \vec{y}_{i-1}, z))].\end{aligned}$$

Since $B$ satisfies $\mathsf{B}\Sigma^0_n$, it follows by Proposition 6.1.2 that this is equivalent in $\mathsf{B}$ to a $\Sigma^0_{n-1}$ formula. Clearly, $(\forall x)(\forall u)(\forall v)[u \leqslant v \wedge \theta(u, x) \rightarrow \theta(v, x)]$ is true in both $M$ an $B$. Also, $b \in M$ belongs to $B$ if and only if there is an $m \in \omega$ such that $M \models \theta(m, a)$. In particular, if $b \in B$ then $M \models \theta(a, b)$ for every nonstandard $a \in M$. Conversely, if $M \models \theta(a, b)$ for some $b \in M$ and every nonstandard $a \in M$, then by underspill (Proposition 6.2.2 (2)) we must have that $M \models \theta(a, m)$ for some $m \in \omega$, and hence $b \in B$. Since $B$ is a $\Sigma^0_{n-1}$ elementary substructure of $M$, it follows that $a \in B$ is nonstandard if and only if $B \models (\forall x)\theta(a, x)$, so $a \in \omega$ if and only if $B \models (\exists x)\neg\theta(a, x)$. □

**Corollary 6.3.6.** *Fix $n \geqslant 2$.*

1. *There exists a model of* $\mathsf{RCA}_0 + \mathsf{I}\Sigma^0_{n-1} + \neg\mathsf{B}\Sigma^0_n$.
2. *There exists a model of* $\mathsf{RCA}_0 + \mathsf{B}\Sigma^0_n + \neg\mathsf{I}\Sigma^0_n$.

*Proof.* For either part, let $M$ be the corresponding model from Theorem 6.3.4 or Theorem 6.3.5. Then, let $\mathcal{M}$ be the $\mathcal{L}_2$ structure with first order part $M$ and second order part all $\Delta^0_1$ definable subsets of $M$. By Theorem 5.10.1, $\mathcal{M} \models \mathsf{RCA}_0$ and has the desired properties. □

## 6.4 Reverse recursion theory

One place induction and bounding arise in reverse mathematics is in its analysis of computability theory, commonly referred to as *reverse recursion theory*. This is usually understood to concern results of classical computability theory, chiefly about the computably enumerable (c.e.) sets and degrees. In a model $\mathcal{M}$ of $\mathsf{PA}^-$, we say a set $S \subseteq M$ is *c.e.* if it is $\Sigma^0_1$ definable in $\mathcal{M}$ (possibly with number parameters). In a model of $\mathcal{M}$ of $\mathsf{RCA}_0$, we can similarly define $S$ being $X$-*c.e.*, for $X \in \mathcal{S}^{\mathcal{M}}$. In fact, as constructions in this part of computability theory are, almost by definition, computable, we may expect the entire discussion to be accommodated in $\mathsf{RCA}_0$. And this is almost the case, except for issues of induction. The main concern is that while a construction may be computable, the verification that it does what it purports to do can be more complex.

We will briefly survey a few results along these lines. We will omit all proofs, which tend to be quite involved in this area.

A central focus of reverse recursion theory is on the priority method, the hallmark technique of modern degree theory. A good introduction can be found in Downey and Hirschfeldt [83], Lerman [194], or Soare [293]. A basic priority construction is organized into stages, and the properties of the set to be constructed are expressed through countably many requirements and assigned relative priorities. There is a plethora of different kinds of priority constructions, of course, but even in the simplest we must take care when we move to a formal system. Consider a finite injury argument. Say our requirements are $\mathcal{R}_0, \mathcal{R}_1, \ldots$, with $\mathcal{R}_i$ having higher priority than $\mathcal{R}_j$ if $i < j$. Typically we will have an argument that each $\mathcal{R}_i$, if it is unsatisfied at a given stage, can "act" (cause us to take some action on its behalf) at a later stage and become satisfied. We can then prove that all requirements are eventually satisfied by induction. Namely, fix $i$ and assume there is a stage $s_0$ so that no $\mathcal{R}_j$ with $j < i$ acts at any stage $s \geqslant s_0$. Now we can appeal to the earlier argument to find a stage $s_1 \geqslant s_0$ at which $\mathcal{R}_i$ acts and becomes satisfied (if it was satisfied already, we take $s_1 = s_0$). Since no $\mathcal{R}_j$ with $j < i$ will act again after stage $s_1$, this means no requirement can injure $\mathcal{R}_i$, and hence it stays satisfied at all stages $s \geqslant s_1$ (and does not act again). This much will be familiar from a first course in computability. What may be less so is the explicit identification of the induction statement here as $\Sigma^0_2$ (assuming that "acting" and "being satisfied" are $\Delta^0_1$ properties, as is usually the case). Thus, on its face, we need $\Sigma^0_2$ induction to formalize this argument.

There are two basic kinds of finite injury arguments, distinguished by whether or not the number of injuries to each requirement is computably bounded. The most famous example of the "bounded" type is the celebrated Friedberg–Muchnik construction of two Turing incomparable c.e. sets, which was historically the very first priority argument. (There, each requirement, once all higher priority requirements have been permanently satisfied, acts at most once. So the number of times $\mathcal{R}i$ will act is at most $2^i$.) The "unbounded" type is typified by the Sacks splitting theorem, that every noncomputable c.e. set can be partitioned into two Turing incomparable c.e. sets. It turns out that both the Friedberg–Muchnik theorem and the Sacks splitting *can* be carried out in $\mathsf{I}\Sigma^0_1$. In fact, in the former case even weaker axioms suffice, while in the latter $\mathsf{I}\Sigma^0_1$ is optimal.

**Theorem 6.4.1 (Chong and Mourad [37]).** $\mathsf{PA}^- + \mathsf{I}\Sigma^0_0 + \mathsf{B}\Sigma^0_1 + \mathsf{Exp}$ *proves the Friedberg–Muchnik theorem.*

**Theorem 6.4.2 (Mytilinaios [230]; Mourad [224]).** *Over* $\mathsf{PA}^- + \mathsf{I}\Sigma^0_0 + \mathsf{B}\Sigma^0_1 + \mathsf{Exp}$, *the following are equivalent.*

1. $\mathsf{I}\Sigma^0_1$.
2. *The Sacks splitting theorem.*

Pushing finite injury constructions of the "bounded" type down into $\mathsf{I}\Sigma^0_1$ is not difficult in general. Suppose we are running such a construction in a model $\mathcal{M}$ of $\mathsf{RCA}_0$. Thus we have requirements and priorities as above (indexed now by elements of $M$), and we also have a $\Sigma^0_1$ definable function $f\colon M \to M$ such that $M$ satisfies that $f(i)$ bounds the number of times $\mathcal{R}_i$ acts during the course of our construction. In this case, for each $i \in M$, the set $I = \{\langle j, n\rangle : j < i \wedge \mathcal{R}_j \text{ acts at least } n \text{ times}\}$ is $\Sigma^0_1$ definable and bounded in $M$ by $b = i \cdot \sum_{j<i} f(j)$.

By bounded $\Sigma^0_1$ comprehension (Proposition 6.2.5), $I$ is $M$-finite, so the formula

$$(\forall \langle j, n\rangle \in I)(\exists s)[\mathcal{R}_j \text{ acts } n \text{ times by the end of stage } s]$$

is an instance of $\mathsf{B}\Sigma^0_1$ in $\mathcal{M}$. Since $\mathcal{M}$ satisfies $\mathsf{B}\Sigma^0_1$ we may fix $s_0$ bounding the existential quantifier. It follows that no $\mathcal{R}_j$ with $j < i$ acts at any stage $s \geqslant s_0$, and we can now proceed as above to conclude that $\mathcal{R}_i$ acts (and is permanently satisfied) at some stage $s_1 \geqslant s_0$.

The above argument does not work for finite injury constructions of the "unbounded" type, and yet many still go through in $\mathsf{I}\Sigma^0_1$. An example relevant to us is the following:

**Theorem 6.4.3 (Hájek and Kučera [133], Mourad [224]).** *Over* $\mathsf{PA}^- + \mathsf{I}\Sigma^0_0 + \mathsf{B}\Sigma^0_1 + \mathsf{Exp}$, *the following are equivalent.*

1. $\mathsf{I}\Sigma^0_1$.
2. *The low basis theorem.*

Note that our proof of the low basis theorem (Theorem 2.8.18) was actually a *forcing* construction (which we will discuss in depth in the next chapter) but can also be presented as a finite injury priority construction with an "unbounded" number of injuries to each requirement (see, e.g., [66], Section 3.4.3).

In general, there are various methods for formalizing these kinds of finite injury constructions, most notably *Shore blocking*, originally developed in $\alpha$-recursion theory by Shore [279], and later adapted for use in arithmetic by Mytilinaios [230]. We will mention an application of this technique in Chapter 9. But not all such constructions are known to be provable in $\mathsf{I}\Sigma^0_1$, and indeed, there are examples that require $\mathsf{B}\Sigma^0_2$. A detailed analysis was undertaken by Groszek and Slaman [130].

For completeness, we mention briefly also *infinite* injury arguments. Again, the most well-known examples here are c.e. set constructions. Recall the following computability theoretic notions.

- A c.e. set $A$ is *maximal* if it is coinfinite and for every c.e. set $B \supseteq A$, either $B =^* \omega$ or $B \setminus A =^* \emptyset$.
- Two sets $A_0$ and $A_1$ form a *minimal pair* if $A_0, A_1 \nleqslant_T \emptyset$ and every $B \leqslant_T A_0, A_1$ satisfies $B \leqslant_T \emptyset$.

The existence of maximal sets and c.e. minimal pairs are both classical applications of the infinity injury method. And, fittingly, the induction needed to prove these results reflects the additional complexity of this method over finite injury.

**Theorem 6.4.4 (Chong and Yang [41]).** *Over* $\mathsf{RCA}_0 + \mathsf{B}\Sigma^0_2$, *the following are equivalent.*

*1.* $\mathsf{I}\Sigma^0_2$.
*2. There exists a maximal set.*

**Theorem 6.4.5 (Chong, Qiang, Slaman, and Yang [38]).** *Over* $\mathsf{RCA}_0 + \mathsf{B}\Sigma^0_2$, *the following are equivalent.*

*1.* $\mathsf{I}\Sigma^0_2$.
*2. There exists a minimal pair of c.e. sets.*

Reverse recursion theory is not limited just to priority constructions. Many other results from computability theory can be, and have been, analyzed in this framework. For example, the limit lemma (Theorem 2.6.5) can be carried out in just $\mathsf{B}\Sigma^0_1$. (See Švejdar [319].) For a much more comprehensive overview, including sketches of some core arguments, we refer to the survey of Chong, Li, and Yang [36].

## 6.5 Hirst's theorem and $\mathsf{B}\Sigma^0_2$

By far the most commonly encountered bounding scheme in reverse mathematics is $\mathsf{B}\Sigma^0_2$. This may be partially attributed to the fact that many constructions in $\mathsf{RCA}_0$ involve $\Sigma^0_2$ and $\Pi^0_2$ formulas (a set being finite or infinite, a function being total, etc.). So, logical manipulations along the lines of Proposition 6.1.2, which are widespread, quickly cause $\mathsf{B}\Sigma^0_2$ to appear. We have also seen in Proposition 6.2.4 that some very elementary number theoretic facts—such as every $\Sigma^0_2$ definable function with domain an initial segment of $\mathbb{N}$ having bounded range—can be equivalent to $\mathsf{B}\Sigma^0_2$. Indeed, this may lead us to think there cannot be many cases where we can do *without* $\mathsf{B}\Sigma^0_2$!

More surprising, however, is that $\mathsf{B}\Sigma^0_2$—a first order principle, solely about properties of numbers—can be equivalent to a second order principle—about sets of numbers. This realization was first made in the following seminal result of Hirst.

**Theorem 6.5.1 (Hirst [154]).** $\mathsf{RCA}_0 \vdash \mathsf{RT}^1 \leftrightarrow \mathsf{B}\Sigma^0_2$.

*Proof.* ($\rightarrow$). We reason in $\mathsf{RCA}_0$ and work with $\mathsf{B}\Pi^0_1$ instead of $\mathsf{B}\Sigma^0_2$ for convenience. Assume $\mathsf{RT}^1$. Let $\varphi(x, y)$ be a $\Pi^0_1$ formula such that for some $z$ we have $(\forall x < z)(\exists y)\varphi(x, y)$. Write $\varphi(x, y)$ as $(\forall u)\psi(u, x, y)$, where $\psi$ is $\Sigma^0_0$. Now define a function $f : \mathbb{N} \to \mathbb{N}$ as follows: for every $a \in \mathbb{N}$,

$$f(a) = \begin{cases} (\mu b < a)(\forall x < z)(\exists y < b)(\forall u < a)\psi(u, x, y) & \text{if } x \text{ exists,} \\ a & \text{otherwise.} \end{cases}$$

Note that $f$ is $\Sigma^0_0$ definable, and so exists.

We claim that $f$ has bounded range. Suppose not. Then by $\Delta^0_1$ comprehension there is a sequence $\langle a_i : i \in \mathbb{N}\rangle$ such that $a_0 < a_1 < \cdots$ and $f(a_0) < f(a_1) < \cdots$. Define $c : \mathbb{N} \to z$ as follows: for every $i$, let

$$c(i) = (\mu x < z)(\forall y < f(a_i))(\exists u < a_{i+1})\neg\psi(u, x, y).$$

Then $c$ exists and is total. Indeed, if there were an $i$ such that $(\forall x < z)(\exists y < f(a_i))(\forall u < a_{i+1})\psi(u, x, y)$ we would have $f(a_{i+1}) \leqslant f(a_i)$ instead of $f(a_i) < f(a_{i+1})$. So $c$ is an instance of $\mathsf{RT}^1$, and we may consequently fix an infinite homogeneous set $H$ for it, say with color $x < z$. By hypothesis, there is a $y$ such that $\varphi(x, y)$. But if we fix $i$ such that $f(a_i) > y$, then $(\exists u < a_{i+1})\neg\psi(u, x, y)$ by definition of $c$, and so in particular $\neg\varphi(x, y)$.

Having proved the claim, let $w$ bound the range of $f$. Thus, for every $a$ we have

$$(\forall x < z)(\exists y < w)(\forall u < a)\psi(u, x, y).$$

Fix $x < z$. Define $d : \mathbb{N} \to w$ as follows: for each $a$, $d(a) = (\mu y < w)(\forall u < a)\psi(u, x, y)$. Then $d$ is an instance of $\mathsf{RT}^1$, and so we may fix an infinite homogeneous set $H$ for it, say with color $y < w$. Thus, there are infinitely many $a$ such that $(\forall u < a)\psi(u, x, y)$, and so it follows that $(\forall u)\psi(u, x, y)$ and therefore that $\varphi(x, y)$. Since $x$ was arbitrary, we conclude that $(\forall x < z)(\exists y < w)\varphi(x, y)$, as desired.

($\leftarrow$). Next, assume $\mathsf{B}\Pi^0_1$ and let $f : \mathbb{N} \to z$ be a given instance of $\mathsf{RT}^1$. Let $\varphi(x, y)$ be the $\Pi^0_1$ formula $(\forall u)[u \geqslant y \to f(u) \neq x]$. Then if $f$ has no infinite homogeneous set, we have

$$(\forall x < z)(\exists y)\varphi(x, y).$$

Fix $w$ as given by $\mathsf{B}\Pi^0_1$ applied to $\varphi$. Clearly, we have $\theta(x, y) \to \theta(x, y^*)$ whenever $y < y^*$, and so in particular

$$(\forall x < z)(\forall u)[u \geqslant w \to f(u) \neq x].$$

But this cannot be since $f$ is a map into $z$. Thus, $f$ must have an infinite homogeneous set and the proof is complete. □

The way $\mathsf{B}\Sigma^0_2$ is used in this proof is interesting. $\mathsf{RT}^1$ admits computable solutions, so there is no sense in which $\mathsf{B}\Sigma^0_2$ is helping with the actual construction of an infinite homogeneous set. Rather, it is used to expose the contradiction in a reduction assumption that no infinite homogeneous set exists.

It is worth noting that Hirst's theorem is unrelated to the issue of nonuniformity in picking an $i$ such that $f(x) = i$ for infinitely many $x$. The same issue occurs with $\mathsf{RT}^1_k$ for any standard $k > 2$, but $\mathsf{RT}^1_k$ is provable in $\mathsf{RCA}_0$. (The same proof works, but the bound $y^*$ can be found directly.) More tellingly, the following principle is completely uniform—it *uniformly* admits computable solutions—yet it exhibits the same behavior as $\mathsf{RT}^1$.

**Definition 6.5.2 (Extraction of a homogeneous set from a limit homogeneous set).** LH is the following statement: for every $c\colon [\mathbb{N}]^2 \to 2$ such that $\lim_y c(x, y) = 1$ for all $x$, there exists an infinite set $H$ such that $c(x, y) = 1$ for all $\langle x, y\rangle \in [H]^2$.

In Definition 8.4.1, we will define a set $L$ to be *limit homogeneous* for a coloring $c\colon [\omega]^2 \to 2$ if it has the above property with respect to all $x \in L$. Every such set has a uniformly $c \oplus L$-computable infinite homogeneous subset (see Proposition 8.4.2). But formalizing this cannot in general be done in $\mathsf{I}\Sigma^0_1$. The proof of the next result is left to Exercise 6.7.6.

**Proposition 6.5.3 (Dzhafarov, Hirschfeldt, and Reitzes [82]).** *Over* $\mathsf{RCA}_0$, LH *is equivalent to* $\mathsf{RT}^1$.

Hirst's theorem makes $\mathsf{B}\Sigma^0_2$ in many cases easier to work with. An example is the following proposition, which reveals yet more guises of $\mathsf{B}\Sigma^0_2$, and which should be compared to Propositions 4.2.10 and 4.3.4.

**Proposition 6.5.4.** *Over* $\mathsf{RCA}_0$, *the following are equivalent.*

1. $\mathsf{B}\Sigma^0_2$.
2. $\mathsf{RT}^1$.
3. IPHP.
4. General-IPHP.
5. FUF.

*Proof.* We argue in $\mathsf{RCA}_0$. That (1) → (2) is one half of Hirst's theorem (Theorem 6.5.1). That (2) → (3) is clear.

To prove (3) → (4), proceed as in the proof of Proposition 4.3.4. Suppose we are given an instance of General-IPHP. This is a sequence $\langle X_i : i \in \mathbb{N}\rangle$ of subsets of $\mathbb{N}$ such that $X_i = \emptyset$ for almost all $i$ and for every $x$ there is a $y \geqslant x$ such that $y \in X_i$ for some $i$. Form the set $S = \{\langle x, \langle i, y\rangle\rangle : y \geqslant x \wedge y \in X_i\}$. Now using $\mathsf{L}\Sigma^0_0$, define $g\colon \mathbb{N} \to \mathbb{N}$ as follows: $g(x) = \langle i, y\rangle$ for the unique $\langle i, y\rangle$ such that $\langle x, \langle i, y\rangle\rangle \in S$ and there is no $\langle x, z\rangle < \langle x, \langle i, y\rangle\rangle$ in $S$. Define $f\colon \mathbb{N} \to \mathbb{N}$ by $f(x) = i$ where $g(x) = \langle i, y\rangle$. By assumption on $\langle X_i : i \in \mathbb{N}\rangle$, the range of $f$ is bounded. Hence, $f$ is an instance of IPHP. By IPHP, there is an $i$ such that $f(x) = i$ for infinitely many $x$. Then $X_i$ is infinite.

To prove (4) → (5), fix an instance of FUF. This is a sequence $\langle F_i : i \in \mathbb{N}\rangle$ of finite sets such that $F_i = \emptyset$ for almost all $i$. If, for every $x$ there were a $y \geqslant x$ such that $y \in F_i$ for some $i$ then $\langle F_i : i \in \mathbb{N}\rangle$ would be an instance of General-IPHP. Since we are assuming General-IPHP, we know that this would imply that for some $i$, $F_i$ is infinite. We conclude that there is an $x$ such that for all $y$, if $y \in F_i$ for some $i$ then $y < x$.

Finally, we prove (5) → (2) (which implies (1) by the other half of Hirst's theorem). Let $f : \mathbb{N} \to z$ be given and assume that no infinite homogeneous set exists. For each $i$, define $F_i = \{x : f(x) = i\}$ if $i < z$ and $F_i = \emptyset$ if $i \geqslant z$. Then $\langle F_i : i \in \mathbb{N}\rangle$ exists, and is an instance of FUF. By FUF, there is a $b$ such that for all $x$, if $x \in F_i$ for some $i$ then $x < b$. But this is impossible, since $f(b) = i$ for some $i < z$, and hence $b \in R_i$. □

Another example of the occurrence of $\mathsf{B}\Sigma^0_2$ happens when a property of individual numbers is phrased instead as a property of an initial segment of the natural numbers. To illustrate this, recall the problem COH from Definition 4.1.6. As a mathematical principle, this states that for every sequence $\langle R_i : i \in \omega\rangle$ of elements of $2^\omega$, there exists an infinite set $S$ such that

$$(\forall i)(\exists w_i)[(\forall x > w_i)[x \in S \to x \in R_i] \vee (\forall x > w_i)[x \in S \to x \in \overline{R_i}]].$$

Consider the following variant introduced by Hirschfeldt and Shore [152].

**Definition 6.5.5 (Strongly cohesive principle).** StCOH is the following statement: for every sequence $\vec{R} = \langle R_i : i \in \omega\rangle$ of elements of $2^\omega$, there exists an infinite set $S$ such that

$$(\forall z)(\exists w)(\forall i < z)[(\forall x > w)[x \in S \to x \in R_i] \vee (\forall x > w)[x \in S \to x \in \overline{R_i}]].$$

As problems, StCOH and COH are equivalent over $\omega$-models. As $\forall\exists$ theorems, where $z$ may be nonstandard, the equivalence is no longer obvious because $\{w_i : i < z\}$ could be unbounded. Indeed, there is provable difference between the principles.

**Proposition 6.5.6 (Hirschfeldt and Shore [152]).** $\mathsf{RCA}_0 \vdash \mathsf{StCOH} \to \mathsf{B}\Sigma^0_2$.

*Proof.* We argue in $\mathsf{RCA}_0$ + StCOH and derive $\mathsf{RT}^1$. Fix $z \in \mathbb{N}$ and an instance $c : \mathbb{N} \to z$ of $\mathsf{RT}^1$. Define an instance $\langle R_i : i \in \mathbb{N}\rangle$ of StCOH as follows: for all $x$,

$$R_i(x) = \begin{cases} 1 & \text{if } c(x) = i, \\ 0 & \text{otherwise.} \end{cases}$$

By StCOH, fix an infinite set $S$ and a $w \in \mathbb{N}$ such that

$$(\forall i < z)[(\forall x > w)[x \in S \to x \in R_i] \vee (\forall x > w)[x \in S \to x \in \overline{R_i}]].$$

Now say $c(w+1) = i$. Since $i < z$, it follows that $(\forall x > w)[x \in S \to x \in R_i]$. Hence $\{x \in R_i : x > w\}$ is an infinite homogeneous set for $c$. □

By contrast, in Theorem 7.7.1 we will see that COH does not imply $\mathsf{B}\Sigma^0_2$ over $\mathsf{RCA}_0$. In fact, we will prove that COH is $\Pi^1_1$ conservative over $\mathsf{I}\Sigma^0_1$. Thus, what may seem like a very casual maneuver in a proof—replacing $(\forall i)(\exists w)$ by $(\forall z)(\exists w)(\forall i < z)$—takes additional inductive power.

We mention one further example which uses a principle introduced by Hirschfeldt and Shore [152]. If $X$ is an infinite set and $\leqslant_X$ a linear ordering of $X$, we say $(X, \leqslant_X)$ is *of order type* $\omega + \omega^*$ if the following hold.

1. $X$ has a $\leqslant_X$-least element $\ell$ and a $\leqslant_X$-greatest element $g$.
2. For all $x \in X$, exactly one of $\{y \in X : y \leqslant_X x\}$ and $\{y \in X : x \leqslant_X y\}$ is infinite.

We say $(X, \leqslant_X)$ is *strongly of order type* $\omega + \omega^*$ if for every finite set

$$\{\ell = x_0 \leqslant_X \cdots \leqslant_X x_k = g\}$$

there exists exactly one $i < k$ such that $\{y \in X : x_i \leqslant_X y \leqslant_X x_{i+1}\}$ is infinite.

**Definition 6.5.7 (Partition principle).** PART is the following statement: if $\leqslant_X$ is a linear ordering of $\mathbb{N}$ such that $(\mathbb{N}, \leqslant_X)$ is of order type $\omega + \omega^*$, then $(\mathbb{N}, \leqslant_X)$ is strongly of order type $\omega + \omega^*$.

It can be seen that PART follows from $\mathsf{RCA}_0 + \mathsf{B}\Sigma^0_2$. (See Exercise 6.7.4.) What is much harder, and indeed, was an open question for some time, is to see that the converse holds as well.

**Theorem 6.5.8 (Chong, Lempp, and Yang [34]).** $\mathsf{RCA}_0 \vdash \mathsf{PART} \leftrightarrow \mathsf{B}\Sigma^0_2$.

The proof, which would take us too far afield in this book, proceeds via an intermediary step. First, a certain special kind of cut is defined, called a *bi-tame cut*. It is then shown that a model $\mathcal{M}$ satisfies $\mathsf{B}\Sigma^0_2$ if and only if $\mathcal{M}$ has no bi-tame cut. This is a characterization very analogous to that of induction in Theorem 6.2.3. Separately, it is then shown that a model $\mathcal{M}$ satisfies PART if and only if $\mathcal{M}$ has no bi-tame cut.

## 6.6 So, why $\Sigma^0_1$ induction?

Having explored more of the universe of fragments of induction, we may well wonder why $\mathsf{I}\Sigma^0_1$ enjoys a privileged position in our investigation of subsystems of second order arithmetic. Even before our discussion, this question arises quite naturally. After all, on the view that $\mathsf{RCA}_0$ is a formalization of computable mathematics, the restriction to $\Delta^0_1$ comprehension is clear, but the restriction to $\Sigma^0_1$ induction less so.

There are two reasonable positions here. On the one hand, we may object to restricting induction at all. We may well commit ourselves to computable methods in our constructions but retain classical reasoning. Then if a property holds of 0, and of $n + 1$ whenever it holds of $n$, we can accept that it holds of all $n$ regardless of its complexity. At the very least, it may seem reasonable to accept something stronger, like $\mathsf{B}\Sigma^0_2$. On the other hand, we may demand restricting induction even further. If, say, we take a constructivist approach, then concluding that a property holds for all $n$ requires us to actually demonstrate this fact, for each individual $n$, constructively.

In this case, the complexity of the property in question matters very much and we may only find it reasonable to consider $\Delta^0_1$ (or perhaps even lower) properties.

Intuitively, the restriction to $\Sigma^0_1$ induction can be seen as a compromise between these two perspectives—that $\mathsf{RCA}_0$ should be a formal system corresponding to computability theory, and that $\mathsf{RCA}_0$ should be a formal system for the foundation of (a large amount of) mathematics.

As to simplicity, we have already seen in Corollary 5.10.2 that $\mathsf{RCA}_0$ is $\Pi^0_1$ conservative over $\mathsf{PRA}$, which is known not to be case with more induction added. It is certainly satisfying that the first order part of our base theory should line up with such an old and well-known metamathematical system. Another measure of simplicity is via ordinal analysis. The *proof theoretic ordinal* of a theory $T$ of arithmetic is the least ordinal number $\alpha$ that $T$ cannot prove to be well ordered. We recommend Pohlers' survey article [249] for more background. The proof theoretic ordinal of $\mathsf{RCA}_0$ is $\omega^\omega$, which is not overly large, and happens to be the same as that of $\mathsf{PRA}$, $\mathsf{PA}^- + \mathsf{I}\Sigma^0_1$, and $\mathsf{WKL}_0$. The fact that theorems can be proven in these systems with restricted induction means that these theorems do not have exceptionally high consistency strength, which is of independent foundational interest.

In terms of formalizing as much of mathematics as possible, we will see throughout this book that $\Sigma^0_1$ induction suffices more often than not. Admittedly, this chapter, particularly the preceding section, presents $\mathsf{B}\Sigma^0_2$ as indispensable to many arguments. But these still make up only a minority within reverse mathematics. Often, even when a classical proof appears on its face to use more than $\Sigma^0_1$ induction, there are ways to temper the proof to make it go through. (We saw examples in Section 6.4, and will see more later, e.g., in the proof of Theorem 8.5.1.) Of course, once we get at least to the level of $\mathsf{ACA}_0$, we have full arithmetical induction anyway, so the question is no longer relevant.

To be sure, there is a rich theory dedicated to the question of what is and is not provable in systems where $\Sigma^0_1$ induction is weakened further, for example to $\mathsf{I}\Sigma^0_0 + \mathsf{B}\Sigma^0_1 + \mathsf{Exp}$, which we saw above. (The latter, when swapped for $\mathsf{I}\Sigma^0_1$ in $\mathsf{RCA}_0$, is sometimes denoted $\mathsf{RCA}^*_0$.) For the purposes of studying "ordinary" mathematical theorems, these systems are quite limited. In Section 6.4, we saw several standard computability theoretic theorems that are equivalent to $\mathsf{I}\Sigma^0_1$ over $\mathsf{PA}^- + \mathsf{I}\Sigma^0_0 + \mathsf{B}\Sigma^0_1 + \mathsf{Exp}$. An even more basic such example is the result that every polynomial over a countable field has only finitely many roots. But for us, perhaps the most convincing along these lines is the following result on definitions by recursion.

**Theorem 6.6.1 (Hirschfeldt and Shore [152]).** *The following are equivalent over* $\mathsf{PA}^- + \mathsf{I}\Sigma^0_0 + \mathsf{B}\Sigma^0_1$.

1. $\mathsf{I}\Sigma^0_1$.
2. *For every $z$, every $m$, and every $\Sigma^0_0$ definable function $f\colon \mathbb{N} \to \mathbb{N}$, there is a function $g\colon m+1 \to \mathbb{N}$ such that $g(0) = z$ and for all $x < m$, $g(x+1) = f(g(x))$.*

Thus $\mathsf{I}\Sigma^0_1$ is needed to carry out even *finite* recursion (using easy-to-define functions). Thus, this level of induction is quite important to maintaining a faithful connection between $\mathsf{RCA}_0$ and computable mathematics. In this sense, then, we see that $\Sigma^0_1$ induction is a kind of sweet spot, neither too weak nor too strong.

One final aspect of the choice of $\Sigma^0_1$ induction is historical. As discussed above, on the whole there are relatively few "ordinary" theorems that are not provable in $\mathsf{RCA}_0$ but are provable with additional induction. This is especially true for principles from analysis and algebra, which were the early focus of the subject. Indeed, the first major counterexample came from a different field—combinatorics, with Hirst's theorem (Theorem 6.5.1). This theorem dates to 1987, by which point in time, of course, the theory $\mathsf{RCA}_0$ was already well established.

## 6.7 Exercises

**Exercise 6.7.1.** Show that the following is provable in $\mathsf{RCA}_0$. If $T \subseteq 2^{<\mathbb{N}}$ is an infinite binary tree, then for each $\ell \in \mathbb{N}$ there exists an extendible $\sigma \in T$ with $|\sigma| = \ell$.

**Exercise 6.7.2.** Let $\Gamma$ be a collection of formulas of $\mathcal{L}_1$ or $\mathcal{L}_2$. The $\Gamma$ *strong induction scheme* ($\mathsf{SI}\Gamma$) is the scheme consisting of all sentences of the form

$$(\forall x)[(\forall y < x)\varphi(y) \to \varphi(x)] \to (\forall x)\varphi(x).$$

for $\varphi(x) \in \Gamma$. Prove that for each $n \geqslant 1$, $\mathsf{RCA}_0 \vdash \mathsf{I}\Sigma^0_n \leftrightarrow \mathsf{SI}\Sigma^0_n \leftrightarrow \mathsf{SI}\Pi^0_n$.

**Exercise 6.7.3.** Show that the following is provable in $\mathsf{RCA}_0$. For every function $f\colon \mathbb{N} \to \mathbb{N}$ and every $w$, there exists a $z$ such that

$$(\forall y < w)[y \in \operatorname{range}(f) \leftrightarrow y \in \operatorname{range}(f \upharpoonright z)].$$

**Exercise 6.7.4.** Show that $\mathsf{RCA}_0 \vdash \mathsf{B}\Sigma^0_2 \to \mathsf{PART}$.

**Exercise 6.7.5 (Corduan, Groszek, and Mileti [56]).** Fix $n > 1$ and suppose $\mathcal{M} \vDash \mathsf{RCA}_0 + \mathsf{B}\Sigma^0_{n-1} + \neg\mathsf{I}\Sigma^0_n$. Let $I \subseteq M$ be a $\Sigma^0_n$-definable proper cut of $\mathcal{M}$. Say $\varphi$ is a $\Sigma^0_{n-2}$ formula such that $a \in I \leftrightarrow \mathcal{M} \vDash (\exists w)(\forall z)\varphi(a,w,z)$ for all $a \in M$. Fix $b \in M \setminus I$ and define $g\colon (M \upharpoonright b) \times M \to M$ by letting $g(a,s)$ be either the $\leqslant^{\mathcal{M}}$-least $w <^{\mathcal{M}} s$ such that $\mathcal{M} \vDash (\forall z < s)\varphi(a,w,z)$, or $s$ if no such $w$ exists.

1. Show that $g$ is $\Delta^0_{n-1}$-definable in $\mathcal{M}$.
2. Show that $\mathcal{M} \vDash g$ is a total function on its domain.
3. Show that if $a \in I$ then $\lim_s g(a,s)$ exists.
4. Show that $a \mapsto \lim_s g(a,s)$ is an unbounded function on $I$.

**Exercise 6.7.6 (Dzhafarov, Hirschfeldt, and Reitzes [82]).** Prove Proposition 6.5.3.

**Exercise 6.7.7.** Prove there is a countable model of $\mathsf{PA}$ with an element that is not $\Sigma^0_n$ definable for any $n \in \omega$. Hence $\bigcup_n K_n(M)$ is a proper subset of $M$. (Much stronger results are possible; see Murawski [228].)

# Chapter 7
# Forcing

Forcing is a profoundly powerful piece of machinery in mathematical logic. It was invented by Cohen in the 1960s as the miraculous ingredient that allowed him to prove that the axiom of choice and the continuum hypothesis are both independent of ZF ([47, 48, 49]). Subsequently, forcing became a household tool in set theory, but not only there.

Shortly after Cohen's breakthroughs, Feferman [102] developed a version for use in arithmetic, which he used to construct a number of examples of exotic subsets of $\omega$, such as a finite tuple of sets each nonarithmetical in the join of the others (see Exercise 7.8.1). This paved the door for the widespread use of forcing in proof theory, model theory, and computability theory. In fact, the rudiments of forcing in computability theory have an even longer history, predating the work of Cohen by a decade. The basic ideas are already present in the finite-extension method, which was developed in the 1950s by Kleene and Post.

Quintessentially, forcing is a framework for constructing objects with prescribed combinatorial properties. The systematic study of the effective properties of these objects was initiated by Jockusch [169] and continues to this day. Unsurprisingly, the technique has immense applications to reverse mathematics, as we explore in this chapter.

## 7.1 A motivating example

The basic premise of forcing is to construct an object $G$ by a sequence of approximations of some kind, called *conditions*, each of which ensures, or *forces*,a particular property about the $G$ obtained in the end. In computability theory and arithmetic, the objects we look to construct are usually subsets of $\omega$, and the properties are usually arithmetical.

A canonical example of this is the construction of a set by a sequence of longer and longer binary strings, the so-called *finite extension method* in computability theory. Here, our conditions are finite strings $\sigma \in 2^{<\omega}$, regarded as initial segments



D. D. Dzhafarov, C. Mummert, *Reverse Mathematics*, Theory and Applications of Computability, https://doi.org/10.1007/978-3-031-11367-3_7

of the characteristic function of the eventual set $G$ we want. We build a sequence $\sigma_0 \preceq \sigma_1 \preceq \cdots$ of conditions and then set $G = \bigcup_i \sigma_i$. During the construction we usually choose $\sigma_i$ carefully to satisfy various properties we want $G$ to have. For example, if we want $G$ to be noncomputable, then we might proceed as follows. We set $\sigma_0 = \langle\rangle$, and having defined $\sigma_e$ for $e \in \omega$, we let $\sigma_{e+1} = \sigma_e {}^\frown (1 - \Phi_e(e))$ if $\Phi_e(e) \downarrow\in \{0, 1\}$, and otherwise we let $\sigma_{e+1} = \sigma_e 0$. In this way, even though we have only determined a small initial segment of $G$ so far, $\sigma_{e+1}$, we already know that $G$ (whatever it will eventually be) will satisfy that $G \neq \Phi_e$. It is in this sense that we can say that $\sigma_{e+1}$ *forces* $G \neq \Phi_e$.

That this is possible is not really surprising. After all, to ensure $G$ differs $\Phi_e$ requires only one bit of information, which is present in a finite initial segment. But the power of forcing comes from the ability to use conditions to similarly force more general properties, including infinitary ones that seemingly do require knowing all of $G$. This is far less obvious.

To illustrate this, let us stay with the example of constructing a noncomputable set $G$ by initial segments. Critical in the finite extension argument is that, for each $e$, we can always find a string forcing $G \neq \Phi_e$, no matter which initial segment of $G$ we may have determined up to that point. For this reason, we say the collection of strings forcing $G \neq \Phi_e$ is *dense*: at every stage of the construction, there is always an opportunity to *meet* this collection (i.e., to extend to some string in it).

We presented the construction in a rather structured way, as is typical in computability arguments: we ensured $\sigma_{e+1}$ forced $\Phi_e \neq G$ for each $e$. We did not actually need to do this: we would have been content for any $\sigma_i$ to force $\Phi_e \neq G$, not necessarily $i = e + 1$. The only thing we actually need to ensure is that the eventual $G$ does not deliberately *avoid* all of the opportunities to force $\Phi_e \neq G$ for each fixed $e$. We call a construction of this kind *generic*, meaning roughly that any property that it is *always* possible to satisfy *will* eventually be satisfied. If one considers the negation of this statement, it is perhaps more intuitively clear that a construction that is not generic is special or "atypical" in a certain sense (see also Figure 7.1). We will make this intuition precise in this chapter.

Conditions do not need to be finite strings, which are "approximations" in perhaps the most obvious sense. A different example is the construction of a path through an infinite computable tree. We have already seen this in action, in the various basis theorems we proved in Section 2.8. The setup there was always the same: given a computable tree $T$, we obtain $G \in [T]$ through a nested sequence of infinite computable subtrees of $T$. Each such subtree "approximates" more of $G$, in the sense that it determines a longer initial segment of $G$, and whittles down the space of possible extensions of this initial segment (i.e., the space of possible elements $G$ of $[T]$ that we could eventually produce). In the parlance of the preceding paragraph, then, the conditions were the computable subtrees of $T$. If we consider, for example, the cone avoidance basis theorem (Theorem 2.8.23), where we were given a noncomputable $C$ and wanted to ensure that $C \not\leq_T G$, we see some of the same issues at work as above. In particular, although we indexed our conditions (subtrees) $T = T_0 \supseteq T_1 \supseteq \cdots$ in such a way that $T_{e+1}$ forced $C \neq \Phi_e^G$, we again only cared that the latter *does get*

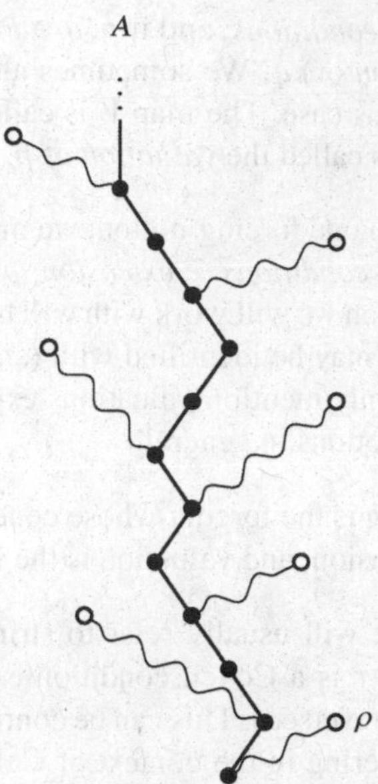


**Figure 7.1.** An illustration of a non-generic set $A \in 2^\omega$, represented by the thick line as a path in $2^{<\omega}$. The solid dots represent initial segments of $A$; the hollow dots, strings $\rho$ forcing a certain property not satisfied by $A$. Thus, $A$ appears to be avoiding all opportunities to force this property, a very specific (hence, "non-generic") behavior.

*forced*, and not at all that it is forced by $T_{e+1}$ or any other specific condition in the construction. So here, too, any $G \in [T]$ obtained *generically* would have worked.

With some intuition in hand, let us pass to a more formal discussion. Throughout this section, it will be helpful to keep in mind the picture of a $G \in \omega^\omega$ being constructed by a sequence of approximations of some kind, as in the previous two examples.

## 7.2 Notions of forcing

We begin by formalizing what, precisely, we mean by "approximation". There are different ways to go about this. Most treatments agree on the intended examples and applications, but possibly differ in certain edge cases. Our approach here is loosely based on one of Shore [283].

**Definition 7.2.1 (Notions of forcing).** A *notion of forcing* (or just *forcing*) is a triple $\mathbb{P} = (P, \leqslant, V)$ where $(P, \leqslant)$ is a partial order and $V$ is a map $P \to 2^{<\omega}$ satisfying the following.

1. If $p, q \in P$ with $p \leqslant q$ then $V(p) \succeq V(q)$.
2. For each $n \in \omega$ and $q \in P$ there is a $p \in P$ with $p \leqslant q$ and $|V(p)| \geqslant n$.

The elements of $P$ are called *conditions*; and if $p, q \in P$ satisfy $p \leqslant q$ then $p$ is said to *extend* (and be an *extension* of) $q$. We sometimes also say $p$ is *stronger* than $q$, and $q$ is *weaker* than $p$, in this case. The map $V$ is called a *valuation map*, and for each $p \in P$, the value $V(p)$ is called the *valuation of* $p$.

When dealing with more than one forcing notion, we may decorate the above terms by $\mathbb{P}$ for clarity, referring to $\mathbb{P}$ *conditions*, $\mathbb{P}$ *extension*, $p \leqslant_{\mathbb{P}} q$, $\mathbb{P}$ *valuation map*, etc.

Every specific forcing notion we will work with will have a natural valuation map, and in these cases the notion may be identified with (and by) the underlying partial order. We will thus usually only mention valuations explicitly in definitions and in proving facts about forcing notions in general.

*Example 7.2.2. Cohen forcing* is the forcing whose conditions are finite strings $\sigma \in 2^{<\omega}$, extension is string extension, and valuation is the identity.

In the context of forcing, we will usually refer to strings as *Cohen conditions*. It is worth emphasizing that if $\tau$ is a Cohen condition extending $\sigma$ then $\tau \leqslant \sigma$ but $\sigma \preceq \tau$ (i.e., the orderings are reversed). This can be confusing, and for this reason we usually stick to the latter ordering in the context of Cohen forcing. This highlights a good question, which is why extension is defined to be $\leqslant$ instead of $\geqslant$ anyway. The reason is that extension is meant to represent that if $q \leqslant p$, then $q$ approximates more of $G$ than $p$, and hence there are fewer possible ways to "complete" $q$ to $G$.

We can make this explicit as follows.

**Definition 7.2.3 (Constructible objects).** Let $\mathbb{P} = (P, \leqslant, V)$ be a notion of forcing. An element $f \in 2^\omega$ is/can be *constructed by* $\mathbb{P}$ if there is a sequence of conditions $p_0 \geqslant p_1 \geqslant \cdots$ such that $\lim_{i\to\infty} |V(p_i)| = \infty$ and $f = \bigcup_{i\in\omega} V(p_i)$. In this case we also say $f$ is *constructed via the sequence* $p_0 \geqslant p_1 \geqslant \cdots$.

We refer to $f$ above as an *object* in general, because depending on the forcing notion we may be viewing it as (the characteristic function) of an object other than just a subset of $\omega$ (see Example 7.2.6). When this is not the case, we will refer to $f$ as a *set* or *real*, as usual.

Every element of $2^\omega$ can be constructed by Cohen forcing. We may well wonder, then, why we need forcing at all and cannot simply stick to binary strings and finite extension arguments. The answer is that different notions of forcing help us better control various aspects of the sets they can construct. To illustrate this, we first give examples of two important forcing notions.

*Example 7.2.4 (Jockusch and Soare [173]).* Let $T$ be an infinite binary tree. *Jockusch–Soare forcing with subtrees of* $T$ is the following notion of forcing.

- The conditions (called *Jockusch–Soare conditions*) are pairs $(U, m)$ where $m \in \omega$ and $U$ is an infinite subtree of $T$ having a unique node of length $m$.
- Extension is defined by $(U^*, m^*) \leqslant (U, m)$ if $U^* \subseteq U$ and $m^* \geqslant m$.
- The valuation map $V$ takes a condition $(U, m)$ to the unique $\sigma \in U$ of length $m$.

Notice that $(T, 0)$ is always a condition, and in fact, it is the $\leqslant$-greatest (i.e., weakest) condition. Likewise if $f$ is any path through $T$, then $(\{\sigma \in 2^{<\omega} : \sigma \prec f\}, n)$ is a condition for every $n$. The objects that can be constructed by Jockusch–Soare forcing with subtrees of $T$ are thus precisely the paths through $T$.

*Example 7.2.5 (Mathias [206]).* Let $X$ be an infinite set. *Mathias forcing within X* is the following notion of forcing.

- The conditions (called *Mathias conditions*) are pairs $(E, I)$, where $E$ is a finite set, $I$ is an infinite subset of $X$, and $E < I$.
- Extension is defined by $(E^*, I^*) \leqslant (E, I)$ if $E \subseteq E^* \subseteq E \cup I$ and $I^* \subseteq I$.
- The valuation map $V$ takes $(E, I)$ to the string $\sigma \in 2^{<\omega}$ of length $\min I$ with $\sigma(n) = 1$ if and only if $n \in E$.

The sets $I$ are called *reservoirs*, and if $(E^*, I^*) \leqslant (E, I)$ and $I \setminus I^*$ is finite then $(E^*, I^*)$ is called a *finite extension* of $(E, I)$.

Here, $(\emptyset, X)$ is the $\leqslant$-greatest condition. The objects that can be constructed by Mathias forcing within $X$ are precisely the subsets of $X$.

In both examples above, when the given tree $T$ or infinite set $X$ is not relevant to a discussion, we may refer simply to *Jockusch–Soare forcing* or *Mathias forcing*. So a *Jockusch–Soare condition* is a condition of *Jockusch–Soare forcing with subtrees of T* for *some* $T$, etc. We include one more illustrative example.

*Example 7.2.6.* Let $\mathbb{P}$ be the notion of forcing whose conditions are functions $\alpha : n \to \omega$ for some $n \in \omega$, with $\alpha^* : n^* \to \omega$ extending $\alpha$ if $n^* \geqslant n$ and $\alpha^*$ extends $\alpha$ as a finite function, and with the valuation of $\alpha : n \to \omega$ being the shortest string $\sigma \in 2^{<\omega}$ such that $\langle x, \alpha(x) \rangle < |\sigma|$ for all $x < |\alpha|$ and $\sigma(y) = 1$ if and only if $y = \langle x, \alpha(x) \rangle$ for some such $x$. The objects that can be constructed by $\mathbb{P}$ are the (characteristic functions of) all functions $f : \omega \to \omega$.

By contrast, not *every* object (set) constructed by Cohen forcing (restricted to $2^{<\omega}$) needs to be a path through a given tree $T$, a subset of a given set $X$, or (the characteristic function) of a function $\omega \to \omega$. More importantly, no single Cohen condition can ensure any of these properties either. For example, outside of trivial cases (e.g., when $T$ has paths extending every finite binary string, or $X =^* \omega$), *every* Cohen condition $\sigma \in 2^{<\omega}$ will have an extension $\tau$ such that no $G \succeq \tau$ is a path through $T$ or a subset of $X$. And, quite possibly, we may need to extend to such a $\tau$ for the sake of satisfying some other property we want $G$ to have.

In the next section, we will see that this situation, of every condition having an extension of a particular kind and our needing to pass to such an extension, is actually crucial in forcing constructions in general. So the issue here is not just an inconvenience.

In the sequel, we will look not just at sequences of conditions but at *filters* of conditions, which are sets of conditions that are consistent in a certain sense.

**Definition 7.2.7 (Filters).** Let $\mathbb{P} = (P, \leqslant, V)$ be a notion of forcing.

1. A *filter* is a subset $F$ of $P$ satisfying the following properties.
   - *(Upward closure):* If $q \in F$ and $q \leqslant p$ then $p \in F$.
   - *(Consistency):* If $p, q \in F$ there exists $r \in F$ such that $r \leqslant p$ and $r \leqslant q$.

   We write $V(F)$ for $\bigcup_{p \in F} V(p)$. If $V(F) \in 2^{\omega}$, we call it the *object* (or *set* or *real*) *determined by* $F$.
2. If $p_0 \geqslant p_1 \geqslant \cdots$ is a sequence of conditions then

$$\{q \in P : (\exists i)[q \geqslant p_i]\}$$

   is a filter, called the filter *generated* by the sequence of the $p_i$.

We will also find the following definition useful.

**Definition 7.2.8.** Let $\mathbb{P} = (P, \leqslant, V)$ be a notion of forcing and $\mathcal{K}$ a property defined for filters.

1. A sequence of conditions $p_0 \geqslant p_1 \geqslant \cdots$ *has the property* $\mathcal{K}$ if the filter generated by this sequence has the property $\mathcal{K}$.
2. $G \in \omega^{\omega}$ *has the property* $\mathcal{K}$ if $G = V(F)$ for some filter $F$ with property $\mathcal{K}$.

Notice that a filter $F$ determines an object if and only if it contains conditions with arbitrarily long valuations. In this case, it is easy to see that $F$ contains a sequence of conditions $p_0 \geqslant p_1 \geqslant \cdots$ with $V(F) = \bigcup_{i \in \omega} V(p_i)$. So, in particular, the object determined by a filter $F$ can be constructed by $\mathbb{P}$. Conversely, if $G$ can be constructed via a sequence of conditions $p_0 \geqslant p_1 \geqslant \cdots$ then $G = V(F)$ for the filter $F$ generated by this sequence. Thus, for the purposes of constructing elements of $\omega^{\omega}$, there is no difference between working with filters or sequences. Filters are more important in set theory, where the forcings tend to be more complicated and (countable) sequences of the kind we are considering do not always suffice. Filters were carried over to computability theory from set theory, but their continued use is mostly a matter of convenience, as it makes stating certain results easier.

*Remark 7.2.9.* It is important to be aware that just because some condition $p \in P$ satisfies $V(p) \preceq G$, this does not mean that $G$ can be constructed via a sequence that contains $p$ (or equivalently, that $p$ is part of a filter that determines $G$). For example, suppose we are dealing with Mathias forcing, and that $p$ is the condition $(\emptyset, 2^{\omega} \setminus \{G\})$. One obvious exception to this is Cohen forcing, where every $G$ can be constructed via the sequence of all its initial segments. There, it is interchangeable to speak of $\sigma$ being an initial segment of $G$ and $\sigma$ being part of a sequence via which $G$ can be constructed. But in general, it is only the latter that properly captures the notion of a condition $p$ being an "initial segment" of $G$. In Mathias forcing, $p$ is part of a sequence via which $G$ can be constructed if and only if $p$ has the form $(E, I)$ where $E \subseteq G \subseteq E \cup I$. In Jockusch–Soare forcing, it is the case if and only if $p$ has the form $(U, m)$ where $G$ is a path through $U$.

We conclude this section by looking at several more examples of forcing notions and the sets they can construct. One common practice is to restrict the domain of a notion of forcing.

*Example 7.2.10.* Fix a set $X$. The following are important variants of Mathias forcing within $X$.

1. If $X$ is computable, then *Mathias forcing within $X$ with computable reservoirs* is the restriction to conditions $(E, I)$ with $I$ computable.
2. If $C \not\leq_T \emptyset$, then *Mathias forcing within $X$ with $C$-cone avoiding reservoirs* is the restriction to conditions $(E, I)$ with $C \not\leq_T I$.
3. *Mathias forcing within $X$ with PA avoiding reservoirs* is the restriction to conditions $(E, I)$ such that $I \not\gg \emptyset$.

In all cases, extension and valuation is as before, within the restricted set of conditions. We can also relativize these notions to any $A \in 2^\omega$, obtaining, e.g., *Mathias forcing within $X$ with $A$-computable reservoirs*, or *Mathias forcing within $X$ with conditions $(E, I)$ such that $C \not\leq_T A \oplus I$.* (Note that in the latter example we do not say "$C$-cone avoiding reservoirs relative to $A$" or anything too cumbersome.)

*Example 7.2.11.* Fix an infinite computable tree $T \subseteq 2^{<\omega}$. *Jockusch–Soare forcing with computable subtrees of $T$* is the restriction to conditions $(U, m)$ where $U$ is computable, with extension and valuation as before. Since the set of paths through an infinite computable tree forms a $\Pi^0_1$ class, this is also sometimes called *forcing with $\Pi^0_1$ classes* (in this case, *$\Pi^0_1$ subclasses of $[T]$*).

A common application of this forcing is to trees $T$ with no computable paths. For every infinite computable subtree $U$ of such a $T$, there is necessarily a maximal $m$ such that $U$ contains a unique node of length $m$, and this $m$ can be found computably from (an index for) $U$. In this case, then, it is customary to specify a condition $(U, m)$ simply by $U$.

We can also obtain new forcing notions from old ones by restricting the extension relation rather than the set of conditions.

*Example 7.2.12.* Consider the forcing notion from Example 7.2.6. Modify the extension relation as follows: $\alpha^* : n^* \to \omega$ extends $\alpha : n \to \omega$ only if $n^* \geqslant n$ and for every $x < |\alpha^*|$ there is a $y < |\alpha|$ such that $\alpha^*(x) \leqslant \alpha(y)$. Then the objects that can be constructed by this forcing are precisely the (characteristic functions of) functions $\omega \to \omega$ with bounded range, i.e., the instances of $\mathsf{IPHP}$ or $\mathsf{RT}^1$.

Another useful method combines two or more forcing notions into one in parallel.

**Definition 7.2.13 (Product forcing).** Let $I \subseteq \omega$ be finite or infinite, and for each $i \in I$, fix a forcing notion $\mathbb{P}_i = (P_i, \leqslant_i, V_i)$. The *product forcing* $\prod_{i \in I} \mathbb{P}_i$ is the notion of forcing defined as follows.

1. The conditions are all tuples $\langle p_i : i \in F \rangle \in \prod_{i \in F} P_i$, where $F$ is a finite initial segment of $I$ (i.e., $F \subseteq I$, and if $i \in F$ and $j \in I$ with $j < i$ then $j \in F$).

2. A condition $\langle p_i^* : i \in F^* \rangle$ extends a condition $\langle p_i : i \in F \rangle$ if $F \subseteq F^*$ and $p_i^* \leqslant_i p_i$ for all $i \in F$.
3. The valuation takes a condition $\langle p_i : i \in F \rangle$ to $\langle V_i(p_i) : i \in F \rangle$, i.e., to the string $\sigma \in 2^{<\omega}$ of length $|F| \cdot m$, where $m = \min\{|V_i(p_i)| : i \in F\}$, and $\sigma(\langle i, x \rangle) = V_i(p_i)(x)$ for all $i \in F$ and all $x < m$.

If $\mathbb{P}_i = \mathbb{P}$ for all $i$, we write $\prod_{i \in I} \mathbb{P}$ for $\prod_{i \in I} \mathbb{P}_i$. If $I$ is finite, the definition is sometimes modified so that $F$ above is always just $I$.

For example, if $\mathbb{C}$ denotes Cohen forcing we can consider the product forcing $\prod_{i \in \omega} \mathbb{C}$. It is easy to see that the sets that can be constructed by $\prod_{i \in \omega} \mathbb{C}$ are all sequences $\langle X_i : i \in \omega \rangle$ of elements $X_i \in 2^{\omega}$. Equivalently, we can view these as instances of $\widehat{\mathsf{RT}^1_2}$ or of $\mathsf{COH}$, if we wish. By contrast, if $\mathbb{P}$ is the forcing from Example 7.2.12 then we can think of the objects that can be constructed by $\prod_{i \in \omega} \mathbb{P}$ as being the instances of $\widehat{\mathsf{IPHP}}$.

## 7.3 Density and genericity

Finding the right partial order and valuation map is the first step in a forcing construction. The next step is to actually force the desired object to have the properties we want by carefully selecting conditions to add to a sequence via which the object is constructed. We cannot do this arbitrarily. For example, if we wish to add a condition from some set $S$, we must not have previously added a condition $p$ that has no extension in $S$! This would seem to require knowing at each "stage" the totality of the sets we will wish to add from in the future, a daunting prospect. We avoid this complication by choosing conditions from *dense* sets.

**Definition 7.3.1 (Density).** Let $\mathbb{P} = (P, \leqslant, V)$ be a notion of forcing.

1. A set $D \subseteq P$ is *dense* if for every $p \in P$ there is a $q \in D$ such that $q \leqslant p$ (i.e., every condition has an extension in $D$).
2. Given $p_0 \in P$, a set $D \subseteq P$ is *dense below* $p_0$ if for every $p \in P$ with $p \leqslant p_0$ there is a $q \in D$ such that $q \leqslant p$.
3. A set $F \subseteq P$ *meets* a set $S \subseteq P$ if $F \cap S \neq \emptyset$; $F$ *avoids* $S$ if there is a $p \in F$ so that no $q \leqslant p$ belongs to $S$.

Thus, in the above parlance, no dense set of conditions can be avoided by any set of conditions, and in particular, if we wish we can always add a condition to our sequence from any given dense set.

We now come to the main definition through which the construction of an object by forcing proceeds.

**Definition 7.3.2 (Genericity).** Let $\mathbb{P} = (P, \leqslant, V)$ be a notion of forcing. Let $\mathcal{D}$ be a collection of dense subsets of $P$. A filter $F$ is $\mathcal{D}$-*generic* (or *generic with respect to* $\mathcal{D}$) if it meets every $D \in \mathcal{D}$.

When $\mathcal{D}$ is understood or implied, we may call a filter or element of $\omega^\omega$ simply *generic* instead of $\mathcal{D}$-generic. When we need to specify $\mathbb{P}$, we may also say a filter or element of $\omega^\omega$ is *$\mathcal{D}$-generic for* $\mathbb{P}$. For some common forcings, like Cohen forcing or Mathias forcing, we may also speak of *Cohen generics* or *Mathias generics*, etc. Using Definition 7.2.8, we obtain also a notion of a descending sequence of conditions being $\mathcal{D}$-generic, as well as of an element of $\omega^\omega$ being $\mathcal{D}$-generic, and we use the same conventions in speaking about these.

The definition of a sequence $p_0 \geqslant p_1 \geqslant \cdots$ being $\mathcal{D}$-generic is weaker than saying that it actually meets (as a set of conditions) every $D \in \mathcal{D}$. This is only equivalent provided each $D$ is closed downward under $\leqslant$, i.e., if $p \in D$ and $q \leqslant p$ implies $q \in D$. This will be the case in virtually all cases of interest to us here, but we maintain the distinction for generality.

The existence of generic filters follows from a result originally proved by Rasiowa and Sikorski [254]. The proof of the version here is quite simple, but it is arguably the most important general result in the theory of forcing.

**Theorem 7.3.3 (Rasiowa and Sikorski [254]).** *Let $\mathbb{P} = (P, \leqslant, V)$ be a notion of forcing, $p$ a condition, and $\mathcal{D}$ a countable family of dense subsets of $P$. There exists a $\mathcal{D}$-generic filter containing $p$.*

*Proof.* Fix $p \in P$. As $\mathcal{D}$ is countable, we may enumerate its members as $D_0, D_1, \ldots$. Let $p_0 = p$, and given $p_i \in P$ for some $i \in \omega$, let $p_{i+1}$ be any extension of $p_i$ in $D_i$, which must exist since $D_i$ is dense. The filter generated by the $p_i$ is $\mathcal{D}$-generic and contains $p$. □

Multiple countable collections of dense sets can be combined, and the Rasiowa–Sikorski theorem can be applied to produce a filter generic with respect to them all. This means that forcing constructions are often modular, divided into parts, each of which specifies a particular collection of dense sets of conditions. A generic filter is then chosen at the end.

*Convention 7.3.4 (Filters determine objects).* By definition, in any forcing notion $(P, \leqslant, V)$ the set of conditions $p$ with $|V(p)| \geqslant n$ is dense for every $n$. We adopt the convention that all generic filters are assumed to be generic also for this collection of dense sets. Therefore, every generic filter $F$ determines a generic object, i.e., $V(F) = \bigcup_{p \in F} V(p)$ is actually an element of $\omega^\omega$.

More generally, we use the adjective *sufficiently generic* to refer to filters generic with respect to all collections of dense sets we have specified explicitly, as well as any that, perhaps, show up frequently enough in a particular discussion for us to take as implied.

**Definition 7.3.5 (Sufficient genericity).** Let $\mathbb{P}$ be a notion of forcing and $\mathcal{K}$ a class of filters. We say *every sufficiently generic filter belongs to* $\mathcal{K}$ if there is a countable collection of dense sets $\mathcal{D}$ so that every $\mathcal{D}$-generic filter belongs to $\mathcal{K}$.

Again, we obtain also a notion of sufficient genericity for descending sequences of conditions and elements of $\omega^\omega$. For example, our initial discussion in Section 7.1

makes it evident that every sufficiently generic set for Cohen forcing is noncomputable. Let us consider some further examples.

*Example 7.3.6.* Recall that a set $Y \in 2^\omega$ is *high* if $Y' \geqslant_T 0''$, or equivalently, if there is a $Y$-computable function that dominates every computable function. Fix any set $X$, and consider Mathias forcing within $X$. For each $e \in \omega$, the collection of conditions $(E, I)$ with the following property is dense: if $\Phi_e$ is a total and increasing function then $p_{E \cup I}(x) \geqslant \Phi_e(x)$ for all $x > |E|$. Indeed, suppose $\Phi_e$ is a total and increasing and $(E, I)$ is any condition, say with $I = \{x_0 < x_1 < \cdots\}$. Let $I^* = \{x_{\Phi_e(i)} : i > |E|\}$. Then $(E, I^*)$ is an extension of $(E, I)$ of the desired kind. Thus, if $G$ is a sufficiently generic set then $p_G$ will dominate every computable function, and therefore $G$ will be high. (Formally, we could let $D_e$ be the set of all conditions $(E, I)$ as above, and then formulate our claim for all $\{D_e : e \in \omega\}$-generic sets.)

*Example 7.3.7.* Let $C$ be a noncomputable set. Consider Mathias forcing with $C$-cone avoiding reservoirs. For each $e \in \omega$, the collection of conditions $(E, I)$ with the following property is dense: there is an $x \in \omega$ such that either $\Phi_e^E(x) \downarrow \neq C(x)$ or $\Phi_e^{E \cup F}(x) \uparrow$ for all finite $F \subseteq I$. (Indeed, suppose $(E, I)$ has no extension with this property. By the failure of the second possibility, for each $x$ there is a finite $F \subseteq I$ such that $\Phi_e^{E \cup F}(x) \downarrow$, and by the failure of the first possibility, the value of this computation must be $C(x)$. Thus, $I$ can compute $C$, contrary to the fact that $(E, I)$ is $C$-cone avoiding.) If $G$ is a sufficiently generic set, it follows that $G \not\geqslant_T C$.

*Example 7.3.8.* Consider Mathias forcing (within $\omega$) with PA avoiding reservoirs $I$. Then for each $e \in \omega$, the collection of conditions $(E, I)$ satisfying the following property is dense: there is an $x$ such that either $\Phi_e^E(x) \downarrow = \Phi_x(x) \downarrow$ or $\Phi_e^{E \cup F}(x) \downarrow = y \rightarrow y \notin \{0, 1\}$ for all finite $F \subseteq I$. (If not, there would be a condition $(E, I)$ such that $I$ computes a 2-valued DNC function, which would mean $I \gg \emptyset$.) If $G$ is a sufficiently generic set, it follows that $G$ computes no 2-valued DNC function, and hence that $G \not\gg \emptyset$.

It is easy to check that, in each of the above examples, the extension $(E^*, I^*)$ of a given condition $(E, I)$ in the relevant dense set satisfies $I^* \leqslant_T I$. Hence, the constructions can be combined. For example, if we force with $C$-cone avoiding conditions, we can combine Examples 7.3.6 and 7.3.7 to obtain a generic $G$ that is both high and does not compute $C$.

We can also obtain computability theoretic facts about various instance-solution problems, as in Chapter 4.

*Example 7.3.9.* Take any instance of the problem $\mathsf{COH}$ defined in Definition 4.1.6. This is a sequence $\vec{R} = \langle R_i : i \in \omega \rangle$ of sets. Fix any set $X$, and consider Mathias forcing within $X$. For each $i \in \omega$, the set of conditions $(E, I)$ satisfying the following property is dense: either $I \subseteq R_i$ or $I \subseteq \overline{R_i}$. Likewise, for each $i$, the set of condition $(E, I)$ with $|E| > i$ is dense. It follows that any sufficiently generic set $G$ is $\vec{R}$-cohesive and infinite, i.e., a $\mathsf{COH}$-solution to $\vec{R}$.

In this proof, the extension $(E^*, I^*)$ of a given condition $(E, I)$ with $I \subseteq R_i$ or $I \subseteq \overline{R_i}$ satisfies $I^* \leqslant_T \vec{R} \oplus I$. Suppose $\vec{R} \not\gg \emptyset$ and consider Mathias forcing with reservoirs

$I$ such that $\vec{R} \oplus I \not\gg \emptyset$. We can then relativize Example 7.3.8 to $\vec{R}$, and combine it with Example 7.3.9 to obtain a set $G$ such that $G$ is $\vec{R}$-cohesive and $\vec{R} \oplus G \not\gg \emptyset$. Thus we have a slicker (and more flexible) proof of Proposition 3.8.5 that COH admits PA avoidance.

We conclude this section by justifying the intuition, alluded to in Section 7.1, that generic sets are "typical" in a certain sense. That sense is topological. (This is not a result we will need in the rest of our discussion, but it is of independent interest.)

**Definition 7.3.10.** Let $\mathbb{P} = (P, \leqslant, V)$ be a notion of forcing.

1. For $p \in P$, $[[p]]$ is the set of all $f \in \omega^\omega$ that can be constructed by $\mathbb{P}$ via a sequence containing $p$.
2. $\mathcal{T}_\mathbb{P} \subseteq \omega^\omega$ is the space of all sets that can be constructed by $\mathbb{P}$, with topology generated by basic open sets of the form $[[p]]$ for $p \in P$.

Note that if $\mathbb{C}$ denotes Cohen forcing then $\mathcal{T}_\mathbb{C}$ is just $2^\omega$ with the usual Cantor space topology, with $[[\sigma]]$ for $\sigma \in 2^{<\omega}$ referring to the usual basic open set.

If $q \leqslant p$ then $[[q]] \subseteq [[p]]$. This is because any sequence containing $q$ and used to construct some $f \in [[q]]$ can be truncated above $q$, and $q$ replaced by $p$, to produce another sequence constructing $f$. From this, it is readily observed that if $D \subseteq P$ is dense (in the sense of forcing) then $\bigcup_{p \in D}[[p]] \subseteq \mathcal{T}_\mathbb{P}$ is dense open (in the sense of the $\mathbb{P}$ topology).

We will need one technical assumption on our forcings.

**Definition 7.3.11 (Filter valuation property).** A notion of forcing $\mathbb{P} = (P, \leqslant, V)$ satisfies the *filter valuation property (FVP)* if for each $f \in \omega^\omega$, the set of condition $p$ with $f \in [[p]]$ is a filter, or equivalently, if for all $p, q \in P$ with $f \in [[p]] \cap [[q]]$, there is an $r \in P$ extending both $p$ and $q$ with $f \in [[r]]$.

It is easy to see that, in addition to the obvious Cohen forcing, all examples of forcing notions we considered in the preceding section satisfy FVP, as will all other forcings we consider in the sequel.

Recall now that a property of the members of a topological space is called *generic* if it holds almost everywhere in the sense of category. That is, a property is generic if it the elements of which it holds form a comeager set (a set containing an intersection of dense open sets). The result here is that this notion of genericity accords nicely with the one from forcing: namely, any property enjoyed by all sufficiently generic objects for a forcing satisfying FVP is generic in the topological sense.

**Proposition 7.3.12.** *Let $\mathbb{P} = (P, \leqslant, V)$ be a notion of forcing satisfying FVP. Then for any collection $\mathcal{D}$ of dense subsets of $P$ the set of $\mathcal{D}$-generic functions is comeager in the $\mathbb{P}$ topology.*

*Proof.* We may assume that for each $n$, the set of all $p \in P$ with $|V(p)| \geqslant n$ is an element of $\mathcal{D}$. For each $D \in \mathcal{D}$, let $U_D = \bigcup_{p \in D}[[p]]$, which is a dense open set in the $\mathbb{P}$ topology, as discussed above. We claim that every element of $\bigcap_{D \in \mathcal{D}} U_D$ is $\mathcal{D}$-generic, which yields the result. Fix $f \in \bigcap_{D \in \mathcal{D}} U_D$ and let $F$ be the set of $p \in P$

with $f \in [[p]]$. Since $\mathbb{P}$ satisfies FVP, $F$ is a filter. Moreover, $F$ meets every $D \in \mathcal{D}$ by choice of $f$, so $F$ is $\mathcal{D}$-generic. By our assumption on $\mathcal{D}$, $F$ contains conditions with arbitrarily long valuations, so $f = V(F)$. □

**Corollary 7.3.13 (Jockusch [169]).** *The collection of Cohen generic sets is comeager in the usual Cantor space topology.*

## 7.4 The forcing relation

There are times when the properties we wish to ensure of our eventual generic are difficult to express in the manner illustrated by our examples up to this point, as explicit dense sets defined in terms of the forcing notion. We therefore introduce a more systematic approach for doing so. To this end, we need to connect forcing with the language we use to talk about generic objects.

Formally, our *forcing language* is $\mathcal{L}_2$ augmented by a new constant symbol $\mathsf{G}$ intended to denote the generic object. We denote this language by $\mathcal{L}_2(\mathsf{G})$. Every formula here is thus of the form $\varphi(\mathsf{G}, \vec{X}, \vec{x})$ for some formula $\varphi(X, \vec{X}, \vec{x})$ of $\mathcal{L}_2$. When we wish to say that $\mathsf{G}$ does not occur in a formula we state explicitly that it is a formula of $\mathcal{L}_2$. Otherwise, we are free to follow the same conventions and notations in $\mathcal{L}_2(\mathsf{G})$ as we did in $\mathcal{L}_2$. In particular, we say a formula of $\mathcal{L}_2(\mathsf{G})$ is $\Sigma^0_n$ or $\Pi^0_n$ if it has the form $\varphi(\mathsf{G})$ for some $\Sigma^0_n$ or $\Pi^0_n$ formula $\varphi(X)$ of $\mathcal{L}_2$.

In what follows, let $\mathcal{M}^\omega$ be the full $\omega$-model $(\omega, \mathcal{P}(\omega))$ and say a sentence of $\mathcal{L}_2$ is *true* if it is satisfied by $\mathcal{M}^\omega$.

**Definition 7.4.1 (Forcing relation).** Let $\mathbb{P} = (P, \leqslant, V)$ be a notion of forcing. Let $p$ be any condition, and $\varphi$ an arithmetical sentence of $\mathcal{L}_2(\mathsf{G})$. We define the relation of *p forcing* $\varphi$, written $p \Vdash \varphi$, as follows.

1. If $\varphi$ is an atomic sentence of $\mathcal{L}_2$ then $p \Vdash \varphi$ if $\varphi$ is true.
2. If $\varphi$ is $t \in \mathsf{G}$ for a term $t$ then $p \Vdash \varphi$ if $t^{\mathcal{M}^\omega} < |V(p)|$ and $V(p)(t^{\mathcal{M}^\omega}) = 1$.
3. If $\varphi$ is $(\exists x < t)\psi(x)$ for a term $t$ then $p \Vdash \psi$ if there is an $m \in \omega$ such that $m <^{\mathcal{M}^\omega} t^{\mathcal{M}^\omega}$ and $p \Vdash \psi(m)$.
4. If $\varphi$ is $\psi_0 \vee \psi_1$ then $p \Vdash \varphi$ if $p \Vdash \psi_0$ or $p \Vdash \psi_1$.
5. If $\varphi$ is $\neg\psi$ then $p \Vdash \varphi$ if $q \nVdash \psi$ for all $q \leqslant p$.
6. If $\varphi$ is $(\exists x)\psi(x)$ then $p \Vdash \varphi$ if there is an $m \in \omega$ such that $p \Vdash \psi(m)$.

We say *p decides* $\varphi$ if $p \Vdash \varphi$ or $p \Vdash \neg\varphi$.

We could also start with a collection $\mathcal{B}$ of subsets of $\omega$ and formulate our discussion over $\mathcal{L}_2(\mathcal{B})$ (as defined in Definition 5.1.3) instead of $\mathcal{L}_2$. Thus our forcing language would be $\mathcal{L}_2(\mathcal{B})(\mathsf{G})$, and the definition of $\Vdash$ would have part (1) modified to apply to atomic formulas $\mathcal{L}_2(\mathcal{B})$. We will not explicitly phrase things in this generality, but everything should be understood as "relativizing" in this sense.

The forcing relation behaves a bit like the satisfaction predicate for structures, but not in all ways. For starters, a condition need not decide a formula $\varphi$, and so it may not force the tautology $\varphi \vee \neg\varphi$ (see Exercise 7.8.4). What we are calling the forcing relation here is properly called *strong forcing*, and should be compared with *weak forcing*, which is more common in set theory. The definition of the latter, which we denote by $\Vdash_w$, differs in its treatment of disjunctions and quantifiers.

- $p \Vdash_w \psi_0 \vee \psi_1$ if for every $q \leqslant p$ there is $r \leqslant q$ such that $r \Vdash_w \psi_0$ or $r \Vdash_w \psi_1$.
- $p \Vdash_w (\exists x)\psi(x)$ if for every $q \leqslant p$ there is $r \leqslant q$ and $n$ such that $r \Vdash_w \psi(n)$.

With these modifications, it is not difficult to see that if $p \Vdash \varphi$ then $p \Vdash_w \varphi$, but not necessarily the other way around. Notably, every condition *does* weakly force every tautology. However, it is also true that the two relations agree up to double negation: $p \Vdash_w \varphi$ if and only if $p \Vdash \neg\neg\varphi$ (see Exercise 7.8.5). Since semantically $\varphi$ and $\neg\neg\varphi$ are equivalent, the choice of which forcing to use is actually inconsequential for deciding properties about generic objects. But strong forcing enjoys an important advantage over weak forcing for computability theory applications, which is that it has a simpler definition. As we describe in the next section, this means that constructions employing strong forcing can be carried out more effectively.

We collect some basic facts about the forcing relation.

**Proposition 7.4.2.** *Let $\mathbb{P} = (P, \leqslant, V)$ be a notion of forcing, $p$ a condition, and $\varphi$ and $\psi$ be arithmetical sentences of $\mathcal{L}_2(\mathsf{G})$.*

1. *If $p \Vdash \varphi$ then $p \Vdash \neg\neg\varphi$.*
2. *If $p \Vdash \neg\neg\varphi$ then some $q \leqslant p$ forces $\varphi$.*
3. *If $p \Vdash \varphi \wedge \psi$ then some $q \leqslant p$ forces both $\varphi$ and $\psi$.*
4. *$p$ cannot force both $\varphi$ an $\neg\varphi$.*
5. *If $p \Vdash \varphi$ and $q \leqslant p$ then $q \Vdash \varphi$.*
6. *If $\varphi$ is a sentence of $\mathcal{L}_2$ and $p$ decides $\varphi$, then $p \Vdash \varphi$ if and only if $\varphi$ is true.*
7. *The collection of conditions that decide $\varphi$ is dense.*

*Proof.* Part 4 follows immediately from the definition of forcing negations. Part 5 is proved by induction on the complexity of $\varphi$, using the monotonicity of the valuation map in the case where $\varphi$ is atomic. The other parts are left to Exercise 7.8.6. □

The last part above is what connects the forcing relation to genericity. The relevant definition is the following.

**Definition 7.4.3 ($n$-genericity).** Let $\mathbb{P} = (P, \leqslant, V)$ be a notion of forcing and $n \in \omega$. A filter $F$ is *$n$-generic* if every $\Sigma^0_n$ sentence $\varphi$ of $\mathcal{L}_2(\mathsf{G})$ is decided by some $p \in F$.

The notion also relativizes. If we work with $\mathcal{L}_2(\mathcal{B})(\mathsf{G})$ for some collection $\mathcal{B}$ of subsets of $\omega$, we get a corresponding notion of $n$-generic *relative to $\mathcal{B}$*, or *$n$-$\mathcal{B}$-generic* for short. We say *$n$-$B$-generic* instead of $n$-$\{B\}$-generic.

We can relate this genericity to our previous one (Definition 7.3.2) as follows. Enumerate as $\varphi_0, \varphi_1, \ldots$ all $\Sigma^0_n$ sentences of $\mathcal{L}_2(\mathsf{G})$. Then, for each $e$, define

$$D_e = \{p \in P : p \Vdash \varphi_e \text{ or } p \Vdash \neg\varphi_e\},$$

and let $\mathcal{D} = \{D_e : e \in \omega\}$. Then $\mathcal{D}$ is a collection of dense sets by Proposition 7.4.2 (7), and being $n$₇generic is precisely the same as being $\mathcal{D}$-generic. In particular, we have the following as an immediate consequence of the Rasiowa–Sikorski theorem.

**Theorem 7.4.4 (Existence of $n$-generics).** *Let $\mathbb{P}$ be a notion of forcing, $p$ a condition, and $n \in \omega$. There exists an $n$-generic filter containing $p$.*

(In the relativized case, we get the existence of an $n$-$\mathcal{B}$-generic so long as $\mathcal{B}$ is a countable collection of subsets of $\omega$.)

Clearly, every $(n+1)$-generic is $n$-generic, for all $n$x. We also have the following basic facts.

**Proposition 7.4.5.** *Let $\mathbb{P} = (P, \leqslant, V)$ be a notion of forcing, $n \in \omega$, and $F$ an $n$-generic filter. If $\varphi$ is a $\Pi^0_n$ sentence $\varphi$ of $\mathcal{L}_2(\mathsf{G})$ then some $p \in F$ decides $\varphi$.*

*Proof.* Fix a $\Sigma^0_n$ sentence $\psi$ so that $\varphi$ is $\neg\psi$. Then some $p \in F$ must decide $\psi$. If $p$ forces $\neg\psi$ then this is $\varphi$. If, instead, $p$ forces $\psi$ then by Proposition 7.4.2 $p$ forces $\neg\neg\psi$, which is $\neg\varphi$. □

With this in hand, we can state and prove the central result of this section, which relates forcing and truth.

**Theorem 7.4.6 (Forcing and truth).** *Let $\mathbb{P} = (P, \leqslant, V)$ be a notion of forcing, $n \in \omega$, $F$ an $n$-generic filter determining the object $G$, and $\varphi(X)$ a $\Sigma^0_n$ or $\Pi^0_n$ formula of $\mathcal{L}_2$. Then $\varphi(G)$ is true if and only if some $p \in F$ forces $\varphi(\mathsf{G})$.*

*Proof.* We proceed by induction on the complexity of $\varphi$. If $\varphi$ is an atomic sentence of $\mathcal{L}_2$, then it is either true, and then forced by all conditions, or false, and then forced by none, and this does not depend on $G$. Next, suppose $\varphi$ is atomic and of the form $t \in X$ for some term $t$. This is satisfied by $G$ if and only if $G(t^{\mathcal{M}^\omega}) = 1$, hence if and only if there is a $p \in F$ with $|V(p)| > t^{\mathcal{M}^\omega}$ and $V(p)(t^{\mathcal{M}^\omega}) = 1$, which precisely means that $p \Vdash \varphi(\mathsf{G})$.

It is clear that if $\varphi$ is $\psi_0 \vee \psi_1$ and the theorem holds of each of $\psi_0$ and $\psi_1$ then it also holds of $\varphi$. Suppose now that $\varphi$ is $\neg\psi$ and that the theorem holds of $\psi$. Notice that $\psi$ is $\Pi^0_n$ or $\Sigma^0_n$, depending as $\varphi$ is $\Sigma^0_n$ or $\Pi^0_n$, respectively. Since $F$ is $n$-$\mathcal{B}$-generic, some condition in $F$ decides $\psi(\mathsf{G})$ (in the former case, by Proposition 7.4.5). Now $\varphi(G)$ is true if and only if $\psi(G)$ is false, which is the case if and only if no condition in $F$ forces $\psi(\mathsf{G})$ by hypothesis. Thus, $\varphi(G)$ holds if and only if a condition in $F$ forces $\neg\psi(\mathsf{G}) = \varphi(\mathsf{G})$. Thus, the theorem holds of $\varphi$.

Next, suppose $\varphi(X)$ is $(\exists x < t)\psi(X, x)$ for some term $t$. We have that $\varphi(G)$ holds if and only if $\psi(G, m)$ holds for some $m < t^{\mathcal{M}^\omega}$. By hypothesis, this is the case if and only if there is a $p \in F$ and an $m < t^{\mathcal{M}^\omega}$ such that $p \Vdash \psi(G, m)$, which means exactly that there is a $p \in F$ that forces $(\exists x < t)\psi(\mathsf{G}, x) = \varphi(\mathsf{G})$. The case of $\varphi(X) = (\exists x)\psi(X, x)$ is handled similarly. This completes the proof. □

Being able to write down the properties we want to decide about our generic object in the forcing language is convenient. In many cases, it eliminates the need to think too much about the particular forcing notion we are working with. We simply take a sufficiently generic object and appeal to Theorem 7.4.6. But in other situations, particularly when we care about effectivity of the generic object, we wish to find an $n$-generic for as small an $n$ as possible, and in these cases we must be more careful.

*Example 7.4.7.* Consider Cohen forcing, and take the sentence

$$(\forall e)(\exists x)[\Phi_e(x)\uparrow \vee \Phi_e(x)\downarrow\neq \mathsf{G}(x)].$$

This is $\Pi^0_3$, and so as remarked above, every 3-generic filter contains a condition that decides it. But we know the negation of this sentence cannot be forced: for all $e \in \omega$ and all conditions $\sigma$, if $x \geqslant |\sigma|$ and $\Phi_e(x)\downarrow$ there is a $\tau \succeq \sigma$ with $\Phi_e(x) \neq \tau(x)$. Hence, by Theorem 7.4.6, every 3-generic set is noncomputable. One way to improve this is to instead take, for each $e \in \omega$, the sentence

$$(\exists x)[\Phi_e(x)\uparrow \vee \Phi_e(x)\downarrow\neq \dot{\mathsf{G}}(x)].$$

This is $\Sigma^0_2$, hence every 2-generic filter contains a condition that decides it, and indeed, by the same argument as above, forces it. We thus conclude that already every 2-generic set is noncomputable. But now consider the sentence

$$(\exists x)[\Phi_e(x)\downarrow\neq \mathsf{G}(x)].$$

This is $\Sigma^0_1$, hence every 1-generic filter contains a condition that decides it. It is no longer the case that the sentence is necessarily forced, but if its negation is forced then our argument actually shows that so is $(\exists x)[\Phi_e(x)\uparrow]$. Hence, in any case, every 1-generic set $G$ still satisfies that $G \neq \Phi_e$ for all $e$, and so is noncomputable.

*Example 7.4.8.* Consider Mathias forcing (within $\omega$), and for each $e \in \omega$, take the sentence

$$(\exists m)(\forall x > m)[\Phi_e(x)\uparrow \vee p_{\mathsf{G}}(x) \geqslant \Phi_e(x)].$$

This is $\Sigma^0_2$, and so by the argument in Example 7.3.6, every 2-generic filter must contain a condition that forces it. Hence, every Mathias 2-generic set is high. But here, this cannot in general be improved to 1-generics (see Exercise 7.8.11).

## 7.5 Effective forcing

In computability theory, forcing constructions are particularly useful when we can gauge their effectiveness. In broad terms, this allows us to conclude that not only can sufficiently generic objects always be found, but that they can be found Turing below certain oracles. To this end, we first need to discuss effective notions of forcing.

*Remark 7.5.1.* In this section, we assume all forcing notions to be countable, and represented by subsets of $\omega$. There can of course be different representations in this sense. For example, in Mathias forcing with computable reservoirs we could identify a condition $(E, I)$ with the pair $\langle e, i\rangle$, where $e$ is a canonical index for $E$, and $i$ is either a $\Delta^0_1$ index for $I$ or a $\Sigma^0_1$ index for $I$ (i.e., $I = \Phi_i$ or $I = W_i$). These kinds of distinctions will typically not affect anything in our discussion, but when they do, we will be explicit about them. Otherwise, we will usually refer simply to conditions, extensions, etc., when of course we really mean the underlying sets of codes, as per Convention 3.4.5.

**Definition 7.5.2.** Fix $A \in 2^\omega$. A notion of forcing $\mathbb{P} = (P, \leqslant, V)$ is *A-computable* if $P$ is (represented by) an $A$-computable subset of $\omega$, $\leqslant$ is an $A$-computable relation on $A$, and $V$ is an $A$-computable map $P \to 2^{<\omega}$.

Thus all forcings here are computable in the join of $P$, $\leqslant$, and $V$. Obviously, Cohen forcing is computable. We give two other examples.

*Example 7.5.3.* Let $X$ be an infinite computable set, and consider Mathias forcing within $X$ with computable reservoirs (Example 7.2.10). The set of conditions is then $\Pi^0_2$. The extension relation is $\Pi^0_1$ relative to pairs of conditions, and the valuation map is computable in the conditions. It follows that this forcing is $\varnothing''$-computable.

*Example 7.5.4.* Let $T \subseteq 2^{<\omega}$ be a computable infinite tree, and consider Jockusch–Soare with computable subtrees of $T$ (Example 7.2.11). The set of conditions here is $\Pi^0_1$, as is the extension relation. The valuation map is computable in the conditions. This forcing is therefore $\varnothing'$-computable.

We now move on to effective forms of density.

**Definition 7.5.5 (Effective density).** Fix $A \in 2^\omega$ and let $\mathbb{P} = (P, \leqslant, V)$ be a notion of forcing.

1. A subset $D$ of $P$ is *A-effectively dense* if there is a partial $A$-computable function $f\colon \omega \to \omega$ such that $f(p)\downarrow$ for every $p \in P$ and $f(p) \leqslant p$ and $f(p) \in D$.
2. A collection $\{D_i : i \in \omega\}$ of subsets of $P$ is *uniformly A-effectively dense* (or, for short, $D_i$ is *uniformly A-effectively dense* for each $i$) if there is a partial $A$-computable function $f\colon \omega \times \omega \to \omega$ such that $f(p, i)\downarrow$ for every $p \in P$ and $i \in \omega$, and $f(p, i) \leqslant p$ and $f(p, i) \in D_i$.

If $\mathbb{P}$ is $A$-computable then for each $i$ the set $\{p \in P : |V(p)| \geqslant i\}$ is $A$-effectively dense, and in fact, the collection of all these sets, across all $i$, is uniformly $A$-effectively dense. We can use this to get the following effective version of the Rasiowa–Sikorski theorem.

**Theorem 7.5.6 (Existence of effective generics).** *Fix $A \in 2^\omega$, let $\mathbb{P} = (P, \leqslant, V)$ be a notion of forcing, $p$ a condition, and $\mathcal{D} = \{D_i : i \in \omega\}$ a uniformly A-effectively dense collection of subsets of $P$. There exists an A-computable $\mathcal{D}$-generic sequence $p_0 \geqslant p_1 \geqslant \cdots$ such that $p_0 = p$ and $p_{i+1}$ meets $D_i$.*

*Proof.* Fix a partial $A$-computable function $f$ witnessing that $\mathcal{D}$ is uniformly $A$-effectively dense. Let $p_0 = p$, and given $p_i$ for some $i \in \omega$, let $p_{i+1} = f(p_i, i)$. □

If we have two or more collections of subsets of $P$, each uniformly $X$-effectively dense, then we can interweave these to produce another such collection. In this way, Theorem 7.5.6 can be quite a powerful tool. Let us illustrate this for each of Jockusch–Soare forcing, Cohen forcing, and Mathias forcing in turn. We begin by revisiting the low basis theorem (Theorem 2.8.18).

*Example 7.5.7.* Let $T$ be an infinite computable tree $T \subseteq 2^{<\omega}$ and consider Jockusch–Soare forcing with computable subtrees of $T$. Notice that the set $L_i$ of all conditions $(U, n)$ such that $n \geqslant i$ is uniformly $\emptyset'$-effectively dense for each $i$. Now, for each $e$ let $J_e$ to be the set of conditions $(U, n)$ such that either $\Phi_e^\sigma(e)\uparrow$ for all $\sigma \in U$, or else $\Phi_e^\sigma(e)\downarrow$ for the unique $\sigma \in U$ of length $n$. Then $J_e$ is uniformly $\emptyset'$-effectively dense for each $e$, which is the iterative step in the proof of the low basis theorem that we saw earlier.

Let $\mathcal{D} = \{D_i : i \in \omega\}$, where $D_{2i} = L_i$ and $D_{2i+1} = L_i$ for all $i$. By Theorem 7.5.6, we can find a $\emptyset'$-computable sequence of conditions $(U_0, n_0) \geqslant (U_1, n_1) \geqslant \cdots$ such that $(U_{2e+1}, n_{2e+1}) \in J_e$ for all $e$, and such that $f = \bigcup_{e \in \omega} V((U_e, n_e)) \in 2^{<\omega}$. As we know, $f \in [T]$. We claim that $f$ is low. Indeed, we have that $e \in f'$ if and only if $\Phi_e^\sigma(e)\downarrow$ for the unique $\sigma \in U_{2e+1}$ of length $n_{2e+1}$. Since the generic sequence is $\emptyset'$-computable and $e$ is arbitrary, $f' \leqslant_T \emptyset'$.

For another application, recall that a set $X$ has hyperimmune free degree if every $X$-computable function is dominated by a computable function (Definition 2.8.19 and Theorem 2.8.21).

**Proposition 7.5.8.** *Let $T \subseteq 2^{<\omega}$ be an infinite computable tree, and consider Jockusch–Soare forcing with computable subtrees of $T$. For each $e$, let $D_e$ be the set of conditions that decide the sentence $(\forall x)[\Phi_e^G(x)\downarrow]$ and let $\mathcal{D} = \{D_e : e \in \omega\}$.*

1. *$\mathcal{D}$ is uniformly $\emptyset''$-effectively dense.*
2. *If $f \in 2^\omega$ is $\mathcal{D}$-generic then $f$ has hyperimmune free degree.*

*Proof.* We leave part (1) to Exercise 7.8.2. For part (2), fix $e$ and a condition $(U, m)$ having $f$ as a path through it that decides the $\Pi_2^0$ sentence $(\forall x)[\Phi_e^G(x)\downarrow]$. If $(U, m)$ forces the negation, then $\Phi_e^f$ cannot be total by Theorem 7.4.6. So suppose instead that $(U, m) \Vdash (\forall x)[\Phi_e^G(x)\downarrow]$. Then for each $x$ the tree $U_x = \{\sigma \in U : \Phi_e^\sigma(x)\uparrow\}$ is finite, else $(U_x, m)$ would be an extension of $(U, m)$ forcing $\Phi_e^G(x)\uparrow$, contradicting monotonicity of the forcing relation (Proposition 7.4.2).

Now define a function $g$ as follows: for each $x$, find an $\ell$ so that $\Phi_e^\sigma(x)\downarrow$ for every $\sigma \in U$ of length $\ell$, and let $g(x) = \max\{\Phi_e^\sigma(x) : \sigma \in U \wedge |\sigma| = \ell\}$. Since $f$ is a path through $U$, it follows that $g(x) \geqslant \Phi_e^f(x)$ for all $x$. Since $e$ was arbitrary, it follows that every $f$-computable function is dominated by a computable function, as desired. □

We obtain the following well-known refinement of the hyperimmune free basis theorem (Theorem 2.8.22).

**Corollary 7.5.9 (Jockusch and Soare [173]).** *Every infinite computable tree $T \subseteq 2^{<\omega}$ has a low$_2$ path of hyperimmune free degree.*

*Proof.* By Theorem 7.5.6 and Proposition 7.5.8, using the fact that for all sets $A$ and $B$, $A'' \leqslant_T B$ if and only if $\{e \in \omega : (\forall x)[\Phi_e^{\mathsf{G}}(x)\downarrow]\} \leqslant_T B$. □

Next, let us turn to Cohen forcing, which as usual exhibits some of the nicest behavior. In particular, here we can very simply calibrate the complexity of the forcing relation itself. The proof of the following is left to the reader (Exercise 7.8.7).

**Lemma 7.5.10 (Jockusch [169]).** *Fix $n \geqslant 1$ along with a $\Sigma_n^0$ sentences $\varphi$ of $\mathcal{L}_2(\mathsf{G})$ and a $\Pi_n^0$ sentence $\psi$ of $\mathcal{L}_2(\mathsf{G})$. Let $\sigma$ be a Cohen condition.*

1. *The relation $\sigma \Vdash \varphi$ is $\Sigma_n^0$ definable, uniformly in an index for $\varphi$ and $\sigma$.*
2. *The relation $\sigma \Vdash \psi$ is $\Pi_n^0$ definable, uniformly in an index for $\psi$ and $\sigma$.*

This has the following consequence, showing that Cohen forcing enjoys an alternate characterization of $n$-genericity that can sometimes be easier to work with. In fact, this is often the way the notion is defined for this forcing.

**Proposition 7.5.11 (Jockusch [169]).** *A filter $F$ is $n$-generic for Cohen forcing if and only if $F$ meets or avoids every set of Cohen conditions definable by a $\Sigma_n^0$ formula of $\mathcal{L}_2$ (with no parameters).*

*Proof.* For the "only if" direction, fix a $\Sigma_n^0$ formula $\varphi(x)$ of $\mathcal{L}_2$ and suppose this defines a set $S$ of Cohen conditions. Let $\psi(X)$ be the formula $(\exists\sigma)[\varphi(\sigma) \wedge \sigma \preceq X]$. Then $\psi(\mathsf{G})$ is a $\Sigma_n^0$ formula of $\mathcal{L}_2(\mathsf{G})$. Consider any $\tau \in F$ that decides $\psi(\mathsf{G})$. If $\tau \Vdash \psi(\mathsf{G})$, then by Theorem 7.4.6, $\psi(G)$ holds for the real $G$ determined by $F$. This means some $\sigma \preceq G$ belongs to $S$, and this $\sigma$ must belong to $F$ (as remarked in Remark 7.2.9). Thus, $F$ meets $S$. Suppose next that $\tau \Vdash \neg\psi(\mathsf{G})$. Then no extension of $\tau$ can belong to $S$, because any such extension would have a further extension forcing $\psi(\mathsf{G})$ (and therefore would force both $\psi(\mathsf{G})$ and $\neg\psi(\mathsf{G})$, which cannot be). Indeed, suppose $\tau^* \succeq \tau$ belongs to $S$, consider any $n$-generic filter $F^*$ containing $\tau^*$, and let $G^*$ be the object determined by $F^*$. Then $\psi(G^*)$ holds, as witnessed by $\tau^*$, so some $\rho \succeq \tau^*$ in $F^*$ must forces $\psi(\mathsf{G})$. We conclude that $\tau$ witnesses that $F$ avoids $S$.

For the "if" direction, fix any $\Sigma_n^0$ formula $\varphi$ of $\mathcal{L}_2(\mathsf{G})$. By Lemma 7.5.10, the set $\{\sigma \in 2^{<\omega} : \sigma \Vdash \varphi\}$ is $\Sigma_n^0$ with the same parameters. If $F$ meets this set then it contains a condition forcing $\varphi$, and if $F$ avoids it then it contains a condition that forces $\neg\varphi$. □

A less technical application of Lemma 7.5.10 is the following, yielding even more effective bounds on Cohen generics than Theorem 7.5.6.

**Proposition 7.5.12 (Jockusch [169]).** *Fix $n \geqslant 1$. In Cohen forcing, we have the following.*

1. *If $G \in 2^\omega$ is $n$-generic then $G^{(n)} \leqslant_T G \oplus \varnothing^{(n)}$.*
2. *There exists a low$_n$ $n$-generic set.*

*Proof.* Let $\varphi_0, \varphi_1, \ldots$ be a computable listing of all $\Sigma^0_1$ sentences of $\mathcal{L}_2(\mathsf{G})$ such that $\varphi_{2e}$ is the sentence $e \in \mathsf{G}^{(n)}$ for every $e$. If $G$ is $n$-generic then for each $i \in \omega$ there must be a $\sigma \preceq G$ that decides $\varphi_i$. By Lemma 7.5.10, $\varnothing^{(n)}$ can determine whether a given condition forces this sentence, and of course, $G$ can determine whether a given Cohen condition is an initial segment of it. Hence, $G \oplus \varnothing^{(n)}$ can find $\sigma$ and determine whether it forces $\varphi_i$ or not. By Theorem 7.4.6, for each $e$ we have $e \in G^{(n)}$ if and only if some $\sigma \preceq G$ forces $\varphi_{2e}$. Thus, $G^{(n)} \leqslant_{\mathrm{T}} G \oplus \varnothing^{(n)}$, and this gives us part (1). For part (2), define $D_i = \{\sigma \in 2^{<\omega} : \sigma \Vdash \varphi_i\}$ for each $i$. Then $\mathcal{D} = \{D_i : i \in \omega\}$ is uniformly $\varnothing^{(n)}$-effectively dense by Lemma 7.5.10. Hence, by Theorem 7.5.6 there is a $\mathcal{D}$-generic $G \leqslant_{\mathrm{T}} \varnothing^{(n)}$, which is of course $n$-generic. By part (1), we have $G^{(n)} \leqslant_{\mathrm{T}} G \oplus \varnothing^{(n)} \leqslant_{\mathrm{T}} \varnothing^{(n)}$, as wanted. □

For comparison, as we have already seen in Example 7.3.6, there is no analogous result to part (2) for Mathias forcing. If $n \geqslant 2$, then in basically every flavor of Mathias forcing (with computable reservoirs, with cone avoiding reservoirs, etc.), every $n$-generic set is high (and so not low$_n$).

We stay with Mathias forcing now, and conclude this section with a series of results will be very useful to us in Chapter 8.

**Lemma 7.5.13.** *Let $\vec{R} = \langle R_i : i \in \omega\rangle$ be an instance of* COH, *and consider Mathias forcing with $\vec{R}$-computable reservoirs. For each $i \in \omega$, let $M_i$ be the set of all conditions $(E, I)$ with either $I \subseteq R_i$ or $I \subseteq \overline{R_i}$. Then for every $X \gg \vec{R}'$, the collection $\{M_i : i \in \omega\}$ uniformly $X$-effectively dense.*

*Proof.* Fix a condition $(E, I)$ and $i \in \omega$. Each of the statements "$I \cap R_i$ is infinite" and "$I \cap \overline{R_i}$ is infinite" is $\Pi^0_2(\vec{R})$, hence $\Pi^0_1(\vec{R}')$. Moreover, at least one of the two is true. By Theorem 2.8.25 (specifically, part (4), as explained there) $X$ can uniformly select one of the two which is true. Define $I^* = I \cap R_i$ if $X$ selects the first statement, and $I^* = I \cap \overline{R_i}$ if it selects the second. Let $E^* = E$. Then $(E^*, I^*)$ is an extension of $(E, I)$ in $M_i$. □

**Lemma 7.5.14.** *Fix $A \in 2^\omega$; and consider Mathias forcing with $A$-computable reservoirs.*

1. *For each $i$, let $L_i$ be the set of conditions $(E, I)$ with $|E| \geqslant i$. Then $\{L_i : i \in \omega\}$ is uniformly effectively dense.*
2. *For each $e \in \omega$, let $J_e$ be the set of conditions that decide the sentence $e \in \mathsf{G}'$. Then $\{J_e : e \in \omega\}$ is uniformly $A'$-effectively dense.*

*Proof.* For (1), given $i$ and $(E, I)$ we can pass to $(E \cup \min I, I \smallsetminus \{\min I\}) \in L_i$. An index for this extension is uniformly computable form an index for $(E, I)$. For (2), fix a condition $(E, I)$ and $e \in \omega$. Now $A'$ can determine, uniformly in $e$, whether there is an $F \subseteq I$ such that $\Phi_e^{E \cup F}(e)\downarrow$ (with use bounded by max $F$). If so, let $E^* = E \cup F$ and $I^* = \{x \in I : x > F\}$. Otherwise, let $E^* = E$ and $I^* = I$. In either case, $(E^*, I^*)$ is an extension of $(E, I)$ in $J_e$. □

**Theorem 7.5.15 (Cholak, Jockusch, and Slaman [33]).** *Let $\vec{R} = \langle R_i : i \in \omega\rangle$ be an instance* COH. *For every $X \gg R'$, there is a solution $G$ to $\vec{R}$ satisfying $G' \leqslant_{\mathrm{T}} X$.*

*Proof.* For each $i$, define $D_{3i} = M_i$, $D_{3i+1} = L_i$, and $D_{3i+2} = J_i$, where $M_i$, $L_i$, and $J_i$ are as defined in the lemmas. Let $\mathcal{D} = \{D_i : i \in \omega\}$. As shown above, $\mathcal{D}$ is uniformly $X$-effectively dense. By Theorem 7.5.6, there is an $X$-computable generic sequence $(E_0, I_0) \geqslant (E_0, I_1) \geqslant \cdots$ with $(E_{3e+2}, I_{3e+2}) \in J_e$ for all $e$. Note that meeting all the $L_i$ implies that the valuations of these conditions have unbounded length, so the sequence determines an object, $G$. As in Example 7.3.9, $G$ is $\vec{R}$-cohesive. And for each $e$ we have $e \in G'$ if and only if $(E_{3e+2}, I_{3e+2}) \Vdash e \in \mathsf{G}'$, which is to say, if and only if $\Phi_e^{E_{3e+2}}(e)\downarrow$. Hence, $G' \leqslant_{\mathrm{T}} X$, as desired. □

**Corollary 7.5.16 (Cholak, Jockusch, and Slaman [33]).** *Every* $\mathsf{COH}$*-instance* $\vec{R} = \langle R_i : i \in \omega \rangle$ *has a solution that is* low$_2$ *relative to* $\vec{R}$.

*Proof.* Relativizing Proposition 2.8.14 to $\vec{R}'$, we get an infinite $\vec{R}'$-computable tree $T$ such that if $f$ is a path through $T$ then $\vec{R}' \oplus f \gg \vec{R}'$ (in fact, $f \gg \vec{R}'$). By the low basis theorem, relativized to $\vec{R}'$, there is some such $f$ satisfying $(\vec{R}' \oplus f)' \leqslant_{\mathrm{T}} \vec{R}''$. Now by Theorem 7.5.15 with $X = \vec{R}' \oplus f$, there is a $\mathsf{COH}$-solution $G$ to $\vec{R}$ such that $G' \leqslant_{\mathrm{T}} \vec{R}' \oplus f$. Hence, we have $G'' \leqslant_{\mathrm{T}} (\vec{R}' \oplus f)' \leqslant_{\mathrm{T}} \vec{R}''$, as was to be shown. □

## 7.6 Forcing in models

The core use of forcing in reverse mathematics is to establish bounds on the instances and solutions of various problems, often for the purposes of building models witnessing separations between mathematical theorems. The framework described up to this point works for working over the full standard model, but we can modify it to work over models in general.

**Definition 7.6.1.** Let $\mathcal{M}$ be an $\mathcal{L}_2$ structure. A *notion of forcing in* $\mathcal{M}$ is a triple $\mathbb{P} = (P, \leqslant, V)$ where $(P, \leqslant)$ is a partial order and $V$ is a map $P \to 2^{<M}$ satisfying the following:

1. If $p, q \in P$ with $p \leqslant q$ then $V(q) \preceq^{\mathcal{M}} V(p)$.
2. For each $n \in M$ and $q \in P$ there is a $p \in P$ such that $p \leqslant q$ and $n \leqslant^{\mathcal{M}} |V(p)|$.

This definition therefore agrees with our earlier one (Definition 7.2.1) for the cases where $\mathcal{M}$ is an $\omega$-model.

One very useful example of the more general definition is when we wish to work over a Scott set. We will employ the following forcing in Chapters 8 and 9.

*Example 7.6.2.* Let $\mathcal{S}$ be a Scott set . (Recall that this means $\mathcal{S}$ is an $\omega$-model such that every infinite binary tree $T \in \mathcal{S}$ has an infinite path $f \in \mathcal{S}$.) For $X \in \mathcal{S}$, we can define *Mathias forcing within X with reservoirs in* $\mathcal{S}$ to be the restriction to Mathias conditions $(E, I)$ such that $I \subseteq X$ and $I \in \mathcal{S}$.

We can formulate notions of density and (generic) filters as before. Given a filter $F$ such that for every $n$ there is a $p \in F$ with $|V(p)| \geqslant^{\mathcal{M}} n$ we can also define $G = \bigcup_{p \in F} V(p)$, the (generic) object determined by $F$. Notice that

$$G(i) = b = (\exists\sigma \in 2^{<M})(\exists p \in F)[V(p) = \sigma \wedge i <^{M} |\sigma| \wedge \sigma(i) = b]$$
$$= (\forall\sigma \in 2^{<M})(\forall p \in F)[V(p) = \sigma \wedge i <^{M} |\sigma| \wedge \sigma(i) = b].$$

Hence, if $P$ and $V$ are subsets of $M$ then $G$ is $\Delta^0_1$ definable in $\mathcal{M}$ from $F$, $P$, and $V$.

With appropriate modifications, most of the principal results from the previous section lift to the current setting, chief among them the following version of the Rasiowa–Sikorski theorem (Theorem 7.3.3).

**Theorem 7.6.3.** *Let $\mathcal{M}$ be a countable $\mathcal{L}_2$ structure, $\mathbb{P} = (P, \leqslant, V)$ a notion of forcing in $\mathcal{M}$, $p$ a condition, and $\mathcal{D} = \{D_i : i \in M\}$ a collection of dense subsets of $P$. There exists a $\mathcal{D}$-generic filter containing $p$.*

As before, $F$ above can always be chosen to determine an object. Thus, we freely follow Convention 7.3.4, appropriately re-interpreted.

We can also adapt the definition of the forcing relation to this more general setting, although we need to take some more care here.

**Definition 7.6.4 (Forcing in a model).** Let $\mathcal{M}$ be an $\mathcal{L}_2$ structure and $\mathbb{P} = (P, \leqslant, V)$ a notion of forcing in $\mathcal{M}$. Let $p$ be any condition, and $\varphi$ an arithmetical sentence of $\mathcal{L}_2(\mathsf{G})$. We define the relation of *$p$ forcing $\varphi$ in $\mathcal{M}$*, written $p \Vdash^{\mathcal{M}} \varphi$, as follows.

1. If $\varphi$ is an atomic sentence of $\mathcal{L}_2$, then $p \Vdash^{\mathcal{M}} \varphi$ if $\mathcal{M} \vDash \varphi$.
2. If $\varphi$ is $t \in \mathsf{G}$ for a term $t$ then $p \Vdash^{\mathcal{M}} \varphi$ if $t^{\mathcal{M}} <^{\mathcal{M}} |V(p)|$ and $V(p)(t^{\mathcal{M}}) = 1$.
3. If $\varphi$ is $(\exists x < t)\psi(x)$ for a term $t$ then $p \Vdash^{\mathcal{M}} \psi$ if there is an $a \in M$ such that $a <^{\mathcal{M}} t^{\mathcal{M}}$ and $p \Vdash^{\mathcal{M}} \psi(a)$.
4. If $\varphi$ is $\psi_0 \vee \psi_1$, then $p \Vdash^{\mathcal{M}} \varphi$ if $p \Vdash^{\mathcal{M}} \psi_0$ or $p \Vdash^{\mathcal{M}} \psi_1$.
5. If $\varphi$ is $\neg\psi$, then $p \Vdash^{\mathcal{M}} \varphi$ if $q \nVdash^{\mathcal{M}} \psi$ for all $q \leqslant p$.
6. If $\varphi$ is $(\exists x)\psi(x)$, then $p \Vdash^{\mathcal{M}} \varphi$ if there is an $a \in M$ such that $p \Vdash^{\mathcal{M}} \psi(a)$.

We say *$p$ decides $\varphi$ in $\mathcal{M}$* if $p \Vdash^{\mathcal{M}} \varphi$ or $p \Vdash^{\mathcal{M}} \neg\varphi$.

If $\mathcal{M}$ is an $\omega$-model then the above agrees with Definition 7.4.1. This is because every $\omega$-model is an $\omega$-submodel of $\mathcal{M}^{\omega}$, so by Theorem 5.9.3, $\omega$-models satisfy the same arithmetical formulas. (We say arithmetical formulas are *absolute* across $\omega$-models.) Thus if $\mathcal{M}$ is *any* $\omega$-model then $\Vdash^{\mathcal{M}}$ is just $\Vdash$.

With Definition 7.6.4 in hand, we can define $n$-genericity (Definition 7.4.3) much as before.

**Definition 7.6.5 ($n$-generic filters).** Let $\mathcal{M}$ be a countable $\mathcal{L}_2$ structure, $\mathbb{P} = (P, \leqslant, V)$ a notion of forcing in $\mathcal{M}$, and $n \in \omega$. A filter $F$ is *$n$-generic in $\mathcal{M}$* if every $\Sigma^0_n$ sentence $\varphi$ of $\mathcal{L}_2(\mathsf{G})$ is decided in $\mathcal{M}$ by some $p \in F$.

The principal results from the previous sections now carry over *mutatis mutandis*, most notably the algebraic properties of the forcing relation (Proposition 7.4.2) and the connection between forcing and truth (Theorem 7.4.6).

**Theorem 7.6.6.** *Let $\mathcal{M}$ be a countable $\mathcal{L}_2$ structure, $\mathbb{P} = (P, \leqslant, V)$ a notion of forcing in $\mathcal{M}$, $n \in \omega$, $F$ an $n$-generic filter determining an object $G \in 2^{M}$, and $\varphi(X)$ a $\Sigma^0_n$ or $\Pi^0_n$ formula of $\mathcal{L}_2$. Then $\mathcal{M}[G] \vDash \varphi(G)$ if and only if some $p \in F$ forces $\varphi(\mathsf{G})$ in $\mathcal{M}$.*

Here, $\mathcal{M}[G]$ is as in Definition 5.9.5.

## 7.7 Harrington's theorem and conservation

An important application of the generalization of forcing to general models is to conservation results. We introduced the basic framework for this in the previous chapter, with Corollaries 5.10.2 and 5.10.7. Each of these earlier results is obtained by definitional extensions. Given a model, we expand its second order part by adding suitably definable sets. As a consequence, formulas involving these new sets can be immediately transformed into formulas not involving them, whence properties like comprehension and induction transfer from the original model to the expanded one. Now, we will instead add sets by forcing. The tension then is that the more generic a set is, the less definable it is over the original model.

The point of this section is to prove Harrington's theorem, which says that $\mathsf{WKL}_0$ has the same first order part as $\mathsf{RCA}_0$. First, though, let us look at the analogous result for $\mathsf{COH}$, which is simpler and can serve as a warm-up. We will see this result generalized in Theorem 8.7.23.

**Theorem 7.7.1 (Cholak, Jockusch, and Slaman [33]).** $\mathsf{RCA}_0 + \mathsf{COH}$ *is* $\Pi^1_1$ *conservative over* $\mathsf{RCA}_0$.

*Proof.* Fix an $\mathcal{L}_2$ structure $\mathcal{M}$ and $\vec{R} \in \mathcal{S}^{\mathcal{M}}$ such that

$$\mathcal{M} \vDash \mathsf{RCA}_0 + \text{"}\vec{R} \text{ is an instance of } \mathsf{COH}\text{"}.$$

By Corollary 5.9.6, it suffices to produce a set $G$ such that $\mathcal{M}[G] \vDash \mathsf{I}\Sigma^0_1$ and

$$\mathcal{M}[G] \vDash \text{"}G \text{ is an infinite } \vec{R}\text{-cohesive set"}.$$

We use Mathias forcing in $\mathcal{M}$. Conditions will thus be pairs $(E, I)$ such that $E, I \in \mathcal{S}^{\mathcal{M}}$, $E$ is $\mathcal{M}$-finite, $I$ is $\mathcal{M}$-infinite, and $\mathcal{M} \vDash E < I$. For any sufficiently generic $G$ for this forcing we will certainly have, in $\mathcal{M}[G]$, that $G$ is an infinite $\vec{R}$-cohesive set, by the same argument as in Example 7.3.9. So, we have only to verify that $\mathcal{M}[G] \vDash \mathsf{I}\Sigma^0_1$. Seeking a contradiction, say $\varphi(X, x)$ is a $\Sigma^0_1$ formula of $\mathcal{L}_2$ such that

$$\mathcal{M}[G] \vDash \varphi(G, 0) \wedge (\forall x)[\varphi(G, x) \to \varphi(G, x+1)] \wedge \neg\varphi(G, a)$$

for some $a \in M$. By Theorem 7.6.6, there is a Mathias condition $(E, I)$ such that $\mathcal{M}[G] \vDash E \subseteq G \subseteq E \cup I$ and such that $(E, I)$ forces

$$\varphi(\mathsf{G}, 0) \wedge (\forall x)[\varphi(\mathsf{G}, x) \to \varphi(\mathsf{G}, x+1)] \wedge \neg\varphi(\mathsf{G}, a) \tag{7.1}$$

in $\mathcal{M}$. Fix a $\Sigma^0_0$ formula $\psi(X, x, y)$ such that $\varphi(X, x)$ is equivalent in $\mathcal{M}$ to $(\exists y)\psi(X, x, y)$. Let $S$ be the set of all $b <^{\mathcal{M}} a$ such that there is a $y$ and a finite set $F \subseteq I$ with $\mathcal{M} \vDash \psi(E \cup F, b, y)$. Thus, $S$ is $\Sigma^0_1$-definable in $\mathcal{M}$, and so belongs to $\mathcal{S}^{\mathcal{M}}$ by bounded $\Sigma^0_1$ comprehension. Now, by definition of the valuation map for Mathias forcing, we see that for all $b <^{\mathcal{M}} a$, if $b \in S$ then $(E, I)$ has an extension that forces $\varphi(\mathsf{G}, b)$, and if $b \notin S$ then $(E, I) \Vdash \neg\varphi(G, b)$. By assumption,

$0 \in S$ and $a \notin S$. Hence, by $\mathsf{L}\Sigma^0_1$ in $\mathcal{M}$, we can fix a $b \leqslant^{\mathcal{M}} a$ such that $b-1 \in S$ and $b \notin S$. It follows that there is an extension of $(E, I)$ that forces $\varphi(\mathsf{G}, b-1) \wedge \neg\varphi(\mathsf{G}, b)$. But this extension must still force (7.1) in $\mathcal{M}$, so we have a contradiction. □

*Remark 7.7.2.* A crucial point in the above proof, which is common across forcing constructions, is that the complexity of the formula we are interested in forcing matches the complexity of forcing that formula. Sometimes, we get this automatically from the definition of forcing, as above. But if the formula is more complicated, or the forcing partial order is more complicated, this need not be the case. There are different strategies for dealing with such situations. We may be able to overcome the obstacle simply by organizing the proof in a careful way (as in Theorem 7.7.3, which we are about to look at). In other cases, we may need instead to force a more general formula (as in Lemma 8.5.7 below), or to restrict our set of conditions somehow (as in Propositions 8.7.4 and 8.7.5).

We now turn to the main result of this section, which is the following seminal result of Harrington.

**Theorem 7.7.3 (Harrington).** $\mathsf{WKL}_0$ *is* $\Pi^1_1$ *conservative over* $\mathsf{RCA}_0$.

Again, we give a model extension argument. Given an instance $T$ of $\mathsf{WKL}$ in a model $\mathcal{M}$ of $\mathsf{RCA}_0$, how do we pick a path $G$ through $T$ so that $\mathcal{M}[G]$ satisfies $\Sigma^0_1$ induction? (It is not difficult to see that we cannot pick a path arbitrarily. See Exercise 7.8.8.) The rough idea is to obtain $G$ by the low basis theorem, but formalizing this is quite delicate.

We will use Jockusch–Soare forcing inside a model $\mathcal{M}$ of $\mathsf{RCA}_0$. Our conditions will thus be pairs $(U, m)$ such that $\mathcal{M}$ satisfies that $U$ is an infinite tree with a unique string of length $m \in M$, with $(U^*, m^*) \leqslant (U, m)$ if $m \leqslant^{\mathcal{M}} m^*$ and $\mathcal{M}$ satisfies that $U^* \subseteq U$, and with the valuation of $(U, m)$ being its unique element of length $m$. We need the following technical lemma.

**Lemma 7.7.4.** *Fix a countable* $\mathcal{L}_2$ *structure* $\mathcal{M}$ *and a* $T \in \mathcal{S}^{\mathcal{M}}$ *such that*

$$\mathcal{M} \vDash \mathsf{RCA}_0 + \text{“}T \text{ is an infinite binary tree”}.$$

*Let* $\varphi(\vec{z})$ *be a* $\Sigma^0_0$ *formula of* $\mathcal{L}_2(\mathsf{G})$. *There exists a* $\Sigma^0_0$ *formula* $\varphi^{\exists}(x, \vec{z})$ *such that for every tuple* $\vec{b}$ *of elements of* $M$ *and every Jockusch–Soare condition* $(U, m)$ *in* $\mathcal{M}$ *the following hold.*

1. *If* $\mathcal{M} \vDash \varphi^{\exists}(\sigma, \vec{b})$ *for some* $\sigma \in 2^{<M}$ *then* $\mathcal{M} \vDash \varphi^{\exists}(\tau, \vec{b})$ *for every* $\tau \succeq^{\mathcal{M}} \sigma$.
2. *If* $\mathcal{M} \vDash \varphi^{\exists}(\sigma, \vec{b})$ *for the unique* $\sigma \in U$ *of length* $m$ *then* $(U, m) \Vdash^{\mathcal{M}} \varphi(\vec{b})$.
3. *If* $(U, m) \Vdash^{\mathcal{M}} \varphi(\vec{b})$ *then there is some* $(U^*, m^*) \leqslant (U, m)$ *in* $\mathcal{M}$ *such that* $\mathcal{M} \vDash \varphi^{\exists}(\sigma, \vec{b})$ *holds for the unique* $\sigma \in U^*$ *of length* $m^*$.

*Proof.* We define $\varphi^{\exists}$ along with an auxiliary formula $\varphi^{\forall}(x, \vec{z})$ such that for every tuple $\vec{b}$ of elements of $M$ and every condition $(U, m)$ in $\mathcal{M}$ the following hold.

4. If $\mathcal{M} \vDash \varphi^{\forall}(\sigma, \vec{b})$ for some $\sigma \in 2^{<M}$ then $\mathcal{M} \vDash \varphi^{\forall}(\rho, \vec{b})$ for ever $\rho \preceq^{\mathcal{M}} \sigma$.

5. If $\mathcal{M} \models \varphi^{\forall}(\sigma, \vec{b})$ for all $\sigma \in U$ then there is some $(U^*, m^*) \leqslant (U, m)$ in $\mathcal{M}$ such that $(U^*, m^*) \Vdash^{\mathcal{M}} \varphi(\vec{b})$.
6. If $(U, m) \Vdash^{\mathcal{M}} \varphi(\vec{b})$ then $\mathcal{M} \models \varphi^{\forall}(\sigma, \vec{b})$ for all $\sigma \in U$.

View $\varphi(\vec{z})$ as $\varphi(\mathsf{G}, \vec{z})$ for a formula $\varphi(X, \vec{z})$ of $\mathcal{L}_2$. We proceed by induction on the complexity of $\varphi$. First, suppose $\varphi$ is atomic. If $X$ does not occur in $\varphi$, then for every tuple $\vec{b}$, $\varphi(\vec{b})$ is a sentence of $\mathcal{L}_2$ which is either true in $\mathcal{M}$ and every condition forces it, or false in $\mathcal{M}$ and every condition forces its negation. In this case, we let $\varphi^{\exists} = \varphi^{\forall} = \varphi$.

If $X$ does occur in $\varphi$, then $\varphi$ is $t \in X$ for some term $t$. We let

$$\varphi^{\exists}(x, \vec{z}) \equiv t < |x| \wedge x(t) = 1$$

and

$$\varphi^{\forall}(x, \vec{z}) \equiv t < |x| \rightarrow x(t) = 1.$$

We now complete the induction. If $\varphi \equiv \neg\psi$ we let $\varphi^{\exists} \equiv \neg\psi^{\forall}$ and $\varphi^{\forall} = \neg\psi^{\exists}$. If $\varphi = \psi_0 \vee \psi_1$ we let $\varphi^{\exists} \equiv \psi_0^{\exists} \vee \psi_1^{\exists}$ and $\varphi^{\forall} \equiv \psi_0^{\forall} \vee \psi_1^{\forall}$. And if $\varphi(X, \vec{z}) = (\exists y < t)\psi(X, y, \vec{z})$, we let $\varphi^{\exists}(x, \vec{z}) \equiv (\exists y < t)\psi^{\exists}(x, y, \vec{z})$ and $\varphi^{\forall}(x, \vec{z}) \equiv (\exists y < t)\psi^{\forall}(x, y, \vec{z})$.

Verifying properties (1)–(6) is straightforward in all these cases, with the exception of property (5) in the very last case. So suppose that for some tuple $\vec{b}$ of elements of $M$ and some condition $(U, m)$, $\mathcal{M} \models \varphi^{\forall}(\sigma, \vec{b})$ for all $\sigma \in U$. We claim that for some $a <^{\mathcal{M}} t^{\mathcal{M}}$, the set of all $\sigma \in U$ such that $\mathcal{M} \models \psi^{\forall}(\sigma, a, \vec{b})$ is $\mathcal{M}$-infinite. If we have this, let $\widehat{U}$ be the set of all these $\sigma$. Then $\widehat{U}$ is a tree in $\mathcal{M}$ by property (4), so $(\widehat{U}, m)$ is an extension of $(U, m)$. Now by induction there is an extension $(U^*, m^*)$ of $(\widehat{U}, m)$, and hence of $(U, m)$, that forces $\psi^{\forall}(\mathsf{G}, a, \vec{b})$. But then this extension also forces $(\exists y < t)\psi(\mathsf{G}, y, \vec{b})$, which is $\varphi(\mathsf{G}, \vec{b})$, so we have (5) for $\varphi^{\forall}$. This means that proving the claim finishes the proof.

Seeking a contradiction, suppose the claim is false. Then for every $a <^{\mathcal{M}} t^{\mathcal{M}}$, the set of $\sigma \in U$ such that $\mathcal{M} \models \psi^{\forall}(\sigma, a, \vec{b})$ is $\mathcal{M}$-finite. In particular, for every $a <^{\mathcal{M}} t^{\mathcal{M}}$ there is a level $\ell \in M$ such that $\mathcal{M} \models \neg\psi^{\forall}(\sigma, a, \vec{b})$ for every $\sigma \in U$ with $|\sigma| = \ell$. Notice that by hypothesis on $\psi^{\forall}$ and property (4), this means also that $\mathcal{M} \models \neg\psi^{\forall}(\tau, a, \vec{b})$ for every $\tau$ with $|\tau| > \ell$. Now since $\psi$ is $\Sigma^0_0$, we can apply $\mathsf{B}\Sigma^0_0$ in $\mathcal{M}$ to find an $m \in M$ such that for every $a <^{\mathcal{M}} t^{\mathcal{M}}$ there exists such an $\ell$ with $\ell <^{\mathcal{M}} m$. But then $\mathcal{M} \models \neg\psi^{\forall}(\tau, a, \vec{b})$ for every $\sigma \in U$ with $|\sigma| = m$ and every $a <^{\mathcal{M}} t^{\mathcal{M}}$, which cannot be. □

*Proof (of Theorem 7.7.3).* Fix a countable $\mathcal{L}_2$ structure $\mathcal{M}$ and a $T \in \mathcal{S}^{\mathcal{M}}$ such that

$$\mathcal{M} \models \mathsf{RCA}_0 + \text{“}T \text{ is an infinite binary tree”}.$$

Consider again Jockusch–Soare forcing with subtrees of $T$ inside $\mathcal{M}$. Let $G$ be 2-$\mathcal{S}^{\mathcal{M}}$-generic in $\mathcal{M}$. We then claim that $\mathcal{M}[G] \models \mathsf{RCA}_0 +$ “$G$ is a path through $T$”. From here, the result follows by Corollary 5.9.6. That $\mathcal{M}[G]$ satisfies $\Delta^0_1$ comprehension follows by definition, while the fact that $G$ is a path through $T$ in $\mathcal{M}[G]$ follows by genericity. So it is enough to show that $\mathcal{M}[G]$ satisfies $\Sigma^0_1$ induction.

Seeking a contradiction, suppose $\varphi(X,x)$ is a $\Sigma^0_1$ formula of $\mathcal{L}_2$ such that

$$\mathcal{M}[G] \vDash \varphi(G,0) \wedge (\forall x)[\varphi(G,x) \rightarrow \varphi(G,x+1)] \wedge \neg\varphi(G,a)$$

for some $a \in M$. By Theorem 7.6.6, there is a condition $(U,m)$ such that $G$ is a path through $U$ and

$$(U,m) \Vdash^{\mathcal{M}} \varphi(\mathsf{G},0) \wedge (\forall x)[\varphi(\mathsf{G},x) \rightarrow \varphi(\mathsf{G},x+1)] \wedge \neg\varphi(\mathsf{G},a). \qquad (7.2)$$

Write $\varphi(X,x)$ as $(\exists y)\psi(X,x,y)$, where $\psi$ is $\Sigma^0_0$. Let $\psi^{\exists}$ be as given by Lemma 7.7.4. Then for each $b \in M$, the set $U_b = \{\sigma \in U : \mathcal{M} \vDash (\forall y \leqslant |\sigma|)\neg\psi^{\exists}(\sigma,b,y)\}$ is $\mathcal{M}$-infinite if and only if, in $\mathcal{M}$, there is an extension $(U^*,m^*)$ of $(U,m)$ that forces $\neg\varphi(\mathsf{G},b)$. So if we let $A$ be the set of all $b \leqslant^{\mathcal{M}} a$ for which $U_b$ is $\mathcal{M}$-infinite, then $a \in A$ and $0 \notin A$. Now $A \in \mathcal{S}^{\mathcal{M}}$ by bounded $\Pi^0_1$ comprehension, and so it has a least element, $b >^{\mathcal{M}} 0$. It follows that there is a $(U^*,m^*)$ such that $\mathcal{M}$ satisfies that $(U^*,m^*)$ is an extension of $(U,m)$ forcing $\varphi(\mathsf{G},b-1) \wedge \neg\varphi(\mathsf{G},b)$, contradicting (7.2). □

Using a more refined argument, it is possible to obtain a version of Harrington's theorem for $\mathsf{I}\Sigma^0_2$ instead of $\mathsf{I}\Sigma^0_1$.

**Theorem 7.7.5 (Cholak, Jockusch, and Slaman [33]).** *Every countable model of* $\mathsf{RCA}_0 + \mathsf{I}\Sigma^0_2$ *is an* $\omega$*-submodel of a countable model of* $\mathsf{WKL}_0 + \mathsf{I}\Sigma^0_2$.

We omit the proof here. As a corollary, $\mathsf{WKL}_0 + \mathsf{I}\Sigma^0_2$ is $\Pi^1_1$ conservative over $\mathsf{RCA}_0 + \mathsf{I}\Sigma^0_2$. This fact, which is of independent interest, actually holds at all levels of the arithmetical hierarchy.

**Theorem 7.7.6 (Hájek [132]; Avigad [7]).** *For all* $n \geqslant 1$, $\mathsf{WKL}_0 + \mathsf{I}\Sigma^0_n$ *is* $\Pi^1_1$ *conservative over* $\mathsf{RCA}_0 + \mathsf{I}\Sigma^0_n$.

## 7.8 Exercises

**Exercise 7.8.1.** Let $G$ be Cohen 1-generic.

1. Write $G = G_0 \oplus G_1$. Show that $G_0$ and $G_1$ form a *minimal pair*, meaning that for any set $X$ computable from both $G_0$ and $G_1$, $X$ is computable.
2. Write $G = \bigoplus_{i \in \omega} G_i$. Show that for all $i$, $G_i \nleq_{\mathrm{T}} \bigoplus_{j \in \omega, j \neq i} G_j$.

**Exercise 7.8.2.** Prove part (1) of Proposition 7.5.8.

**Exercise 7.8.3 (Yu [329]).** Let $G$ be Cohen $n$-generic and write $G = G_0 \oplus G_1$. Show that $G_0$ is Cohen $n$-generic and that $G_1$ is Cohen $n$-generic relative to $G_0$. (This is analogous to van Lambalgen's theorem (Theorem 9.4.17) from algorithmic randomness.)

**Exercise 7.8.4.** Give an example of a notion of forcing, a sentence $\varphi$ of the forcing language, and a condition that does not force $\varphi \vee \neg\varphi$.

**Exercise 7.8.5.** Show that in any forcing notion, if $\varphi$ is any sentence in the forcing language and $p$ any condition, then $p \Vdash_w \varphi$ if and only if $p \Vdash \neg\neg\varphi$.

**Exercise 7.8.6.** Complete the proof of Proposition 7.4.2.

**Exercise 7.8.7.** Prove Lemma 7.5.10.

**Exercise 7.8.8.** This exercise explains why the path in Harrington's theorem (Theorem 7.7.3) had to be chosen so carefully. Fix $n > 1$, let $\mathcal{M}$ be a model of $\mathsf{RCA}_0 + \neg\mathsf{I}\Sigma^0_n$. Let $T = 2^{<M}$, so that $T \in \mathcal{S}^{\mathcal{M}}$ and $\mathcal{M}$ satisfies that $T$ is an infinite binary tree. Show that there is a path $f$ through $T$ such that $\mathcal{M}[f] \vDash \neg\mathsf{I}\Sigma^0_1$. (Hint: Use the fact from Theorem 6.2.3 that $\mathcal{M}$ has a $\Sigma^0_n$ definable proper cut.)

**Exercise 7.8.9.** Let $(E, I)$ be Mathias conditions, let $\varphi(X)$ be a $\Sigma^0_1$ formula of $\mathcal{L}_2$. Show that if $(E, I) \Vdash \varphi(\mathsf{G})$ then $\varphi(S)$ holds for every set $S$ with $E \subseteq S \subseteq E \cup I$.

**Exercise 7.8.10.** Let $(E^*, I^*) \leqslant (E, I)$ be Mathias conditions and let $\varphi$ be a $\Sigma^0_1$ formula of $\mathcal{L}_2(\mathsf{G})$. Show that if $(E^*, I^*) \Vdash \varphi$ then so does $(E^*, I \setminus \max E^*)$. So in Mathias forcing, $\Sigma^0_1$ formulas can always be forced by finite extensions. (Hint: Induction on complexity.)

**Exercise 7.8.11.** Show that in Mathias forcing with computable reservoirs there exist low 1-generics. (Hint: Use Exercise 7.8.10.)

# Part III
# Combinatorics

# Chapter 8
# Ramsey's theorem

Ramsey proved his eponymous theorem in his 1929 paper "On a problem of formal logic" [253]. The focus was actually not on combinatorics at all, but rather on the *Entscheidungsproblem*, which was still open at the time. (The dramatic results of Church and Turing showing that the problem is unsolvable were still a few years away.) Ramsey used his theorem to study a special case of the problem. More precisely, he used the finitary Ramsey's theorem (Definition 3.3.6), though in fact he proved the infinite version first and then deduced the finitary one from it. (This is still a common proof.) In the intervening hundred years, both versions have had a vast impact and legacy. Our interest will be confined to the infinite version.

In broad terms, Ramsey's theorem can be thought of as saying that some amount of order is necessary in any configuration of objects. Understanding this order—how it is organized and how it can be described—has been the focus of much interest in combinatorics and logic, not least because it is so intrinsically captivating. In computability theory specifically, it has spawned a long and fruitful line of research.

The earliest foray into the effective aspects of Ramsey's theorem was by Specker, who showed that Ramsey's theorem does not hold computably.

**Theorem 8.0.1 (Specker [300]).** *For every $n \geqslant 2$, $\mathsf{RT}^2$ omits computable solutions.*

Historically, this was an important result, showing that a purely combinatorial result could have interesting logical (and in particular, computability theoretic) content. The mantle was then picked up by Jockusch in his seminal paper "Ramsey's theorem and recursion theory" [168], in which he not only significantly extended Specker's result, but brought to bear the full machinery of the time to give a remarkably deep analysis of Ramsey's theorem from the point of view of computability and complexity. This, it can be said, founded and firmly established computable combinatorics as a viable field.

In this and the next chapter, we survey the reverse mathematics of combinatorics, starting now with Ramsey's theorem itself. Note that we have collected a number of results about Ramsey's theorem already, in Chapters 3, 4 and 6. For example, we have Hirst's theorem (Theorem 6.5.1) that $\mathsf{RT}^1$ is equivalent over $\mathsf{RCA}_0$ to $\mathsf{B}\Sigma^0_2$. We have also seen, in Corollary 4.5.10, that $\mathsf{RT}^n \equiv_\omega \mathsf{RT}^n_k$ for every $k \geqslant 2$. The proof



D. D. Dzhafarov, C. Mummert, *Reverse Mathematics*, Theory and Applications of Computability, https://doi.org/10.1007/978-3-031-11367-3_8

can be easily formalized to show that for all (standard) $k \geqslant 2$, $\mathsf{RCA}_0 \vdash \mathsf{RT}^n_2 \leftrightarrow \mathsf{RT}^n_k$. (The point being, that when $k$ is standard the induction in the proof can be carried out externally, rather than in $\mathsf{RCA}_0$.) Of course, it is also trivially the case that $\mathsf{RCA}_0 \vdash (\forall n)\mathsf{RT}^n_1$.

For completeness, we mention also that $\mathsf{RCA}_0$ proves the finitary Ramsey's theorem (FRT) defined in Definition 3.3.6. A proof in $\mathsf{PA}^- + \mathsf{I}\Sigma^0_1$ is given by Hájek and Pudlák [134, Chapter II.1]. Since FRT can be expressed as an arithmetical, and hence $\Pi^1_1$, statement, provability in $\mathsf{RCA}_0$ follows by Corollary 5.10.2.

We begin our discussion by understanding the complexity of homogeneous sets in terms of the arithmetical hierarchy. This will also give us some preliminary results in terms of subsystems of second order arithmetic.

## 8.1 Upper bounds

Our first result is the following important theorem of Jockusch.

**Theorem 8.1.1 (Jockusch [168]).** *For all $n \geqslant 1$, $\mathsf{RT}^n$ admits $\Pi^0_n$ solutions.*

For $n = 1$ this is trivial. For $n \geqslant 2$ it follows by taking $I = \omega$ in the following lemma.

**Lemma 8.1.2 (Jockusch [168]).** *Fix $n \geqslant 2$, $k \geqslant 1$, and an infinite set $I$. Then every $c\colon [I]^2 \to k$ has an infinite homogeneous set $H \subseteq I$ that is $\Pi^0_n(c \oplus I)$.*

To prove it, we introduce an auxiliary notion that is quite helpful in certain kinds of constructions of homogeneous sets.

**Definition 8.1.3.** Fix $n \geqslant 2$, $k \geqslant 1$, and a coloring $c\colon [\omega]^n \to k$. A set $F$ is *pre-homogeneous* for $c$ if for each $\vec{x} \in [F]^{n-1}$ there is an $i < k$ such that $c(\vec{x}, y) = i$ for all $y \in F$ with $\vec{x} < y$.

We leave the following to the reader and proceed to the proof of the main result.

**Lemma 8.1.4.** *Fix $n \geqslant 2$, $k \geqslant 1$, and a coloring $c\colon [\omega]^n \to k$. If $P$ is an infinite pre-homogeneous set for $c$ then $c$ has a $(c \oplus P)$-computable infinite homogeneous set $H \subseteq P$.*

*Proof (of Lemma 8.1.2).* By induction on $n$.

***Inductive basis.*** Assume $n = 2$. As usual, we restrict to computable instances for simplicity, with the full result following by relativization. So fix $k \geqslant 2$, an infinite computable set $I$, and a computable coloring $c\colon [I]^2 \to k$. We begin by constructing a $\Pi^0_2$ infinite set $P \subseteq I$ which is pre-homogeneous for $c$. We use a $\varnothing'$ oracle to enumerate elements of $I$ into $\overline{P}$ by stages. At each stage $s$, we define a finite nonempty set $F_s$ of elements of $I$ not yet enumerated into $\overline{P}$, as well as a function $d_s\colon F_s \to k$. We enumerate at most finitely many elements into $\overline{P}$ at each stage.

***Construction.***

*Stage* $s = 0$*:* Let $F_0 = \{\min I\}$ and $d_0(\min I) = 0$.

*Stage* $s > 0$*:* Assume inductively that $F_{s-1}$ and $d_{s-1}$ have been defined. Fix the largest $z \in F_{s-1}$ for which the following holds: there exists $w > F_{s-1}$ in $I$ such that $w$ has not yet been enumerated into $\overline{P}$ and $c(x, w) = d_{s-1}(x)$ for all $x < z$ in $F_{s-1}$. Note that $z$ exists, since the latter condition is always satisfied by $z = \min F_{s-1}$. Fix a corresponding $w$ that minimizes $c(z, w)$, and let $F_s = \{x \in F_s : x \leqslant z\} \cup \{w\}$. Then, define

$$d_s(x) = \begin{cases} d_{s-1}(x) & \text{if } x < z, \\ c(z, w) & \text{if } x = z, \\ 0 & \text{if } x = w. \end{cases}$$

Finally, enumerate all numbers $x \leqslant s$ in $I \setminus F_s$ into $\overline{P}$.

***Verification.*** As the construction is computable in $\varnothing'$, it follows that $\overline{P}$ is $\Sigma^0_1(\varnothing')$. Hence, $P$ is $\Pi^0_2$, as desired. We prove the following claims.

*Claim 1: A number* $x \in I$ *belongs to* $P$ *if and only if* $x \in F_s$ *for all sufficiently large* $s$. By construction, $x \in I$ is enumerated into $\overline{P}$ at a stage $s$ if and only if $x \leqslant s$ and $x \notin F_s$.

*Claim 2: If* $x \in P$ *and* $x \in F_s$ *for some* $s$ *then* $x \in F_t$ *for all* $t \geqslant s$. If $x \in F_s \setminus F_{s+1}$ then it must be that the numbers $z$ and $w$ found at stage $s + 1$ of the construction satisfy $z < x < w$. But then after stage $s + 1$, only numbers larger than $w$ are ever eligible to be part of $P$.

*Claim 3: For each* $x \in P$, $\lim_s d_s(x)$ *exists.* Fix $x \in P$, and assume the claim for all $y < x$ in $P$. By the first claim, we can find $s_0 \geqslant x$ such that for all $s \geqslant s_0$, if $y \leqslant x$ belongs to $P$ then $y \in F_s$, and if $y < x$ is in $P$ then $d_s(y) = d_{s_0}(y)$. Note that since $s \geqslant x$, every $y < x$ in $I \setminus P$ is enumerated into $\overline{P}$ by stage $s_0$, so no such $y$ can belong to $F_s$ for any $s \geqslant s_0$. Now the only reason we could have $d_s(x) \neq d_{s_0}(x)$ for some $s > s_0$ is if $x$ is the number $z$ found at stage $s$ of the construction. By construction, and our hypothesis on the $y < x$, this means there are infinitely many $w \in I$ such that $c(y, w) = d_{s_0}(y)$ for each $y < x$ in $P$, but $c(x, w) \neq d_{s_0}(x)$ for almost all such $w$. It follows that the value of $d_s(x)$ is changed in at most finitely many stages $s > s_0$.

*Claim 4: P is pre-homogeneous.* Fix $x \in P$, and let $s_0$ be the least stage $s$ so that $x \in F_s$. By minimality of $s_0$, either $s_0 = 0$ and $x = \min I$, or $s_0 > 0$ and $x$ is the number $w$ found at stage $s_0$ of the construction. In any case, $x$ is the largest element of $F_{s_0}$. By Claim 2, if $s > s_0$ then $x \in F_s$. Thus, the number $z$ found at any such stage $s$ of the construction must satisfy $x \leqslant z$. Moreover, as noted in the preceding claim, the number of such stages $s$ at which $x = z$ must be finite. Let $s_1 > s_0$ be the final stage at which $x = z$, or $s_0 + 1$ if there is no such stage. Thus, $F_{s_1}$ contains precisely one element larger than $x$ and $d_{s_1}(x) = d_s(x)$ for all $s \geqslant s_1$. Moreover, at every stage $s \geqslant s_1$ an element $w$ is found and added to $F_s$, and this number must satisfy $c(x, z) = d_s(x) = d_{s_1}(x)$.

*Claim 5: $P$ is infinite.* The proof is similar to that of Claim 4. Suppose $x \in P$. We exhibit a larger element of $P$. Let $s_1$ be as in Claim 4, i.e., the last stage $s$ such that $x$ is the number $z$ found at this stage. Thus $F_{s_1}$ contains precisely one element $w > x$, and we have $d_{s_1}(x) = d_s(x)$ for all $s \geqslant s_1$, and $d_{s_1}(x) = c(x, w)$ by definition. Hence, $w$ is never enumerated into $\overline{P}$ and so belongs to $P$.

The verification is complete.

To complete the proof of the $n = 2$ case, define $H_i = \{x \in P : \lim_s d_s(x) = i\}$ for each $i < k$. Thus, each $H_i$ is homogeneous for $c$ with color $i$. By construction, for each $x \in P$ the value of $d_s(x)$ can only increase with $s$. Hence, for each $i$ we have that $x \in \bigcup_{j \leqslant i} H_j$ if and only if $d_s(x) \leqslant i$ for all $s$. Since $d \leqslant_T \varnothing'$, this means $\bigcup_{j \leqslant i} H_j$ is $\Pi^0_1(\varnothing')$ and therefore $\Pi^0_2$. Let $i < k$ be least such that $H_i$ is infinite, and fix $m$ so that no $x > m$ has $\lim_s d_s(x) < i$. Then $H = \{x \in \bigcup_{j \leqslant i} H_j : x > m\}$ is a $\Pi^0_2$ infinite homogeneous set for $c$.

***Inductive step.*** Assume $n > 2$ and the result holds for $n - 1$. Fix $k \geqslant 2$, $I \subseteq \omega$, and a coloring $c \colon [I]^n \to k$. By the low basis theorem (Theorem 2.8.18) relative to $(c \oplus I)'$, fix $X \gg (c \oplus I)'$ with $X' \leqslant_T (c \oplus I)''$. It then suffices to produce an infinite homogeneous set $H$ for $c$ which is $\Pi^0_{n-1}(X)$. For, by Post's theorem (Theorem 2.6.2), we have that

$$\Pi^0_{n-1}(X) = \Pi^0_{n-2}(X') \subseteq \Pi^0_{n-2}((c \oplus I)'') = \Pi^0_n(c \oplus I)$$

as classes of sets. We first define an $X$-computable set $I^* = \{x_0 < x_1 < \cdots\} \subseteq I$ inductively as follows. Let $x_0 < \cdots < x_{n-2}$ be the least $n - 1$ elements of $I$ and define $I_{n-1} = \{x \in I : x > x_{n-2}\}$. Suppose next that for some $s \geqslant n - 1$ we have defined $x_0 < \cdots < x_{s-1}$ along with a $(c \oplus I)$-computable infinite set $I_s \subseteq I$ with $x_s < I_s$. Since $s \geqslant n - 1$, we may fix $m \geqslant 1$ so that there are $m$ elements of $[\{x_0, \ldots, x_{s-1}\}]^{n-1}$ containing $x_{s-1}$. Enumerate these as $\vec{x}_0, \ldots, \vec{x}_{m-1}$. Now define $J_0 = I_s$ and suppose that for some $\ell < m$ we have defined a $(c \oplus I)$-computable infinite set $J_\ell \subseteq I_s$. Since $X \gg (c \oplus I)'$, it follows by Theorem 2.8.25 that $X$ can uniformly compute a color $i_\ell < k$ such that there are infinitely many $y \in J_\ell$ with $c(\vec{x}_\ell, y) = i_\ell$. Let $J_{\ell+1} = \{y \in J_\ell : c(\vec{x}_\ell, y) = i_\ell\}$. Finally, let $x_s$ be the least element of $J_m$, and let $I_{s+1} = \{x \in J_m : x > x_s\}$. Thus, for each $\ell < m$ we have $c(\vec{x}_\ell, x_s) = i_\ell$ and $c(\vec{x}_\ell, y) = i_\ell$ for all $y \in I_s$.

By construction, if $\vec{x} \in [I^*]^{n-1}$ then $c(\vec{x}, y)$ is the same for all $y > \vec{x}$ in $I^*$. We may thus define a $(c \oplus I^*)$-computable, and hence $X$-computable, coloring $c^* \colon [I^*]^{n-1} \to k$ where $c^*(\vec{x}) = c(\vec{x}, y)$ for the least $y > \vec{x}$ in $I^*$. By induction hypothesis, there is a $\Pi^0_{n-1}(X)$ infinite homogeneous set $H \subseteq I^* \subseteq I$ for $c^*$. Clearly, this is also homogeneous for $c$. The proof is complete. □

The fact that Ramsey's theorem admits arithmetical solutions should lead us to believe that the theorem is provable in $\mathsf{ACA}_0$. Indeed, for each fixed exponent, this is the case. Formalizing the kind of argument we just gave is tricky, however. Instead, we give a separate argument that is more direct and less delicate.

**Proposition 8.1.5.** *For all $n \geqslant 1$, $\mathsf{ACA}_0 \vdash \mathsf{RT}^n$.*

*Proof.* For $n = 1$, this follows by Hirst's theorem (Theorem 6.5.1) and the fact that $\mathsf{B}\Sigma^0_2$ is provable in $\mathsf{ACA}_0$. For $n \geqslant 2$, we recall the proof of Proposition 4.5.12 which gave an explicit construction of an infinite homogeneous set for a given $k$-coloring. This can be readily formalized in $\mathsf{ACA}_0$. Note that the proof featured an induction: for each $m < n$, we used $\mathsf{RT}^{(m)}$ to prove $\mathsf{RT}^{m+1}$. But as $n$ here is a fixed standard number, and the induction is finite (proceeding only up to $m = n - 1$), we are not actually using induction within $\mathsf{ACA}_0$ here. □

What of the full Ramsey's theorem, $\mathsf{RT}$? It is still a problem that admits arithmetical solutions, but both Theorem 8.1.1 and the proof of the preceding proposition seem to suggest that the complexity of the solutions increases with the exponent. And we will soon see that this is unavoidable. (So, e.g., there is no $m \in \omega$ so that every computable instance of $\mathsf{RT}$ has a $\varnothing^{(m)}$-computable solution.) From here, the fact that $\mathsf{ACA}_0$ does not prove $\mathsf{RT}$ follows by Theorem 5.6.7. (Another way to see this is described in Exercise 8.10.8.) The evident reason is that $\mathsf{ACA}_0$ cannot prove $(\forall n)(\forall X)[X^{(n)} \text{ exists}]$. That much is the role of the of the subsystem $\mathsf{ACA}_0'$ from Definition 5.6.5, discussed in detail in Section 5.6.1. And this turns out to capture the strength of $\mathsf{RT}$ precisely.

**Theorem 8.1.6 (McAloon [207]).** $\mathsf{RCA}_0 \vdash \mathsf{RT} \leftrightarrow \mathsf{ACA}_0'$.

*Proof.* First, we show $\mathsf{ACA}_0' \vdash \mathsf{RT}$. Arguing in $\mathsf{ACA}_0'$, fix $n \geqslant 2$, $k \geqslant 1$, and $c \colon [\mathbb{N}]^n \to k$. Let $\alpha_0$ be an increasing string of length $n - 1$. Let $T$ be the set of all $\alpha \preceq \alpha_0$ as well as all increasing $\alpha \in \mathbb{N}^{<\mathbb{N}}$ satisfying the following conditions.

- $\alpha_0 \preceq \alpha$.
- $\mathrm{range}(\alpha)$ is pre-homogeneous for $c$.
- For each $j < \ell$, there are infinitely many $y$ such that $c(\vec{x}, y) = c(\vec{x}, \alpha(j))$ for every $\vec{x} \in [\mathrm{range}(\alpha \upharpoonright j)]^{(n-1)}$, and $\alpha(j)$ is the least such $y$.

$T$ is clearly a tree, and it is finitely branching since for each $\alpha \in T$ and each $i < k$ there can be at most one $x$ such that $\alpha x \in T$. Note that the defining conditions above are uniformly arithmetical in $n$, and so $T$ is $\Delta^0_{n+t}$-definable, for some (standard) number $t$. Thus, $T$ exists.

We claim that $T$ is infinite. Since $T$ is finitely branching, we must show that for each $\ell$ there is an $\alpha \in T$ with $|\alpha| = \ell$. We proceed by induction on $\ell$. If $\ell \leqslant n - 1$, we have $\alpha_0$ in $T$ along with its initial segments. Suppose next that $\ell = n$. There is a single tuple $\vec{x} \in [\mathrm{range}(\alpha_0)]^{n-1}$, and by $\mathsf{RT}^1_k$, there is an $i < k$ such that $c(\vec{x}, y) = i$ for infinitely many $y$. Fix some such $i$, and the least corresponding $y$. Then $\alpha_0 y \in T$. Finally, suppose $\ell > n$, and assume the result holds for $\ell - 1$. Fix any $\alpha \in T$ of length $\ell - 1$ and let $x = \alpha(|\alpha| - 1)$, so that $\alpha = \alpha^\# x$. Let $I$ be the set of all $y > x$ such that $c(\vec{x}, y) = c(\vec{x}, x)$ for all $\vec{x} \in [\mathrm{range}(\alpha^\#)]^{(n-1)}$, which is infinite by assumption. Now enumerate the tuples $\vec{x} \in [\mathrm{range}(\alpha)]^{(n-1)}$ that contain $x$ as $\vec{x}_0, \ldots, \vec{x}_{m-1}$. Define $p \colon m \to k$ by recursion, as follows: for each $s < m$, let $p(s)$ be the least $i < k$ such that there are infinitely many $y \in I$ such that $c(\vec{x}_t, y) = p(t)$ for all $t < s$ and

$c(\vec{x}_s, y) = i$. Let $I^*$ be the set of all $y \in I$ such that $c(\vec{x}_s, y) = p(s)$ for all $s < m$, which is infinite by construction. Let $x^* = \min I^*$; then it is easily seen that $\alpha x^* \in T$. The claim is proved.

Since $T$ is a finitely branching, infinite subtree of $\mathbb{N}^{<\mathbb{N}}$, we may fix a path $f \in [T]$. It is clear that range$(f)$ is an infinite pre-homogeneous set for $c$. For each $x \in$ range$(f)$, let $d(x) = c(x, y)$ for the least $y > x$ in range$(f)$, which gives us an instance $d$: range$(f) \to k$ of $\mathsf{RT}^1_k$. As in Lemma 8.1.4, it now follows that if $H \subseteq$ range$(f)$ is homogeneous for $d$ then $H$ is homogeneous for $c$. This completes the proof that $\mathsf{ACA}_0' \vdash \mathsf{RT}$.

Now assume $\mathsf{RCA}_0 + \mathsf{RT}$. We $\mathsf{RT} \to \mathsf{RT}^3$; in Corollary 8.2.6 below, we will see that $\mathsf{RT}^3 \to \mathsf{ACA}_0$. Thus, we may argue in $\mathsf{ACA}_0 + \mathsf{RT}$. To derive $\mathsf{ACA}_0'$, fix a set $X$ and $n \geqslant 1$. W must show that $X^{(n)}$ exists. For an arbitrary set $Z$ and $s_0 \in \omega$, define

$$Z^{(0)}_{s_0} = Z_0 \upharpoonright s_0,$$

and for $i \geqslant 1$ and $s_0, \ldots, s_i \in \omega$, define

$$Z^{(i)}_{s_0,\ldots,s_i} = \{e < s_0 : (\exists s < s_1)[\Phi_{e,s}^{Z^{(i-1)}_{s_1,\ldots,s_i}}(e)\downarrow]\}.$$

Here we refer to a coding of Turing functionals in second order arithmetic, as discussed in Section 5.5.3. In particular, by arithmetical comprehension, $Z^{(i)}_{s_0,\ldots,s_i}$ exists for all $i$ and $s_0, \ldots, s_i$.

We now define a coloring $c \colon [\omega]^{2n+1} \to n+2$, which is an instance of $\mathsf{RT}$. Given $s_0 < s_1 < \cdots < s_n < t_1 < \cdots < t_n$, first check if there is an $i \leqslant n$ such that

$$X^{(i)}_{s_0,s_1,\ldots,s_i} \neq X^{(i)}_{s_0,t_1,\ldots,t_i}.$$

If so, set $c(s_0, s_1, \ldots, s_n, t_1, \ldots, t_n) = i$ for the least such $i$. And otherwise, set $c(s_0, s_1, \ldots, s_n, t_1, \ldots, t_n) = n + 1$. Clearly, $c$ exists, and we may thus apply $\mathsf{RT}$ to obtain an infinite homogeneous set $H$ for it.

We claim that for every $i \leqslant n$ and all $s_0 < \cdots < s_i$ in $H$,

$$X^{(i)}_{s_0,\ldots,s_i} = X^{(i)} \upharpoonright s_0.$$

Notice that the statement of the claim is arithmetical, so since we are working in $\mathsf{ACA}_0$ we can prove it by induction. For $i = 0$, this is clear. So suppose $1 \leqslant i \leqslant n$ and the result holds for $i - 1$. Thus for all $t_1 < \cdots < t_i$ in $H$ we have that

$$X^{(i-1)}_{t_1,\ldots,t_i} = X^{(i-1)} \upharpoonright t_1.$$

Fix $s_0 < \cdots < s_i$ in $H$. Fix $u$ so that for all $t > u$ and all $e < s_0$, we have that

$$\Phi_e^{X^{(i-1)}}(e)\downarrow \leftrightarrow (\exists s < t)[\Phi_e^{X^{(i-1)} \upharpoonright t}(e)\downarrow].$$

Then for any tuple $t_1 < \cdots < t_i$ in $H$ with $t_1$ larger than $s_0$ and $u$, we have

$$\begin{aligned} X^{(i)}_{s_0,t_1,\ldots,t_i} &= \{e < s_0 : (\exists s < t_1)[\Phi^{X^{(i-1)}_{t_1,\ldots,t_i}}_{e,s}(e)\downarrow]\} \\ &= \{e < s_0 : (\exists s < t_1)[\Phi^{X^{(i-1)}\upharpoonright t_1}_{e,s}(e)\downarrow]\} \\ &= X^{(i)} \upharpoonright s_0. \end{aligned}$$

But $t_1, \ldots, t_i$ were arbitrary. So if we take any $t_{i+1} < \cdots < t_n < u_1 < \cdots < u_n$ in $H$ with $t_{i+1}$ larger than $t_i$, then

$$X^{(i)}_{s_0,t_1,\ldots,t_i} = X^{(i)} \upharpoonright s_0 = X^{(i)}_{s_0,u_1,\ldots,u_i},$$

and consequently $c(s_0, t_1, \ldots, t_n, u_1, \ldots, u_n) \neq i$. Since $H$ is homogeneous, we must have $c(s_0, s_1, \ldots, s_n, t_1, \ldots, t_n) \neq i$, and therefore

$$X^{(i)}_{s_0,s_1,\ldots,s_i} = X^{(i)}_{s_0,t_1,\ldots,t_i} = X^{(i)} \upharpoonright s_0,$$

as desired.

With the claim in hand, we proceed as follows. Let $p \colon \omega \to H$ be the principal function of $H$. Then, for each $i \leqslant n$, define

$$X_i = \bigcup_{j\in\omega} X^{(i)}_{p(j),\ldots,p(j+i)},$$

which exists by arithmetical comprehension. As we showed, $X^{(i)}_{p(j),\ldots,p(j+i)} = X^{(i)} \upharpoonright p(j)$ for every $j$, and therefore $X_i = X^{(i)}$. Thus, $\langle X_0, \ldots, X_n\rangle$ is a sequence witnessing that $X^{(n)}$ exists. The proof is complete. □

## 8.2 Lower bounds

We now turn to lower bounds. Our first task is to show that the $\Pi^0_n$ bound from Theorem 8.1.1 cannot be improved. This also yields a significant strengthening of Specker's theorem (Theorem 8.0.1). The following lemma will be useful here and subsequently.

**Lemma 8.2.1.** *Fix $n, k \geqslant 1$, and let $c^* \colon [\omega]^n \to k$ be a $\varnothing'$-computable coloring. There exists a computable $c \colon [\omega]^{n+1} \to k$ such that every infinite homogeneous set for $c$ is homogeneous for $c^*$.*

*Proof.* Let $c$ be a computable approximation $c^*$ as given by the limit lemma. That is, $c$ is a map $[\omega]^{n+1} \to k$ such that $\lim_s c(\vec{x}, s) = c^*(\vec{x})$ for every tuple $\vec{x} \in [\omega]^n$. It is easy to see that this has the desired property. □

**Theorem 8.2.2 (Jockusch [168]).** *For each $n \geqslant 2$, $\mathsf{RT}^n_2$ omits $\Sigma^0_n$ solutions.*

*Proof.* We aim to show that for every set $A$ there is an $A$-computable instance $c$ of $\mathsf{RT}^n_2$ with no $\Sigma^0_n(A)$ solution. It suffices to show $c$ has no $\Delta^0_n(A)$ solution since every

$\Sigma^0_n(A)$ infinite set has a $\Delta^0_n(A)$ infinite subset. The proof is now by induction on $n \geqslant 2$.

***Inductive basis.*** Fix $n = 2$. We construct a computable coloring $c\colon [\omega]^2 \to 2$ with no infinite $\Delta^0_2$ solution. The result follows by relativization. We aim to ensure that for each $e$, $\Phi_e(\varnothing')$ is not (the characteristic function of) an infinite homogeneous set for $c$. We thus assume throughout the proof that if $\Phi_e(\varnothing')(x)\downarrow$ then $\Phi_e(\varnothing')(x) < 2$ and also $\Phi_e(\varnothing')(y)\downarrow$ for all $y < x$. For each $e$, we also fix a computable approximation $\{D_{e,s} : s \in \omega\}$ to $\Phi_e(\varnothing')$. Thus $D_{e,s}$ is a finite subset of $\omega \upharpoonright s$, and for each $x < s$ we have $x \in D_{e,s}$ if and only if $\Phi_e(\varnothing')(x)[s]\downarrow= 1$. In particular, if $\Phi_e(\varnothing')(x) = 1$ then $x \in D_{e,s}$ for all sufficiently large $s$, and if $\Phi_e(\varnothing')(x) = 0$ then $x \notin D_{e,s}$ for all sufficiently large $s$.

We construct $c$ by stages. At stage $s \geqslant 0$, we define $c(x, s)$ for all $x < s$. Fix $s$, and assume $c$ is defined on all pairs $(x, y)$ with $x < y < s$. For each $e < s$ in order, proceed as follows. Compute the set $D_{e,s}$, and if it contains at least $2e + 2$ many elements, fix the least two that are not claimed (to be defined below) by any $e^* < e$. Say these numbers are $x < y$, and say that $e$ *claims* these numbers. Then, define $c(x, s) = 0$ and $c(y, s) = 1$. If $D_{e,s}$ does not contain at least $2e + 2$ many elements, do nothing; in this case, $e$ claims no numbers. Now if $e + 1 < s$, proceed to $e + 1$ (meaning, repeat the above with $e + 1$ in place of $e$). If instead $e + 1 = s$, define $c(x, s) = 0$ for any $x$ not claimed by any $e^* < s$, and move to stage $s + 1$. This completes the construction.

We clearly end up with a computable coloring $c\colon [\omega]^2 \to 2$. To verify that $c$ has no $\Delta^0_2$ infinite homogeneous set, fix any $e$ such that $\Phi_e(\varnothing')$ is total and (the set it defines) contains at least $2e + 2$ many elements. Let $x_0 < \cdots < x_{2e+1}$ be the least $2e+2$ many such elements Now fix $s_0$ so that for all $s \geqslant s_0$ and all $x \leqslant x_{2e+1}$ we have that $D_{e,s}(x) = \Phi_e(\varnothing')(x)$. Thus for all $s \geqslant s_0$, the least $2e + 2$ elements of $D_{e,s}$ are precisely $x_0 < \cdots < x_{2e+1}$. Since each $e^* < e$ claims at most two numbers at every stage, collectively these $e^*$ can claim at most $2e$ many numbers. Thus, at every stage $s \geqslant \max\{e, s_0\}$, $e$ will be able to claim $x_i < x_j$ for some $i < j < 2e + 2$, and we will thus define $c(x_i, s) \neq c(x_j, s)$. Notice that $i$ and $j$ here may depend on $s$! But if $x_0, \ldots, x_{2e+1}$ are *all* part of an infinite homogeneous set for $c$, it follows that this set cannot contain any $s \geqslant \max\{e, s_0\}$ and so must be finite. Thus, $\Phi_e(\varnothing')$ cannot be an infinite homogeneous set for $c$, as was to be shown.

***Inductive step.*** Fix $n > 2$ and assume the result holds for $n - 1$. Then we may fix a $\varnothing'$-computable coloring $c^*\colon [\omega]^{n-1} \to 2$ with no $\Delta^0_{n-1}(\varnothing')$ infinite homogeneous set. By Lemma 8.2.1, we may fix a computable $c\colon [\omega]^n \to 2$ such that every infinite homogeneous set for $c$ is also homogeneous for $c^*$. In particular, $c$ has no $\Delta^0_{n-1}(\varnothing')$ infinite homogeneous set. But $\Delta^0_{n-1}(\varnothing') = \Delta^0_n$, so $c$ witnesses the desired result. □

The following consequence is immediate.

**Corollary 8.2.3.** *For $n \geqslant 2$, $\mathsf{RCA}_0 \nvdash \mathsf{RT}^n_2$.*

In fact, for $n \geqslant 3$ this result can be significantly improved. Indeed, we can show that Ramsey's theorem can, in fact, encode a lot of specific information.

**Theorem 8.2.4 (Jockusch [168]).** *Fix a set $A$. For every $n \geqslant 2$, there exists an $A$-computable coloring $c\colon [\omega]^n \to 2$ such that if $H$ is any infinite homogeneous set for $c$ then $A^{(n-2)} \leqslant_T A \oplus H$.*

*Proof.* The case $n = 2$ is trivial, so we may assume $n \geqslant 3$. We take $A = \varnothing$ for simplicity. The full theorem easily follows by relativization.

First, we claim that there is an increasing $\varnothing^{(n-2)}$-computable function $f\colon \omega \to \omega$ such that if $g\colon \omega \to \omega$ is any function that dominates $f$ (meaning, $f(x) < g(x)$ for almost all $x \in \omega$) then $\varnothing^{(n-2)} \leqslant_T g$. Namely, let

$$f(x) = \sup\{y \in \omega : (\exists e < x)[\Phi_e^{\varnothing^{(n-3)}}(x) \downarrow= y\}.$$

Clearly, $f \leqslant_T \varnothing^{(n-2)}$. Moreover, $f$ dominates every partial function computable in $\varnothing^{(n-3)}$, and any such function computes $\varnothing^{(n-2)}$. (See Exercise 8.10.2.) Since any function that dominates $f$ will also have this property, the claim is proved.

We proceed as follows. Define $c^*\colon [\omega]^2 \to 2$ by $c^*(x, y) = 1$ if and only if $y > f(x)$. For every $x$ we have $c^*(x, y) = 1$ for almost all $y$. Hence, all infinite homogeneous sets for $c^*$ have color 1. And since $f$ is increasing, it follows that if $H = \{h_0 < h_1 < \cdots\}$ is homogeneous for $c^*$ then $h_{x+1} > f(h_x) > f(x)$ for all $x$. Thus, the principal function of $H \smallsetminus h_0$ dominates $f$, meaning $\varnothing^{(n-2)} \leqslant_T H$.

Now, $c^*$ is computable from $f$ and hence from $\varnothing^{(n-2)}$. If we iterate Lemma 8.2.1 $n - 2$ times, we obtain a computable $c\colon [\omega]^n \to 2$ such that every infinite homogeneous set $H$ for $c$ is homogeneous for $c^*$. By construction, this is the desired coloring. □

In the parlance of Section 3.6, we get the following consequence.

**Corollary 8.2.5.** *For each $n \geqslant 3$, $\mathsf{RT}^n_2$ codes the jump (i.e., $\mathsf{TJ} \leqslant_c \mathsf{RT}^n_2$).*

*Proof.* Relativize the proof Theorem 8.2.4. □

**Corollary 8.2.6.** *For all $n \geqslant 3$ and $k \geqslant 2$, $\mathsf{RCA}_0 \vdash \mathsf{ACA}_0 \leftrightarrow \mathsf{RT}^n_2 \leftrightarrow \mathsf{RT}^n$.*

Whether Corollary 8.2.5 (and Corollary 8.2.6) also holds for $n = 2$ was left as an open question by Jockusch in [168]. Surprisingly, the answer turns out to be *no*, as we discuss in the next section. Before that, one final remark is that using the results of this section we can now prove Proposition 4.6.11. Recall that this said the following: for each $n \geqslant 3$, we have $\mathsf{RT} \equiv_\omega \mathsf{RT}^n_2$, but $\mathsf{RT}$ neither admits computable solutions nor is it computably reducible to $m$ applications of $\mathsf{RT}^n_2$ for any $m$.

*Proof (of Proposition 4.6.11).* Fix $n \geqslant 3$. First, we show $\mathsf{RT} \leqslant_\omega \mathsf{RT}^n_2$. Fix an $\omega$-model $\mathcal{S}$ of $\mathsf{RT}^n_2$. Fix any $X \in \mathcal{C}$. Then for each $m \geqslant 1$, by taking $A = X^{(m-1)}$ in Theorem 8.2.4, it follows by induction that $X^{(m)} \in \mathcal{S}$. Now consider an arbitrary instance $c$ of $\mathsf{RT}$ in $\mathcal{S}$. This is a coloring $c\colon [\omega]^m \to k$ for some $m, k \geqslant 1$. By Theorem 8.1.1, $c$ has a $c^{(m)}$-computable infinite homogeneous set, and so it has a solution in $\mathcal{S}$. We conclude $\mathcal{S}$ is a model of $\mathsf{RT}$, as was to be shown.

Next, fix $m$ and suppose $\mathsf{RT}$ is computably reducible to $m$ applications of $\mathsf{RT}^n_2$. For each instance $d$ of $\mathsf{RT}^n_2$, fix a $d^{(n)}$-computable solution, $H_d$. Using Theorem 8.2.2, fix a computable instance $c\colon [\omega]^{mn+1} \to 2$ of RT with no $\varnothing^{(mn)}$-computable solution. Now there must be a sequence of instances $d_0, \ldots, d_{m-1}$ of $\mathsf{RT}^n_2$ such that $d_0 \leqslant_{\mathrm{T}} c$ and for $i > 0$, $d_i \leqslant_{\mathrm{T}} c \oplus H_{d_0} \oplus \cdots \oplus H_{d_{i-1}}$, and such that $c \oplus H_{d_0} \oplus \cdots \oplus H_{d_{m-1}}$ computes an RT-solution to $c$. But $c \oplus H_{d_0} \oplus \cdots \oplus H_{d_{m-1}} \leqslant_{\mathrm{T}} \varnothing^{(mn)}$, and $\varnothing^{(mn)}$ computes no solution to $c$, so we have reached a contradiction. □

## 8.3 Seetapun's theorem

Two decades after Jockusch blew open the investigation into the effective content of Ramsey's theorem, the main outstanding question concerning the computability theoretic strength of $\mathsf{RT}^2$ was finally answered. The solution, showing that $\mathsf{RT}^2$ does *not* code the jump, was obtained in the early 1990s by Seetapun, while still a graduate student at UC Berkeley. His result has spurred on much of the work in computable combinatorics over the past thirty years.

**Theorem 8.3.1 (Seetapun; see [275]).** $\mathsf{RT}^2$ *admits cone avoidance.*

**Corollary 8.3.2.** *Over* $\mathsf{RCA}_0$, $\mathsf{RT}^2$ *does not imply* $\mathsf{ACA}_0$.

In this section, we present Seetapun's original argument, which is in many ways more direct but more combinatorially involved. In Section 8.5, we will give a second, more computability theoretic proof, as an application of the Cholak–Jockusch–Slaman decomposition.

Before proceeding, we make one general remark that will make the rest of our discussion simpler and more natural.

*Remark 8.3.3 (Colorings on general domains).* Although we formulated Ramsey's theorem in Definition 3.2.5 in terms of colorings on $\omega$, we can also consider the more general version, where an instance is a pair $\langle X, c\rangle$ for an infinite set $X$ and coloring $c\colon [X]^n \to k$, and a solution is an infinite homogeneous set for $c$ contained in $X$. For definiteness, let us denote this problem, and its associated $\forall\exists$ theorem form, by General-$\mathsf{RT}^n_k$. Then the following proposition justifies our moving freely between it and the original version in most situations.

**Proposition 8.3.4.**

1. *For all* $n, k \geqslant 1$, $\mathsf{RT}^n_k \equiv_{\mathrm{W}}$ General-$\mathsf{RT}^n_k$.
2. $\mathsf{RCA}_0 \vdash (\forall n)(\forall k)[\mathsf{RT}^n_k \leftrightarrow \text{General-}\mathsf{RT}^n_k]$.

*Proof.* We prove (1), and leave (2) to the reader. Evidently, $\mathsf{RT}^n_k$ is a subproblem of General-$\mathsf{RT}^n_k$, so trivially $\mathsf{RT}^n_k \leqslant_{\mathrm{W}}$ General-$\mathsf{RT}^n_k$. In the other direction, fix an instance $\langle X, c\rangle$ of General-$\mathsf{RT}^n_k$. Let $p\colon \omega \to X$ be the principal function of $X$, and define a coloring $\widehat{c}\colon [\omega]^n \to k$ as follows: for $x < y$, let $\widehat{c}(x, y) = c(p(x), p(y))$. Note that $\widehat{c}$ is uniformly computable from $\langle X, c\rangle$. Now suppose $\widehat{H}$ is any infinite homogeneous set for $\widehat{c}$. Then $H = \{p(x) : x \in \widehat{H}\}$ is an infinite subset of $X$, uniformly computable from $X \oplus H$. It is easy to see that $H$ is homogeneous for $c$. □

Here, and in other constructions of homogeneous sets for colorings, we will use the following elaboration of Mathias forcing.

**Definition 8.3.5.** For each $k \geqslant 1$, *$k$-fold Mathias forcing* is the following notion of forcing.

1. The conditions are tuple $(E_0, \ldots, E_{k-1}, I)$ such that for each $i < k$:
   - $E_i$ is a finite set,
   - $I$ is an infinite set,
   - $E_i < I$.
2. Extension is defined by $(E_0^*, \ldots, E_{k-1}^*, I^*) \leqslant (E_0, E_1, I)$ if
   - for each $i < k$, $E_i \subseteq E_i^* \subseteq E_i \cup I$,
   - $I^* \subseteq I$.
3. The valuation of $(E_0, \ldots, E_{k-1}, I)$ is the string $\sigma \in 2^{<\omega}$ of length $k \cdot \min I$, where, for each $i < k$ and $x < I$, $\sigma(kx + i) = 1$ if and only if $x \in E_i$.

So, if $(E_0, \ldots, E_{k-1}, I)$ is a condition then for each $i$, $(E_i, I)$ is a Mathias condition, and indeed, so is $(E_0 \oplus \cdots \oplus E_{k-1}, I)$. A generic filter here thus determines an object of the form $G = H_0 \oplus \cdots \oplus H_{k-1}$.

We can restrict this forcing in various ways as with ordinary Mathias forcing, e.g., by asking for the reservoirs $I$ be to be computable or low or cone avoiding. For convenience, we add the symbol $\mathtt{H}_i$ to our forcing language to refer to the component $H_i$ of the generic object. Note that this does not actually alter the forcing language. We are merely defining $x \in \mathtt{H}_i$ to be an abbreviation for $x \in \mathtt{G} \wedge (\exists d < x)[x = kd+i]$.

*Remark 8.3.6.* The way this forcing is used in proofs of Ramsey's theorem is to add, e.g., the following clauses to the definition of a condition $(E_0, \ldots, E_{k-1}, I)$ for a given coloring $c\colon [\omega]^2 \to k$.

- $c(x, y) = i$ for all $\langle x, y\rangle \in [E_i]^2$.
- $c(x, y) = i$ for all $x \in E_i$ and $y \in I$.

In this case, the generic object $G = H_0 \oplus \cdots \oplus H_{k-1}$ satisfies that each $H_i$ is homogeneous for $c$ with color $i$. (However, $H_i$ may not be infinite).

An important fact used in the proof of Seetapun's theorem, and as we will see, other computability theoretic constructions, is the following. In some sense it is just a logical observation, but its utility is such that we present it as a lemma in its own right.

**Lemma 8.3.7 (Lachlan's disjunction).** *Fix $k \geqslant 1$, let $A_0, \ldots, A_{k-1}$ be sets, and let $\{\mathcal{R}_e : e \in \omega\}$ be a countable family of requirements. If, for all tuples $e_0, \ldots, e_{k-1} \in \omega$ there is an $i < k$ such that $A_i$ satisfies $\mathcal{R}_{e_i}$, then there is an $i < k$ such that $A_i$ satisfies $\mathcal{R}_e$ for all $e \in \omega$.*

By considering the contrapositive, the proof of the above fact is trivial. But the result is very useful indeed. A typical use is the construction a homogeneous set for a $k$-coloring satisfying some collection of requirements as above. In this case, a common strategy is to build, for each $i < k$, a homogeneous set $H_i$ with color $i$. We can almost never ensure that *each* $H_i$ satisfies all the requirements, but by playing the different homogeneous sets off against each other, we can often show that for each tuple $e_0, \ldots, e_{k-1}$, there is at least one $i$ such that $H_i$ satisfies the $e_i$th requirement. If one thinks of this construction as happening dynamically, where each $H_i$ is built up either by stages, or in the case of a forcing construction, condition by condition, then it is usually not possible to know such an $i$ "in advance". Lachlan's disjunction ensures that we do not need to. Instead, the construction can complete and the appropriate $i$ found "at the end".

Let us move on to proving Seetapun's theorem, where we will see a specific illustration of using Lachlan's disjunction. As mentioned, this uses a very clever combinatorial set-up.

*Proof (of Theorem 8.3.1).* By Corollary 4.5.10 and Theorem 4.6.9, we have that $\mathsf{RT}^2 \leqslant_\omega \mathsf{RT}^2_2$. Hence, it suffices to prove the result for $\mathsf{RT}^2_2$. As usual, we deal with computable instances, with the general result following by relativization. So fix $C \nleq_{\mathrm{T}} \emptyset$ and let $c\colon [\omega]^2 \to 2$ be a computable coloring. Seeking a contradiction, suppose $c$ has no infinite homogeneous set $H$ such that $C \nleq_{\mathrm{T}} H$.

We force with 2-fold Mathias conditions with the additional clauses mentioned in Remark 8.3.6 and with $C$-cone avoiding reservoirs. (In the general case, where $c$ is not necessarily computable, we use reservoirs $I$ with $C \nleq_{\mathrm{T}} c \oplus I$.) Our assumption about the homogeneous sets of $c$ implies that if $H_0 \oplus H_1$ is a sufficiently generic object for this forcing then $H_0$ and $H_1$ are both infinite. Indeed, fix any condition $(E_0, E_1, I)$. Fix $i < 2$. If there were no $x \in I$ such that $c(x, y) = i$ for infinitely many $y \in I$, then $I$ would have an infinite $I$-computable (hence $C$-cone avoiding) subset that is homogeneous for $c$ with color $1 - i$. Since this is impossible, there must indeed be such an $x$. Setting $E^*_i = E_i \cup \{x\}$, $E^*_{1-i} = E_{1-i}$, and $I^* = \{y \in I : y > x\}$ yields an extension $(E^*_0, E^*_1, I^*)$ of $(E_0, E_1, I)$ with $|E^*_i| > |E_i|$. Since $i$ was arbitrary, it follows that for each $n$, the set of conditions $(E_0, E_1, I)$ with $|E_i| > n$ for each $i$ is dense.

It thus suffices to show that for a sufficiently generic object $H_0 \oplus H_1$ there is an $i < 2$ such that $H_i$ satisfies the following requirements for every $e \in \omega$:

$$\mathcal{R}_e\colon (\exists w)\neg[\Phi^{H_i}_{e_i}(w) \downarrow= C(w)].$$

As discussed above, we will not try to ascertain this $i$ right away. Instead, we will satisfy the following modified requirements for all $e_0, e_1 \in \omega$:

$$\mathcal{R}^*_{e_0,e_1}\colon (\exists i < 2)(\exists w)\neg(\Phi^{H_i}_{e_i}(w) \downarrow= C(w)].$$

From here, the fact that either $H_0$ or $H_1$ satisfies all the requirements $\mathcal{R}_e$ follows by Lachlan's disjunction.

So fix $e_0, e_1 \in \omega$ along with a condition $(E_0, E_1, I)$. We claim that there is an extension $(E_0^*, E_1^*, I^*)$ forcing that there is an $i < 2$ such that

$$(\exists w)\neg[\Phi_{e_i}^{\mathrm{H}_i}(w) \downarrow= C(w)]. \tag{8.1}$$

Since $(E_0, E_1, I)$ is arbitrary, it follows that the set of conditions forcing (8.1) for some $i < 2$ is dense. By genericity, this implies that every sufficiently generic $H_0 \oplus H_1$ satisfies $\mathcal{R}^*_{e_0,e_1}$, as desired.

To prove the claim, we first need a series of definitions. For each $i < 2$, a nonempty finite set $F$ is an *i-blob* if the following hold.

- $F \subseteq I$ (so $E_i < F$).
- $c(x, y) = i$ for all $x, y \in F$.
- There is a $w > E_i$ and for each $j < 2$ a subset $F^j \subseteq F$ such that $\Phi_{e_i}^{E_i \cup F^j}(w) \downarrow= j$, with use bounded by $\max F^j$.

Thus, in particular, the sets $F^0$ and $F^1$ satisfy $\Phi_{e_0}^{E_i \cup F^0}(w) \downarrow\neq \Phi_{e_0}^{E_i \cup F^1}(w) \downarrow$, and so one of these two computations differs from $C(w)$.

A *Seetapun sequence* is an infinite sequence $F_0, F_1, \ldots$ of 0-blobs with $F_s < F_{s+1}$ for all $s$. The *Seetapun tree* determined by such a sequence is the set of all $\alpha \in \omega^{<\omega}$ satisfying the following.

- $\alpha(s) \in F_s$ for all $s < |\alpha|$.
- There is no 1-blob $F \subseteq \mathrm{range}(\alpha^\#)$.

The Seetapun tree is indeed a tree, and it is finitely branching. Figure 8.1 provides a visual. The clause that the 1-blob $F$ be contained in $\mathrm{range}(\alpha^\#)$, rather than $\mathrm{range}(\alpha)$, is for convenience. It means that every terminal $\alpha \in T$ has a 1-blob in its range.

We proceed to the construction of the extension $(E_0^*, E_1^*, I^*)$. There are two cases to consider.

*Case 1: For some i < 2, there exists an infinite C-cone avoiding subset of I containing no i-blob.* Let the subset of $I$ in question be $I^*$, and let $E_0^* = E_0$ and $E_1^* = E_1$. Clearly, $(E_0^*, E_1^*, I^*)$ extends $(E_0, E_1, I)$, and we claim that it forces (8.1). Consider any $w > E_0^* = E_0$. By assumption, there do not exist $F^0, F^1 \subseteq I^*$ homogeneous for $c$ with color $i$ such that $\Phi_{e_i}^{E_i \cup F^j}(w) \downarrow= j$ for each $j < 2$, else the union of these sets would be an $i$-blob. Hence, if $\Phi_{e_i}^{E_i \cup F}(w) \downarrow< 2$ for some $F \subseteq I^*$ homogeneous for $c$ with color $i$, then the value of the computation depends only on $w$, and not on $F$. Denote this value by $v_w$. Thus, $v_w$ is defined if and only if some $F \subseteq I^*$ as above exists. Note that we can search for such an $F$ computably in $I^*$, meaning that $v_w$, if defined, can be found uniformly $I^*$-computably from $w$. It follows that if $v_w$ is defined for *every* $w$, then the sequence $\langle v_w : w \in \omega\rangle$ is $I^*$-computable and hence does not compute $C$. This leads to the conclusion that there must be a $w$ such that either $v_w$ is undefined or $v_w \neq C(w)$. But if some extension $(\widehat{E}_0, \widehat{E}_1, \widehat{I})$ of $(E_0^*, E_1^*, I^*)$ forced $\Phi_{e_i}^{\mathrm{H}_i}(w) \downarrow= C(w)$, then $\widehat{E}_i \setminus E_i^*$ would be precisely an $F \subseteq I^*$ witnessing that $v_w$ is defined and equal to $C(w)$. Hence, $(E_0^*, E_1^*, I^*)$ must force $\neg(\Phi_{e_i}^{\mathrm{H}_i}(w) \downarrow= C(w))$, as wanted.

*Case 2: Otherwise.* We first use the failure of Case 1 to construct a Seetapun sequence $F_0, F_1, \ldots$. Let $F_0$ be any 0-blob contained in $I$, which exists by hypothesis, and suppose that we have defined $F_s$ for some $s \in \omega$. Let $F_{s+1}$ be any 0-blob contained in $\{y \in I : y > F_s\}$, which again exists by hypothesis. Clearly, this Seetapun sequence is computable in $I$.

It follows that the Seetapun tree corresponding to this sequence is also $I$-computable. As $T$ is finitely branching, then if it is infinite may fix a $C$-cone avoiding path $P$ through it, using Theorem 2.8.23. Now $F_s < F_{s+1}$ for all $s$, so also $P(s) < P(s+1)$. Hence, range$(P)$ is computable from $P$, and therefore is in particular $C$-cone avoiding. But by definition of the Seetapun tree, range$(P)$ contains no 1-blob, which contradicts Case 1 not holding.

We conclude that $T$ is finite. There is thus an $m \in \omega$ such that every $\alpha \in T$ has length at most $m$. Thus, the only elements of our Seetapun sequence needed to construct $T$ are the $F_s$ for $s < m$. By considering the elements of $\bigcup_{s<m} F_s$ in turn, we can thin out $I$ to an $I$-computable infinite set $J \subseteq I$ with the property that for every $x \in \bigcup_{s<m} F_s$ there is a color $i_x < 2$ such that $c(x, y) = i_x$ for every $y \in J$. More specifically, enumerate the elements of $\bigcup_{s<m} F_s$ as $x_0, \ldots, x_{v-1}$. Define $J_0 = I$, and suppose inductively that we have defined $J_u$ for some $u < v$. Let $i_{x_u}$ be the least $i < 2$ such that there are infinitely many $y \in J_u$ with $c(x_u, y) = i$. Then let $J_{u+1}$ be the set of all such $y$. Finally, take $J = J_v$.

Notice that for any finite set $F \subseteq \bigcup_{s<m} F_s$, if $F$ is homogeneous for $c$ with color $i$ and every $x \in F_s$ has $i_x = i$, then setting $\widehat{E}_i = E_i \cup F$, $\widehat{E}_{1-i} = E_{1-i}$, and $\widehat{I} = \{y \in J : y > F\}$ produces an extension $(\widehat{E}_0, \widehat{E}_1, \widehat{I})$ of $(E_0, E_1, I)$.

We now consider two subcases.

*Subcase 2a: There exists $s < m$ such that $i_x = 0$ for all $x \in F_s$.* Fix such an $s$. By definition, there is a $j < 2$ such that $\Phi_{e_0}^{E_0 \cup F_s^j}(w) \downarrow \neq C(w)$, with the use of the computation bounded by $\max F_s^j$. Let $E_0^* = E_0 \cup F_s^j$, let $E_1^* = E_1$, and let $I^* = \{y \in J : y > F_s^j\}$. Then $(E_0^*, E_1^*, I^*)$ is the desired extension of $(E_0, E_1, I)$.

*Subcase 2b: Otherwise.* For each $s < m$, fix $x_s \in F_s$ with $i_{x_s} = 1$. Let $\alpha$ be the string of length $m$ with $\alpha(s) = x_s$ for each $s < m$. Then either $\alpha \notin T$ or $\alpha$ is terminal in $T$. Either way, range$(\alpha)$ contains some 1-blob, $F$. By definition, there is a $j < 2$ such that $\Phi_{e_1}^{E_1 \cup F^j}(w) \downarrow \neq C(w)$, with the use of the computation bounded by $\max F^j$. Let $E_0^* = E_0$, let $E_1^* = E_1 \cup F^j$, and let $I^* = \{y \in J : y > F^j\}$. Then $(E_0^*, E_1^*, I^*)$ is the desired extension of $(E_0, E_1, I)$. □

## 8.4 Stability and cohesiveness

The difference between $\mathsf{RT}^2$ and $\mathsf{RT}^n$ for $n \geqslant 3$ exposed by Seetapun's theorem created an incentive to understand Ramsey's theorem for pairs more deeply and in new ways. One such approach that proved extremely fruitful was invented in the seminal work of Cholak, Jockusch, and Slaman [33]. The idea is to decompose

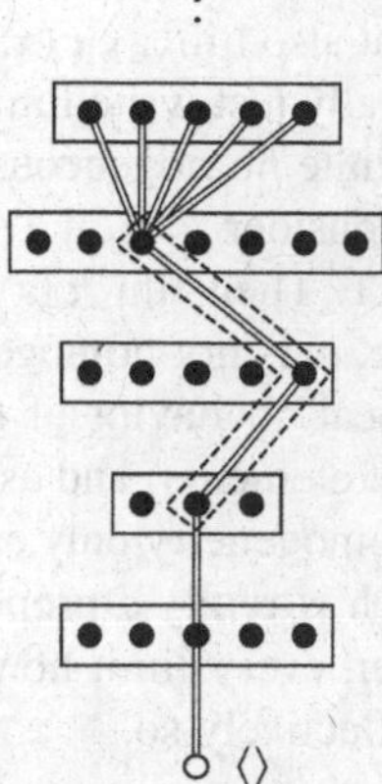

**Figure 8.1.** An illustration of a Seetapun tree. The bottom node is the root, $\langle\rangle$. Each horizontal rectangle above this is a 0-blob, with its elements represented by the dots inside. The vertical arrangement of 0-blobs represents a Seetapun sequence. Highlighted are five different nodes in the Seetapun tree, represented as paths via doubled lines. The nodes agree on their first four values. The dashed outline represents a 1-blob in the (common portion of the) ranges of these nodes. Hence, these nodes are maximal in the tree (i.e., they have length 5 and have no extensions in the tree of length 6).

$\mathsf{RT}^2$ into two "simpler" principles, each of which is some sort of elaboration on Ramsey's theorem for *singletons* ($\mathsf{RT}^1$), which is a much simpler theorem to analyze. In this section we state these two principles, and in the next section we prove the Cholak–Jockusch–Slaman decomposition and discuss its significance.

### 8.4.1 Stability

The first way to "simplify" Ramsey's theorem for singletons is to restrict to so-called *stable* colorings.

**Definition 8.4.1.** Fix $k \geqslant 1$, an infinite set $X$, and a coloring $c\colon [X]^2 \to k$.

1. For $x \in X$, an infinite set $Y \subseteq X$, and $i < k$, we write $\lim_{y \in Y} c(x, y) = i$ if $c(x, y) = i$ for almost all (i.e., all but finitely many) $y \in Y$.
2. A set $L \subseteq X$ is *limit homogeneous* for $c$ if there is an $i < k$ such that $\lim_{y \in L} c(x, y) = i$ for all $x \in L$.
3. $c$ is *stable* if for each $x \in X$ there is an $i < k$, called the *limit color of $x$ under $c$*, such that $\lim_{y \in X} c(x, y) = i$.
4. If $c$ is stable, the coloring $d\colon X \to k$ defined by $d(x) = \lim_{y \in X} c(x, y)$ is called the *coloring of singletons induced by $c$*.

By analogy with homogeneous sets, we may also say the limit homogeneous set $L$ in the above definition is limit homogeneous *with color $i$*. Note that if $c$ is stable and

$L \subseteq X$ is an infinite limit homogeneous set for $c$ with color $i$ then we not only have $\lim_{y \in L} c(x, y) = i$ for all $x \in L$ but also $\lim_{y \in X} c(x, y) = i$. When $X = \omega$ this agrees with Definition 2.6.4, so we typically just write $\lim_y$ in place of $\lim_{y \in \omega}$.

It is easy to see that every infinite homogeneous set is limit homogeneous. The converse is false. For example, consider $c\colon [\omega]^2 \to 2$ defined by $c(x, x+1) = 0$ and $c(x, y) = 1$ for all $y > x + 1$. Then $\lim_y c(x, y) = 1$ for all $x$, so $\omega$ is limit homogeneous for $c$. But of course, $\omega$ is not homogeneous for $c$. To amplify on this, homogeneity cares about the "local" behavior of the coloring (i.e., what color is assigned to each particular pair of elements) and as such is still a property of pairs, even for stable colorings. *Limit* homogeneity only cares about the "global" behavior (i.e., the limit colors), and as such is really a property of singletons. (We will say more about this shortly.) However, every limit homogeneous set can be "thinned" out to a homogeneous one, and effectively so.

**Proposition 8.4.2.** *Fix $k \geqslant 1$ and a coloring $c\colon [\omega]^2 \to k$. If $L$ is an infinite limit homogeneous set for $c$ then $c$ has a $(c \oplus L)$-computable infinite homogeneous set $H \subseteq L$.*

*Proof.* Enumerate the elements of $L$ as $x_0 < x_1 < \cdots$. Fix $i < k$ such that $L$ is limit homogeneous with color $i$. Let $n_0 = 0$, and suppose that for some $s \geqslant 0$ we have defined numbers $n_0 < \cdots < n_s$. Since $\lim_y c(x, y) = i$ for all $x \in L$, there exists a number $n > n_s$ such that $c(x_{n_t}, x_n) = i$ for all $t \leqslant s$. Let $n_{s+1}$ be the least such $n$. Then by induction, for all $t < s$ we then have that $c(x_{n_t}, x_{n_s}) = i$. Thus, $H = \{x_{n_s} : s \in \omega\}$ is an infinite homogeneous subset of $L$, and clearly $H$ is computable from $c \oplus L$. □

*Remark 8.4.3 (Complexity of determining limit colors).* Given a coloring $c\colon [\omega]^2 \to k$, determining whether an element $x$ has a limit color is uniformly $\Sigma_2^{0,c}$ in $x$, since

$$\lim_y c(x, y) \text{ exists} \leftrightarrow (\exists i < k)(\exists z > x)(\forall y \geqslant z)[c(x, y) = i].$$

But if we know that $x$ *has* a limit color, then determining its value is uniformly $\Delta_2^{0,c}$ in $x$, as

$$\begin{aligned} \lim_y c(x, y) = i &\leftrightarrow (\exists z > x)(\forall y \geqslant z)[c(x, y) = i] \\ &\leftrightarrow (\forall z > x)(\exists y \geqslant z)[c(x, y) = i]. \end{aligned}$$

**Proposition 8.4.4.** *Fix $k \geqslant 1$ and a stable coloring $c\colon [\omega]^2 \to k$. Let $d$ be the coloring of singletons induced by $c$. Then $d$ is uniformly $c'$-computable, and its infinite homogeneous sets are precisely the limit homogeneous sets of $c$.*

We leave the proof of the proposition to the reader. Building homogeneous sets for colorings of *singletons* is much easier than for colorings of pairs. Thus, Propositions 8.4.2 and 8.4.4 can be thought of as saying that for stable colorings of pairs, we can trade the combinatorial complexity of building homogeneous sets for the computational complexity of finding limit colors.

Using the limit lemma, we can obtain a kind of converse to the preceding proposition.

**Proposition 8.4.5.** *Fix $k \geqslant 1$, a set $A$, and an $A'$-computable coloring $d\colon \omega \to k$. Then there exists a uniformly $A$-computable stable coloring $c\colon [\omega]^2 \to k$ so that $d$ is the coloring of singletons induced by $c$.*

*Proof.* By the limit lemma relativized to $A$, we may fix an $A$-computable approximation to $d$. This is a function $\widehat{d}\colon \omega^2 \to \omega$ such that for every $x$ and every sufficiently large $y$, we have $\widehat{d}(x,y) = d(x)$. Define $c\colon [\omega]^2 \to k$ as follows: for $x < y$, let $c(x,y) = \widehat{d}(x,y)$ if $\widehat{d}(x,y) < k$, and otherwise let $c(x,y) = 0$. By the uniformity of the limit lemma, $c$ is uniformly $A$-computable. Also, $\lim_y c(x,y) = \lim_y \widehat{d}(x,y) = d(x)$, as desired. □

From the computational point of view, we conclude that finding limit homogeneous sets for $A$-computable stable colorings of pairs is *the same* as finding homogeneous sets for $A'$-computable colorings of singletons. Frequently, it is convenient to think of the latter as an actual $\Delta^{0,A}_2$ finite partition $\langle P_0, \ldots, P_{k-1}\rangle$ of $\omega$. On that view, we are just looking at infinite subsets of the $P_i$. In particular, if $k = 2$, finding an infinite limit homogeneous set of a given $A$-computable stable coloring is the same as finding an infinite subset of a given $\Delta^{0,A}_2$ set or of its complement.

We can formulate the discussion in terms of $\forall\exists$ theorems and problems.

**Definition 8.4.6 (Stable Ramsey's theorem).**

1. For $k \geqslant 1$, *the stable Ramsey's theorem for $k$-colorings* ($\mathsf{SRT}^2_k$) is the following statement: every stable coloring $c\colon [\omega]^n \to k$ has an infinite homogeneous set.
2. For $k \geqslant 1$, *the $\Delta^0_2$ $k$-partition subset principle* ($\mathsf{D}^2_k$) is the following statement: every stable coloring $c\colon [\omega]^n \to k$ has an infinite limit homogeneous set.

$\mathsf{SRT}^2$ and $\mathsf{D}^2$ are defined as $(\forall k)\mathsf{SRT}^2_k$ and $(\forall k)\mathsf{D}^2_k$.

Each of these has a problem form obtained from the theorem form in the standard way. As usual, we will shift perspectives between the forms freely depending on context. For example, one immediate corollary of Proposition 8.4.4 is the following result.

**Corollary 8.4.7.** $\mathsf{SRT}^2$ *admits $\Delta^0_2$ solutions.*

As in Remark 8.3.3 and Proposition 8.3.4, we can apply the above principles also to colorings defined on other infinite subsets of the natural numbers in Weihrauch reductions and proofs over $\mathsf{RCA}_0$.

How do $\mathsf{SRT}^2_k$ and $\mathsf{D}^2_k$ compare under our various measures for studying different problems and theorems? For starters, the following proposition is an immediate consequence of Proposition 8.4.2.

**Proposition 8.4.8.** *For each $k \in \omega$, $\mathsf{SRT}^2_k \equiv_{\mathrm{c}} \mathsf{D}^2_k$.*

By contrast, it turns out that $\mathsf{SRT}^2_2 \nleqslant_{\mathrm{W}} \mathsf{D}^2_2$ and $\mathsf{SRT}^2_2 \nleqslant_{\mathrm{sc}} \mathsf{D}^2_2$, as we show in Theorem 9.1.18. For now, the most relevant reducibility to consider is provability under $\mathsf{RCA}_0$. Obviously, $\mathsf{RCA}_0 \vdash (\forall k)[\mathsf{SRT}^2_k \to \mathsf{D}^2_k]$. In the other direction, it may seem straightforward to formalize Proposition 8.4.2 in second order arithmetic. But there is a hitch. Suppose that, in an arbitrary model $\mathcal{M}$ of $\mathsf{RCA}_0$, we have a $c$ that $\mathcal{M}$ thinks is a stable 2-coloring of pairs. Suppose we have numbers $x_0, \ldots, x_{a-1} \in M$, each with the same limit color $i < 2$. How do we know there is a number $x >^{\mathcal{M}} x_{a-1}$ such that $c(x_b, x) = i$ for all $b <^{\mathcal{M}} a$? The natural justification—for each $b <^{\mathcal{M}} a$, choose $s_b > x_b$ so that $c(x_b, y) = i$ for all $y \geqslant^{\mathcal{M}} s_b$, and then let $x = \max\{s_b : b \leqslant^{\mathcal{M}} a\}$—actually uses $\mathsf{B}\Sigma^0_2$. Since $\mathcal{M}$ may not satisfy $\mathsf{B}\Sigma^0_2$, we need first of all to show that this follows from $\mathsf{D}^2_2$ itself.

**Theorem 8.4.9 (Chong, Lempp, and Yang [34]).** $\mathsf{RCA}_0 \vdash \mathsf{D}^2_2 \to \mathsf{B}\Sigma^0_2$.

*Proof.* Recall the principle $\mathsf{PART}$ from Definition 6.5.7. By Theorem 6.5.8, this is equivalent to $\mathsf{B}\Sigma^0_2$. It thus suffices to show that $\mathsf{D}^2_2 \to \mathsf{PART}$.

We argue in $\mathsf{RCA}_0$. Let $\leqslant_L$ be a linear ordering of $\mathbb{N}$ such that $(\mathbb{N}, \leqslant_L)$ is of order type $\omega + \omega^*$. To show that it is strongly of order type $\omega + \omega^*$, let $\{\ell = x_0 \leqslant_L \cdots \leqslant_L x_k = g\}$ be a finite set, where $\ell$ and $g$ are the least and greatest elements under $\leqslant_L$. And seeking a contradiction, suppose that for each $i < k$ the set $\{y \in \mathbb{N} : x_i \leqslant_L y \leqslant_L x_{i+1}\}$ is finite. We claim that for each $i < k$,

$$\{y \in \mathbb{N} : x_i \leqslant_L y\} \text{ is infinite.} \tag{8.2}$$

This is obvious for $i = 0$, and we should like to prove it for all $i$ by induction. However, we cannot do so directly since it is a $\Pi^0_2$ statement. Before addressing this, note that once this claim is proved we have our contradiction: since $x_{k+1} = g$ we have that

$$\{y \in \mathbb{N} : x_k \leqslant_L y\} = \{y \in \mathbb{N} : x_k \leqslant_L y \leqslant_L x_{k+1}\}.$$

To prove (8.2), define $c\colon [\mathbb{N}]^2 \to 2$ as follows: for all $x < y$, let $c(x, y) = 1$ if $x \leqslant_L y$, and let $c(x, y) = 0$ if $y \leqslant_L x$. Note that $c$ is stable. Indeed, fix $x$. By assumption, exactly one of $\{y \in \mathbb{N} : y \leqslant_L x\}$ and $\{y \in \mathbb{N} : x \leqslant_L y\}$ is infinite. In the former case, $c(x, y) = 0$ for almost all $y$. In the latter, $c(x, y) = 1$ for almost all $y$.

We can thus apply $\mathsf{D}^2_2$ to obtain an infinite limit homogeneous set $L$ for $c$. If $L$ has color 0, then for every $x \in \mathbb{N}$ we have

$$\{y \in \mathbb{N} : x \leqslant_L y\} \text{ is infinite} \leftrightarrow (\forall y \in L)[y \nleqslant_L x],$$

while if $L$ has color 1 then similarly

$$\{y \in \mathbb{N} : x \leqslant_L y\} \text{ is infinite} \leftrightarrow (\exists y \in L)[x \leqslant_L y].$$

It follows in particular that (8.2) is equivalent to either a $\Sigma^0_1$ of $\Pi^0_1$ formula in $i$, and therefore can be proved by induction. □

*Remark 8.4.10.* By Theorem 6.5.1, $\mathsf{B}\Sigma^0_2$ is equivalent over $\mathsf{RCA}_0$ to $\mathsf{RT}^1$, and it is easy to see that $\mathsf{RT}^1$ is a consequence of $\mathsf{D}^2$. Thus the emphasis in the above theorem is on the fact that we can obtain $\mathsf{B}\Sigma^0_2$ even when the number of colors in the instances of $\mathsf{D}^2$ is fixed. It is much more straightforward to show separately that $\mathsf{SRT}^2_2$ implies $\mathsf{RT}^1$ over $\mathsf{RCA}_0$ (Exercise 8.10.3).

**Corollary 8.4.11.** $\mathsf{RCA}_0 \vdash (\forall k)[\mathsf{SRT}^2_k \leftrightarrow \mathsf{D}^2_k]$.

*Proof.* As noted above, $\mathsf{RCA}_0 \vdash (\forall k)[\mathsf{SRT}^2_k \rightarrow \mathsf{D}^2_k]$. Obviously, $\mathsf{RCA}_0 \vdash \mathsf{D}^2_1 \rightarrow \mathsf{SRT}^2_1$. If $k > 2$ then $\mathsf{D}^2_k \rightarrow \mathsf{D}^2_2$, so $\mathsf{RCA}_0 + \mathsf{D}^2_k \vdash \mathsf{B}\Sigma^0_2$ by the preceding proposition. Now the argument of Proposition 8.4.2 can be carried out. □

The main import of these results is that, in most situations, we can use $\mathsf{SRT}^2_k$ and $\mathsf{D}^2_k$ interchangeably. This is useful because, as we will see, constructing limit homogeneous sets is easier than constructing homogeneous ones. This makes sense: as discussed earlier, for limit homogeneous sets we do not care about "local" behavior, so there are fewer things to ensure. In particular, working with $\mathsf{D}^2_k$ or $\mathsf{D}^2$ is usually more convenient than directly with $\mathsf{SRT}^2_k$ or $\mathsf{SRT}^2$.

### 8.4.2 Cohesiveness

We now move on to the second "simplification" of $\mathsf{RT}^2$, which is the principle $\mathsf{COH}$ from Definition 4.1.6. This is far less obviously related to Ramsey's theorem, but as a first pass we can look at the following definition and result.

**Definition 8.4.12.** Let $\mathsf{P}$ be an instance solution problem all of whose solutions are subsets of $\omega$. Then $\mathsf{P}$ *with finite errors*, denoted $\mathsf{P}^{\mathrm{fe}}$, is the problem whose instances are the same as those of $\mathsf{P}$, and the $\mathsf{P}^{\mathrm{fe}}$-solutions to any such instance $X$ are all sets $Z =^* Y$ for some $\mathsf{P}$-solution $Y$ to $X$.

Thus, for example, consider $(\mathsf{RT}^1_2)^{\mathrm{fe}}$. A solution to a given coloring $c\colon \omega \rightarrow 2$ is now any infinite set which is *almost* homogeneous, i.e., homogeneous up to finitely many modifications. $\mathsf{COH}$, it turns out, is essentially the parallelized version (as defined in Definition 3.1.5) of this principle, i.e., $\widehat{(\mathsf{RT}^1_2)^{\mathrm{fe}}}$. (A note of caution: for a problem $\mathsf{P}$ as above, $\widehat{\mathsf{P}}^{\mathrm{fe}}$ and $\widehat{\mathsf{P}^{\mathrm{fe}}}$ are not the same!) This connection is not *a priori* obvious, even if it may seem to be. By considering characteristic functions, the instances of $\mathsf{COH}$ and $\widehat{(\mathsf{RT}^1_2)^{\mathrm{fe}}}$ are seen to be the same. Given a sequence of colorings $\omega \rightarrow 2$, an $\widehat{(\mathsf{RT}^1_2)^{\mathrm{fe}}}$-solution is a sequence of almost homogeneous sets, none of which needs to have any relation to any of the others. A solution to $\mathsf{COH}$, on the other hand, is a *single* set that is almost homogeneous for all the colorings in the instance simultaneously.

**Theorem 8.4.13 (Jockusch and Stephan [167]).** *Fix a set $A$. The following are equivalent for a set $X$.*

1. *$X' \gg A'$.*
2. *$X$ computes a solution to every $A$-computable instance of* $\mathsf{COH}$.
3. *$X$ computes a solution to every $A$-computable instance of* $\widehat{(\mathsf{RT}^1_2)^{\mathrm{fe}}}$.

*Proof.* (1) $\rightarrow$ (2): Fix $X' \gg A'$ and an $A$-computable instance $\langle R_i : i \in \omega \rangle$ of $\mathsf{COH}$. For each $\sigma \in 2^{<\omega}$, we define $R_\sigma$ inductively as follows: if $\sigma = \langle\rangle$, we let

$R_\sigma = \omega$; if $\sigma = \tau 1$ then $R_\sigma = R_\tau \cap R_{|\tau|}$; if $\sigma = \tau 0$ then $R_\sigma = R_\tau \cap \overline{R_{|\tau|}}$. Note that if $R_\tau$ is infinite then so is at least one of $R_{\tau 0}$ and $R_{\tau 1}$. Moreover, there is at least one string $\sigma$ of every length such that $R_\sigma$ is infinite. Since $X' \gg A'$, it follows by Theorem 2.8.25 that $X'$ computes a function $f\colon \omega \to 2^{<\omega}$ such that for each $n$, $|f(n)| = n$ and $R_{f(n)}$ is infinite. Fix an $X$-computable limit approximation to $f$, which we may take to be a function $\widehat{f}\colon \omega^2 \to 2^{<\omega}$ such that $|\widehat{f}(n,s)| = n$ for all $n$ and $s$, and $\widehat{f}(n,s) = f(n)$ for all sufficiently large $s$. We may further assume that if $n < m$ then $\widehat{f}(m,s) \preceq \widehat{f}(n,s)$ for all $s$.

We now define a sequence of numbers $x_0 < x_1 < \cdots$. For $n \in \omega$, search for the least $s > n$ and the least $x > \max\{x_m : m < n\}$ such that $x \in R_{\widehat{f}(n,s)}$, and let $x_n = x$. Note that the search for $x_n$ must succeed since for $R_{\widehat{f}(n,s)} = R_{f(n)}$ for all sufficiently large $s$ and $R_{f(n)}$ is infinite. Let $S = \{x_n : n \in \omega\}$. Then $S$ is $X$-computable and infinite, and we claim it is cohesive. Indeed, fix $i$ and let $n$ be such that $R_{\widehat{f}(i+1,s)} = R_{f(i+1)}$ for all $s \geqslant n$. Then all the $x_m$ for $m \geqslant n$ belong to $R_{f(i+1)}$ and so either to $R_i$ (if $f(i+1)(i) = 1$) or to $\overline{R_i}$ (if $f(i+1)(i) = 0$). Hence, either $S \subseteq^* R_i$ or $S \subseteq^* \overline{R_i}$, respectively.

(2) $\to$ (3): Immediate, as $\widehat{(\mathsf{RT}^1_2)^{\mathrm{fe}}}$ is a subproblem of COH.

(3) $\to$ (1): We define an $A$-computable instance $\langle R_i : i \in \omega\rangle$ of $(\mathsf{RT}^1_2)^{\mathrm{fe}}$ as follows. Fix $i$. Using a fixed $A$-computable approximation to $A'$, we can approximate $\Phi_i^{A'}(i)$, letting $\Phi_i^{A'}(i)[s]$ denote its value (if any) at stage $s$ of this approximation. For each $s$, if $\Phi_i^{A'}(i)[s] \downarrow= 0$ let $R_i(s) = 1$. Otherwise, let $R_i(s) = 0$. Now suppose $S \leqslant_{\mathrm{T}} X$ is an infinite cohesive set for $\langle R_i : i \in \omega\rangle$. We define an $X'$-computable function $f$ as follows: given $i$, first use $X'$ to compute whether $S \subseteq^* R_i$ or $S \subseteq^* \overline{R_i}$, and then set $f(i) = 1$ in the first case and $f(i) = 0$ in the second. For each $i$, if $\Phi_i^{A'}(i) \downarrow= 0$ then for all sufficiently large $s$ we have $\Phi_i^{A'}(i)[s] \downarrow= 0$, and hence $R_i(s) = 1$. In this case, we thus have $f(i) = 1 \neq \Phi_i^{A'}(i)$. If $\Phi_i^{A'}(i) \downarrow\neq 0$ then we instead have $R_i(s) = 0$ for all sufficiently large $s$, hence $f(i) = 0$. We conclude that $f$ is DNC relative to $A'$, which means that $X' \gg A'$, as wanted. □

The proof of the last implication actually exhibits an $A$-computable instance of COH, every solution to whichcomputes a solution to every other $A$-computable instance. In the parlance introduced in Exercise 4.8.1, this yields:

**Corollary 8.4.14.** COH *admits universal instances.*

This gives us a nice degree theoretic characterization of the $\omega$-models of COH. Namely, it follows that an $\omega$-model $\mathcal{M}$ satisfies COH if and only if, for every $A \in \mathcal{S}^{\mathcal{M}}$, there is an $X \in \mathcal{S}^{\mathcal{M}}$ with $X' \gg A'$.

Observing the uniformity in the proof of Theorem 8.4.13 also yields the following more direct relationship.

**Corollary 8.4.15.** COH $\equiv_{\mathrm{W}} \widehat{(\mathsf{RT}^1_2)^{\mathrm{fe}}}$.

All told, we find that COH, like $\mathsf{SRT}^2_2$ and $\mathsf{D}^2_2$, is some kind of variation on Ramsey's theorem for singletons. Unlike $\mathsf{SRT}^2_2$ and $\mathsf{D}^2_2$, however, it is not at all evident that $\mathsf{RT}^2_2$ implies COH (over $\mathsf{RCA}_0$, or in any other sense). In fact, it does, and in a strong way.

**Theorem 8.4.16 (Cholak, Jockusch, and Slaman [33]).** COH *is uniformly identity reducible to* $\mathsf{RT}^2_2$.

*Proof.* Let $\langle R_i : i \in \omega \rangle$ be an instance of COH. First, we may assume that for all $x < y$ there is an $i$ such that $R_i(x) \neq R_i(y)$, i.e., $R_i$ contains one of $x$ and $y$ but not the other. (Indeed, we can immediately define a new family $\langle R^*_i : i \in \omega \rangle$, with $R^*_{2i} = R_i$ and $R^*_{2i+1} = \{i\}$, which has the desired property. This new family is uniformly computable from the first, and any infinite cohesive set for it is also cohesive for the original.) We now define a coloring $c \colon [\omega]^2 \to 2$ as follows: given $x < y$, find the least $i \in \omega$ such that $R_i(x) \neq R_i(y)$, and then output $R_i(x)$. Note that $c$ is uniformly computable from $\langle R_i : i \in \omega \rangle$. Let $H$ be any $\mathsf{RT}^2_2$-solution to $c$. We claim that $H$ is an infinite cohesive set for $\langle R_i : i \in \omega \rangle$. We proceed by induction, showing that for each $i$, either $H \subseteq^* R_i$ or $H \subseteq^* \overline{R_i}$. Fix $i$, and assume the result is true for all $j < i$. Let $b$ be so that for each $j < i$, either every $x > b$ in $H$ is in $R_j$ or every $x > b$ in $H$ is in $\overline{R_j}$. Now if $H$ intersects both $R_i$ and $\overline{R_i}$ infinitely, we can fix $b < x < y < z$ in $H$ so that $R_i(x) \neq R_i(y) \neq R_i(z)$. By assumption, $R_j(x) = R_j(y) = R_j(z)$ for every $j < i$. But then by definition of $c$, we must have $c(x, y) \neq c(y, z)$, which is impossible since $x, y, z \in H$ and $H$ is homogeneous for $c$. We conclude that either $H \cap R_i$ or $H \cap \overline{R_i}$ is finite, i.e., $H \subseteq^* R_i$ or $H \subseteq^* \overline{R_i}$, as desired. □

We conclude with the following proposition, which brings our conversation full circle back to stability.

**Proposition 8.4.17.** *Fix $k \geqslant 1$ and a coloring $c \colon [\omega]^2 \to k$. There exists an instance of* COH *such that if $S$ is any solution to this instance then $c \upharpoonright [S]^2$ is stable.*

*Proof.* For each $x$ and $i < k$, let $R_{kx+i} = \{y > x : c(x, y) = i\}$, thereby obtaining an instance $\langle R_n : n \in \omega \rangle$ of COH. Suppose $S$ is an infinite cohesive set for this family. We claim the stronger fact that $\lim_{y \in S} c(x, y)$ exists for *every* $x \in \omega$. Indeed, fix $x$. For each $n$, either $S \subseteq^* R_n$ or $S \subseteq^* \overline{R_n}$, hence by definition there must be an $i < k$ such that $S \subseteq^* R_{kx+i}$. Thus $c(x, y) = i$ for almost all $y \in S$, meaning $\lim_{y \in S} c(x, y) = i$. □

## 8.5 The Cholak–Jockusch–Slaman decomposition

We can now put together all the pieces from the previous section and prove that $\mathsf{RT}^2$ can be decomposed, or split, into the two "simpler" principles $\mathsf{SRT}^2$ and COH. We will then look at several representative applications.

The *Cholak–Jockusch–Slaman decomposition* is the following seminal result.

**Theorem 8.5.1 (Cholak, Jockusch, and Slaman [33]).** $\mathsf{RCA}_0 \vdash (\forall k \geqslant 2)[\mathsf{RT}^2_k \leftrightarrow \mathsf{SRT}^2_k + \mathsf{COH}]$.

The proof proceeds by formalizing our earlier arguments. But as may be expected, the main issues we run into are ones of induction. For ease of presentation, we break the proof into two lemmas.

**Lemma 8.5.2.** $\mathsf{RCA}_0 \vdash (\forall k \geqslant 2)[\mathsf{RT}^2_k \to \mathsf{SRT}^2_k + \mathsf{COH}]$.

*Proof (Jockusch and Lempp, unpublished; Mileti [211]).* Obviously, $\mathsf{RCA}_0$ proves $(\forall k)[\mathsf{RT}^2_k \to \mathsf{SRT}^2_k]$. We next show that $\mathsf{RT}^2_2 \to \mathsf{COH}$. To this end, we would like to formalize the proof of Theorem 8.4.16. The only issue there is with the induction at the end. In $\mathsf{RCA}_0$, we cannot prove by induction for each $i$ that either $H \subseteq^* R_i$ or $H \subseteq^* \overline{R_i}$, as the latter is a $\Sigma^0_2$ formula. As is often the case, we can fix this by being more careful. We now argue in $\mathsf{RCA}_0$. Recall that the coloring $c$ that $H$ is homogeneous for is defined as follows: for all $x < y$, if $i$ is least such that $R_i$ contains one of $x$ and $y$ but not the other, then $c(x, y) = R_i(x)$.

Say $H$ is homogeneous with color $v < 2$. Let $p \colon \mathbb{N} \to H$ be the principal function of $H$, which exists by $\Delta^0_1$ comprehension. We aim to prove the following for each $i$: for each finite set $F \subseteq \mathbb{N}$, if $R_i(p(x)) = 1 - v$ and $R_i(p(x+1)) = v$ for all $x \in F$, then $|F| < 2^i$. Note that this is now a $\Pi^0_1$ formula, so we *can* prove it by induction. For $i = 0$, this is immediate. For if $x \in \mathbb{N}$ satisfied $R_0(p(x)) = 1 - v$ and $R_0(p(x+1)) = v$ then by definition of $c$ we would have $c(p(x), p(x + 1)) = R_0(p(x)) = 1 - v$, a contradiction. So fix $i > 0$ and assume the result is true for all $j < i$. For each $x \in F$, let $j_x$ be the least $j$ such that $R_j(p(x)) \neq R_j(p(x+1))$. As $R_i(p(x)) \neq R_i(p(x+1))$ we have $j_x \leqslant i$, but as $p(x) < p(x + 1)$ and $R_i(p(x)) \neq v$ we must in fact have $j_x < i$. For each $j < i$ let $F_j = \{x \in F : j_x = j\}$. Then $F = \bigcup_{j<i} F_j$, and from the inductive hypothesis it follows that

$$|F| = \sum_{j<i} |F_j| < \sum_{j<i} 2^j = 2^i - 1.$$

This proves the claim.

It remains to verify, using the claim, that $H$ is cohesive for $\langle R_i : i \in \mathbb{N}\rangle$. Suppose not and fix an $i$ such that both $H \cap R_i$ and $H \cap \overline{R_i}$ are infinite. Then there exist finite sets $F$ of arbitrary size satisfying $R_i(p(x)) = 1 - v$ and $R_i(p(x + 1)) = v$ for all $x \in F$. This is impossible by the claim. □

**Lemma 8.5.3.** $\mathsf{RCA}_0 \vdash (\forall k \geqslant 2)[\mathsf{SRT}^2_k + \mathsf{COH} \to \mathsf{RT}^2_k]$.

*Proof.* The first step is to formalize Proposition 8.4.17. We argue in $\mathsf{RCA}_0 + \mathsf{SRT}^2_2 + \mathsf{COH}$. Fix $k \geqslant 1$ and $c \colon [\mathbb{N}]^2 \to k$. Define the family $\langle R_n : n \in \mathbb{N}\rangle$ as in Proposition 8.4.17, and apply $\mathsf{COH}$ to find an infinite set $S$ cohesive for this family. As before, we know that for each $n$, either $S \subseteq^* R_n$ or $S \subseteq^* \overline{R_n}$. Fix $x$, and suppose towards a contradiction that for each $i < k$ we had $S \subseteq^* \overline{R_{kx+i}}$. Then for each $i < k$ there is a $b$ such that $y > b \to y \notin R_{kx+i}$. By $\mathsf{B}\Sigma^0_2$, which is a consequence of $\mathsf{SRT}^2_2$, we may fix a $b$ so that for $y > b \to (\forall i < k)[y \notin R_{kx+i}]$. By definition, this means $c(x, b+1) \neq i$ for all $i < k$, which cannot be. So, we may fix $i$ such that $S \subseteq^* R_{kx+i}$, whence as before we conclude $c(x, y) = i$ for almost all $y \in S$. Thus, $c \upharpoonright [S]^2$ is stable. Applying $\mathsf{SRT}^2_2$, we obtain an infinite homogeneous set $H \subseteq S$ for $c \upharpoonright [S]^2$. Then $H$ is also homogeneous for $c$, as desired. □

We can also formulate a version of this result for $\leqslant_W$. Recall the compositional product, $*$, from Definition 4.5.11.

**Theorem 8.5.4 (Cholak, Jockusch, and Slaman [33]).** *For all $k \geqslant 1$, $\mathsf{RT}^2_k \leqslant_W \mathsf{SRT}^2_k * \mathsf{COH}$.*

*Proof.* For completeness, we spell out the details. Given $c\colon [\omega]^2 \to k$, let $\langle R_n : n \in \omega\rangle$ be the instance of $\mathsf{COH}$ defined in Proposition 8.4.17. Let $\Gamma$ be the Turing functional that, given a pair of sets $\langle A, B\rangle$ as an oracle, outputs the set $\{\langle x, y, i\rangle \in A : x, y \in B\}$. In particular, if $S$ is any infinite cohesive set for $\langle R_n : n \in \omega\rangle$ then $\Gamma(c, S) = c \upharpoonright [S]^2$, which is an instance of $\mathsf{SRT}^2_k$. The pair $\langle\langle R_n : n \in \omega\rangle, \Gamma\rangle$ is thus an instance of $\mathsf{SRT}^2_k * \mathsf{COH}$, uniformly computable from $c$, and any solution to this instance is an infinite homogeneous set $H \subseteq S$ for $c \upharpoonright [S]^2$ and hence for $c$. □

Curiously, even though $\mathsf{SRT}^2_2$ is a subproblem of $\mathsf{RT}^2_2$, and $\mathsf{COH} \leqslant_W \mathsf{RT}^2_2$ by Theorem 8.4.16, we cannot improve the above to an equivalence. As shown by Dzhafarov, Goh, Hirschfeldt, Patey, and Pauly [81], $\mathsf{COH} * \mathsf{SRT}^2_2 \nleqslant_W \mathsf{RT}^2_2$.

The main interest in the Cholak–Jockusch–Slaman decomposition, however, is in second order arithmetic, and ergo in Theorem 8.5.1. It is here that the majority of applications come from.

### 8.5.1 A different proof of Seetapun's theorem

The Cholak–Jockusch–Slaman decomposition can be applied to obtain results about $\mathsf{RT}^2$ by establishing them separately for $\mathsf{SRT}^2$ and for $\mathsf{COH}$. The main advantage here is that such proofs are more modular and thus more readily adaptable to other situations. A good example is Seetapun's theorem, which we now re-prove using this approach. We will see several other examples in the next section, as well as in the exercises.

The basic outline for proving cone avoidance of $\mathsf{RT}^2$ breaks down into the following two steps.

1. Prove that $\mathsf{COH}$ admits cone avoidance.
2. Prove that $\mathsf{SRT}^2$ admits cone avoidance.

To see how these fit together, let us look at computable instances for simplicity. Given a computable instance $c$ of $\mathsf{RT}^2$, we apply cone avoidance of $\mathsf{COH}$ to obtain a cohesive set $S$ that does not compute a given noncomputable set $C$. The restriction of $c$ to this cohesive set is stable and $S$-computable. Hence, by cone avoidance of $\mathsf{SRT}^2_2$ (now relative to $S$) we can find an infinite homogeneous set $H \subseteq S$ for $c \upharpoonright [S]^2$ such that $C \nleqslant_T S \oplus H$. And of course, $H$ is homogeneous for $c$.

As remarked earlier, it is much easier to work with $\mathsf{D}^2$ in place of $\mathsf{SRT}^2$. But another simplification is to go even further, and work with $\mathsf{RT}^1$ instead. By Proposition 8.4.4, we can pass from any $\mathsf{D}^2$-instance to an $\mathsf{RT}^1$-instance with the same solutions. However, we cannot simply replace $\mathsf{SRT}^2$ by $\mathsf{RT}^1$ in (2). The problem is that the $\mathsf{RT}^1$

instance is only computable from the jump of the $\mathsf{D}^2$-instance. In our example above, $d$ would thus be $S'$-computable, so if $C = \varnothing'$, say, then we could very well have $C \leqslant_T d$. In this case, cone avoidance of $\mathsf{RT}^1$ would not by itself suffice to conclude that $d$ has a solution $H$ with $C \nleqslant_T S \oplus H$. So instead, we prove the following:

2′. $\mathsf{RT}^1$ admits strong cone avoidance.

Here, we recall the notion of *strong* cone avoidance from Definition 3.6.13. This is a stronger property than (ordinary) cone avoidance, but ensuring it is worth it for the benefit of being able to work with a coloring of singletons instead of pairs. Note that $\mathsf{RT}^1$ admits strong cone avoidance if and only if $\mathsf{D}^2$ does. So certainly (2′) implies (2). It is worth noting that we cannot prove strong cone avoidance for $\mathsf{SRT}^2$ directly, even if we wanted to.

**Proposition 8.5.5.** $\mathsf{SRT}^2_2$ *does not admit strong cone avoidance.*

*Proof.* Fix a $\varnothing'$-computable increasing $f$ such that $\varnothing'$ is computable from any function that dominates $f$ (Exercise 8.10.2.) Define $c\colon [\omega]^2 \to 2$ by

$$c(x, y) = \begin{cases} 0 & \text{if } y \leqslant f(x), \\ 1 & \text{otherwise}, \end{cases}$$

for all $x < y$. Thus, for each $x$ and all sufficiently large $y > x$ we have $c(x, y) = 1$, so $f$ is stable. And any infinite homogeneous set for $c$ must have color 1. Let $H$ be any such set, say $H = \{h_0 < h_1 < \cdots\}$. Then for all $x$ we have $h_{x+1} > f(h_x) \geqslant f(x)$, so the $H$-computable function $x \mapsto h_{x+1}$ dominates $f$. We conclude $\varnothing' \leqslant_T H$. □

Thus, even though $\mathsf{SRT}^2$ and $\mathsf{D}^2$ are interchangeable in most contexts, they are different in certain aspects.

We now proceed with the proof. First, we state the two key lemmas we will need.

**Lemma 8.5.6 (Cholak, Jockusch, and Slaman [33]).** $\mathsf{COH}$ *admits cone avoidance.*

**Lemma 8.5.7 (Dzhafarov and Jockusch [86]).** $\mathsf{RT}^1$ *admits strong cone avoidance.*

Next, let us see in detail how these combine to yield Theorem 8.3.1.

*Proof (of Theorem 8.3.1 from Lemmas 8.5.6 and 8.5.7).* By Corollary 4.6.12, we know that $\mathsf{RT}^2 \equiv_\omega \mathsf{RT}^2_2$, so it suffices to prove cone avoidance for the latter. So fix sets $A$ and $C$ with $C \nleqslant_T A$ and an $A$-computable coloring $c\colon [\omega]^2 \to 2$. Define an instance $\{R_x : x \in \omega\}$ of $\mathsf{COH}$ by $R_x = \{y > x : c(x, y) = 0\}$ for all $x \in \omega$. By Lemma 8.5.6, fix an infinite cohesive set $S$ for this family of sets satisfying $C \nleqslant_T A \oplus S$. By Proposition 8.4.17, $c \upharpoonright [S]^2$ is stable. Applying Lemma 8.5.7, we may fix an infinite limit homogeneous set $L \subseteq S$ for $c$ satisfying $C \nleqslant_T A \oplus S \oplus L$. Now by Proposition 8.4.2, we may fix an infinite $c \oplus L$-computable (and hence $A \oplus L$-computable) homogeneous set $H \subseteq L$ for $c \upharpoonright [S]^2$ (and therefore $c$). We have $C \nleqslant_T A \oplus H$, as desired. □

It remains to prove the lemmas in turn. The first is a simple application of facts we have already collected earlier in our discussion.

*Proof (of Lemma 8.5.6).* Fix sets $A$ and $C$ with $C \not\leqslant_T A$, and let $\langle R_i : i \in \omega \rangle$ be any $A$-computable instance of COH. Consider Mathias forcing with conditions $(E, I)$ such that $I \subseteq \omega$ and $C \not\leqslant_T A \oplus I$. By Example 7.3.7, relativized to $A$, every sufficiently generic set $G$ satisfies $C \not\leqslant_T A \oplus G$, and by Example 7.3.9, every sufficiently generic set $G$ is cohesive for the $R_i$. Thus, any sufficiently Mathias generic set $G$ is a solution witnessing cone avoidance of COH. □

*Proof (of Lemma 8.5.7).* Fix sets $A$ and $C$ with $C \not\leqslant_T A$, along with a coloring $c \colon \omega \to k$ for some $k$. (Note that there is no hypothesis that $c$ be $A$-computable, or effective in any other way.) We aim to produce an infinite homogeneous set $H$ for $c$ such that $C \not\leqslant_T A \oplus H$.

First, we define an infinite set $X \subseteq \omega$ with $C \not\leqslant_T A \oplus X$ and a set $K \subseteq \{0, \ldots, k-1\}$, as follows. Define $X_0 = \omega$ and $K_0 = \{0, \ldots, k-1\}$, and suppose we have defined $X_i$ and $K_i$ for some $i < k$. If there is an infinite set $X^* \subseteq X_i$ such that $C \not\leqslant_T A \oplus X^*$ and $c(x) \neq i$ for all $x \in X^*$, then let $X_{i+1} = X^*$ and let $K_{i+1} = K_i \setminus \{i\}$. Otherwise, let $X_{i+1} = X_i$ and $K_{i+1} = K_i$. Finally, set $X = X_k$ and $K = K_k$. By construction, for all $i < k$, if $i \in K$ then $c(x) = i$ for infinitely may $x \in X$, and if $i \notin K$ then $c(x) \neq i$ for all $x \in X$. For ease of notation, assume $K = \{0, \ldots, \ell - 1\}$ for some $\ell \leqslant k$.

We force with $\ell$-fold Mathias conditions $(E_0, \ldots, E_{\ell-1}, I)$ within $X$ with the following additional clauses.

- For each $i < \ell$, $c(x) = i$ for all $x \in E_i$.
- $C \not\leqslant_T A \oplus I$.

By choice of $X$ and $K$, it is easy to see that for each $i < \ell$ and each $n \in \omega$, the set of conditions $(E_0, \ldots, E_{\ell-1}, I)$ with $|E_i| > n$ is dense. Hence, any sufficiently generic object for this forcing will have the form $G = H_0 \oplus \cdots \oplus H_{\ell-1}$, where each $H_i$ is an infinite homogeneous set for $c$ with color $i$.

Next, fix $\ell$ indices $e_0, \ldots, e_{\ell-1} \in \omega$. We claim that the set of conditions forcing that there is an $i < \ell$ such that

$$(\exists w)\neg[\Phi_{e_i}^{\mathrm{H}_i}(w) \downarrow = C(w)] \tag{8.3}$$

is dense. Fix any condition $(E_0, \ldots, E_{\ell-1}, I)$. We exhibit an extension forcing (8.3). Consider the class $\mathcal{C}$ of all sets $Z = Z_0 \oplus \cdots \oplus Z_{\ell-1}$ as follows.

- $Z_0, \ldots, Z_{\ell-1}$ partition $I$.
- There is no $i < \ell$ and $w \geqslant \min I$ such that there exists $F_0, F_1 \subseteq Z_i$ with $\Phi_{e_i}^{A \oplus (E_i \cup F_0)}(w) \downarrow \neq \Phi_{e_i}^{A \oplus (E_i \cup F_1)}(w) \downarrow$ (with uses bounded by $2 \max F_0$ and $2 \max F_1$, respectively).

Notice that $\mathcal{C}$ is a $\Pi_1^0(A \oplus I)$ class. We investigate two cases.

*Case 1:* $\mathcal{C} \neq \emptyset$. By the cone avoidance basis theorem (Theorem 2.8.23) relative to $A \oplus I$, we may fix $Z = Z_0 \oplus \cdots \oplus Z_{\ell-1} \in \mathcal{C}$ such that $C \not\leqslant_T A \oplus Z$. Since $I$ is infinite and the $Z_i$ partition $I$, there must be an $i < \ell$ such that $Z_i$ is infinite. Fix such an $i$. Then the condition $(E_0^*, \ldots, E_{\ell-1}^*, I^*) = (E_0, \ldots, E_{\ell-1}, Z_i)$ is an extension of $(E_0, \ldots, E_{\ell-1}, I)$ forcing that there is a $w \geqslant \min I$ for which $\Phi_{e_i}^{A \oplus \mathrm{H}_i}(w) \uparrow$, which implies (8.3).

*Case 2:* $C = \emptyset$. In this case, by compactness, we may fix a level $n \in \omega$ such that if $Z_0, \ldots, Z_{\ell-1}$ partition $I$ then there is an $i < \ell$, a $w \geqslant \max I$, and sets $F_0, F_1 \subseteq Z_i \upharpoonright n$ such that $\Phi_{e_i}^{A \oplus (E_i \cup F_0)}(w) \downarrow \neq \Phi_{e_i}^{A \oplus F_1}(w) \downarrow$. Then there is $b < 2$ such that $\Phi_{e_i}^{A \oplus (E_i \cup F_b)}(w) \neq C(w)$. In this case, for $j \neq i$, set $E_j^* = E_j$, set $E_i^* = E_i \cup F_b$, and set $I^* = \{x \in I : x > F_b\}$. Then $(E_0^*, \ldots, E_{\ell-1}^*, I^*)$ is an extension of $(E_0, \ldots, E_{\ell-1}, I)$ forcing that there is a $w \geqslant \min I$ for which $\Phi_{e_i}^{A \oplus \mathrm{H}_i}(w) \downarrow \neq C(w)$.

By Lachlan's disjunction, we conclude that if $G = H_0 \oplus \cdots \oplus H_{\ell-1}$ is sufficiently generic then there is an $i < \ell$ such that $C \nleqslant_{\mathrm{T}} A \oplus H_i$. Combined with our earlier observation that $H_i$ is homogeneous for $c$ completes the proof. □

### 8.5.2 Other applications

Another application, broadly following the same outline as the alternative proof of Seetapun's theorem in the previous section, is the following.

**Theorem 8.5.8 (Cholak, Jockusch, and Slaman [33]).** *Fix $A \in 2^\omega$ and $X \gg A'$. Every $A$-computable instance of* $\mathsf{RT}^2$ *has a solution $H$ satisfying $(A \oplus H)' \leqslant_{\mathrm{T}} X$.*

This has the following corollary, which complements Seetapun's theorem

**Corollary 8.5.9.** $\mathsf{RT}^2$ *admits low$_2$ solutions.*

*Proof.* Fix $A$ and by the low basis theorem (Theorem 2.8.18) relative to $A'$, choose $X \gg A'$ with $X' \leqslant_{\mathrm{T}} A''$. Applying Theorem 8.5.8, every $A$-computable instance $c$ of $\mathsf{RT}^2$ has a solution $H$ such that $(A \oplus H)' \leqslant_{\mathrm{T}} X$. Hence $(A \oplus H)'' \leqslant_{\mathrm{T}} X' \leqslant_{\mathrm{T}} A''$.□

Of course, by Theorem 8.2.2, $\mathsf{RT}^2$ omits $\Sigma_2^0$—hence $\Delta_2^0$, and so certainly low—solutions. So the above cannot be improved from low$_2$ to low. Since no low$_2$ set can compute $\emptyset'$, Corollary 8.5.9 establishes certain kinds of cone avoidance for $\mathsf{RT}^2$. In fact, it is possible to add in full cone avoidance.

**Theorem 8.5.10 (Dzhafarov and Jockusch [86]).** *Fix $A \in 2^\omega$ and $C \nleqslant_{\mathrm{T}} A$. Every* $\mathsf{RT}^2$ *instance has a solution $H$ such that $(A \oplus H)'' \leqslant_{\mathrm{T}} A''$ and $C \nleqslant_{\mathrm{T}} A \oplus H$.*

We omit the proof, which uses techniques similar to those we will see below.

Let us move on to the proof of Theorem 8.5.8. Using the Cholak–Jockusch–Slaman decomposition, we split the proof into a version for $\mathsf{COH}$ and a version for $\mathsf{SRT}_2^2$. We begin with the former.

**Lemma 8.5.11.** *Fix $A \in 2^\omega$ and $X \gg A'$. Every $A$-computable instance of* $\mathsf{COH}$ *has a solution $G$ satisfying $(A \oplus G)' \leqslant_{\mathrm{T}} X$.*

*Proof.* We prove the result for $A = \emptyset$. The general case easily follows by relativization. Fix $X \gg \emptyset'$ and a computable instance $\vec{R} = \{R_i : i \in \omega\}$ of $\mathsf{COH}$. Consider Mathias forcing with computable reservoirs. We prove two claims.

*Claim 1: For each $e \in \omega$, the set of conditions deciding $e \in \mathsf{G}'$ is uniformly $X$-effectively dense.* Indeed, fix any condition $(E, I)$. Then $\emptyset'$ (and hence $X$) can tell if there is a finite $F \subseteq E$ such that $\Phi_e^{E \cup F}(e)\downarrow$ (with use bounded by $\max F$, as usual). If so, $\emptyset'$ can find the least such $F$, and then $(E^*, I^*) = (E \cup F, \{y \in I : y > F\})$ is an extension of $(E, I)$ forcing $e \in \mathsf{G}'$. On the other hand, if there is no such $F$ then $(E, I)$ already forces $\neg(e \in \mathsf{G})'$.

*Claim 2: For each $i \in \omega$, the set of conditions forcing $\mathsf{G} \subseteq^* R_i \vee \mathsf{G} \subseteq^* \overline{R_i}$ is uniformly $X$-effectively dense.* Fix any condition $(E, I)$. Note that "$I \cap R_i$ is infinite" and 1"$I \cap \overline{R_i}$ is infinite" are both $\Pi^0_1(\emptyset')$ statements, at least one of which is true. By Theorem 2.8.25 (4), relativized to $\emptyset'$, $X$ can uniformly computably identify one of these two statements which is true. If it is the former, we let $(E^*, I^*) = (E, I \cap R_i)$, which forces $\mathsf{G} \subseteq^* R_i$. If it is the latter, we let $(E^*, I^*) = (E, I \cap \overline{R_i})$, which forces $\mathsf{G} \subseteq^* \overline{R_i}$.

To complete the proof, note that for each $n$, the collection of conditions $(E, I)$ with $|E| = n$ is uniformly computably dense (even though the forcing is only $\emptyset''$-computable). Hence, by Theorem 7.5.6 (as in Example 7.5.7) there is an $X$-computable generic sequence $(E_0, I_0) \geqslant (E_1, I_1) \geqslant \cdots$ such that $G = \bigcup_s E_s \in \omega^\omega$ and for all $s$, if $s = 3e+1$ then $(E_s, I_s)$ decides $e \in \mathsf{G}'$, and if $s = 3e+2$ then $(E_s, I_s)$ forces $\mathsf{G} \subseteq^* R_i \vee \mathsf{G} \subseteq^* \overline{R_i}$. It follows that $G$ is $\vec{R}$-cohesive. And for each $e$, $e \in G'$ if and only if $\Phi_e^{E_{3e+1}}(e)\downarrow$ (with use bounded by $E_{3e+1}$). This can be checked by $X$ since it knows the entire sequence of conditions, so $G' \leqslant_T X$, as desired. □

Now let us prove a version of Theorem 8.5.8 for $\mathsf{SRT}^2_2$. We emphasize we are looking at $\mathsf{SRT}^2_2$, not $\mathsf{SRT}^2$.

**Lemma 8.5.12.** *Fix $A \in 2^\omega$ and $X \gg A'$. Every $A$-computable instance of $\mathsf{SRT}^2_2$ has a solution $H$ satisfying $(A \oplus H)' \leqslant_T X$.*

*Proof.* Again, we prove just the unrelativized version. Fix a computable stable coloring $c\colon [\omega]^2 \to 2$. If $c$ has a low infinite homogeneous set, we may take this to be $H$. Then $H' \leqslant_T \emptyset' \leqslant_T X$, so we are done. So suppose $c$ has no low infinite homogeneous set. For ease of notation, for each $i < 2$ define $P_i = \{x : \lim_y c(x, y) = i\}$. By Remark 8.4.3, $P_0$ and $P_1$ are both $\emptyset'$-computable. Our assumption implies that neither has a low infinite subset, as this would be a low limit homogeneous set for $c$, which could then be thinned to a low homogeneous set using Proposition 8.4.2.

We produce an infinite subset $G$ of either $P_0$ or $P_1$ with $G' \leqslant_T X$. We force with 2-fold Mathias conditions $(E_0, E_1, I)$ such that $E_i \subseteq P_i$ for each $i < 2$ and such that $I$ is low. A sufficiently generic filter thus yields limit homogeneous sets $G_0$ and $G_1$ for $c$ with colors 0 and 1, respectively. We use $\mathsf{G}_0$ and $\mathsf{G}_1$ as names for these objects in our forcing language, though these are just abbreviations for definitions using the parameter $\mathsf{G}$, as in Remark 8.3.6.

The two key density facts are the following.

*Claim 1: For each $n \in \omega$ and each $i < 2$, the set of conditions forcing $(\exists x \geqslant n)[x \in \mathsf{G}_i]$ is uniformly $X$-effectively dense.* Fix $n$ and $i$ and a condition $(E_0, E_1, I)$. Since $P_{1-i}$ has no infinite low subset, it must be that $I \cap P_i$ is infinite. In particular, we can $I$-computably (and hence certainly $X$-computably) find an $x \geqslant n$ in $I \cap P_i$. Let $E_i^* = E_i \cup \{x\}$, $E_{1-i}^* = E_{1-i}$, and $I^* = \{y \in I : y > x\}$. Then $(E_0^*, E_1^*, I^*)$ is an extension of $(E_0, E_1, I)$ forcing $x \in \mathsf{G}_i$.

*Claim 2: For all $e_0, e_1 \in \omega$, the set of conditions that decide, for some $i < 2$, whether $e_i \in \mathsf{G}_i'$ is uniformly $X$-effectively dense.* Consider the class $C$ of all sets $Z = Z_0 \oplus Z_1$ as follows.

- $Z_0 \cup Z_1 = I$.
- For each $i < 2$ and every $F \subseteq Z_i$, $\Phi_{e_i}^{E_i \cup F}(e_i) \uparrow$.

Then $C$ is a $\Pi_1^0(I)$ class. Clearly, $I$ can uniformly compute a tree $T \subseteq 2^{<\omega}$ such that $C = [T]$. Since $I$ is low, this means that $\varnothing'$ can uniformly compute whether or not $C \neq \varnothing$, since this is the same as whether or not $T$ is infinite. Suppose first that $C = \varnothing$. Then in particular, $(I \cap P_0) \oplus (I \cap P_1) \notin C$. Since $(I \cap P_0) \oplus (I \cap P_1) \leqslant_T \varnothing'$, it follows that $\varnothing'$ (and hence $X$) can uniformly search for an $i < 2$ and a finite set $F \subseteq I \cap P_i$ such that $\Phi_{e_i}^{E_i \cup F}(e_i) \downarrow$, and this search must succeed. For the least $i$ such that some such $F$ is found, let $E_i^* = E_i \cup F$, let $E_{1-i}^* = E_{1-i}$, and let $I^* = \{y \in I : y > F\}$. Then $(E_0^*, E_1^*, I^*)$ is an extension of $(E_0, E_1, I)$ forcing $\Phi_{e_i}^{\mathsf{G}_i}(e_i) \downarrow$. Suppose next that $C \neq \varnothing$. Since $I$ is low, $\varnothing'$ (and hence $X$) can uniformly compute an element $Z \in C$ such that $(I \oplus Z)' \leqslant_T I' \equiv_T \varnothing'$. Thus, each of the statements "$Z_0$ is infinite" and "$Z_1$ is infinite" are uniformly $\Pi_1^0(\varnothing')$, and one of them is true. So by Theorem 2.8.25 (4), relativized to $\varnothing'$, $X$ can uniformly computably find an $i < 2$ such that $Z_i$ is infinite. In this case, we let $(E_0^*, E_1^*, I^*) = (E_0, E_1, Z_i)$, which is an extension of $(E_0, E_1, I)$ forcing $\Phi_{e_i}^{\mathsf{G}_i}(e_i) \uparrow$.

Now apply Theorem 7.5.6 to construct an $X$-computable sequence of conditions $(E_0^0, E_1^0, I^0) \geqslant (E_0^1, E_1^1, I^1) \geqslant \cdots$ such that the following hold for all $s$: if $s = 3n + i$, $i < 2$, then $(E_0^s, E_1^s, I^s) \Vdash (\exists x \geqslant n)[x \in \mathsf{G}_i]$; if $s = 3\langle e_0, e_1\rangle + 2$, then $(E_0^s, E_1^s, I^s)$ decides, for some $i < 2$, whether $e_i \in \mathsf{G}_i'$.

It is clear that for each $i < 2$, $G_i = \bigcup_s E_i^s$ is an infinite subset of $P_i$.

Now by Lachlan's disjunction (Lemma 8.3.7), there is an $i < 2$ such that for all $e \in \omega$ there is an $s$ such that $(E_0^s, E_1^s, I^s)$ decides whether $e \in \mathsf{G}_i'$. Fix this $i$. Since $X$ knows the sequence of conditions, given $e$, it can search for (and find) an $s$ such that either $\Phi_{e_i}^{E_i^s}(e_i) \downarrow$ or, for all finite $F \subseteq I^s$, $\Phi_{e_i}^{E_i^s \cup F}(e_i) \uparrow$. In this way, $X$ can conclude whether $e \in G_i'$ or not. Thus, $G_i' \leqslant_T X$.

Finally, using Proposition 8.4.2, we thin $G_i$ to an infinite set $H \subseteq G_i$ which is computable from $G_i$ and homogeneous for $c$. □

We can now prove the main theorem.

*Proof (of Theorem 8.5.8).* We wish to show that for every $k \geqslant 2$, every $A \in 2^\omega$, and every $X \gg A'$, every $A$-computable instance $c$ of $\mathsf{RT}_k^2$ has a solution $H$ such that $(A \oplus H)' \leqslant_T X$. We proceed by induction on $k$. We prove the base case, $k = 2$, for $A = \varnothing$. The full case is obtained by relativization. Using density of $\gg$ (Proposition 2.8.26), fix $X_0$ such that $X \gg X_0 \gg \varnothing'$. Fix a computable instance $c$ of $\mathsf{RT}_2^2$. Define a computable instance $\vec{R} = \langle R_x : x \in \omega\rangle$ of $\mathsf{COH}$ by setting $R_x = \{y > x : c(x, y) = 0\}$ for all $x$. By Lemma 8.5.11, there is an infinite $\vec{R}$-cohesive set $G$ with $G' \leqslant_T X_0$. As in Proposition 8.4.17, $c \upharpoonright [G]^2$ is stable. Since $c \upharpoonright [G]^2$ is $G$-computable and $X \gg X_0 \geqslant_T G'$, it follows by Lemma 8.5.12, realtivized to $G$, that there is an infinite homogeneous set $H \subseteq G$ for $c \upharpoonright [G]^2$ with $(G \oplus H)' \leqslant_T X$. Since $H$ is also homogeneous for $c$, we are done.

Now fix $k > 2$ and assume the result holds for $k - 1$. Again, we take $A = \emptyset$ for simplicity. Fix a computable instance $c$ of $\mathsf{RT}^2_k$. Using the density of $\gg$ (Proposition 2.8.26), fix $X_0$ such that $X \gg X_0 \gg \emptyset'$. Define a coloring $c_0 \colon [\omega]^2 \to k - 1$ as follows: for all $\vec{x} \in [\omega]^2$,

$$c_0(\vec{x}) = \begin{cases} c(\vec{x}) & \text{if } c(\vec{x}) < k - 2, \\ k - 1 & \text{otherwise.} \end{cases}$$

Since $X_0 \gg \emptyset'$ and $c_0$ is a computable instance of $\mathsf{RT}^2_{k-1}$, we may apply the inductive hypothesis to obtain an infinite homogeneous set $H_0$ for $c_0$ such that $H_0' \leqslant_T X_0$. If $H_0$ has color $i < k - 2$ for $c_0$, then $H_0$ is also homogeneous for $c$ and so we can take $H = H_0$ and have $H' \leqslant_T X_0 \leqslant_T X$. Otherwise, $c(\vec{x})$ is either $k - 2$ or $k - 1$ for all $\vec{x} \in H_0$. In this case, define a coloring $c_1 \colon [H_0]^2 \to 2$ as follows: for all $\vec{x} \in [H_0]^2$, $c_1(\vec{x}) = c(\vec{x}) - k + 2$. Since $X \gg H_1'$ and $c_1$ is an $H_0$-computable instance of $\mathsf{RT}^2_2$, we may apply the inductive hypothesis, relativized to $H_0$, to obtain an infinite homogeneous set $H \subseteq H_0$ for $c_1$ such that $(H_0 \oplus H)' \leqslant_T X$. Now $H$ is also homogeneous for $c$ and we have $H' \leqslant_T X$, as desired. □

## 8.6 Liu's theorem

Having seen that Ramsey's theorem for pairs is strictly weaker than arithmetical comprehension, the next logical point of comparison is with weak König's lemma. For starters, we can already show that $\mathsf{WKL}_0$ does not prove $\mathsf{RT}^2_2$. Indeed, there is an $\omega$-model satisfying $\mathsf{WKL}$ contained entirely in the low sets, but by Theorem 8.1.1, $\mathsf{RT}^2_2$ has a computable instance with no $\Delta^0_2$ (let alone low) solution. Thus, $\mathsf{RT}^2_2 \nleqslant_\omega \mathsf{WKL}$. What about in the other direction, does $\mathsf{RCA}_0 \vdash \mathsf{RT}^2 \to \mathsf{WKL}$? Following Seetapun's proof this question quickly gained prominence as a major problem in computable combinatorics, particularly as more and more attempts at solving it proved unsuccessful. A correct proof was finally found in 2011, by Liu.

**Theorem 8.6.1 (Liu).** $\mathsf{RT}^2$ *admits PA avoidance.*

**Corollary 8.6.2.** $\mathsf{WKL} \nleqslant_\omega \mathsf{RT}^2_2$. *Hence also* $\mathsf{RCA}_0 \nvdash \mathsf{RT}^2 \to \mathsf{WKL}$ *(and so the two principles are incomparable over* $\mathsf{RCA}_0$.*)*

The story here bears an almost uncanny resemblance to the resolution of another longstanding open problem in computability theorem—Post's problem, by Friedberg and Muchnik—in that the breakthrough did not come from a senior researcher, long established in the field, but rather a newcomer. Indeed, like Friedberg and Muchnik, Liu was still an undergraduate student at the time his proof was published. And what a proof! As we will see, it has many novel elements that really set it apart from the arguments we have seen up to this point. At the same time, the proof still employs (a variant of) Mathias forcing and follows the Cholak–Jockusch–Slaman decomposition, so we will not be wholly on unfamiliar ground.

The proof breaks into two halves, following the general outline of the proof of Seetapun's theorem that we saw in Section 8.5.1. We will utilize the notion of *strong* PA avoidance from Definition 3.6.13.

**Proposition 8.6.3.** $\mathsf{COH}$ *admits PA avoidance.*

**Proposition 8.6.4.** $\mathsf{RT}^1_2$ *admits strong PA avoidance.*

We have already proved the former result in Proposition 3.8.5, where we dealt with $\mathsf{COH}$ in the guise of $\mathsf{SeqCompact}_{2^\omega}$. (Recall that $\mathsf{COH} \equiv_c \mathsf{SeqCompact}_{2^\omega}$, by Proposition 4.1.7.) Thus, the remainder of this section is dedicated to proving Proposition 8.6.4.

## 8.6.1 Preliminaries

To prove Proposition 8.6.4, we must show that for all sets $A, C$ with $A \not\gg C$, every instance $c\colon \omega \to 2$ of $\mathsf{RT}^1_2$ (computable from $A$ or not) has a solution $G$ such that $A \oplus G \not\gg C$. For ease of notation, we work with sets and subsets rather than colorings, per Proposition 8.4.4. That is, we show that every set $P$ has an infinite subset $G$ in it or its complement such that $A \oplus G \not\gg C$. (Formally, $P = \{x : c(x) = 0\}$, but we will not mention $c$ again.) We may assume that $P$ is not computable, as otherwise either $P$ or $\overline{P}$ is infinite and can serve as $G$. As a result, we can take $A = C = \emptyset$. The only property of $\emptyset$ we will invoke is that $\emptyset \not\gg \emptyset$, and hence that it does not compute $P$. Thus, the full result easily follows by relativization.

**Definition 8.6.5.**

1. A *pre-condition* is a sequence $p = (E_0, \ldots, E_{n-1}, C)$ where the $E_i$ are finite sets, called the *finite parts* of the pre-condition, and $C$ is a $\Pi^0_1$ class such that $E_i < X_i$ for every $i < n$ and every $X_0 \oplus \cdots \oplus X_{n-1} \in C$. We call $n$ the *length* of the pre-condition.
2. A finite part $E_i$ of the pre-condition $(E_0, \ldots, E_{n-1}, C)$ is *acceptable* if there exists $X_0 \oplus \cdots X_{n-1} \in C$ such that $X_i \cap A$ and $X_i \cap \overline{P}$ are both infinite.
3. We call $(E_0, \ldots, E_{n-1}, C)$ a *condition* if for some $N \in \omega$, called the *bound* of the condition, whenever $X_0 \oplus \cdots \oplus X_{n-1}$ belongs to $C$ then every $x > N$ belongs to $\bigcup_{i<n} X_i$.
4. A pre-condition $(\widehat{E}_0, \ldots, \widehat{E}_{m-1}, \widehat{C})$ *extends* $(E_0, \ldots, E_{n-1}, C)$ if $m \geqslant n$ and there exists a surjective map $c\colon m \to n$ such that for all $i < m$ and all $Y_0 \oplus \cdots \oplus Y_{m-1} \in \widehat{C}$ there exists $X_0 \oplus \cdots \oplus X_{n-1} \in C$ such that $E_{c(i)} \subseteq \widehat{E}_i \subseteq E_{c(i)} \cup X_{c(i)}$. In this case, we say that $c$ *witnesses* this extension, and that $\widehat{E}_i$ is a *child* of $E_{c(i)}$.

5. We call $(\widehat{E}_0, \ldots, \widehat{E}_{m-1}, \widehat{C})$ a *finite extension* of $(E_0, \ldots, E_{n-1}, C)$ if $m = n$, the witness $c$ is the identity, and for every $Y_0 \oplus \cdots \oplus Y_{n-1} \in \widehat{C}$ there exists $X_0 \oplus \cdots \oplus X_{n-1} \in C$ such that $Y_i =^* X_i$ for all $i < n$.
6. A set $G$ *satisfies* $(E_0, \ldots, E_{n-1}, C)$ *via* $i < n$ if there exists some $i < n$ and some $X_0 \oplus \cdots \oplus X_{n-1} \in C$ such that $G$ satisfies $(E_i, X_i)$. In this case, we also say that $G$ satisfies $(E_0, \ldots, E_{n-1}, C)$ *via* $i$.

We wish to construct a set $G$ meeting the requirements

$$\mathcal{P}_e \colon G \cap P \text{ and } G \cap \overline{P} \text{ each contain an element} > e$$

for all $e \in \omega$, as well as

$$\mathcal{R}_{e_0,e_1} \colon \Phi_{e_0}^{G \cap P} \text{ or } \Phi_{e_1}^{G \cap \overline{P}} \text{ is not a 2-valued DNC function}$$

for all $e_0, e_1 \in \omega$. Such a $G$ will thus have infinite intersection with both $P$ and $\overline{P}$, and by Lachlan's disjunction (Lemma 8.3.7), either $\Phi_{e_0}^{G \cap P}$ will not be a 2-valued DNC function for all $e_0$, or $\Phi_{e_1}^{G \cap \overline{P}}$ will not be a 2-valued DNC function for all $e_1$.

The proof of the proposition will follow from the following two lemmas.

**Lemma 8.6.6.** *Every condition $p$ has an acceptable part. Thus, given $e \in \omega$, $p$ has a finite extension $q$ such that if $E$ is a child of an acceptable part of $p$ then $E \cap P$ and $E \cap \overline{P}$ both contain an element $> e$.*

**Lemma 8.6.7.** *Let $p$ be a condition, and let $e_0, e_1 \in \omega$ be given. There exists an extension $q$ of $p$ such that for all $G$ satisfying $q$, either $\Phi_{e_0}^{G \cap P}$ is not a 2-valued DNC function or $\Phi_{e_1}^{G \cap \overline{P}}$ is not a 2-valued DNC function.*

*Proof (of Proposition 8.6.4).* First, we inductively define a sequence of conditions

$$p_0, p_1, \ldots$$

with $p_{i+1}$ extending $p_i$ for all $i \in \omega$. We begin with the condition $p_0 = (\emptyset, \{\omega\})$, and assume we are given $p_i$ for some $i \in \omega$. If $i = 2e$, let $p_{i+1}$ be the finite extension obtained by applying Lemma 8.6.6 to $p_i$. If $i = 2\langle e_0, e_1 \rangle + 1$, let $p_{i+1}$ be the extension obtained by applying Lemma 8.6.7 to $p_s$.

Now let $T$ be the set of all the acceptable parts of the conditions $p_s$, ordered by setting $E \leqslant \widehat{E}$ for $E, \widehat{E} \in T$ just in case there exist $i_0 < \cdots < i_{k-1}$ and $E_0, \ldots, E_{k-1}$ such that $E_0 = E$, $E_{k-1} = \widehat{E}$, each $E_j$ is an acceptable part of $p_{i_j}$, and each $E_{j+1}$ is a child of $E_j$. Under this ordering, $T$ forms a finitely branching tree, because if $\widehat{E}$ is an acceptable part of some condition and a child of a finite part $E$ of another condition, then $E$ is also an acceptable part. Furthermore, $T$ is infinite, since every $p_i$ has an acceptable part by Lemma 8.6.6.

So let $\{G_0, G_1, \ldots\}$ with $G_0 \leqslant G_1 \leqslant \cdots$ be any infinite path through this tree, and set $G = \bigcup_{i \in \omega} G_i$. It is then readily seen that $G$ satisfies $p_{2\langle e_0, e_1 \rangle+1}$ for all $e_0, e_1 \in \omega$, and hence that it meets $R_{e_0,e_1}$. Similarly, $G$ satisfies $p_{2e}$ for all $e$, and hence meets $\mathcal{P}_e$. □

### 8.6.2 Proof of Lemma 8.6.6

We require one preliminary lemma.

**Lemma 8.6.8.** *Let $p = (E_0, \ldots, E_{n-1}, C)$ be a condition, and let $e \in \omega$ be given. There exists $X_0 \oplus \cdots \oplus X_{n-1} \in C$, $i < n$, and $x, y > e$ with $x \in X_i \cap P$ and $y \in X_i \cap \overline{P}$.*

*Proof.* Suppose not. Since $C$ is a $\Pi^0_1$ class and $P$ is noncomputable, we may choose some $X_0 \oplus \cdots \oplus X_{n-1} \in C$ that does not compute $P$. By assumption, for each $i < n$ either $X_i \subseteq^* P$ or $X_i \subseteq^* \overline{P}$. Let $S = \{i < n : X_i \subseteq^* P\}$, noting that $\{i < n : i \notin S_0\} = \{i < n : X_i \subseteq^* \overline{P}\}$. Since $p$ is a condition, we have $\bigcup_{i<n} X_i =^* \omega$. So, given $x$ larger than $e$ and the bound of $p$, we have that $x \in P$ if and only if $x \in X_i$ for some $i \in S$, meaning $P \leqslant_T X_0 \oplus \cdots \oplus X_{n-1}$, a contradiction. □

Now the proof of Lemma 8.6.6 follows.

*Proof (of Lemma 8.6.6).* Fix a condition $p = (E_0, \ldots, E_{n-1}, C)$ and an $e \in \omega$. For each $s \in \omega$, we define finite sets $F^s_0, \ldots, F^s_{n-1}$ such that the class $C_s$ of all $X_0 \oplus \cdots \oplus X_{n-1} \in C$ with

$$F_i \neq \emptyset \to X_i \upharpoonright \max F_i + 1 = F_i$$

for all $i < n$ is nonempty. Let $F^0_0 = \cdots = F^0_{n-1} = \emptyset$, so that $C_0 = C \neq \emptyset$, and assume inductively that $F^s_0, \ldots, F^s_{n-1}$ have been defined for some $s$. Then $(E_0, \ldots, E_{n-1}, C_s)$ is a condition, so applying the previous lemma with

$$e = \max(E_0 \cup F^s_0) \cup \cdots \cup (E_{n-1} \cup F^s_{n-1}),$$

we may fix $X_0 \oplus \cdots \oplus X_{n-1} \in C_s$, $i < n$, and $x \in X_i \cap P$ and $y \in X_i \cap \overline{P}$. Let

$$F^{s+1}_j = \begin{cases} F^s_j & \text{if } j \neq i \\ F^s_j \cup \{z \in X_i : z \leqslant x, y\} & \text{if } j = i, \end{cases}$$

noting that $X_i \upharpoonright \max F^{s+1}_i = F^{s+1}_i$ by definition, so $X_0 \oplus \cdots \oplus X_{n-1} \in C_{s+1} \neq \emptyset$. If we now let $F_i = \bigcup_s F^s_i$ for each $i$, then $F_0 \oplus \cdots \oplus F_{n-1}$ must belong to $C$ by compactness. Furthermore, for all $s$ and $i$, if $F^{s+1}_i \neq F^s_i$ then $F^{s+1}_i - F^s_i$ intersects both $P$ and $\overline{P}$. Since at each stage $F^{s+1}_i \neq F^s_i$ for some $i$, there must be some $i$ for which this is the case at infinitely many $s$. Then $F_i$ has infinite intersection with both $P$ and $\overline{P}$, meaning $E_i$ is acceptable. □

### 8.6.3 Proof of Lemma 8.6.7

We first require a crucial definition.

**Definition 8.6.9.** Let $p = (E_0, \ldots, E_{n-1}, C)$ be a condition, and let $e_0, e_1 \in \omega$ be given.

1. A *bit assignment* is a finite partial function $v\colon \omega \to 2$. A bit assignment is *correct* if $\mathrm{dom}(v) \neq \emptyset$ and $\Phi_x(x) \downarrow= v(x)$ for all $x \in \mathrm{dom}(v)$.
2. Given a bit assignment $v$, let $S_v$ be the $\Pi^0_1$ class of all sets of the form

$$X_{0,0} \oplus X_{0,1} \oplus \cdots \oplus X_{n-1,0} \oplus X_{n-1,1}$$

such that $(X_{0,0} \cup X_{0,1}) \oplus \cdots \oplus (X_{n-1,0} \cup X_{n-1,1}) \in C$ and for all $i < n$ and all finite sets $F$,

- if $F$ satisfies $(E_i \cap P, X_{i,0})$ and $\Phi^F_{e_0}(x) \downarrow\in \{0,1\}$ for some $x \in \mathrm{dom}(v)$, then $\Phi^F_{e_j}(x) \neq v(x)$,
- if $F$ satisfies $(E_i \cap \overline{P}, X_{i,1})$ and $\Phi^F_{e_1}(x) \downarrow\in \{0,1\}$ for some $x \in \mathrm{dom}(v)$, then $\Phi^F_{e_j}(x) \neq v(x)$.

3. We say *$p$ forces agreement on $e_0, e_1$* if there exists a correct bit assignment $v$ such that $S_v = \emptyset$.
4. Given bit assignments $v_0, \ldots, v_{s-1}$, let $\mathrm{Cross}(S_{v_0}, \ldots, S_{v_{s-1}})$ denote the class of all sets of the form

$$\bigoplus_{i<n} \bigoplus_{j<2} \bigoplus_{k<l<s} X^k_{i,j} \cap X^l_{i,j}$$

where, for each $k < s$,

$$X^k_{0,0} \oplus X^k_{0,1} \oplus \cdots \oplus X^k_{n-1,0} \oplus X^k_{n-1,1} \in S_{v_k}.$$

5. For a set $I$ of indices $i < n$, we let $S_{v,I}$ be defined as $S_v$ above but with the definition applying only to those $i \in I$. We say $p$ forces agreement on $e_0, e_1$ *inside $I$* if there is a correct bit assignment $v$ such that $S_{v,I} = \emptyset$, and define $\mathrm{Cross}(S_{v_0,I}, \ldots, S_{v_{s-1},I})$ as above but with $S_{v_k}$ replaced by $S_{v_k,I}$.

**Lemma 8.6.10.** *Let $p = (E_0, \ldots, E_{n-1}, C)$ be a condition, and let $v_0, \ldots, v_{2n}$ be bit assignments. Then*

$$(E_0, \ldots, E_0, \ldots, E_{n-1}, \ldots, E_{n-1}, \mathrm{Cross}(S_{v_0}, \ldots, S_{v_{2n}})),$$

*where each $E_i$ appears $2\binom{2n+1}{2}$ times is a condition extending $p$ with the same bound. For any set $I$ of indices $i < n$, the same result holds if $\mathrm{Cross}(S_{v_0}, \ldots, S_{v_{2n}})$ is replaced by $\mathrm{Cross}(S_{v_0,I}, \ldots, S_{v_{2n},I})$.*

*Proof.* It is not difficult to see that

$$(E_0, \ldots, E_0, \ldots, E_{n-1}, \ldots, E_{n-1}, \mathrm{Cross}(S_{v_0}, \ldots, S_{v_{2n}}))$$

is a pre-condition extending $p$, as witnessed by $c\colon 2n\binom{2n+1}{2} \to n$ given by $c(2i\binom{2n+1}{2} + j) = i$ for each $i < n$ and $j < 2\binom{2n+1}{2}$. To see that is in fact a condition, let $N$ be the bound of $p$, and fix any $x > N$ along with

$$X^k_{0,0} \oplus X^k_{1,0} \oplus \cdots \oplus X^k_{0,n-1} \oplus X^k_{1,n-1} \in S_{v_k}$$

for each $k < 2n + 1$. Since $(X^k_{0,0} \cup X^k_{1,0}) \oplus \cdots \oplus (X^k_{0,n-1} \cup X^k_{1,n-1}) \in C$ for each such $k$, there exists some $i_{k,x} < n$ and $j_{k,x} < 2$ such that $x \in X^k_{j_{k,x},i_{k,x}}$. But since there are only $2n$ many pairs $(j_{k,x}, i_{k,x})$, there must exist $k \neq l$ such that $(j_{k,x}, i_{k,x}) = (j_{l,x}, i_{l,x})$. Hence, $x \in X^k_{i_{k,x},j_{k,x}} \cap X^l_{i_{k,x},j_{k,x}}$, and thus

$$x \in \bigoplus_{i<n} \bigoplus_{j<2} \bigoplus_{k<l<2n+1} X^k_{i,j} \cap X^l_{i,j} \in \mathrm{Cross}(S_{v_0}, \ldots, S_{v_{2n}}).$$

**Lemma 8.6.11.** *Let $p = (E_0, \ldots, E_{n-1}, C)$ be a condition, and let $e_0, e_1 \in \omega$ be given. If $p$ does not force agreement on $e_0, e_1$ then there exist bit assignments $v_0, \ldots, v_{2n}$ such that $S_{v_i} \neq \emptyset$ for each $i < 2n + 1$, and for each $i < j < 2n + 1$ there is some $x \in \mathrm{dom}(v_i) \cap \mathrm{dom}(v_j)$ with $v_i(x) \neq v_j(x)$. For any set $I$ of indices $i < n$, the same result holds if forcing agreement is replaced by forcing agreement inside $I$ and $S_{v_i}$ is replaced by $S_{v_i,I}$.*

*Proof.* We construct, for each $\sigma \in 2^{<\omega}$, an element $x_\sigma \in \omega$ with certain properties described below. Given a bit assignment $v$ not defined on $x_{\sigma \upharpoonright i}$ for any $i < |\sigma|$, let

$$v_\sigma = v \cup \bigcup_{i<|\sigma|} \{\langle x_{\sigma \upharpoonright i}, \sigma(i)\rangle\}.$$

Each $x_\sigma$ will satisfy that $\Phi_{x_\sigma}(x_\sigma)$ does not converge to 0 or 1, and if $v$ is any correct bit assignment not defined on $x_{\sigma \upharpoonright i}$ for any $i$ then each of $S_{v_\sigma \cup \{\langle x_\sigma, 0\rangle\}}$ and $S_{v_\sigma \cup \{\langle x_\sigma, 1\rangle\}}$ is nonempty. Let $y$ be any number such that $\Phi_y(y) \downarrow \in \{0, 1\}$.

To begin, we claim that there must exist an $x \in \omega$ such that $\Phi_x(x)$ does not converge to 0 or 1, and for each correct bit assignment $v$ not defined on $x$, each of $S_{v \cup \{\langle x, 0\rangle\}}$ and $S_{v \cup \{\langle x, 1\rangle\}}$ is nonempty. If not, we could compute a 2-valued DNC function $f$ as follows: given $x$, we wait until either $\Phi_x(x) \downarrow \in \{0, 1\}$, in which case we let $f(x) = 1 - \Phi_x(x)$, or until we find a correct bit assignment $v$ not defined on $x$ and a $j \in \{0, 1\}$ such that the $\Pi^0_1$ class $S_{v \cup \{\langle x, j\rangle\}}$ is empty, in which case we let $f(x) = j$. (Note that if $\Phi_x(x)$ converged and was equal to $j$ then $v \cup \{\langle x, j\rangle\}$ would be a correct bit assignment, so $S_{v \cup \{\langle x, j\rangle\}}$ could not be empty by assumption.) In fact, the same argument shows that there must be infinitely many $x$ with this property, so let $x_\lambda$ be any such $x > y$.

Now suppose $x_\sigma$ has been defined for some $\sigma \in 2^{<\omega}$, and fix $k \in \{0, 1\}$. We claim that there must exist an $x \in \omega$ such that $\Phi_x(x)$ does not converge to 0 or 1, and for each correct bit assignment $v$ not defined on $x$ or on $x_{\sigma \upharpoonright i}$ for any $i < |\sigma|$, each of $S_{v_\sigma \cup \{\langle x_\sigma, k\rangle, \langle x, 0\rangle\}}$ and $S_{v_\sigma \cup \{\langle x_\sigma, k\rangle, \langle x, 1\rangle\}}$ is nonempty. If not, we could compute a 2-valued DNC function $f$ as follows: given $x$, we wait until either $\Phi_x(x) \downarrow \in \{0, 1\}$, in which case we let $f(x) = 1 - \Phi_x(x)$, or until we find a correct bit assignment $v$ not defined on $x$ or on $x_{\sigma \upharpoonright i}$ for any $i$ and a $j \in \{0, 1\}$ such that the $\Pi^0_1$ class $S_{v_\sigma \cup \{\langle x_\sigma, k\rangle, \langle x, j\rangle\}}$ is empty, in which case we let $f(x) = j$. (Note that if $\Phi_x(x)$ converged and was equal to $j$ then $w = v \cup \{\langle x, j\rangle\}$ would be a correct bit assignment not defined on $x_{\sigma \upharpoonright i}$ for any $i$, so

$$S_{v_\sigma \cup \{\langle x_\sigma, k\rangle, \langle x, j\rangle\}} = S_{w_\sigma \cup \{\langle x_\sigma, k\rangle\}}$$

could not be empty by choice of $x_\sigma$.) The same argument shows that there must be infinitely many $x$ with this property, so let $x_{\sigma k}$ be any such $x > y$.

Having defined $x_\sigma$ for all $\sigma$, let $v$ be any correct bit assignment not defined on any of them (which exists since we chose all $x_\sigma$ to be larger than $y$, so $v = \{\langle y, \Phi_y(y)\rangle\}$ would work). Let $\sigma_0, \dots, \sigma_{2n}$ be any pairwise incompatible elements of $2^{<\omega}$ of the same length, and for $i < 2n+1$, let $v_i = v_{\sigma_i}$. By construction, each $S_{v_i}$ is nonempty. Furthermore, given $i < j < 2n+1$, if $\tau$ denotes the longest common initial segment of $\sigma_i$ and $\sigma_j$, then

$$v_i(x_\tau) = \sigma_i(|\tau|) \neq \sigma_j(|\tau|) = v_j(x_\tau),$$

as desired. □

Given a condition $p$, let $I_p$ be the set of all $i < n$ such that there exists a $G$ satisfying $(E_0, \dots, E_{n-1}, C)$ via $i$ and either $\Phi_{e_0}^{G\cap P}$ or $\Phi_{e_1}^{G\cap \overline{P}}$ is a 2-valued DNC function. Lemma 8.6.7 is an immediate consequence of the following:

**Lemma 8.6.12.** *Let $p$ be a condition so that $I_p \neq \emptyset$, and let $e_0, e_1 \in \omega$ be given. If $p$ forces agreement on $e_0, e_1$ inside $I_p$ then $p$ has a finite extension $q$ such that $|I_q| < |I_p|$. Otherwise, $p$ has an extension $q$ such that $I_q = \emptyset$.*

*Proof.* Let $p = (E_0, \dots, E_{n-1}, C)$, and suppose first that $p$ forces agreement on $e_0, e_1$ inside $I_p$. Then there exists a correct bit assignment $v$ such that $S_{v,I_p} = \emptyset$. Fixing any $X_0 \oplus \cdots \oplus X_{n-1} \in C$, we then have that

$$(X_0 \cap P) \oplus (X_0 \cap \overline{P}) \oplus \cdots \oplus (X_n - 1 \cap P) \oplus (X_n - 1 \cap \overline{P})$$

does not belong to $S_{v,I_p}$. Hence, for some $i \in I_p$, there either exists a finite set $F$ satisfying $(E_i \cap P, X_i \cap P)$ such that $\Phi_{e_0}^F$ agrees with $v$ on some $x \in \mathrm{dom}(v)$, or else there exists a finite set $F$ satisfying $(E_i \cap \overline{P}, X_i \cap \overline{P})$ such that $\Phi_{e_1}^F$ agrees with $v$ on some $x \in \mathrm{dom}(v)$. Say the former possibility holds, the latter being analogous. We then fix some such $F$ and $x$ and define

$$E_j = \begin{cases} E_j & \text{if } j \neq i \\ F & \text{if } j = i, \end{cases}$$

and let $Q$ consist of all sets of the form

$$\bigoplus_{j<i} Y_j \oplus \{y \in Y_i : y > \varphi_{e_0}^F(x)\} \oplus \bigoplus_{i<j<n} Y_j$$

where $Y_0 \oplus \cdots \oplus Y_{n-1} \in C$. Now for any $G$ satisfying $(E_0, \dots, E_{n-1}, Q)$ via $i$, $F \subseteq G \cap P$ so if $\Phi_{e_0}^{G\cap P}$ is total and $\{0,1\}$-valued then it equals $\Phi_x(x)$ and is therefore not 2-valued DNC. Thus, $i \notin I_q$. Since clearly $I_q \subseteq I_p$, this implies that $|I_q| < |I_p|$.

Now suppose $p$ does not force agreement on $e_0, e_1$ inside $I_p$. Then, using Lemma 8.6.11, fix bit assignments $v_0, \dots, v_{2n}$ such that $S_{v_i,I_p} \neq \emptyset$ for each $i < 2n+1$, and for each $i < j < 2n+1$ there is some $x \in \mathrm{dom}(v_i) \cap \mathrm{dom}(v_j)$ with $v_i(x) \neq v_j(x)$. Let $q$ be

$$(E_0, \ldots, E_0, \ldots, E_{n-1}, \ldots, E_{n-1}, \mathrm{Cross}(S_{v_0}, \ldots, S_{v_{s-1}})),$$

where each $E_i$ appears $2\binom{2n+1}{2}$ times, which by Lemma 8.6.10 is an extension of $p$. Let $c\colon 2n\binom{2n+1}{2} \to n$ witness this extension. Now consider any set $G$ satisfying $q$, say via $i < n$. Then $G$ satisfies $(E_{c(i)}, X^k_{c(i),j} \cap X^k_{c(i),j})$ for some $k, l < 2n+1, j < 2$,

$$X^k_{0,0} \oplus X^k_{0,1} \oplus \cdots \oplus X^k_{n-1,0} \oplus X^k_{n-1,1} \in S_{v_k, I_p},$$

and

$$X^l_{0,0} \oplus X^l_{0,1} \oplus \cdots \oplus X^l_{n-1,0} \oplus X^l_{n-1,1} \in S_{v_l, I_p}.$$

Fix $x$ so that $v_k(x) \neq v_l(x)$, and suppose $j = 0$ for simplicity, the case where $j = 1$ being analogous. Then if $\Phi^{G \cap P}_{e_0}(x) \downarrow\in \{0,1\}$ we have $\Phi^{G \cap P}_{e_0}(x) \neq v_k(x)$ since $G \cap P$ satisfies $(E_{c(i)} \cap P, X^k_{c(i),0})$, and so $\Phi^{G \cap P}_{e_0}(x) = v_l(x)$ since $v_k$ and $v_l$ are $\{0,1\}$-valued. But $\Phi^{G \cap P}_{e_0}(x) \neq v_l(x)$ since $G \cap P$ satisfies $(E_{c(i)} \cap P, X^l_{c(i),0})$, which is a contradiction. Thus, $\Phi^{G \cap P}_{e_0}$ is not total. We conclude $I_q = \emptyset$, as desired. □

*Proof (of Lemma 8.6.7).* If $I_p = \emptyset$, there is nothing to prove. Otherwise, let $p_0 = p$, and having defined $p_s$ for some $s \in \omega$ with $I_{p_s} \neq \emptyset$, let $p_{s+1}$ be an extension of $p_s$ with $|I_{p_{s+1}}| < |I_{p_s}|$, as given by the previous lemma. By that lemma, there exists some $s$ such that $I_{p_s} = \emptyset$, and we let this be $q$. □

## 8.7 The first order part of RT

We now turn briefly to the first order strength of Ramsey's theorem. For $n = 1$ and $n \geqslant 3$, we already a near complete picture. $\mathsf{RCA}_0 \vdash \mathsf{RT}^1_k$ for each specific $k$, so its first part lies below $\mathsf{I}\Sigma^0_1$ by Corollary 5.10.2 (and so is trivial, from our perspective). $\mathsf{RT}^1$ is equivalent to $\mathsf{B}\Sigma^0_2$ by Hirst's theorem (Theorem 6.5.1), and so this is also its first order part. For $n \geqslant 3$, $\mathsf{RT}^n$, and also $\mathsf{RT}^n_k$ for each specific $k \geqslant 2$, is equivalent to $\mathsf{ACA}_0$, so its first order part is just PA by Corollary 5.10.7. For completeness, we can also note that $\mathsf{RCA}_0 \vdash (\forall n)\mathsf{RT}^n_1$, so here too the first order part is clear. As with the second order strength studied above, then, the interesting (and more complicated) case is $n = 2$. Unsurprisingly, the first order part of Ramsey's theorem for pairs, like its second order part, has been the subject of a great deal of research.

### 8.7.1 Two versus arbitrarily many colors

On the technical side, our main goal for this chapter is to present a proof that the first order part of $\mathsf{RT}^2_2$ is strictly weaker than that of $\mathsf{RT}^2$ (even $\mathsf{SRT}^2$). Thus, in particular, $\mathsf{RT}^2_2$ and $\mathsf{RT}^2$ are not equivalent over $\mathsf{RCA}_0$. The two key theorems are the following.

**Theorem 8.7.1 (Cholak, Jockusch, and Slaman [33]).** $\mathsf{WKL}_0 + \mathsf{RT}^2_2 + \mathsf{I}\Sigma^0_2$ *is* $\Pi^1_1$ *conservative over* $\mathsf{WKL}_0 + \mathsf{I}\Sigma^0_2$.

**Theorem 8.7.2 (Cholak, Jockusch, and Slaman [33]).** $\mathsf{RCA}_0 \vdash \mathsf{SRT}^2 \to \mathsf{B}\Sigma^0_3$.

Since $\mathsf{B}\Sigma^0_3$ is strictly stronger than $\mathsf{I}\Sigma^0_2$ over $\mathsf{RCA}_0$ (and hence over $\mathsf{WKL}_0$, by Harrington's theorem, Theorem 7.7.3), the following corollary is immediate.

**Corollary 8.7.3.** *For each* $k \geqslant 1$, $\mathsf{RCA}_0 \nvdash \mathsf{RT}^2_k \to \mathsf{SRT}^2$.

The proof of Theorem 8.7.1 utilizes the Cholak–Jockusch–Slaman decomposition, via the following two propositions.

**Proposition 8.7.4 (Cholak, Jockusch, and Slaman [33]).** *Every countable model of* $\mathsf{WKL}_0 + \mathsf{I}\Sigma^0_2$ *is an* $\omega$*-submodel of a model of* $\mathsf{WKL}_0 + \mathsf{COH} + \mathsf{I}\Sigma^0_2$.

**Proposition 8.7.5 (Cholak, Jockusch, and Slaman [33]).** *Every countable model of* $\mathsf{WKL}_0 + \mathsf{I}\Sigma^0_2$ *is an* $\omega$*-submodel of a model of* $\mathsf{WKL}_0 + \mathsf{SRT}^2_2 + \mathsf{I}\Sigma^0_2$.

With these in hand, Theorem 8.7.1 easily follows.

*Proof (of Theorem 8.7.1).* By Theorem 5.9.4, it suffices to show that every model of $\mathsf{RCA}_0 + \mathsf{I}\Sigma^0_2$ is an $\omega$-submodel of $\mathsf{RCA}_0 + \mathsf{RT}^2_2 + \mathsf{I}\Sigma^0_2$. Fix a countable model $\mathcal{M}$ satisfying $\mathsf{RCA}_0 + \mathsf{I}\Sigma^0_2$ and suppose $c \in \mathcal{S}^{\mathcal{M}}$ is a coloring $[M]^2 \to 2$. By Corollary 5.9.6, it suffices to produce a set $G$ such that $\mathcal{M}[G]$ satisfies $\mathsf{I}\Sigma^0_2$ and the fact that $G$ is an infinite homogeneous set for $c$. We proceed as in Lemma 8.5.3, defining an instance $\vec{R}$ of $\mathsf{COH}$ in $\mathcal{S}^{\mathcal{M}}$ such that $c \upharpoonright [X]^2$ is stable for every solution $X$ to this instance. Apply Proposition 8.7.4 to find $\mathcal{M}^* \supseteq_\omega \mathcal{M}$ satisfying $\mathsf{RCA}_0 + \mathsf{COH} + \mathsf{I}\Sigma^0_2$. Since $\vec{R} \in \mathcal{S}^{\mathcal{M}} \subseteq \mathcal{S}^{\mathcal{M}^*}$, we can choose a $\mathsf{COH}$-solution $X$ to $\vec{R}$ in $\mathcal{S}^{\mathcal{M}^*}$. Thus, $c \upharpoonright [X]^2$ is an instance of $\mathsf{SRT}^2_2$ in $\mathcal{S}^{\mathcal{M}^*}$. Apply Proposition 8.7.5 to find $\mathcal{M}^{**} \supseteq_\omega \mathcal{M}^*$ satisfying $\mathsf{RCA}_0 + \mathsf{SRT}^2_2 + \mathsf{I}\Sigma^0_2$. Then $c \upharpoonright [X]^2 \in \mathcal{S}^{\mathcal{M}^{**}}$ so we can choose an $\mathsf{SRT}^2_2$-solution $G \subseteq X$ to $c \upharpoonright [X]^2$. Since $\mathcal{M}^{**}$ and $\mathcal{M}$ have the same first order part, it follows that $\mathcal{M}[G]$ satisfies $\mathsf{I}\Sigma^0_2$. Obviously, $G$ is an $\mathsf{RT}^2_2$-solution to $c$ in $\mathcal{M}[G]$, so we are done. □

We delay the proofs of Propositions 8.7.4 and 8.7.5 to the next sections. For now, we turn to Theorem 8.7.2. The proof has some elements in common with that of Hirst's theorem, but in other ways it is quite different.

*Proof (of Theorem 8.7.2).* We argue in $\mathsf{RCA}_0 + \mathsf{SRT}^2$ and derive $\mathsf{B}\Pi^0_2$ (which is equivalent to $\mathsf{B}\Sigma^0_3$ by Theorem 6.1.3). By Theorem 8.4.9 (or Exercise 8.10.3), we may assume $\mathsf{B}\Pi^0_1$. Seeking a contradiction, suppose $\mathsf{B}\Pi^0_2$ is false. Thus there is a $\Pi^0_2$ formula $\varphi(x, y)$ and a $z$ such that the following two facts are true.

1. $(\forall x < z)(\exists y)\varphi(x, y)$.
2. $(\forall w)(\exists x < z)(\forall y < w)\neg\varphi(x, y)$.

For ease of terminology, say $w$ is *bad* for $x < z$ if $(\forall y < w)\neg\varphi(x, y)$. Thus, (2) says that every $w$ is bad for some $x < z$.

Let $\psi(u, v, x, y)$ be a $\Sigma^0_0$ formula such that $\varphi(x, y) \leftrightarrow (\forall u)(\exists v)\psi(u, v, x, y)$. Define a map $r\colon [\mathbb{N}]^2 \times z \to \mathbb{N}$ as follows: given $w < t$ and $x < z$, let $r(w, t, x)$ be the largest number $u^* < t$ such that

$$(\exists y < w)(\forall u < u^*)(\exists v < t)\psi(u, v, x, y).$$

Since $u^*$ can, in principle, be 0, it follows that $r$ is total. Next, define a coloring $c\colon [\mathbb{N}]^2 \to z$, as follows: given $w < t$, let $c(w, t)$ be the least $x < z$ such that $r(w, t, x) \leqslant r(w, t, x^*)$ for all $x^* < z$.

The key properties about $r$ are the following. Fix $w$ and $x < z$, and suppose first that $w$ is bad for $x$. Thus, $(\forall y < w)(\exists u)(\forall v)\neg\psi(u, v, x, y)$. By $\mathsf{B}\Pi^0_1$, there is a $\widehat{u}$ such that $(\forall y < w)(\exists u < \widehat{u})(\forall v)\neg\psi(u, v, x, y)$, and by $\mathsf{L}\Pi^0_1$ we may assume $\widehat{u}$ is least with this property. Let $u^* = \widehat{u} - 1$. By minimality of $\widehat{u}$, we have $(\exists y < w)(\forall u < u^*)(\exists v)\psi(u, v, x, y)$, so by $\mathsf{B}\Pi^0_1$ we may choose an $s$ such that $(\exists y < w)(\forall u < u^*)(\exists v < s)\psi(u, v, x, y)$. Then for all $t \geqslant s$ we have $r(w, t, x) = u^*$. In particular, $\lim_t r(w, t, x)$ exists.

Now suppose $w$ is not bad for $x$, so that $(\exists y < w)(\forall u)(\exists v)\psi(u, v, x, y)$. Fix any witness $y < w$. Let $m$ be arbitrary, and apply $\mathsf{B}\Pi^0_1$ and $\mathsf{L}\Pi^0_1$ to find a the least $s$ such that $(\forall u < m)(\exists v < s)\psi(u, v, x, y)$. Then for all $t \geqslant s$, we have $r(w, t, x) \geqslant m$. We conclude that $\lim_t r(w, t, x) = \infty$.

We now combine these two facts to prove that $c$ is stable. Fix $w$, along with any $r^*$ such that $\lim_t r(w, t, x) = r^*$ for some $x < z$. (This exists, since $w$ is bad for at least one $x < z$.) Now consider the sentence

$$(\forall x < z)(\exists s)(\forall t \geqslant s)[r(w, t, x) = r(w, s, x) \vee r(w, t, x) > r^*].$$

By our remarks above about $r$, this sentence is true. Hence, by $\mathsf{B}\Pi^0_1$, we may fix an $s_0$ so that

$$(\forall x < z)(\exists s < s_0)(\forall t \geqslant s)[r(w, t, x) = r(w, s, x) \vee r(w, t, x) > r^*].$$

So for every $x < z$, either $w$ is bad for $x$ and $r(w, t, x)$ reaches its limit by $s_0$, or this limit is always larger than $r^*$ past $s_0$ (irrespective of whether $w$ is bad for $x$ or not). By choice of $r^*$, it follows that it $t \geqslant s_0$ then $c(w, t) = x$ for some $x$ satisfying the first alternative, and hence that $c(w, t)$ is the same (namely, $c(w, t) = x$ for the least $x$ that minimizes $\lim_t r(w, t, x)$).

To complete the proof we claim that for each $x < z$ there are at most finitely may $w \in \mathbb{N}$ with $\lim_t c(w, t) = x$. Clearly, this is impossible in the presence of $\mathsf{SRT}^2$, so this yields the desired contradiction. Fix $x$. By (1), there is $y$ such that $(\forall u)(\exists v)\psi(u, v, x, y)$. Fix any $w > y$. Then $w$ is not bad for $x$, hence $\lim_t r(w, t, x) = \infty$. In particular, we cannot have $\lim_t c(w, t) = x$. The proof is complete. □

### 8.7.2 Proof of Proposition 8.7.4

Throughout this section we work over a fixed countable model $\mathcal{M}$ of $\mathsf{WKL} + \mathsf{I}\Sigma^0_2$. By Theorem 7.7.5, every countable model of $\mathsf{RCA}_0 + \mathsf{I}\Sigma^0_2$ is an $\omega$-submodel of a countable model of $\mathsf{WKL}_0 + \mathsf{I}\Sigma^0_2$. So it suffices to show that for a given instance $\vec{R} \in \mathcal{S}^{\mathcal{M}}$ of $\mathsf{COH}$ there is a $G$ such that $\mathcal{M}[G] \models \mathsf{I}\Sigma^0_2 +$ "$G$ is an infinite $\vec{R}$-cohesive set".

We will add $G$ by an elaboration on Mathias conditions in $\mathcal{M}$, in the sense of Definition 7.6.1. Recall that Mathias conditions in this context are pairs $(E, I)$, where $E, I \in \mathcal{S}^{\mathcal{M}}$, $E$ is $\mathcal{M}$-finite, $I$ is $\mathcal{M}$-infinite, and $E <^{\mathcal{M}} I$. Since $\mathcal{M}$ is fixed, we omit saying "Mathias condition *in* $\mathcal{M}$", "forces *in* $\mathcal{M}$", etc., and say simply "Mathias condition", "forces", etc.

We will need the following definition.

**Definition 8.7.6.** A *negation set* is an $\mathcal{M}$-finite collection $\Gamma$ of $\Pi^0_1$ formulas of $\mathcal{L}_2$ of the form $\psi(X, \vec{x})$, i.e., having one free set variable and one or more free number variables.

Here and throughout, we abbreviate $\vec{d}$ being an $\mathcal{M}$-finite tuple of elements of $M$ by $\vec{d} \in M$. The idea of the proof is the following. We construct a Mathias generic set $G$, and satisfy $\mathsf{I}\Sigma^0_2$ in $\mathcal{M}[G]$ by considering each $\Sigma^0_2$ formula of $\mathcal{L}_2$ in turn. Such a formula may have $G$ as a parameter, and so has the form $(\exists \vec{x})\psi(G, \vec{x}, y)$, where $\psi(X, \vec{x}, y)$ is a $\Pi^0_1$ formula with parameters from $\mathcal{M}$. In general, there are two ways induction can hold in $\mathcal{M}[G]$ for a formula $\varphi(y)$:

- $(\forall y)\varphi(y)$ holds in the model.
- There is a $b \in M$ such that $(\forall y < b)\varphi(y) \wedge \neg\varphi(b)$ holds.

Thus, we should like to show that for every $\Pi^0_1$ formula $\psi(X, \vec{x}, y)$ and every $b$, it is dense either to force $(\exists \vec{x})\psi(\mathtt{G}, \vec{x}, b)$ or to force $\neg(\exists \vec{x})\psi(\mathtt{G}, \vec{x}, b)$. However, as discussed in Remark 7.7.2, we need the complexity of forcing these statement to match that of the formulas themselves. The problem is that forcing the negation of a $\Sigma^0_2$ statement is too complicated (it is not $\Pi^0_2$). This is where negation sets come in. If we cannot force $(\exists \vec{x})\psi(\mathtt{G}, \vec{x}, b)$ for some $b$, we add $\psi(X, \vec{x}, b)$ to a running negation set. We then argue that for every $\vec{d} \in M$, it is dense to force $\neg\psi(\mathtt{G}, \vec{d}, b)$, thereby ensuring that $\neg(\exists \vec{x})\psi(G, \vec{x}, b)$ holds in $\mathcal{M}[G]$.

Let us pass to the details. Given a negation set $\Gamma$, we label each Mathias condition as either $\Gamma$-*large* or $\Gamma$-*small*, in a way that will be defined below. The specifics of this are not needed to state the key lemmas and derive Proposition 8.7.4 from them, so we do that first. The following technical lemma spells out the key properties of largeness and smallness that we will need.

**Lemma 8.7.7.** *Let* $\Gamma$ *be a negation set.*

1. *If* $(E, I)$ *and* $(E^*, I^*)$ *are Mathias conditions,* $(E, I)$ *is* $\Gamma$*-large, and* $(E^*, I^*)$ *is a finite extension of* $(E, I)$ *(i.e.,* $(E^*, I^*) \leqslant (E, I)$ *and* $I \setminus I^*$ *is finite), then* $(E^*, I^*)$ *is* $\Gamma$*-large.*

2. *If $(E, I)$ is a $\Gamma$-large Mathias condition and $\psi(X, \vec{x}) \in \Gamma$, then $(E, I)$ does not force $\psi(\mathsf{G}, \vec{d})$ for any $\vec{d} \in M$.*
3. *If $(E, I)$ is a $\Gamma$-large Mathias condition, $a \in M$, and for each $j < a$, $E_j$ and $I_j$ are sets in $\mathcal{S}^{\mathcal{M}}$ such that $E_j$ is $\mathcal{M}$-finite, $E \subseteq E_j \subseteq E_j \cup I$, and $I = \bigcup_{j<a} I_j$, then there is a $j < a$ such that $(E_j, I_j)$ is $\Gamma$-large.*

The argument will be organized using an auxiliary notion of forcing that pairs Mathias conditions with negation sets.

**Definition 8.7.8.** We define the following notion of forcing.

1. The conditions are triples $(\Gamma, E, I)$, where $\Gamma$ is a negation set and $(E, I)$ is a $\Gamma$-large Mathias condition.
2. Extension is defined by $(\Gamma^*, E^*, I^*) \leqslant (\Gamma, E, I)$ if $\Gamma \subseteq \Gamma^*$ and $(E^*, I^*)$ extends $(E, I)$ as Mathias conditions.
3. The valuation of $(\Gamma, E, I)$ is the valuation of $(E, I)$ as a Mathias condition.

As we will see below, the set of conditions here is nonempty. Because the valuation map ignores the negation set, a generic for this forcing is thus also a Mathias generic (but not conversely). By the same token, $(\Gamma, E, I)$ forces a formula if and only if the Mathias condition $(E, I)$ forces it.

The following lemmas help make formal our discussion above concerning preserving $\mathsf{I}\Sigma^0_2$ in $\mathcal{M}[G]$. We will use the following notation for convenience. Given $\vec{d} \in M$ with $\vec{d} = \langle d_0, \ldots, d_{a-1} \rangle$ for some $a \in M$, and a formula $\psi(X, \vec{x})$ with $\vec{x} = \langle x_0, \ldots, x_{n-1} \rangle$ for some $n \in \omega$, then writing $\psi(X, \vec{d})$, means that $a >^{\mathcal{M}} n$ and refers to the formula where, for each $i < n$, $x_i$ has been replaced by $d_i$.

**Lemma 8.7.9.** *Let $\psi(X, \vec{x}, y)$ be a $\Pi^0_1$ formula of $\mathcal{L}_2$. For every $a \in M$, the set of conditions $(\Gamma, E, I)$ satisfying one of the following two properties is dense.*

1. *For every $b \leqslant^{\mathcal{M}} a$, $(E, I) \Vdash^{\mathcal{M}} (\exists \vec{x})\psi(\mathsf{G}, \vec{x}, b)$.*
2. *There is a $b \leqslant^{\mathcal{M}} a$ such that for every $c <^{\mathcal{M}} b$, $(E, I) \Vdash^{\mathcal{M}} (\exists \vec{x})\psi(\mathsf{G}, \vec{x}, c)$ and $\psi(X, \vec{x}, b) \in \Gamma$.*

**Lemma 8.7.10.** *Let $(\Gamma, E, I)$ be a condition and $\psi(X, \vec{x}) \in \Gamma$. For every $\vec{d} \in M$, the set of conditions $(\Gamma^*, E^*, I^*)$ forcing $\neg\psi(\mathsf{G}, \vec{d})$ is dense below $(\Gamma, E, I)$.*

Proposition 8.7.4 now follows easily.

*Proof (of Proposition 8.7.4).* Let $G$ be sufficiently generic for the forcing in Definition 8.7.8. To see that $\mathcal{M} \vDash \mathsf{I}\Sigma^0_2$, let $\psi(X, \vec{x}, y)$ be a $\Pi^0_1$ formula of $\mathcal{L}_2$ with parameters from $\mathcal{M}$. By Lemma 8.7.9, either $\mathcal{M}[G] \vDash (\exists \vec{x})\psi(G, \vec{x}, b)$ for every $b \in M$, or there is a $b \in M$ such that $\mathcal{M}[G] \vDash (\exists \vec{x})\psi(G, \vec{x}, c)$ for every $c <^{\mathcal{M}} b$ and, by Lemma 8.7.10, $\mathcal{M}[G] \vDash \neg\psi(G, \vec{d}, b)$ for every tuple $\vec{d}$ of elements of $M$, meaning $\mathcal{M}[G] \vDash \neg(\exists \vec{x})\psi(G, \vec{x}, c)$. Thus, induction holds. To see that $G$ is cohesive for a given instance of $\mathsf{COH}$ in $\mathcal{S}^{\mathcal{M}}$, fix any set $R \in \mathcal{S}^{\mathcal{M}}$. By Lemma 8.7.7, if $(\Gamma, E, I)$ is a condition then so is either $(\Gamma, E, I \cap R)$ or $(\Gamma, E, I \cap \overline{R})$. Hence, the set of conditions $(\Gamma, E, I)$ with $I \subseteq^* R$ or $I \subseteq^* \overline{R}$ is dense, so either $G \subseteq^* R$ or $G \subseteq^* \overline{R}$. (Thus, $G$ is actually simultaneously cohesive for *every* instance of $\mathsf{COH}$ in $\mathcal{S}^{\mathcal{M}}$.) □

So, it remains to prove the lemmas, and to this end we of course first need to make precise what it means for a condition to be $\Gamma$-large or $\Gamma$-small.

**Definition 8.7.11.** Let $\Gamma = \{\psi_i(X, \vec{x}) : i < m\}$ be a negation set.

1. A Mathias condition $(E, I)$ is $\Gamma$-*small* if there exist
   - $\ell \in M$,
   - $\mathcal{M}$-finite sequences
     - $\langle m_i : i <^{\mathcal{M}} \ell \rangle$, of elements of $M$,
     - $\langle \vec{d}_i : i <^{\mathcal{M}} \ell \rangle$, of $\mathcal{M}$-finite tuples of elements of $M$,
     - $\langle E_i : i <^{\mathcal{M}} \ell \rangle$, of $\mathcal{M}$-finite subsets of $M$,

     each coded by an element of $M$,
   - an $\mathcal{M}$-finite sequence $\langle I_i : i <^{\mathcal{M}} \ell \rangle$ of elements of $\mathcal{S}^{\mathcal{M}}$, coded as an element of $\mathcal{S}^{\mathcal{M}}$,

   such that the following hold in $\mathcal{M}$:

   a. $I = \bigcup_{i<\ell} I_i$,
   b. for each $i < \ell$, $E \subseteq E_i \subseteq E \cup I$ and either $I_i$ is bounded by $\max \vec{d}_i$ or $\psi_{m_i}(E_i \cup F, \vec{d}_i)$ for every finite set $F \subseteq I_i$.
2. A Mathias condition $(E, I)$ that is not $\Gamma$-small is $\Gamma$-*large*.

*Proof (of Lemma 8.7.7).* We leave the proofs of (1) and (2) to the exercises (Exercise 8.10.9).

For (3), fix $\Gamma$. Let $\varphi(X, E, I, x)$ be the formula asserting: that there exist $\ell < x$ and (codes for) finite sequences $\langle m_i : i < \ell \rangle$, $\langle \vec{d}_i : i < \ell \rangle$, and $\langle E_i : i < \ell \rangle$ smaller than $x$; that $X = \langle I_i : i < \ell \rangle$; and that clauses (a) and (b) in Definition 8.7.11 hold. Then a Mathias condition $(E, I)$ is $\Gamma$-small precisely if $\mathcal{M} \vDash (\exists x)(\exists X)\varphi(X, E, I, x)$. Notice that clauses (a) and (b) are $\Pi^0_1$, and therefore so is $\varphi$. By Exercise 5.13.16, there is a $\Sigma^0_0$ formula $\theta(X, E, I, x)$ such that in $\mathcal{M}$, a set satisfies $\varphi(X, E, I, x)$ if and only if $(\forall k)\theta(X \upharpoonright k, E, I, x)$. Given $a \in M$, let $T_a$ be the set of all $\mathcal{M}$-finite binary sequences $\sigma$ such that $(\forall k \leqslant |\sigma|)\theta(\sigma \upharpoonright k, E, I, a)$. Then $\{T_a : a \in M\}$ belongs to $\mathcal{S}^{\mathcal{M}}$. Moreover, each $T_a$ is a binary tree and $\varphi(X, E, I, a)$ holds in $\mathcal{M}$ if and only if $X$ is an $\mathcal{M}$-infinite path through $T_a$. Since $\mathcal{M}$ satisfies WKL, $T$ has such a path if and only if $T_a$ is infinite. Thus, $(\exists X)\varphi(X, E, I, x)$ is equivalent in $\mathcal{M}$ to a $\Pi^0_1$ formula.

Now, fix a $\Gamma$-large Mathias condition $(E, I)$, $a \in M$, and sets $E_j$ and $I_j$ for $j <^{\mathcal{M}} a$ as in the statement of the lemma. We must show that there is a $j$ such that $(E_j, I_j)$ is a $\Gamma$-large Mathias condition. Seeking a contradiction, suppose that for each $j <^{\mathcal{M}} a$, either $I_j$ is $\mathcal{M}$-finite or $(E_j, I_j)$ is $\Gamma$-small. Then $\mathcal{M}$ satisfies

$$(\forall j < a)(\exists x)[(\forall y)[y \in I_j \to y < x] \vee (\exists X)\varphi(X, E_j, I_j, x)].$$

By the discussion above, the matrix of this formula is equivalent in $\mathcal{M}$ to a $\Pi^0_1$ formula. Since $\mathcal{M}$ satisfies $\mathsf{I}\Sigma^0_2$, it also satisfies $\mathsf{B}\Pi^0_1$. Thus, we may fix a $b \in M$ such that

$$(\forall j < a)(\exists x < b)[(\forall y)[y \in I_j \to y < x] \vee (\exists X)\varphi(X, E_j, I_j, x)].$$

Let $\theta(z)$ be the formula

$$(\exists X)(\forall j < z)(\exists x < b)[(\forall y)[y \in I_j \to y < x] \vee \varphi(X^{[j]}, E_j, I_j, x)].$$

By Proposition 6.1.2, the formula

$$(\forall j < z)(\exists x < b)[(\forall y)[y \in I_j \to y < x] \vee \varphi(X^{[j]}, E_j, I_j, x)]$$

is equivalent in $\mathcal{M}$ to a $\Pi^0_1$ formula, and therefore so is $\theta(z)$. We may thus use $\mathsf{I}\Pi^0_1$ in $\mathcal{M}$ to prove that $(\forall z \leqslant a)\theta(z)$ holds. Trivially, $\mathcal{M} \vDash \theta(0)$. So fix a nonzero $z \leqslant^{\mathcal{M}} a$, and assume $\mathcal{M} \vDash \theta(z-1)$, as witnessed by $X \in \mathcal{S}^{\mathcal{M}}$. If $I_{z-1}$ is $\mathcal{M}$-finite, we let $X_z = X \cup (\{z-1\} \times \{I_{z-1}\})$, so that $X_z^{[z-1]} = \langle I_{z-1} \rangle$, and then $X_z$ witnesses that $\mathcal{M} \vDash \theta(z)$. If $I_{z-1}$ is $\mathcal{M}$-infinite, then by hypothesis we can fix $X_{z-1} \in \mathcal{S}^{\mathcal{M}}$ such that $(\exists x < b)\varphi(X_{z-1}, E_{z-1}, I_{z-1}, x)$. If we let $X_z = X \cup (\{z-1\} \times X_{z-1})$, so that $X_z^{[z-1]} = X_{z-1}$, then $X_z$ witnesses that $\mathcal{M} \vDash \theta(z)$.

To complete the proof, fix $X_a \in \mathcal{S}^{\mathcal{M}}$ witnessing that $\mathcal{M} \vDash \theta(a)$. By bounded $\Pi^0_1$ comprehension (using $X_a$ as a parameter), we can form the set of all pairs $\langle j, x_j \rangle$ with $j <^{\mathcal{M}} a$ and $x_j <^{\mathcal{M}} b$ such that $\mathcal{M} \vDash \varphi(X_a^{[j]}, E_j, I_j, x_j)$. Now by $\mathsf{L}\Pi^0_1$, for each $j <^{\mathcal{M}} a$ we can fix an $\ell_j <^{\mathcal{M}} x_j$ and (codes for) sequences $\langle m_{j,i} : i <^{\mathcal{M}} \ell_j \rangle$, $\langle \vec{d}_{j,i} : i <^{\mathcal{M}} \ell_j \rangle$, and $\langle E_{j,i} : i <^{\mathcal{M}} \ell_j \rangle$ smaller than $x_j$ such that clauses (a) and (b) in Definition 8.7.11 hold with the sequence of sets coded by $X_a^{[j]}$. Let $\ell = \sum_{j <^{\mathcal{M}} a} \ell_j$, and let $\ell_{-1} = 0$ for definiteness. Define a sequence $\langle m_k : k <^{\mathcal{M}} \ell \rangle$ as follows: for $k <^{\mathcal{M}} \ell$, fix $j <^{\mathcal{M}} a$ and $i <^{\mathcal{M}} \ell_j$ such that $k = \ell_{j-1} + i$, and let $m_k = m_{j,i}$. (So, $m_0 = m_{0,0}, m_1 = m_{0,1}, \ldots, m_{\ell_0} = m_{1,0}, m_{\ell_0+1} = m_{1,1}, \ldots$, etc.) Define $\langle \vec{d}_k : k <^{\mathcal{M}} \ell \rangle$ and $\langle E_k : k <^{\mathcal{M}} \ell \rangle$ analogously. Also, for each $j <^{\mathcal{M}} a$ we can decode $X_a^{[j]}$ as $\langle I_{j,i} : i <^{\mathcal{M}} \ell_j \rangle$, and so analogously define a sequence of sets $\langle I_k : k <^{\mathcal{M}} a \rangle$. As $\mathcal{M}$ satisfies $\mathsf{B}\Sigma^0_2$, an $\mathcal{M}$-finite union of $\mathcal{M}$-finite sets is $\mathcal{M}$-finite, by Proposition 6.5.4. Thus, each of these sequences is $\mathcal{M}$-finite. But now it is easy to see that $\ell$, $\langle m_k : k <^{\mathcal{M}} \ell \rangle$, $\langle \vec{d}_k : k <^{\mathcal{M}} \ell \rangle$, $\langle E_k : k <^{\mathcal{M}} \ell \rangle$, and $\langle I_k : k <^{\mathcal{M}} a \rangle$ witness that $(E, I)$ is $\Gamma$-small, which is a contradiction. □

*Proof (of Lemma 8.7.9).* Fix a $\Pi^0_1$ formula $\psi(X, \vec{x}, y)$, $a \in M$, and a condition $(\Gamma, E, I)$. We must find an extension $(\Gamma^*, E^*, I^*)$ satisfying either item (1) or (2) in the statement of the lemma.

Let $\varphi(\ell, z)$ be the formula asserting that $z \leqslant a$ and there exist (codes for) finite sequences $\langle m_i : i <^{\mathcal{M}} \ell \rangle$, $\langle \vec{d}_i : i <^{\mathcal{M}} \ell \rangle$, $\langle E_i : i <^{\mathcal{M}} \ell \rangle$, $\langle I_i : i <^{\mathcal{M}} \ell \rangle$, and $\langle e_i : i < \ell \rangle$ such that one of the following holds.

- Clauses (a) and (b) in Definition 8.7.11 hold with $\ell$ and the sequences $\langle m_i : i <^{\mathcal{M}} \ell \rangle$, $\langle \vec{d}_i : i <^{\mathcal{M}} \ell \rangle$, $\langle E_i : i <^{\mathcal{M}} \ell \rangle$, and $\langle I_i : i <^{\mathcal{M}} \ell \rangle$ and clauses (a) and (b) hold.
- For each $i < \ell$, $e_i = \langle \vec{e}_{i,y} \in M : y \leqslant z \rangle$ and $\psi(E_i \cup F, \vec{e}_{i,y}, y)$ holds for every $y \leqslant z$ and every finite sets $F \subseteq I_i$.

Just as in the proof of the preceding lemma, $\varphi$ is equivalent in $\mathcal{M}$ to a $\Pi^0_1$ formula. Thus, $(\exists \ell)\varphi(\ell, z)$ is $\Sigma^0_2$. We consider two cases.

*Case 1:* $\mathcal{M} \vDash (\exists \ell)\varphi(\ell, a)$. Fix a witnessing $\ell$ and corresponding sequences $\langle m_i : i <^{\mathcal{M}} \ell\rangle$, $\langle \vec{d}_i : i <^{\mathcal{M}} \ell\rangle$, $\langle E_i : i <^{\mathcal{M}} \ell\rangle$, $\langle I_i : i <^{\mathcal{M}} \ell\rangle$, and $\langle e_i : i < \ell\rangle$. Since $(E, I)$ is $\Gamma$-large, it follows by Lemma 8.7.7 (3) that $(E_i, I_i)$ is $\Gamma$-large for some $i <^{\mathcal{M}} \ell$. This means the first case in the definition of $\varphi$ above does not hold for $i$, so the second must hold. Thus, $(E_i, I_i)$ forces $\psi(\mathsf{G}, \vec{e}_{i,b}, b)$, and therefore in particular, $(\exists \vec{x})\psi(\mathsf{G}, \vec{x}, b)$, for every $b \leqslant^{\mathcal{M}} a$. In this case, we can therefore take $(\Gamma^*, E^*, I^*) = (\Gamma, E_i, I_i)$.

*Case 2: Otherwise.* So $\mathcal{M} \vDash (\exists z \leqslant a)\neg(\exists \ell)\varphi(\ell, z)$. Since $\mathcal{M}$ satisfies $\mathsf{I}\Sigma^0_2$ (and therefore $\mathsf{L}\Pi^0_2$), we can fix the least $b \leqslant a$ such that $\mathcal{M} \vDash \neg(\exists \ell)\varphi(\ell, b)$. Since no $\ell$ and no sequences witness that first case in the definition of $\varphi$ holds, this means in particular that $(E, I)$ is $\Gamma \cup \{\psi(X, \vec{x}, b)\}$-large. If $b = 0$, then we can take $(\Gamma^*, E^*, I^*) = (\Gamma \cup \{\psi(X, \vec{x}, b)\}, E, I)$. So suppose $b >^{\mathcal{M}} 0$. Fix $\ell$ such that $\mathcal{M} \vDash \varphi(\ell, b-1)$ and fix the corresponding sequences $\langle m_i : i <^{\mathcal{M}} \ell\rangle$, $\langle \vec{d}_i : i <^{\mathcal{M}} \ell\rangle$, $\langle E_i : i <^{\mathcal{M}} \ell\rangle$, $\langle I_i : i <^{\mathcal{M}} \ell\rangle$, and $\langle e_i : i < \ell\rangle$. By Lemma 8.7.7 (3), $(E_i, I_i)$ is $\Gamma$-large for some $i <^{\mathcal{M}} \ell$. As in the preceding case, it thus follows that $(E_i, I_i)$ forces $(\exists \vec{x})\psi(\mathsf{G}, \vec{x}, c)$ for every $c <^{\mathcal{M}} b$. Hence, $(\Gamma^*, E^*, I^*) = (\Gamma \cup \{\psi(X, \vec{x}, b)\}, E_i, I_i)$ can serve as the desired extension. □

*Proof (of Lemma 8.7.10).* Fix a condition $(\Gamma, E, I)$, a formula $\psi(X, \vec{x})$ in $\Gamma$, and $\vec{d} \in M$. Let $(\Gamma^*, E^*, I^*)$ be any extension of $(\Gamma, E, I)$. We must exhibit an extension of this condition that forces $\neg\varphi(\mathsf{G}, \vec{d})$. By Lemma 8.7.10 (2), $(E^*, I^*)$ does not force $\psi(\mathsf{G}, \vec{d})$. Since $\psi(X, \vec{x})$ is $\Pi^0_1$, it is really $\neg\theta(X, \vec{x})$ for some $\Sigma^0_1$ formula $\theta$. Thus, the fact that $(E^*, I^*)$ does not force $\psi(\mathsf{G}, \vec{d})$ means some $(E^{**}, I^{**}) \leqslant (E^*, I^*)$ forces $\theta(\mathsf{G}, \vec{d})$. By Exercise 7.8.10, this means that $(E^{**}, I^* \setminus \max E^{**})$ forces $\theta(\mathsf{G}, \vec{d})$. By Proposition 7.4.2, $(E^{**}, I^* \setminus \max E^{**})$ also forces $\neg\neg\theta(\mathsf{G}, \vec{d})$, which is $\neg\varphi(\mathsf{G}, \vec{d})$. By Lemma 8.7.10 (1), the finite extension $(E^{**}, I^* \setminus \max E^{**})$ is $\Gamma^*$-large. Hence, $(\Gamma^*, E^{**}, I^* \setminus \max E^{**})$ is the desired extension. □

### 8.7.3 Proof of Proposition 8.7.5

We again fix a countable model $\mathcal{M}$ of $\mathsf{WKL} + \mathsf{I}\Sigma^0_2$, and this time also fix an instance $c \in \mathcal{S}^{\mathcal{M}}$ of $\mathsf{SRT}^2_2$. We show that there is a $G$ such that $\mathcal{M}[G] \vDash \mathsf{I}\Sigma^0_2$ + "$G$ is an infinite limit homogeneous set for $c$". The argument is very similar to that in the preceding question, but with modifications. We begin with a generalization of Mathias forcing. Let us say here that a formula $\gamma(x)$ of $\mathcal{L}_2$ is $\Delta^0_2$ *in* $\mathcal{M}$ if it is $\Sigma^0_2$ and equivalent in $\mathcal{M}$ to a $\Pi^0_2$ formula.

**Definition 8.7.12.** *Mathias pseudo-forcing* is the following notion of forcing:

- The conditions are pairs $(E, \gamma)$, where $E \in \mathcal{S}^{\mathcal{M}}$ is $\mathcal{M}$-finite and $\gamma$ is a $\Delta^0_2$ formula in $\mathcal{M}$ such that $\mathcal{M} \vDash (\forall x)[\gamma(x) \to x > E] \wedge (\forall x)(\exists y)[y > x \wedge \gamma(y)]$.

- Extension is defined by $(E^*, \gamma^*) \leqslant (E, \gamma)$ if $\mathcal{M} \vDash E \subseteq E^* \wedge (\forall x)[x \in E^* \rightarrow x \in E \vee \gamma(x)] \wedge (\forall x)[\gamma^*(x) \rightarrow \gamma(x)]$.
- The valuation of $(E, \gamma)$ is the string $\sigma \in 2^{<\mathcal{S}^{\mathcal{M}}}$ of length the $\leqslant^{\mathcal{M}}$-least element $a \in \mathcal{S}^{\mathcal{M}}$ such that $\mathcal{M} \vDash \gamma(a)$ (which exists, since $\mathcal{M} \vDash \mathsf{L}\Sigma^0_2$) with $\mathcal{M} \vDash (\forall x)[\sigma(x) = 1 \leftrightarrow x \in E]$.

We will call the conditions in Mathias pseudo-forcing *Mathias pseudo-conditions*, although it should be noted that this is a *bona fide* forcing notion in $\mathcal{M}$ in the sense of Definition 7.6.1. Every Mathias condition is clearly a Mathias pseudo-condition. Conversely, a Mathias pseudo-condition $(E, \gamma)$ is a Mathias condition only if $\{a \in \mathcal{S}^{\mathcal{M}} : \mathcal{M} \vDash \gamma(a)\}$ belongs to $\mathcal{S}^{\mathcal{M}}$ (which it need not). For ease of notation, we write $(E, \text{“}I\text{”})$ if $(E, \gamma)$ is a Mathias pseudo-condition and $I$ is the set defined by $\gamma$. We can then also write things like $a \in \text{“}I\text{”}$, $\text{“}I\text{”} \subseteq S$ for $S \in \mathcal{S}^{\mathcal{M}}$, etc., as shorthands for $\mathcal{M} \vDash \gamma(a)$, $\mathcal{M} \vDash (\forall x)[\gamma(x) \rightarrow x \in S]$, etc.

For each $i < 2$, let $P_i = \{a \in M : \mathcal{M} \vDash \lim_y c(x, y) = i\}$, and note that $P_0$ and $P_1$ are $\Delta^0_2$-definable in $\mathcal{M}$ by Remark 8.4.3. Say a Mathias pseudo-condition $(E, \text{“}I\text{”})$ is *$i$-acceptable* if $\mathcal{M} \vDash E \subseteq \text{“}P_i\text{”}$. A generic $G$ for forcing with $i$-acceptable Mathias conditions is thus a subset of $P_i$, meaning a limit homogeneous set for $c$ with color $i$. Below, we will first restrict to 0-acceptable Mathias conditions in an attempt to make $G \subseteq P_0$, and failing that, we will switch to 1-acceptable Mathias conditions and build $G \subseteq P_1$. We need to suitably adapt the definition of smallness and largeness to 0-acceptable and 1-acceptable Mathias conditions and pseudo-conditions, and prove analogs of Lemmas 8.7.7, 8.7.9 and 8.7.10. We do this slightly differently in the 0-acceptable and 1-acceptable case.

First, we restrict to 0-acceptable Mathias conditions and pseudo-conditions and define smallness and largeness.

**Definition 8.7.13.** Let $\Gamma$ be a negation set.

1. A Mathias condition $(E, I)$ is *$\Gamma$-small$_0$* if it is 0-acceptable and $\Gamma$-small as in Definition 8.7.11, but with the modification that sequence $\langle E_i : i <^{\mathcal{M}} \ell\rangle$ additionally satisfies $\mathcal{M} \vDash (\forall i < \ell)[E_i \subseteq \text{“}P_0\text{”}]$.
2. A Mathias pseudo-condition $(E, \text{“}I\text{”})$ is *$\Gamma$-small$_0$* if it is 0-acceptable and $\Gamma$-small$_0$ as defined for Mathias conditions, but with the following modification. Instead of an $\mathcal{M}$-finite sequence $\langle I_i : i <^{\mathcal{M}} \ell\rangle$ of elements of $\mathcal{S}^{\mathcal{M}}$ such that $\mathcal{M} \vDash I = \bigcup_{i<\ell} I_i$, there is a $\Delta^0_2$ formula $\gamma(i, x)$ such that, writing “$I_i$” for $\{a \in M : \mathcal{M} \vDash \gamma(i, a)\}$, we have $\mathcal{M} \vDash \text{“}I\text{”} = \bigcup_{i<\ell} \text{“}I_i\text{”}$. The rest of the definition is then unchanged, except that $I$ is replaced by “$I$” and $I_i$ by “$I_i$” throughout.

Thus, the only distinction between the definition for Mathias conditions and for Mathias pseudo-conditions is whether the partitions in the definition need to be actual sets in $\mathcal{S}^{\mathcal{M}}$ or merely $\Delta^0_2$-definable in $\mathcal{M}$. Although Mathias conditions are pseudo-conditions, and so in principle could be $\Gamma$-small$_0$ in either sense above, it will always be clear which is meant. Whenever we specify a Mathias condition, $\Gamma$-small$_0$ is to be interpreted as (1), and whenever we specify a Mathias pseudo-condition, $\Gamma$-small$_0$ is to be interpreted as (2).

Let us now consider Lemma 8.7.7. Part (1) is unproblematic and goes through with basically the same proof. The same is true of part (3), only we note that the additional clause above that $\mathcal{M} \models (\forall i < \ell)[E_i \subseteq \text{“}P_0\text{”}]$ is equivalent to a $\Sigma^0_2$ formula (and so does not increase the complexity of being $\Gamma$-small$_0$). This is because $\mathsf{B}\Pi^0_1$ holds in $\mathcal{M}$, and therefore $\mathcal{M} \models E_i \subseteq \text{“}P_0\text{”} \leftrightarrow (\exists w)(\forall x \in E_i)(\forall y)[y > w \to c(x, y) = 0]$. As to part (2), this is more subtle, and requires an additional hypothesis to go through here.

For every negation set $\Gamma$, whenever a Mathias condition $(E, I)$ is $\Gamma$-small$_0$ then so is the Mathias pseudo-condition $(E, I \cap \text{“}P_0\text{”})$. (8.4)

**Lemma 8.7.14.** *Let $\Gamma$ be a negation set and assume* (8.4) *holds. If $(E, I)$ is a $\Gamma$-large$_0$ Mathias condition and $\psi(X, \vec{x}) \in \Gamma$, then $(E, I)$ does not force $\psi(\mathsf{G}, \vec{d})$ (in the partial order of* 0*-acceptable Mathias conditions) for any $\vec{d} \in M$.*

*Proof.* Suppose towards a contradiction that $(E, I)$ forces $\psi(\mathsf{G}, \vec{d})$. Since $\psi$ is $\Pi^0_1$ and we have restricted our forcing to just 0-acceptable Mathias conditions, it follows from the definition of forcing and Exercise 7.8.10 that $\psi(E \cup F, \vec{d})$ holds for every $\mathcal{M}$-finite set $F \subseteq I$ such that $\mathcal{M} \models F \subseteq \text{“}P_0\text{”}$. But then it is readily seen that $(E, I \cap \text{“}P_0\text{”})$ is actually $\Gamma$-small (by the same proof as in the original lemma). □

We can now formulate an analog of the auxiliary forcing notion in Definition 8.7.8. Namely, conditions are triples $(\Gamma, E, I)$, where $\Gamma$ is a negation set and $(E, I)$ is a $\Gamma$-large$_0$ Mathias condition. Extension and valuation is defined as before. With this, Lemmas 8.7.9 and 8.7.10 now carry over with obvious modifications, using Lemma 8.7.14 in place of Lemma 8.7.7 (2) where needed. (It is worth emphasizing that these all involve only the auxiliary forcing notion, and hence ultimately only Mathias conditions. Mathias pseudo-conditions are only used in stating condition (8.4) and proving Lemma 8.7.14.) All the pieces can now be assembled as in the previous section.

*Proof (of Proposition 8.7.5 under the hypothesis* (8.4)*).* We may assume $c$ has no solution in $\mathcal{M}$, as otherwise there is nothing to do. Thus, for every $\mathcal{M}$-infinite set $S \in \mathcal{S}^{\mathcal{M}}$ it must be that $\mathcal{M} \models (\forall x)(\exists y)[y > x \wedge y \in S \cap \text{“}P_0\text{”}]$, as otherwise for some $b \in M$, $\{a \in S : a > b\}$ would be an $\mathcal{M}$-infinite subset of $P_1$ in $\mathcal{S}^{\mathcal{M}}$, and hence a solution to $c$ in $\mathcal{M}$. It follows that for each $a \in M$, the set of conditions $(\Gamma, E, I)$ such that $\mathcal{M} \models (\exists x)[x \in E \wedge x > a]$ is dense, and a generic $G$ for this forcing is consequently an $\mathcal{M}$-infinite subset of $P_0$. All that remains is to verify that $\mathcal{M}[G]$ satisfies $\mathsf{I}\Sigma^0_2$, and this is done exactly as in the proof of Proposition 8.7.4 in the previous section, but using the analogs of Lemmas 8.7.7, 8.7.9 and 8.7.10 discussed above. □

To complete the proof, we must handle the case that (8.4) fails. Let us fix a counterexample, then, which is a negation set $\widehat{\Gamma}$ and Mathias condition $(\widehat{E}, \widehat{I})$. Thus, $(\widehat{E}, \widehat{I})$ is $\widehat{\Gamma}$-large$_0$ but the Mathias pseudo-condition $(\widehat{E}, \widehat{I} \cap \text{“}P_0\text{”})$ is $\widehat{\Gamma}$-small$_0$. All definitions below will be made with respect to $\widehat{\Gamma}$ and $(\widehat{E}, \widehat{I})$. We start again with smallness, this time for 1-acceptable Mathias pseudo-conditions.

**Definition 8.7.15.** Let $\Gamma \supseteq \widehat{\Gamma}$ be a negation set.

1. A Mathias condition $(E, I)$ is $\Gamma$-*small*$_1$ if it is 1-acceptable, $I \subseteq \widehat{I}$, and $(E, I)$ is $\Gamma$-small as in Definition 8.7.11, but with clause (b) modified to say that either $I_i$ is bounded by $\max \vec{d}_i$, or $\psi_{m_i}(E_i \cup F, \vec{d}_i)$ for every finite set $F \subseteq I_i$, or $\psi_{m_i}(\widehat{E} \cup F, \vec{d}_i)$ for every finite set $F \subseteq I_i$.
2. A Mathias pseudo-condition $(E, \text{“}I\text{”})$ is $\Gamma$-*small*$_1$ if it is 1-acceptable, $\mathcal{M} \vDash \text{“}I\text{”} \subseteq \text{“}\widehat{I}\text{”}$, and $(E, \text{“}I\text{”})$ is $\Gamma$-small$_1$ as defined for Mathias conditions, but with the following modification. Instead of an $\mathcal{M}$-finite sequence $\langle I_i : i <^{\mathcal{M}} \ell \rangle$ of elements of $\mathcal{S}^{\mathcal{M}}$ such that $\mathcal{M} \vDash I = \bigcup_{i<\ell} I_i$, there is a $\Delta^0_2$ formula $\gamma(i, x)$ such that, writing “$I_i$” for $\{a \in M : \mathcal{M} \vDash \gamma(i, a)\}$, we have $\mathcal{M} \vDash \text{“}I\text{”} = \bigcup_{i<\ell} \text{“}I_i\text{”}$. The rest of the definition is then unchanged, except that $I$ is replaced by “$I$” and $I_i$ by “$I_i$” throughout.

Looking next at Lemma 8.7.7, parts (1) and (3) again lift without issue, just as on the 0-acceptable side. In part (3), a Mathias condition being $\Gamma$-small$_1$ is $\Sigma^0_2$-definable by the same argument as for $\Gamma$-small$_0$. Part (2) now carries over without any additional hypotheses.

**Lemma 8.7.16.** *If $(E, I)$ is a $\Gamma$-large$_1$ Mathias condition and $\psi(X, \vec{x}) \in \Gamma$, then $(E, I)$ does not force $\psi(\mathsf{G}, \vec{d})$ (in the partial order of 1-acceptable Mathias conditions) for any $\vec{d} \in M$.*

*Proof.* We show that if $(E, I)$ is $\Gamma$-large$_1$ then so is the Mathias pseudo-condition $(E, I \cap \text{“}P_1\text{”})$. The conclusion then follows just like in Lemma 8.7.14. Seeking a contradiction, suppose $(E, I \cap \text{“}P_1\text{”})$ is $\widehat{\Gamma}$-small$_1$. Since $I \subseteq \widehat{I}$ and $(\widehat{E}, \widehat{I} \cap \text{“}P_0\text{”})$ is $\Gamma$-small$_0$, it is readily seen from the definitions that $(E, I \cap \text{“}P_0\text{”})$ is $\Gamma$-small$_1$. Since $\Gamma \supseteq \widehat{\Gamma}$, this means that $(E, I \cap \text{“}P_0\text{”})$ is actually $\widehat{\Gamma}$-small$_1$. Since $(I \cap P_0) \cup (I \cap P_1) = I$, the idea now is to combine the witnesses for $(E, I \cap \text{“}P_0\text{”})$ and $(E, I \cap \text{“}P_1\text{”})$ to conclude that $(E, I)$ is $\widehat{\Gamma}$-small$_1$. We can easily fix $a \in M$ that bounds all the (codes for) sequences of elements of $M$, tuples of elements of $M$, and $\mathcal{M}$-finite subsets of $M$ that witness that $(E, I \cap \text{“}P_0\text{”})$ and $(E, I \cap \text{“}P_1\text{”})$ are $\Gamma$-small$_1$. However, for the witnessing partitions of reservoirs, this is more complicated. Since $(E, I \cap \text{“}P_0\text{”})$ and $(E, I \cap \text{“}P_1\text{”})$ are Mathias pseudo-conditions, these partitions are given by formulas. By contrast, $(E, I)$ is a Mathias condition, so a witnessing partition for it needs to be an actual element of $\mathcal{S}^{\mathcal{M}}$. This can be handled as follows. As in the proof of Lemma 8.7.7 (3), there is a $\Sigma^0_0$ formula $\theta(X, Y, Z, x)$ such that a Mathias condition $(E^*, I^*)$ is $\Gamma$-small$_1$ precisely if $\mathcal{M} \vDash (\exists x)(\exists X)(\forall k)\theta(X \upharpoonright k, E^*, I^*, x)$. Let $T$ be the set of all $\mathcal{M}$-finite binary sequences $\sigma$ such that $(\forall k \leqslant |\sigma|)\theta(\sigma \upharpoonright k, E, I, a)$. By choice of $a$, $(E, I)$ will be $\Gamma$-small$_1$ if we can show that there is an infinite path $X$ through $T$. Since $\mathcal{M} \vDash \mathsf{WKL}$, it suffices to show that for every $b \in M$ there is a $\sigma \in T$ with $|\sigma| = b$. Let $\ell_0$ and $\ell_1$ be such that the witnessing partitions of $I \cap P_0$ and $I \cap P_1$ have sizes $\ell_0$ and $\ell_1$, respectively, and let $\gamma_0$ and $\gamma_1$ be the formulas defining these partitions. Define $\sigma \in 2^b$ as follows: given $\langle j, x \rangle < b$, if $j <^{\mathcal{M}} \ell_0$ let $\sigma(\langle j, x \rangle) = 1$ if and only if $\mathcal{M} \vDash \gamma_0(j, x)$, and if $\ell_0 \leqslant^{\mathcal{M}} j$ let $\sigma(\langle j, x \rangle) = 1$ if and only if $\mathcal{M} \vDash \gamma_1(j - \ell_0, x)$. Then $\sigma$ is an initial segment of the characteristic function of the partition of $I$ obtained by merging the partitions (of $I \cap P_0$ and $I \cap P_1$) defined by $\gamma_0$ and $\gamma_1$. Clearly, $\sigma \in T$. □

Now, define the auxiliary notion of forcing whose conditions are triples $(\Gamma, E, I)$ such that $\Gamma \supseteq \widehat{\Gamma}$ and $(E, I)$ is a $\Gamma$-large$_1$ Mathias condition. Lemmas 8.7.9 and 8.7.10 then carry over, *mutatis mutandis*.

*Proof (of Proposition 8.7.5).* Assume $(\widehat{\Gamma}, \widehat{E}, \widehat{I})$ is a counterexample to (8.4). Again, we assume $c$ has no solution in $\mathcal{M}$, so that a generic $G$ for our forcing notion is an $\mathcal{M}$-infinite subset of $P_1$. $\mathcal{M}[G]$ satisfies $\mathsf{I}\Sigma^0_2$ as before. □

### 8.7.4 What else is known?

The first order part of Ramsey's theorem is now quite well understood. In fact, for the version with arbitrarily many colors ($\mathsf{RT}^2$), it is known exactly. Cholak, Jockusch, and Slaman [33] established an analogue of Theorem 8.7.1 one level up in the arithmetical hierarchy.

**Theorem 8.7.17 (Cholak, Jockusch, and Slaman [33]).** $\mathsf{WKL}_0 + \mathsf{RT}^2 + \mathsf{I}\Sigma^0_3$ *is* $\Pi^1_1$ *conservative over* $\mathsf{RCA}_0 + \mathsf{I}\Sigma^0_3$.

Once again, the proof proceeds by separate model extension arguments for $\mathsf{COH}$ and $\mathsf{SRT}^2$. Belanger [13], building on earlier results of Hájek (Theorem 7.7.6), then obtained a strengthening of the $\mathsf{COH}$ result, showing that $\mathsf{WKL}_0 + \mathsf{COH}$ is $\Pi^1_1$ conservative over $\mathsf{RCA}_0 + \mathsf{B}\Sigma^0_3$. Finally, Slaman and Yokoyama [292] established the same conservation level also for $\mathsf{SRT}^2$. Thus, we obtain:

**Theorem 8.7.18 (Slaman and Yokoyama [292]).** $\mathsf{WKL}_0 + \mathsf{RT}^2$ *is* $\Pi^1_1$ *conservative over* $\mathsf{RCA}_0 + \mathsf{B}\Sigma^0_3$.

Combining this with the fact that $\mathsf{SRT}^2$ implies $\mathsf{B}\Sigma^0_3$ (Theorem 8.7.2), it follows that the first order part of $\mathsf{RT}^2$ (and indeed, $\mathsf{SRT}^2$ as well as $\mathsf{WKL} + \mathsf{RT}^2$) is $\mathsf{B}\Sigma^0_3$.

When the number of colors is fixed (and at least 2), Theorem 8.7.1 gives $\mathsf{I}\Sigma^0_2$ as an upper bound, while Hirst's theorem (and the fact that $\mathsf{RT}^2_2 \to \mathsf{RT}^1$ over $\mathsf{RCA}_0$) gives $\mathsf{B}\Sigma^0_2$ as a lower bound. The first order part in this case, while not exactly pinned down, is known to lie closer to the latter.

**Theorem 8.7.19 (Chong, Slaman, and Yang [40]).** $\mathsf{RCA}_0 + \mathsf{RT}^2_2 \nvdash \mathsf{I}\Sigma^0_2$.

**Theorem 8.7.20 (Patey and Yokoyama [246]).** $\mathsf{WKL}_0 + \mathsf{RT}^2_2$ *is* $\Sigma^0_3$ *conservative over* $\mathsf{RCA}_0 + \mathsf{B}\Sigma^0_2$.

Whether the first order part of $\mathsf{RT}^2_2$ is *exactly* $\mathsf{B}\Sigma^0_2$ remains open.

*Question 8.7.21.* Is $\mathsf{RCA}_0 + \mathsf{RT}^2_2$ is $\Pi^1_1$ conservative over $\mathsf{RCA}_0 + \mathsf{B}\Sigma^0_2$?

But $\Sigma^0_3$ formulas already include a large segment of natural arithmetical statements, in particular, all consistency statements, and all $\forall\exists$ theorems of arithmetic. Indeed, one consequence of Theorem 8.7.20 is that $\mathsf{RT}^2_2$ is "finitistically reducible", in the sense discussed in the introduction, a particularly striking fact given that Ramsey's theorem (in the form we are discussing here) seems so intrinsically to be about infinite sets. (The interested reader may enjoy looking at the March 2016 edition of *Quanta Magazine*, where a popular account of Theorem 8.7.20 and its implications is presented.)

It is also still open whether, like for the versions with arbitrarily many colors, the first order strength of $\mathsf{RT}^2_2$ is the same as that of $\mathsf{SRT}^2_2$. More generally, elucidating the precise relationship between Ramsey's theorem for pairs and the stable Ramsey's theorem for pairs has been, and is still, one of the central themes in computable combinatorics and the reverse mathematics of combinatorial problems. We dedicate the next section to exploring this question in more detail.

We conclude this section with one more result, concerning the first order strength of $\mathsf{COH}$. In Theorem 7.7.1 we proved that $\mathsf{COH}$ is $\Pi^1_1$ conservative over $\mathsf{RCA}_0$, mostly as a warm-up for Harrington's theorem (Theorem 7.7.3). We cannot improve this to $\Pi^1_2$, of course, since $\mathsf{COH}$ is itself a $\Pi^1_2$ statement. But as it turns out, we can improve it to a *restricted class of* $\Pi^1_2$ *statements*.

**Definition 8.7.22 (Hirschfeldt and Shore [152]).** A $\mathcal{L}_2$ theory $T_1$ is *restricted* $\Pi^1_2$ *conservative* over an $\mathcal{L}_2$ theory $T_2$ if every sentence of the form

$$(\forall X)[\varphi(X) \to (\exists Y)\psi(X,Y)],$$

where $\varphi$ is arithmetical and $\psi$ is $\Sigma^0_3$ that is provable in $T_1$ is provable in $T_2$.

We will prove the following result.

**Theorem 8.7.23 (Hirschfeldt and Shore [152]).** $\mathsf{RCA}_0 + \mathsf{COH}$ *is restricted* $\Pi^1_2$ *conservative over* $\mathsf{RCA}_0$.

Note that $\mathsf{COH}$ itself has the form $(\forall X)[\varphi(X) \to (\exists Y)\psi(X,Y)]$ where $\varphi$ is arithmetical and $\psi$ is $\Pi^0_3$. To prove the theorem, we begin with the following lemma.

**Lemma 8.7.24.** *Let $\mathcal{M}$ be a countable model* $\mathsf{RCA}_0$ *and suppose $\psi$ is a $\Sigma^0_3$ formula such that $\mathcal{M} \vDash (\forall Y)[\neg\psi(A,Y)]$ for some $A \in \mathcal{S}^{\mathcal{M}}$. Then if $G$ is sufficiently generic for Mathias forcing in $\mathcal{M}$, $\mathcal{M}[G] \vDash (\forall Y)[\neg\psi(A,Y)]$.*

*Proof.* We prove the contrapositive. Fix $G$, and suppose there is a $B \in \mathcal{M}[G]$ such that $\mathcal{M}[G] \vDash \psi(A,B)$. As $B$ is $\Delta^0_1$-definable from $G$ and parameters in $\mathcal{M}$, we can fix an $e \in \omega$ such that $B = \Phi_e(G \oplus C)$ for some $C \in \mathcal{S}^{\mathcal{M}}$. Consider the formula

$$\Phi_e(Z \oplus W) \text{ is total, } \{0,1\}\text{-valued, and } \psi(X, \Phi_e(Z \oplus W)). \tag{8.5}$$

Over $\mathsf{RCA}_0$, this is equivalent to a $\Pi^0_2$ formula. Let $\theta(X,W,Z,x,y)$ be a $\Sigma^0_0$ formula so that $\mathsf{RCA}_0$ proves that (8.5) is equivalent to $(\forall x)(\exists y)\theta(X,W,Z,x,y)$. Since $G$ is generic, there is a condition $(E, I)$ such that $\mathcal{M}[G] \models E \subseteq G \subseteq E \cup I$ and $(E, I)$ forces $(\forall x)(\exists y)\theta(A, C, \mathsf{G}, x, y)$ in $\mathcal{M}$. So for every $b \in M$, every extension $(E^*, I^*)$ of $(E, I)$ has a further extension forcing $(\exists y)\theta(A, C, \mathsf{G}, b, y)$. By Exercise 7.8.10, the latter can always be achieved by an $\mathcal{M}$-finite extension of $(E^*, I^*)$. Hence, by recursion in $\mathcal{M}$, we can define a sequence $\langle E_b : b \in M\rangle$ of $\mathcal{M}$-finite sets with $E = E_0$ and satisfying the following for every $b \in M$.

- $E_b \subseteq E_{b+1} \cup E_b \cup I$.
- $E_{b+1} \setminus E_b$ is $\mathcal{M}$-finite.
- $E_{b+1}$ forces $(\exists y)\theta(A, C, \mathsf{G}, x, y)$.

As $\mathcal{M} \models \mathsf{RCA}_0$, the sequence $\langle E_b : b \in M\rangle$ is in $\mathcal{S}^{\mathcal{M}}$, and consequently so is $H = \bigcup_{b \in M} E_b$. By Exercise 7.8.9, $\mathcal{M} \models (\forall x)(\exists y)\theta(A, C, H, x, y)$. In other words, $\mathcal{M}$ satisfies that $\Phi_e(H \oplus C)$ defines a set and that $\psi(A, \Phi_e(H \oplus C))$ holds. But $\Phi_e(H \oplus C) \in \mathcal{S}^{\mathcal{M}}$, so $\mathcal{M} \models (\exists Y)\psi(A, Y)$, as was to be shown. □

*Proof (of Theorem 8.7.23).* Fix an arithmetical formula $\varphi$ and a $\Sigma^0_3$ formula $\psi$, and suppose $\mathcal{M}$ is a countable model of $\mathsf{RCA}_0$ satisfying $\neg(\forall X)[\varphi(X) \to (\exists Y)\psi(X, Y)]$. We can fix $A \in \mathcal{S}^{\mathcal{M}}$ such that $\mathcal{M} \models \varphi(A) \wedge (\forall Y)[\neg\psi(A, Y)]$.

Let $\mathcal{N}_0 = \mathcal{M}$, and for $i \in \omega$, let $G_i$ be sufficiently generic for Mathias forcing in $\mathcal{N}_i$ and let $\mathcal{N}_{i+1} = \mathcal{N}_i[G_i]$. Note that $\mathcal{N}_i$ is an $\omega$-submodel of $\mathcal{N}_{i+1}$ for all $i$. Hence, all the $\mathcal{N}_i$ have the same first order part as $\mathcal{M}$, as does $\mathcal{N} = \bigcup_{i \in \omega} \mathcal{N}_i$. By the proof of Theorem 7.7.1, $\mathcal{N}_i \models \mathsf{RCA}_0$ for all $i$, and moreover, if $\vec{R}$ is an instance of COH in $\mathcal{N}_i$ then $G_i$ is $\vec{R}$-cohesive. It follows that $\mathcal{N} \models \mathsf{RCA}_0 + \mathsf{COH}$.

We claim that $\mathcal{N} \models \neg(\forall X)[\varphi(X) \to (\exists Y)\psi(X, Y)]$, which proves what we want. Since $\mathcal{M}$ is an $\omega$-submodel of $\mathcal{N}$, $A \in \mathcal{S}^{\mathcal{N}}$, and since $\varphi$ is arithmetical we have $\mathcal{N} \models \varphi(A)$. By the lemma, it follows by (external) induction on $i$ that $\mathcal{N}_i \models (\forall Y)[\neg\psi(A, Y)]$. Since every set in $\mathcal{S}^{\mathcal{N}}$ belongs to $\mathcal{S}^{\mathcal{N}_i}$ for some $i$, we must have $\mathcal{N} \models (\forall Y)[\neg\psi(A, Y)]$ as well. The proof is complete. □

## 8.8 The $\mathsf{SRT}^2_2$ vs. COH problem

The Cholak–Jockusch–Slaman decomposition raises an immediate question, which is whether it is actually *proper*. That is, we should like to know if either of $\mathsf{SRT}^2_2$ or COH already implies $\mathsf{RT}^2_2$ over $\mathsf{RCA}_0$—by itself. Of course, this is equivalent to asking if either of $\mathsf{SRT}^2$ or COH implies the other. This has come to be called by some the $\mathsf{SRT}^2_2$ *vs.* COH *problem*.

We already know part of the answer. By Theorem 7.7.1, COH is $\Pi^1_1$-conservative over $\mathsf{RCA}_0$, whereas $\mathsf{RT}^2_2$ is not since it implies $\mathsf{B}\Sigma^0_2$.

**Corollary 8.8.1.** $\mathsf{RCA}_0 \nvdash \mathsf{COH} \to \mathsf{RT}^2_2$ *(or* $\mathsf{SRT}^2_2$*).*

Hirschfeldt, Jockusch, Kjos-Hanssen, Lempp, and Slaman [149] showed it is also possible to prove this using a computability theoretic argument instead of a proof theoretic one. This gives a stronger separation, witnessed by an $\omega$-model.

**Theorem 8.8.2 (Hirschfeldt, et al. [149]).** $\mathsf{RT}^2_2 \not\leqslant_\omega \mathsf{COH}$.

*Proof.* The proof makes use of the principle DNR, introduced in Definition 4.3.9. The first step is to show that $\mathsf{DNR} \leqslant_\omega \mathsf{RT}^2_2$, and this is left to the exercises (Exercise 8.10.4). We show that $\mathsf{DNR} \not\leqslant_\omega \mathsf{COH}$, which gives the theorem. Let $C$ be the class of all sets $Y$ that compute no DNC function. Then $C$ is closed downward under $\leqslant_{\mathrm{T}}$, and by definition, DNR does not admit preservation of $C$ (in the sense of Definition 3.6.11). By Theorem 4.6.13, it therefore suffices to show that COH admits preservation of $C$. That is, we must show that for every set $A$ that computes no DNC function, every $A$-computable instance of COH has a solution $Y$ such that $A \oplus Y$ still computes no DNC function. We prove this in the case $A = \emptyset$. The full result follows easily by relativization. Let $\vec{R} = \langle R_i : i \in \omega \rangle$ be a computable instance of COH. We force with Mathias conditions with computable reservoirs. As discussed in Example 7.3.9, any sufficiently generic set $G$ for this forcing will be $\vec{R}$-cohesive. We claim no such $G$ computes a DNC function. To this end, it clearly suffices to show that for each Turing functional $\Gamma$, the set of conditions forcing, for some $e \in \omega$, either that $\Gamma^G(e) \downarrow = \Phi_e(e) \downarrow$, or that $\Gamma^G(e) \uparrow$, is dense. To see this, fix $\Gamma$ and any condition $(E, I)$. If there is an $e$ and a finite set $F \subseteq I$ such that $\Gamma^{E \cup F}(e) \downarrow = \Phi_e(e) \downarrow$, then $(E \cup F, \{x \in I : x > F\})$ is an extension of $(E, I)$ forcing $\Gamma^G(e) \downarrow = \Phi_e(e) \downarrow$. If, instead, there is an $e$ such that $\Gamma^{E \cup F}(e) \uparrow$ for all finite $F \subseteq I$, then $(E, I)$ itself forces $\Gamma^G(e) \uparrow$. So assume neither is true. We derive a contradiction. By assumption, for each $e$ there is a finite set $F \subseteq I$ such that $\Gamma^{E \cup F}(e) \downarrow$, and for this $e$, if $\Phi_e(e) \downarrow$ then it must be the case that $\Gamma^{E \cup F}(e) \neq \Phi_e(e)$. But then since $I$ is computable we can compute a DNC function $f$ as follows: given $e$, search for the least $F \subseteq I$ such that $\Gamma^{E \cup F}(e) \downarrow$, and let $f(e) = \Gamma^{E \cup F}(e)$. Since no DNC function can be computable, the proof is complete. $\square$

What about $\mathsf{SRT}^2_2$: does it imply $\mathsf{RT}^2_2$, or equivalently, COH? The question was first asked by Cholak, Jockusch, and Slaman [33]. Over the next decade and a half, it saw a spectacular amount of interest. Particularly after the proof of Liu's theorem (Theorem 8.6.1), it became *the* question in the reverse mathematics of combinatorics. To be sure, there are some obvious and immediate differences between $\mathsf{SRT}^2_2$ and $\mathsf{RT}^2_2$. For instance, by Corollary 8.4.7 and Theorem 8.2.2, $\mathsf{SRT}^2_2$ admits $\Delta^0_2$ solutions whereas $\mathsf{RT}^2_2$ omits $\Delta^0_2$ solutions. This yields:

**Corollary 8.8.3.** $\mathsf{RT}^2_2 \not\leqslant_{\mathrm{c}} \mathsf{SRT}^2_2$.

But by Theorem 8.4.13, COH also admits $\Delta^0_2$ solutions, so this does not settle whether or not COH is computably reducible to $\mathsf{SRT}^2_2$. And of course, $\leqslant_{\mathrm{c}}$ measures only one-time computable transformations, not repeated applications as might be found in an $\omega$-model reduction or a proof in $\mathsf{RCA}_0$.

Several ideas were proffered for separating $\mathsf{SRT}^2_2$ from $\mathsf{RT}^2_2$, focused on isolating a computability theoretic property, enjoyed by $\mathsf{SRT}^2_2$ and failed by $\mathsf{RT}^2_2$, that could be iterated to produce an $\omega$-model satisfying the former and not the latter. Cholak, Jockusch, and Slaman [33] asked whether $\mathsf{SRT}^2_2$ might admit low solutions, which when combined with Theorem 8.2.2, would yield the desired $\omega$-model. Unfortunately, this possibility was quickly quashed.

**Theorem 8.8.4 (Downey, Hirschfeldt, Lempp, and Solomon [75]).** $\mathsf{SRT}^2_2$ *omits low solutions.*

We will discuss the proof of this result below. Attempts were then made to show that for some $n > 1$, $\mathsf{SRT}^2_2$ admits low$_n$ $\Delta^0_2$ solutions, which can also be used to produce an $\omega$-model separation. This remains open, but ultimately was not used to answer the main question. Hirschfeldt, Jockusch, Kjos-Hanssen, Lempp, and Slaman [149] obtained a partial result, showing that $\mathsf{SRT}^2_2$ admits incomplete $\Delta^0_2$ solutions (i.e., every $A$-computable instance of $\mathsf{SRT}^2_2$ has a solution $H$ with $H \oplus A <_T A'$). But this cannot be iterated, and so does not yield an $\omega$-model witnessing a separation. More complicated degree theoretic properties were later considered by Kach and Solomon [174], who also proposed a framework for organizing and studying such properties, and the resulting $\omega$-models, more generally. Alas, the specific properties they looked at did not yield a separation either.

In the end, the solution was finally obtained, in some sense, by going back to the very beginning. Given a model $\mathcal{M}$ and set $X \in \mathcal{S}^{\mathcal{M}}$, say $X$ is *low in* $\mathcal{M}$ if every $\Delta^0_2$ formula having at most $X$ as a set parameter is equivalent in $\mathcal{M}$ to a $\Delta^0_2$ formula with no set parameters. (Here, "$\Delta^0_2$ formula" is being used as in Section 8.7.3.)

**Theorem 8.8.5 (Chong, Slaman, and Yang [39]).** *There exists a countable model* $\mathcal{M}$ *satisfying* $\mathsf{RCA}_0 + \mathsf{SRT}^2_2$ *such that every* $X \in \mathcal{S}^{\mathcal{M}}$ *is low in* $\mathcal{M}$.

By contrast, it is not difficult to formalize the $n = 2$ case in the proof of Theorem 8.2.2 in $\mathsf{RCA}_0 + \mathsf{B}\Sigma^0_2$. That is, $\mathsf{RCA}_0 + \mathsf{B}\Sigma^0_2$ proves that $\mathsf{RT}^2_2$ omits $\Delta^0_2$ (and hence also low) solutions. It follows that the model $\mathcal{M}$ above cannot also satisfy $\mathsf{RT}^2_2$. And so we have the answer:

**Corollary 8.8.6 (Chong, Slaman, and Yang[39]).** $\mathsf{RCA}_0 \nvdash \mathsf{SRT}^2_2 \to \mathsf{RT}^2_2$.

The proof is remarkable in a couple of ways. Most visibly, it is not an $\omega$-model separation, unlike the earlier attempts. Indeed, $\mathcal{M}$ above is designed specifically so as to allow something that would be impossible if it were standard, namely, for Theorem 8.8.4 to fail and $\mathsf{SRT}^2_2$ to admit low solutions. Thus, this is not a purely proof theoretic result either (like the conservation theorems we have seen), as there is an important interplay between the first order and second-order parts.

Effectively, Theorem 8.8.5 shows that the original attempt at separating $\mathsf{SRT}^2_2$ and $\mathsf{RT}^2_2$, using low sets, does work, just over a nonstandard universe. This means that the proof of Theorem 8.8.4 requires more induction to formalize than is available in the model $\mathcal{M}$. The proof there is an infinite injury priority argument, and as pointed out e.g. in [218], it can be carried out in $\mathsf{RCA}_0 + \mathsf{I}\Sigma^0_2$. That fits, as Chong, Slaman, and Yang [39] show that their model $\mathcal{M}$ satisfies $\neg\mathsf{I}\Sigma^0_2$. On the other hand, $\mathcal{M}$ certainly satisfies $\mathsf{B}\Sigma^0_2$, since this follows from $\mathsf{SRT}^2_2$. So, we get an interesting reverse recursion theory result: *any* proof of Theorem 8.8.4 necessarily makes use of at least some induction above $\mathsf{B}\Sigma^0_2$. In particular, per our discussion in Section 6.4, the theorem cannot be proved by any (typical) finite injury argument.

The question thus naturally turns to $\omega$-models. This has seen another long string of attempts at a solution, including revivals of some of the earlier ideas that preceded the Chong–Slaman–Yang result. These were joined by new efforts that focused on finer reducibilities as a way to "work up" to an $\omega$-model separation. In light of Corollary 8.8.3, that $\mathsf{RT}^2_2 \not\leqslant_{\mathrm{c}} \mathsf{SRT}^2_2$, these looked to compare $\mathsf{COH}$ and $\mathsf{SRT}^2_2$ directly. Dzhafarov [79, 80] showed that $\mathsf{COH} \not\leqslant_{\mathrm{W}} \mathsf{SRT}^2$ and $\mathsf{COH} \not\leqslant_{\mathrm{sc}} \mathsf{SRT}^2_2$. Later, Dzhafarov, Patey, Solomon, and Westrick [90] extended the latter result by showing that $\mathsf{COH} \not\leqslant_{\mathrm{sc}} \mathsf{SRT}^2$. (As we will see in Section 9.1, $\mathsf{SRT}^2$ is strictly stronger than $\mathsf{SRT}^2_2$ under both $\leqslant_{\mathrm{W}}$ and $\leqslant_{\mathrm{sc}}$. This is analogous to Propositions 4.3.7 and 4.4.6). We will say more about these partial results at the end of this section. Later on, we will encounter some of the techniques developed by these results, which have since found applications also to other problems. (One example is the *tree labeling method* discussed in Sections 9.1 and 9.2.)

The $\omega$-model separation was found at long last in 2019, once again by rather different methods than those that had been tried up to that point.

**Theorem 8.8.7 (Monin and Patey [218]).** *For all $A, C \in 2^\omega$ with $A' \not\gg C$, every $A$-computable instance of* $\mathsf{SRT}^2_2$ *has a solution $H$ so that $(A \oplus H)' \not\gg C$.*

As noted following Theorem 8.4.13 above, every $\omega$-model of $\mathsf{COH}$ must contain a set $X$ with $X' \gg \varnothing'$. But by iterating and dovetailing the above theorem (taking $C = \varnothing$), it is possible to obtain an $\omega$-model of $\mathsf{SRT}^2_2$ that contains no such $X$. Hence, we obtain:

**Corollary 8.8.8 (Monin and Patey [218]).** $\mathsf{COH} \not\leqslant_\omega \mathsf{SRT}^2_2$ *(and so* $\mathsf{RT}^2_2 \not\leqslant_\omega \mathsf{SRT}^2_2$*). Hence also* $\mathsf{RCA}_0 \nvdash \mathsf{SRT}^2 \to \mathsf{COH}$ *(and so* $\mathsf{RCA}_0 \nvdash \mathsf{SRT}^2 \to \mathsf{RT}^2_2$*).*

Recall that $\mathsf{SRT}^2_2$ does not imply $\mathsf{SRT}^2$ over $\mathsf{RCA}_0$ by Corollary 8.7.3, so in fact the second part above is a stronger conclusion than Corollary 8.8.6.

The statement of Theorem 8.8.7 is essentially that of Liu's theorem, but one jump up. (Accordingly, some authors call the property expressed there *jump PA avoidance*.) Liu's proof may seem tantalizingly close to being directly modifiable to give this result, but this is somewhat illusory. Indeed, there are significant technical obstacles, most notably because the relevant formulas to force are now one quantifier more complex. The innovation of the Monin–Patey proof is the development of a suitable notion of largeness that keeps the complexity of forcing these formulas down (which, as per in Remark 7.7.2, is crucial). This is the key breakthrough that was missing from earlier attempts and sets their argument apart.

Let us conclude this section by looking at what else can be asked about $\mathsf{SRT}^2_2$ and $\mathsf{COH}$. The following definition has come up independently in several works (e.g., [79, 148, 243]) but was first isolated and named by Monin and Patey (in work unrelated to their proof of Theorem 8.8.7).

**Definition 8.8.9 (Monin and Patey [217]).** Let $\mathsf{P}$ and $\mathsf{Q}$ be problems.

1. $\mathsf{P}$ is *omnisciently computably reducible* to $\mathsf{Q}$, written $\mathsf{P} \leqslant_{\mathrm{oc}} \mathsf{Q}$, if for every $\mathsf{P}$-instance $X$ there is a $\mathsf{Q}$-instance $\widehat{X}$ such that if $\widehat{Y}$ is any $\mathsf{Q}$-solution to $\widehat{X}$ then $X \oplus \widehat{Y}$ computes a $\mathsf{P}$-solution to $X$.

2. P is *strongly omnisciently computably reducible* to Q, written P $\leqslant_{\text{soc}}$ Q, if for every P-instance $X$ there is a Q-instance $\widehat{X}$ such that if $\widehat{Y}$ is any Q-solution to $\widehat{X}$ then $\widehat{Y}$ computes a P-solution to $X$.

The emphasis here is on the fact that the Q-instance $\widehat{X}$ need *not* be computable from the P-instance $X$, or be effective in any other way. Clearly, $\leqslant_{\text{oc}}$ implies $\leqslant_{\text{soc}}$ which implies $\leqslant_{\text{sc}}$, and $\leqslant_{\text{oc}}$ also implies $\leqslant_{\text{c}}$. These reducibilities thus eliminate some of the computational dependence between two problems, and so in a sense, allow for their comparison on more purely combinatorial terms: is one problem "combinatorially" reducible to another? A good example that illustrates this is the following, which is immediate from Propositions 8.4.4 and 8.4.5.

**Proposition 8.8.10.** *For all $k \geqslant 1$, $\mathsf{D}^2_k \equiv_{\text{soc}} \mathsf{RT}^1_k$.*

Indeed, to someone uninterested in computability theoretic considerations, finding a limit homogeneous set for a stable coloring of pairs is no different than finding a homogeneous set for a coloring of singletons.

These reducibilities are thus quite interesting in the specific case of the $\mathsf{SRT}^2_2$ vs. COH problem. Several things here are known. For one, we have the following considerable strengthening of the aforementioned result that COH $\nleqslant_{\text{sc}}$ $\mathsf{SRT}^2$.

**Theorem 8.8.11 (Dzhafarov, Patey, Solomon, and Westrick [90]).** COH $\nleqslant_{\text{soc}}$ $\mathsf{SRT}^2$.

(See also Theorems 9.1.1 and 9.1.10 below, which are related.) This is quite different from the original question, whether COH $\leqslant_\omega$ $\mathsf{SRT}^2_2$. But it gives new insight into the relationship between the two principles, especially when combined with the following.

**Theorem 8.8.12 (Cholak, Dzhafarov, Hirschfeldt, and Patey [28]).** COH $\leqslant_{\text{oc}}$ $\mathsf{SRT}^2_2$.

*Proof.* Let $\vec{R} = \langle R_0, R_1, \ldots \rangle$ be a given instance of COH. We use the notation $R_\sigma$ for $\sigma \in 2^{<\omega}$, defined in the proof of Theorem 8.4.13. Let $c\colon [\omega]^2 \to 2$ be the following coloring: for $x < y$, set

$$c(x, y) = \begin{cases} 0 & \text{if } (\exists \sigma \in 2^{<\omega})(\exists z \geqslant y)[|\sigma| = x \wedge z \in R_\sigma \text{ and } R_\sigma \text{ is finite}], \\ 1 & \text{otherwise.} \end{cases}$$

We claim that $\lim_y c(x, y) = 1$ for all $x$, and so in particular that $c$ is an instance of $\mathsf{SRT}^2_2$. Given $x$, fix the least $m_x \geqslant x$ such that for all $\sigma$ of length $x$, if $R_\sigma$ is finite then $R_\sigma \leqslant m_x$. Then $c(x, y) = 1$ for all $y > m_x$.

Let $H = \{h_0 < h_1 < \cdots\}$ be any infinite homogeneous set for $c$, necessarily with color 1. The minimality of $m_x$ above implies that, for each $x$, $c(x, y) = 0$ for all $x < y \leqslant m_x$ (so the value of $c(x, y)$ changes at most once). From here it is readily seen that $h_{x+1} > m_{h_x} \geqslant m_x$ for all $x$.

We now define a sequence of binary strings $\sigma_0 \preceq \sigma_1 \preceq \cdots$ computably from $\vec{R} \oplus H$. Let $\sigma_0 = \langle\rangle$, and assume inductively that $\sigma_{x-1}$ has been defined for some $x > 0$ and that $R_{\sigma_{x-1}}$ is infinite. At least one of $R_{\sigma_{x-1}} \cap R_x$ and $R_{\sigma_{x-1}} \cap \overline{R_x}$ is infinite, so we can search for and find a $z > h_{x+1}$ in one of these two intersections. Let $\sigma_x$ be $\sigma_{x-1}0$ or $\sigma_{x-1}1$, depending on which of the two we find the least such $z$ in. Now, $z > m_x$, so $R_{\sigma_x}$ is infinite.

Now let $y_0 = \min R_{\sigma_0}$, and having defined $y_{x-1}$ for some $x > 0$ let $y_x$ be the least element of $R_{\sigma_x}$ larger than $y_{x-1}$. Then $S = \{y_x : x \in \omega\}$ is an infinite $\vec{R}$-cohesive set computable from the sequence $\sigma_0 \preceq \sigma_1 \preceq \cdots$ and hence from $\vec{R} \oplus H$. □

The above proof exploits the sparsity of homogeneous sets, much like the proof of Proposition 8.5.5. As there, the same argument would not work for $\mathsf{D}^2$ (or equivalently, $\mathsf{RT}^1$, since we are comparing under omniscient reductions), because $\omega$ is limit homogeneous for $c$. And so we are led naturally to the following key question.

*Question 8.8.13.* Is it the case that $\mathsf{COH} \leqslant_{\mathrm{oc}} \mathsf{D}^2$ (or equivalently, $\mathsf{RT}^1$)?

This is the distillation of the $\mathsf{SRT}^2_2$ vs. COH problem to its most combinatorial form, and will likely require rather different methods to tackle.

## 8.9 Summary: Ramsey's theorem and the "big five"

Having studied the computability theoretic properties of Ramsey's theorem in some detail, we are now prepared to locate the theorem and its restrictions within the hierarchy of subsystems of second order arithmetic.

**Theorem 8.9.1.**

(Ramsey's theorem for singletons):

1. *For each* $k \geqslant 1$, $\mathsf{RCA}_0 \vdash \mathsf{RT}^1_k$.
2. $\mathsf{RCA}_0 \vdash \mathsf{RT}^1 \leftrightarrow \mathsf{B}\Sigma^0_2$.

(Ramsey's theorem for pairs):

3. $\mathsf{RCA}_0 \vdash \mathsf{RT}^2_1$.
4. $\mathsf{ACA}_0 \vdash \mathsf{RT}^2$.
5. *Over* $\mathsf{RCA}_0$, $\mathsf{RT}^2$ *and* WKL *are incomparable.*
6. *For all* $k \geqslant 1$, $\mathsf{RCA}_0 \vdash \mathsf{RT}^2 \to \mathsf{RT}^2_k$.
7. $\mathsf{RCA}_0 \nvdash \mathsf{RT}^2_k \to \mathsf{SRT}^2$.
8. *For all* $k \geqslant 2$, $\mathsf{RCA}_0 \vdash \mathsf{RT}^2_k \leftrightarrow \mathsf{SRT}^2_k + \mathsf{COH}$.
9. $\mathsf{RCA}_0 \vdash \mathsf{RT}^2 \leftrightarrow \mathsf{SRT}^2 + \mathsf{COH}$.
10. $\mathsf{RCA}_0 \nvdash \mathsf{COH} \to \mathsf{SRT}^2_2$ *and* $\mathsf{RCA}_0 \nvdash \mathsf{SRT}^2 \to \mathsf{COH}$.

(Ramsey's theorem for exponent $n \geqslant 3$):

*11.* $\mathsf{RCA}_0 \vdash \mathsf{RT}^n_1$.
*12. For all* $k \geqslant 2$, $\mathsf{RCA}_0 \vdash \mathsf{RT}^n_k \leftrightarrow \mathsf{RT}^n \leftrightarrow \mathsf{ACA}_0$.

*Proof.* Parts (1) is trivial. Part (2) is Hirst's theorem (Theorem 6.5.1). Part (3) is trivial. Part (4) is proved in Proposition 8.1.5. Part (5) follows by Liu's theorem (Theorem 8.6.1) and the preceding discussion. Part (6) is clear. Part (7) is Corollary 8.7.3. Parts (8) and (9) are in Theorem 8.5.1. The first half of part (10) is Corollary 8.8.1, and the second is the Chong–Slaman–Yang theorem and Monin–Patey theorem (Theorems 8.8.5 and 8.8.7). Finally, (11) is trivial, and (12) is Corollary 8.2.6. □

The diagram in Figure 8.2 gives us our first snapshot of the rich and complicated tapestry of relationships between combinatorial principles. This has come to be called the *reverse mathematics zoo*, and will be the subject of Section 9.12 below.

## 8.10 Exercises

**Exercise 8.10.1 (Jockusch [168]).** Fix $A \gg \varnothing'$ and $n, k \geqslant 1$. Show that every computable $c\colon [\omega]^n \to k$ has an infinite $A$-computable pre-homogeneous set.

**Exercise 8.10.2.** Show that if $f\colon \omega \to \omega$ is total and dominates every partial computable function then $\varnothing' \leqslant_{\mathrm{T}} f$. $\varnothing^{(n-2)}$.

**Exercise 8.10.3.** Show that $\mathsf{RCA}_0 \vdash \mathsf{SRT}^2_2 \to \mathsf{RT}^1$.

**Exercise 8.10.4 (Hirschfeldt, Jockusch, Kjos-Hanssen, Lempp, and Solomon [149]).** Show that $\mathsf{DNR} \leqslant_{\mathrm{c}} \mathsf{D}^2_2$. (Hint: First, using an argument similar to the $n = 2$ case of Theorem 8.2.2, show that there is a set $P \leqslant_{\mathrm{T}} \varnothing'$ such that for all $e$, if $W_e$ is a subset of $P$ or $\overline{P}$ then $|W_e| < 3e + 2$. Next, suppose $L$ is an infinite subset of $P$ or $\overline{P}$, and let $g$ be the $L$-computable function such that for all $e$, $W_{g(e)}$ consists of the least $3e + 2$ many elements of $L$. Show that $W_{g(e)} \neq W_e$ for all $e$. Finally, apply Exercise 2.9.16.)

**Exercise 8.10.5.** Show that $\mathsf{RCA}_0 + \mathsf{COH}$ proves the following: if $S$ is an infinite set and $\vec{R}$ is a family of sets, then there exists an infinite $\vec{R}$-cohesive set $X \subseteq S$.

**Exercise 8.10.6.** Use Theorem 8.4.13 to show that $\mathsf{COH}$ omits solutions of hyperimmune free degree.

**Exercise 8.10.7 (Hirschfeldt and Shore [152]).** Let $\mathsf{CRT}^2_2$ be the statement that for every coloring $c\colon [\omega]^2 \to 2$ there is an infinite set $S$ such that $c \upharpoonright [S]^2$ is stable. Show that $\mathsf{RCA}_0 + \mathsf{B}\Sigma^0_2 \vdash \mathsf{CRT}^2_2 \leftrightarrow \mathsf{COH}$ (and so $\mathsf{COH} \equiv_\omega \mathsf{CRT}^2_2$).

**Exercise 8.10.8.** The *Paris–Harrington theorem* ($\mathsf{PH}$) is the following result strengthening the finitary Ramsey's theorem ($\mathsf{FRT}$) from Definition 3.3.6: for all $n, k, m \geqslant 1$

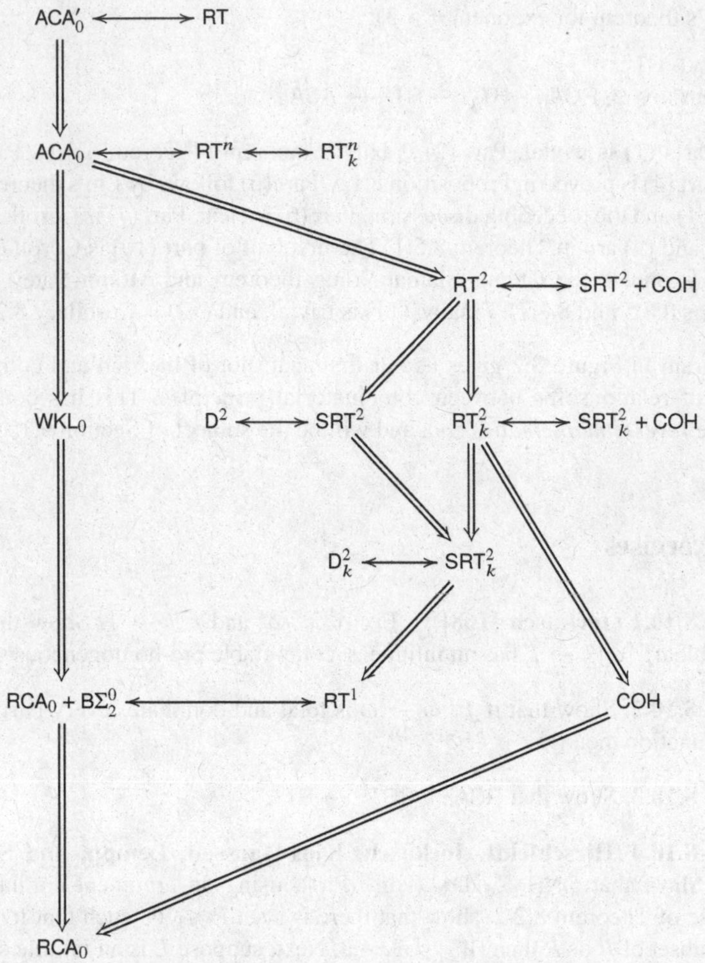


**Figure 8.2.** The location of versions of Ramsey's theorem below $\mathsf{ACA}_0'$. Here, $n \geqslant 3$ and $k \geqslant 2$ are arbitrary. Arrows denote implications over $\mathsf{RCA}_0$; double arrows are implications that cannot be reversed. No additional arrows can be added.

with $m \geqslant n$, there is an $N \in \omega$ such that for every finite set $X$ with $|X| \geqslant N$, every $c\colon [X]^n \to k$ has a homogeneous set $H \subseteq X$ with $|X| \geqslant \max\{m, \min X\}$. Show that $\mathsf{RCA}_0 \vdash \mathsf{RT} \to \mathsf{PH}$. It is a famous result of Paris and Harrington [238] that PH is not provable in PA. (For a proof of this fact, see Kaye [176, Section 14.3].) Conclude by Corollary 5.10.7 that RT is not provable in $\mathsf{ACA}_0$.

**Exercise 8.10.9.** Complete the proof of Lemma 8.7.7 by proving parts (1) and (2).

# Chapter 9
# Other combinatorial principles

Seetapun's theorem (Theorem 8.3.1) and the follow-up work of Cholak, Jockusch, and Slaman [33] led to the realization that not only *are* there natural principles defying the "big five" phenomenon in reverse mathematics, but that there may be many of them. The interest in finding more examples, and understanding why their strength is so different from that of most other theorems studied before, has grown into a massive research program. Many dozens of other such principles have now been identified. Curiously enough, most of them (though not all) have turned out to come from combinatorics and, indeed, are related to Ramsey's theorem for pairs in one way or another.

The goal of this chapter is to give a partial survey the results of this investigation, highlighting recent developments and connections to currently open problems. Another overview is given by Hirschfeldt [147]. The overlap between that treatment and ours is minimal, and we feel the two accounts are quite complementary.

## 9.1 Finer results about RT

We have seen that Ramsey's theorem exhibits different complexity bounds based on the exponent, but not based on the number of colors. However, in Chapter 4, we saw that versions of Ramsey's theorem for different numbers of colors with the same exponent could be separated using finer reducibilities. In this section, we continue this investigation and obtain very precise results about the principles $\mathsf{RT}^n_k$.

### 9.1.1 Ramsey's theorem for singletons

We begin by looking at $\mathsf{RT}^1_k$. In Proposition 4.3.7 (Exercise 4.8.4) we saw that if $k > j$ then $\mathsf{RT}^1_k \nleq_{\mathrm{W}} \mathsf{RT}^1_j$. In Proposition 4.4.6, we also mentioned (without proof)

 
D. D. Dzhafarov, C. Mummert, *Reverse Mathematics*, Theory and Applications of Computability, https://doi.org/10.1007/978-3-031-11367-3_9

an analogous result for $\leqslant_{sc}$: if $k > j$ then $\mathsf{RT}^1_k \nleqslant_{sc} \mathsf{RT}^1_j$. We now prove this result, in the following considerably stronger form.

**Theorem 9.1.1 (Dzhafarov, Patey, Solomon, and Westrick [90]).** *For all $k > j$, $\mathsf{RT}^1_k \nleqslant_{sc} \mathsf{SRT}^2_j$.*

We must exhibit an instance $c$ of $\mathsf{RT}^1_k$, and for each Turing functional $\Phi$ for which $\Phi^c$ is an instance of $\mathsf{SRT}^2_j$, an infinite homogeneous set $H$ for $\Phi^c$ that computes no infinite homogeneous set for $c$. This is achieved via a cardinality argument, similar to Proposition 4.3.7. But there are two complicating factors. First, $H$ must now avoid computing an infinite homogeneous set for $c$ via all possible Turing functionals, not just a single one. And second, more problematically, $H$ must be homogeneous for a stable coloring of pairs, which is harder to ensure than being homogeneous set for a coloring of singletons, or indeed, being limit homogeneous for a stable coloring of pairs. More precisely, because we need to control the internal structure of $H$ (the colors between the elements of $H$), we always need $\Phi^c$ to already be defined on the elements we wish to add to $H$, and this in turn requires more of $c$ to already be defined. This causes a tension with diagonalizing the infinite homogeneous sets of $c$.

Before discussing how to overcome this obstacle, let us define the components of the proof, state the main technical lemma we will need, and then see how Theorem 9.1.1 follows. We use separate forcing constructions for the construction of our $\mathsf{RT}^1_k$ instance $c$ and our $\mathsf{SRT}^2_j$ solution $H$. For $c$, we use Cohen forcing with conditions $\sigma \in k^{<\omega}$; for $H$, we use Mathias forcing. However, the two constructions are necessarily intertwined. For the rest of this section, let $k > j$ be fixed.

**Definition 9.1.2.** Fix $\ell < j$ and a Turing functional $\Phi$. A Cohen condition $\sigma \in k^{<\omega}$ is $(\Phi, \ell)$-*compatible* with a Mathias condition $(E, I)$ if the following hold.

- $\sigma \Vdash \Phi^{\mathsf{G}}$ is a stable coloring $[\omega]^2 \to j$ for some $j > \ell$.
- $\sigma \Vdash \Phi^{\mathsf{G}}(x, y) = \ell$ for all $x, y \in E$.
- $\sigma \Vdash (\forall y \geqslant \min I)[\Phi^{\mathsf{G}}(x, y) = j]$ for all $x \in E$.

**Definition 9.1.3.** We define the following notion of forcing.

1. Conditions are finite tuples $p$ consisting of a Cohen condition $\sigma^p \in k^{<\omega}$, a finite set $S^p$ of Turing functionals, and for each $\Phi \in S^p$, finite sets $E^p_{\Phi,0}, \dots, E^p_{\Phi,j-1}$ and an infinite set $I^p_\Phi$, such that the following hold.
   - $\sigma^p \Vdash \Phi^{\mathsf{G}}$ is a stable coloring $[\omega]^2 \to j$.
   - For each $\ell < j$, $(E^p_{\Phi,\ell}, I^p_\Phi)$ is a Mathias condition.
   - For each $\ell < j$, $\sigma^p$ is $(\Phi, \ell)$-compatible with $(E^p_{\Phi,\ell}, I^p_\Phi)$.
2. Extension is defined by $q \leqslant p$ if $\sigma^q \succeq \sigma^p$, $S^q \supseteq S^p$, and for each $\Phi \in S^p$ and $\ell < j$, $(E^q_{\Phi,\ell}, I^q_\Phi) \leqslant (E^p_{\Phi,\ell}, I^p_\Phi)$ as Mathias conditions.
3. The valuation of a condition $p$ is the join of the valuation of $\sigma^p$ as a Cohen condition with the valuations, for each $\Phi \in S^p$, of the Mathias conditions $(E^p_{\Phi,0}, I^p_\Phi), \dots, (E^p_{\Phi,j-1}, I^p_\Phi)$.

Observe that for each Turing functional $\Phi$, the set of conditions $p$ having $\Phi \in S^p$, or else having $\sigma^p$ force that $\Phi^{\mathsf{G}}$ is not a stable coloring $[\omega]^2 \to j$, is dense. Hence, a generic for the above forcing consists of an instance $c^G$ of $\mathsf{RT}^1_k$, and for each $\Phi$ such that $\Phi^{c^G}$ is an instance of $\mathsf{SRT}^2_j$, a $j$-tuple of homogeneous sets $H^G_{\Phi,0}, \dots, H^G_{\Phi,j-1}$. By genericity, and the definition of compatibility, it is easily seen that there is at least one $\ell < j$ such that $H^G_{\Phi,\ell}$ is infinite.

We will only deal with the forcing relation with respect to Cohen conditions, and so continue to use the symbol $\mathsf{G}$ for the name of the generic $\mathsf{RT}^1_k$ instance $c^G$ in the $k^{<\omega}$ forcing language. Our main objective will be to prove the following density lemma.

**Lemma 9.1.4.** *Fix $i < k$, $\ell < j$, and Turing functionals $\Phi$ and $\Psi$. The set of conditions $p$ satisfying one of the following properties is dense.*

1. *For all $x \in I^p_\Phi$, $\sigma^p \Vdash \neg(\lim_y \Phi^{\mathsf{G}}(x, y) = \ell)$.*
2. *For all finite sets $F \subseteq I^p_\Phi$ and all sufficiently large $x \in \omega$, $\Psi^{E^p_{\Phi,\ell} \cup F}(x) \simeq 0$.*
3. *For each $x \in I^p_\Phi$ there exists a unique number $w^x \in \omega$ such that for all $\tau \succeq \sigma^p$, if $\tau(w^x) = i$ then $\tau \Vdash \neg(\lim_y \Phi^{\mathsf{G}}(x, y) = \ell)$.*
4. *There exists a $w \in \omega$ such that $\sigma^p(w) = i$ and $\Psi^{E^p_{\Phi,\ell}}(w) \downarrow= 1$.*

The theorem now easily follows.

*Proof (of Theorem 9.1.1).* Choose a sufficiently generic filter $\mathcal{F}$ for the forcing in Definition 9.1.3, and let $c^G$ be as above. Suppose $\Phi^{c^G}$ is an instance of $\mathsf{SRT}^2_j$, and let $H^G_{\Phi,0}, \dots, H^G_{\Phi,j-1}$ again be as above. We will show that there is an $\ell < j$ such that $H^G_{\Phi,\ell}$ is infinite and $\Psi^{H^G_{\Phi,\ell}}$ is not an infinite homogeneous set for $c$, for any Turing functional $\Psi$.

Seeking a contradiction, suppose otherwise. Let $C$ be the (necessarily nonempty) set of all $\ell < j$ such that $H^G_{\Phi,\ell}$ is infinite. We may thus fix a condition $p_0 \in \mathcal{F}$ such that, for each $x \in I^{p_0}_\Phi$ and each $\ell \notin C$, $\sigma^{p_0}$ forces $\neg(\lim_y \Phi^{\mathsf{G}}(x, y) = \ell)$. Now, for each $\ell \in C$, fix a Turing functional $\Psi_\ell$ via which $H^G_{\Phi,\ell}$ computes an infinite homogeneous set for $c^G$, say with color $i_\ell < k$. Since $j < k$, we may fix an $i < k$ such that $i \neq i_\ell$ for all $\ell < j$.

Apply Lemma 9.1.4 repeatedly with $i$, $\Phi$, and $\Psi_\ell$ for each $\ell \in C$ in turn, to obtain a condition $p \leqslant p_0$ in $\mathcal{F}$. By assumption, properties (2) and (4) cannot hold for any $\ell \in C$, since $E^p_{\Phi,\ell} \subseteq H^G_{\Phi,\ell} \subseteq I^p_\Phi$, $\sigma^p \preceq c^G$, and $\Psi$ maps $H^G_{\Phi,\ell}$ onto an infinite homogeneous set for $c^G$ with color different from $i$. And by genericity, property (1) cannot hold for any $\ell \in C$, since $H^G_{\Phi,\ell}$ is homogeneous (and hence limit homogeneous) for $\Phi^{c^G}$ with color $\ell$. Hence, $p$ witnesses that (3) holds for each $\ell \in C$.

It follows that for each $\ell \in C$, each number $x \in I^p_\Phi$ corresponds to a unique number $w^x_\ell \in \omega$ such that if $\tau$ is any extension of $\sigma^p$ with $\tau(w^x_\ell) = i$ then $\tau \Vdash \neg(\lim_y \Phi^{\mathsf{G}}(x, y) = \ell)$. The uniqueness means we can choose an $x$ such that $w^x_\ell \geqslant |\sigma^p|$ for each $\ell \in C$. Then we can find a $\tau \succeq \sigma^p$ with $\tau(w^x_\ell) = i$ for all $\ell \in C$. (It may be that

$w^x_\ell = w^x_{\ell^*}$ for some $\ell \neq \ell^*$, but this does not matter.) So $\tau$ forces $\neg(\lim_y \Phi^G(x, y) = \ell)$ for all $\ell \in C$. But $x \in I^p_\Phi \subseteq I^{p_0}_\Phi$ and $\tau \geq \sigma^{p_0}$, so $\tau \Vdash \neg(\lim_y \Phi^G(x, y) = \ell)$ for all $\ell < j$, which cannot be. □

We move on to prove Lemma 9.1.4. Let us fix $i < k$, $\ell < j$, and the functionals $\Phi$ and $\Psi$ for the remainder of this discussion. We may assume $\Psi$ is $\{0, 1\}$-valued, so that if $\Psi^Z$ is total for some $Z$ then we can view it as a set.

**Definition 9.1.5.** Fix $n \in \omega$ and a Mathias condition $(E, I)$. We define $T(n, E, I) \subseteq I^{<\omega}$ as follows.

- $\langle\rangle \in T(n, E, I)$.
- A nonempty string $\alpha \in I^{<\omega}$ belongs to $T(n, E, I)$ if and only if $\alpha$ is increasing and

$$(\forall F \subseteq \mathrm{range}(\alpha^\#))(\forall x \geqslant n)[\Psi^{E \cup F}(x) \simeq 0].$$

The following basic facts about $T(n, E, I)$ are straightforward to check. The key point to remember in, e.g., (2) or (3), is that since $(E, I)$ is a Mathias condition, $I$ is an infinite set.

**Lemma 9.1.6.** *Fix $n \in \omega$, a Mathias condition $(E, I)$, and let $T = T(n, E, I)$. Then $T$ has the following properties.*

1. *$T$ is closed under initial segments, so $T$ is a tree.*
2. *If $\alpha \in T$ is not terminal, then $\alpha x \in T$ for all $x \in I$ with $x >$ range$(\alpha)$.*
3. *If $\alpha \in T$ is terminal, then there is an $F \subseteq$ range$(\alpha)$ and an $x \geqslant n$ such that $\Psi^{E \cup F}(x) \downarrow = 1$.*
4. *If $T$ is not well founded and $I^* =$ range$(f)$ for some $f \in [T]$ then for all sets $Z$ with $E \subseteq Z \subseteq E \cup I^*$, either $\Psi^Z$ is not total or $\Psi^Z$ is a finite set.*

The main combinatorial construct we need is a *labeling* of the nodes in this tree, which we use as a guide for adding elements to the homogeneous sets we are building. The method used in the proof of Lemma 9.1.4 is accordingly called the *tree labeling method*. It was originally developed by Dzhafarov [80], but then extended for the proof of the present theorem. We will mention other applications of this method later.

**Definition 9.1.7 (Labeling and labeled subtree).** Fix $n \in \omega$, a Mathias condition $(E, I)$, and let $T = T(n, E, I)$. Suppose $T$ is well founded.

1. For each terminal $\alpha \in T$, the least $F \subseteq \mathrm{range}(\alpha)$ and $x \geqslant n$ as in Lemma 9.1.6 (3) are denoted $F^\alpha$ and $w^\alpha$, respectively.
2. The *label* of a node $\alpha \in T$ is defined inductively as follows.
   - If $\alpha$ is terminal, its label is $w^\alpha$.
   - Suppose $\alpha$ is not terminal, and every $\beta \succeq \alpha \in T$ has been labeled. If infinitely many immediate successors of $\alpha$ in $T$ have a common label $w$, the label of $\alpha$ is the least such $w$. Otherwise, the label of $\alpha$ is $\infty$.

3. The *labeled subtree* of $T$ is the tree $T^L = T^L(n, E, I)$ defined inductively as follows.
   - $\langle\rangle \in T^L$.
   - Suppose $\alpha \in T^L$ and $\alpha$ is not terminal in $T$. If $\alpha$ has infinitely many immediate successors in $T$ with the same label as its own (either numerical, or $\infty$), then each such successor is in $T^L$. Otherwise, $\alpha$ has immediate successors with infinitely many numerical labels $w$, and for each such $w$, the least successor of $\alpha$ with label $w$ is in $T^L$.

We again collect some basic facts about this definition.

**Lemma 9.1.8.** *Fix $n \in \omega$, a Mathias condition $(E, I)$, and let $T = T(n, E, I)$. Suppose $T$ is well-founded and let $T^L = T^L(n, E, I)$.*

1. *$\alpha \in T^L$ is terminal in $T^L$ if and only if it is terminal in $T$.*
2. *If $\alpha \in T^L$ has label $w \in \omega$, then so does every $\beta \geq \alpha$ in $T^L$, and $w \geqslant n$.*
3. *Each nonterminal $\alpha \in T^L$ has infinitely many immediate successors in $T^L$, and these either all have the same numerical label $w$, in which case so does $\alpha$, or they all have label $\infty$, in which case so does $\alpha$, or they each have a different numerical label, in which case $\alpha$ has label $\infty$.*
4. *If $\alpha \in T$ is terminal then its label is some $w \in \omega$, and there is an $F \subseteq \operatorname{range}(\alpha)$ such that $\Psi^{E\cup F}(w) \downarrow= 1$.*

We will also need the following less obvious fact, which will be the main tool used for building homogeneous sets in the proof of Lemma 9.1.4. We say a Cohen condition $\sigma \in k^{<\omega}$ forces that a finite set $F$ is *homogeneous and limit homogeneous for $\Phi^G$ with color $\ell$* if $\sigma \Vdash \Phi^G(x, y) = \ell$, for all $x, y \in F$, and $\sigma \Vdash \lim_y \Phi^G(x, y) = \ell$, for all $x \in F$.

**Lemma 9.1.9.** *Fix $n \in \omega$, a Mathias condition $(E, I)$, and let $T = T(n, E, I)$. Suppose $T$ is well-founded and let $T^L = T^L(n, E, I)$. Suppose $\sigma \in k^{<\omega}$ is $(\Phi, \ell)$-compatible with $(E, I)$. Then one of the following possibilities holds:*

1. *There is a $\sigma^* \succeq \sigma$ and an infinite set $I^* \subseteq I$ such that*
$$\sigma \Vdash \neg(\lim_y \Phi^G(x, y) = \ell)$$
*for each $x \in I^*$.*
2. *If $\alpha \in T^L$ and $\sigma$ forces that $E \cup \operatorname{range}(\alpha)$ is homogeneous and limit homogeneous for $\Phi^G$ with color $\ell$, then there is a $\sigma^* \succeq \sigma$ and an $\alpha^* \succeq \alpha$ such that $\alpha^*$ is terminal in $T^L$ and $\sigma^*$ forces that $E \cup \operatorname{range}(\alpha^*)$ is homogeneous and limit homogeneous for $\Phi^G$ with color $\ell$.*

*Proof.* Suppose $\sigma \in k^{<\omega}$ is $(\Phi, \ell)$-compatible with $(E, I)$, and that no $\sigma^*$ and $I^*$ as in part (1) exist. Let $\alpha \in T^L$ be as in (2). We define a finite sequence $\sigma_0 \preceq \sigma_1 \preceq \cdots$ of Cohen conditions and a finite sequence $\alpha_0 \preceq \alpha_1 \preceq \cdots$ of elements of $T^L$. Let $\sigma_0 = \sigma$ and $\alpha_0 = \alpha$. Fix $s > 0$, and assume inductively that we have defined $\sigma_s$ and

$\alpha_s$, and that $\sigma_s$ forces that $\alpha_s$ is homogeneous and limit homogeneous for $\Phi^G$ with color $\ell$. If $\alpha_s$ is terminal, we do nothing. Otherwise, fix $b$ such that $\sigma_s \Vdash (\forall y \geqslant b)[\Phi^G(x,y) = \ell]$ for all $x \in \text{range}(\alpha_s)$ and let $I^* = \{x \in I : x \geqslant b \wedge \alpha_s x \in T^L\}$. Then $I^*$ is infinite by definition of $T^L$. If there were no $x \in I^*$ and no $\tau \succeq \sigma_s$ forcing that $\lim_y \Phi^G(x,y) = \ell$ then (1) would apply with $\sigma^* = \sigma_s$ and $I^*$, a contradiction. So we may fix some such $\tau$ and $x$. Let $\sigma_{s+1} = \tau$ and let $\alpha_{s+1} = \alpha_s x$. Clearly, the inductive hypotheses are preserved. This completes the construction. Letting $\sigma^*$ be the final element defined in the sequence $\sigma_0, \sigma_1, \ldots$, and $\alpha^*$ the final element defined in the sequence $\alpha_0, \alpha_1, \ldots$, we obtain the conclusion of (2). □

We can now prove our main lemma, completing the proof of Theorem 9.1.1.

*Proof (of Lemma 9.1.4).* Fix a Cohen condition $\sigma \in k^{<\omega}$ and a Mathias condition $(E, I)$ for which $\sigma$ is $(\Phi, \ell)$-compatible. We prove there is a $\sigma^* \preceq \sigma$ and an $(E^*, I^*) \leqslant (E, I)$ such that $\sigma^*$ is $(\Phi, \ell)$-compatible with $(E^*, I^*)$ and one of the following properties holds.

1. For all $x \in I^*$, $\sigma^* \Vdash \neg(\lim_y \Phi^G(x,y) = \ell)$.
2. For all finite sets $F \subseteq I^*$ and all sufficiently large $x \in \omega$, $\Psi^{E^* \cup F}(x) \simeq 0$.
3. For each $x \in I^*$ there exists a unique number $w^x \in \omega$ such that for all $\tau \succeq \sigma^*$, if $\tau(w^x) = i$ then $\tau \Vdash \neg(\lim_y \Phi^G(x,y) = \ell)$.
4. There exists a $w \in \omega$ such that $\sigma^*(w) = i$ and $\Psi^{E^*}(w) \downarrow= 1$.

These are just like (1)–(4) in the statement of Lemma 9.1.4, but this presentation is notationally simpler. And since $\sigma$ and $(E, I)$ are arbitrary, this will gives us the lemma. We break into cases.

*Case 1: There is a $\sigma^* \succeq \sigma$ and an infinite set $I^* \subseteq I$ such that $\sigma^* \Vdash \neg(\lim_y \Phi^G(x,y) = \ell)$ for each $x \in I^*$.* We let $E^* = E$. Then $\sigma^*$ and $(E^*, I^*)$ satisfy (1) above.

*Case 2: Otherwise.* Let $n = \max\{|\sigma|, E\}$ and let $T = T(n, E, I)$. If $T$ is not well-founded, pick any $f \in [T]$ and let $I^* = \text{range}(f)$. In this case, let $\sigma^* = \sigma$ and $E^* = E$. Then by Lemma 9.1.6 (4), it follows that $\sigma^*$ and $(E^*, I^*)$ satisfy (2) above. So assume next that $T$ is well-founded. Let $T^L = T^L(n, E, I)$. We break into subcases based on the label of the root, $\langle\rangle$.

*Subcase 2a: $\langle\rangle$ has label $w \in \omega$.* Let $\sigma_0$ be any extension of $\sigma$ with $\sigma_0(w) = i$, which exists since $w \geqslant n \geqslant |\sigma|$. Now since $\sigma_0$ is $(\Phi, \ell)$-compatible with $(E, I)$ since $\sigma$ is, and therefore $\sigma_0$ forces that $E \cup \text{range}(\langle\rangle) = E$ is homogeneous and limit homogeneous for $\Phi^G$ with color $\ell$. Since Case 1 does not hold, Lemma 9.1.9 yields a $\sigma^* \succeq \sigma_0$ and a terminal $\alpha^* \in T^L$ such that $\sigma^*$ forces that $E \cup \text{range}(\alpha^*)$ is homogeneous and limit homogeneous for $\Phi^G$ with color $\ell$. By Lemma 9.1.8 (2), the label of $\alpha^*$ is $w$, so by Lemma 9.1.8 (4), there is an $F \subseteq \text{range}(\alpha^*)$ such that $\Psi^{E \cup F}(w) \downarrow= 1$. Let $E^* = E \cup F$ and $I^* = \{x \in I : x > E^*\}$. Then $\sigma^*$ is $(\Phi, \ell)$-compatible with $(E^*, I^*)$ and (4) holds.

*Subcase 2b: $\langle\rangle$ has label $\infty$.* Fix a $\preceq$-maximal $\alpha \in T^L$ with label $\infty$ for which there is a $\sigma_0 \succeq \sigma$ forcing that range$(\alpha)$ is homogeneous and limit homogeneous for $\Phi^G$ with color $\ell$. (Note that such an $\alpha$ exists because $T$ is well-founded.) Fix $b$ so that

$\sigma_0 \Vdash (\forall y \geqslant b)[\Phi^G(x, y) = \ell]$, for all $x \in \text{range}(\alpha)$. By Lemma 9.1.8 (4), $\alpha$ is not terminal in $T^L$, so $I^* = \{x \in I : x \geqslant b \wedge \alpha x \in T^L\}$ is infinite. We claim that $\alpha x$ has numerical label, for each $x \in I^*$. If not, then $\alpha x$ would have label $\infty$ for each such $x$, and so the reason for $\alpha$ being maximal would be that no extension of $\sigma$ forces $\lim_y \Phi^G(x, y) = \ell$. But then $\sigma$ would force $\neg(\lim_y \Phi^G(x, y) = \ell)$ for all $x \in I^*$, which cannot be since we are not in Case 1. So the claim holds. For each $x \in I^*$, let $w^x \in \omega$ be the label of $\alpha x$. By Lemma 9.1.8 (3), each $w^x$ is unique. Now if, for each $x \in I^*$, every $\tau \succeq \sigma_0$ with $\tau(w^x) = i$ forces $\neg(\lim_y \Phi^G(x, y) = \ell)$, we can set $E^* = E$, and then $\sigma^*$ and $(E^*, I^*)$ satisfy (3) above. So assume not, and fix a witnessing $x \in I^*$ and corresponding $\tau$. Since $x \geqslant b$, we have $\tau$ forcing that range$(\alpha x)$ is homogeneous and limit homogeneous for $\Phi^G$. Lemma 9.1.9 now yields a $\sigma^* \succeq \tau$ and a terminal $\alpha^* \in T^L$ extending $\alpha x$ such that $\sigma^*$ forces that $E \cup \text{range}(\alpha^*)$ is homogeneous and limit homogeneous for $\Phi^G$ with color $\ell$. By Lemma 9.1.8 (2), the label of $\alpha^*$ is $w^x$, so by Lemma 9.1.8 (4), there is an $F \subseteq \text{range}(\alpha^*)$ such that $\Psi^{E \cup F}(w^x) \downarrow= 1$. Let $E^* = E \cup F$ and $I^* = \{x \in I : x > E^*\}$. Then $\sigma^*$ is $(\Phi, \ell)$-compatible with $(E^*, I^*)$ and (4) holds. □

Theorem 9.1.1 can be further strengthened using the notion of strong omniscient computable reducibility, which was introduced in Definition 8.8.9.

**Theorem 9.1.10 (Dzhafarov, Patey, Solomon, and Westrick [90]).** *For all $k > j$, $\mathsf{RT}^1_k \nleqslant_{\text{soc}} \mathsf{SRT}^2_j$.*

We omit the proof, as it requires ideas from set theory that are somewhat outside the scope of this text. Basically, because we now need to deal with arbitrary instances of $\mathsf{SRT}^2_j$, rather than merely those computable from $c^G$, we lose the ability to directly "talk about" these instances in our forcing language. The workaround is to make $c^G$ be generic over a suitable countable model of ZFC, diagonalize all $\mathsf{RT}^1_j$ instances in this model (which we *can* name), and then apply an absoluteness argument to extend the result to $\mathsf{RT}^1_j$ instances in general. The combinatorics underlying this argument, however, are exactly the same, and the reader who has understood these combinatorics and is familiar with basic forcing arguments in set theory should have little difficulty following the proof in [90].

### 9.1.2 Ramsey's theorem for higher exponents

We now move on to consider $\mathsf{RT}^n_k$, for $n \geqslant 2$. Again, in terms of the analysis in Chapter 8, all such versions are equivalent. Hover, under finer reducibilities, it is possible to show that if $k \neq j$ then $\mathsf{RT}^n_k$ and $\mathsf{RT}^n_j$ are actually quite far apart.

**Theorem 9.1.11 (Patey [243]).** *For all $n \geqslant 2$ and $k > j \geqslant 1$, $\mathsf{RT}^n_k \nleqslant_{\text{c}} \mathsf{RT}^n_j$.*

Dorais, Dzhafarov, Hirst, Mileti, and Shafer [72] showed earlier that $\mathsf{RT}^n_k \nleqslant_{\text{sW}} \mathsf{RT}^n_j$, and then Hirschfeldt and Jockusch [148] and Brattka and Rakotoniaina [20] independently showed that $\mathsf{RT}^n_k \nleqslant_{\text{W}} \mathsf{RT}^n_j$. Each of these results used rather different methods. Patey's theorem above greatly extends these results, and has a remarkably elegant proof, which we now present.

Recall that a set $S$ is *hyperimmune* if its principal function is not dominated by any computable function. Every hyperimmune set is, in particular, *immune*, meaning it has no computable infinite subset. The key definition and lemma we need for the proof are the following.

**Definition 9.1.12 (Patey [243]).** Fix $k \geqslant m \geqslant 1$. A problem $\mathsf{P}$ *admits preservation of m among k hyperimmunities* if for every $A \in 2^{\omega}$ and every sequence $S_0, \ldots, S_{k-1}$ of $A$-hyperimmune sets, every $A$-computable instance $X$ of $\mathsf{P}$ has a solution $Y$ such that at least $m$ many of the sets $S_0, \ldots, S_{k-1}$ are $(A \oplus Y)$-hyperimmune.

**Lemma 9.1.13 (Patey [243]).** *For all $k > j \geqslant 1$, $\mathsf{RT}^2_j$ admits preservation of* 2 *among k hyperimmunities.*

Let us see how the lemma implies Theorem 9.1.11.

*Proof (of Theorem 9.1.11).* Since $\mathsf{RT}^n_1$ is computably true and, for all $k \geqslant 2$, $\mathsf{RT}^n_k$ is not (by Theorem 8.2.2), the result is obvious if $j = 1$. So, we may assume $j \geqslant 2$. We first prove the following claim.

*Claim. There exists a $\Delta^0_n$ sequence $\langle S_0, \ldots, S_{k-1} \rangle$ whose members partition $\omega$ such that for every computable $d\colon [\omega]^n \to j$ there exits an infinite homogeneous set $H$ for $d$ and distinct numbers $i_0, i_1 < k$ with the property that every infinite $H$-computable set intersects both $S_{i_0}$ and $S_{i_1}$.*

The proof is by induction on $n$. First, suppose $n = 2$. By Exercise 2.9.15, there is a $\Delta^0_2$ sequence of sets $\langle S_0, \ldots, S_{k-1} \rangle$ whose members partition $\omega$ and for each $i < k$, $\overline{S_i}$ is hyperimmune. Consider any computable $d\colon [\omega]^2 \to j$. By Lemma 9.1.13, $d$ has an infinite homogeneous set $H$ such that for some $i_0, i_1 < k$, $\overline{S_{i_0}}$ and $\overline{S_{i_1}}$ are $H$-hyperimmune, and so in particular $H$-immune. This means every $H$-computable infinite set intersects both $S_{i_0}$ and $S_{i_1}$, as was to be shown.

Clearly, the above relativizes. So fix $n > 2$, and assume that the result holds for $n - 1$ and that this also relativizes. We prove the result for $n$ in unrelativized form, but the proof will easily be seen to relativize as well. Fix $A \gg \varnothing'$ with $A' \leqslant_{\mathrm{T}} \varnothing''$, and apply the inductive hypothesis relative to $A$ to obtain a $\Delta^0_{n-1}(A)$ sequence $\langle S_0, \ldots, S_{k-1} \rangle$. By Post's theorem (Theorem 2.6.2) have

$$\Delta^0_{n-1}(A) = \Delta^0_{n-2}(A') \subseteq \Delta^0_{n-2}(\varnothing'') = \Delta^0_n,$$

so $\langle S_0, \ldots, S_{k-1} \rangle$ is $\Delta^0_n$. We claim that this is the desired partition. Fix any computable $d\colon [\omega]^n \to j$. By Exercise 8.10.1, $d$ has an $A$-computable infinite pre-homogeneous set $P$. Define $d^*\colon [P]^{n-1} \to j$ as follows: for all $\vec{x} \in [P]^{(n-1)}$, $d^*(\vec{x}) = d(\vec{x}, y)$ for the least $y \in P$ with $y > \vec{x}$ (which is the same value for any $y \in P$ with $y > \vec{x}$). Then $d^*$ is $P$-computable, and hence $A$-computable. So, by assumption, there is an infinite set $H \subseteq P$ homogeneous for $d^*$ and distinct numbers $i_0, i_1 < k$ such that every $H$-computable infinite set intersects both $S_{i_0}$ and $S_{i_1}$. But $H$ is clearly also homogeneous for $d$. This proves the claim.

To complete the proof, we exhibit a computable instance $c$ of $\mathsf{RT}^n_k$ such that every $c$-computable (hence computable) instance $d$ of $\mathsf{RT}^n_j$ has a solution computing no solution to $c$. Fix a $\Delta^0_n$ sequence $\langle S_0, \ldots, S_{k-1} \rangle$ as in the claim. By iterating Proposition 8.4.5 $n-1$ times, we obtain a computable coloring $c \colon [\omega]^n \to k$ whose infinite homogeneous sets are precisely the infinite subsets of the $S_i$. But by choice of $\langle S_0, \ldots, S_{k-1} \rangle$, every computable $d \colon [\omega]^n \to j$ has an infinite homogeneous set $H$ such that no $H$-computable infinite set is contained in any $S_i$. □

Let us return to fill in the proof of Lemma 9.1.13. The proof uses the Cholak–Jockusch–Slaman decomposition by establishing analogous results for $\mathsf{COH}$ and for $\mathsf{D}^2_j$. Let us begin with the former. Notice that preserving 2 among $k$ hyperimmunities, which is what we wish to establish for $\mathsf{RT}^2_j$, is not a property (like lowness or cone avoidance, for example) that iterates well. Thus, for $\mathsf{COH}$, we actually prove a stronger preservation property.

**Definition 9.1.14 (Patey [243]).** Fix $k \geqslant m \geqslant 1$. A problem $\mathsf{P}$ *admits preservation of hyperimmunity* if for every $A \in 2^\omega$ and every sequence $S_0, S_1, \ldots$ of $A$-hyperimmune sets, every $A$-computable instance $X$ of $\mathsf{P}$ has a solution $Y$ such that the sets $S_0, S_1, \ldots$ is $(A \oplus Y)$-hyperimmune.

**Lemma 9.1.15 (Patey [243]).** $\mathsf{COH}$ *admits preservation of hyperimmunity.*

*Proof.* We prove the unrelativized version. Let $\vec{R} = \langle R_i : i \in \omega \rangle$ be a computable instance of $\mathsf{COH}$ and let $S_0, S_1, \ldots$ be a sequence of hyperimmune sets. Let $G$ be sufficiently generic for Mathias forcing with computable reservoirs. Then $G$ is an infinite $\vec{R}$-cohesive set, and we claim that each of the sets $S_0, S_1, \ldots$ is $G$-hyperimmune. To see this, consider an arbitrary condition $(E, I)$ and arbitrary $e, i \in \omega$. We claim there is an extension $(E^*, I^*)$ of $(E, I)$ forcing that $\Phi^{\mathrm{G}}_e$ does not dominate $p_{B_i}$. Define a partial computable $f \colon \omega \to \omega$ as follows: given $x \in \omega$, let $F$ be the least subset of $I$, if it exists, such that $\Phi^{E \cup F}(x) \downarrow$ (with use bounded by $\max F$), and set $f(x) = \Phi^{E \cup F}(x)$; otherwise, $f(x) \uparrow$.

Now, if $f$ is not total, then $(E, I)$ forces that $\Phi^{\mathrm{G}}_e$ is not total, so we can just set $(E^*, I^*) = (E, I)$. Otherwise, $f$ is a computable function, and since $S_i$ is hyperimmune there must be an $x$ such that $f(x) < p_{B_i}(x)$. In this case, fix $F \subseteq I$ such that $\Phi^{E \cup F}(x) \downarrow < p_{B_i}(x)$, and let $E^* = E \cup F$ and $I^* = \{y \in Iy > \max F\}$. Now $(E^*, I^*)$ is the desired extension. By genericity, no $G$-computable function can dominate $p_{B_i}$ for any $i$, hence each of the sets $B_0, B_1, \ldots$ is $G$-hyperimmune, which is what we wanted to show. □

Now let us consider the situation for $\mathsf{D}^2_j$. As in the alternative proof of Seetapun's theorem given in Section 8.5.1, it is more convenient to work with $\mathsf{RT}^1_j$, and so we prove a "strong" version of the property we wish to preserve. The relevant definition is the following, building on Definition 9.1.12.

**Definition 9.1.16 (Patey [243]).** Fix $k \geqslant m \geqslant 1$ and let $\mathsf{P}$ be a problem. Then $\mathsf{P}$ admits *strong preservation of $m$ among $k$ hyperimmunities* if the above holds for every instance $X$ of $\mathsf{P}$ (computable from $A$ or not).

**Lemma 9.1.17 (Patey [243]).** *For all $k > j \geqslant 1$, $\mathsf{RT}^1_j$ admits strong preservation of 2 among k hyperimmunities.*

*Proof.* The proof is by induction on $j$. If $j = 1$, then the result holds trivially. Indeed, there is only one instance of $\mathsf{RT}^1_1$, the constant coloring $x \mapsto 0$, and this has $\omega$ as a solution. And if $S_0, \ldots, S_{k-1}$ are $A$-hyperimmune then of course they are also $(A \oplus \omega)$-hyperimmune. So fix $j > 1$, and assume the result holds for $j - 1$. Fix $A \in 2^\omega$, a sequence of $A$-hyperimmune sets $S_0, \ldots, S_{k-1}$, and a coloring $c\colon \omega \to j$ (not necessarily $A$-computable).

*Case 1: There is an infinite set $I$, an $\ell < j$, and distinct numbers $i_0, i_1 < k$ such that $c(x) \neq \ell$ for all $x \in I$ and $S_{i_0}$ and $S_{i_1}$ are $(A \oplus X)$-hyperimmune.* By relabeling, we may assume $\ell = j - 1$. Say $I = \{m_0 < m_1 < \cdots\}$. Define $c^*\colon \omega \to j - 1$ as follows: for all $x < y$, $c^*(x) = c(m_x)$. Then $c^* \leqslant_T c \oplus I \leqslant_T A \oplus I$. By inductive hypothesis relative to $A \oplus I$ and using the sequence $S_{i_0}, S_{i_1}$, there is an infinite homogeneous set $H^*$ for $c^*$ such that two among the sets $S_{i_0}$ and $S_{i_1}$ (i.e., both of them) are $(A \oplus I \oplus H^*)$-hyperimmune. Let $H = \{m_x : x \in H^*\}$, which now is an infinite homogeneous set for $c$. Then $H \leqslant_T I \oplus H^*$, so also $S_{i_0}$ and $S_{i_1}$ are $(A \oplus H)$-hyperimmune.

*Case 2: Otherwise.* In this case, we obtain $H$ by forcing. We force with $j$-fold Mathias conditions $(E_0, \ldots, E_{j-1}, I)$ such that each of the sets $S_0, \ldots, S_{k-1}$ is $(A \oplus I)$-hyperimmune, and such that for each $\ell < j$ and $x \in E_\ell$, $c(x) = \ell$. Thus in particular, $(\emptyset, \ldots, \emptyset, \omega)$ is a condition, and a sufficiently generic yields sets $G_0, \ldots, G_{j-1}$ such that $H_\ell$ is homogeneous for $c$ with color $\ell$. Let $\mathsf{G}_0, \ldots, \mathsf{G}_{j-1}$ be names in our forcing language for these generic sets. (As usual, these are really abbreviations for definitions using the symbol $\mathsf{G}$.) We prove two density claims about this notion.

*Claim 1: For each $m \in \omega$ and $\ell < j$, the set of conditions forcing $(\exists x)[x \geqslant m \wedge x \in \mathsf{G}_\ell]$ is dense.* Fix a condition $(E_0, \ldots, E_{j-1}, I)$. Since we are not in Case 1, for each $\ell < j$ there must exist infinitely many $x \in I$ such that $c(x) = \ell$, because $S_0, \ldots, S_{k-1}$ are $(A \oplus I)$-hyperimmune. So given $m$ and $\ell$, fix $x \geqslant m$ in $I$ with $c(x) = \ell$, let $E^*_\ell = E_\ell \cup \{x\}$, $E^*_{\ell^*} = E_{\ell^*}$ for all $\ell^* \neq \ell$, and let $I^* = \{y \in I : y > x\}$. Then $(E^*_0, \ldots, E^*_{j-1}, I^*)$ is an extension of $(E_0, \ldots, E_{j-1}, I)$ witnessing the claim.

*Claim 2: For each $i < k$ and all $e_0, \ldots, e_{j-1} \in \omega$, the collection of conditions forcing*

$$(\exists j < \ell)[\Phi^{A \oplus \mathsf{G}_\ell}_{e_\ell} \text{ does not dominate } p_{S_i}]$$

*is dense.* Fix a condition $(E_0, \ldots, E_{j-1}, I)$. We exhibit an extension that forces, for some $j < \ell$, either that $\Phi^{A \oplus \mathsf{G}_\ell}_{e_\ell}$ is not total, or that there is an $x$ such that $\Phi^{A \oplus \mathsf{G}_\ell}_{e_\ell}(x) \downarrow < p_{S_i}(x)$. For each $x \in \omega$, let $C_x$ be the class of all sets $Z = Z_0 \oplus \cdots \oplus Z_{j-1}$ as follows.

- $Z_0, \ldots, Z_{\ell-1}$ partition $I$.
- For all $\ell < j$ and all finite sets $F \subseteq Z_\ell$, $\Phi^{A \oplus (E_\ell \cup F)}_{e_\ell}(x) \uparrow$.

Notice that this is a $\Pi^0_1(A \oplus I)$ class, with index as such uniform in $x$.

First, suppose there is an $x \in \omega$ such that $C_x \neq \emptyset$. In that case, using the hyperimmune free basis theorem (Theorem 2.8.22) relative to $A \oplus I$, we can find some $Z = Z_0 \oplus \cdots \oplus Z_{j-1} \in C_x$ of hyperimmune free degree relative to $A \oplus I$. This means that every $(A \oplus I \oplus Z)$-computable function is dominated by an $(A \oplus I)$-computable function. Since $S_i$ is $(A \oplus I)$-hyperimmune, it follows that it is also $(A \oplus I \oplus Z)$-hyperimmune, and so in particular $(A \oplus Z_\ell)$-hyperimmune for each $\ell < j$. Fix $\ell < j$ such that $Z_\ell$ is infinite, and set $(E^*_0, \ldots, E^*_{j-1}, I^*) = (E_0, \ldots, E_{j-1}, Z_\ell)$. Then $(E^*_0, \ldots, E^*_{j-1}, I^*)$ is an extension of $(E_0, \ldots, E_{j-1}, I)$ forcing that $\Phi^{A \oplus G_\ell}_{e_\ell}$ is not total.

So suppose next that $C_x = \emptyset$ for every $x$. Let $f \colon \omega \to \omega$ be the function defined as follows. Given $x \in \omega$, it follows by definition of $C_x$ that there exists an $m$ such that for every partition $F_0, \ldots, F_{j-1}$ of $I \upharpoonright m$ (i.e., of a finite set into finite sets) there is an $\ell < j$ and a finite set $F \subseteq F_\ell$ such that $\Phi^{A \oplus (E_\ell \cup F)}_{e_\ell}(x) \downarrow$. Moreover, $m$ can be found $(A \oplus I)$-computably, and so can the values of all of the resulting computation of the form $\Phi^{A \oplus (E_\ell \cup F)}_{e_\ell}(x) \downarrow$ for $F$ contained in some part of some partition of $F_0, \ldots, F_{j-1}$ of $I \upharpoonright m$. Then $f(x)$ is the supremum of all these computations. This make $f$ an $(A \oplus I)$-computable function. Since $S_i$ is $(A \oplus I)$-hyperimmune, we can fix an $x$ such that $f(x) < p_{S_i}(x)$. Now for each $\ell < j$, let $F_\ell = \{x \in I : x < m \wedge c(x) = \ell\}$, so that $F_0, \ldots, F_{j-1}$ partition $I \upharpoonright m$. Fix $\ell < j$ such that there is an $F \subseteq F_\ell$ for which $\Phi^{A \oplus (E_\ell \cup F)}_{e_\ell}(x) \downarrow$, so that $\Phi^{A \oplus (E_\ell \cup F)}_{e_\ell}(x) \leqslant f(x)$ by definition of $f$. Let $E^*_\ell = E_\ell \cup F$, $E^*_{\ell^*} = E_{\ell^*}$ for all $\ell^* \neq \ell$, and let $I^* = \{y \in I : y > \max F\}$. Then $(E^*_0, \ldots, E^*_{j-1}, I^*)$ is an extension of $(E_0, \ldots, E_{j-1}, I)$ forcing that $\Phi^{A \oplus G_\ell}_{e_\ell}(x) < p_{S_i}(x)$.

Having proved our claims, let $G_0, \ldots, G_{\ell-1}$ be sufficiently generic for our notion of forcing. By the first claim, for each $\ell < j$, $G_\ell$ is an infinite homogeneous set for $c$ with color $\ell$. By the second claim, for each $i < k$ and all $e_0, \ldots, e_{j-1} \in \omega$ there is a $j < \ell$ such that $\Phi^{A \oplus G_\ell}_{e_\ell}$ does not dominate $p_{S_i}$. By Lachlan's disjunction (Lemma 8.3.7) it follows that for each $i < k$, there is an $\ell = \ell_i < j$ such that $\Phi^{A \oplus G_\ell}_e$ does not dominate $p_{S_i}$ for any $e \in \omega$, meaning $S_i$ is $(A \oplus G_\ell)$-hyperimmune. But since $k > j$, there must exist distinct $i_0, i_1 < k$ with $\ell_{i_0} = \ell_{i_1}$. So let $H = G_{\ell_{i_0}}$. Then both $S_{i_0}$ and $S_{i_1}$ are $(A \oplus H)$-hyperimmune, which is what was to be shown. □

We are now ready to put everything together to prove Lemma 9.1.13.

*Proof (of Lemma 9.1.13).* Fix $k \geqslant 2$, $j \geqslant 1$, $A \in 2^\omega$, a sequence $S_0, \ldots, S_{k-1}$ of $A$-hyperimmune sets, and an $A$-computable coloring $c \colon [\omega]^2 \to k$. For each $x$ and $i < k$, let $R_{kx+i} = \{y > x : c(x, y) = i\}$, and let $\vec{R}$ be the family of all these sets. Then $\vec{R}$ is an $A$-computable instance of $\mathsf{COH}$. By Lemma 9.1.15, there exists an infinite $\vec{R}$-cohesive set $G$ such that the sets $S_0, \ldots, S_{k-1}$ are $(A \oplus G)$-hyperimmune. As in Proposition 8.4.17, $c \upharpoonright [G]^2$ is a stable coloring of pairs, so we can define $c^* \colon G \to 2$ by $c^*(x) = \lim_{y \in G} c(x, y)$ for all $x \in G$. By Lemma 9.1.17, relative to $A \oplus G$, there is an infinite homogeneous set $H^* H$ for $c^*$ and distinct numbers $i_0, i_1 < k$ such that $S_{i_0}$ and $S_{i_1}$ are $(A \oplus G \oplus H^*)$-hyperimmune. By definition, $H^*$ is limit homogeneous for $c \upharpoonright [G]^2$, and so by Proposition 8.4.2 there is an infinite homogeneous set $H \subseteq H^*$ for $c \upharpoonright [G]^2$ which is computable from $(c \upharpoonright [G]^2) \oplus H^* \leqslant_{\mathrm{T}} c \oplus G \oplus H^* \leqslant_{\mathrm{T}} A \oplus G \oplus H^*$. So $S_{i_0}$ and $S_{i_1}$ are obviously also $(A \oplus H)$-hyperimmune. Since $H$ is homogeneous also for $c$ (not just $c \upharpoonright [G]^2$), the proof is complete. □

### 9.1.3 Homogeneity vs. limit homogeneity

For the final topic of this section, we look at $\mathsf{SRT}^2_2$ and its variant $\mathsf{D}^2_2$. We saw in Proposition 8.4.8 and Corollary 8.4.11 that $\mathsf{SRT}^2_2 \equiv_c \mathsf{D}^2_2$ and that $\mathsf{RCA}_0 \vdash \mathsf{SRT}^2_2 \leftrightarrow \mathsf{D}^2_2$. With respect to the reducibilities studied in Chapter 4, the next ones to consider are $\leqslant_W$ and $\leqslant_{sc}$. And as it turns out, these are able to distinguish the two principles.

**Theorem 9.1.18 (Dzhafarov [80]).**

*1.* $\mathsf{SRT}^2_2 \nleqslant_W \mathsf{D}^2$.
*2.* $\mathsf{SRT}^2_2 \nleqslant_{sc} \mathsf{D}^2$.

One way to interpret this is that the proof of Proposition 8.4.2 (the result that every limit homogeneous set can be computably thinned to a homogeneous one) cannot be made fully uniform, nor can it avoid making essential use of the initial coloring.

*Proof (of Theorem 9.1.18).* We begin with part (1). Fix Turing functionals $\Phi$ and $\Psi$. We build a stable coloring $c\colon [\omega]^2 \to 2$ such that if $\Phi^c$ is a stable coloring $[\omega]^2 \to k$ for some $k \geqslant 2$, then $\Phi^c$ has an infinite limit homogeneous set $L$ such that $\Psi^{c \oplus L}$ is not homogeneous for $c$. In fact, it is possible to build $c$ to be computable, but we will give a forcing argument, as this makes the argument shorter.

Conditions are triples $\langle n, \sigma, f\rangle$ such that $n \in \omega$, $\sigma$ is a function $[n]^2 \to 2$, and $f$ is a function $n \to 2$. A condition $\langle n^*, \sigma^*, f^*\rangle$ extends $\langle n, \sigma, f\rangle$ if $n^* \geqslant n$, $\sigma^* \supseteq \sigma$ and $f^* \supseteq f$ as functions, and for all $x < n$ and all $y$ with $n \leqslant y < n^*$, $\sigma^*(x, y) = f(x)$. The valuation of $\langle n, \sigma, f\rangle$ is simply $\sigma$, as a finite set of codes of ordered triples $\langle x, y, i\rangle$ for $x < y$ and $i < 2$. Every sufficiently generic filter thus determines a stable coloring $G\colon [\omega]^2 \to 2$: for all $x$, $\lim_y G(x, y) = f(x)$ for any condition $\langle n, \sigma, f\rangle$ in the filter with $x < n$.

Given a condition $\langle n, \sigma, f\rangle$ and $i < 2$, let $c_{\sigma,f,i}\colon [\omega]^2 \to 2$ be the coloring defined as follows: for $x < y$,

$$c_{\sigma,f,i}(x, y) = \begin{cases} \sigma(x, y) & \text{if } x, y < n, \\ f(x) & \text{if } x < n \wedge n \leqslant y, \\ i & \text{if } n \leqslant x, y. \end{cases}$$

Our desired coloring $c$ will either be a generic $G$ as above, or else some $c_{\sigma,f,i}$.

*Case 1: There is a condition $\langle \sigma, n, f\rangle$, an $i < 2$, and an infinite low set $I \subseteq \omega$ with the following property: there is no finite set $F \subseteq I$ and no numbers $v > u \geqslant n$ such that $\Psi^{c_{\sigma,f,i} \oplus F}(u) \downarrow = \Psi^{c_{\sigma,f,i} \oplus F}(v) \downarrow = 1$.* In this case, take $c = c_{\sigma,f,i}$. If $\Phi^c$ is a stable coloring, let $L$ be any infinite limit homogeneous set for it contained in $I$. Then by assumption, $\Psi^{c \oplus L}$ cannot be an infinite set.

*Case 2: Otherwise.* If there is no condition forcing that $\Phi^G$ is a stable coloring, we may take $c$ to be any generic $G$, and we are done. Thus, let us fix a condition $\langle n_0, \sigma_0, f_0\rangle$ forcing that $\Phi^G$ is a stable coloring $[\omega]^2 \to k$ for some $k \geqslant 2$. Fix a $\subseteq$-minimal nonempty set $C \subseteq \{0, 1, \dots, k-1\}$ with the property that there is a condition $\langle n_1, \sigma_1, f_1\rangle \leqslant \langle n_0, \sigma_0, f_0\rangle$ and an infinite low set $I \subseteq \omega$ such that $\langle n_1, \sigma_1, f_1\rangle$ forces the following.

- For each $x \in I$, $\lim_y \Phi^G(x, y) \in C$.
- For each $j \in C$ there are infinitely many $x \in I$ with $\lim_y \Phi^G(x, y) = j$.

(Since $I$ is low, it can be defined and hence talked about in our forcing language.) Say $C = \{k_0 < \cdots k_{m-1}\}$, $m > 0$. Consider the class $\mathcal{C}$ of all sets $Z = Z_0 \oplus \cdots \oplus Z_{m-1}$ as follows.

- For each $\ell < m$, there is no finite set $F \subseteq Z_\ell$ and no numbers $v > u \geqslant n_1$ such that $\Psi^{c_{\sigma_1,f_1,0} \oplus F}(u) \downarrow = \Psi^{c_{\sigma_1,f_1,0} \oplus F}(v) \downarrow = 1$.
- $Z_0, \ldots, Z_{m-1}$ partition $I$.

Notice that $\mathcal{C}$ is a $\Pi^0_1(I)$ class. We investigate two subcases.

*Subcase 2a:* $\mathcal{C} \neq \emptyset$. Apply the low basis theorem to find an element $Z = Z_0 \oplus \cdots \oplus Z_{k-1}$ of $\mathcal{C}$ which is low over $I$, and hence low. Fix $\ell < k$ such that $Z_\ell$ is infinite. Now $\langle n_1, \sigma_1, f_1\rangle$, $i = 0$, and $Z_\ell$ witness that we are in Case 1, a contradiction.

*Subcase 2b:* $\mathcal{C} = \emptyset$. By compactness, there is an $s \in \omega$ such that for any partition $Z_0, \ldots, Z_{m-1}$ of $I$ there is a $\ell < m$ and a finite $F \subseteq Z_\ell \upharpoonright s$ such that $\Psi^{c_{\sigma_1,f_1,0} \oplus F}(u) \downarrow = \Psi^{c_{\sigma_1,f_1,0} \oplus F}(v) \downarrow = 1$ for some numbers $v > u \geqslant n_1$. By our use conventions, this also means that $u, v < s$. Let $n_2 = s$, $\sigma_2 = c_{\sigma_1,f_1,0} \upharpoonright [n_2]^2$, and let $f_2 \colon n_2 \to 2$ be defined as follows: for $x < n_1$, $f_2(x) = f_1(x)$, and for $n_1 \leqslant x < n_2$, $f_2(x) = 1$. Note that $\langle n_2, \sigma_2, f_2\rangle \leqslant \langle n_1, \sigma_1, f_1\rangle$. Let $G$ be the stable coloring determined by any sufficiently generic filter containing the condition $\langle n_2, \sigma_2, f_2\rangle$, and let $c = G$. Then $\Phi^c$ is a stable coloring $[\omega]^2 \to k$. By choice of $I$, the set $L_\ell = \{x \in I : \lim_y \Phi^c(x, y) = k_\ell\}$ is infinite for each $\ell < m$, and $L_0, \ldots, L_{m-1}$ partition $I$. Since $L_0 \oplus \cdots \oplus L_{m-1} \notin \mathcal{C}$, we may fix a $\ell < m$ and a finite $F \subseteq L_\ell \upharpoonright s$ such that $\Psi^{c_{\sigma_1,f_1,0} \oplus F}(u) \downarrow = \Psi^{c_{\sigma_1,f_1,0} \oplus F}(v) \downarrow = 1$ for some numbers $v, u$ with $n_1 \leqslant u < v < s$. Since $c$ agrees with $c_{\sigma_1,f_1,0}$ below $s$, which also bounds the use of these computations, it follows that $\Psi^{c \oplus F}(u) \downarrow = \Psi^{c \oplus F}(v) \downarrow = 1$. But by definition, $c(u, v) = 0$ and $\lim_y \Phi^c(u) = \ell_2(u) = 1$. Let $L = F \cup \{x \in L_\ell : x > s\}$. Then $L$ is an infinite limit homogeneous set for $\Phi^c$, and $\Psi^{c \oplus L}$, assuming it is the characteristic function of a set, contains $u$ and $v$ and so cannot be homogeneous for $c$.

This completes the proof of part (1). Part (2) can be proved using the same notion of forcing and a similar diagonalization. □

## 9.2 Partial and linear orders

In this section, we move from colorings to partial and linear orderings. Throughout, "partial order" and "linear order" will refer to orderings of subsets of $\omega$ (or, if we are arguing in a model of $\mathsf{RCA}_0$, of $\mathbb{N}$). We call such an order *finite* or *infinite* if its domain is a finite or infinite set, respectively. For completeness, let us mention that if $(P, \leqslant_P)$ is some partial order, we write: $x <_P y$ if $x \leqslant_P y$ and $x \neq y$; and $x \mid_P y$

if $x \nleqslant_P y$ and $y \nleqslant_P x$. We reserve $\leqslant$ for the usual linear ordering of the natural numbers. Similarly, "least" and "greatest" will always be meant with respect to $\leqslant$. For general partial orders $\leqslant_P$, we will say "$\leqslant_P$-least" and "$\leqslant_P$-greatest", etc.

Recall that if $(P, \leqslant_P)$ is a partial order, then a subset $S$ of $P$ is a *chain* if for all $x, y \in S$, either $x \leqslant_P y$ or $y \leqslant_P x$, and it is an *antichain* if for all $x, y \in S$, $x \mid_P y$. The principles we study here are motivated by a combinatorial result known as Dilworth's theorem. This states that in a finite partial order $(P, \leqslant_P)$, the size of the largest antichain is equal to the size of the smallest partition of $P$ into chains. One version of this for the countably infinite setting is the following.

**Definition 9.2.1 (Chain/antichain principle).** CAC is the following statement: every infinite partial order $(P, \leqslant_P)$ has an infinite chain or antichain.

A related version for linear orders can be obtained using the following definition.

**Definition 9.2.2.** Let $(L, \leqslant_L)$ be a linear order. A subset $S$ of $L$ is:

1. an *ascending sequence* if for all $x, y \in S$, if $x \leqslant y$ then $x \leqslant_L y$,
2. a *descending sequence* if for all $x, y \in S$, if $x \leqslant y$ then $y \leqslant_L x$.

In the parlance of order theory, an ascending sequence is in particular a suborder of order type (i.e., isomorphic to) $\omega$, while a descending sequence is a suborder of order type $\omega^*$ (the reverse of the natural ordering on $\omega$). One point of caution, however, is that these are not equivalent. A suborder of type $\omega$, for example, is simply a subset $P^*$ of $P$ such that every $x \in P^*$ has only finitely many $\leqslant_P$-predecessors in $P^*$. But there is no reason the ordering needs to also respect the natural ordering. We will return to this somewhat subtle distinction at the end of this chapter.

**Definition 9.2.3 (Ascending/descending sequence principle).** ADS is the following statement: every infinite linear order has an infinite ascending sequence or descending sequence.

The computability theoretic content of partial and linear orders has been studied extensively since at least the 1970s and 1980s. In reverse mathematics specifically, it began with the principles CAC and ADS. These were formulated by Hirschfeldt and Shore [152] in their seminal paper on combinatorial principles weaker than Ramsey's theorem for pairs. Notably, except for Ramsey's theorem and logical principles like induction, bounding, etc., these were two of the very first principles found to lie outside the "big five" classification.

### 9.2.1 Equivalences and bounds

We begin with some basic relationships between CAC, ADS, and Ramsey's theorem. First, notice that a partial order $(P, \leqslant_P)$ naturally induces a 2-coloring of $[P]^2$. Namely, one color can be used for $\leqslant_P$-comparable pairs, another for $\leqslant_P$-incomparable pairs. A homogeneous set is then obviously either a chain or an antichain. This witnesses that CAC is identity reducible to $\mathsf{RT}^2_2$ and also the following:

**Proposition 9.2.4 (Hirschfeldt and Shore [152]).** $\mathsf{RCA}_0 \vdash \mathsf{RT}^2_2 \to \mathsf{CAC}$.

A linear order $(L, \leqslant_L)$ also naturally induces a 2-coloring of $[P]^2$, and this can be used to show that ADS is a consequence of $\mathsf{RT}^2_2$ much as above. We will return to this shortly. However, ADS is also a consequence of CAC.

**Proposition 9.2.5 (Hirschfeldt and Shore [152]).** $\mathsf{RCA}_0 \vdash \mathsf{CAC} \to \mathsf{ADS}$.

*Proof.* We prove that ADS is identity reducible to CAC. The result is just a formalization of this argument. Fix a linear order $(L, \leqslant_L)$. Define a partial ordering $\leqslant_P$ of $L$ as follows: for $x < y$ in $L$, set $x \leqslant_P y$ if $x \leqslant_L y$, and set $x \mid_P y$ if $y \leqslant_L x$. Thus, $\leqslant_P$ tests whether or not $x$ and $y$ are ordered the same or opposite way as they under the natural ordering. Clearly, $\leqslant_P$ is computable from $\leqslant_L$. It is easy to check that $\leqslant_P$ is indeed a partial order. Reflexivity and antisymmetry are obvious. For transitivity, if $x \leqslant_P y$ and $y \leqslant_P z$ then it must be that $x \leqslant y \leqslant z$ and $x \leqslant_L y \leqslant_L z$, so also $x \leqslant z$ and $x \leqslant_L z$, and therefore $x \leqslant_P z$. Now, let $S = \{x_0 < x_1 < \cdots\} \subseteq L$ be an infinite chain or antichain for $\leqslant_P$. If $S$ is a chain then we must have $x_0 <_P x_1 <_P \cdots$ and therefore also $x_0 <_L x_1 <_L \cdots$. Thus, $S$ is an ascending sequence for $\leqslant_L$. If $S$ is an antichain, then we must have $x_0 >_L x_1 >_L \cdots$, so $S$ is an infinite descending sequence for $\leqslant_L$. □

Testing whether a pair of numbers is or is not ordered the same way as under the natural ordering turns out to be very useful, and is encountered repeatedly in the context of CAC and ADS. We could use it, for example, to give an alternative proof of Proposition 9.2.4. Namely, given a partial order $(P, \leqslant_P)$, define a 3-coloring $c_{(P,\leqslant_P)}$ of $[P]^2$ as follows: for $x < y$ in $P$, let

$$c_{(P,\leqslant_P)}(x,y) = \begin{cases} 0 & \text{if } y \leqslant_P x, \\ 1 & \text{if } x \leqslant_P y, \\ 2 & \text{if } x \mid_P y. \end{cases}$$

An infinite homogeneous set for $c_{(P,\leqslant_P)}$ is again a CAC-solution to $\leqslant_P$. And if we instead start with a linear order $(L, \leqslant_L)$, then $c_{(L,\leqslant_L)}$ will be a 2-coloring (the color 2 will never be used), and an infinite homogeneous set for $c_{(L,\leqslant_L)}$ will then be an ADS-solution to $\leqslant_L$.

It turns out that such colorings can be used to completely characterize CAC and ADS in terms of restrictions of Ramsey's theorem for pairs. The key observation is that these colorings behave like partial/linear orders, in the following sense.

**Definition 9.2.6.** Fix $k \geqslant 2$ and a coloring $c\colon [\omega]^2 \to k$.

1. $c$ is *semi-transitive* if for all but at most one $i < k$, if $c(x, y) = c(y, z) = i$ then $c(x, z) = i$.
2. $c$ is *transitive* if for all $i < k$, if $c(x, y) = c(y, z) = i$ then $c(x, z) = i$.

If $(P, \leqslant_P)$ is a partial order, then the coloring $c_{(P,\leqslant_P)}$ is easily seen to be semi-transitive (the only color $i < 3$ that need not satisfy the transitive property is $i = 2$). So if $(L, \leqslant_L)$ is a linear order, the coloring $c_{(L,\leqslant_L)}$ is transitive.

**Proposition 9.2.7 (Hirschfeldt and Shore [152]).** *The following are provable in* $\mathsf{RCA}_0$.

1. CAC *is equivalent to the restriction of* $\mathsf{RT}^2_2$ *to semi-transitive colorings.*
2. ADS *is equivalent to the restriction of* $\mathsf{RT}^2_2$ *to transitive colorings.*

*Proof.* The proof of (1) is divided into two parts: first, that the restriction of $\mathsf{RT}^2_2$ to semi-transitive colorings is identity reducible to CAC; and second, that CAC is identity reducible to the restriction of $\mathsf{RT}^2_3$ to semi-transitive colorings. It is straightforward to formalize these arguments over $\mathsf{RCA}_0$, and equally straightforward (using basically the same proof as Corollary 4.6.12) to show that for all $k \geqslant 2$, the restrictions of $\mathsf{RT}^2_2$ and $\mathsf{RT}^2_k$ to semi-transitive colorings are equivalent over $\mathsf{RCA}_0$. To prove the reductions, first fix a semi-transitive coloring $c\colon [\omega]^2 \to 2$. Say $c(x, y) = c(y, z) = 1 \to c(x, z) = 1$, for all $x < y < z$. Define a $c$-computable partial ordering $\leqslant_P$ of $\omega$ as follows: for $x < y$, set $x \leqslant_P y$ if $c(x, y) = 1$. Now every infinite chain for $\leqslant_P$ is an infinite homogeneous set for $c$ with color 1, and every infinite antichain for $\leqslant_P$ is an infinite homogeneous set for $c$ with color 0. Next, let $(P, \leqslant_P)$ be an instance of CAC. The coloring $c_{(P,\leqslant_P)}\colon [P]^2 \to 3$ defined above is computable from $(P, \leqslant_P)$. And as we noted, it is semi-transitive, and every infinite homogeneous set for it is an infinite chain or antichain for $\leqslant_P$.

Now, to prove (2), we prove that each of ADS and the restriction of $\mathsf{RT}^2_2$ to transitive colorings is identity reducible to the other. Fix a transitive coloring $c\colon [\omega]^2 \to 2$. We construct a $c$-computable linear ordering $\leqslant_L$ by stages. At stage $s \in \omega$, we define $\leqslant_L$ on $\omega \upharpoonright s + 1$. Thus, at stage 0, we just define $0 \leqslant_L 0$. Now take $s > 0$ and assume that $\leqslant_L$ has been defined on $\omega \upharpoonright s$. If there is an $x < s$ such that $c(x, s) = 1$, choose the $\leqslant_L$-largest such $x$ and define $y <_L s$ for all $y \leqslant_L x$. Define $s <_L y$ for all other $y < s$. If no such $x$ exists, meaning $c(x, s) = 0$ for all $x < s$, then define $s <_L x$ for all $x < s$. In either case, define $s \leqslant_L s$. This completes the construction.

It is clear from that $\leqslant_L$ is indeed a linear order. (New elements are always added to the linear order either above all previously added elements, below all previously added elements, or between two previously ordered elements.) Consider any ADS-solution $S$ for $\leqslant_L$, and suppose first that $S$ is ascending. We claim that $S$ is homogeneous for $c$ with color 1. If not, fix the least $y \in S$ such that there is a $w$ in $S$ with $y < w$ (and hence also $y <_L w$) and $c(y, w) = 0$. Let $s$ be the least such $w$ for our fixed $y$. For us to have ordered $y <_L s$ at stage $s$ of the construction, there must be an $x < s$ such that $y \leqslant_L x$ and $c(x, s) = 1$. Obviously, $y \neq x$. If $x < y$ then for us to have ordered $y \leqslant_L x$ we must have $c(x, y) = 0$. But then since $c(y, s) = 0$, transitivity of $c$ implies $c(x, s) = 0$, a contradiction. So it must be that $y < x$. If $c(y, x) = 1$ then since $c(x, s) = 1$, transitivity of $c$ implies $c(y, s) = 1$, a contradiction. Thus, we must have $c(y, x) = 0$. But now $x$ can serve as a smaller witness $w$ above than $s$, contradicting the choice of $s$.

To complete the proof, suppose $S$ is descending. Then it follows readily from the construction that $S$ is homogeneous for $c$ with color 0. For the converse, let $(L, \leqslant_L)$ be an instance of ADS. Then the coloring $c_{(L,\leqslant_L)}\colon [L]^2 \to 2$ is computable from $(L, \leqslant_L)$ and every infinite homogeneous set for it is an infinite ascending or descending sequence for $\leqslant_L$. □

A basic question now is how "close" ADS and CAC are to each other, and of course, to $\mathsf{RT}^2_2$. For starters, these principles are not computably true. In the case of CAC, Herrmann [142] showed that there is a computable partial ordering of $\omega$ with no infinite $\Sigma^0_2$ chain or antichain. In the case of ADS, Tennenbaum (see Rosenstein [260]) and Denisov (see Goncharov and Nurtazin [127]) independently constructed a computable linear order with no computable suborder of order type $\omega$ or $\omega^*$. In our parlance, this implies there is a computable instance of each of CAC and ADS with no computable solution. The following provides a stronger lower bound.

**Theorem 9.2.8 (Hirschfeldt and Shore [152]).** $\mathsf{RCA}_0 \vdash \mathsf{ADS} \to \mathsf{COH}$.

*Proof.* Arguing in $\mathsf{RCA}_0$, fix $\vec{R} = \langle R_i : i \in \omega\rangle$, an instance of COH. For each $x \in \mathbb{N}$, let $\sigma_x \in 2^{<\mathbb{N}}$ be the string of length $x$ defined by: $\sigma_x(i) = R_i(x)$ for all $i < x$. (So the set of $\sigma_x$ exists, as does the map $x \mapsto \sigma_x$.) Define a linear ordering $\leqslant_L$ of $\mathbb{N}$ as follows: $x \leqslant_L y$ if and only if $\sigma_x$ lexicographically precedes $\sigma_y$, i.e., $\sigma_x \preceq \sigma_y$ or

$$(\exists i < x)(\forall j < i)[\sigma_x(j) = \sigma_y(j) \land \sigma_x(i) < \sigma_y(i)].$$

Clearly, $\leqslant_L$ exists and as such is an instance of ADS. Fix any ADS-solution, $S$, to this instance, and suppose first that it is ascending. We claim that $S$ is $\vec{R}$-cohesive. Fix $i$ and consider the set

$$M = \{\sigma \in 2^{i+1} : (\exists y \in S)[\sigma \text{ lexicographically precedes } \sigma_y]\}.$$

This is $\Sigma^0_1$-definable and bounded, and so exists by $\mathsf{I}\Sigma^0_1$. Hence, we can fix the lexicographically largest $\sigma \in M$, say with witness $y \in S$. Since $S$ is ascending, $\sigma_y$, and therefore also $\sigma$, must lexicographically precede $\sigma_x$ for all $x > y$ in $S$. Then for all sufficiently large $x \in S$, we necessarily have $\sigma \preceq \sigma_x$, as otherwise $\sigma_x \upharpoonright i+1$ would be an element of $M$ (with witness $x$) lexicographically larger than $\sigma$. Hence, for all such $x$, we have by definition that $R_i(x) = \sigma_x(i) = \sigma(i)$, so either $S \subseteq^* R_i$ or $S \subseteq^* \overline{R_i}$. Since $i$ was arbitrary, we conclude that $S$ is $\vec{R}$-cohesive. This completes the proof in the case that $S$ is ascending. If $S$ is descending, we interchange "lexicographically precedes" with "lexicographically succeeds" and "lexicographically largest" with "lexicographically least", and then the proof is the same. □

Note that, in terms of finer reducibilities, what we have actually shown is that COH is identity reducible to ADS.

### 9.2.2 Stable partial and linear orders

As with Ramsey's theorem for pairs, we can try to understand CAC and ADS better by dividing each into simpler forms, analogously to the Cholak–Jockusch–Slaman decomposition. We have already remarked that this idea has proved fruitful for many principles, but here is our first example of how this is done outside of Ramsey's theorem proper. The starting point is an analogue of "limit color" in the setting of partial/linear orders. For a partial order $(P, \leqslant_P)$, recall the induced coloring $c_{(P,\leqslant_P)}$ discussed in the previous section. If *this* coloring is stable, then considering its limit colors leads to the following definitions.

**Definition 9.2.9.** Let $(P, \leqslant_P)$ be an infinite partial order. Then $x \in P$ is:

1. $\leqslant_P$-*small* if $x \leqslant_P y$ for almost all $y \in P$,
2. $\leqslant_P$-*large* if $y \leqslant_P x$ for almost all $y \in P$,
3. $\leqslant_P$-*isolated* if $x \mid_P y$ for almost all $y \in P$.

Given $S \subseteq P$, say $x \in S$ is $\leqslant_P$-small, $\leqslant_P$-large, or $\leqslant_P$-isolated *in* $S$ if $x \leqslant_P y$, $y \leqslant_P x$, or $x \mid_P y$, respectively, for almost all $y \in S$.

When $\leqslant_P$ is fixed, we may call $x$ as above just *small*, *large*, or *isolated*.

The obvious idea is to call a partial or linear order *stable* just in case every element is small, large, or isolated (i.e., precisely if the induced coloring above is stable). For partial orders, we also consider an alternative, stronger notion of stability.

**Definition 9.2.10 (Stable partial and linear orders).**

1. An infinite partial order is *weakly stable* if every element is small, large, or isolated.
2. An infinite partial order is *stable* if either every element is small or isolated, or every element is large or isolated.
3. An infinite linear order is *stable* if every element is either small or large.
4. We define the following principles:
   - WSCAC is the restriction of CAC to weakly stable partial orders.
   - SCAC is the restriction of CAC to stable partial orders.
   - SADS is the restriction of ADS to stable linear orders.

Therefore, a stable linear order is one of order type $\omega + \omega^*$, as defined in Section 6.5.

Unfortunately, the above terminology can be a bit confusing. A stable linear order is necessarily weakly stable as a partial order, but it is not necessarily stable as a partial order. Aesthetically, it would maybe make more sense to re-name "weakly stable" by "stable", and "stable" by "strongly stable". But the above terminology is well established in the literature, and actually, for most intents and purposes the distinction is inconsequential. Most importantly, we have the following theorem due to Jockusch, Kastermans, Lempp, Lerman, and Solomon [171].

**Theorem 9.2.11.** $\mathsf{RCA}_0 \vdash \mathsf{SCAC} \leftrightarrow \mathsf{WSCAC}$.

*Proof.* Obviously, WSCAC $\to$ SCAC (every stable partial order is weakly stable.) To show that SCAC $\to$ WSCAC, we argue in $\mathsf{RCA}_0$. Let $(P, \leqslant_P)$ be an infinite weakly stable partial order. Define a new partial ordering $\leqslant_P^*$ of $P$ as follows: for all $x, y$ in $P$, set $x \leqslant_P^* y$ if and only if $x \leqslant_P y$ and $x \leqslant y$.

It is easy to check that this is indeed a partial order. Note that if $x \in P$ is $\leqslant_P$-small then it is also $\leqslant_P^*$-small. And if $x$ is $\leqslant_P$-large or $\leqslant_P$-isolated, then it is $\leqslant_P^*$-isolated. Thus, by weak stability of $\leqslant_P$, every $x \in P$ is either $\leqslant_P^*$-small or $\leqslant_P^*$-isolated, so

$(P, \leqslant_P^*)$ is stable. Apply SCAC to find an infinite $S \subseteq P$ which is either a chain or antichain for $\leqslant_P^*$. In the former case, $S$ is also clearly a chain for $\leqslant_P$. In the latter case, $S$ must consist entirely of $\leqslant_P$-large or $\leqslant_P$-isolated elements. Hence, $(S, \leqslant_P)$ is a stable partial order. Apply SCAC to find an infinite $S^* \subseteq S \subseteq P$ which is either a chain or antichain for $\leqslant_P$, and the proof is complete. □

For now, then, we will stick to SCAC, as it tends to be slightly easier to work with. But we will revisit WSCAC, and its relationship to SCAC, at the end of this chapter, when we consider finer reducibilities.

**Proposition 9.2.12 (Hirschfeldt and Shore [152]).** *The following are provable in* $\mathsf{RCA}_0$.

1. SCAC *is equivalent to the restriction of* $\mathsf{SRT}^2_2$ *to semi-transitive colorings.*
2. SADS *is equivalent to the restriction of* $\mathsf{SRT}^2_2$ *to transitive colorings.*
3. $\mathsf{RCA}_0 \vdash \mathsf{SRT}^2_2 \to \mathsf{SCAC} \to \mathsf{SADS} \to \mathsf{B}\Sigma^0_2$.

*Proof.* For (1) and (2), the proof is identical to that of Proposition 9.2.7. All that is needed is to observe that if $(P, \leqslant_P)$ is a stable partial order then the induced coloring $c_{(P,\leqslant_P)}$ is a stable. Our (stronger) notion of stability also implies that $c_{(P,\leqslant_P)}$ is a 2-coloring (in fact, it uses only the colors 0 and 2 or 1 and 2). Part (3) is obvious except for the last part, that $\mathsf{SADS} \to \mathsf{B}\Sigma^0_2$. This proceeds by showing that SADS implies the principle PART from Definition 6.5.7; the details are left to the exercises. □

Like $\mathsf{SRT}^2_2$, each of SCAC and SADS admits $\Delta^0_2$ solutions. This follows from parts (1) and (2) above, and noting that the equivalences there are actually computable reductions (in fact, identity reductions). But it can also be seen directly. First, note that given a stable partial or linear order, determining whether an element of the domain is small, large, or (if the order is partial) isolated, is computable in the jump of the order. This is just like determining an element's limit color under a stable coloring of pairs (see Remark 8.4.3).

Hence, the jump can compute an analogue of an infinite limit homogeneous set, i.e., an infinite set of elements, all of which are small, or all of which are large, or all of which are isolated. Just like in Proposition 8.4.2, this limit homogeneous set can then be computably thinned to an actual solution for the order—a chain or antichain in the case of SCAC, and an ascending or descending sequence in the case of SADS. See Exercise 9.13.2 to work this out in detail.

To obtain decompositions, we now need to formulate suitable analogues of COH. It is not necessarily obvious how to do this, since COH is not *prima facie* any kind of restriction of Ramsey's theorem. However, we can formulate these analogues in a canonical way (if not particular interesting, from a combinatorial perspective), as in the equivalence given by Exercise 8.10.7. We can simply let the "cohesive version" be the statement that each instance is stable on some infinite subdomain. As it turns out, this is not a new principle in the case of CAC.

**Proposition 9.2.13 (Hirschfeldt and Shore [152]).** *Over* $\mathsf{RCA}_0$, $\mathsf{ADS}$ *is equivalent to the statement that for every infinite partial order* $(P, \leqslant_P)$ *there is an infinite set* $S \subseteq P$ *such that* $(S, \leqslant_P)$ *is stable.*

*Proof.* We argue in $\mathsf{RCA}_0$. Let an infinite partial order $(P, \leqslant_P)$ be given. By Exercise 9.13.1, there exists a linear order $\leqslant_L$ of $P$ extending $\leqslant_P$ (so $x \leqslant_P y \to x \leqslant_L y$, for all $x, y \in P$). By $\mathsf{ADS}$, we may fix an infinite ascending or descending sequence $S \subseteq P$ for $\leqslant_L$. Now define a family $\vec{R} = \{R_x : x \in \omega\}$ of subsets of $S$ by setting, for each $x \in S$,

$$R_x = \{y \in S : x < y \wedge x \mid_P y\}.$$

Since $\mathsf{ADS}$ implies $\mathsf{COH}$, we may invoke Exercise 8.10.5 to find an infinite $\vec{R}$-cohesive set $X \subseteq S$. Let $\leqslant_X$ be the restriction of $\leqslant_P$ to $X$. Then $(X, \leqslant_X)$ is stable. Indeed, consider any $x \in X$. If $X \subseteq^* R_x$ then $x \mid_P y$ for almost all $y \in X$. Hence, $x$ is $\leqslant_X$-isolated. On the other hand, suppose $X \subseteq^* \overline{R_x}$, which means $x \leqslant_P y$ or $y \leqslant_P x$ for almost all $y$. If $S$ is ascending then $x <_L y$ for all $y > x$ in $S$, so since $\leqslant_L$ extends $\leqslant_P$, we must have $x <_P y$ for almost all $y \in X$. Thus, $x$ is $\leqslant_X$-small. And if $S$ is descending, then we analogously get that $x$ is $\leqslant_X$-large. Since this depends only on $S$ and not on $x$, we conclude that either every element of $X$ is $\leqslant_X$-isolated or $\leqslant_X$-small, or every element of $X$ is $\leqslant_X$-isolated of $\leqslant_X$-large. □

For $\mathsf{ADS}$, we formulate a separate principle.

**Definition 9.2.14.** $\mathsf{CADS}$ is the following statement: for every infinite linear order $(L, \leqslant_L)$ there is an infinite set $S \subseteq L$ such that $(S, \leqslant_L)$ is stable.

The following proposition summarizes the basic relationships between the "cohesive versions". Of note is that by parts (3) and (4), $\mathsf{CADS}$ and $\mathsf{COH}$ are equivalent modulo $\mathsf{B}\Sigma^0_2$ (and hence, each is also equivalent in the same way to the principle $\mathsf{CRT}^2_2$ defined in Exercise 8.10.7.)

**Proposition 9.2.15 (Hirschfeldt and Shore [152]).**

*1.* $\mathsf{RCA}_0 \vdash \mathsf{CAC} \leftrightarrow \mathsf{SCAC} + \mathsf{ADS}$.
2. $\mathsf{RCA}_0 \vdash \mathsf{ADS} \leftrightarrow \mathsf{SADS} + \mathsf{CADS}$.
*3.* $\mathsf{RCA}_0 \vdash \mathsf{ADS} \to \mathsf{COH} \to \mathsf{CADS}$.
*4.* $\mathsf{RCA}_0 + \mathsf{B}\Sigma^0_2 \vdash \mathsf{CADS} \to \mathsf{COH}$.

*Proof.* Parts (1) and (2) are immediate by Proposition 9.2.13 and the definitions. The first implication in part (3) is Theorem 9.2.8. The proof of the second implication is similar to that of Proposition 9.2.13.

Finally, for (4), we proceed as in the proof of Theorem 9.2.8. Given an instance $\vec{R} = \langle R_i : i \in \omega \rangle$, define a linear order $\leqslant_L$ as there. Applying $\mathsf{CADS}$ (instead of $\mathsf{ADS}$, this time), we get an infinite set $S$ such that $\leqslant_L$ is stable on $S$. If there are only finitely many $\leqslant_L$-small or finitely many $\leqslant_L$-large elements in $S$, then using $\mathsf{B}\Sigma^0_2$ we can thin $S$ to an infinite ascending or descending sequence for $\leqslant_L$. (See Exercise 9.13.2.) We can then argue as in Theorem 9.2.8 that this is cohesive for $\vec{R}$.

So assume there are infinitely many $\leqslant_L$-small and $\leqslant_L$-large elements in $S$. By $\mathsf{I}\Sigma^0_1$, $S$ has a $\leqslant_L$-least element, $0_L$, as well as a $\leqslant_L$-largest element, $1_L$. Now fix $i$, and let

$$M = \{\sigma \in 2^{i+1} : (\exists y \in S)[\sigma = \sigma_y \upharpoonright i+1]\}.$$

(Recall that $\sigma_y$ is the string of length $y$ such that $\sigma_y(i) = R_i(y)$ for all $i < y$.) The set $M$ exists by bounded $\Sigma^0_1$ comprehension. Say $M$ has $b$ many elements. So $b \leqslant 2^{i+1}$, but also $b > 0$: since $S$ is infinite, we may take any $y > i$ in $S$, and then $\sigma = \sigma_y \upharpoonright i+1$ belongs to $M$. For each $\sigma \in M$, let $y_\sigma$ be the least $y \in S$ such that $\sigma = \sigma_y \upharpoonright i+1$, which also exists by $\mathsf{I}\Sigma^0_1$. Let $y_0 <_L \cdots <_L y_{b-1}$ list all elements $S$ of the form $y_\sigma$ for $\sigma \in M$. For completeness, let $y_{-1} = 0_L$ and $y_b = 1_L$. (Note that it could be that $y_{-1} = y_0$ or $y_b = y_{b-1}$.)

We next define a coloring $c \colon \mathbb{N} \to b+1$ as follows: given $x \in \mathbb{N}$, let $c(x)$ be the least $j < b+1$ such that $x \leqslant_L y_j$. Since $x \leqslant_L y_b$, the existence of $c(x)$ follows by $\mathsf{I}\Sigma^0_1$, so $c$ is well-defined. By $\mathsf{RT}^1$ (which follows from $\mathsf{B}\Sigma^0_2$) we may fix $j < b+1$ such that $c(x) = j$ for infinitely many $x \in \mathbb{N}$. By definition of $c$, all such $x$ satisfy $y_{j-1} <_L x \leqslant_L y_j$. Hence, $y_{j-1}$ is $\leqslant_L$-small in $S$ and $y_j$ is $\leqslant_L$-large in $S$, and we actually have that $c(x) = j$ for almost all $x \in S$. In particular, almost all $x \in S$ satisfy $x > i$ and $y_{j-1} <_L x <_L y_j$.

We claim that for any such $x$, either $\sigma_{y_{j-1}} \preceq \sigma_x$ or $\sigma_x \preceq \sigma_{y_j}$. Indeed, by definition of $\leqslant_L$, $\sigma_{y_{j-1}}$ lexicographically precedes $\sigma_x$, which lexicographically precedes $\sigma_{y_j}$. And since $x > i$, $\sigma = \sigma_x \upharpoonright i+1 \in M$. So if $\sigma_{y_{j-1}} \npreceq \sigma_x$ and $\sigma_x \npreceq \sigma_{y_j}$, then we would have that $y_{j-1} <_L y_\sigma <_L y_j$. But this is impossible by how the numbering $y_0, \ldots, y_b$ was chosen. So the claim holds.

Now by the preceding claim, either $\sigma_{y_{j-1}} \preceq \sigma_x$ for almost all $x \in S$, or $\sigma_x \preceq \sigma_{y_j}$ for almost all $x \in S$. In the former case, we have that $R_i(x) = \sigma_x(i) = \sigma_{j-1}(i)$ for almost all $x \in S$, and in the latter we have that $R_i(x) = \sigma_x(i) = \sigma_j(i)$ for almost all $x \in S$. Thus, either $S \subseteq^* R_i$ or $S \subseteq^* \overline{R_i}$. Since $i$ was arbitrary, we conclude that $S$ is $\vec{R}$-cohesive. □

### 9.2.3 Separations over $\mathsf{RCA}_0$

Having established implications between CAC, ADS, and their stable and cohesive variants, let us turn to nonimplications. To begin, we have the following.

**Proposition 9.2.16 (Hirschfeldt and Shore [152]).** $\mathsf{RCA}_0 \nvdash \mathsf{COH} \to \mathsf{SADS}$.

*Proof.* On the one hand, $\mathsf{SADS} \to \mathsf{B}\Sigma^0_2$ by Proposition 9.2.12. On the other, COH is $\Pi^1_1$ conservative over $\mathsf{RCA}_0$ by Theorem 7.7.1 and therefore does not imply $\mathsf{B}\Sigma^0_2$. Hence, COH does not imply SADS over $\mathsf{RCA}_0$. □

In fact, Wang [322], has established also the stronger fact that $\mathsf{SADS} \nleqslant_\omega \mathsf{COH}$. The proof, which we omit, uses a preservation property based on definability (see Definition 9.5.5). We will see a somewhat similar idea used in the proof of Theorem 9.2.22 below.

Another pair of principles we can separate immediately is SADS and WKL. Of course, SADS does not imply WKL over $\mathsf{RCA}_0$ since even $\mathsf{RT}^2_2$ does not. The failure of the converse follows from the fact that SADS implies $\mathsf{B}\Sigma^0_2$ while, by Harrington's theorem (Theorem 7.7.3), WKL is $\Pi^1_1$ conservative over $\mathsf{RCA}_0$. We can improve this with an $\omega$-model separation.

**Proposition 9.2.17 (Csima and Mileti [59]).** SADS *omits hyperimmune free solutions.*

*Proof.* The aforementioned result of Tennenbaum and Denisov, that there is a computable linear order with no computable suborder of order type $\omega$ or $\omega^*$, is actually of order type $\omega + \omega^*$. It is easy to computably thin the domain of this order to produce a computable instance $(L, \leqslant_L)$ of SADS with no computable solution. We claim that, in fact, every solution to $(L, \leqslant_L)$ is hyperimmune. Suppose not. Let $S \subseteq L$ be any solution, and let $f$ be a computable function dominating $p_S$. Then for all $x$ we have $p_S(x) \leqslant f(x) \leqslant p_S(f(x))$, so in particular, $f(x)$ is $\leqslant_L$-small or $\leqslant_L$-large depending as $S$ is ascending or descending. But then, using $f$, we can find an infinite computable set either of $\leqslant_L$-small or $\leqslant_L$-large elements, and this can then be computably thinned using Exercise 9.13.2 to a computable solution to $(L, \leqslant_L)$. □

**Corollary 9.2.18.** SADS $\nleqslant_\omega$ WKL. *Therefore,* WKL *and* SADS *are incomparable over* $\mathsf{RCA}_0$.

*Proof.* By Exercise 4.8.13, there is an $\omega$-model of WKL consisting entirely of sets that have hyperimmune free degree. By the preceding proposition, SADS is false in this model. □

Although each of $\mathsf{SRT}^2_2$, SCAC, and SADS admit $\Delta^0_2$ solutions (as noted above), a surprising point of difference is that SCAC and SADS actually also admit low solutions. Recall that, by Theorem 8.8.4, this is not the case for $\mathsf{SRT}^2_2$.

**Theorem 9.2.19 (Hirschfeldt and Shore [152]).** SCAC *admits low solutions.*

*Proof.* We prove the result for computable stable partial orders. The full result follows by relativization. So suppose $(P, \leqslant_P)$ is computable and stable, say with every element small or isolated. (The case where every element is large or isolated is symmetric.) If there is a computable infinite antichain for $\leqslant_P$ then we are done, so suppose otherwise. Call a string $\alpha \in \omega^{<\omega}$ a *precondition* if the following hold.

- For all $i < j < |\alpha|$, $\alpha(i) < \alpha(j)$.
- For all $i < j < |\alpha|$, $\alpha(i) <_P \alpha(j)$.

Now, consider the notion of forcing whose conditions are strings $\alpha \in P^{<\omega}$ such that $\alpha$ is a precondition and also $\alpha(|\alpha| - 1)$ is $\leqslant_P$-small (and hence so is $\alpha(i)$ for all $i < |\alpha|$). Extension is as usual for strings, and the valuation of a condition $\alpha$ is simply its range as a subset of $\omega$. (Note that this is a $\varnothing'$-computable forcing.) A sufficiently generic set $G$ is thus a chain for $\leqslant_P$ (in fact, an ascending sequence). Clearly, for each $n$, the set of conditions $\alpha$ with $|\alpha| > n$ is uniformly $(P, \leqslant_P)$-effectively dense, as otherwise almost every element of $P$ would be $\leqslant_P$-isolated, which would then contradict our assumption by Exercise 9.13.2. We next argue that for each $e \in \omega$, the set of conditions that decide whether $e \in \mathsf{G}'$ is uniformly $\varnothing'$-effectively dense. To see this, fix $e$ and a condition $\alpha$. We search for a condition $\alpha^* \geq \alpha$ such that one of the following holds.

1. $\Phi_e^{\alpha^*}(e) \downarrow$.
2. $\Phi_e^{\beta}(e) \uparrow$ for every precondition $\beta \geq \alpha$.

Since the set of preconditions (as opposed to conditions) is computable, the search can be performed using $\varnothing'$. We claim it must succeed. Assume not. For every condition $\alpha^* \geq \alpha$ there is a precondition $\beta \geq \alpha^*$ such that $\Phi_e^{\beta}(e) \downarrow$, else (2) would hold. Moreover, no such $\beta$ can be a condition, else (1) would hold, so $\beta(|\beta| - 1)$ must be $\leqslant_P$-isolated. Let $S^*$ be the set of all numbers of the form $\beta(|\beta| - 1)$ for $\beta$ a precondition extending $\alpha$ with $\Phi_e^{\beta}(e) \downarrow$. Then $S^*$ is an infinite c.e. set of $\leqslant_P$-isolated elements, and hence has an infinite computable subset $S$. Now by Exercise 9.13.2, there is an infinite computable antichain for $\leqslant_P$ contained in $S$, which contradicts our assumption. So the claim holds, and we must find the desired $\alpha^*$. If (1) holds, then $\alpha^* \Vdash e \in \mathsf{G}'$. If (2) holds, then $\alpha^* \Vdash e \notin \mathsf{G}$, as every conditions is a precondition. This proves that the set of conditions that decide whether $e \in \mathsf{G}'$ is uniformly $\varnothing'$-effectively dense. As usual, we can now take a generic $G$ according to Theorem 7.5.6 and conclude that $G' \leqslant_{\mathrm{T}} \varnothing'$. □

**Corollary 9.2.20.** $\mathsf{SRT}^2_2 \nleqslant_\omega \mathsf{SCAC}$. *Hence also* $\mathsf{RCA}_0 \nvdash \mathsf{SCAC} \to \mathsf{SRT}^2_2$.

Next, recall the problem DNR, which we defined in Definition 4.3.9 and saw again in the proof of Theorem 8.8.2.

**Theorem 9.2.21 (Hirschfeldt and Shore [152]).** $\mathsf{DNR} \nleqslant_\omega \mathsf{CAC}$. *Hence also* $\mathsf{RCA}_0 \nvdash \mathsf{CAC} \to \mathsf{DNR}$.

*Proof.* Let $C$ be the class of all sets $Y$ such that $Y$ computes no DNC function. Clearly, $C$ is closed downward under $\leqslant_{\mathrm{T}}$. We show that CAC admits preservation of $C$. Of course, by definition, DNR does not admit preservation of $C$ (with witness $\varnothing$). The result then follows by Theorem 4.6.13.

We begin by showing that SCAC admits preservation of $C$. That is, given a set $A$ computing no DNC function, and an $A$-computable instance of SCAC, i.e., an $A$-computable infinite stable partial order $(P, \leqslant_P)$, we exhibit an infinite chain or antichain for $(P, \leqslant_P)$ whose join with $A$ computes no DNC function. As usual, we take $A = \varnothing$ for ease of notation, with the full result following by relativization. First, if $\leqslant_P$ has an infinite computable antichain, then we are done, so assume not. In particular, there are infinitely many $\leqslant_P$-small elements in $P$. Say every element of $P$ is either $\leqslant_P$-small or $\leqslant_P$-isolated, the other case being symmetric. Consider the notion of forcing from Theorem 9.2.19, as well as the notion of precondition defined there. We claim that for every Turing functional $\Gamma$, the set of conditions $\alpha$ forcing, for some $e$, either that $\Gamma^{\mathsf{G}}(e) \uparrow$ or that $\Gamma^{\mathsf{G}}(e) \downarrow = \Phi_e(e) \downarrow$, is dense. Any sufficiently generic $G$ such that $\Gamma^G$ is total will consequently satisfy that there is an $e$ such that $\Gamma^G(e) \downarrow = \Phi_e(e) \downarrow$, so $\Gamma^G$ will not be a DNC function.

To prove the claim, fix a Turing functional $\Gamma$ and a condition $\alpha$. If there is an $e$ and a condition $\alpha^* \geq \alpha$ with $\Gamma^{\alpha^*}(e) \downarrow = \Phi_e(e) \downarrow$, then $\alpha^*$ forces $\Gamma^G(e) \downarrow = \Phi_e(e) \downarrow$. If, instead, there is an $e$ and a condition $\alpha^* \geq \alpha$ such that $\Gamma^\beta(e) \uparrow$ for every precondition $\beta \geq \alpha^*$, then $\alpha^*$ forces $\Gamma^G(e) \uparrow$ (since every condition is a precondition). So assume neither possibility holds; we derive a contradiction. By assumption, for every $e$ there is a precondition $\beta \geq \alpha$ such that $\Gamma^\beta(e) \downarrow$. Moreover, if $\Phi_e(e) \downarrow$ and $\Gamma^\beta(e) = \Phi_e(e)$, then $\beta$ cannot be a condition, so $\beta(|\beta| - 1)$ must be $\leqslant_P$-isolated.

Now, if there were only finitely many $e$ such that $\Gamma^\beta(e) \downarrow = \Phi_e(e) \downarrow$ for some precondition $\beta \geq \alpha$, then we could compute a function $f$ with $f(e) \neq \Phi_e(e)$ for almost all $e$, as follows: given $e$, search for the least precondition $\beta \geq \alpha$ such that $\Gamma^\beta(e) \downarrow$ and set $f(e) = \Gamma^\beta(e)$. Since the set of preconditions is computable, so would be $f$. But then a finite modification of $f$ would be a computable DNC function, which is impossible. So there must be infinitely many $e$ such that $\Gamma^\beta(e) \downarrow = \Phi_e(e) \downarrow$ for some precondition $\beta \geq \alpha$. And by our use conventions, we must have $|\beta| \geqslant e$ if $\Gamma^\beta(e) \downarrow$. It follows that there is an infinite computable set $S \subseteq P$ of numbers of the form $\beta(|\beta| - 1)$ for some precondition $\beta \geq \alpha$ with $\Gamma^\beta(e) \downarrow = \Phi_e(e) \downarrow$. But then $S$ is an infinite computable set of $\leqslant_P$-isolated elements of $P$, which can be thinned to an infinite computable antichain for $\leqslant_P$ by Exercise 9.13.2. This contradicts our assumption that $\leqslant_P$ has no such antichain. This completes the proof that SCAC admits preservation of $C$.

Now, since SADS $\leqslant_\omega$ SCAC, it follows that SADS admits preservation of $C$. We also recall, from the proof of Theorem 8.8.2, that COH admits preservation of $C$. Hence, by Exercise 4.8.12, SADS + COH admits preservation of $C$. Since SADS+COH $\equiv_\omega$ ADS by Proposition 9.2.15, it follows that ADS admits preservation of $C$, and hence so does SCAC + ADS. And since CAC $\equiv_\omega$ SCAC + ADS, also by Proposition 9.2.15, we conclude at last that CAC admits preservation of $C$. □

The final separation we consider in this section is between CAC and ADS. This was left open by Hirschfeldt and Shore [152], and formed a major question in reverse mathematics for a number of years. The eventual resolution actually established the following stronger fact.

**Theorem 9.2.22 (Lerman, Solomon, and Towsner [195]).** SCAC $\nleq_\omega$ ADS. *Hence also* $\mathsf{RCA}_0 \nvdash$ ADS $\rightarrow$ SCAC.

The proof in [195] is a two-part forcing argument. A *ground forcing* is used to create a suitably complicated instance of SCAC. Then, an *iterated forcing* is used to add solutions to instances of ADS that do not compute solutions to this SCAC instance. (A nice introduction to this technique for more general applications is given by Patey [245].) Here, we give a somewhat simplified proof due to Patey [243] which uses preservation properties in the sense of Definition 3.6.11. We will need the following combinatorial notion.

**Definition 9.2.23 (Patey [243]).**

1. A formula $\varphi(X, Y)$ of sets is *essential* if for every $n \in \omega$ there is a finite $E > n$ such that for every $m \in \omega$, $\varphi(E, F)$ holds for some finite $F > m$.

2. Fix $A, P_0, P_1 \in 2^\omega$. Then $P_0, P_1$ are *dependently $A$-hyperimmune* if for every $\Sigma_1^{0,A}$ essential formula $\varphi(X,Y)$, $\varphi(F_0, F_1)$ holds for some finite sets $F_0 \subseteq \overline{P_0}$ and $F_1 \subseteq \overline{P_1}$.
3. A problem P *admits preservation of dependent hyperimmunity* if for every $A \in 2^\omega$, and all dependently $A$-hyperimmune sets $P_0, P_1$, every $A$-computable instance $X$ of P has a solution $Y$ such that $P_0, P_1$ are dependently $(A \oplus Y)$-hyperimmune.

Clearly, if $P_0, P_1$ are dependently $A$-hyperimmune, they are dependently $A^*$-hyperimmune for all $A^* \leqslant_T A$. So the class $\mathcal{C}_{P_0,P_1}$ of all $A$ such that $P_0, P_1$ are dependently $A$-hyperimmune is closed downwards under $\leqslant_T$, and a problem admits preservation of dependent hyperimmunity if and only if, for all $P_0, P_1 \in 2^\omega$, it admits preservation of $\mathcal{C}_{P_0,P_1}$.

The two main facts from which the theorem follows are the following.

**Lemma 9.2.24 (Patey [243]).** ADS *admits preservation of dependent hyperimmunity.*

**Lemma 9.2.25 (Patey [243]).** SCAC *does not admit preservation of dependent hyperimmunity.*

*Proof (of Theorem 9.2.22; Patey [243]).* Fix $P_0, P_1 \in 2^\omega$ such that SADS does not admit preservation of $\mathcal{C}_{P_0,P_1}$. Since ADS admits preservation of dependent hyperimmunity, it admits preservation of $\mathcal{C}_{P_0,P_1}$. From here, the conclusion follows by Theorem 4.6.13. □

Let us now prove Lemmas 9.2.24 and 9.2.25, beginning with the former.

*Proof (of Lemma 9.2.24).* As usual, we prove the result in the unrelativized setting for simplicity. Let $(L, \leqslant_L)$ be a computable linear order, and let $P_0, P_1$ be a pair of dependently $\emptyset$-hyperimmune sets. We must show that there is an infinite ascending or descending sequence $G$ for $\leqslant_L$ such that $P_0, P_1$ are dependently $G$-hyperimmune. First, suppose there is an infinite set $I$, whose elements are either all $\leqslant_L$-small in $I$ or all $\leqslant_L$-large in $I$, and such that $P_0, P_1$ are dependently $I$-hyperimmune. By Exercise 9.13.2, $I$ can be computably thinned to a solution $S$ for $(L, \leqslant_L)$, and $P_0, P_1$ are dependently $S$-hyperimmune since $S \leqslant_T I$. In this case, we can thus take $G = S$.

Assume next that no $I$ as above exists. We force with 2-fold Mathias conditions $(E_0, E_1, I)$ satisfying the following additional properties.

- $E_0$ is $\leqslant_L$-ascending and $x \leqslant_L y$ for all $x \in E_0$ and $y \in I$.
- $E_1$ is $\leqslant_L$-descending and $x \geqslant_L y$ for all $x \in E_1$ and $y \in I$.
- $x \leqslant_L y$ for all $x \in E_0$ and all $y \in E_1$.
- $P_0, P_1$ are dependently $I$-hyperimmune.

(Trivially, $(\emptyset, \emptyset, \omega)$ is a condition.) A sufficiently generic filter yields objects $G_0$ and $G_1$, where $G_0$ is a $\leqslant_L$-ascending sequence and $G_1$ is a $\leqslant_L$-descending sequence (and $x \leqslant_L y$ for all $x \in G_0$ and all $y \in G_1$). It is easy to see, from our assumption, that both $G_0$ and $G_1$ are infinite sets. We claim that for some $i < 2$, $P_0, P_1$ are dependently $G_i$-hyperimmune, so we can then take $G = G_i$ to complete the proof.

Let $\mathsf{G}_0$ and $\mathsf{G}_1$ be names for $G_0$ and $G_1$ in the forcing language. The way we prove the claim is to show, for every pair $\varphi_0(\mathsf{G}_0, X, Y)$ and $\varphi_1(\mathsf{G}_1, X, Y)$ of $\Sigma^0_1$ formulas of $\mathcal{L}_2(\mathsf{G})$, that the set of conditions forcing the following is dense: for some $i < 2$, if $\varphi_i(\mathsf{G}_i, X, Y)$ is essential, then

$$(\exists \text{ finite } F_0 \subseteq \overline{P}_0)(\exists \text{ finite } F_1 \subseteq \overline{P}_1)[\varphi_{e_i}(\mathsf{G}_i, F_0, F_1)]. \tag{9.1}$$

The claim then follows by genericity and Lachlan's disjunction.

So fix $\varphi_0(\mathsf{G}_0, X, Y)$ and $\varphi_1(\mathsf{G}_1, X, Y)$ as above and let $(E_0, E_1, I)$ be any condition. By passing to an extension if necessary, we can assume $(E_0, E_1, I)$ forces that both $\varphi_0(\mathsf{G}_0, X, Y)$ and $\varphi_1(\mathsf{G}_1, X, Y)$ are essential (otherwise we are done). Let $\psi(X, Y)$ be the formula asserting there are finite sets $D_0, D_1 \subseteq I$ such that the following hold.

- $D_0$ is $\leqslant_L$-ascending.
- $D_1$ is $\leqslant_L$-descending.
- $x \leqslant_L y$ for all $x \in D_0$ and all $y \in D_1$.
- for each $i < 2$, there exist finite sets $F_{i,0} \subseteq X$ and $F_{i,1} \subseteq Y$ such that $\varphi_{e_i}(E_i \cup D_i, F_{i,0}, F_{i,1})$ holds.

Then $\psi$ is $\Sigma^{0,I}_1$, and we claim it is essential. For each $n \in \omega$ we exhibit a finite set $F_0$ such that for each $m \in \omega$ there is a finite set $F_1$ for which $\psi(F_0, F_1)$ holds. Fix $n$. Take a generic filter containing $(E_0, E_1, I)$, and let $G_0$ a $G_1$ be the objects determined by it. By genericity, $\varphi_0(G_0, X, Y)$ and $\varphi_0(G_1, X, Y)$ are essential. Hence, we can fix finite sets $F_{0,0}, F_{1,0} > n$ such that for every $m$, there exist finite sets $F_{0,1}, F_{1,1} > y$ for which $\varphi_0(G_0, F_{0,0}, F_{0,1})$ and $\varphi_0(G_1, F_{1,0}, F_{1,1})$ hold. Let $F_0 = F_{0,0} \cup F_{1,0}$. Now fix $m \in \omega$. By choice of $F_{0,0}$ and $F_{1,0}$, we can fix finite sets $F_{0,1}, F_{1,1} > m$ such that $\varphi_0(G_0, F_{0,0}, F_{0,1})$ and $\varphi_0(G_1, F_{1,0}, F_{1,1})$ hold. Let $F_1 = F_{1,0} \cup F_{1,1}$. Now since $\varphi_0$ and $\varphi_1$ are existential formulas, there is a $k \in \omega$ such that $\varphi_0(G_0 \upharpoonright k, F_{0,0}, F_{0,1})$ and $\varphi_0(G_1 \upharpoonright k, F_{1,0}, F_{1,1})$ hold. Without loss of generality, $k > E_0, E_1$. Hence, for each $i < k$, $G_i \upharpoonright k = E_i \cup D_i$ for some $D_i \subseteq I$. Since $G_0$ is $\leqslant_L$-ascending, so is $D_0$. Similarly, $D_1$ is $\leqslant_L$-descending. And since $x \leqslant_L y$ for all $x \in G_0$ and all $y \in D_1$, the same is true of all $x \in D_0$ and all $y \in D_1$. Thus, $\psi(F_0, F_1)$ holds, as claimed.

Since $P_0, P_1$ are dependently $I$-hyperimmune, it follows that there exists $F_0 \subseteq \overline{P_0}$ and $F_1 \subseteq \overline{P_1}$ for which $\psi(F_0, F_1)$ holds. Fix the witnessing sets $D_0, D_1$ and, for each $i < 2$, $F_{i,0}$ and $F_{i,1}$. Then $F_{i,0} \subseteq \overline{P_0}$ and $F_{i,1} \subseteq \overline{P_1}$. If there are infinitely many $y \in I$ such that $x \leqslant_L y$ for all $x \in D_0$ (or equivalently, for the $\leqslant_L$-largest element of $D_0$), then set $E^*_0 = E_0 \cup D_0$, $E^*_1 = E_1$, and $I^* = \{y \in I : y > D_0 \wedge (\forall x \in D_0)[x \leqslant_L y]\}$. Otherwise, since $x \leqslant_L y$ for all $x \in D_0$ and all $y \in D_1$, it follows that there are infinitely many $y \in I$ such that $x \geqslant_L y$ for all $x \in D_1$. In this case, set $E^*_0 = E_0$, $E^*_1 = E_1 \cup D_1$, and $I^* = \{y \in I : y > D_1 \wedge (\forall x \in D_1)[x \geqslant_L y]\}$. In either case, $(E^*_0, E^*_1, I^*)$ is an extension of $(E_0, E_1, I)$ forcing (9.1). □

*Proof (of Lemma 9.2.25).* Recall by Proposition 9.2.12 that $\mathsf{SCAC}$ is $\omega$-model equivalent to the restriction of $\mathsf{SRT}^2_2$ to semi-transitive colorings. We show that the latter problem does not admit preservation of dependent hyperimmunities. To this end, we show that there exists a computable stable semi-transitive coloring $c \colon [\omega]^2 \to 2$ such that the sets $P_0 = \{x \in \omega : \lim_y c(x, y) = 0\}$ and

$P_1 = \{x \in \omega : \lim_y c(x, y) = 1\}$ are dependently $\varnothing$-hyperimmune. To see that this suffices, suppose $H$ is any infinite homogeneous set for $c$. Let $\varphi(X, Y)$ be the $\Sigma_1^{0,H}$ formula $X \neq \varnothing \wedge Y \neq \varnothing \wedge X \subseteq H \wedge Y \subseteq H$. Since $H$ is infinite, $\varphi$ is easily seen to be essential. But clearly, there do not exist $F_0 \subseteq \overline{P_0}$ and $F_1 \subseteq \overline{P_1}$ such that $\varphi(F_0, F_1)$ holds, since either $H \cap P_0 = \varnothing$ or $H \cap P_1 = \varnothing$. Hence, $P_0, P_1$ are not dependently $H$-hyperimmune.

The construction is a finite injury argument. Fix an effective enumeration $\varphi_0, \varphi_1, \ldots$ of all $\Sigma_1^0$ formulas. For each $e \in \omega$, we aim to satisfy the following requirement:

$$\mathcal{R}_e : \ \varphi_e \text{ is essential} \rightarrow (\exists F_0 \subseteq P_0)(\exists F_1 \subseteq P_1)[\varphi_e(F_0, F_1)].$$

We actually construct the sets $P_0$ and $P_1$ first, and then define $c$ appropriately after. We proceed by stages. At stage $s$, we define approximations $P_{0,s}$ and $P_{1,s}$ to $P_0$ and $P_1$, respectively, with $P_{0,s} \cup P_{1,s} = \omega \upharpoonright s$. For each $e$, we also define a number $m_{e,s} \leqslant s$.

***Construction.*** To begin, set $P_{0,0} = P_{1,0} = \varnothing$ and $m_{e,0} = 0$ for all $e$. Declare all requirements *active*. Next, fix $s \in \omega$ and suppose $P_{0,s}$ and $P_{1,s}$ have been defined, along with $m_{e,s}$ for each $e$.

Choose the least $e < s$, if it exists, such that $\mathcal{R}_e$ is active and there exist finite sets $F_0 < F_1$ contained in $[m_s, s]$ for which $\varphi_e(F_0, F_1)$ holds (with witness bounded by $s$). In this case, say that $\mathcal{R}_e$ *acts* at stage $s+1$, and declare it *inactive*. Let

$$P_{0,s+1} = (P_{0,s} \cap [0, m_{e,s})) \cup [m_{e,s}, \max F_0]$$

and

$$P_{1,s+1} = (P_{1,s} \cap [0, m_{e,s})) \cup (\max F_0, s].$$

Since $P_{0,s} \cup P_{1,s} = \omega \upharpoonright s$ by assumption, we have that $P_{0,s+1} \cup P_{1,s+1} = \omega \upharpoonright s+1$. For $e^* \leqslant e$ set $m_{e^*,s+1} = m_{e^*,s}$. For $e^* > e$, set $m_{e^*,e} = s+1$, and declare $\mathcal{R}_{e^*}$ active.

If no requirement acts at stage $s+1$, simply set $P_{0,s+1} = P_{0,s} \cup \{s\}$ and $P_{1,s+1} = P_{1,s}$, and $m_{e,s+1} = m_e$ for all $e$. This completes the construction.

***Verification.*** It is easy to verify, by induction on $e$, that every requirement acts at most finitely many stages. Indeed, for $\mathcal{R}_e$ to act at some stage, it must be active at the beginning of the stage, and at this point it is declared inactive. The the only reason it can act again is because some $\mathcal{R}_{e^*}$ for $e^* < e$ acts at a later stage, at which point $\mathcal{R}_e$ is again declared active. By the same token, every requirement is active or inactive at cofinitely many stages. Notice, too, that if a requirement is inactive at cofinitely many stages then it is satisfied.

Now fix any $n \in \omega$. We claim that for each $i < 2$, $P_{i,s} \upharpoonright n$ is the same for all sufficiently large $s$. This is trivially true if no requirement ever acts at any stage after $n$. Otherwise, choose any stage $s_0 \geqslant n$ at which some requirement, say $\mathcal{R}_{e_0}$, acts. Let $s_1 \geqslant s_0$ be such that no requirement $\mathcal{R}_e$ with $e \leqslant e_0$ acts again at any stage $s \geqslant s_1$. Then for any such $s$, the only requirements $\mathcal{R}_e$ that can act satisfy $m_{e,s} \geqslant m_{e,s_0} \geqslant n$, and hence $P_{i,s} \upharpoonright n = P_{i,s_1} \upharpoonright n$. This proves the claim. Now for each $i < 2$, set $P_i = \{x \in \omega : (\forall^\infty s)[x \in P_{i,s}]\}$ for each $i$. By construction, every $x \in \omega$ belongs to exactly one of $P_0$ or $P_1$.

We next claim that each $\mathcal{R}_e$ is satisfied. We proceed by induction on $e$. Assume all $\mathcal{R}_{e^*}$ for $e^* < e$ are satisfied, and let $s_0$ be a stage such that no such $\mathcal{R}_{e^*}$ acts at stage $s_0$ or after. Seeking a contradiction, suppose $\mathcal{R}_e$ is not satisfied. Then $\mathcal{R}_e$ must be active at every stage $s \geqslant s_0$. Also, $\varphi_e$ must be essential, else $\mathcal{R}_e$ would be satisfied trivially. So, there exist finite sets $F_0, F_1$ such that $m_{e,s_0} < F_0 < F_1$ and $\varphi_e(F_0, F_1)$ holds. Choose any $s > \max(s_0, \max F_1)$ and large enough to bound the existential quantifier in $\varphi_e(F_0, F_1)$. By construction, $m_{e,s} = m_{e,s_0}$, so $F_0$ and $F_1$ are contained in $[m_e, s)$. But then $\mathcal{R}_e$ acts at stage $s$, becoming inactive, a contradiction.

To complete the proof, we define a computable stable semi-transitive coloring $c$ so that for each $i < 2$, $P_i = \{x \in \omega : \lim_y c(x, y) = i\}$. Given $x < s$, let $c(x, s) = i$ for the unique $i$ such that $x \in P_{i,s}$. Since the construction is computable, so is $c$. Since, for each $x$ and each $i < 2$, $P_{i,s} \upharpoonright x + 1$ is the same for all sufficiently large $s$, it follows that $\lim_s c(x, s)$ exists. Hence, $c$ is stable, and $P_0$ and $P_1$ are the desired sets.

It remains only to show that $c$ is semi-transitive. Fix $x < y < z$ such that $c(x, y) = c(y, z) = 0$. We claim that $c(x, z) = 0$. Seeking a contradiction, suppose not, so that $c(x, z) = 1$. Then $x \in P_{0,y}$ and $x \in P_{1,z}$, so we can fix the largest $s < z$ so that $x \in P_{0,s}$. In particular, $y \leqslant s$ and $x \in P_{1,s+1}$. By construction, $P_{1,s+1} = (P_{1,s} \cap [0, m_{e,s})) \cup (\max F_0, s]$ for some finite set $F_0$ found during the action of some strategy $\mathcal{R}_e$ at stage $s+1$ of the construction. We must thus have that $m_{e,s} < x$, and since $x < y \leqslant s$, it follows that also $y \in P_{1,s+1}$. But $y \in P_{0,z}$, so we can fix the largest stage $s^*$ with $s + 1 \leqslant s^* < z$ such that $y \in P_{1,s^*}$. By construction, $P_{0,s^*+1} = (P_{0,s^*} \cap [0, m_{e^*,s^*})) \cup [m_{e^*,s^*}, \max F_0^*]$ for some finite set $F_0^*$ found during the action of some strategy $\mathcal{R}_{e^*}$ at stage $s^* + 1$ of the construction. If $e^* \geqslant e$ then $m_{e^*,s^*} \geqslant m_{e^*,s+1} = s+1 > y$, so $y$, being an element of $P_{1,s^*}$, could not be an element of $P_{0,s^*+1}$. Thus, $e^* < e$, and hence $m_{e^*,s^*} \leqslant m_{e,s^*} = m_{e,s} < x < y \leqslant \max F_0^*$. It follows that $x \in P_{0,s^*+1}$, which contradicts the choice of $s$ since $s^* + 1 > s$. □

Of the implications proved above, the only one remaining unaddressed is COH $\rightarrow$ CADS in Proposition 9.2.15, and this is because whether or not CADS implies COH remains open.

*Question 9.2.26.* Is it the case that $\mathsf{RCA}_0 \vdash \mathsf{CADS} \rightarrow \mathsf{COH}$?

But recall that CADS does imply COH over $\mathsf{RCA}_0 + \mathsf{B}\Sigma^0_2$ (Proposition 9.2.15 (4)), so a negative answer to the question would necessarily need to involve nonstandard models. We summarize everything in Figure 9.1.

### 9.2.4 Variants under finer reducibilities

We conclude this section with a brief discussion of some related results, mostly omitting proofs. The main focus here is on technical variations of the principles CAC and ADS.

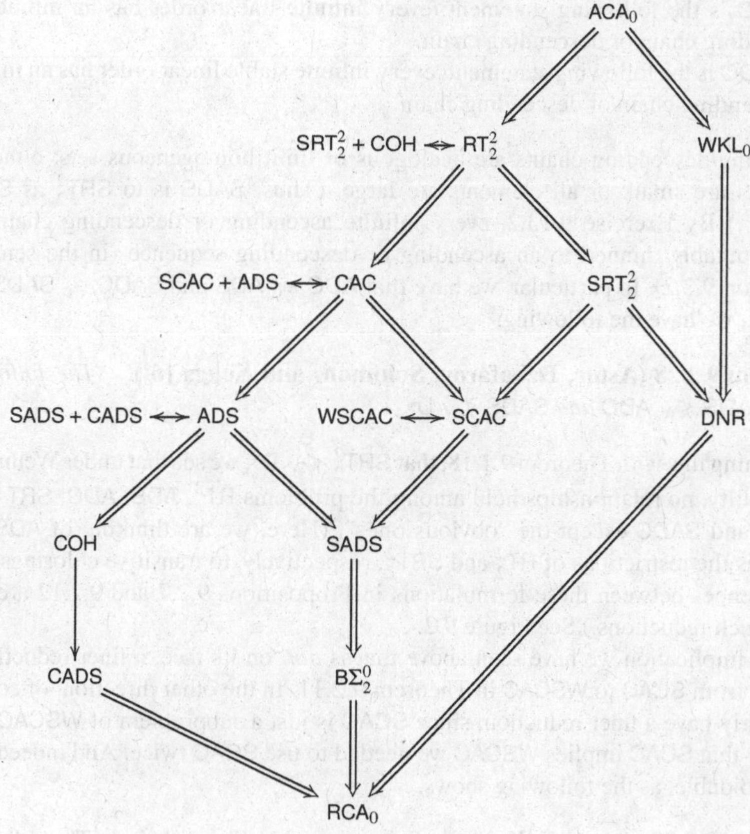


**Figure 9.1.** The location of CAC, ADS, and their stable and cohesive variants alongside $\mathsf{RT}^2_2$. Arrows denote implications over $\mathsf{RCA}_0$; double arrows are implications that cannot be reversed. Except for a potential arrow from CADS to COH, no additional arrows can be added.

A quick inspection of the proofs shows that most of the implications illustrated in Figure 9.1 are actually formalizations in $\mathsf{RCA}_0$ of finer reducibilities. Indeed, almost all of them are Weihrauch reductions. Astor, Dzhafarov, Solomon, and Suggs [6] observed that, under $\leqslant_{\mathrm{W}}$, some of the principles we have considered can break apart. More precisely, they noticed that the definition of ascending or descending sequence in the formulation of ADS can be computably weakened, as follows.

**Definition 9.2.27 (Astor, Dzhafarov, Solomon, and Suggs [6]).**

1. Let $(L, \leqslant_L)$ be an infinite partial order. A subset $S$ of $L$ is

   - an *ascending chain* if every $x \in S$ is $\leqslant_L$-small in $S$,
   - a *descending chain* if every $x \in S$ is $\leqslant_L$-large in $S$.

2. ADC is the following statement: every infinite linear order has an infinite ascending chain or descending chain.
3. SADC is the following statement: every infinite stable linear order has an infinite ascending chain or descending chain.

Ascending/descending chains are analogous of limit homogeneous sets: either all elements are small, or all elements are large. (Thus, SADS is to $\mathsf{SRT}^2_2$ as SADC is to $\mathsf{D}^2$.) By Exercise 9.13.2, every infinite ascending or descending chain can be computably thinned to an ascending or descending sequence (in the sense of Definition 9.2.2). In particular, we have that ADC $\equiv_c$ ADS and SADC $\equiv_c$ SADS. By contrast, we have the following.

**Theorem 9.2.28 (Astor, Dzhafarov, Solomon, and Suggs [6]).** *The following hold:* SADS $\nleqslant_W$ ADC *and* SADS $\nleqslant_W$ $\mathsf{D}^2$.

Combining this with Theorem 9.1.18, that $\mathsf{SRT}^2_2 \nleqslant_W \mathsf{D}^2$, we see that under Weihrauch reducibility, no relationships hold among the problems $\mathsf{RT}^2$, ADS, ADC, $\mathsf{SRT}^2$, $\mathsf{D}^2$, SADS, and SADC except the "obvious ones". (Here, we are thinking of ADS and SADS as the restrictions of $\mathsf{RT}^2_2$ and $\mathsf{SRT}^2_2$, respectively, to transitive colorings. The equivalences between these formulations in Propositions 9.2.7 and 9.2.12 are also Weihrauch reductions.) See Figure 9.2.

One implication we have seen above that is *not*, on its face, a finer reduction is the one from SCAC to WSCAC in Theorem 9.2.11. In the other direction, of course, we clearly have a finer reduction, since SCAC is just a subproblem of WSCAC. But to show that SCAC implies WSCAC we needed to use SCAC twice. And indeed, this is unavoidable, as the following shows.

**Theorem 9.2.29 (Astor, Dzhafarov, Solomon, and Suggs [6]).** *The following holds:* WSCAC $\nleqslant_c$ SCAC.

As pointed out in [6], it may be tempting to ascribe this nonreduction to the fact that WSCAC allows for three "limit behaviors" (small, large, and isolated) while SCAC allows only for two (small and isolated, or large and isolated). In this way, we may be reminded of Theorem 9.1.11, that $\mathsf{RT}^2_3 \nleqslant_c \mathsf{RT}^2_2$. But curiously, WSCAC is computably reducible (in fact, Weihrauch reducible) to $\mathsf{SRT}^2_2$. (See Exercise 9.13.3.) So, cardinality cannot be all that is going on. And indeed, while the original proof of Theorem 9.2.28 was a direct and somewhat blunt forcing argument, a subsequent alternative proof due to Patey [241] helped better elucidate some of the computability theoretic underpinnings of the separation. The idea of Patey's argument uses preservation properties similar to those used in the proof of Theorem 9.2.22.

With partial orders, too, one can consider some variations detectable under finer reducibilities. One very interesting example is the following.

**Definition 9.2.30 (Hughes [163]; Towsner [311]).**

1. Let $(P, \leqslant_P)$ be a partial order. A set $S \subseteq P$ is *ω-ordered* if $x \leqslant_P y$ implies $x \leqslant y$, for all $x, y \in P$.

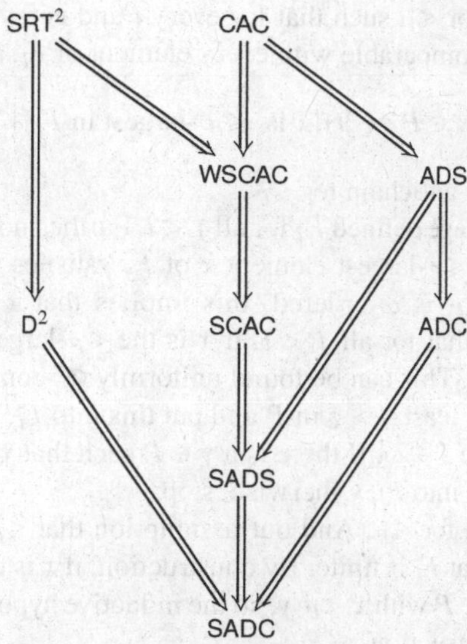


**Figure 9.2.** Relationships between CAC, ADS, ADC, and their stable forms under Weihrauch reducibility. All arrows (reductions) are actually strong Weihrauch reductions, and no additional arrows can be added.

2. $\mathsf{CAC}^{\mathrm{ord}}$ is the following statement: every infinite $\omega$-ordered partial order has an infinite chain or antichain.
3. $\mathsf{SCAC}^{\mathrm{ord}}$ is the following statement: every infinite $\omega$-ordered stable partial order has an infinite $\omega$-ordered chain or antichain.

The motivation for looking at $\omega$-ordered partial orders is that, in virtually all applications—such as when a partial order is defined as part of a proof involving CAC—it ends up being $\omega$-ordered. In this respect, partial orders that are not $\omega$-ordered show up less often, and as such are a bit less natural. So we might view the comparison of $\mathsf{CAC}^{\mathrm{ord}}$ to CAC as a comparison of "natural uses" of CAC to "all possible uses in principle".

The following result may be interpreted as saying that, under provability over $\mathsf{RCA}_0$, the "natural cases" are all there is.

**Proposition 9.2.31 (Hughes [163]; Towsner [311]).** $\mathsf{RCA}_0$ *proves* $\mathsf{CAC} \leftrightarrow \mathsf{CAC}^{\mathrm{ord}}$.

By contrast, under computable reducibility there is a detectable difference.

**Theorem 9.2.32 (Hughes [163]).** *The problem* $\mathsf{CAC}^{\mathrm{ord}}$ *admits* $\Delta^0_2$ *solutions. Hence,* $\mathsf{CAC} \nleqslant_{\mathrm{c}} \mathsf{CAC}^{\mathrm{ord}}$.

*Proof.* We prove the result in the unrelativized case. Fix an infinite computable $\omega$-ordered partial order $(P, \leqslant_P)$. If this has an infinite computable chain, we are done. So suppose otherwise. We build a $\varnothing'$-computable sequence of nonempty finite chains $F_0 < F_1 < \cdots$ for $\leqslant_P$ such that for every $i$ and every $j < i$, the $\leqslant_P$-largest element of $F_j$ is $\leqslant_P$-incomparable with every element of $F_i$. In particular, the set

$$\{x \in P : (\exists i)[x \text{ is } \leqslant_P \text{-largest in } F_i]\}$$

will be a $\varnothing'$-computable antichain for $\leqslant_P$.

Fix $i$ and assume we have defined $F_j$ for all $j < i$. Furthermore, assume inductively that for each $j < i$, the $\leqslant_P$-largest element $x$ of $F_j$ satisfies that there is no $y \in P$ with $x <_P y$. Since $\leqslant_P$ is $\omega$-ordered, this implies that $x$ is isolated in $P$. Fix $z > \max \bigcup_{j<i} F_j$ such that for all $j < i$, if $x$ is the $\leqslant_P$-largest element of $F_j$ then $x \mid_P y$ for all $y > z$ in $P$. This can be found uniformly $\varnothing'$-computably. We define $F_i$ as follows. Let $x_0$ be the least $y > z$ in $P$ and put this into $F_i$. Now suppose we have defined $x_s \in F_i$ for some $s \in \omega$. If there is a $y \in P$ such that $x_s <_P y$, let $x_{s+1}$ be the least such $y$ and put this into $F_i$. Otherwise, stop.

Clearly, $F_i$ is a chain for $\leqslant_P$. And our assumption that $\leqslant_P$ has no infinite computable chain implies that $F_i$ is finite. By construction, if $x$ is the $\leqslant_P$-largest element of $F_i$ then there is no $y \in P$ with $x <_P y$, so the inductive hypotheses are maintained. This completes the construction.

To complete the proof, recall the aforementioned result of Herrmann [142], that CAC omits $\Delta^0_2$ solutions. Hence, CAC $\nleqslant_{\mathrm{c}}$ CAC$^{\mathrm{ord}}$, as desired. □

Surprisingly, the above does not go through for stable partial orders.

**Theorem 9.2.33 (Hughes [163]).** SCAC $\equiv_{\mathrm{c}}$ SCAC$^{\mathrm{ord}}$.

*Proof.* Fix an infinite partial order $(P, \leqslant_P)$. For simplicity, assume it is computable. The general case follows by relativization. Also, assume that every element of $P$ is either $\leqslant_P$-small or $\leqslant_P$-isolated. The case where every element is either $\leqslant_P$-large or $\leqslant_P$-isolated is symmetric. Now define a computable partial ordering $\leqslant_Q$ of $P$ as follows: for $x, y$ in $P$, set $x \leqslant_Q y$ if and only if $x \leqslant y$ and $x \leqslant_P y$. Thus, $(P, \leqslant_Q)$ is $\omega$-ordered. Let $S \subseteq P$ be an SCAC$^{\mathrm{ord}}$-solution to $(P, \leqslant_Q)$. If $S$ is a chain for $\leqslant_Q$, then it is also clearly a chain for $\leqslant_P$, and thus serves as a SCAC-solution to $(P, \leqslant_P)$. So suppose instead that $S$ is an antichain for $\leqslant_Q$. By definition of $\leqslant_Q$, if $x$ is any element of $S$ then we must have $x \nleqslant_P y$ for all $y > x$ in $S$. Since $\leqslant_P$ is stable, this means every $x \in S$ is $\leqslant_P$-isolated. Using Exercise 9.13.2, we can consequently computably thin $S$ out to an infinite antichain for $\leqslant_P$. □

However, the principles SCAC and SCAC$^{\mathrm{ord}}$ can be separated by finer reducibilities.

**Theorem 9.2.34 (Hughes [163]).**

*1.* SCAC $\nleqslant_{\mathrm{W}}$ SCAC$^{\mathrm{ord}}$.
2. SCAC $\nleqslant_{\mathrm{sc}}$ SCAC$^{\mathrm{ord}}$.

The proofs of both results are forcing arguments. Part 2 employs a combinatorial elaboration on the tree labeling method, which we saw used in the proof of Theorem 9.1.1.

## 9.3 Polarized Ramsey's theorem

In what follows, it is helpful to recall Definition 3.2.4, and the convention following it. Namely, for a set $X$, $[X]^n$ denotes the collection of all finite subsets of $X$ of size $n$, not just increasing $n$-tuples of elements of $X$, even though we usually treat the two as interchangeable. Here, the distinction is important.

**Definition 9.3.1.** Fix $n, k \geqslant 1$ and $c\colon [\omega]^n \to k$. A tuple $(H_0, \ldots, H_{n-1})$ of subsets of $\omega$ is:

1. *p-homogeneous* for $c$ if $c$ is constant on all finite sets $F = \{x_0, \ldots, x_{n-1}\}$ with $x_0 \in H_0, \ldots, x_{n-1} \in H_{n-1}$ and $|F| = n$,
2. *increasing p-homogeneous* for $c$ if $c$ is constant on all finite sets $F = \{x_0, \ldots, x_{n-1}\}$ with $x_0 \in H_0, \ldots, x_{n-1} \in H_{n-1}$ and $x_0 < \cdots < x_{n-1}$.

A p-homogeneous or increasing p-homogeneous set $(H_0, \ldots, H_{n-1})$ is *infinite* if each of $H_0, \ldots, H_{n-1}$ is infinite.

Alternatively, if we wanted to stick with just considering colorings defined on increasing tuples, then $(H_0, \ldots, H_{n-1})$ would be $p$-homogeneous if $c$ were constant on all Cartesian products of $H_0, \ldots, H_{n-1}$, in any order. And $(H_0, \ldots, H_{n-1})$ would be increasing $p$-homogeneous just if $c$ were constant on $H_0 \times \cdots \times H_{n-1}$.

We now have the following "polarized" analogue of Definition 3.2.5.

**Definition 9.3.2 (Polarized and increasing polarized** $\mathsf{RT}$**).**

1. For $n, k \geqslant 1$, $\mathsf{PT}^n_k$ is the following statement: every $c\colon [\omega]^n \to k$ has an infinite p-homogeneous set.
2. For $n, k \geqslant 1$, $\mathsf{IPT}^n_k$ is the following statement: every $c\colon [\omega]^n \to k$ has an infinite increasing p-homogeneous set.
3. For $n \geqslant 1$, $\mathsf{PT}^n$ and $\mathsf{IPT}^n$ are $(\forall k)\mathsf{PT}^n_k$ and $(\forall k)\mathsf{IPT}^n_k$, respectively.
4. $\mathsf{PT}$ and $\mathsf{IPT}$ are $(\forall n)\mathsf{PT}^n$ and $(\forall n)\mathsf{IPT}^n$, respectively.

As with Ramsey's theorem, $\mathsf{RT}^n_1$ is trivially provable in $\mathsf{RCA}_0$, and for all $k \geqslant 2$, $\mathsf{RT}^n_k$ is equivalent over $\mathsf{RCA}_0$ to $\mathsf{RT}^n_2$, so we usually just stick to the latter when discussing implications and equivalences.

Various kinds of polarized partition results have been studied extensively in combinatorics (see, e.g., Chvátal [46]; Erdős and Hajnál [95], and Erdős, Hajnál, and Milner [96, 97]). The specific principles $\mathsf{PT}$ and $\mathsf{IPT}$, inspired by these results, were first formulated by Dzhafarov and Hirst [83] specifically in the context of reverse mathematics.

Note that the parts of a p-homogeneous set need not be disjoint. Thus, if we have an infinite homogeneous set $H$ for a given coloring of $[\omega]^n$, then setting $H_0 = \cdots = H_{n-1} = H$ yields an infinite p-homogeneous set $\langle H_0, \ldots, H_{n-1}\rangle$. Therefore, we have $\mathsf{PT}^n_k \leqslant_{\mathrm{sc}} \mathsf{RT}^n_k$. Also, every p-homogeneous set is clearly increasing p-homogeneous, so $\mathsf{IPT}^n_k$ is a subproblem of $\mathsf{PT}^n_k$. By the same token, $\mathsf{RCA}_0 \vdash \mathsf{RT}^n_2 \to \mathsf{PT}^n_2 \to \mathsf{IPT}^n_2$. (And all the same is true if we quantify out the colors and/or the exponents.) For $n \geqslant 3$, these implications reverse.

**Proposition 9.3.3 (Dzhafarov and Hirst [83]).** *Fix* $n \geqslant 3$.

1. $\mathsf{RCA}_0 \vdash \mathsf{RT}^n_2 \leftrightarrow \mathsf{PT}^n_2 \leftrightarrow \mathsf{IPT}^n_2$.
2. $\mathsf{RCA}_0 \vdash \mathsf{RT}^n \leftrightarrow \mathsf{PT}^n \leftrightarrow \mathsf{IPT}^n$.
3. $\mathsf{RCA}_0 \vdash \mathsf{RT} \leftrightarrow \mathsf{PT} \leftrightarrow \mathsf{IPT}$.

The reversals consist in showing that $\mathsf{IPT}^3_2 \to \mathsf{ACA}_0$ and $\mathsf{IPT} \to \mathsf{ACA}'_0$ and appealing to the equivalences of these with $\mathsf{RT}^3_2$ and $\mathsf{RT}$ (Theorem 8.1.6 and Corollary 8.2.6), respectively.

More interestingly, we also have an equivalence in the case of $n = 2$, at least between Ramsey's theorem and the polarized Ramsey's theorem. This is somewhat surprising, as the solutions for these two principles are otherwise quite different. Consider the following example from [83]: for numbers $x < y$, let $c(x, y)$ be 0 or 1 depending as $x$ and $y$ have different or like parity. Then $(H_0, H_1)$, where $H_0$ is the set of even numbers and $H_1$ the set of odds, is an infinite p-homogeneous set for $c$, with color 0. But no infinite set can be homogeneous for $p$ with color 0. This highlights part of the combinatorial difference between the two principles. As we will see, their equivalence over $\mathsf{RCA}_0$ is rather indirect.

**Theorem 9.3.4 (Dzhafarov and Hirst [83]).** $\mathsf{RCA}_0 \vdash \mathsf{RT}^2_2 \leftrightarrow \mathsf{PT}^2_2$.

*Proof.* As noted, $\mathsf{RT}^2_2 \to \mathsf{PT}^2_2$. For the converse, we argue in $\mathsf{RCA}_0$ and show that $\mathsf{PT}^2_k$ implies each of $\mathsf{D}^2_2$ and $\mathsf{ADS}$. Since $\mathsf{ADS} \to \mathsf{COH}$ by Theorem 9.2.8, it follows that $\mathsf{PT}^2_2 \to \mathsf{RT}^2_2$ by Theorem 8.5.1 (and the equivalence of $\mathsf{D}^2_2$ with $\mathsf{SRT}^2_2$, Corollary 8.4.11).

First, fix a stable coloring $c\colon [\mathbb{N}]^2 \to 2$. Regarding this as an instance of $\mathsf{PT}^2_2$, let $(H_0, H_1)$ be an infinite p-homogeneous set for $c$, say with color $i < 2$. We claim that $H_0$ is limit homogeneous for $c$ with color $i$. Indeed, fix $x \in H_0$. Then for all $y \in H_1$ different from $x$ we have that $c(x, y) = i$. But as $H_1$ is infinite, it must be that $i$ is the limit color of $x$.

Now let $\leqslant_L$ be a linear ordering of $\mathbb{N}$. Let $c\colon [\mathbb{N}]^2 \to 2$ be the induced coloring of pairs: that is, for all $x < y$,

$$c(x, y) = \begin{cases} 0 & \text{if } y <_L x, \\ 1 & \text{if } x <_L y. \end{cases}$$

Let $(H_0, H_1)$ be an infinite p-homogeneous set for $c$, say with color $i < 2$. By primitive recursion, define a sequence $x_0 < x_1 < \cdots$ as follows: $x_0 = \min H_0$; given $x_{2j}$ for some $j \geqslant 0$, let $x_{2j+1}$ be the least $x > x_{2j+1}$ in $H_1$; given $x_{2j+1}$, let $x_{2j+2}$ be the least $x > x_{2j+1}$ in $H_0$. By definition, $c(x_k, x_{k+1}) = i$ for all $k$. Hence, if $i = 1$ then $x_0 <_L x_1 <_L \cdots$, and if $i = 0$ then $x_0 >_L x_1 >_L \cdots$. Thus, $\{x_0, x_1, \ldots\}$ is either an ascending or descending sequence for $\leqslant_L$. □

Let us break this down a bit more. The proof shows that each of $\mathsf{D}^2_2$ and $\mathsf{ADS}$ is computably reducible to $\mathsf{PT}^2_2$, and we also have $\mathsf{SRT}^2_2 \leqslant_{\mathrm{c}} \mathsf{D}^2_2$ and $\mathsf{COH} \leqslant_{\mathrm{c}} \mathsf{ADS}$ by Proposition 8.4.8 and Theorem 9.2.8. In terms of the compositional product from

Definition 4.5.11, Theorem 8.5.4 says that $\mathsf{RT}^2_2 \leqslant_c \mathsf{SRT}^2_2 * \mathsf{COH}$. Thus, what we really showed above proof is that $\mathsf{RT}^2_2 \leqslant_c \mathsf{PT}^2_2 * \mathsf{PT}^2_2$. This raises the following question.

*Question 9.3.5.* Is it the case that $\mathsf{RT}^2_2 \leqslant_c \mathsf{PT}^2_2$?

Another open question is whether Theorem 9.3.4 can be strengthened by replacing $\mathsf{PT}^2_2$ with $\mathsf{IPT}^2_2$.

*Question 9.3.6.* Is it the case that $\mathsf{RCA}_0 \vdash \mathsf{IPT}^2_2 \to \mathsf{RT}^2_2$?

We can get a lower bound on the strength of $\mathsf{IPT}^2_k$ by considering the stable versions of each of $\mathsf{PT}^2_k$ and $\mathsf{IPT}^2_k$. Let $\mathsf{SPT}^2_k$ and $\mathsf{SIPT}^2_k$ be the restrictions of each of these principles, respectively, to stable colorings.

**Proposition 9.3.7 (Dzhafarov and Hirst [83]).**

*1.* $\mathsf{RCA}_0 \vdash \mathsf{IPT}^2_2 \to \mathsf{SRT}^2_2$.
2. $\mathsf{RCA}_0 \vdash \mathsf{SRT}^2_2 \leftrightarrow \mathsf{SPT}^2_2 \leftrightarrow \mathsf{SIPT}^2_2$.

*Proof.* Again, the case $k = 1$ is trivial, so assume $k \geqslant 2$. The argument that $\mathsf{PT}^2_k \to \mathsf{SRT}^2_k$ in Theorem 9.3.4 actually shows that $\mathsf{SIPT}^2_k \to \mathsf{SRT}^2_k$. Clearly, $\mathsf{IPT}^2_k \to \mathsf{SIPT}^2_k$, so we have (1). Also, $\mathsf{SRT}^2_k \to \mathsf{SPT}^2_k \to \mathsf{SIPT}^2_k$, since each principle in this chain of implications is a subproblem of the previous. Thus we also have (2).□

It turns out that part (2) also holds under Weihrauch reducibility, but not under any finer reducibility notions.

**Theorem 9.3.8 (Nichols [232]).**

*1.* $\mathsf{SRT}^2_2 \equiv_W \mathsf{SPT}^2_2 \equiv_W \mathsf{SIPT}^2_2$.
2. $\mathsf{SRT}^2_2 \nleqslant_{sc} \mathsf{SPT}^2_2$.
*3.* $\mathsf{SPT}^2_2 \nleqslant_{sc} \mathsf{SIPT}^2_2$.
*4.* $\mathsf{SIPT}^2_2 \nleqslant_{sc} \mathsf{D}^2_2$.

This complements Theorems 9.1.18 and 9.2.34. Collectively, these results suggest that in the pantheon of reducibilities we work with, $\leqslant_{sc}$ is best at distinguishing results on the basis of combinatorial (as opposed to computability theoretic) properties. Parts (2)–(4) are proved by a forcing argument using the tree labeling method. Part (3) has a particularly intricate combinatorial core, reflective of how close $\mathsf{SPT}^2_2$ and $\mathsf{SIPT}^2_2$ are to one another.

Finally, we present a result due to Patey [240] showing that Proposition 9.3.7 (1) is a strict implication. For this, we let $\mathsf{DNR}(\varnothing')$ be the principle asserting that for every set $X$, there exists a function which is DNC relative to $X'$. The following somewhat more careful formulation shows how this statement is formalized in $\mathsf{RCA}_0$.

**Definition 9.3.9.** $\mathsf{DNR}(\varnothing')$ is the following statement: for every set $X$ there exists a function $f\colon \mathbb{N} \to \mathbb{N}$ such that for all $e \in \omega$, if $\sigma \in 2^{<\omega}$ is such that $\Phi^\sigma_e(e)\downarrow$ and $\sigma(i) = 1$ if and only if $\Phi_i(i)\downarrow$ for all $i < |\sigma|$, then $f(e) \neq \Phi^\sigma_e(e)$.

**Theorem 9.3.10 (Patey [240]).** $\mathsf{RCA}_0 \vdash \mathsf{IPT}^2_2 \to \mathsf{DNR}(\varnothing')$.

*Proof.* We argue in $\mathsf{RCA}_0$. First, we can formalize the $S^m_n$ theorem (Theorem 2.4.10), in the manner of Section 5.5.3. Now fix $X$. Using the $S^m_n$ theorem and the limit lemma (which is provable in $\mathsf{I}\Sigma^0_1$, as noted in Section 6.4) we can conclude that there is a function $g\colon \mathbb{N}^2 \to 2$ (of the form $\Phi^X_i$, for some $i \in \mathbb{N}$) such that for all $e \in \mathbb{N}$, $\lim_s g(e,s)$ exists if and only if there exists a $\sigma \in 2^{<\mathbb{N}}$ is such that $\Phi^\sigma_e(e) \downarrow = \lim_s g(e,s)$ and $\sigma(i) = 1$ if and only if $\Phi_i(i) \downarrow$ for all $i < |\sigma|$, and in this case $\lim_s g(e,s) = \Phi^\sigma_e(e)$. Therefore, it suffices to exhibit $f\colon \mathbb{N} \to \mathbb{N}$ such that $f(e) \neq \lim_s g(e,s)$ if the latter exists.

The argument now is somewhat similar to Theorem 8.2.2. We define an instance $c\colon [\mathbb{N}]^2 \to 2$ of $\mathsf{IPT}^2_2$ by induction. At stage $s > 0$, we define $c(x,s)$ for all $x < s$. Let $s > 0$ be given. For each $e < s$, decode $g(e,s) \in \mathbb{N}$ as a tuple of numbers smaller than $s$ of length $2e+2$. Let $D_{e,s}$ be the finite set of these numbers.

Each $e < s$ now *claims* a pair of elements of $D_{e,s}$. Namely, $e$ claims the least two elements of $D_{e,s}$ that are not claimed by any $e^* < e$. (Note that $e^*$ can only claim elements of $D_{e^*,s}$, but $D_{e^*,s}$ and $D_{e,s}$ can intersect.) Since $D_{e,s}$ has size $2e+2$, there are guaranteed to be at least two elements for $e$ to claim. Say $e$ claims $x < y$, and set $c(x,s) = 0$ and $c(y,s) = 1$. Finally, fix any $x < s$ not claimed by any $e < s$ and set $c(x,s) = 0$. This completes the construction.

It is clear that $c$ exists and that it is an instance of $\mathsf{IPT}^2_2$. Let $\langle H_0, H_q \rangle$ be an infinite increasing p-homogeneous set for $c$. Define $f\colon \mathbb{N} \to \mathbb{N}$ as follows: for all $e$, let $f(e)$ code the least $2e+2$ many elements of $H_0$. Seeking a contradiction, suppose $\lim_s g(e,s)$ exists and $f(e) = \lim_s g(e,s)$. Fix $s_0$ so that $g(e,s) = g(e,s_0)$ for all $s \geqslant s_0$. Then for all $s \geqslant s_0$, $D_{e,s}$ equals the first $2e+2$ many elements of $H_0$. Hence, by construction, for all $s \geqslant s_0$ there exist $x, y \in D_{e,s} \subseteq H_0$ such that $c(x,s) \neq c(y,s)$. But since $H_1$ is infinite we can choose such an $s \geqslant s_0$ in $H_1$, and then we have a contradiction. □

**Corollary 9.3.11 (Patey [240]).** $\mathsf{RCA}_0 \nvdash \mathsf{SRT}^2_2 \to \mathsf{IPT}^2_2$

*Proof.* Consider the model $\mathcal{M}$ of Chong, Slaman, and Yang [39] from Theorem 8.8.5. This model satisfies $\mathsf{RCA}_0 + \mathsf{SRT}^2_2$, plus every $X \in \mathcal{S}^{\mathcal{M}}$ is low in $\mathcal{M}$. But clearly no $f$ as in the definition of $\mathsf{DNR}(\varnothing')$ can be low in $\mathcal{M}$. So, $\mathsf{IPT}^2_2$ must fail in $\mathcal{M}$. □

The Monin–Patey theorem (Corollary 8.8.8) also suggests the following question.

*Question 9.3.12.* Is it the case that $\mathsf{IPT}^2_2 \leqslant_\omega \mathsf{SRT}^2_2$?

## 9.4 Rainbow Ramsey's theorem

The principle $\mathsf{DNR}(\varnothing')$ defined at the end of the preceding section turns out to have a characterization as a variant of Ramsey's theorem. Thus, its appearance above in connection with $\mathsf{IPT}^2_2$ is not entirely random. It is actually a surprising fact that computability theoretic properties can be equivalent to purely combinatorial ones in our framework. One occurrence of this that we have already seen is the degree characterization of the solutions of COH, in Theorem 8.4.13. We will see several others in Section 9.10 below.

Our interest here will be in the following principle known as the *rainbow Ramsey's theorem*, which was first studied in reverse mathematics by Csima and Mileti [59].

**Definition 9.4.1 (Rainbow Ramsey's theorem).** For $n, k \geqslant 1$, $\mathsf{RRT}^n_k$ is the following statement: for all $c\colon [\omega]^n \to \omega$, if $|c^{-1}(z)| \leqslant k$ for all $z \in \omega$, then there exists an infinite set $R \subseteq \omega$ such that $d$ is injective on $[R]^n$.

The set $R$ above is called a *rainbow* for the coloring $f$. A function $c$ with the property that $|c^{-1}(z)| \leqslant k$ for all $z$ is sometimes called *k-bounded* in the literature, but we will not use this term to avoid confusion with earlier terms like "$k$-valued" and "computably bounded".

The rainbow Ramsey's theorem in some sense says the opposite of Ramsey's theorem: instead of looking for a set on which the coloring uses just one color, it looks for a set on which it uses each color *at most once*. However, though it may not be obvious at first, the latter property is actually a consequence of homogeneity.

**Proposition 9.4.2 (Galvin, unpublished).**

- *For all $n, k \geqslant 1$, $\mathsf{RRT}^n_k$ is identity reducible to $\mathsf{RT}^n_k$.*
- $\mathsf{RCA}_0 \vdash (\forall n)(\forall k)[\mathsf{RT}^n_k \to \mathsf{RRT}^n_k]$.

*Proof.* We prove (1). Let $c\colon [\omega]^n \to \omega$ be an instance of $\mathsf{RRT}^n_k$. Define $d\colon [\omega]^n \to k$ as follows: for all $\vec{x} \in [\omega]^n$,

$$d(\vec{x}) = |\{\vec{y} \leqslant \max \vec{x} : \vec{y} \neq \vec{x} \wedge c(\vec{y}) = c(\vec{x})\}|.$$

As there are at most $k+1$ many tuples of any given color under $c$, we have $d(\vec{x}) < k$. Now let $H$ be an infinite homogeneous set for $d$ and suppose $\vec{y}, \vec{x} \in [H]^n$, say with $\vec{y} \leqslant \max \vec{x}$. Then if we had $\vec{y} \neq \vec{x}$ and $c(\vec{y}) = c(\vec{x})$ we would have by definition that $d(\vec{x}) \geqslant d(\vec{y}) + 1$, which cannot be. We conclude that $c$ is injective on $[H]^n$. $\square$

Csima and Mileti [59] undertook an in-depth analysis of the computability theoretic and proof theoretic content of the rainbow Ramsey's theorem. The previous lemma immediately implies that $\mathsf{RRT}^n_k$ inherits the upper bounds on its strength from $\mathsf{RT}^n_k$. Interestingly, in terms of the arithmetical hierarchy, it also has the same lower bounds. Thus, we have the following analogue of Theorems 8.1.1 and 8.2.2.

**Theorem 9.4.3 (Csima and Mileti [59]).** *Fix $n, k \geqslant 1$.*

1. *For each $n, k \geqslant 1$, $\mathsf{RRT}^n_k$ admits $\Pi^0_n$ solutions and $\mathsf{ACA}_0 \vdash \mathsf{RRT}^n_k$. Hence, also $\mathsf{ACA}_0' \vdash (\forall n)(\forall k)\mathsf{RRT}^n_k$.*
2. *For each $n \geqslant 1$ and $k \geqslant 2$, $\mathsf{RRT}^n_k$ omits $\Sigma^0_n$ solutions. Hence, $\mathsf{ACA}_0 \nvdash (\forall n)(\forall k)\mathsf{RRT}^n_k$.*

However, in other ways the rainbow Ramsey's theorem turns out to be notably weaker than Ramsey's theorem. The following result of Wang resolved a longstanding question about whether an analogue of Jockusch's Theorem 8.2.4 holds in this setting (i.e., whether we can code the jump).

**Theorem 9.4.4 (Wang [321]).** $(\forall n)(\forall k)\mathsf{RRT}^n_k$ *admits strong cone avoidance. Hence,* $\mathsf{TJ} \nleq_\omega (\forall n)(\forall k)\mathsf{RRT}^n_k$ *and* $\mathsf{RCA}_0 \nvdash (\forall n)(\forall k)\mathsf{RRT}^n_k \to \mathsf{ACA}_0$.

Even though the $n = 2$ case of the rainbow Ramsey's theorem is no longer as special as it is for Ramsey's theorem, it is nonetheless of special interest. Arguably the most striking result of Csima and Mileti is an unexpected connection between $\mathsf{RRT}^2_k$ and algorithmic randomness. We pause here to go over some basic notions from this subject. We will need these again in Section 9.11.

We begin by recalling *Cantor space measure* (or *"fair coin" measure*), $\mu$. As usual, this is first defined on basic open sets: for $\sigma \in 2^{<\omega}$, the measure of $[[\sigma]]$ is $\mu([[\sigma]]) = 2^{-|\sigma|}$. Now, consider an arbitrary open set, $\mathcal{U} \subseteq 2^\omega$. This can be written (not uniquely) in the form $\bigcup_{\sigma\in U}[[\sigma]]$, where $U \subseteq 2^{<\omega}$ is *prefix free* (i.e., if $\sigma < \tau \in U$ then $\sigma \notin U$). Then $\mu(\mathcal{U})$ is defined to be $\sum_{\sigma\in U} 2^{-|\sigma|}$, and it can be shown that this value does not depend on the choice of $U$. (See Exercise 9.13.5). From here, the definition of $\mu$ is extended to other subsets of $2^\omega$ as in the usual definition of Lebesgue measure on $\mathbb{R}$. First, every set $C$ is assigned an *outer measure* $\mu^*(C)$, equal to the infimum over all open sets $\mathcal{U} \supseteq C$ of $\mu(\mathcal{U})$. Second, every such $C$ is assigned an *inner measure* $\mu_*(C) = 1 - \mu^*(2^\omega \setminus C)$. If $\mu^*(C) = \mu_*(C)$ then $C$ is called *measurable*, and $\mu(C)$ is defined to be $\mu^*(C)$.

We can now pass to one of the central notions of algorithmic randomness. In what follows, say a sequence of $\langle \mathcal{U}_n : n \in \omega\rangle$ of subsets of $2^\omega$ is *uniformly* $\Sigma^0_1$ if there is a uniformly c.e. sequence $\langle U_n : n \in \omega\rangle$ of subsets of $2^{<\omega}$ such that $\mathcal{U}_n = [[U]]_n$ for all $n$.

**Definition 9.4.5 (Martin-Löf [205]).**

1. A *Martin-Löf test* is a uniformly $\Sigma^0_1$ sequence $\langle \mathcal{U}_n : n \in \omega\rangle$ of subsets of $2^\omega$ such that $\mu(\mathcal{U}_n) \leqslant 2^{-n}$ for all $n$.
2. A set $X$ *passes* a Martin-Löf test $\langle \mathcal{U}_n : n \in \omega\rangle$ if $X \notin \bigcap_n [[U_n]]$.
3. A set $X$ is *1-random* (or *Martin-Löf random*) if it passes every Martin-Löf test.
4. For $n \geqslant 1$, a set $X$ is *n-random* if it is 1-random relative to $\varnothing^{(n-1)}$.

The intuition here is that if $X \in 2^\omega$ is random, there should be nothing special that we can say about it. That is, $X$ should not satisfy any property that can be effectively described and is not enjoyed by all (or almost all, in the sense of measure) sets. For example, an $X$ whose every other bit is 0, or whose elements as a set form an arithmetical progression, or whose odd and even halves compute one another—intuitively, none of these sets should be random. (See Exercise 9.13.4.) A Martin-Löf test is thus a test against some such property, with the measure condition on the members of the test ensuring that it captures only measure 0 many sets. A 1-random set, being one that passes every test, should thus be random in the intuitive sense.

We leave to the exercises the following well-known and important facts.

**Theorem 9.4.6 (Martin-Löf [205]).** *There exists a Martin-Löf test $\langle \mathcal{U}_n : n \in \omega \rangle$, called a* universal Martin-Löf test, *such that $X$ is 1-random if and only if it passes $\langle \mathcal{U}_n : n \in \omega \rangle$.*

**Corollary 9.4.7.** *There exists a nonempty $\Pi^0_1$ class all of whose elements are 1-random.*

Of course, there is much more that can be said here, and there are many other definitions of randomness that have been considered (some equivalent to the above, some not). Our discussion will be confined to a handful of applications, so we refer the reader to other books—most notably those of Downey and Hirschfeldt [83] and Nies [233]—for a thorough discussion, including a comparison of randomness notions, connections with other computability theoretic properties, and historical background. For an overview of recent developments in the subject, see [106].

Let us now return to our discussion of the rainbow Ramsey's theorem. The hallmark result of [59] is the following.

**Theorem 9.4.8 (Csima and Mileti [59]).** *Fix $k \geqslant 1$ and $A \in 2^\omega$. If $c: [\omega]^2 \to k$ is an $A$-computable instance of $\mathsf{RRT}^2_k$ and $X$ is 1-random relative to $A'$ then $A \oplus X$ computes an $\mathsf{RRT}^2_k$-solution to $c$.*

Essentially, this says that if we are given an instance of $\mathsf{RRT}^2_k$ and guess numbers *randomly enough*, then we will obtain a solution to this instance. Intuitively, this makes sense. Instances of $\mathsf{RRT}^2_k$ are *almost* injective to begin with (on all of $[\omega]^2$), and given that $k$ is fixed, the probability of picking two numbers colored the same should be low. Still, it is remarkable that this can be made precise and that it is true.

To better understand Theorem 9.4.8, and $\mathsf{RRT}^2_k$ in general, we need several auxiliary combinatorial notions. The first may be regarded as a weak form of stability. (For other definitions of *stable* instances of $\mathsf{RRT}^2_k$, see Patey [240].)

**Definition 9.4.9.** Fix $k \geqslant 1$. An instance $c: [\omega]^2 \to \omega$ of $\mathsf{RRT}^2_k$ is *normal* if for all $x_0 < x_1$ and $y_0 < y_1$, if $x_1 \neq y_1$ then $c(x_0, x_1) \neq c(y_0, y_1)$.

**Proposition 9.4.10 (Csima and Mileti [59]).** *For all $n, k \geqslant 1$, if $c: [\omega]^n \to \omega$ is an instance of $\mathsf{RRT}^n_k$ there exists an infinite $c$-computable set $I \subseteq \omega$ such that $c \upharpoonright [I]^n$ is normal.*

We leave the proof to the exercises.

**Definition 9.4.11.** Fix $n, k \geqslant 1$ and let $c: [\omega]^2 \to \omega$ be an instance of $\mathsf{RRT}^n_k$.

1. If $F \subseteq \omega$ is finite, then $\mathrm{Viab}_c(F) = \{x > F : c \upharpoonright [F \cup \{x\}]^2 \text{ is injective}\}$.
2. $F \subseteq \omega$ is *admissible* for $c$ if $F$ is finite and $\mathrm{Viab}_c(F)$ is infinite.

The following result says that if a finite set $F$ looks like an initial segment of a solution to $c$ (because $c \upharpoonright [F]^2$ is injective) and $F$ has infinitely many *one*-point extensions that still look like an initial segment to a solution, then in fact, $F$ *can* be extended to a solution. This makes $\mathsf{RRT}^2$-solutions quite special. (The analogous result for Ramsey's theorem is very much false.)

**Proposition 9.4.12 (Csima and Mileti [59]).** *For all $k \geqslant 1$, if $c\colon [\omega]^2 \to \omega$ is a normal instance of* $\mathsf{RRT}^2_k$ *and $F \subseteq \omega$ is admissible for $c$ then*

$$|\{x \in \mathrm{Viab}_c(F) : F \cup \{x\} \text{ is not admissible for } c\}| \leqslant k \cdot |F|.$$

*Proof.* If $k = 1$ there is nothing to prove, so assume $k \geqslant 2$. Let $n = k \cdot |F|$. Seeking a contradiction, suppose there are at least $n+1$ many $x \in \mathrm{Viab}_c(F)$ such that $F \cup \{x\}$ is not admissible for $c$. Let $x_0, \ldots, x_n$ witness this fact. Thus $\mathrm{Viab}_c(F \cup \{x_i\})$ is finite for each $i \leqslant n$. Since $F$ is admissible for $c$, we can choose a $z > \bigcup_{i \leqslant n} \mathrm{Viab}_c(F \cup \{x_i\})$ in the infinite set $\mathrm{Viab}_c(F)$.

Fix $i \leqslant n$. By choice of $z$, $c$ is not injective on $[F \cup \{x_i, z\}]^2$. But $x_i, z \in \mathrm{Viab}_c(F)$, so $c$ is injective on both $[F \cup \{x_i\}]^2$ and on $[F \cup \{z\}]^2$. It follows that one of the witnesses to noninjectivity of $c$ on $[F \cup \{x_i, z\}]^2$ must be $(x_i, z)$. And since $c$ is normal, the other witness must be $(y, z)$ for some $y \in F$.

We conclude that for each $i \leqslant n$ there exists a $y_i \in F$ such that $c(x_i, z) = c(y_i, z)$. The assignment $i \mapsto y_i$ defines a map $k \cdot |F| \to |F|$. If the preimage of each $y \in F$ under this map had size smaller than $k$, then the size of the full preimage of $F$ would have size smaller than $k \cdot |F|$. So we can fix a $y \in F$ whose preimage $G \subseteq \{0, 1, \ldots, n\}$ has size at least $k$. Thus, $y_i = y$ for all $i \in G$. But then for all $i \in G$ we have $c(x_i, z) = c(y, z)$. Since all the $x_i$ for $i \in G$ are distinct, and $y$ is distinct from each of these, we have at least $k + 1$ many pairs all with the same color under $c$. This contradicts the fact that $c$ is an instance of $\mathsf{RRT}^2_k$. □

We now come to the aforementioned characterization of $\mathsf{RRT}^2_2$ in terms of DNC functions relative to $\varnothing'$.

**Theorem 9.4.13 (Miller, unpublished).** *Fix $k \geqslant 2$.*

1. $\mathsf{DNR}(\varnothing') \equiv_{\mathrm{W}} \mathsf{RRT}^2_k$.
2. $\mathsf{RCA}_0 \vdash \mathsf{DNR}(\varnothing') \leftrightarrow \mathsf{RRT}^2_k$.

*Proof.* We prove (1). The proof uses Propositions 9.4.10 and 9.4.12, and each of these can be formalized in $\mathsf{RCA}_0$ easily enough. It is then straightforward to formalize the argument and obtain (2).

($\mathsf{DNR}(\varnothing') \leqslant_{\mathrm{W}} \mathsf{RRT}^2_k$). An instance of a $\mathsf{DNR}(\varnothing')$ is a set, $A$, and a solution is a function DNC relative to $A'$. We prove the result for $A = \varnothing$. The general case then follows by relativization. It therefore suffices to build a computable instance $c$ of $\mathsf{RRT}^2_k$, every solution to which computes a function DNC relative to $\varnothing'$. In fact, the $c$ we build will be an instance of $\mathsf{RRT}^2_2$.

We construct $c$ by stages, very similarly to the coloring in the proof of Theorem 9.3.10. As there, let $g\colon \omega^2 \to 2$ be a computable function such that for all $e$, if $\Phi^{\varnothing'}_e(e)\downarrow$ then $\lim_s g(e, s) \downarrow= \Phi^{\varnothing'}_e(e)$. At stage $s > 0$, define $c(x, s)$ for all $x < s$. For each $e < s$, decode $g(e, s)$ as a tuple of numbers smaller than $s$ of length $2e + 2$ and let $D_{e,s}$ be the set of these numbers. Each $e < s$ now *claims* the least two elements $x, y$ of $D_{e,s}$ not claimed by any $e^* < e$ and sets $c(x, s) = c(y, s) = \langle e, s\rangle$. At the end, for any $x < s$ not claimed by any $e < s$ let $c(x, s) = \langle x + s, s\rangle$. This completes the construction.

It is easy to see that $|c^{-1}(z)| \leqslant 2$ for all $z \in \omega$. Let $R = \{r_0 < r_1 < \cdots\}$ be any $\mathsf{RRT}^2_2$-solution to $c$. Define $f \leqslant_T R$ as follows: for all $e$, let $f(e)$ be a code for the least $2e+2$ many elements of $R$. We claim that $f$ is DNC relative to $\varnothing'$. If not, $\Phi_e(e)\downarrow= f(e)$ for some $e$. Fix $s$ larger than all elements in the tuple coded by $f(e)$ and large enough so that $g(e,t) = \Phi_e(e)$ for all $t \geqslant s$. Since $R$ is infinite, we can without loss of generality take $s \in R$. Then at stage $s$ of the construction, $D_{e,s}$ consists of the least $2e+2$ many elements of $R$, and we set $c(x,s) = c(y,s)$ for some $x, y \in D_{e,s}$. Since $(x,s), (y,s) \in [R]^2$, this contradicts that $f \upharpoonright [R]^2$ is injective.

($\mathsf{RRT}^2_k \leqslant_W \mathsf{DNR}(\varnothing')$). Fix $k \geqslant 1$ and let $c\colon \omega \to \omega$ be an instance of $\mathsf{RRT}^2_k$. Again, we deal with the unrelativized case, so assume $c$ is computable. By Proposition 9.4.10 there is an infinite computable $I \subseteq \omega$ such that $c \upharpoonright [I]^2$ is stable. By passing everything through a computable bijection, we may assume $I = \omega$ for ease of notation.

Let $f$ be DNC relative to $\varnothing'$. We define an infinite $f$-computable set $R = \{r_0 < r_1 < \cdots\}$ such that for each $i$, $R_i = \{r_j : j < i\}$ is admissible for $c$. Notice that this implies that $c$ is injective on $[R]^2$. The conclusion is trivially true for $R_0 = \varnothing$. Assume, then, that for some $i \in \omega$ we have defined $r_j$ for all $j < i$ and that $R_i$ is admissible for $c$. Say $\mathrm{Viab}_c(R_i) = \{x_0 < x_1 < \cdots\}$, and without loss of generality, assume $x_0 > R_i$. Define

$$W = \{m \in \omega : R_i \cup \{x_m\} \text{ is not admissible for } c\}.$$

By Proposition 9.4.12, $|W| \leqslant k \cdot |R_i| = ki$. Notice that $\mathrm{Viab}_c(R_i)$ is uniformly computable in $R_i$, so $W$ is uniformly $\Sigma^0_2$ in $i$ and hence uniformly $\Sigma^{0,\varnothing'}_1$.

So, for each $\ell < ki$, we can uniformly computably find an index $e_\ell$ such that, if the $\ell$th number enumerated into $W$ is $m$, then $\Phi^{\varnothing'}_{e_\ell}(x)\downarrow= m(\ell)$ for all $x$, where we interpret $m$ as a code for a string of length $ki$. Now, let $m = \langle f(e_\ell) : \ell < ki\rangle$. Then for all $\ell < ki$ we have $m(\ell) = f(e_\ell) \neq \Phi^{\varnothing'}_{e_\ell}(e_\ell)$, so $m$ cannot belong to $W$. Hence $\mathrm{Viab}_c(R_i) \cup \{x_m\}$ is admissible for $c$, and we define $r_i = x_m$. □

The connection between the rainbow Ramsey's theorem for pairs had not been observed in, say, Ramsey's theorem, and so initially appeared somewhat mysterious. Miller's theorem (Theorem 9.4.13) clarifies this connection in a number of ways. For one, it yields Theorem 9.4.8 by relativizing to $\varnothing'$ the following well-known result of Kučera. Most proofs of this result in textbooks use Kolmogorov complexity, so for convenience, we include a proof here that uses tests.

**Theorem 9.4.14 (Kučera [192]).** *Every 1-random set X computes a DNC function.*

*Proof.* For each $n \in \omega$, define

$$\mathcal{U}_n = \{X \in 2^\omega : (\exists e > n)[\Phi_e(e)\downarrow= X \upharpoonright e]\},$$

where $X \upharpoonright e$ is interpreted as a code for a string of length $e$. Thus $\mathcal{U}_n = \bigcup_{e>n}[[V_n]]$, where

$$V_e = \{\sigma \in 2^e : \Phi_e(e)\downarrow= \sigma \in 2^e\} = \begin{cases} \sigma & \text{if } \Phi_e(e)\downarrow= \sigma \in 2^e, \\ \varnothing & \text{otherwise.} \end{cases}$$

From here it is easy to see that $\langle \mathcal{U}_n : n \in \omega \rangle$ is a uniformly $\Sigma^0_1$ sequence and that for each $n$,

$$\mu(\mathcal{U}_n) \leqslant \sum_{n>e} \mu([[V_n]]) \leqslant \sum_{n>e} 2^{-e} = 2^{-n}.$$

It follows that $\langle \mathcal{U}_n : n \in \omega \rangle$ is a Martin-Löf test. If $X$ is 1-random, there is consequently an $n$ such that $X \notin \mathcal{U}_n$. By construction, this means that for all $e > n$, if $\Phi_e(e) \downarrow$ then $\Phi_e(e) \neq X \upharpoonright e$.

Define $f \leqslant_T X$ by letting $f(e)$ be a code for $X \upharpoonright e$ for all $e$. Then $f(e) \neq \Phi_e(e)$ for all $e > n$, so a finite modification of $f$ is DNC. In particular, $X$ computes a DNC function. □

Some other results about the strength of $\mathsf{RRT}^2_2$, originally proved in [59] by different arguments, admit simplified proofs thanks to Theorem 9.4.13. One example is the following theorem. The proof we give here is obtained by combining separate results of Monin [216] and Patey [240].

**Theorem 9.4.15 (Csima and Mileti [59]).** *The problem* $\mathsf{RRT}^2_2$ *omits solutions of hyperimmune free degree.*

*Proof (Monin and Patey, see [240]).* We prove the theorem for computable instances of $\mathsf{RRT}^2_2$. The full result follows by relativization. In light of Theorem 9.4.13, it is enough to show that if $f$ is DNC relative to $\varnothing'$, then $f$ does not have hyperimmune free degree. Then the computable coloring $c$ constructed in the proof that $\mathsf{RRT}^2_k \leqslant_W \mathsf{DNR}(\varnothing')$ will have no solution of hyperimmune free degree, completing the result.

Seeking a contradiction, suppose otherwise. By Theorem 2.8.21, this means every $f$-computable function is dominated by a computable function. Consider the class

$$\mathcal{U} = \{X \in 2^\omega : (\exists e)[\Phi^{\varnothing'}_e(e) \downarrow = X(e)]\}.$$

Clearly, $X \in \mathcal{U}$ if and only if it is not DNC relative to $\varnothing'$. Hence, $f \notin \mathcal{U}$. Now, it is not difficult to check that $\mathcal{U}$ can be written as $\bigcup_n C_n$, where each $C_n$ is a $\Pi^0_1$ class, and whose index as such is uniformly computable in $n$. Let $T_0, T_1, \ldots$ be a uniformly computable sequence of subtrees of $2^{<\omega}$ with $C_n = [T_n]$ for all $n$. Thus, $f \in \bigcap_n [[\{\sigma \in 2^{<\omega} : \sigma \notin T_n\}]]$.

Define a function $g \leqslant_T f$ as follows: given $n$, let $g(n)$ be the least $k \in \omega$ such that $f \upharpoonright k \notin T_n$. By assumption, $g$ is dominated by some computable function, $h$. Let

$$\mathcal{D} = \bigcap_n [[\{\sigma \in 2^{h(n)} : \sigma \notin T_n\}]].$$

Then $f \in \mathcal{D}$ by definition of $g$ and $h$. Since $[[\{\sigma \in 2^{h(n)} : \sigma \notin T_n\}]]$ is a clopen set in the Cantor space (being a finite union of basic open sets) , $\mathcal{D}$ is actually a $\Pi^0_1$ class. And since $\mathcal{D}$ contains $f$, it is nonempty. Hence, we can fix a $\varnothing'$-computable path $p \in \mathcal{D}$. But $\mathcal{D} \subseteq 2^\omega \setminus \mathcal{U}$, so $p$ is also DNC relative to $\varnothing'$. Obviously, this is impossible. □

**Corollary 9.4.16 (Csima and Mileti [59]).** $\mathsf{RRT}^2_2 \nleqslant_\omega \mathsf{WKL}$. *Hence, also* $\mathsf{RCA}_0 \nvdash \mathsf{WKL} \to \mathsf{RRT}^2_2$.

*Proof.* Let $\mathcal{S}$ be an $\omega$-model of WKL consisting entirely of sets that have hyperimmune free degree (Exercise 4.8.13). By Lemma Theorem 9.4.15, fix a computable instance $c$ of $\mathsf{RRT}^2_2$ with no solution of hyperimmune free degree. Then $c$ witnesses that $\mathcal{S} \nvDash \mathsf{RRT}^2_2$. □

Overall, $\mathsf{RRT}^2_2$ turns out to be a very weak principle. By Theorem 9.4.13, it implies DNR over $\mathsf{RCA}_0$. However, we now show that it implies neither SADS nor COH. There are two technical ingredients to this. The first is the following famous result of van Lambalgen.

**Theorem 9.4.17 (van Lambalgen [317]).** *Fix $n \geqslant 1$. A set $X = X_0 \oplus X_1$ is $n$-random if and only if $X_0$ is $n$-random and $X_1$ is $n$-random relative to $X_0$.*

For a proof, see, e.g., Downey and Hirschfeldt [83, Theorems 6.9.1 and 6.9.2]. The second technical ingredient is the following. Here we will only it for $n = 2$, which is also the version proved in Csima and Mileti [59]. But the proof is the same for any $n$, and we will use the more general version in Section 9.11.

**Theorem 9.4.18 (Folklore).** *Fix $n \geqslant 1$ and let $X$ be any $n$-random set. There exists an $\omega$-model $\mathcal{S}$ with the following properties.*

1. *Every set in $\mathcal{S}$ is $X$-computable.*
2. *For each $Y \in \mathcal{S}$ there is a $Y^* \in \mathcal{S}$ which is $n$-random relative to $Y$.*

*Proof.* Fix $X$. Write $X = \bigoplus_i X_i$ and define

$$\mathcal{S} = \{Y \in 2^\omega : (\exists k)[Y \leqslant_{\mathrm{T}} \bigoplus_{i<k} X_i]\}.$$

Clearly, this is an $\omega$-model and every $Y \in \mathcal{S}$ is $X$-computable. Fix any such $Y$, say with $Y \leqslant_{\mathrm{T}} \bigoplus_{i<k} X_i$. Since

$$X = (\bigoplus_{i<k} X_i \oplus X_k) \oplus \bigoplus_{i>k} X_i$$

is $n$-random, it follows by van Lambalgen's theorem that $\bigoplus_{i<k} X_i \oplus X_k$ is $n$-random. Hence, again by van Lambalgen's theorem, $X_k$ is $n$-random relative to $\bigoplus_{i<k} X_i$, and therefore relative to $Y$. Since $X_k \in \mathcal{S}$, the proof is complete. □

**Corollary 9.4.19 (Csima and Mileti [59]).** *Let $X$ be 2-random. Then there exists an $\omega$-model of $\mathsf{RRT}^2_2$ consisting entirely of $X$-computable sets.*

*Proof.* By Theorem 9.4.18 with $n = 2$ and Theorem 9.4.8. □

We now get a couple of separation results in rapid succession.

**Corollary 9.4.20 (Csima and Mileti [59]).** $\mathsf{RRT}^3_2 \not\leqslant_\omega \mathsf{RRT}^2_2$. *Hence, also* $\mathsf{RCA}_0 \nvdash \mathsf{RRT}^2_2 \to \mathsf{RRT}^3_2$.

*Proof.* By Corollary 9.4.7 relative to $\varnothing'$, there exists a $\Pi^{0,\varnothing'}_1$ class all of whose elements are 2-random. Hence, there exists a $\Delta^0_2$ 2-random set. Let $\mathcal{S}$ be a model of $\mathsf{RRT}^2_2$ consisting entirely of $X$-computable sets, as given by Corollary 9.4.19. In particular, every element of $\mathcal{S}$ is $\Delta^0_3$. By Theorem 9.4.3, there is a computable instance of $\mathsf{RRT}^3_2$ having no $\Sigma^0_3$ (hence, no $\Delta^0_3$) solution. This coloring witnesses that $\mathcal{S} \nvDash \mathsf{RRT}^3_2$. □

A similar argument allows us to show that $\mathsf{RRT}^2_2$ does not imply $\mathsf{SADS}$. Here, we need one technical result, whose proof we omit.

**Theorem 9.4.21 (Mileti [211]).** *If $S \in 2^\omega$ is hyperimmune then*

$$\mu(\{X \in 2^\omega : X \text{ computes an infinite subset of } S)\} = 0.$$

**Corollary 9.4.22 (Csima and Mileti [59]).** $\mathsf{SADS} \not\leqslant_\omega \mathsf{RRT}^2_2$. *Hence, also* $\mathsf{RCA}_0 \nvdash \mathsf{RRT}^2_2 \to \mathsf{SADS}$.

*Proof.* Let $(L, \leqslant_L)$ be the computable $\mathsf{SADS}$-instance given by Proposition 9.2.17. Thus, every solution to this instance is hyperimmune. Let $A \subseteq L$ be the set of all $\leqslant_L$-small elements of $L$, and $B \subseteq L$ the set of all $\leqslant_L$-large elements of $L$. So also each of $A$ and $B$ must be hyperimmune, and every solution to $(L, \leqslant_L)$ is a subset of $A$ or $B$. It follows that

$$\mu(\{X \in 2^\omega : X \text{ computes an } \mathsf{SADS}\text{-solution to } (L, \leqslant_L)\}) = 0.$$

But the measure of all 2-random sets is 1, so we can choose some such $X$ that computes no solution to $(L, \leqslant_L)$. By Corollary 9.4.19, let $\mathcal{S}$ be an $\omega$-model of $\mathsf{RRT}^2_2$ consisting entirely of $X$-computable sets. Then $(L, \leqslant_L)$ witnesses that $\mathcal{S} \nvDash \mathsf{SADS}$. □

The weakness of $\mathsf{RRT}^2_2$ is further confirmed by looking at its first order part. We do this briefly, without giving any proofs, which tend to be rather involved. The first order consequences of $\mathsf{RRT}^2_2$ were first considered by Conidis and Slaman [55], who established the following conservation result.

**Theorem 9.4.23 (Conidis and Slaman [55]).** *The system* $\mathsf{RCA}_0 + \mathsf{RRT}^2_2$ *is* $\Pi^1_1$-*conservative over* $\mathsf{RCA}_0 + \mathsf{B}\Sigma^0_2$.

This was subsequently expanded as follows.

**Theorem 9.4.24 (Slaman, unpublished).** $\mathsf{RCA}_0 \nvdash \mathsf{RRT}^2_2 \to \mathsf{B}\Sigma^0_2$.

In fact, both results hold for the stronger principle 2-$\mathsf{RAN}$ in place of $\mathsf{RRT}^2_2$, which is the formal analogue of the existence of a 2-random set. This implies $\mathsf{RRT}^2_2$ by formalizing Theorem 9.4.8 in $\mathsf{RCA}_0$. (For a thorough investigation of this principle in second order arithmetic, including alternative formalizations, see Avigad, Dean, and Rute [9].)

However, $\mathsf{RRT}^2_2$ does not have entirely trivial first order part either. The following principle was introduced by Seetapun and Slaman [275].

**Definition 9.4.25.** Let $\Gamma$ be a collection of formulas of $\mathcal{L}_2$. The $\Gamma$ *cardinality scheme* ($\mathsf{C}\Gamma$) is the scheme consisting of all sentences of the form, for $\varphi \in \Gamma$: if $\varphi(x, y)$ defines an injective function then it has unbounded range.

**Theorem 9.4.26 (Conidis and Slaman [55]).** $\mathsf{RCA}_0 \vdash \mathsf{RRT}^2_2 \to \mathsf{C}\Sigma^0_2$.

Seetapun and Slaman [275] showed that $\mathsf{C}\Sigma^0_2$ is not provable in $\mathsf{RCA}_0$. But beyond this, it is very weak indeed.For example, Slaman (unpublished) showed that for all $n$, $\mathsf{I}\Sigma^0_n + \bigwedge_m \mathsf{C}\Sigma^0_m$ does not imply $\mathsf{B}\Sigma^0_{n+1}$ over $\mathsf{RCA}_0$.

We conclude this section with an observation and an open-ended question. The proof of Proposition 9.4.2, that $\mathsf{RT}^n_k$ implies $\mathsf{RRT}^n_k$, suggests more generally that any kind of structure satisfying a version of Ramsey's theorem should also satisfy a version of the rainbow Ramsey's theorem. We will explore a number of other combinatorial relatives of Ramsey's theorem in the rest of this chapter, and for basically all of them no rainbow version has yet been studied. We can thus ask the following.

*Question 9.4.27.* What other variants of Ramsey's theorem admit a rainbow version? How does it relate to the rainbow Ramsey's theorem? How does it relate to algorithmic randomness?

## 9.5 Erdős–Moser theorem

The next variant of Ramsey's theorem we consider is the so-called *Erdős–Moser theorem*, also known as the *tournament principle*. Many a sports fan has lamented the fact that the relation of (one team) *winning against* (another team) is not transitive. If only! But if we can free ourselves of the emotion associated with such events in real life, we can notice some interesting related mathematics questions. For example, given $N \in \omega$, how many teams would have to play in a round-robin tournament (assuming no ties) to guarantee that there are at least $N$ many teams among which winning *is* transitive (i.e., if Team A beats Team B, and Team B beats Team C, then Team A also beats Team C)? In combinatorics, this and related problems have received a great deal of interest. The name of what has come to be called the Erdős–Moser theorem in reverse mathematics derives from a pair of influential papers by Erdős and Moser [99, 100].

Recall the definition of a transitive coloring from Definition 9.2.6.

**Definition 9.5.1 (Erdős–Moser principle).** $\mathsf{EM}$ is the following statement: for every coloring $c \colon [\omega]^2 \to 2$ there exists an infinite set $T \subseteq \omega$ such that $c \upharpoonright [T]^2$ is transitive.

A tournament is perhaps most easily seen as a directed graph, with nodes representing teams (or players), and arrows indicating who won against whom. A 2-coloring represents the same information, but highlights the immediate connection of EM to Ramsey's theorem. More precisely, EM is just a subproblem of $\mathsf{RT}^2_2$. As such, it inherits admitting $\Pi^0_2$ solutions. Kach, Lerman, Solomon, and Weber (unpublished; see [195]) showed that it omits $\Sigma^0_2$ solutions. So, with respect to the arithmetical hierarchy, EM behaves just like $\mathsf{RT}^2_2$. Dzhafarov, Kach, Lerman, and Solomon (unpublished; see [195]) also showed that EM omits solutions of hyperimmune free degree, which is also like $\mathsf{RT}^2_2$.

Over $\mathsf{RCA}_0$, things are more interesting.

**Proposition 9.5.2 (Folklore).** $\mathsf{RCA}_0 \vdash \mathsf{RT}^2_2 \leftrightarrow \mathsf{EM} + \mathsf{ADS}$.

*Proof.* For the nontrivial direction, fix an instance $c\colon [\mathbb{N}]^2 \to 2$ of $\mathsf{RT}^2_2$. Apply EM to find an infinite set $T$ such that $c \upharpoonright [T]^2$ is transitive. By Proposition 9.2.7, we can view $c \upharpoonright [T]^2$ as an instance of ADS. Let $H \subseteq T$ be any ADS-solution to this instance. Then $c \upharpoonright [T]^2$, and hence also $c$, is constant on $[H]^2$. Thus, $H$ is homogeneous for $c$ and we are done. □

The previous result gives a new decomposition of $\mathsf{RT}^2_2$, akin to the Cholak–Jockusch–Slaman decomposition (Theorem 8.5.1). (Note that, as in Theorem 8.5.4, what the proof actually shows is that $\mathsf{RT}^2_2 \leqslant_{\mathrm{W}} \mathsf{EM} * \mathsf{ADS}$.) This decomposition also turns out to be proper. We already know that ADS does not imply $\mathsf{RT}^2_2$ (Theorems 9.2.21 and 9.2.22), so it does not imply EM. The question whether EM implies ADS was open for many years.

**Theorem 9.5.3 (Lerman, Solomon, and Towsner [195]).** $\mathsf{ADS} \nleqslant_\omega \mathsf{EM}$. *Hence, also* $\mathsf{RCA}_0 \nvdash \mathsf{EM} \to \mathsf{ADS}$.

The proof in [195] actually showed that $\mathsf{SRT}^2_2 \nleqslant_\omega \mathsf{EM}$. This was done by an iterated forcing construction, similar in style to the original separation of CAC from ADS (Theorem 9.2.22), but requiring a deep analysis of a very different set of combinatorics. Subsequently, both Wang [322] and Patey [243] were able to improve on this to give even stronger separations. Recall the notion of a problem admitting preservation of hyperimmunity from Definition 9.1.14.

**Theorem 9.5.4 (Patey [243]).** EM *admits preservation of hyperimmunity but* SADS *does not.*

In fact, Patey [243] showed that SADS does not even preserve two hyperimmune sets (i.e., 2 among 2 hyperimmunities, in the parlance of Definition 9.1.12). Hence, by Theorem 4.6.13, we have $\mathsf{SADS} \nleqslant_\omega \mathsf{EM}$. Wang [322] used a different preservation property to obtain the same result.

**Definition 9.5.5 (Wang [322]).** Let $\Gamma$ denote one of $\Sigma^0_n$, $\Pi^0_n$, or $\Delta^0_n$ for some $n \in \omega$. A problem P *admits preservation of* $\Gamma$ *definitions* if for every $A \in 2^\omega$ and every $Z \in 2^\omega$ which is properly $\Gamma$ relative to $A$, every $A$-computable instance $X$ of P has a solution $Y$ such that $Z$ is properly $\Gamma$ relative to $A \oplus Y$.

The use of "properly" above is important (see Definition 2.6.1) as otherwise the definition is trivial.

**Theorem 9.5.6 (Wang [322]).** EM *admits preservation of $\Delta^0_2$ definitions but* SADS *does not.*

Actually, Wang [322] showed that a large number of principles admit preservation of $\Delta^0_2$ definitions, including COH and WKL. So by Exercise 4.8.12, SADS is not $\omega$-model reducible even to, say, WKL + COH + EM.

SADS is one of the weakest principles we have studied so far, so the fact that EM does not imply it tells us that the latter is somewhat weak as well. However, EM does have some strength. In addition to Theorem 9.5.2, we have a version of Theorem 9.3.10 for EM.

**Theorem 9.5.7 (Wang; see [175]).** $\mathsf{RCA}_0 \vdash \mathsf{EM} \to \mathsf{DNR}(\emptyset')$.

On the first order side, we also get an implication to $\mathsf{B}\Sigma^0_2$.

**Theorem 9.5.8 (Kreuzer [187]).** $\mathsf{RCA}_0 \vdash \mathsf{EM} \to \mathsf{B}\Sigma^0_2$.

*Proof.* We use Hirst's theorem (Theorem 6.5.1) and prove that $\mathsf{EM} \to \mathsf{RT}^1$. Arguing in $\mathsf{RCA}_0$, fix $c \colon \mathbb{N} \to k$, $k \geqslant 1$. Define $d \colon [\mathbb{N}]^2 \to 2$ by letting $d(x,y)$ be 1 or 0 depending as $c(x) = c(y)$ or $c(x) \neq c(y)$, respectively. Apply EM to find an infinite set $T$ such that $c \upharpoonright [T]^2$ is transitive. For all $x < y < z$ in $T$, if $c(x) \neq c(y)$ and $c(y) \neq c(z)$ then $d(x,y) = d(y,z) = 0$. Hence, we must have $d(x,z) = 0$ by transitivity, so $c(x) \neq c(z)$. It follows that $c$ colors the elements of $T$ in intervals: if $c(x) = i < k$ and $c(y) \neq i$ for some $y > x$ then $c(z) \neq i$ for all $z \geqslant y$. By bounded $\Sigma^0_1$ comprehension, we can form the set $F = \{i < k : (\exists x \in T)[c(x) = i]\}$. Let $W = \{x \in T : c(x) \in F \wedge (\forall y < x)[y \in T \to c(y) \neq c(x)]\}$. Then $W$ exists, and it must be finite. (If not, fix $x_0 < \cdots < x_k$ in $W$. By definition, $c$ is injective on $\{x_\ell : \ell \leqslant k\}$. But then this is an injection of a set of size $k+1$ into a set of size $k$, contradicting Proposition 6.2.7.) Let $x$ be the largest element of $W$. Then by our observation, we have $c(y) = c(x)$ for all $y \geqslant x$ in $T$. In particular, $\{y \in T : y \geqslant x\}$ is an infinite homogeneous set for $c$. □

## 9.6 The Chubb–Hirst–McNicholl tree theorem

A different variant of Ramsey's theorem is obtained by changing the structure on which colorings are defined. Especially interesting here are trees.

**Definition 9.6.1.** Fix numbers $n, k \geqslant 1$ and a set $T \subseteq 2^{<\omega}$.

1. $[T]^n = \{\langle \sigma_0, \ldots, \sigma_{n-1} \rangle \in T^n : \sigma_0 \prec \cdots \prec \sigma_{n-1}\}$.
2. A *$k$-coloring of* $[T]^n$ is a map $c \colon [T]^n \to k$.
3. A set $S \subseteq T$ is *homogeneous* for $c \colon [T]^n \to k$ if $c$ is constant on $[S]^n$.

We follow all the same conventions for the above definition as we do for the analogous definition for colorings of (subsets of) $[\omega]^n$ (Definition 3.2.4).

We will shortly justify why we are restricting to colorings of *comparable* nodes here. First, we define the kinds of sets we want to be homogeneous.

**Definition 9.6.2.** A set $T \subseteq 2^{<\omega}$ is *isomorphic to* $2^{<\omega}$, written $T \cong 2^{<\omega}$, if $(T, \preceq)$ and $(2^{<\omega}, T)$ are isomorphic structures (i.e., there is a bijection $f\colon T \to 2^{<\omega}$ such that for all $\sigma, \tau \in T$, $\sigma \preceq \tau$ if and only if $f(\sigma) \preceq f(\tau)$).

Intuitively, the homogeneous object should "look like" the object used to define the coloring. Here this is $2^{<\omega}$, but it is exactly the same in the case of Ramsey's theorem. There, a homogeneous object is not just any set, but an infinite set, i.e., one which, together with $\leqslant$, is isomorphic to $(\omega, \leqslant)$.

We can now state the following tree version of Ramsey's theorem, originally introduced by Chubb, Hirst, and McNicholl [45].

**Definition 9.6.3 (Chubb–Hirst–McNicholl tree theorem).**

1. For $n, k \geqslant 1$, $\mathsf{TT}^n_k$ is the following statement: every $c\colon [2^{<\omega}]^n \to k$ has a homogeneous set $T \cong 2^{<\omega}$.
2. For $n \geqslant 1$, $\mathsf{TT}^n$ is $(\forall k)\mathsf{TT}^n_k$.
3. $\mathsf{TT}$ is $(\forall n)\mathsf{TT}^n$.

In the reverse mathematics literature, TT is sometimes called "Ramsey's theorem on trees", but more commonly simply the "tree theorem". We will refer to it here either by its abbreviation or as the "Chubb–Hirst–McNicholl tree theorem", so as to distinguish it from another principle, *Milliken's tree theorem*, which we discuss below.

The reason $[T]^n$ is defined to consist only of $n$-tuples of comparable nodes of $T$, rather than arbitrary size-$n$ subsets of $T$, is that the latter would render the tree theorem false. Consider, e.g., the coloring that assigns a pair of nodes $(\sigma, \tau)$ the color 0 if $\sigma \mid \tau$, and the color 1 otherwise. This coloring is not constant on the (comparable and incomparable) pairs of any $T \cong 2^{<\omega}$. We will return to this issue at the end of the section.

Perhaps the most obvious first question in any investigation of TT is how it compares to RT. It is easy to see that for all $n, k \geqslant 1$, $\mathsf{RT}^n_k \leqslant_{\mathrm{sW}} \mathsf{TT}^n_k$. Indeed, fix $c\colon [\omega]^n \to k$ and define $d\colon [2^{<\omega}]^n \to k$ as follows: for all $\sigma_0 \prec \cdots \prec \sigma_{n-1}$ in $2^{<\omega}$, let

$$d(\sigma_0, \ldots, \sigma_{n-1}) = c(|\sigma_0|, \ldots, |\sigma_{n-1}|).$$

Note that $(|\sigma_0|, \ldots, |\sigma_{n-1}|) \in [\omega]^n$ since $|\sigma_0| < \cdots < |\sigma_{n-1}|$. (The coloring $d$ is called the *level coloring* induced by $c$.) Now let $T \cong 2^{<\omega}$ be any homogeneous set for $d$, and let $\langle \sigma_i : i \in \omega \rangle$ be the sequence of lexicographically least elements of $T$ Then $H = \{|\sigma_i| : i \in \omega\}$ is an infinite homogeneous set for $c$.

The argument can be formalized to also show that $\mathsf{TT}^n_k \to \mathsf{RT}^n_k$ over $\mathsf{RCA}_0$. (The only point requiring a bit of justification is the existence of the sequence of $\sigma_i$, and this is Exercise 9.13.13.) Thus $\mathsf{TT}^n_k$ inherits all the lower complexity bounds of Ramsey's theorem. Chubb, Hirst, and McNicholl [45] showed that, in terms of the arithmetical and jump hierarchies, it also has the same upper bounds, and indeed behaves very similarly. However, because of the tree structure, some of the proofs—like that of (1) below—are surprisingly more complicated.

**Theorem 9.6.4 (Chubb, Hirst, and McNicholl [45]).**

1. *For all* $n \geqslant 1$, $\mathsf{TT}^n$ *admits* $\Pi^0_n$ *solutions and* $\mathsf{ACA}_0 \vdash \mathsf{TT}^n$.
2. *For all* $n \geqslant 2$, $\mathsf{TT}^n_2$ *omits* $\Sigma^0_n$ *solutions and* $\mathsf{RCA}_0 \vdash \mathsf{ACA}'_0 \leftrightarrow \mathsf{TT}$.
3. *For all* $n \geqslant 3$, $\mathsf{TT}^n_2$ *codes the jump and* $\mathsf{RCA}_0 \vdash \mathsf{ACA} \leftrightarrow \mathsf{TT}^n_2 \leftrightarrow \mathsf{TT}^n$.
4. $\mathsf{RCA}_0 \vdash (\forall n, k \geqslant 1)[\mathsf{TT}^n_k \to \mathsf{RT}^n_k]$.
5. $\mathsf{RCA}_0 \vdash \mathsf{I}\Sigma^0_2 \to \mathsf{TT}^1 \to \mathsf{B}\Sigma^0_2$.

Historically, the first evidence that the Chubb–Hirst–McNicholl tree theorem behaves differently from Ramsey's theorem was the following result showing that Hirst's theorem does not lift to the tree setting. The proof utilizes a very clever model theoretic argument to translate a failure of induction into a combinatorial advantage.

**Theorem 9.6.5 (Corduan, Groszek, and Mileti [56]).** $\mathsf{RCA}_0 \nvdash \mathsf{B}\Sigma^0_2 \to \mathsf{TT}^1$.

*Proof.* The main insight is that there exists a computable map $f\colon \omega \times 2^{<\omega} \to 2$ with the following property: for each $n \in \omega$ and $e < n$, if $\Phi_e$ is a computable set $T \cong 2^{<\omega}$ then there exist incomparable nodes $\sigma_0, \sigma_1 \in T$ such that for each $i < 2$, $f(n, \tau) = i$ for all $\tau \in T$ extending $\sigma_i$. This is proved by a finite injury argument with a computably bounded number of injuries (see Exercise 9.13.15). As discussed in Section 6.4, this can be formalized in $\mathsf{RCA}_0$.

We will use this to construct a model of $\mathsf{RCA}_0 + \mathsf{B}\Sigma^0_2 + \neg\mathsf{TT}^1$. Start with model $\mathcal{M}$ of $\mathsf{RCA}_0 + \mathsf{B}\Sigma^0_2 + \neg\mathsf{I}\Sigma^0_2$. Fix a $b \in M$ and a function $g\colon (M \upharpoonright b) \times M \to M$ as given by Exercise 6.7.5. So there is a proper cut $I \subseteq M$ such that $\mathcal{M} \vDash \lim_s g(a, s)$ exists for all $a \in I$, and the values of these limits are unbounded in $M$. Also, $g$ is $\Delta^0_1$-definable in $\mathcal{M}$. Let $A$ be the join of the set parameters in this definition, so that $g$ is actually $\Delta^0_1(A)$-definable (with no other set parameters).

Let $f$ be as in the observation above, relativized to $A$. Notably, $f$ is also $\Delta^0_1(A)$-definable in $\mathcal{M}$. We define a $\Delta^0_1(A)$-definable instance $c$ of $\mathsf{TT}^1$ as follows. Given $\sigma \in 2^{<M}$, let $c(\sigma)$ be the (code of) the sequence

$$\langle f(g(a, |\sigma|), \sigma) : a <^M b\rangle.$$

Thus, $c$ is a $2^b$-coloring.

Now, consider the structure $\mathcal{M}^*$ obtained from $\mathcal{M}$ by restricting the second order part to those $S \in \mathcal{S}^{\mathcal{M}}$ that are $\Delta^0_1(A)$ in $\mathcal{M}$ with no other parameters. This is an $\omega$-submodel of $\mathcal{M}$, so by Theorem 5.9.3, every $\Pi^1_1$ sentence true in $\mathcal{M}$ is also true in $\mathcal{M}^*$. In particular, $\mathcal{M}^* \vDash \mathsf{RCA}_0 + \mathsf{B}\Sigma^0_2$. Also, the defining properties of both $g$ and $f$ are $\Pi^1_1$, and so hold in $\mathcal{M}^*$. We claim that $c$ has no solution in $\mathcal{S}^{\mathcal{M}^*}$. Since $c \in \mathcal{S}^{\mathcal{M}^*}$ it follows that $\mathcal{M}^* \nvDash \mathsf{TT}^1$, which completes the proof.

To prove the claim, consider any $T$ in $\mathcal{S}^{\mathcal{M}^*}$. By definition, $T$ is $A$-computable (in the sense of $\mathcal{M}^*$); hence, there is an $e \in M$ such that $T = \Phi^A_e$ in $\mathcal{M}^*$. Since the values of $\lim_s g(a,s)$ are unbounded on $I$, we can choose $a \in I$ with $\lim_s g(a,s) = n >^M e$. If $T \cong 2^{<M}$, choose $\sigma_0, \sigma_1 \in T$ as in the definition of $f$ for this $n$ and $e$. Also, fix $s_0$ so that $g(a,s) = n$ for all $s \geqslant^M s_0$, and for each $i < 2$, fix $\tau_i \succeq \sigma_i$ in $T$ with $|\tau_i| \geqslant s_0$. Then for each $i < 2$ we have $f(g(a,|\tau_i|),\tau_i) = f(n,\tau_i) = i$, so $f(g(a,|\tau_0|),\tau_0) \neq f(g(a,|\tau_1|),\tau_1)$. But then $c(\tau_0) \neq c(\tau_1)$, so $T$ is not homogeneous for $c$. Since $T$ was an arbitrary element of $\mathcal{S}^{\mathcal{M}^*}$, this proves our claim. □

The reader may wish to ponder why, exactly, the above argument cannot be replicated for $\mathsf{RT}^1$ in place of $\mathsf{TT}^1$. (Of course, we know $\mathsf{B}\Sigma^0_2$ *does* imply $\mathsf{RT}^1$.) Can the argument be adapted to work for $\mathsf{RT}^2_2$?

Using different model theoretic techniques, it is possible to obtain further results concerning the strength of $\mathsf{TT}^1$. The following theorem complements the previous one to establish that both implications in part (5) of Theorem 9.6.4 are strict.

**Theorem 9.6.6 (Chong, Li, Wang, and Yang [35]).** $\mathsf{RCA}_0 \nvdash \mathsf{TT}^1 \to \mathsf{I}\Sigma^0_2$.

And the following provides an extension of Theorem 8.7.20.

**Theorem 9.6.7 (Chong, Wang, and Yang [43]).** $\mathsf{WKL}+\mathsf{RT}^2_2+\mathsf{TT}^1$ *is* $\Pi^0_3$*-conservative over* $\mathsf{RCA}_0 + \mathsf{B}\Sigma^0_2$.

On the second order side, parts (1) and (4) of Theorem 9.6.4 immediately raise the tantalizing possibility that $\mathsf{TT}^2_2$ might lie strictly between $\mathsf{ACA}_0$ and $\mathsf{RT}^2_2$. This is compelling, since $\mathsf{RT}^2_2$ is the strongest of the principles we have seen so far that lie strictly below arithmetical comprehension. The first step in "dethroning" $\mathsf{RT}^2_2$ from this position was the following separation.

**Theorem 9.6.8 (Patey [242]).** $\mathsf{TT}^2_2 \nleq_\omega \mathsf{RT}^2_2$. *Hence, also* $\mathsf{RCA}_0 \nvdash \mathsf{RT}^2_2 \to \mathsf{TT}^2_2$.

The second step was to establish a version of Seetapun's theorem for trees. This was obtained somewhat later.

**Theorem 9.6.9 (Dzhafarov and Patey [89]).** $\mathsf{TT}^2_2$ *admits cone avoidance. Hence, also* $\mathsf{RCA}_0 \nvdash \mathsf{TT}^2_2 \to \mathsf{ACA}_0$.

And later still, it was even shown that a version of Liu's theorem (Theorem 8.6.1) for trees holds as well.

**Theorem 9.6.10 (Chong, Li, Liu, and Yang [42]).** $\mathsf{TT}^2_2$ *admits PA avoidance. Hence, also* $\mathsf{RCA}_0 \nvdash \mathsf{TT}^2_2 \to \mathsf{WKL}$.

Theorem 9.6.8 is obtained using a carefully designed preservation property based on the fact that, in any $T \cong 2^{<\omega}$, it is possible to do "different things" above each of a given pair of incomparable nodes. (Essentially, this is also the basis for the combinatorial ingredient in the proof of Theorem 9.6.5 above.) The proofs of Theorems 9.6.9 and 9.6.10 both rely on an analogue of the Cholak–Jockusch–Slaman decomposition for trees, and in particular, on suitably formulated stable and cohesive versions of $\mathsf{TT}^2_2$. (Dzhafarov, Hirst, and Lakins [84] showed that there are actually several possible candidates for what a stable form of $\mathsf{TT}^2_2$ should look like, and each of these produces its own cohesive version. See Exercise 9.13.14.) In the case of Theorem 9.6.9, strong cone avoidance for $\mathsf{TT}^1_2$ is established, and then cone avoidance for a cohesive form of $\mathsf{TT}^2_2$. The combinatorial core of the argument features a use of so-called *forcing with bushy trees*, originally developed by Kumabe [189, 190] and used widely in the study of the Turing degrees and in algorithmic randomness. (For a nice survey of this method, see Khan and Miller [178].) The proof of Theorem 9.6.10 relies on more intricate preservation properties of the stable and cohesive forms of $\mathsf{TT}^2_2$, and in particular, gives a new and somewhat more modular proof of Liu's theorem.

## 9.7 Milliken's tree theorem

There is another kind of "Ramsey's theorem on trees", of which TT is actually just a special case. This is Milliken's tree theorem, originally proved by Milliken in [214], and later refined in [215]. A slightly more modern proof appears in the book of Todorčević [310]. While the Chubb–Hirst–McNicholl tree theorem (TT) of the preceding section was invented directly in the reverse mathematics literature, Milliken's tree theorem has been studied extensively in descriptive set theory and combinatorics. At first glance, it may perhaps seem as an oddly technical result, but in fact it is incredibly elegant. Importantly, it is also a natural generalization of a number of partition results, including Ramsey's theorem and many others. Such generalizations are a main object of study in *structural Ramsey theory* (see Nešetřil and Rödl [231]).

To state Milliken's tree theorem, we begin with preliminary definitions.

**Definition 9.7.1.** Fix $T \subseteq \omega^{<\omega}$.

1. $T$ is *rooted* if there is an $\alpha \in T$ such that $\alpha \preceq \beta$ for all $\beta \in T$.
2. $T$ is *meet-closed* if for all $\alpha, \beta \in T$, the longest common initial segment of $\alpha$ and $\beta$ belongs to $T$.
3. For $\alpha \in T$, the *level of $\alpha$ in $T$*, denoted $\mathrm{lvl}_T(\alpha)$, is $|\{\tau \in T : \beta \prec \alpha\}|$.
4. For each $n \in \omega$, $T(n) = \{\alpha \in T : \mathrm{lvl}_T(\alpha) = n\}$.
5. The *height* of $T$, denoted $\mathrm{ht}(T)$, is $\sup\{n > 0 : T(n-1) \neq \emptyset\}$.
6. For $k \in \omega$, $\alpha \in T$ is *$k$-branching in $T$* if there exist exactly $k$ many distinct $\beta \succ \alpha$ in $T$ with $\mathrm{lvl}_T(\beta) = \mathrm{lvl}_T(\alpha) + 1$.
7. $\alpha \in T$ is a *leaf* of $T$ it is 0-branching in $T$.
8. $T$ is *finitely branching* if every $\alpha \in T$ is $k$-branching for some $k \in \omega$.

So, structurally, rooted meet-closed sets "look like" finitely-branching trees, even though they do not have to be trees at all in our usual sense. (In fact, in the literature these sets are often referred to as "trees" for simplicity, but we will not do so here to avoid any possibility of confusion.)

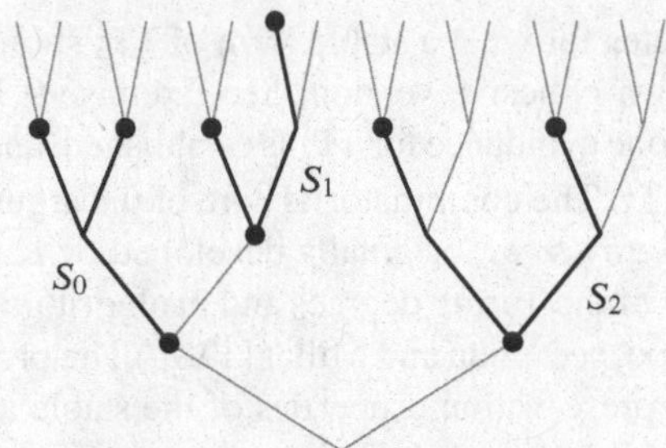


**Figure 9.3.** An illustration of several subsets, $S_0$, $S_1$, and $S_2$, of $T = 2^{\leqslant 4}$. $T$ is represented by thin gray lines. The elements of each $S_i$ are the solid nodes connected by heavy black lines. $S_0$ is not a strong subtree of $T$ because it is not meet closed. $S_1$ is not a strong subtree because it fails condition (1) in Definition 9.7.2. Only $S_2$ is a strong subtree of $T$; $S_2 \in \mathcal{S}_2(T)$. Each of the $S_i$ is isomorphic to $2^{<2}$ as structures under $\preceq$.

**Definition 9.7.2.** Let $T \subseteq \omega^{<\omega}$ be rooted, meet-closed, and finitely branching. A set $S \subseteq T$ is a *strong subtree* of $T$ if it is rooted, meet-closed, and the following are true.

1. There exists $f \colon \mathrm{ht}(S) \to \mathrm{ht}(T)$ so that $\mathrm{lvl}_T(\alpha) = f(\mathrm{lvl}_S(\alpha))$ for all $\alpha \in S$.
2. For all $\alpha \in S$ and $k \in \omega$, if $\alpha$ is $k$-branching in $T$ and $\mathrm{lvl}_T(\alpha) + 1 < \mathrm{ht}(S)$ then $\alpha$ is $k$-branching in $S$.

For $\eta \in \omega \cup \{\omega\}$, $\mathcal{S}_\eta(T)$ is the set of all strong subtrees of $T$ of height $\eta$.

The defining property of the function $f$ above can be restated as follows: nodes at the same level in $S$ must come from the same level in $T$. See Figure 9.3. Notice that if $T \in \mathcal{S}_\omega(2^{<\omega})$ then in particular $T \cong 2^{<\omega}$.

**Definition 9.7.3 (Milliken's tree theorem).**

1. For $n, k \geqslant 1$, $\mathsf{MTT}^n_k$ is the following statement: if $T \subseteq \omega^{<\omega}$ is rooted, meet-closed, and finitely branching, with height $\omega$ and no leaves, then for every $c \colon \mathcal{S}_n(T) \to k$ there exists $S \in \mathcal{S}_\omega(T)$ such that $c$ is constant on $\mathcal{S}_n(S)$.
2. For $n \geqslant 1$, $\mathsf{MTT}^n$ is $(\forall k)\mathsf{MTT}^n_k$.
3. $\mathsf{MTT}$ is $(\forall n)\mathsf{MTT}^n$.

It is not difficult to see that $\mathsf{MTT}^n_k$ implies $\mathsf{TT}^n_k$ for all $n$ and $k$ (see Exercise 9.13.17). Hence, in particular, $\mathsf{MTT}^n_k$ implies $\mathsf{RT}^n_k$ and most of the principles below $\mathsf{ACA}_0$ that we have seen so far. This reflects the aforementioned fact that, combinatorially, it is a common generalization of these and many other partition results.

The problem of finding the complexity of $\mathsf{MTT}$ from the point of view of computability theory and reverse mathematics was first posed by Dobrinen [68] and by Dobrinen, Laflamme, and Sauer [69]. One motivation for this has to do with how Milliken's tree theorem is proved. Namely, every proof of $\mathsf{MTT}$ actually proves a seemingly stronger *product form* of Milliken's tree theorem (which we define below). It is natural to wonder if this is necessarily so, and of course reverse mathematics lends itself well to this analysis. Indeed, we will see that the answer is yes. First, we need some additional definitions.

**Definition 9.7.4.** Fix $d \geqslant 1$ and let $T_0, \ldots, T_{d-1} \subseteq \omega^{<\omega}$ be rooted, meet-closed, and finitely branching. For $\eta \in \omega \cup \{\omega\}$, $\mathcal{S}_\eta(T_0, \ldots, T_{d-1})$ is the set of all $(S_0, \ldots, S_{d-1})$ such that $S_i \in \mathcal{S}_\eta(T_i)$ for all $i < d$, and the function $f$ witnessing this (from Definition 9.7.2 (1)) is the same for all $i$.

**Definition 9.7.5 (Product form of Milliken's tree theorem).**

1. For $n, k \geqslant 1$, $\mathsf{PMTT}^n_k$ is the following statement: fix $d \geqslant 1$ and let $T_0, \ldots, T_{d-1} \subseteq \omega^{<\omega}$ be rooted, meet-closed, and finitely branching, with height $\omega$ and no leaves. Then for every $c \colon \mathcal{S}_n(T_0, \ldots, T_{d-1}) \to k$ there exists $\langle S_0, \ldots, S_{d-1} \rangle \in \mathcal{S}_\omega(T_0, \ldots, T_{d-1})$ such that $c$ is constant on the set $\mathcal{S}_n(S_0, \ldots, S_{d-1})$.
2. For $n \geqslant 1$, $\mathsf{PMTT}^n$ is $(\forall k)\mathsf{PMTT}^n_k$.
3. $\mathsf{PMTT}$ is $(\forall n)\mathsf{PMTT}^n$.

Thus, Milliken's tree theorem is just the product form of Milliken's tree theorem with $d = 1$. As pointed out in [214], the principle $\mathsf{PMTT}^1$ was first proved by Laver (unpublished), and then Pincus [248] showed it to be a consequence of an earlier combinatorial theorem due to Halpern and Laüchli [135]. As a result, $\mathsf{PMTT}^1$ is commonly referred to either as the *Halpern–Laüchli theorem* or the *Halpern–Laüchli–Laver–Pincus (HLLP) theorem.*

It may look like $\mathsf{PMTT}$ is just the parallelization of $\mathsf{MTT}$ (in the sense of Definition 3.1.5), but the requirement in the definition of $\mathcal{S}_\eta(T_0, \ldots, T_{d-1})$ of a common level function significantly complicates the combinatorics. Indeed, already proving $\mathsf{PMTT}^1$ seems to encompass most of the combinatorial complexity of (the full) Milliken's tree theorem. This is quite a contrast from, say, $\mathsf{RT}^1$ and (the full) Ramsey's theorem. However, it turns out that in terms of *computational* complexity the situations are the same. The first part of the following result was independently noted by Simpson [285, Theorem 15].

**Theorem 9.7.6 (Anglès d'Auriac, Cholak, Dzhafarov, Monin, and Patey [61]).** $\mathsf{PMTT}^1$ *admits computable solutions. In fact, for every* $A \in 2^{<\omega}$, *every* $A$*-computable instance of* $\mathsf{PMTT}^1$ *theorem has an* $A$*-computable solution whose index can be found uniformly arithmetically in* $A$.

Much like in standard proofs of Ramsey's theorem (e.g., Lemma 8.1.2) $\mathsf{PMTT}$ can be proved by induction. The inductive basis is $\mathsf{PMTT}^1$, and the inductive step, that $\mathsf{PMTT}^{n+1}$ follows from $\mathsf{PMTT}^n$, is obtained by applying $\mathsf{PMTT}^1$ again. This immediately yields part (1) of the following corollary. Combining this with earlier results then yields parts (2) and (3).

**Corollary 9.7.7 (Anglès d'Auriac, Cholak, Dzhafarov, Monin, and Patey [61]).**

1. *For all* $n, k \geqslant 1$, $\mathsf{ACA}_0 \vdash \mathsf{PMTT}^n_k$.
2. *For all* $n \geqslant 3$, $\mathsf{RCA}_0 \vdash \mathsf{ACA} \leftrightarrow \mathsf{PMT}^n \leftrightarrow \mathsf{MTT}^n \leftrightarrow \mathsf{TT}^n \leftrightarrow \mathsf{RT}^n$.
3. $\mathsf{RCA}_0 \vdash \mathsf{ACA}' \leftrightarrow \mathsf{PMTT} \leftrightarrow \mathsf{MTT} \leftrightarrow \mathsf{TT} \leftrightarrow \mathsf{RT}$. *In fact,* $\mathsf{PMTT} \equiv_{\mathrm{W}} \mathsf{MTT}$.

Part (3) helps explain why a proof of $\mathsf{MTT}$ that is not also a proof of $\mathsf{PMTT}$ is difficult to find. By the metrics of reverse mathematics, the two principles have the same combinatorial cores.

Once again, the $n = 2$ situation turns out to be different.

**Theorem 9.7.8 (Anglès d'Auriac, Cholak, Dzhafarov, Monin, and Patey [61]).** $\mathsf{PMTT}^2$ *admits cone avoidance. Hence, also* $\mathsf{RCA}_0 \nvdash \mathsf{PMTT}^2 \to \mathsf{ACA}_0$.

The proof follows the by-now familiar pattern. First, we formulate a stable version of $\mathsf{PMTT}^2$ and show that every $A$-computable instance of the stable version yields an $A'$-computable instance of $\mathsf{PMTT}^1$ with the same solutions (see Exercise 9.13.18). This is exactly analogous to the relationship between $\mathsf{D}^2$ and $\mathsf{RT}^1$ given by Propositions 8.4.4 and 8.4.5. Second, a cohesive version is formulated, which is simply the following: for all $k \geqslant 1$ and all instances $c\colon \mathcal{S}_2(T_0,\ldots,T_{d-1}) \to k$ of $\mathsf{PMTT}^2_k$ there exists $\langle S_0,\ldots,S_{d-1}\rangle \in \mathcal{S}_\omega(T_0,\ldots,T_{d-1})$ such that $c \upharpoonright \mathcal{S}_2(S_0,\ldots,S_{d-1})$ is stable. The main work then falls to showing that this cohesive version admits cone avoidance, and that $\mathsf{PMTT}^1$ admits strong cone avoidance.

A number of questions remain open around $\mathsf{PMTT}^2$ and $\mathsf{MTT}^2$. In particular, the root of Dobrinen's question about the strength of Milliken's tree theorem remains open for colorings of strong subtrees of height 2.

*Question 9.7.9.* Is it the case that $\mathsf{RCA}_0 \vdash \mathsf{MTT}^2 \to \mathsf{PMTT}^2$? How do $\mathsf{PMTT}^2$ and $\mathsf{MTT}^2$ compare under reducibilities finer than $\leqslant_c$?

Likewise the precise relationship between $\mathsf{MTT}^2$ and $\mathsf{TT}^2$ is unknown.

*Question 9.7.10.* Is it the case that $\mathsf{RCA}_0 \vdash \mathsf{TT}^2 \to \mathsf{MTT}^2$?

Finally, while we have an analogue of Seetapun's theorem for Milliken's tree theorem, it is open whether the analogue of Liu's theorem holds as well.

*Question 9.7.11.* Over $\mathsf{RCA}_0$, does $\mathsf{MTT}^2$ (or even $\mathsf{PMTT}^2$) imply $\mathsf{WKL}$?

## 9.8 Thin set and free set theorems

In this section, we consider the *thin set theorem* and *free set theorem*, both of which are interesting consequences of Ramsey's theorem. These were first described in the context of reverse mathematics by Friedman [110] (see also Simpson [287]) and later expounded in an open questions survey by Friedman and Simpson [116]. The first deep dive into these questions came a bit later, by Cholak, Giusto, Hirst, and Jockusch [32].

**Definition 9.8.1.** Fix $n \geqslant 1$, $X, Y \in 2^\omega$, and a coloring $c\colon [X]^n \to Y$.

1. A set $Z \subseteq X$ is *thin* for $c$ if $c$ is not surjective.
2. A set $Z \subseteq X$ is *free* for $c$ if for all $\vec{x} \in [Z]^n$, if $f(\vec{x}) \in Z$ then $f(\vec{x}) \in \vec{x}$.

**Definition 9.8.2 (Thin set theorem, free set theorem).** Fix $n \geqslant 1$.

1. $\mathsf{TS}^n_\omega$ is the following statement: every $c\colon [\omega]^n \to \omega$ has an infinite thin set.
2. $\mathsf{FS}^n_\omega$ is the following statement: every $c\colon [\omega]^n \to \omega$ has an infinite free set.

We note that some texts, e.g., [32] and [147], use the alternate abbreviations $\mathsf{TS}^n$, $\mathsf{FS}^n$, $\mathsf{TS}(n)$, and $\mathsf{FS}(n)$. As we will see, the above notation can be more easily generalized in a manner that is consistent with other principles. In the case of the free set theorem, it also makes it easier to discern from the notation $\mathsf{FS}(X)$ that will be defined in the next section.

One example of a "free set" in mathematics is a set $X$ of linearly independent vectors in a vector space. This has the property that if $x_0, \ldots, x_{n-1} \in X$ and $c(x_0, \ldots, x_{n-1})$ is a linear combination of these vectors that is also in $X$, then necessarily $c(x_0, \ldots, x_{n-1}) \in \{x_0, \ldots, x_{n-1}\}$, by linear independence. $\mathsf{FS}$ can thus be seen as some kind of combinatorial distillation of this and similar notions of "independence". (For some other examples, see [32]. A related discussion also appears in Knight [182].) In contrast, $\mathsf{TS}$ may appear to be somewhat less motivated. The two nonetheless have much in common.

**Theorem 9.8.3 (Friedman [110]; also Cholak, Giusto, Hirst, and Jockusch [32]).** *Fix $n \geqslant 1$.*

1. $\mathsf{RCA}_0 \vdash \mathsf{RT}^n_2 \to \mathsf{FS}^n_\omega \to \mathsf{TS}^n_\omega$.
2. $\mathsf{RCA}_0 \vdash \mathsf{FS}^{n+1}_\omega \to \mathsf{FS}^n_\omega$.
3. $\mathsf{RCA}_0 \vdash \mathsf{TS}^{n+1}_\omega \to \mathsf{TS}^n_\omega$.

*Proof.* We prove the implication $\mathsf{RT}^n_2 \to \mathsf{FS}^n_\omega$ in part (1) and leave the other implications for the exercises. We argue in $\mathsf{RCA}_0 + \mathsf{RT}^n_2$. Fix $c \colon [\mathbb{N}]^n \to \mathbb{N}$, an instance of $\mathsf{FS}^n_\omega$.

Given $\vec{x} \in [\mathbb{N}]^n$, let $k(\vec{x})$ denote the least $k < n$, if it exists, such that $c(\vec{x}) < \vec{x}(k)$. Now by recursion, define $\vec{x}^0 = \vec{x}$, and having defined $\vec{x}^m \in [\mathbb{N}]^n$ for some $m$, if $i(\vec{x}^m)$ is defined then define $\vec{x}^{m+1} \in [\mathbb{N}]^n$ with

$$\vec{x}^{m+1}(k) = \begin{cases} \vec{x}^m(k) & \text{if } k \neq k(\vec{x}^m), \\ c(\vec{x}^m) & \text{if } k = k(\vec{x}^m) \end{cases}$$

for all $k < n$. Hence, $\vec{x}^{m+1}(k(\vec{x}^m)) < \vec{x}^m(k(\vec{x}^m))$. Let $m(\vec{x})$ be the least $m$ such that $k(\vec{x}^m)$ is not defined or $k(\vec{x}^m) \neq k(\vec{x})$. Note that the set of $m$ for which $k(\vec{x}^m)$ is defined and equal to $k(\vec{x})$ is $\Sigma^0_0$-definable and therefore exists. If this set is empty, then $m(\vec{x}) = 0$. Otherwise, $m(\vec{x})$ exists by $\mathsf{L}\Sigma^0_0$.

We now define an instance $d$ of $\mathsf{RT}^n_{2n+2}$, which is a consequence of $\mathsf{RT}^n_2$, as follows: given $\vec{x} \in [\mathbb{N}]^n$, let

$$d(\vec{x}) = \begin{cases} 0 & \text{if } c(\vec{x}) \in \vec{x}, \\ 1 & \text{if } c(\vec{x}) > \vec{x}, \\ 2k + b & \text{if } c(\vec{x}) \notin \vec{x} \wedge c(\vec{x}) \not> \vec{x} \wedge k = k(\vec{x}) \wedge m(\vec{x}) \equiv b \pmod 2. \end{cases}$$

Note that $d$ is well-defined: if $c(\vec{x}) \notin \vec{x}$ and $c(\vec{x}) \not> \vec{x}$ then $k(\vec{x})$ is defined and so is $m(\vec{x})$. Let $H$ be any infinite homogeneous set for $d$, say with color $i < 2n + 2$. We break into cases according to the value of $i$.

*Case 1: $i = 0$.* In this case, $H$ is a free set for $c$.

*Case 2: $i = 1$.* We define an infinite set $X = \{x_0 < x_1 < \cdots\} \subseteq H$ recursively as follows. Let $x_0 = \min H$. Now fix $s > 0$ and assume we have defined $x_t$ for all $t < s$. Let $x_s$ be the least $x$ larger than $x_{s-1}$ and $c(\vec{x})$ for every $\vec{x} \in [\{x_t : t < s\}^n]$. (Since the latter set is finite, the range of $c$ restricted to it is finite by Proposition 6.2.4, and so $x$ exists because $H$ is infinite.) This completes the construction. Now fix any $\vec{x} \in [X]^n$. We claim that $c(\vec{x}) \notin X$, whence it follows that $X$ is free for $c$. Suppose not. Say $c(\vec{x}) = x_s$. Since $X$ is a subset of $H$, it is still homogeneous for $d$ with color 1, and so $c(\vec{x}) > \vec{x}$. It follows that $\vec{x} \in [\{x_t : t < s\}]^n$. But by construction, $x_s$ is larger than $c(\vec{y})$ for all $\vec{y} \in [\{x_t : t < s\}]^n$. In particular, $x_s > c(\vec{x}) = x_s$, a contradiction.

*Case 3: $i > 1$.* We claim that $H$ is free for $c$. To see this, consider any $\vec{x} \in [H]^n$ and suppose $c(\vec{x}) \in H$. Since we are in Case 3, $k(\vec{x})$ and $m(\vec{x})$ are defined. And since $\vec{x} \in [H]^n$ and $c(\vec{x}) \in H$, we must also have $\vec{x}^1 \in [H]^n$. By homogeneity of $H$, it follows that $d(\vec{x}) = d(\vec{x}^1)$, which means that $k(\vec{x}) = k(\vec{x}^1)$ and $m(\vec{x}) \equiv m(\vec{x}^1) \pmod 2$. But $(\vec{x}^1)^m = \vec{x}^{m+1}$ for all $m$, so clearly $m(\vec{x}) = m(\vec{x}^1) + 1$. Hence, no $\vec{x} \in [H]^n$ can have $c(\vec{x}) \in H$. □

**Corollary 9.8.4.**

1. $\mathsf{RCA}_0 \vdash \mathsf{FS}^1_\omega$ *and for all $n \geqslant 2$,* $\mathsf{ACA}_0 \vdash \mathsf{FS}^2_\omega$.
2. $\mathsf{ACA}_0' \vdash \mathsf{FS}_\omega$.

Note that the proof above actually shows that $\mathsf{FS}^n_\omega \leqslant_c \mathsf{RT}^n_{2n+2}$. The following question is open:

*Question 9.8.5.* Is it the case that $\mathsf{FS}^n_\omega \leqslant_c \mathsf{RT}^n_j$ for some $j < 2n + 2$?

Friedman [110] showed separately that $\mathsf{FS}^n_\omega \leqslant_\omega \mathsf{RT}^{n+1}_{n+2}$. Since $\mathsf{RT}^3_2 \to \mathsf{ACA} \to \mathsf{RT}^{n+1}_{n+2}$, this yields an alternative proof of the above theorem for $n \geqslant 3$.

A minor modification to the proof of Theorem 9.8.3 can be used to show that the solution produced there can be chosen to be $\Pi^0_n$. By a separate argument similar to the proof of Theorem 8.2.2, it is not difficult to build a computable instance of $\mathsf{TS}^n_\omega$ with no $\Delta^0_n$ solution. Hence, the free set theorem and consequently also the thin set theorem enjoy the same arithmetical bounds familiar to us from Ramsey's theorem and, by now, many other principles.

**Theorem 9.8.6 (Cholak, Giusto, Hirst, and Jockusch [32]).**

1. *For every $n \geqslant 1$, $\mathsf{FS}^n_\omega$ admits $\Pi^0_n$ solutions.*
2. *For every $n \geqslant 1$, $\mathsf{TS}^n_\omega$ omits $\Sigma^0_n$ solutions.*

Surprisingly, the free set theorem is much more closely related to the *rainbow* Ramsey's theorem than to Ramsey's theorem itself. The first result showing this was by Wang, which has an immediate and important corollary.

**Theorem 9.8.7 (Wang [321]).** *For all $n \geqslant 1$, $\mathsf{RRT}^n_2$ is uniformly identity reducible to $\mathsf{FS}^n_\omega$ and $\mathsf{RCA}_0 \vdash \mathsf{FS}^n_\omega \to \mathsf{RRT}^n_2$.*

**Corollary 9.8.8.** $(\forall n)\mathsf{FS}^n_\omega$ *admits strong cone avoidance. Hence,* $\mathsf{TJ} \not\leqslant_\omega \mathsf{FS}^n_\omega$ *and* $\mathsf{RCA}_0 \nvdash \mathsf{FS}^n_\omega \to \mathsf{ACA}_0$.

*Proof.* Immediate by Theorem 9.4.4. □

Later, Patey obtained a partial reversal to Wang's theorem.

**Theorem 9.8.9 (Patey [240]).** *For all* $n \geqslant 1$, $\mathsf{RCA}_0 \vdash \mathsf{RRT}^{2n+1}_2 \to \mathsf{FS}^2_\omega$.

**Corollary 9.8.10.** $(\forall n)\mathsf{RRT}^n_2 \equiv_\omega (\forall n)\mathsf{FS}^n_\omega$.

It remains open whether this equivalence extends to arbitrary models.

*Question 9.8.11.* Is it the case that $\mathsf{RCA}_0 \vdash (\forall n)\mathsf{RRT}^n_2 \leftrightarrow (\forall n)\mathsf{FS}^n_\omega$?

It also remains open how close the free set theorem and thin set theorem are to one another. The combinatorics of the two theorems are very similar, which has thus far frustrated all attempts to separate them.

*Question 9.8.12.* Is it the case that $\mathsf{RCA}_0 \vdash (\forall n)[\mathsf{TS}^n_\omega \to \mathsf{FS}^n_\omega]$? For each $n$, does there exist an $m \geqslant n$ such that $\mathsf{RCA}_0 \vdash \mathsf{TS}^m_\omega \to \mathsf{FS}^n_\omega$?

In light of Theorem 9.8.9, the following result may be viewed as a partial step towards a positive answer to the second question.

**Theorem 9.8.13 (Patey [240]).** $\mathsf{RCA}_0 \vdash \mathsf{TS}^2_\omega \to \mathsf{RRT}^2_2$.

This is currently the only nontrivial lower bound on the strength of $\mathsf{TS}^2_\omega$ in terms of implications over $\mathsf{RCA}_0$, extending a prior result by Rice [257] that $\mathsf{TS}^2_\omega$ implies $\mathsf{DNR}$.

There is a natural generalization of $\mathsf{TS}^n_\omega$ that was originally considered by Dorais, Dzhafarov, Hirst, Mileti, and Shafer [72].

**Definition 9.8.14 (Thin set theorem for finite colorings).** For $n \geqslant 1$ and $k \geqslant 2$, $\mathsf{TS}^n_k$ is the following statement: every $c \colon [\omega]^n \to \omega$ has an infinite thin set.

The indexing here works opposite to that of Ramsey's theorem: if $k > j$, then it is $\mathsf{TS}^n_k$ that is *prima facie* weaker than $\mathsf{TS}^n_j$.

**Proposition 9.8.15 (Dorais, Dzhafarov, Hirst, Mileti, and Shafer [72]).** *Fix* $n \geqslant 1$ *and* $k > j \geqslant 2$.

1. $\mathsf{TS}^n_k$ *is uniformly identity reducible to* $\mathsf{TS}^n_j$.
2. $\mathsf{TS}^n_\omega$ *is uniformly identity reducible to* $\mathsf{TS}^n_k$.
3. $\mathsf{RCA}_0 \vdash \mathsf{TS}^n_\omega \to \mathsf{TS}^n_j \to \mathsf{TS}^n_k$.

The implications are strict.

**Theorem 9.8.16 (Patey [243]).** *For all* $n \geqslant 1$ *and* $k > j \geqslant 2$, $\mathsf{TS}^n_j \not\leqslant_\omega \mathsf{TS}^n_k$. *Hence, also* $\mathsf{RCA}_0 \nvdash \mathsf{TS}^n_k \to \mathsf{TS}^n_j$.

Compare this with Proposition 4.3.7 and Theorems 9.1.1 and 9.1.11 for Ramsey's theorem, where the principles $\mathsf{RT}^n_k$ for different values of $k$ are equivalent over $\mathsf{RCA}_0$ but can be separated under finer reducibilities.

Notice that $\mathsf{TS}^n_2$ is *exactly* $\mathsf{RT}^n_2$, so Theorem 9.8.16 actually gives us a strictly descending sequence of principles weaker than $\mathsf{RT}^n_2$,

$$\mathsf{RT}^n_2 = \mathsf{TS}^n_2 \to \mathsf{TS}^n_3 \to \mathsf{TS}^n_4 \to \cdots \to \mathsf{TS}^n_\omega.$$

This reveals something curious. On one end of the above spectrum, the thin set theorem behaves like Ramsey's theorem (because the two coincide). On the other end, it behaves quite differently. For example, for $n \geqslant 3$ we know $\mathsf{TS}^n_2 = \mathsf{RT}^n_2$ is equivalent to $\mathsf{ACA}_0$, but even $(\forall n)\mathsf{TS}^n_\omega$ is strictly weaker. For any such behavior, we can ask where the demarcation occurs, whether between $k = 2$ and $k > 3$, between $k \in \omega$ and $\omega$, or—the most interesting possibility—somewhere in-between. It turns out that coding the jump is an example of this latter case.

**Proposition 9.8.17 (Dorais, Dzhafarov, Hirst, Mileti, and Shafer [72]).** *For all* $n \geqslant 1$, $\mathsf{RCA}_0 \vdash \mathsf{TS}^{n+2}_{2^n} \to \mathsf{ACA}_0$.

*Proof.* We argue in $\mathsf{RCA}_0 + \mathsf{TS}^{n+2}_{2^n}$. Fix an injection $f : \mathbb{N} \to \mathbb{N}$. We show range$(f)$ exists. For each $k < n$, define $c_k : \mathbb{N}^{n+2} \to 2$ as follows: given $\vec{x} \in [\mathbb{N}]^{n+2}$, let

$$c_k(\vec{x}) = \begin{cases} 1 & \text{if } (\exists z)[\vec{x}(k) < y < \vec{x}(k+1) \wedge f(y) < \vec{x}(0)], \\ 0 & \text{otherwise.} \end{cases}$$

Now define $c : \mathbb{N}^{n+2} \to 2^n$ by $c(\vec{x}) = \langle c_k(\vec{x}) : k < n\rangle$. Apply $\mathsf{TS}^{n+2}_{2^n}$ to find an infinite thin set $X$ for $c$. Say $\langle b_k : k < n\rangle \in 2^n$ is such that $c(\vec{x}) \neq \langle b_k : k < n\rangle$ for all $\vec{x} \in [X]^{n+2}$. Fix the largest $k_0 < n$ such that for all $k < k_0$ and all $m$, there exists $\vec{x} \in [X \cap [m, \infty)]^{n+2}$ with $c_k(\vec{x}) = b_k$. (Note that $k_0$ exists because $n$ is standard.) We claim that $b_{k_0} = 1$. Indeed, consider any $\vec{x} = \langle x_0, \ldots, x_{n+1}\rangle \in [X]^{n+2}$ with $c_k(\vec{x}) = b_k$ for all $k < k_0$. Since $f$ is injective, the set $F = \{y \in \mathbb{N} : f(y) < \vec{x}(0)\}$ is finite by Proposition 6.2.7. Choose $x^*_{k_0} < x^*_{k_0+1} < \cdots < x^*_{n+1}$ in $X$ with $x^*_{k_0}$ larger than $x_{k_0}$ and $\max F$. For $k < k_0$ let $x^*_k = x_k$, and let $\vec{x}^* = \langle x^*_0, \ldots, x^*_{n+1}\rangle \in [X]^{n+2}$. Then $c_k(\vec{x}^*) = b_k$ for all $k < k_0$, and $c_{k_0}(\vec{x}^*) = 0$. By maximality of $k_0$ we cannot have $c_{k_0}(\vec{x}^*) = b_{k_0}$, hence $b_{k_0} = 1$ as claimed. With this in hand, we can give a $\Pi^0_1$ definition of the range of $f$, as follows. A number $z \in \mathbb{N}$ belongs to range$(f)$ if and only if for all $\vec{x} \in [X]^{n+2}$ with $z < \vec{x}(0)$ and $c_k(\vec{x}) = b_k$ for all $k < k_0$ there is a $y \leqslant \vec{x}(k_0)$ such that $f(y) = z$. Indeed, if for some such $\vec{x}$ we had that $f(y) = z$ for a $y > \vec{x}(k_0)$ then we could choose $x^*_{k_0+1} < \cdots < x^*_{n+1}$ in $X$ with $y < x^*_{k_0+1}$, let $x^*_k = x_k$ for all $k \leqslant k_0$ and $\vec{x}^* = \langle x^*_0, \ldots, x^*_{n+1}\rangle \in [X]^{n+2}$, and obtain that $c_{k_0}(\vec{x}^*) = 1 = b_{k_0}$, a contradiction. □

By contrast, we have the following result related to Corollary 9.8.8.

**Theorem 9.8.18 (Wang [321]).** *For each* $n \geqslant 1$ *there is a* $k \geqslant 2$ *such that* $\mathsf{TS}^n_k$ *admits strong cone avoidance. In particular,* $\mathsf{TJ} \nleq_\omega \mathsf{TS}^n_k$ *and* $\mathsf{RCA}_0 \nvdash \mathsf{TS}^n_k \to \mathsf{ACA}_0$.

By Proposition 9.8.17, the $k$ in Wang's theorem must be larger than $2^n$. Wang himself showed that if $k$ is at least the $n$th *Schröder number*, $S_n$, then this is large enough. The Schröder numbers are defined by the following recurrence relation:

$$S_0 = 1,$$
$$S_{n+1} = S_n + \sum_{k \leqslant n} S_k S_{n-k-1}.$$

It can be checked that $S_n > 2^n$ for all sufficiently large $n$, so Wang asked whether the number of colors $k$ above can be characterized exactly. Amazingly, the answer is yes! The characterization is in terms of the *Catalan numbers*, $d_n$, which are defined using the recurrence

$$d_0 = 1,$$
$$d_{n+1} = \sum_{k \leqslant n} d_k d_{n-k-1}.$$

**Theorem 9.8.19 (Cholak and Patey [30]).** *Fix $n \geqslant 1$.*

1. $\mathsf{TS}^n_k$ *admits strong cone avoidance if and only if $k > d_n$.*
2. $\mathsf{TS}^n_k$ *admits cone avoidance if and only if $k > d_{n-1}$.*
3. *If $k \leqslant d_{n-1}$ then* $\mathsf{TS}^n_k$ *codes the jump and* $\mathsf{RCA}_0 \vdash \mathsf{TS}^n_k \rightarrow \mathsf{ACA}_0$.

For $n = 2$, this phenomenon is even more intriguing, since $\mathsf{TS}^2_2 = \mathsf{RT}^2_2$ implies so many other principles while as already mentioned, $\mathsf{TS}^2_\omega$ seems to imply so few. There are many potential questions to investigate. For example, recently, Liu and Patey [199] showed that $\mathsf{EM} \nleqslant_\omega \mathsf{TS}^2_4$, and hence $\mathsf{RCA}_0 \nvdash \mathsf{TS}^2_4 \rightarrow \mathsf{EM}$. The question of whether this is sharp is open.

*Question 9.8.20.* Is it the case that $\mathsf{RCA}_0 \vdash \mathsf{TS}^2_3 \rightarrow \mathsf{EM}$?

## 9.9 Hindman's theorem

We now turn to a partition result that behaves rather differently from Ramsey's theorem and, indeed, is in many ways even more surprising. This is *Hindman's theorem*, first proved by Hindman [143] as the resolution to a longstanding conjecture in combinatorics. As with Ramsey's theorem, there are a number of different proofs of this result. Hindman's original proof was a complicated combinatorial argument. Later, a simpler combinatorial proof was discovered by Baumgartner [12]. Galvin and Glazer (see [50]) gave a proof using idempotent ultrafilters. Blass, Hirst, and Simpson [17] found a way to make Hindman's proof more effective. They used this to establish what is still the best upper bound on the strength of Hindman's theorem in the context of reverse mathematics. We will see this result in Section 9.9.2. There, we will use an elegant and greatly simplified proof of Hindman's theorem due to Towsner [311].

We begin with definitions and the statement of Hindman's theorem.

**Definition 9.9.1.** For $X \subseteq \omega$,

$$\mathrm{FS}(X) = \{x \in \omega : (\exists n \geqslant 1)(\exists y_0 < \cdots < y_{n-1} \in X)[x = y_0 + \cdots + y_{n-1}]\}.$$

Worth stressing is that $\mathrm{FS}(X)$ is the set of all *nonempty* finite sums of *distinct* elements of $X$. We also repeat the caution from the previous section that this notion has nothing to do with the free set theorem studied there.

**Definition 9.9.2 (Hindman's theorem).** HT is the following statement: for every $k \geqslant 1$ and every $c\colon \omega \to k$, there is an infinite set $I \subseteq \omega$ such that $c$ is constant on $\mathrm{FS}(I)$.

The set of finite sums from an infinite set is called an *IP set*, so Hindman's theorem can also be stated as follows: every coloring of the integers is constant on some IP set. Despite this being a common terminology, we will avoid it here in favor of the more explicit FS notation.

In spite of appearances, most similarities between Hindman's theorem and Ramsey's theorem are superficial. Their internal combinatorics seem quite different. For example, as we will see in Section 10.6, Hindman's theorem has a close relationship with topological dynamics; no "dynamical" proof of Ramsey's theorem is known. Indeed, even the following "finite version" of Hindman's theorem, proved some forty years earlier, required considerable techniques.

**Theorem 9.9.3 (Rado [252]).** *For every $n \in \omega$ and every $c\colon \omega \to k$, $k \geqslant 1$, there is a set $F \subseteq \omega$ with $|F| \geqslant n$ such that $c$ is constant on* $\mathrm{FS}(F)$.

As a taste, let us prove this result for $n = 3$ and $k = 2$. Although the argument is elementary, it is quite a bit less trivial than the analogous result for Ramsey's theorem. Indeed, for colorings of $n$-tuples with $n \geqslant 3$, the latter is vacuous, and for colorings of singletons it is trivial. For colorings of pairs, the proof is easy and well-known: any six numbers contain among them three such that a given coloring of pairs is constant on all the pairs formed from these three numbers (a fact that can be summarized as $R^2_2(3) = 6$, using the notation for Ramsey numbers introduced in Section 3.3).

*Proof (of Theorem 9.9.3 for $n = 3$ and $k = 2$).* Without loss of generality, $c(0) = 0$. We may then also assume there are infinitely many $x \in \omega$ with $c(x) = 0$, since otherwise the result is trivial. Now, if there exists positive numbers $x < y$ such that $c(x) = c(y) = c(x + y) = 0$, we can take $F = \{0, x, y\}$. So assume not. Fix positive numbers $x_0 < x_1 < \cdots < x_5$ such that $c(x_i) = 0$ for all $i \leqslant 5$ and such that $x_{i+1} - x_i$ is different for each $i < 5$. Let $d_i = x_{i+1} - x_i$. Since $x_i + d_i = x_{i+1}$ for all $i < 4$, our assumption implies that $c(d_i) = 1$. More generally, since $x_i + d_i + d_{i+1} + \cdots d_{i+j} = x_{i+j+1}$ for all $i, j$ with $i + j < 4$, we have that $c(d_i + d_{i+1} + \cdots d_{i+j}) = 1$. Notice also that we cannot have $c(d_0 + d_3) = c(d_1 + d_4) = c(d_0 + d_1 + d_3 + d_4) = 0$.

So, if $c(d_0 + d_3) = 1$, we can take $F = \{d_0, d_1 + d_2, d_3\}$. If $c(d_1 + d_4) = 1$, we can take $F = \{d_1, d_2 + d_3, d_4\}$. And if $c(d_0 + d_1 + d_3 + d_4) = 0$ we can take $F = \{d_0 + d_1, d_2, d_3 + d_4\}$. □

This proof is a bit cheeky, since we allow ourselves the use of 0. Still, it illustrates some of the subtle ways in which Hindman's theorem is combinatorially more complicated than Ramsey's theorem. In general, the additional relations imposed by sums are quite significant.

### 9.9.1 Apartness, gaps, and finite unions

An important property of HT-solutions is that they always contain subsequences of numbers that are very "spread apart" in a certain technical sense. As we will see, one use of these is to encode information into such solutions. To make this precise, let us begin with some preliminary definitions and results. We have already discussed a number of ways of coding finite sets by numbers. Here, we will employ binary representations, as expressed in the following lemma. The proof is left to Exercise 9.13.10.

**Lemma 9.9.4 (Binary representation of the integers).** *The following is provable in* $\mathsf{RCA}_0$. *For every $x \in \mathbb{N}$, there exists a unique $\sigma \in \mathbb{N}^{<\mathbb{N}}$ with $\sigma(i) < \sigma(j)$ for all $i < j < |\sigma|$ and such that $x = \sum_{i<|\sigma|} 2^{\sigma(i)}$.*

Let us fix some notation to reflect this coding.

**Definition 9.9.5.** The following definitions are made in $\mathsf{RCA}_0$.

1. Let $b\colon \mathbb{N} \to \mathcal{P}_{\text{fin}}(\mathbb{N})$ be the function which assigns to each $x \in \mathbb{N}$ the range of the unique $\sigma \in \mathbb{N}^{<\mathbb{N}}$ given by Lemma 9.9.4.
2. The function $b$ has an inverse, $b^{-1}\colon \mathcal{P}_{\text{fin}}(\mathbb{N}) \to \mathbb{N}$, defined by $b^{-1}(F) = 2^{x_0} + \cdots + 2^{x_{n-1}}$ for all $F = \{x_0 < \cdots < x_{n-1}\} \in \mathcal{P}_{\text{fin}}(\mathbb{N})$.
3. For $x \in \mathbb{N}$, let $\lambda(x) = \min b(x)$ and $\mu(x) = \max b(x)$.

The justification for the maps $b$ and $b^{-1}$ actually being inverses (and hence also for our notation) follows by Lemma 9.9.4. The following basic facts concerning $\lambda$ and $\mu$ are easy to verify in $\mathsf{RCA}_0$.

- For all $x, y \in \mathbb{N}$, if $\lambda(x) = \lambda(y)$ then $\lambda(x + y) \geqslant \lambda(x) + 1$.
- For all $x, y \in \mathbb{N}$, if $\lambda(x) < \lambda(y)$ then $\lambda(x + y) = \lambda(x)$.
- For all nonempty finite sets $F < E$, $\mu(b^{-1}(F)) < \lambda(b^{-1}(E))$.

The next lemma is slightly more subtle. There is a simple inductive proof, but it requires quantifying over infinite sets and hence requires induction for $\Pi^1_1$ formulas. We give a more careful argument that avoids this.

**Lemma 9.9.6.** *The following is provable in* $\mathsf{RCA}_0 + \mathsf{B}\Sigma^0_2$*: for every $m \in \mathbb{N}$ and every infinite $I \subseteq \mathbb{N}$, there exists $x \in \mathrm{FS}(I)$ with $\lambda(x) \geqslant m$.*

*Proof.* Fix $k$ and $I$, and suppose every $x \in \mathrm{FS}(I)$ satisfies $\lambda(x) < m$. By $\mathsf{B}\Sigma^0_2$, in the guise of $\mathsf{RT}^1$, there exists an $\ell < m$ such that $\lambda(x) = \ell$ for infinitely many $x \in \mathrm{FS}(I)$. In particular, the set

$$F = \{\ell < m : (\exists x_0, x_1 \in \mathrm{FS}(I))[x_0 < x_1 \wedge \lambda(x_0) = \lambda(x_1) = \ell]\}$$

is nonempty. Note that $F$ exists by bounded $\Sigma^0_1$ comprehension. We claim that $\ell < m$ belongs to $F$ if and only if there are infinitely many $x \in \mathrm{FS}(I)$ with $\lambda(x) = \ell$. The "if" part is clear. For the "only if", suppose $\ell < k$ belongs to $F$ but there are only finitely many $x$ with $\lambda(x) = \ell$. Then we can fix $x_0 < x_1$ in $\mathrm{FS}(I)$ with $\lambda(x_0) = \lambda(x_1) = \ell$, and such that $\lambda(x) \neq \ell$ for all $x > x_0$ in $\mathrm{FS}(I)$. Let $x_2 = x_1 + (x_0 + x_1)$. We have that $\lambda(x_0 + x_1) > \ell$ and hence $\lambda(x_2) = \ell$. But $x_2 > x_1$, so we have a contradiction to the maximality of $x_1$. Thus, the claim holds. To complete the proof, fix the largest $\ell < m$ in $F$. By the claim, we can fix a sequence $x_0 < x_1 < \cdots$ of elements of $\mathrm{FS}(I)$ with $\lambda(x_i) = \ell$ for all $i$. For each $i$, let $y_i = x_{2i} + x_{2i+1}$. Then $y_0 < y_1 < \cdots$ is a sequence of elements of $\mathrm{FS}(I)$ with $\lambda(y_i) > \ell$ for all $i$. By a further application of $\mathsf{B}\Sigma^0_2$, there is an $\ell^* > \ell$ such that $\lambda(y_i) = \ell^*$ for infinitely many $i$. But then $\ell^* \in F$, contradicting the maximality of $\ell$. We conclude that there is an $x \in \mathrm{FS}(I)$ with $\lambda(x) \geqslant m$, as was to be shown. □

With this in hand, we can state the "apartness" property we mentioned, and prove that it is enjoyed by HT-solutions.

**Definition 9.9.7.**

1. A set $X \subseteq \omega$ satisfies the *apartness property* if for all $x < y$ in $X$, $\mu(x) < \lambda(y)$.
2. HT *with apartness* is the statement of HT, augmented to include that the solution set $I$ satisfies the apartness property.

**Corollary 9.9.8.**

1. *For every infinite set $I \subseteq \omega$ there is an infinite $I$-computable subset of* $\mathrm{FS}(I)$ *satisfying the apartness property.*
2. $\mathsf{RCA}_0 + \mathsf{B}\Sigma^0_2$ *proves that for every infinite set $I \subseteq \mathbb{N}$ there is an infinite subset of* $\mathrm{FS}(I)$ *satisfying the apartness property.*

*Proof.* Define a sequence of elements $x_0 < x_1 < \cdots$ of FS()$I$) recursively as follows. Let $x_0 = \min I$, and given $x_i$ for some $i \in \mathbb{N}$, let $x_{i+1}$ be the least element of $\mathrm{FS}(I)$ such that $x_i > x_{i+1}$ and $\mu(x_i) < \lambda(x_{i+1})$. The existence of $x_{i+1}$ follows by Lemma 9.9.6. □

**Theorem 9.9.9.**

1. $\mathsf{RCA}_0$ *proves that* HT *is equivalent to* HT *with the apartness property.*
2. HT *is Weihrauch equivalent to* HT *with the apartness property.*

*Proof.* For (1), we argue in $\mathsf{RCA}_0 + \mathsf{HT}$. Fix an instance $c \colon \mathbb{N} \to k$ of HT. Apply HT to get a solution, $I$. Now as $\mathsf{RT}^1$ is a subproblem of HT, we have at our disposal $\mathsf{B}\Sigma^0_2$. Hence, we can apply Corollary 9.9.8 to find an infinite set $J \subseteq \mathrm{FS}(I)$ satisfying the apartness property. We have $\mathrm{FS}(J) \subseteq \mathrm{FS}(I)$, hence $c$ is constant on $J$. Part (2) is similar. □

The key point is that apartness really is natural to the combinatorics of Hindman's theorem. We will see several applications of it, both in this section and again in Section 9.9.3. To begin, we will use apartness in an essential way to prove the following:

**Theorem 9.9.10 (Blass, Hirst, and Simpson [17]).** $\mathsf{RCA}_0 \vdash \mathsf{HT} \to \mathsf{ACA}_0$.

As usual, the proof will be a formalization of the fact that HT codes the jump. However, as there are some small subtleties in the proof concerning induction, we will directly give the formal version. First, some definitions.

**Definition 9.9.11.** The following definitions are made in $\mathsf{RCA}_0$, for a fixed function $f \colon \mathbb{N} \to \mathbb{N}$. Let $b \colon \mathbb{N} \to \mathcal{P}_{\text{fin}}(\mathbb{N})$ be the map of Definition 9.9.5.

1. A *gap in* $x \in \omega$ is a pair $\{u, v\} \subseteq b(n)$ such that $u < v$ and there is no $w \in b(x)$ with $u < w < v$.
2. A *short gap in* $x$ (with respect to $f$) is a gap $\{u, v\}$ in $x$ such that there exists $n < u$ in the range of $f$ with $f(s) \neq n$ for all $s < v$.
3. A *very short gap in* $x$ (with respect to $f$) is a gap $\{u, v\}$ in $x$ such that there exists $n < u$ with $f(s) = n$ for some $s < \mu(x)$ but $f(s) \neq n$ for all $s < v$.
4. For $x \in \mathbb{N}$, $\mathrm{SG}(x)$ is the number of short gaps in $n$.
5. For $x \in \mathbb{N}$, $\mathrm{VSG}(x)$ is the number of very short gaps in $n$.

*Proof (of Theorem 9.9.10).* We argue in $\mathsf{RCA}_0 + \mathsf{HT}$. Fix an injective function $f \colon \mathbb{N} \to \mathbb{N}$. We claim that the range of $f$ exists. In what follows, all short gaps and very short gaps will be meant with respect to this $f$.

Define $c \colon \mathbb{N} \to 2$ by $c(x) = 0$ if $\mathrm{VSG}(x)$ is even, and $c(x) = 1$ if $\mathrm{VSG}(x)$ is odd. Notice that $c$ exists by $\Delta^0_1$ comprehension, as the definition of $\mathrm{VSG}(x)$ is uniformly $\Sigma^0_0$ in $x$. Let $I$ be a solution to $c$ as given by HT with apartness.

Consider any $x < y$ in $\mathrm{FS}(I)$. Since $I$ satisfies the apartness property, it follows that $\mu(x) < \lambda(y)$. Hence, the gaps in $x + y$ are precisely the gaps in $x$, the pair $\{\mu(x), \lambda(y)\}$, and the gaps in $y$. Also, every short gap in $x$ or $y$ is short in $x+y$. Thus, $\mathrm{SG}(x+y)$ is either $\mathrm{SG}(x)+\mathrm{SG}(y)+1$ or $\mathrm{SG}(x)+\mathrm{SG}(y)$, depending as $\{\mu(x), \lambda(y)\}$ is or is not short in $x + y$. We claim that the former case is impossible, i.e., that $\{\mu(x), \lambda(y)\}$ is never short in $x + y$. This will complete the proof, since then for all $n \in \mathbb{N}$ we will have

$$n \in \mathrm{range}(f) \leftrightarrow (\forall x, y \in \mathrm{FS}(x))[x < y \wedge n < \mu(x) \to (\exists s < \lambda(y)[f(s) = n]],$$

giving a $\Pi^0_1$ definition of the range of $f$.

To prove the claim, we show that $\mathrm{SG}(x)$ is even for all $x \in \mathrm{FS}(x)$. Thus, we cannot have $\mathrm{SG}(x + y) = \mathrm{SG}(x) + \mathrm{SG}(y) + 1$ above, and the claim follows. So fix $x$. Let $F = \{n < \mu(x) : (\exists s)[f(s) = n]\}$, which exists by bounded $\Sigma^0_1$ comprehension. By $\mathsf{B}\Sigma^0_2$, which we have on account of having assumed HT, we can fix $s_0 > x$ such that for all $n \in F$, $f(s) = n$ for some $s \leqslant s_0$. And because $I$ satisfies the apartness property, we can fix a $y > s$ in $I$. Let us count which of the gaps in $x + y$ are very short. Every very short gap in $y$ is very short in $x + y$ since $\mu(y) = \mu(x + y)$. By choice of $y$, $\{\mu(x), \lambda(y)\}$ is not very short (or even short) in $x + y$. But by contrast, every short gap in $x$ is very short in $x + y$ since $\mu(x) < \lambda(y) < \mu(y)$. Of course, no gap in $x$ that is not short can be very short in $x + y$. Thus, we have

$$\mathrm{VSG}(x + y) = \mathrm{SG}(x) + \mathrm{VSG}(y).$$

But since $y, x+y \in \mathrm{FS}(I)$, we have $c(y) = c(x+y)$, and so $\mathrm{VSG}(y)$ and $\mathrm{VSG}(x+y)$ have the same parity. We conclude that $\mathrm{SG}(x)$ is even. □

We conclude this section by mentioning an oft-stated variant of Hindman's theorem, which is sometimes easier to work with. We call this version the *finite unions theorem*, and list it under the abbreviation FUT below.

**Definition 9.9.12.** Fix $X \subseteq \omega$.

1. $\mathcal{P}_{\text{fin}}(X)$ is the set of nonempty finite subsets of $X$.
2. For $\mathcal{U} \subseteq \mathcal{P}_{\text{fin}}(X)$,

$$\mathrm{FU}(\mathcal{U}) = \{E \in \mathcal{P}_{\text{fin}}(X) : (\exists n \geqslant 1)(\exists F_0, \ldots, F_{n-1} \in \mathcal{U})[E = \bigcup_{i<n} F_i]\}.$$

3. $\mathcal{U} \subseteq \mathcal{P}_{\text{fin}}(X)$ satisfies the *apartness property* if for all $E, F \in \mathcal{U}$, either $E < F$ or $F < E$.

As with HT, we are looking at *nonempty* finite unions here.

**Definition 9.9.13 (Finite unions theorem).** FUT is the following statement: for every $k \geqslant 1$ and every $c\colon \mathcal{P}_{\text{fin}}(\omega) \to k$, there is an infinite set $\mathcal{U} \subseteq \mathcal{P}_{\text{fin}}(\omega)$ with the apartness property such that $c$ is constant on $\mathrm{FU}(\mathcal{S})$.

It is straightforward to formalize this definition in $\mathsf{RCA}_0$. One observation is that, with unions of sets, there is no way to get an analogue of Corollary 9.9.8, so the apartness property is explicitly part of the definition here.

**Proposition 9.9.14.**

*1.* $\mathsf{RCA}_0 \vdash \mathsf{HT} \leftrightarrow \mathsf{FUT}$.
*2.* $\mathsf{HT} \equiv_{\mathrm{W}} \mathsf{FUT}$.

*Proof.* We argue in $\mathsf{RCA}_0$ and prove (1). To show that HT $\to$ FUT, assume HT and let $c\colon \mathcal{P}_{\text{fin}}(\mathbb{N}) \to k$ be given. Let $b\colon \mathbb{N} \to \mathcal{P}_{\text{fin}}(N)$ be the map defined in Definition 9.9.5. Define $d\colon \mathbb{N} \to k$ by $d(x) = c(b(x))$ for all $x$, and let $I$ be an infinite set so that $c$ is constant on $\mathrm{FS}(I)$. By Theorem 9.9.9, we may assume $I$ satisfies the apartness property. Let $\mathcal{U} = \{b(x) : x \in I\}$. Then $\mathcal{U}$ satisfies the apartness property and $c$ is constant on $\mathrm{FU}(\mathcal{U})$.

For the converse, assume FUT and let $c\colon \mathbb{N} \to k$ be given. Let $b^{-1}$ be the inverse map of Definition 9.9.5, and define $d\colon \mathcal{P}_{\text{fin}}(\mathbb{N}) \to k$ by $d(F) = c(b^{-1}(F))$ for all $F \in \mathcal{P}_{\text{fin}}(\mathbb{N})$. Apply FUT to get an infinite set $\mathcal{U} \subseteq \mathcal{P}_{\text{fin}}(\omega)$ with the apartness property such that $d$ is constant on $\mathrm{FU}(\mathcal{U})$. Let $I = \{b^{-1}(F) : F \in \mathcal{U}\}$. If $E, F$ are finite sets with $E < F$ then $b^{-1}(E \cup F) = b^{-1}(E) + b^{-1}(F)$. So by $\Sigma^0_1$ induction on $n$, if $F_0, \ldots, F_{n-1}$ are finite sets such that, for each $i < n-1$, either $F_i < F_{i+1}$ or $F_{i+1} < F_i$, then $b^{-1}(\bigcup_{i<n} F_i) = \sum_{i<n} b^{-1}(F_i)$. It follows that $c$ is constant on $\mathrm{FS}(I)$. □

## 9.9.2 Towsner's simple proof

Hindman's original proof of HT used a complex but purely combinatorial argument. Subsequently, several "high powered" proofs of Hindman's theorem were obtained. One method uses ultrafilters and the topology of the Stone–Čech compactification of $\omega$; see Hindman and Strauss [145]. Another method uses the Auslander–Ellis theorem of topological dynamics; see Furstenberg [120]. We sketch a generalization of this dynamical proof in Theorem 10.6.15. These methods are difficult to formalize in second-order arithmetic although, as we will see, some progress has been made on formalizing the proofs via ultrafilters.

The best known upper bound on HT, which we prove in this section, involves the stronger system $\mathsf{ACA}_0^+$ from Definition 5.6.8. Blass, Hirst, and Simpson obtained the following theorem through an intricate analysis of Hindman's original, purely combinatorial proof.

**Theorem 9.9.15 (Blass, Hirst, and Simpson [17]).** $\mathsf{ACA}_0^+ \vdash \mathsf{HT}$.

The question of whether this bound is optimal is open.

*Question 9.9.16.* Is it the case that $\mathsf{RCA}_0 + \mathsf{HT} \vdash \mathsf{ACA}^+$?

Indeed, it is not even known whether Theorem 9.9.10, which shows that HT codes the jump, can be improved to code the double jump.

*Question 9.9.17.* Does there exists a computable instance of HT every solution to which computes $\varnothing''$?

On the flip side, it is still possible that HT could be provable in $\mathsf{ACA}_0$.

*Question 9.9.18.* Is it the case that $\mathsf{ACA}_0 \vdash \mathsf{HT}$?

By Theorem 5.6.7, a positive answer here would require HT to admit $\Delta^0_n$ solutions for some fixed $n \in \omega$. The strongest result here, brand new at the time of this writing, is that in this case $n$ would have to be at least 4.

**Theorem 9.9.19 (Yuke [332]).** *There exists a computable instance of* HT *with no* $\varnothing''$*-computable solution.*

Let us turn to Theorem 9.9.15. We present a newer proof due to Towsner [311], who called it a "simple proof of Hindman's theorem". And indeed, computability and reverse math aside, Towsner's is arguably the most elementary proof of Hindman's theorem to date. Keeping track of the effectivity in the argument yields the following.

**Theorem 9.9.20.** *For every $A \in 2^\omega$ and every $X \gg A^{(\omega)}$, every $A$-computable instance of* HT *has an $X$-computable solution.*

The proof is readily formalized in $\mathsf{RCA}_0$. Since $\mathsf{ACA}_0^+$ certainly implies WKL, this yields Theorem 9.9.15.

Towsner actually proved the analogue of Theorem 9.9.20 for FUT, rather than HT directly. These are equivalent, as we know, but we have translated our presentation from finite sets and unions to integers and sums. The setup is otherwise unchanged. In Lemma 9.9.24, we incorporate some expositional elements from another presentation of the proof, due to Anglès d'Auriac [60]. We will use the following notions.

**Definition 9.9.21.** Fix $X \subseteq \omega$ and $c\colon \omega \to k$, $k \geqslant 1$. Let $b\colon \omega \to \mathcal{P}_{\text{fin}}(\omega)$ be the map defined in Definition 9.9.5.

1. For $y \in \omega$, $X \perp y = \{x \in X : b(x) \cap b(y) = \varnothing\}$.
2. For $Y \subseteq \omega$, $X \perp Y = \bigcap_{y \in Y} X \perp y$.
3. $X$ *half-matches* $y \in \omega$ (relative to $c$) if $(\exists x \in X)[c(y) = c(x+y)]$.
4. $X$ *full-matches* $y \in \omega$ (relative to $c$) if $(\exists x \in X)[c(y) = c(x+y) = c(x)]$.
5. $X$ *half-matches* $Y \subseteq \omega$ or *full-matches* $Y \subseteq \omega$ if it, respectively, half-matches or full-matches every $y \in Y$.

To begin, we prove a series of technical lemmas. Each comes in two parts: first, a purely combinatorial fact asserting the existence of certain sets; second, a calculation of an upper bound on the complexity of these sets relative to the givens. The reader who has not yet seen a proof of Hindman's theorem may wish, at first pass, to simply read the combinatorial parts, and only then go back and reread the proofs with a view to the effectivity estimates. Incidentally, the hypotheses of all the lemmas are the same. In particular, we always have a fixed HT-instance, $c$. All references to half-matching and full-matching should be understood as being with respect to this $c$, unless explicitly specified otherwise.

**Lemma 9.9.22.** *Suppose $I \subseteq \omega$ satisfies the apartness property, and let $c\colon \mathrm{FS}(I) \to k$, $k \geqslant 1$ be given.*

*1. For every finite set $F \subseteq I$, one of the following holds.*

*a. There is a finite set $D \subseteq I \perp F$ such that for every $x \in \mathrm{FS}(I \perp (F \cup D))$, $F$ does not half-match $x + y$ for some $y \in \mathrm{FS}(D)$.*

*b. There is an infinite set $J \subseteq I \perp F$ such that $F$ half-matches $\mathrm{FS}(J)$.*

*2. If $c$ and $I$ are computable, then $J$ in (b) can be chosen to be computable.*

*Proof.* For (1), fix $F$ and assume (a) fails. We define a set $J^* = \{x_0, x_1, \dots\} \subseteq \min I \perp F$ inductively so that $F$ half-matches any sum of two or more elements if $J^*$. Let $x_0 = \min I \perp F$. Fix $i > 0$, and assume we have defined $x_j$ for all $j < i$. Let $D$ be the finite set $\{x_j : j < i\}$. Then by assumption, there is an $x \in \mathrm{FS}(I \perp (F \cup D))$ such that $F$ half-matches $x + y$ for all $y \in \mathrm{FS}(D)$. Let $x_i$ be the least such $x$. Now any sum of two or more elements of $D \cup \{x_i\}$ is either a sum of two or more elements of $D$ or has the form $x_i + y$ for some $y \in \mathrm{FS}(D)$. Hence, by induction, it is half-matched by

$F$. This completes the construction. Let $J = \{x_{2i} + x_{2i+1} : i \in \mathbb{N}\}$. Clearly, $J^*$ satisfies the apartness property, and hence so does $J$. And as every elements of $\mathrm{FS}(J)$ is a sum of two or more elements of $J^*$, it is half-matched by $F$. Thus, we have (b).

Clearly, $J$ is computable in $c$ and $I$. So we have (2). □

**Lemma 9.9.23.** *Suppose $I \subseteq \omega$ satisfies the apartness property, and let $c\colon \mathrm{FS}(I) \to k$, $k \geqslant 1$ be given.*

1. *There is a finite set $F \subseteq I$ and an infinite set $J \subseteq I \perp F$ such that $F$ half-matches* $\mathrm{FS}(J)$.
2. *If $c$ and $I$ are computable, then $J$ can be chosen to be computable.*

*Proof.* For (1), assume not. Then for every finite set $F \subseteq I$, alternative (a) holds in the previous lemma. For each $i \leqslant k$, we define a finite set $F_i$ inductively as follows. Let $F_0 = \{\min I\}$ and suppose that for some $i \leqslant k-1$, $F_i$ has been defined. By the previous lemma, there is a finite set $D \subseteq I \perp F_i$ such that for every $x \in \mathrm{FS}(I \perp (F_i \cup D))$, $F_i$ does not half-match $x + y$ for some $y \in \mathrm{FS}(D)$. Let $D_i$ be the least such $D$, and let $F_{i+1} = \mathrm{FS}(F_i \cup D_i)$. This completes the construction.

Now, let $F = F_k$ and $J = I \perp F_k$. We claim that $F$ half matches $\mathrm{FS}(J)$, contrary to our assumption.

To prove the claim, fix any $x \in \mathrm{FS}(J)$. We define numbers $x_0, \ldots, x_k$ by reverse induction on $i \leqslant k$, as follows. Let $x_k = x$, noting that $x_k \in \mathrm{FS}(I \perp F_k)$. Now assume that we have defined $x_i$ for some positive $i \leqslant k$ and that $x_i \in \mathrm{FS}(I \perp F_i)$. By construction, there is a $y \in D_{i-1}$ such that $F_{i-1}$ does not half-match $x_i + y$. Let $y_{i-1}$ be the least such $y$, and let $x_{i-1} = x_i + y_{i-1}$. Notice that $x_{i-1} \in \mathrm{FS}(I \perp F_{i-1})$ since $F_{i-1} \subseteq F_i$ and $D_{i-1} \subseteq I \perp F_{i-1}$.

Fix $i < k$, so that $x_i = x_{i+1} + y_i$. By definition, $F_i$ does not half-match $x_{i+1} + y_i$, so we must have that $c(x_{i+1} + y_i) \neq c(x_{i+1} + y_i + y)$ for all $y \in F_i$. Now, a straightforward induction shows that if $j < i$ then $x_j = x_{i+1} + y_i + y$ for some $y \in F_i$. (Induct on $i - j$.) Thus we have in particular that $c(x_j) \neq c(x_{i+1} + y_i) = c(x_i)$. Since, by Proposition 6.2.7, there is no map from $k$ to any $k^* < k$, it must be that every color is used by $c(x_i)$ for some $i < k$. Thus $c(x) = c(x_i)$ for some $i < k$. But $x_i = x + y$ for some $y \in F_k$, so $x_i$ witnesses that $F_k$ half-matches $x$, as was to be shown.

For (2), note that $J$ here was obtained by finitely many iterations of the previous lemma, using the same $c$ and $I$ at each step. □

**Lemma 9.9.24.** *Suppose $I \subseteq \omega$ satisfies the apartness property, and let $c\colon \mathrm{FS}(I) \to k$, $k \geqslant 1$ be given.*

1. *One of the following holds.*
   a. *There is a finite set $F \subseteq I$ and an infinite set $J \subseteq I \perp F$ such that $F$ full-matches* $\mathrm{FS}(J)$.
   b. *There is an $\ell < k$ and an infinite set $J \subseteq I$ such that $c(x) \neq \ell$ for all* $x \in \mathrm{FS}(J)$.
2. *If $c$ and $I$ are computable, then $J$ in (a) can be chosen to be computable, and $J$ in (b) can be chosen to be $\varnothing^{(3)}$-computable.*

*Proof.* To prove (1), assume (a) fails. For each $i$, we define a finite set $F_i$, an infinite set $I_i \subseteq I$, and a finite coloring $c_i$ of $\mathrm{FS}(I_i)$. We will also ensure that $F_{i+1} \subseteq I_i$ and that $I_{i+1} \subseteq I_i \perp F_{i+1}$. Let $F_0 = \emptyset$, $I_0 = I$, and $c_0 = c \upharpoonright \mathrm{FS}(I_0)$. Now fix $i \in \omega$, and assume $F_i$, $I_i$, and $c_i$ have been defined. Apply Lemma 9.9.23 (1) to $c_i$ and $I_i$ to find a finite set $F_{i+1} \subseteq I_i$ and an infinite set $I_{i+1} \subseteq I_i \perp F_{i+1}$ satisfying the apartness condition such that $F_{i+1}$ half-matches $\mathrm{FS}(I_{i+1})$ with respect to $c_i$. By removing an initial segment if necessary, we may also assume $I_{i+1} > F_{i+1}$, which will be useful later. Define $c_{i+1}$ on $\mathrm{FS}(I_{i+1})$ as follows: given $x \in \mathrm{FS}(I_{i+1})$, choose the least $y \in F_{i+1}$ such that $c_i(x) = c_i(x + y)$ and let $c_{i+1}(x) = \langle y, c_i(x)\rangle$. (Thus, $c$ uses at most $|F_{i+1}| \times k$ many colors.)

As we have assumed that (a) fails, it follows that for each $i$, $\mathrm{FS}(\bigcup_{j \leqslant i} F_j)$ does not full-match $\mathrm{FS}(I_i)$ relative to $c$. Let $x_i \in \mathrm{FS}(I_i)$ be the least witness to this fact. Fix the least $\ell < k$ such that $c(x_i) = \ell$ for infinitely many $i$, and let $i_0 < i_1 < \cdots$ enumerate all positive such $i$. Consider any $n \in \omega$. We show that for each $i < i_n$, there exists $y_i \in F_{i+1}$ such that if $y \in \mathrm{FS}(\{y_i : i < i_n\})$ then $c(x_{i_n} + y) = c(x_{i_n})$. First, given $i$ and $x \in \mathrm{FS}(I_{i+1})$, let $y_i(x)$ be the least $y \in F_{i+1}$ such that $c_i(x) = c_i(x + y)$, as in the definition of $c_{i+1}$ above. Then for each $i < i_n$, let $y_i = y_i(x_{i_n})$.

Let us verify that the above choice works. We claim that for all $m \geqslant 1$ and all $y = y_{j_0} + \cdots + y_{j_{m-1}} \in \mathrm{FS}(\{y_i : i < i_n\})$ with $j_{m-1} < \cdots < j_0 < i_n$, the following holds: for all $j < j_{m-1}$,

$$y_j(x_{i_n} + y_{j_0} + \cdots + y_{j_{m-1}}) = y_j. \tag{9.2}$$

and

$$c_j(x_{i_n} + y_{j_0} + \cdots + y_{j_{m-1}}) = c_j(x_{i_n}). \tag{9.3}$$

Taking $j = 0$ in the second equality gives the desired conclusion since $c_0 = c$. By definition, for each $j \leqslant j_0$ we have

$$c_j(x_{i_n} + y) = (y_{j-1}(x_{i_n} + y), c_{j-1}(x_{i_n} + y)).$$

and

$$c_j(x_{i_n}) = \langle y_{j-1}, c_{j-1}(x_{i_n})\rangle.$$

Therefore, if $c_j(x_{i_n} + y) = c_j(x_{i_n})$ then $y_{j-1}(x_{i_n} + y) = y_{j-1}$ and $c_{j-1}(x_{i_n} + y_{j_0}) = c_{j-1}(x_{i_n})$. It thus suffices to prove that

$$c_{j_{m-1}}(x_{i_n} + y) = c_{j_{m-1}}(x_{i_n}), \tag{9.4}$$

and then the claim follows for $y$ by reverse induction. We proceed by induction on $m \geqslant 1$, verifying that (9.4) holds for $m$ and concluding that so do (9.2) and (9.3). If $m = 1$, then $y = y_{j_0}$ and $c_{j_0}(x_{i_n} + y_{j_0}) = c_{j_0}(x_{i_n})$ by choice of $y_{j_0}$. Thus, the claim holds for $m = 1$. Now suppose the claim holds for some $m \geqslant 1$ and fix $y = y_{j_0} + \cdots + y_{j_{m-1}} + y_{j_m} \in \mathrm{FS}(\{y_i : i < i_n\})$ with $j_m < j_{m-1} < \cdots < j_0$. Since $j_m < j_{m-1}$, we have by hypothesis that

$$y_{j_m}(x_{i_n} + y_{j_0} + \cdots + y_{j_{m-1}}) = y_{j_m}$$

and

$$c_{j_m}(x_{i_n} + y_{j_0} + \cdots + y_{j_{m-1}}) = c_{j_m}(x_{i_n}).$$

Together, these yield

$$c_{j_m}(x_{i_n} + y) = c_{j_m}(x_{i_n} + y_{j_0} + \cdots + y_{j_{m-1}}) = c_{j_m}(x_{i_n}).$$

Thus, the claim holds for $m + 1$.

Now for each $n$, all $y \in y \in \mathrm{FS}(\{y_i : i < i_n\})$ must satisfy $c(y) \neq \ell$. This is because by assumption, $\mathrm{FS}(\bigcup_{j \leqslant i_n} F_j)$ does not full-match $x_{i_n}$, $y_i \in F_{i+1}$ for all $i < i_n$, and $c(x_{i_n}) = c(x_{i_n} + y) = \ell$ for all $y \in \mathrm{FS}(\{y_i : i < i_n\})$. We can now complete the proof. Let $T \subseteq \omega^{<\omega}$ be the following tree: $\alpha \in \omega^{<\omega}$ if and only if $\alpha(i) \in F_{i+1}$ for all $i < |\alpha|$ and $c(y) \neq \ell$ for all $y \in \mathrm{FS}(\mathrm{range}(\alpha))$. Clearly, $T$ is finitely-branching, and our argument above shows that for every $n$, there is an $\alpha \in T$ of length $n$. Hence, $T$ is an infinite tree, and so there exists a path $f \in [T]$. Let $J = \mathrm{range}(f)$. By construction, $c(y) \neq \ell$ for all $y \in \mathrm{FS}(J)$. Thus we have proved that (b) holds.

Now for (2), suppose (a) does not hold with a computable witness. We now use Lemma 9.9.24 (2) to define the sequence of $F_i$, $I_i$, and $c_i$, maintaining that $I_i$ and $c_i$ are computable for all $i$. Thus, the hypothesis that (a) fails for computable sets, even if not for all sets, is enough to always guarantee the existence of $F_{i+1}$ and $I_{i+1}$. It is easily seen that, given (indices for) $F_i$, $I_i$, and $c_i$, finding (indices for) $F_{i+1}$ and $I_{i+1}$ can be found uniformly computably in $\varnothing^{(2)}$. Thus the sequence of all the $F_i$ is $\varnothing^{(2)}$-computable, as is the tree $T$ defined from them. This tree is also $\varnothing^{(2)}$-computably bounded, hence a path $f \in [T]$ can be chosen computable in $\varnothing^{(3)}$. As the construction arranges for $F_{i+1} \subseteq I_i > F_i$ for all $i$, it follows that $F_0 < F_1 < \cdots$ and hence that $f$ is an increasing function. Thus, the desired solution $J = \mathrm{range}(f)$ is computable in $f$, and so also in $\varnothing^{(3)}$, as claimed. □

**Lemma 9.9.25.** *Suppose $I \subseteq \omega$ satisfies the apartness property, and let $c \colon \mathrm{FS}(I) \to k$, $k \geqslant 1$ be given.*

1. *One of the following holds.*
   a. *There is a finite set $F \subseteq I$ and an infinite set $J \subseteq I \perp F$ such that $F$ full-matches* $\mathrm{FS}(J)$.
   b. *There is an infinite set $J \subseteq I$ such that $c$ is constant on* $\mathrm{FS}(J)$.
2. *If $c$ and $I$ are computable, then $J$ in (a) or (b) can be chosen to be $\varnothing^{(3k-1)}$-computable.*

*Proof.* For (1), again assume (a) fails. For each $i < k$, we define an infinite set $I_i \subseteq I$ and a coloring $c_i \colon \mathrm{FS}(I_i) \to k - i$ such that:

$$\text{if } c_i(x) = c_i(y) \text{ for some } x, y \in \mathrm{FS}(J_i) \text{ then also } c(x) = c(y). \tag{9.5}$$

Let $c_0 = c$ and $I_0 = I$, and assume that $c_i$ and $I_i$ have been defined for some $i < k-1$. Apply Lemma 9.9.24 (1) to $c_i$ and $I_i$. Since we have assumed that (a) fails for $c$ and

$I$, it must also fail for $c_i$ and $I_i$ by (9.5). So, there is an $\ell < k - i$ and an infinite set $J \subseteq I_i$ such that $c_i(x) \neq \ell$ for all $x \in \mathrm{FS}(J)$. Let $I_{i+1} = J$. Define $c_{i+1}$ as follows: for $x \in \mathrm{FS}(I_{i+1})$,

$$c_{i+1}(x) = \begin{cases} c_i(x) & \text{if } c_i(x) < \ell, \\ c_i(x) - 1 & \text{otherwise.} \end{cases}$$

So $c_{i+1}$ is a $(k - \ell - 1)$-coloring, and clearly if $c_{i+1}(x) = c_{i+1}(y)$ then $c_i(x) = c_i(y)$ and hence by induction, $c(x) = c(y)$. By construction, $c_{k-1}$ is constant on the infinite set $I_{k-1}$. Hence, $c$ is constant on $\mathrm{FS}(I_{k-1})$, so taking $J = I_{k-1}$ gives us (b).

For (2), we assume (1) does not hold with a $\varnothing^{(3^{k-1})}$-computable witness. Now iterate Lemma 9.9.24 (2) to define the sequence of $I_i$ and $c_i$ above, so that $I_i$ and $c_i$ are computable in $\varnothing^{(3^i)}$. Since $\varnothing^{(3^i)} \leqslant_{\mathrm{T}} \varnothing^{(3^{k-1})}$ for all $i < k$, our assumption suffices to keep the iteration going. And the solution set $J = I_{k-1}$ is computable in $\varnothing^{(3^{k-1})}$, as claimed. □

We are now ready to put all the pieces together. First, let us assemble just the combinatorial halves (the first parts of all the lemmas) to prove Hindman's theorem.

*Proof (of Hindman's theorem).* Fix $c \colon \omega \to k$, $k \geqslant 1$. Applying Corollary 9.9.8, let $I \subseteq \omega$ be an infinite set satisfying the apartness property.

Apply Lemma 9.9.25 to $c$ and $I$. If alternative (b) holds, we are done.

So assume not. For each $i$, we define a finite set $F_i$, an infinite set $I_i \subseteq I$, and a finite coloring $c_i$ of $\mathrm{FS}(I_i)$. Let $F_0 = \varnothing$, $I_i = I$, and $c_0 = c \upharpoonright \mathrm{FS}(I_0)$. Suppose we are given $F_i$, $I_i$, and $c_i$ for some $i$. By hypothesis, there is a finite set $F \subseteq I_i$ and an infinite set $J \subseteq I \perp F$ such that $F$ full-matches $\mathrm{FS}(J)$ relative to $c$. Let $F_{i+1} = F$ and $I_{i+1} = J$. Define $c_{i+1}$ as follows: given $x \in \mathrm{FS}(I_{i+1})$ choose the least $y \in F_{i+1}$ such that $c(x) = c(x + y) = c(y)$, and let $c_{i+1}(x) = \langle y, c_i(x)\rangle$.

Choose the least $\ell < k$ such that for every $i$, there are infinitely many $x \in I_i$ with $c(x) = \ell$. For each $i$, let $x_i$ be the least $x \in I_i$ with $c(x) = \ell$ and such that $x > x_j$ for all $j < i$. Thus we have $x_0 < x_1 < \cdots$. Analogously to the construction in Lemma 9.9.24, for each $i$ we can fix $y_0 < \cdots < y_{i-1}$ with $y_j \in F_{j+1}$ for each $j$, and such that $c(y) = c(x_i + y) = c(x_i + y)$ for all $y \in \mathrm{FS}(\{y_j : j < i\})$. In particular, $c(y) = \ell$ for all such $y$.

Let $T$ be the tree of all $\alpha \in \omega^{<\omega}$ with $\alpha(j) \in F_{j+1}$ for all $j < |\alpha|$ and such that $c(y) = \ell$ for all $y \in \mathrm{FS}(\mathrm{range}(\alpha))$. Then $T$ is infinite, so we can fix $f \in [T]$. Let $J = \mathrm{range}(f)$. Then $c(y) = \ell$ for all $y \in \mathrm{FS}(J)$, as desired. □

Now, we use the effectivity estimates in the lemmas (the second parts) to obtain a complexity bound on the set $J$ in the preceding argument.

*Proof (of Theorem 9.9.20).* Take $A = \varnothing$. The second part of each of the lemmas above relativizes to any set, so also this proof relativizes without issue. So, fix a computable $c \colon \omega \to k$, $k \geqslant 1$. By the effectivity of Corollary 9.9.8, we may pick out an infinite computable set $I \subseteq \omega$ satisfying the apartness property.

We now proceed in a slightly different order than before. Let $F_0 = \varnothing$, $I_0 = I$, and $c_0 = c$. Assume, inductively, that for some $i \in \omega$, we have defined a finite set $F_i$, an infinite set $I_i \subseteq I$, and a finite coloring $c_i \colon \mathrm{FS}(I_i) \to k_i$ for some $k_i \geqslant k$, all with the same properties as before. But assume, in addition, that $c_i$ and $I_i$ are computable in $B = \varnothing^{n_i}$ for some $n_i \in \omega$. (Thus, $k_0 = k$ and $n_0 = 0$.) It is now that we apply Lemma 9.9.25, to $c_i$ and $I_i$. Relativizing part (2) of Lemma 9.9.25 to $B$, we can quantify over infinite $B^{(3^{k_i-1})}$-computable sets to determine whether part (a) or (b) holds in the lemma.

If part (b) holds, then we have an infinite $B^{(3^{k_i-1})}$-computable $J \subseteq I$ such that $c_i$ is constant on $J$. By induction, and the way the $c_j$ are defined for all $j \leqslant i$, it follows that also $c$ is constant on $J$. Hence, we are done, and our solution is arithmetical.

If (a) holds, then we instead have a finite set $F \subseteq I_i$ and a $B^{(3^{k_i-1})}$-computable $J \subseteq I \perp F$ such that $F$ full-matches $\mathrm{FS}(J)$, and these can be used to define $F_{i+1}$, $I_{i+1}$, and $c_{i+1}$ as before. Thus, $c_{i+1}$ will be a $(|F_{i+1}| \times k_i)$-coloring of $\mathrm{FS}(I_{i+1})$. By cutting off an initial segment if necessary, we may assume $I_{i+1} > F_{i+1}$.

Letting $n_{i+1} = (3^{k_i-1}) \cdot n_i$, we have that $c_{i+1}$ and $I_{i+1}$ are $\varnothing^{(n_{i+1})}$-computable. Notice that to quantify over infinite $B^{(5k_i-5)}$-computable sets above requires a $(B^{(3^{k_i-1})})''$ oracle. Hence, indices for each of $F_{i+1}$, $I_{i+1}$, and $c_{i+1}$, as well as the number $n_{i+1}$, can be found uniformly $(B^{(3^{k_i-1})})''$-computable from $F_i$, $I_i$, $c_i$, and $n_i$. This completes the construction.

Now, if part (a) holds for every $i$, then we end up constructing the sequence $F_0 < F_1 < \cdots$ as before. By the preceding paragraph, this sequence is uniformly computable in $\varnothing^{(\omega)}$, hence so is the tree $T$ defined from the $F_i$ in the proof of Hindman's theorem. As this tree is $\varnothing^{(\omega)}$-computable and $\varnothing^{(\omega)}$-computably bounded, any $X \gg \varnothing^{(\omega)}$ computes a path $f \in [T]$ (Exercise 4.8.5). Since, for each $i$, we ensured that $I_i > F_i$, $f$ is an increasing function and so $J = \mathrm{range}(f)$ is computable from $f$ and therefore from $X$. As before, $J$ is an HT-solution to $c$. □

The notions of half-matching and full-matching have been generalized and extended by Anglès d'Auriac, Monin, and Patey (see [60]), who used them to give a set of combinatorial conditions under which Question 9.9.16 (and, relatedly, the question of whether HT admits arithmetical solutions) would have a positive answer.

## 9.9.3 Variants with bounded sums

We now consider two classes of variations of Hindman's theorem, which are quite interesting in their own right, but also help further elucidate some of the connections between the computable and combinatorial content of HT.

**Definition 9.9.26.** Fix $X \subseteq \omega$.

1. For each $n \geqslant 1$, $\mathrm{FS}^{=n}(X)$ denotes the set
$$\{x_0 + \cdots + x_{n-1} : (\forall i < n)[x_i \in X \wedge (\forall j < n)[i \neq j \rightarrow x_i \neq x_j]]\}.$$
2. For each $n \geqslant 1$, $\mathrm{FS}^{\leqslant n}(X) = \bigcup_{1 \leqslant m \leqslant n} \mathrm{FS}^{=m}(X)$.

Thus, $\mathrm{FS}(X) = \bigcup_{n \geqslant 1} \mathrm{FS}^{=n}(X) = \bigcup_{n \geqslant 1} \mathrm{FS}^{\leqslant n}(X)$.

**Definition 9.9.27** ($\mathsf{HT}$ **for bounded and exact sums).** Fix $n, k \geqslant 1$.

1. For all $n, k \geqslant 1$, $\mathsf{HT}^{=n}_k$ is the following statement: every $c\colon \omega \to k$, there is an infinite set $I \subseteq \omega$ such that $c$ is constant on $\mathrm{FS}^{=n}(I)_k$.
2. For all $n, k \geqslant 1$, $\mathsf{HT}^{\leqslant n}_k$ is the following statement: every $c\colon \omega \to k$, there is an infinite set $I \subseteq \omega$ such that $c$ is constant on $\mathrm{FS}^{\leqslant n}(I)$.
3. For all $n \geqslant 1$, $\mathsf{HT}^{=n}$ is the statement $(\forall k \geqslant 1)\ \mathsf{HT}^{=n}_k$ and $\mathsf{HT}^{\leqslant n}$ is the statement $(\forall k \geqslant 1)\ \mathsf{HT}^{\leqslant n}_k$.

For $\mathsf{P}$ any of these statements, $\mathsf{P}$ *with apartness* denotes the same statement, augmented to include that the solution set $I$ satisfies the apartness property.

The original interest here was from Hindman, Leader, and Strauss [144], who observed the following rather surprising fact. Combinatorially, it would seem that $\mathsf{HT}^{\leqslant 2}$ is a much simpler problem than the full version of Hindman's theorem, $\mathsf{HT}$. And yet, it seems just as hard to prove the former as it is the latter. This motivated the authors to pose the following question.

*Question 9.9.28 (Hindman, Leader, and Strauss [144]).* Is there a proof of $\mathsf{HT}^{\leqslant 2}$ that would not also be a proof of $\mathsf{HT}$?

This is reminiscent of the comparison of Milliken's tree theorem and the product form of Milliken's tree theorem in Section 9.7. Indeed, we can quickly recast this into the following reverse mathematics question.

*Question 9.9.29.* Does $\mathsf{HT}^{\leqslant 2}_2 \to \mathsf{HT}$ over $\mathsf{RCA}_0$ (or even $\mathsf{ACA}_0$)?

And irrespective of the answer, we can also ask the following analogue of Question 9.9.16.

*Question 9.9.30.* Is it the case that $\mathsf{ACA}_0 \vdash \mathsf{HT}^{\leqslant 2}_2$?

Both questions remain (tantalizingly!) open. In Section 10.6.2, we discuss an iterated version of Hindman's theorem which is also provable in $\mathsf{ACA}_0^+$, and which is equivalent over $\mathsf{ACA}_0$ to several other principles.

The principle $\mathsf{HT}^{=n}$ is a bit different from $\mathsf{HT}^{\leqslant n}$. For one, observe that for all $n, k \geqslant 1$, $\mathsf{HT}^{=n}$ is identity reducible to $\mathsf{RT}^n$ and $\mathsf{HT}^{=n}_k$ is identity reducible to $\mathsf{RT}^n_k$. (Given an $\mathsf{HT}^{=n}_k$-instance $c$, define an $\mathsf{RT}^n_k$-instance $d$ by $d(x_0, \ldots, x_{n-1}) = c(x_0 + \cdots + x_{n-1})$. Now if $H$ is any infinite homogeneous set for $d$ then $c$ is constant on $\mathrm{FS}^{=n}(H)$.) So the analogue of Question 9.9.29 for $\mathsf{HT}^{=2}$ has a negative answer: we know that $\mathsf{RT}^2$, and hence $\mathsf{HT}^{=2}$, does not imply $\mathsf{ACA}_0$, but that $\mathsf{HT}$ does. We also get an easy positive answer to the analogue of Question 9.9.30 for $\mathsf{HT}^{=n}$, since we know that $\mathsf{ACA}_0 \vdash \mathsf{RT}^n$.

Notice that this argument does not work also for $\mathsf{HT}^{\leqslant n}$. All we can get by applying $\mathsf{RT}^n$ to an instance $c$ of $\mathsf{HT}^{\leqslant n}_k$ is an infinite set $I \subseteq \omega$ such that $c$ is constant on $\mathrm{FS}^{=m}(I)$ for all $m \leqslant n$. But $c$ may take different colors on $\mathrm{FS}^{=m}(I)$ for different values of $m$.

The relationships between the principles in Definition 9.9.27, for different values of $n$ and $k$, are far less obvious than, say, for $\mathsf{RT}^n_k$, and much is still unknown. For example, it is not the case, at least not by the argument we gave before, that P is always equivalent to P with apartness. The following result shows that for $\mathsf{HT}^{\leqslant n}$ we can recover this in part.

**Theorem 9.9.31 (Carlucci, Kołodziejczyk, Lepore, and Zdanowski [26]).** *Fix $n \geqslant 2$ and $k \geqslant 1$.*

1. $\mathsf{RCA}_0 \vdash \mathsf{HT}^{\leqslant n}_{2k} \to \mathsf{HT}^{\leqslant n}_{k}$ *with apartness.*
2. $\mathsf{HT}^{\leqslant n}_{k}$ *with apartness* $\leqslant_{\mathrm{sc}} \mathsf{HT}^{\leqslant n}_{2k}$.

*Proof.* Let us prove (2). Fix a coloring $c\colon \omega \to k$. We define $d\colon \omega \to 2k$, as follows. First, define a maps $i, a\colon \omega \to \omega$ as follows. Given $x \in \omega$, fix $n$ and $a_0, \ldots, a_{n-1} \in \{0,1,2\}$ such that $x = \sum_{i<n} a_i \cdot 3^i$. (Thus, we are expressing $x$ in base 3.) Then, let $i(x)$ be the least $i < n$ such that $a_i > 0$ (or 0 if $x = 0$), and let $a(x) = a_i$. Now, we define $d$: for each $x \in \omega$, let

$$d(x) = \begin{cases} c(x) & \text{if } a(x) = 1, \\ k + c(x) & \text{if } a(x) = 2. \end{cases}$$

Clearly, $d \leqslant_{\mathrm{T}} c$. Let $I \subseteq \omega$ be an $\mathsf{HT}^{\leqslant n}_{2k}$-solution to $d$. Say $d(x) = j < 2k$ for all $x \in I$. If $j < k$, then we must have $a(x) = 1$ for all such $x$, and otherwise we must have $a(x)(x) = 2$. We claim that $i$ is injective on $I$. Suppose not, and say $i(x) = i(y)$. Since $n \geqslant 2$, we have that $c(x+y) = j$ and hence either $a(x) = a(y) = a(x+y) = 1$ or $a(x) = a(y) = a(x+y) = 2$. But this is impossible: if $a(x) = a(y) = 1$ then $a(x+y) = 2$, and if $a(x) = a(y) = 2$ then $a(x+y) = 1$. Thus, the claim holds, and now it is easy to computably thin $I$ to an infinite set satisfying the apartness condition. □

If we add the apartness condition to $\mathsf{HT}^{=n}$, we can also get a converse to the fact that $\mathsf{ACA}_0 \vdash \mathsf{HT}^{=n}$. Thus, our best known lower bound for the full Hindman's theorem is already present in this (comparatively weak) principle.

**Theorem 9.9.32 (Carlucci, Kołodziejczyk, Lepore, and Zdanowski [26]).** *For each $n \geqslant 3$ and $k \geqslant 2$, the following are equivalent over* $\mathsf{RCA}_0$.

1. $\mathsf{ACA}_0$.
2. $\mathsf{HT}^{=n}_{k}$ *with apartness.*

*Proof.* First, we argue in $\mathsf{ACA}_0$. Fix a coloring $c\colon \mathbb{N} \to k$. By Corollary 9.9.8, fix an infinite set $I \subseteq \mathbb{N}$ satisfying the apartness condition. Define $d\colon [I]^3 \to 2$ as in our discussion above: $d(x,y,z) = x + y + z$. Applying $\mathsf{RT}^n_k$, which is provable from $\mathsf{ACA}_0$ (Corollary 8.2.6), we get an infinite homogeneous set $H \subseteq I$ for $d$. Clearly $H$ satisfies the apartness condition (being a subset of $I$), and $c$ is constant on $\mathrm{FS}^{=n}(H)$ by construction.

For the reversal, we argue in $\mathsf{RCA}_0 + \mathsf{HT}^{=n}_2$. Let $f\colon \mathbb{N} \to \mathbb{N}$ be an injective function. We show that range$(f)$ exists. Recall the notion of *gap* from Definition 9.9.11. Say a gap $\{u, v\}$ in a number $x$ is *important* if

$$(\exists y < \lambda(x))[y \in \mathrm{range}(f \restriction [u, v])].$$

We define $c: \mathbb{N} \to 2$ by $c(x) = 0$ if the number of important gaps in $x$ is even, and $c(x) = 1$ if the number of important gaps in $x$ is odd. Being an important gap is $\Sigma^0_0$-definable, so $c$ exists. Fix $I$ infinite and satisfying the apartness property such that $c$ is constant on $\mathrm{FS}^{=n}(I)$, say with value $j < 2$.

We make the following two claims about $I$.

*Claim 1: For each $x \in I$, there exists $x_1 > x$ in $I$ such that $c(x + x_1) = j$.* To see this, fix $x$ and let $z$ be such that

$$(\forall y < \lambda(x))[y \in \mathrm{range}(f) \leftrightarrow y \in \mathrm{range}(f \restriction z)]. \tag{9.6}$$

(See Exercise 6.7.3). Consider any $x_1 < \cdots < x_{n-1}$ in $I$ with $\max\{x, z\} < x_1$. Since $I$ satisfies the apartness property, we have

$$\lambda(x) = \lambda(x + x_1) = \lambda(x + x_1 + \cdots + x_{n-1}),$$

and so by choice of $z$, each of $x$, $x + x_1$, and $x + x_1 + \cdots + x_{n-1}$ have the same number of important gaps. Thus in particular,

$$c(x + x_1) = c(x + x_1 + \cdots + x_{n-1}) = j$$

since $x + x_1 + \cdots + x_{n-1} \in \mathrm{FS}^{=n}(I)$.

*Claim 2: For all $x < x_1$ in $I$ with $c(x + x_1) = j$, we have*

$$(\forall y < \lambda(x))[y \in \mathrm{range}(f) \to y \in \mathrm{range}(f \restriction \mu(x_1))].$$

Let such $x < x_1$ be given, and let $z$ be as in (9.6). Fix $x_2 < \cdots < x_{n-1}$ in $I$ with $\max\{x_1, z\} < x_2$. By choice of $z$, no gap $\{u, v\}$ with $\lambda(x_2) \leqslant u < v$ is important in $x + x_1 + \cdots + x_{n-1}$. So, if $\{\mu(x_1), \lambda(x_2)\}$ were important then there would be exactly one more important gap in $x + x_1 + \cdots + x_{n-1}$ than in $x + x_1$. But $x + x_1 + \cdots x_{n-1} \in \mathrm{FS}^{=n}(I)$, so $c(x + x_1 + \cdots x_{n-1}) = j$ and therefore $c(x + x_1 + \cdots x_{n-1}) = c(x + x_1)$. Thus, $\{\mu(x_1), \lambda(x_2)\}$ cannot be important in $x + x_1 + \cdots + x_{n-1}$, and consequently we must have $z \leqslant \mu(x_1)$.

By the first and second claims, we conclude that $y \in \mathrm{range}(f)$ if and only if

$$(\forall x, x_1 \in I)[y < \lambda(x) \wedge x < x_1 \wedge c(x + x_1) = j \to y \in \mathrm{range}(f \restriction \mu(x_1))].$$

Thus, we have a $\Pi^0_1$ definition of the range of $f$, so the range exists. □

Thus, for $n \geqslant 3$ and $k \geqslant 2$, we have in particular that $\mathsf{HT}^{=n}$ with apartness, $\mathsf{HT}^{=n}_k$ with apartness, $\mathsf{RT}^n$, and $\mathsf{RT}^n_k$ are all equivalent. The situation is therefore much as we might expect naïvely. For example, we indirectly get many familiar facts, like that $\mathsf{HT}^{=n} \to \mathsf{HT}^{=n-1}$, that $\mathsf{HT}^{=n}_k$ and $\mathsf{HT}^{=n}_2$ are equivalent, etc.

On the other hand, the proof of Theorem 9.9.32 is somewhat surprising. More precisely, the coding used is quite different from that used to prove the analogous result for $\mathsf{RT}^n$ (Corollary 8.2.5). And in fact, the same coding can be used to prove the following result, which, given that it is a statement about pairs, is arguably more unexpected.

**Theorem 9.9.33 (Carlucci, Kołodziejczyk, Lepore, and Zdanowski [26]).** $\mathsf{RCA}_0 \vdash \mathsf{HT}_2^{\leqslant 2}$ *with apartness* $\to \mathsf{ACA}_0$. *Hence,* $\mathsf{RCA}_0 \vdash \mathsf{HT}_4^{\leqslant 2} \to \mathsf{ACA}_0$.

The proof is left to Exercise 9.13.12. Using another elaboration on the same idea, it is also possible to prove the following, whose proof we omit.

**Theorem 9.9.34 (Dzhafarov, Jockusch, Solomon, and Westrick [85]).** $\mathsf{RCA}_0 \vdash \mathsf{HT}_3^{\leqslant 3} \to \mathsf{ACA}_0$.

As noted, we do not even know if $\mathsf{HT}_2^{\leqslant 2}$ is provable in $\mathsf{ACA}_0$. Thus, we do not have equivalences above (at least, not yet).

In terms of lower bounds, the preceding results leave open the situation for $\mathsf{HT}_2^{\leqslant 3}$ and $\mathsf{HT}_k^{\leqslant 2}$, $2 \leqslant k < 4$. We show that none of these principles is trivial from the point of view of reverse mathematics by establishing the following lower bound.

**Theorem 9.9.35 (Dzhafarov, Jockusch, Solomon, and Westrick [85]).** $\mathsf{RCA}_0 \vdash \mathsf{HT}_2^{\leqslant 2} \to \mathsf{SRT}_2^2$.

*Proof.* We argue in $\mathsf{RCA}_0 + \mathsf{HT}_2^{\leqslant 2}$ and derive $\mathsf{D}_2^2$. Fix a stable coloring $c\colon [\mathbb{N}]^2 \to 2$. Let $i$ and $a$ be the maps $\mathbb{N} \to \mathbb{N}$ defined in the proof of Theorem 9.9.31. Define $d\colon \mathbb{N} \to 2$ as follows: for all $x \in \mathbb{N}$,

$$d(x) = \begin{cases} c(i(x), x) & \text{if } a(x) = 1, \\ 1 - c(i(x), x) & \text{if } a(x) = 2. \end{cases}$$

Let $I$ be an infinite set such that $d$ is constant on $\mathrm{FS}^{\leqslant 2}(I)$. We will need to prove several claims.

*Claim 1: For all $i \in \omega$ and $a \in \{1, 2\}$, there are at most finitely many $x \in I$ with $i(x) = i$ and $a(x) = a$.* Fix $i$ and $a$, and suppose towards a contradiction that $J = \{x \in I : i(x) = i \wedge a(x) = a\}$ is infinite. Let $\ell$ be the limit color of $i$ under $c$, and let $s_0$ be large enough so that $c(i, z) = \ell$ for all $z \geqslant s_0$. Since $J$ is infinite, we can find $x, y \in J$ with $s_0 \leqslant x < y$. So, $c(i, x) = c(i, x + y) = \ell$. Now, because $i(x) = i(y) = i$ and $a(x) = a(y) = a$, we have $i(x + y) = i$ and $a(x + y) \neq a$. Thus, depending as $a$ is 1 or 2, we either have $d(x) = \ell$ and $d(x + y) = 1 - \ell$, or $d(x) = 1 - \ell$ and $d(x + y) = \ell$. Either way, $d(x) \neq d(x + y)$. But $x$ and $x + y$ both belong to $\mathrm{FS}^{\leqslant 2}(J) \subseteq \mathrm{FS}^{\leqslant 2}(I)$, so we should have $d(x) = d(x + y)$. This is a contradiction.

*Claim 2: For all $i \in \omega$ and $a \in \{1, 2\}$, if $x \in I$ satisfies $i(x) = i$ and $a(x) = a$ then*

$$\lim_s c(i, s) = \begin{cases} d(x) & \textit{if } a = 1, \\ 1 - d(x) & \textit{if } a(x) = 2. \end{cases}$$

Fix $i$, $a$, and $x$. As above, let $\ell$ be the limit color of $i$ under $c$, and let $s_0$ be large enough so that $c(i,z) = \ell$ for all $z \geqslant s_0$. Since $I$ is infinite, it follows by Claim 1 that there is a $y \geqslant s_0$ in $I$ with $i(y) > i$. We thus have $c(i, x+y) = \ell$. And since $i(x) = i$, we also have $i(x+y) = i$. Thus, because $x + y \in \mathrm{FS}^{\leqslant 2}(I)$, we conclude that

$$d(x) = d(x+y) = \begin{cases} \ell & \text{if } a = 1, \\ 1-\ell & \text{if } a = 2. \end{cases}$$

This proves the claim.

To complete the proof, apply $\mathsf{RT}^1_4$ to find a pair $(j,a) \in 2 \times 2$ such that for infinitely many $x \in I$, $d(x) = j$ and $a(x) = a$. By Claim 1, the collection of all $i$ such that $i = i(x)$ for some $x \in I$ with $d(x) = j$ and $a(x) = a$ is infinite. Moreover, as this collection is $\Sigma^0_1$-definable, it follows by Exercise 5.13.5 that there is an infinite set consisting only of such $i$. Call this set $L$. By Claim 2, $L$ is limit homogeneous for $c$.□

Interestingly, we can also get this bound for $\mathsf{HT}^{=2}_2$ with apartness. In fact, here we can prove slightly more. Recall the principle $\mathsf{IPT}^2_2$ from Section 9.3.

**Theorem 9.9.36 (Carlucci, Kołodziejczyk, Lepore, and Zdanowski [26]).** $\mathsf{RCA}_0 \vdash \mathsf{HT}^{=2}_2$ *with apartness* $\rightarrow \mathsf{IPT}^2_2$.

*Proof.* We argue in $\mathsf{RCA}_0$. Fix a coloring $c \colon [\mathbb{N}]^2 \rightarrow 2$ and define $d \colon \mathbb{N} \rightarrow 2$ as follows: for all $x \in \mathbb{N}$,

$$d(x) = \begin{cases} c(\lambda(x), \mu(x)) & \text{if } \lambda(x) \neq \mu(x), \\ 0 & \text{if } \lambda(x) = \mu(x). \end{cases}$$

Let $I$ be an infinite set satisfying the apartness property and such that $d$ is constant on $\mathrm{FS}^{=2}(I)$, say with value $\ell < 2$. Say $I = \{x_0 < x_1 < \cdots\}$, and let $H_0 = \{\lambda(x_{2i} : i \in \mathbb{N}\}$ and $H_1 = \{\mu(x_{2i+1} : i \in \mathbb{N}\}$. We claim that $(H_0, H_1)$ is an increasing $p$-homogeneous set for $c$, in fact with color $\ell$. To see this, choose $x < y$ with $x \in H_0$ and $y \in H_1$. Then $x = \lambda(x_i)$ for some even $i$ and $y = \mu(x_j)$ for some odd $j$. In particular, $i \neq j$. Now if $j \leqslant i$ then also $x_j < x_i$, which implies that $\mu(x_j) < \lambda(x_i)$ by virtue of $I$ satisfying the apartness property. Since this is not the case, we must have $i < j$. Hence, again because $I$ satisfies the apartness property, we also have that $\mu(x_i) < \lambda(x_j)$. So, $x = \lambda(x_i) = \lambda(x_i + x_j)$ and $y = \mu(x_j) = \mu(x_i + x_j)$. Since $x_i + x_j \in \mathrm{FS}^{=2}(I)$, it now follows by definition that

$$c(x,y) = c(\lambda(x_i + x_j), \lambda(x_i + x_j)) = \ell.$$

The only nontrivial principle from Definition 9.9.27 remaining for which we have not established a lower bound above $\mathsf{RCA}_0$ is $\mathsf{HT}^{=2}$ (and $\mathsf{HT}^{=2}_k$, $k \geqslant 2$), without apartness. We turn to this principle in the next section. Before wrapping up this section, though, let us briefly note a few other variants of $\mathsf{HT}$ that have been considered in the literature. First, Carlucci, Kołodziejczyk, Lepore, and Zdanowski [26] also looked at "bounded versions" of the principle $\mathsf{FUT}$. The principles $\mathsf{FUT}^{=n}$, $\mathsf{FUT}^{=n}_k$,

$\mathsf{FUT}^{\leqslant n}$, and $\mathsf{FUT}_k^{\leqslant n}$ can be defined in the obvious way. As in Proposition 9.9.14, each of these is equivalent to the corresponding version of HT, with apartness. Carlucci [25, 24] also considered several "weak yet strong" versions of HT, with the property that they not only imply $\mathsf{ACA}_0$ (like the full HT) but they are easily provable in $\mathsf{ACA}_0$ (and so seem weaker). A contrasting view might be that these give credence to Question 9.9.30 having an affirmative answer. Carlucci [25] also showed that several well-known partition regularity results, including Van der Waerden's theorem, can be expressed in terms of such restrictions of HT.

We summarize the relationships established in this section in Figure 9.4.

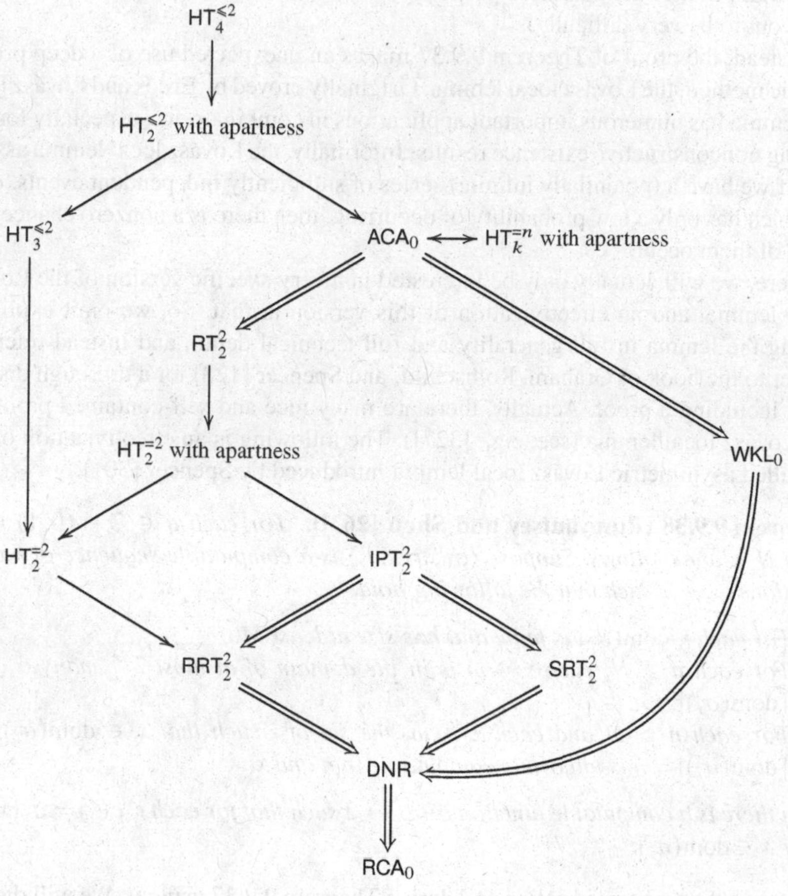


**Figure 9.4.** Relationships between some of the versions of Hindman's theorem for bounded sums. Here, $n \geqslant 3$ and $k \geqslant 3$ are arbitrary. Arrows denote implications over $\mathsf{RCA}_0$; double arrows are implications that cannot be reversed.

### 9.9.4 Applications of the Lovász local lemma

The goal of this section is to present a proof of the following theorem of Csima, Dzhafarov, Hirschfeldt, Jockusch, Solomon, and Westrick [57].

**Theorem 9.9.37.** $\mathsf{HT}_2^{=2}$ *omits computable solutions.*

This turns out to be surprisingly different from the case of the other bounded versions of Hindman's theorem considered above. For all of those, we were able to establish implications to principles above $\mathsf{RCA}_0$ (whether to $\mathsf{ACA}_0$, $\mathsf{SRT}^2_2$, or $\mathsf{IPT}^2_2$) by clever, but nonetheless elementary, combinatorial arguments. For $\mathsf{HT}^{=2}$ (without the assumption of the apartness property!) any kind of direct combinatorial coding turns out to be very difficult.

Instead, the proof of Theorem 9.9.37 makes an unexpected use of a deep probabilistic method, the Lovász local lemma. Originally proved by Erdős and Lovász [98], the lemma has numerous important applications in combinatorics, especially for obtaining nonconstructive existence results. Informally, the Lovász local lemma asserts that if we have a (potentially infinite) series of sufficiently independent events, each of which has only a low probability of occurring, then there is a nonzero chance that none of them occur.

Here, we will actually only be interested in a very specific version of the Lovász local lemma, and an effectivization of this version at that. So, we omit explicitly stating the lemma in full generality and full technical detail, and instead refer the reader to the book of Graham, Rothschild, and Spencer [128] for a thorough discussion, including a proof. Actually, there are many nice and self-contained proofs of the Lovász local lemma (see, e.g., [327]). The following is an effectivization of the so-called asymmetric Lovász local lemma introduced by Spencer [301].

**Theorem 9.9.38 (Rumyantsev and Shen [263]).** *For each $q \in \mathbb{Q} \cap (0,1)$ there is an $N \in \omega$ as follows. Suppose $\langle \sigma_0, \sigma_1, \ldots \rangle$ is a computable sequence of partial functions $\omega \to 2$ such that the following hold.*

1. *For each $i$, $\mathrm{dom}(\sigma_i)$ is finite and has size at least $M$.*
2. *For each $n \geqslant N$, each $x \in \omega$ is in the domain of at most $2^{qn}$ many $\sigma_i$ with $|\mathrm{dom}(\sigma_i)| = n$.*
3. *For each $n \geqslant N$ and each $x \in \omega$, the set of $i$ such that $x \in \mathrm{dom}(\sigma_i)$ and $|\mathrm{dom}(\sigma_i)| = n$ is uniformly computable in $n$ and $x$.*

*Then there is a computable function $c\colon \omega \to 2$ such that for each $i$, $c(x) = \sigma_i(x)$ for some $x \in \mathrm{dom}(\sigma_i)$.*

Let us see the theorem in action and derive Theorem 9.9.37 using it. We will discuss more applications afterward. In what follows, if $F \subseteq \omega$ and $y \in \omega$, write $F + y$ for the set $\{x \in \omega : (\exists w)[w \in F \wedge x = w + y]\}$.

*Proof (of Theorem 9.9.37).* As usual, we prove the result in unrelativized form. Our goal is thus to define a computable coloring $\omega \to 2$ such that if the c.e. set $W_e$ is infinite, then $c$ is not constant on $\mathrm{FS}^{=2}(W_e)$. To achieve the latter, we will ensure that there is a finite set $F \subseteq W_e$ such that $c$ is not constant on $F + y$ for any sufficiently large $y$. Clearly, this yields the desired conclusion. The idea now is to obtain $c$ by a suitable application of Theorem 9.9.38.

Here, we will follow the convention that for every $e$, $W_{e,0} = \emptyset$ and $|W_{e,s+1} \setminus W_{e,s}| \leqslant 1$ for all $s$. (That is, in the computable enumeration of $W_e$, at most one new element is enumerated at each stage.)

Fix any $q \in \mathbb{Q} \cap (0, 1)$. Say, for concreteness, $q = 1/2$. Let $N$ be as given by Theorem 9.9.38 for this $q$. By increasing $N$ if necessary, we may assume that $n \leqslant 2^{qn-1}$ for all $n \geqslant N$. For each $e \in \omega$, let $n_e = N + e$, and for each $k = \langle e, s, y \rangle$ define a finite set $E_k$ as follows. If $|W_{e,s}| = n_e$ and $|W_{e,t}| \neq n$ for any $t < s$, we let $E_k = W_{e,s}$, and otherwise we let $E_k = \emptyset$. Thus, either $|E_k| = n_e$ or $|E_k| = 0$.

We claim that for each $n \geqslant N$, every $x \in \omega$ is in at most $n$ many $E_k$ with $|E_k| = n$. Say $n = n_e$. If $W_e$ does not contain at least $n$ elements, then there are no $E_k$ with $|E_k| = n$, so there is nothing to prove. Otherwise, we can let $s$ be least such that $|W_{e,s}| = n$. Then if $x \in E_k$ and $|E_k| = n$, we must have that $k = \langle e, s, y \rangle$ for some $y$, and that $x = w + y$ for some $w \in W_{e,s}$. But there are only $n$ many elements in $W_{e,s}$, so at most $n$ many different $y$ such that $x - y$ belongs to $W_{e,s}$. This proves the claim.

It is clear that for each $n \geqslant N$ and each $x \in \omega$, the set of $k$ such that $x \in E_k$ and $|E_k| = n$ is uniformly computable in $n$ and $x$. Let $e = n - N$, so that $n = n_e$. Now, given $k$, we have that $x \in E_k$ if and only if $k = \langle e, s, y \rangle$ for some $s$ and $y$ and $x = w + y$ for some $w \in W_{e,s}$.

To complete the proof, we define a computable sequence $\langle \sigma_i : i \in \omega \rangle$ of partial functions $\omega \to 2$ with finite domain. Fix $i$, and assume we have defined $\sigma_j$ for all $j < 2i$. First, find the least $k = \langle e, s, y \rangle$ such that $|W_{e,s}| = n_e$, $|W_{e,t}| \neq n_e$ for any $t < s$, and no $\sigma_j$ with $j < 2i$ has $\mathrm{dom}(\sigma_j) = E_k$. Then, let $\mathrm{dom}(\sigma_{2i}) = \mathrm{dom}(\sigma_{2i+1}) = E_k$, and define $\sigma_{2i}(x) = 0$ and $\sigma_{2i+1}(x) = 1$ for all $x \in E_k$. Thus, for every $k$, if $E_k \neq \emptyset$ then $E_k = \mathrm{dom}(\sigma_i)$ for some $i$.

By construction, the sequence $\langle \sigma_i : i \in \omega \rangle$ satisfies clauses (1) and (3) in the statement of Theorem 9.9.38. Given $n \geqslant N$ and $x$, since $x$ is in at most $n$ many $E_k$ with $|E_k| = n$, it follows that $x$ is in the domain of at most $2n$ many $\sigma_i$ with $|\mathrm{dom}(\sigma_i)| = n$. But $2n \leqslant 2 \cdot 2^{qn-1} = 2^{qn}$ by choice of $N$, so $\langle \sigma_i : i \in \omega \rangle$ also satisfies clause (2).

So, let $c : \omega \to 2$ be the computable function given by Theorem 9.9.38. Fix $i$, and let $k$ be such that $\mathrm{dom}(\sigma_{2i}) = \mathrm{dom}(\sigma_{2i+1}) = E_k$. Then there exist $x_0, x_1 \in E_k$ such that $c(x_0) = \sigma_{2i}(x_0) = 0$ and $c(x_1) = \sigma_{2i+1}(x_1) = 1$. Hence, $c$ is not constant on $E_k$. But for every infinite $W_e$ there is a finite $F \subseteq W_e$ (namely, $F = W_{e,s}$ for the least $s$ for which $|W_{e,s}| = n_e$) and numbers $k, y$ such that $F + y = E_k$. So as mentioned at the outset, $c$ is not constant on $\mathrm{FS}^{=2}(W_e)$. The theorem is proved. □

As noted in [57], the key insight that the Lovász local lemma might be of use in proving Theorem 9.9.37 is the following. First, observe that if $F \subseteq W_e$ is finite and $y$ and $y^*$ are far enough apart, then $F + y$ and $F + y^*$ are disjoint. Thus, how a coloring $c : \omega \to 2$ is defined on $F + y$ is entirely independent of how it is defined on $F + y^*$. Furthermore, the probability that $F + y$ is homogeneous for $c$ is only $2^{-|F|}$, and hence small if $|F|$ is large. If we think of coloring finite sets as events, we can thus recognize the general shape of the premise of the Lovász local lemma, as informally described earlier.

We omit the proof of the following theorem and corollary of Csima, Dzhafarov, Hirschfeldt, Jockusch, Solomon, and Westrick [57], which can be proved by a more careful implementation of Theorem 9.9.38. This extends Theorem 9.9.37 and has a couple of interesting corollaries.

**Theorem 9.9.39 (Csima, et al. [57]).** *There is a computable instance of* $\mathsf{HT}^{=2}_2$ *all of whose solutions are DNC relative to* $\varnothing'$.

**Corollary 9.9.40 (Csima, et al. [57]).**

1. $\mathsf{HT}^{=2}_2$ *omits* $\Sigma^0_2$ *solutions.*
2. $\mathsf{RCA}_0 \vdash \mathsf{HT}^{=2}_2 \to \mathsf{RRT}^2_2$.

*Proof.* For (1), note that solutions to instance of $\mathsf{HT}^{=2}_2$ are closed under infinite subset. Every infinite $\Sigma^0_2$ set has an infinite $\Delta^0_2$ subset, but of course no $\Delta^0_2$ set can be DNC relative to $\varnothing'$. Part (2) is obtained by formalizing the proof of Theorem 9.9.39 in $\mathsf{RCA}_0$ and applying Theorem 9.4.13. □

The Lovász local lemma (or more precisely, the computable version due to Rumyantsev and Shen, Theorem 9.9.38) has also found applications to other problems in reverse mathematics and computable combinatorics, for example in the work of Liu, Monin, and Patey [198] on ordered variable words. Hirschfeldt and Reitzes [151] have also recently employed it in their study of "thin set versions" of Hindman's theorem. And an analysis of constructive properties of other versions of the Lovász local lemma has recently been undertaken by Mourad [223].

## 9.10 Model theoretic and set theoretic principles

Our discussion will now appear to digress from Ramsey's theorem and all things combinatorial, and delve instead into some model theory and set theory. We say "appear" because (as may be inferred from the fact that this section is still included in a chapter on combinatorics) the principles we will discuss nevertheless turn out to be deeply combinatorial, and indeed, even related to some of the consequences of $\mathsf{RT}^2_2$ discussed above.

### 9.10.1 Languages, theories, and models

We begin with some background on effectivizing and formalizing the basic elements of logic and model theory. Complete introductions to *computable model theory*, also known as *computable structure theory*, are given by Ash and Knight [4], Harizanov [136], and Montalbán [221].

Throughout, all formal languages we will consider will be countable. This makes formalizing many notions straightforward, just with $\omega$ used as the domain of all representations. For example, a *language* is formally a sequence of disjoint subsets $\mathcal{V}$, $C$, and for each $k > 0$, $\mathcal{F}_k$ and $\mathcal{R}_k$, of $\omega$ (any of which may be empty). We interpret these in the usual way: $\mathcal{V}$ is the set of *variables*, $C$ the set of *constant symbols*, $\mathcal{F}_k$ the set of *function symbols of arity* $k$, and $\mathcal{R}_k$ the set of *relation symbols of arity* $k$.

By using our coding for finite sequences of numbers, given a language we can in turn represent *terms*, *formulas*, and *deductions* (relative to some fixed standard deductive system) as subsets of $\omega$. In the classical setting, these codings are developed in standard accounts of Gödel's incompleteness theorem. The codings are used for much more general purposes in computable model theory.

The sets of (codes for) terms and formulas are each clearly computable from the language, as is the set of codes for deductions (as we can assume our deductive system is computable). All of this can be directly formalized in $\mathsf{RCA}_0$, and in particular, given a language, $\mathsf{RCA}_0$ can prove that each of these sets exists.

The coding extends to sets of formulas, which are naturally regarded as subsets of $\omega$ via codes. The notions of consistency and completeness of a set of formulas are defined as usual. We distinguish arbitrary sets of formulas from *theories*, which we understand to be sets of formulas closed under deduction. This is a nontrivial distinction, because $\mathsf{RCA}_0$ is not able to form the deductive closure of a set of formula, as the following proposition shows. We leave the proof as an exercise.

**Proposition 9.10.1.** *Over* $\mathsf{RCA}_0$, *the following are equivalent.*

1. $\mathsf{WKL}_0$.
2. *For every language* $\mathcal{L}$, *every consistent set of* $\mathcal{L}$*-sentences is contained in a consistent* $\mathcal{L}$ *theory.*

In computable structure theory, a computable theory (or more generally, an $A$-computable theory, for some $A \in 2^\omega$) is more commonly called a *decidable theory* (an $A$*-decidable theory*).

For structures, we restrict first of all to those with countable domain. Since we are working in a countable language anyway, this is of no great consequence by the downward Löwenheim–Skolem theorem. The domain of a structure can thus be identified with a subset of $\omega$. The rest of the structure then includes a subset of the domain interpreting the constant symbols, along with functions and relations on this domain of appropriate arities interpreting the corresponding function and relation symbols. We also always assume that a structure comes together with its elementary diagram, i.e., the set of all (codes for) sentences with parameters for elements of the domain that are true in the structure. By taking disjoint unions, a structure can again be coded by a single subset of $\omega$, and if this is computable ($A$-computable) the structure is called a *decidable structure* ($A$*-decidable structure*). So in particular, in a decidable structure the satisfaction predicate, $\models$, is computable.

In computable structure theory there is also interest in structures where the elementary diagram is *not* included. The computable ($A$-computable) such structures are then simply called *computable structures* (*A-computable structures*). In these, only the atomic diagram is available. Every decidable structure is computable, but not conversely. For example, the standard model of Peano arithmetic is computable, but if it were decidable then its elementary diagram would be a computable (hence computably axiomatizable) complete consistent extension of Peano arithmetic, contradicting Gödel's first incompleteness theorem. This argument can be extended to yield Tennenbaum's theorem (see [308]), that the only computable model of Peano arithmetic is the standard one.

However, it is possible for a computable structure to have a decidable copy, say, or for every computable model of some theory to in fact be decidable. In general there are many aspects of computable and decidable structures that can be considered. See, e.g., Cholak, Goncharov, Khoussainov, and Shore [29] or Harrison-Trainor [139] for more thorough discussions. Our interest in this section will be in the stronger type of model, with elementary diagram included, as is common in reverse mathematics.

## 9.10.2 The atomic model theorem

We will consider several theorems of classical model theory concerning the omitting and realizing of types. We quickly review the key underlying notions. For a more complete introduction to these topics, consult any standard introduction to model theory, e.g., Marker [202] or Sacks [265]).

**Definition 9.10.2.** Fix a (countable) language $\mathcal{L}$.

1. A formula $\varphi$ is an *atom* of a theory $T$ if for every formula $\psi$ in the same free variables, exactly one of $T \vdash \varphi \to \psi$ or $T \vdash \varphi \to \neg\psi$ holds.
2. A theory $T$ is *atomic* if for every formula $\psi$ such that $T + \psi$ is consistent there is an atom $\varphi$ of $T$ such that $T \vdash \varphi \to \psi$.
3. A set $\Gamma$ of formulas in the same fixed variables is:
   a. a *partial type* of a theory $T$ if $T + \Gamma$ is consistent,
   b. a *(complete) type* of a theory $T$ if it is a partial type not properly contained in any other partial type.
4. A partial type $\Gamma$ is *principal* (or *isolated*) if there is a formula $\varphi$ such that $T + \varphi$ is consistent and $T \vdash \varphi \to \psi$ for all $\psi \in \Gamma$. We say $\varphi$ *isolates* $\Gamma$.
5. A partial type $\Gamma$ of a theory $T$ is:
   a. *realized* in a model $\mathcal{B} \vDash T$ if there is a tuple $\vec{b}$ of elements of $B$ such that $\mathcal{B} \vDash \psi(\vec{b})$ for all $\psi \in \Gamma$,
   b. *omitted* in a model $\mathcal{B} \vDash T$ if it is not realized in $\mathcal{B}$.
6. For a theory $T$, a model $\mathcal{B} \vDash T$ is *atomic* if the only types of $T$ realized in $\mathcal{B}$ are principal.

Throughout this section, we will be interested exclusively in complete theories. If $T$ is complete, then every principal partial type of $T$ is realized in every model of $T$. (Suppose $\Gamma$ is isolated by $\varphi$. Then $T + \varphi$ is consistent, hence by completeness, $T$ must include $(\exists \vec{x})\varphi(\vec{x})$. It follows that if $\mathcal{B}$ is any model of $T$ then $\varphi(\vec{b})$ holds in $\mathcal{B}$ for some tuple $\vec{b}$ of elements of $B$. Hence, $\psi(\vec{b})$ holds in $\mathcal{B}$ for all $\psi \in \Gamma$.) By contrast, the famous *omitting types theorem*, due to Henkin [141] and Orey [236], says that if $\Gamma$ is nonprincipal then it is omitted in some countable model of $T$. (More generally, any countable collection of nonprincipal types can be omitted in some such model.) An atomic model (of a complete theory) is thus one which omits every type it can.

Classically, a theory has an atomic model if and only if it is atomic. Hirschfeldt, Shore, and Slaman [153] showed that the "only if" direction of this equivalence was provable in $\mathsf{RCA}_0$. They called the converse the *atomic model theorem*.

**Definition 9.10.3 (Hirschfeldt, Shore, and Slaman [153]).** The *atomic model theorem* ($\mathsf{AMT}$) asserts: every complete atomic theory has an atomic model.

This principle has its origins in much older work in computable structure theory going back to the 1970s. Millar [212] constructed a complete atomic decidable theory with no computable atomic model, which in particular means that $\mathsf{AMT}$ is not computably true. Goncharov and Nurtazin [127] and Harrington [137] showed that a complete atomic decidable theory has a decidable model if and only if there is a computable listing of the theory's principal types. (Here, by a listing of a class $\mathcal{C} \subseteq 2^\omega$ we mean a sequence $\langle X_i : i \in \omega \rangle$ such that $X_i \in \mathcal{C}$ for each $i$ and each $X \in \mathcal{C}$ is equal to $X_i$ for some (possibly many) $i \in \omega$.) This is a very useful result, which shows by relativization that, to construct an $A$-computable atomic model of a given complete atomic decidable theory, it suffices to produce an $A$-computable enumeration of its principal types. This is actually a combinatorial task, as provided by the following definition and theorem.

**Definition 9.10.4.** Fix a tree $U \subseteq 2^{<\omega}$.

1. $f \in [U]$ is *isolated* in $U$ if there is a $\sigma \in U$ such that $f$ is the only path of $U$ extending $\sigma$. We say $\sigma$ *isolates* $f$ in $U$.
2. The isolated paths through $U$ are *dense in* $U$ if every $\sigma \in T$ is extended by some isolated $f \in [U]$.
3. A *listing* of the isolated paths of $U$ is a sequence $\langle f_i : i \in \omega \rangle$ such that each $f_i$ is an isolated path of $U$ and each isolated path of $U$ is equal to $f_i$ for some (possibly many) $i \in \omega$.

**Theorem 9.10.5 (Csima, Hirschfeldt, Knight, and Soare [58]).**

1. *If $T$ is a complete atomic decidable theory, there is an infinite computable tree $U \subseteq 2^{<\omega}$ with no dead ends and isolated paths dense such that:*
   a. *every path of $U$ computes a type of $T$,*
   b. *every type of $T$ computes a path of $U$,*
   c. *every listing of the isolated paths of $S$ computes a listing of the principal types of $T$.*

2. *If $U \subseteq 2^{<\omega}$ is an infinite computable tree with no dead ends and isolated paths dense, there is a complete atomic decidable theory $T$ such that:*

   a. *every path of $U$ computes a type of $T$,*
   b. *every type of $T$ computes a path of $U$,*
   c. *every listing of the principal types of $T$ computes a listing of the isolated paths of $U$.*

Thus, to understand the degrees of atomic models of computable atomic decidable theories we can study the degrees of listings of isolated paths of trees $U$ as above. Csima, Hirschfeldt, Knight, and Soare [58] called the degrees of such sets *prime bounding*, in reference to *prime models*. (Prime models are those that elementarily embed into every model of the same complete theory. These coincide with atomic models for countable theories.) Thus, $X$ has prime bounding degree if and only if it computes a solution to every computable instance of AMT. In [58], several characterizations of the prime bounding degrees were provided, with the following being a notable example.

**Theorem 9.10.6 (Csima, Hirschfeldt, Knight, and Soare [58]).** *If $X \leqslant_T \emptyset'$ then $X$ has prime bounding degree if and only if it is nonlow$_2$.*

Since AMT is not computably true, we get a lower bound in the form of non-provability in $\mathsf{RCA}_0$. The following upper bound complements Theorem 9.10.5 in providing a combinatorial connection for AMT.

**Theorem 9.10.7 (Hirschfeldt, Shore, and Slaman [153]).** $\mathsf{RCA}_0$ *proves that* SADS *implies* AMT.

The proof is a fairly involved priority construction, requiring the method of Shore blocking (discussed in Section 6.4) to formalize in $\mathsf{RCA}_0$. For a nice self-contained presentation of the argument, see [147]. The main takeaway of the result, beyond the surprising relationship to linear orders, is that AMT is actually a relatively weak principle. In fact, it is strictly weaker than SADS.

**Theorem 9.10.8 (Hirschfeldt, Shore, and Slaman [153]).** AMT $\nleqslant_\omega$ SADS. *Hence, also* $\mathsf{RCA}_0 \nvdash \mathsf{AMT} \to \mathsf{SADS}$.

By Theorem 9.2.19, SADS admits low solutions, and hence so does AMT. In particular, there is an $\omega$-model of AMT consisting entirely of low sets. However, by Theorem 9.10.6, no such model can consist of sets all below a *single* fixed low set. (Since every $\omega$-model contains every complete atomic decidable theory, any set computing every member of an $\omega$-model of AMT has prime bounding degree.) On the other hand, by Theorem 4.6.16 there *is* such a $\omega$-model of WKL, so in particular, AMT $\nleqslant_\omega$ WKL.

On the first order side of things, we have the following result concerning the notion of restricted $\Pi^1_2$ conservativity from Section 8.7.4.

**Theorem 9.10.9 (Hirschfeldt, Shore, and Slaman [153]).** AMT *is restricted* $\Pi^1_2$ *conservative over* $\mathsf{RCA}_0$.

The proof is similar to that of Theorem 8.7.23, but using Cohen forcing instead of Mathias forcing. (The analogue of Lemma 8.7.24 is that, if $\mathcal{M}$ is a model of $\mathsf{RCA}_0$ and $\psi$ is a $\Sigma^0_3$ formula such that $\mathcal{M} \vDash (\forall Y)[\neg\psi(A, Y)]$ for some $A \in \mathcal{S}^{\mathcal{M}}$, then the same holds in $\mathcal{M}[G]$ for $G$ sufficiently generic for Cohen forcing in $\mathcal{M}$.) It follows, for one, that AMT does not imply $\mathsf{B}\Sigma^0_2$, which in combination with Theorem 9.10.8 means it does not imply any other principles in Figure 9.1.

But AMT does imply some new principles. Millar [212] showed that the classical omitting types theorem admits computable solutions. The following stronger version, where (complete) types are replaced by partial types, turns out to be more interesting.

**Definition 9.10.10 (Hirschfeldt, Shore, and Slaman [153]).** The *omitting partial types principle* (OPT) is the following statement: for every complete theory $T$ and every enumerable collection $C$ of partial types of $T$ there is a model of $T$ omitting all nonprincipal elements of $C$.

**Proposition 9.10.11 (Hirschfeldt, Shore, and Slaman [153]).** $\mathsf{RCA}_0 \vdash \mathsf{AMT} \to \mathsf{OPT}$ *but* $\mathsf{AMT} \nleq_\omega \mathsf{OPT}$.

The striking fact proved in [153] and used in separating OPT from AMT, is that OPT has a characterization purely in computability theoretic terms.

**Theorem 9.10.12 (Hirschfeldt, Shore, and Slaman [153]).** $\mathsf{RCA}_0 \vdash \mathsf{OPT} \leftrightarrow (\forall X)(\exists Y)[Y$ *is hyperimmune relative to* $X]$.

This is reminiscent of Miller's Theorem 9.4.13, which characterized the rainbow Ramsey's theorem in terms of functions DNC relative to $\varnothing'$, only here we have an even more ubiquitous computability theoretic property. Any principle that omits solutions of hyperimmune-degree (e.g., COH, $\mathsf{RRT}^2_2$, EM) therefore implies OPT, at least over $\omega$-models. In contrast, any principle that admits solutions of hyperimmune free degree does not imply OPT. Notably, $\mathsf{OPT} \nleq_\omega \mathsf{WKL}$.

The way this can be used to build an $\omega$-model of OPT in which AMT fails is as follows. Fix a $\mathrm{low}_2$ c.e. set $A$. Using Sacks' theorem that the c.e. degrees are dense (meaning, if $X <_\mathrm{T} Y$ are c.e., then there exists a c.e. set $Z$ such that $X <_\mathrm{T} Z <_\mathrm{T} Y$), we can build a sequence of c.e. sets, $A_0 <_\mathrm{T} A_1 <_\mathrm{T} \cdots <_\mathrm{T} A$. Then for each $i$ we have $A_i <_\mathrm{T} A_{i+1} <_\mathrm{T} A \leqslant_\mathrm{T} \varnothing' \leqslant_\mathrm{T} A_i'$, so by Theorem 2.8.20, relativized to $A_i$, it follows that $A_{i+1}$ is hyperimmune relative to $A_i$. Hence, $\mathcal{S} = \{Y \in 2^\omega : (\exists i)[Y \leqslant_\mathrm{T} A_i]\}$ is an $\omega$-model of OPT. On the other hand, by Theorem 9.10.6, no $\mathrm{low}_2$ set below $\varnothing'$ can bound a model of AMT. (A similar argument can be used to separate OPT from virtually every principle we have discussed so far.)

In the use of Sacks' density theorem above, we have a rare application in reverse mathematics of a classical computability theoretic result proved by an infinite injury argument. We will see another in the proof of Corollary 9.10.24 below. There is a more direct separation that avoids injury altogether, due to Patey [244], who showed that EM does not imply AMT over $\mathsf{RCA}_0$. As remarked in Section 9.5, EM omits solutions of hyperimmune free degree, and the proof of this formalizes to show that EM implies OPT.

Along with OPT, Hirschfeldt, Shore, and Slaman [153] introduced a weak version of AMT known as the *atomic model theorem with subenumerable types*. Say two partial types $\Gamma$ and $\Gamma^*$ are *equivalent* if for all formulas $\varphi$, $\Gamma \vdash \varphi \leftrightarrow \Gamma^* \vdash \varphi$. A *subenumeration* of the types of a theory $T$ is a listing $\langle \Gamma_i : i \in \omega \rangle$ of a class of partial types of $T$ such that for every type $\Gamma$ of $T$ there is an $i$ such that $\Gamma$ and $\Gamma_i$ are equivalent. Thus a subenumeration is a listing (up to equivalence) of the types of $\Gamma$, "hidden" among some other partial types. We say the types of a theory are *subenumerable* if such a subenumeration exists.

**Definition 9.10.13 (Hirschfeldt, Shore, and Slaman [153]).** The *atomic model theorem with subenumerable types* (AST) is the following statement: every complete theory whose types are subenumerable has an atomic model.

As stated, this is only interesting in the context of reverse mathematics, since classically there is no content to a subenumeration existing or not. Alternatively, if we regard the associated problem form, then the subenumeration should be understood as being part of the instance. A *computable* instance of AST is therefore a complete atomic decidable theory, together with a computable subenumeration of the types of the theory. In particular, in such a theory every type is actually computable. By Theorem 9.10.5, this means that finding solutions to computable instances of AST is the same as finding listings of isolated paths of infinite computable trees $U \subseteq 2^{<\omega}$ with no dead ends, isolated paths dense, and all paths computable.

Goncharov and Nurtazin [127] showed that AST is not computably true. On the other hand, we have the following result.

**Theorem 9.10.14 (Hirschfeldt [146]).** *Let $U \subseteq 2^{<\omega}$ be an infinite computable tree with no dead ends, isolated paths dense, and all paths computable. Then every noncomputable set computes a listing of the isolated paths of $U$.*

*Proof.* Fix any noncomputable set, $X$. Let $\sigma_0, \sigma_1, \ldots$ be a computable enumeration of the elements of $U$. For each $n$, we define a path $f_n \in [U]$ as follows. Let $\tau_{n,0} = \sigma_n$ and suppose we have defined $\tau_{n,s} \in U$ for some $s \in \omega$. Since $U$ has no dead ends, either $\tau_{n,s}0$ or $\tau_{n,s}1$ must belong to $U$. If both do, call $s$ a *coding stage*. Let $k$ be the number of coding stages prior to $s$, and let $\tau_{n,s+1} = \tau_{n,s} \frown X(k)$. If $s$ is not a coding stage, let $\tau_{n,s+1}$ be whichever of $\tau_{n,s}0$ or $\tau_{n,s}1$ belongs to $U$. Let $f_n = \bigcup_s \tau_{n,s}$. Clearly, $\langle f_n : n \in \omega \rangle$ is uniformly computable in $X$ and each $f_n$ is a path through $U$. Moreover, every isolated path through $T$ must be some $f_n$. Indeed, if $f \in [U]$ is isolated by $\sigma_n$ then $f_n = f$. We claim that, in fact, each $f_n$ is isolated in $U$, whence it follows that $\langle f_n : n \in \omega \rangle$ is an $X$-computable listing of the isolated paths of $U$, as desired. Suppose not, and fix $n$ to the contrary. Since $f_n \in [U]$, it is computable by assumption on $U$. Hence, the sequence $\langle \tau_{n,s} : s \in \omega \rangle$ is computable, and so is the set of coding stages in the construction of $f_n$. Now, $f_n$ is not isolated, so the set of coding stages is infinite. Enumerate these as $s_0 < s_1 < \cdots$. Then for all $k \in \omega$ we have $X(k) = \tau_{n,s_k}(k)$, which means that $X$ is computable, a contradiction. □

Relativizing this argument, and formalizing it in $\mathsf{RCA}_0$, yields a computability theoretic characterization of AST.

**Theorem 9.10.15 (Hirschfeldt, Shore, and Slaman [153]).** *The following are equivalent over* $\mathsf{RCA}_0$.

1. AST.
2. $(\forall X)(\exists Y)[Y \not\leqslant_T X]$.

This shows that AST is essentially the weakest *possible* principle that is not computably true, i.e., which does not hold in REC. (The only conceivably weaker such principles would have to be ones that hold in topped models, which are arguably not natural.)

Since no hyperimmune set can be computable, it is clear that AST $\leqslant_\omega$ OPT. This can be formalized in $\mathsf{RCA}_0$ to get an implication.

**Proposition 9.10.16 (Hirschfeldt, Shore, and Slaman [153]).** $\mathsf{RCA}_0 \vdash \mathsf{OPT} \rightarrow \mathsf{AST}$.

On the other hand, using Theorems 9.10.12 and 9.10.15 one can construct an $\omega$-model of AST in which OPT fails (Exercise 9.13.19.)

There are other classes of models with interesting computability theoretic and reverse mathematics properties. Harris [138] undertook a similar study of *saturated models*, while Hirschfeldt, Lange, and Shore [150] studied *homogeneous models*.

### 9.10.3 The finite intersection principle

We now switch fields, and move from model theory to set theory. Our interest will be in a family of principles well-known from classical investigations of the axiom of choice. There has been extensive study, stretching back over a century, of various equivalents and consequences of the axiom of choice, relative to choice-free axiomatizations of set theory (e.g., ZF). The most famous of these, of course, are Zorn's lemma and the well-ordering principle. But there are myriad others, from nearly every corner of mathematics, and more still if one looks at equivalents of various weaker forms of choice. A monumental catalog of such principles has been put together by Rubin and Rubin [261, 262], later expanded by Howard and Rubin [162].

The study of what happens to these equivalences in $\mathsf{Z}_2$ was initiated by Dzhafarov and Mummert [87, 88]. To be sure, to even express most of the principles studied alongside choice into $\mathcal{L}_2$ requires a significant miniaturization: principles concerning arbitrary sets now concern only sets of numbers. But this makes the situation all the more interesting. Stripped of the power that comes from being able to speak about sets in general, regardless of type or complexity, we are left with just the underlying combinatorics.

The principles studied in [88] concern various intersection principles, most famous among them the *finite intersection principle*. This (classically) asserts that every family of nonempty sets has a $\subseteq$-maximal subfamily with the so-called *finite intersection property*, meaning that any finite subfamily has nonempty intersection.

The finite intersection principle, and its variants for intersections of fixed size, are each equivalent to the axiom of choice over ZF (see [262], Chapter 4). In [88], these principles were miniaturized to second order arithmetic as follows.

**Definition 9.10.17.** Let $\vec{X} = \langle X_i : i \in \omega \rangle$ and $\vec{Y} = \langle Y_i : i \in \omega \rangle$ be families of sets.

1. We write $X \in \vec{X}$ if $X = X_i$ for some $i \in \omega$.
2. $\vec{X}$ is *nontrivial* if $X_i \neq \emptyset$ for some $i$.
3. $\vec{Y} = \langle Y_i : i \in \omega \rangle$ is a *subfamily* of $\vec{X}$ if $Y_i \in \vec{X}$ for every $i$.
4. For $n \geqslant 1$, $\vec{Y}$ satisfies the *n intersection property*, or *n-i.p.*, if for every finite set $F \subseteq \omega$ of size $n$, $\bigcap_{i \in F} Y_i \neq \emptyset$.
5. $\vec{Y}$ satisfies the *finite intersection property*, or *f.i.p.*, if for every nonempty finite set $F \subseteq \omega$, $\bigcap_{i \in F} Y_i \neq \emptyset$.
6. If $\vec{Y}$ is subfamily of $\vec{X}$ satisfying $n$-i.p. or f.i.p., then $\vec{Y}$ is a *maximal* subfamily with this property if for every other subfamily $\vec{Z}$ of $\vec{X}$ with the same property, if $\vec{Y}$ is a subfamily of $\vec{Z}$ then $\vec{Z}$ is a subfamily of $\vec{Y}$.

**Definition 9.10.18 (Dzhafarov and Mummert [88]).**
1. For $n \geqslant 1$, the *n intersection principle* ($n$IP) is the following statement: every nontrivial family of sets has a maximal subfamily satisfying $n$-i.p.
2. The *finite intersection principle* (FIP) is the following statement: every nontrivial family of sets has a maximal subfamily satisfying f.i.p.

The basic relationship between FIP and $n$IP is the following.

**Proposition 9.10.19 (Dzhafarov and Mummert [88]).** *For all $m > n \geqslant 2$,*

$$\mathsf{RCA}_0 \vdash \mathsf{FIP} \rightarrow m\mathsf{IP} \rightarrow n\mathsf{IP}.$$

Because of the maximality condition, this does require a bit of argument. See Exercise 9.13.21.

Most surprising about the strength of FIP is that it is closely related to the principles AMT and OPT of the previous section. This is in spite of the fact that the classical versions of the model theoretic principles are all provable in ZF, and are thus unrelated to the axiom of choice in any way.

Dzhafarov and Mummert [88] showed that FIP follows from a principle called $\Pi^0_1$G, introduced by Hirschfeldt, Shore, and Slaman [153]. This is a kind of strong genericity principle, known by results of Conidis [51] to be $\omega$-model equivalent to AMT. It follows that over $\omega$-models, FIP is implied by AMT. Now, Hirschfeldt, Shore, and Slaman [153] showed that to formalize the equivalence between $\Pi^0_1$G and AMT requires $\mathsf{I}\Sigma^0_2$. But it turns out that this is unnecessary for the implication to FIP.

**Theorem 9.10.20 (Day, Dzhafarov, and Miller, unpublished; Hirschfeldt and Greenberg, unpublished).** $\mathsf{RCA}_0 \vdash \mathsf{AMT} \rightarrow \mathsf{FIP}$.

Dzhafarov and Mummert [88] showed $\mathsf{AMT} \nleqslant_\omega \mathsf{FIP}$, so also $\mathsf{RCA}_0 \nvdash \mathsf{FIP} \rightarrow \mathsf{AMT}$. But by contrast they also showed the following.

**Theorem 9.10.21 (Dzhafarov and Mummert [88]).** $\mathsf{RCA}_0 \vdash \mathsf{2IP} \rightarrow \mathsf{OPT}$.

*Proof (Downey, Diamondstone, Greenberg and Turetsky [65]).* We exhibit a computable instance $\vec{X} = \langle X_i : i \in \omega \rangle$ of 2IP, every solution to which computes a hyperimmune function. Formalizing this in $\mathsf{RCA}_0$ and applying Theorem 9.10.12 yields the result.

We build $\vec{X}$ by stages. When we say that we "make" two sets $X_i$ and $X_j$ intersect at some stage $s$, we mean the following.

- Add $2\langle i, s\rangle$ to $X_i$ and $2\langle j, s\rangle$ to $X_j$.
- Add $2\langle i, j, s\rangle + 1$ to both $X_i$ and $X_j$.

It follows that if $i \neq j$ then $X_i \cap X_j$ will always be contained in the set of odd numbers. At stage 0, we make each of the following pairs of sets intersect.

- $X_{a(e,x)}$ and $X_{a(e,x^*)}$, for all $e$ and all $x \neq x^*$.
- $X_{a(e,x)}$ and $X_{a(e^*,x^*)}$, for all $e \neq e^*$ and all $x$ and $x^*$.

At stage $s$, consider all $e, x \leqslant s$. If $\Phi_e(e)[s] \downarrow$ make $X_{b(e)}$ intersect $X_{a(e^*,x)}$, for all $e \neq e^*$ and all $x$. If also $\Phi_e(x)[s] \downarrow$, make $X_{b(e)}$ intersect $X_{a(e,x)}$. This completes the construction. Clearly, $\vec{X}$ is computable.

Let $\vec{Y} = \langle Y_i : i \in \omega \rangle$ be any maximal subfamily of $\vec{X}$ satisfying 2-i.p. Each set in $\vec{Y}$ is nonempty, so given $i$ we can $\vec{Y}$-computably find the least $s$ such that $2\langle j, s\rangle \in Y_i$ for some $j$. That is, we can $\vec{Y}$-computably find the unique $j$ such that $Y_i = X_j$. Now, by construction and maximality, if $X_{b(e)} \in \vec{Y}$ then $X_{a(e,x)} \in \vec{Y}$ if and only if $\Phi_e(x) \downarrow$, whereas if $X_{b(e)} \notin \vec{Y}$ then $X_{a(e,x)} \in \vec{Y}$ for all $x$.

We use this fact to define a $\vec{Y}$-computable function $f$, as follows. Fix $x$. Let $i(x)$ be the least $j \in \omega$ such that for each $e \leqslant x$, at least one of the following holds.

1. There is an $i \leqslant j$ with $Y_i = X_{b(e)}$.
2. There is an $i \leqslant j$ with $Y_i = X_{a(e,x+1)}$.

So if (1) holds then $\Phi_e(e) \downarrow$, and if (1) and (2) both hold then $\Phi_e(x+1) \downarrow$ and so, by our use conventions, also $\Phi_e(x) \downarrow$. Let $f(x)$ be least number larger than $\Phi_e(e)$ if (1) holds for $x = e$; and larger than $\Phi_e(x)$ for all $e < x$ for which both (1) and (2) hold. Clearly, the map $x \mapsto i(x)$ is $\vec{Y}$-computable, as is determining for each $e \leqslant x$ which of (1) or (2) hold, so $f$ is $\vec{Y}$-computable.

We claim that $f$ is not dominated by any computable function. To see this, suppose $\Phi_e$ is total. Then $X_{b(e)}$ belongs to $\vec{Y}$, as does $X_{a(e,x)}$ for all $x$. Fix the least $j$ such that $Y_j = X_{b(e)}$. If there is no $i < j$ such that $Y_i = X_{a(e,x)}$ for some $x > e$ then in the definition of $f$ we must have $i(e) \geqslant j$, and therefore $f(e) > \Phi_e(e)$. Otherwise, fix the largest $x > e$ such that $Y_i = X_{a(e,x)}$ for some $i < j$. In particular, there is no $i \leqslant j$ with $Y_i = X_{a(e,x+1)}$. Hence, we must have $i(x) \geqslant j$, and therefore $f(x) > \Phi_e(x)$. □

Thus, the set theoretic principles FIP and 2IP are sandwiched between the model theoretic principles AMT and OPT. Dzhafarov and Mummert [88] left open the question of whether OPT is actually equivalent to FIP (or 2IP). A number of degree theoretic upper bounds on the complexity of FIP were established in [88]. Downey,

Diamondstone, Greenberg, and Turetsky [65] observed that all of them are enjoyed by Cohen 1-generics, which led them to the following: if $X \leqslant_T \varnothing'$, then $X$ computes a solution to every computable instance of $\mathsf{FIP}$ if and only if it is 1-generic. (As it turns out, $\mathsf{FIP}$ admits a universal instance, so this is the same as asking about a single computable instance of $\mathsf{FIP}$. See Exercise 9.13.20.) Remarkably, Cholak, Downey, and Igusa [27] showed that 1-genericity also characterizes the strength of $\mathsf{FIP}$ outside the $\varnothing'$-computable degrees.

**Definition 9.10.22 (Existence of 1-generics).** $\mathsf{1\text{-}GEN}$ is the following statement: $(\forall X)(\exists Y)[Y$ is Cohen 1-generic relative to $X]$.

**Theorem 9.10.23 (Cholak, Downey, and Igusa [27]).**

1. *For all* $n \geqslant 2$, $\mathsf{FIP} \equiv_c n\mathsf{IP} \equiv_c \mathsf{1\text{-}GEN}$.
2. $\mathsf{RCA}_0 \vdash \mathsf{FIP} \leftrightarrow \mathsf{1\text{-}GEN}$
3. *For* $n \geqslant 2$, $\mathsf{RCA}_0 + \mathsf{I}\Sigma^0_2 \vdash n\mathsf{IP} \leftrightarrow \mathsf{1\text{-}GEN}$.

Thus, yet again we see a computability theoretic characterization of a noncomputability theoretic principle. Using the characterization, it is possible to answer the question from [88] and separate $\mathsf{FIP}$ from $\mathsf{OPT}$. The proof uses a surprisingly strong degree theoretic fact.

**Corollary 9.10.24.** $\mathsf{OPT} \nleqslant_\omega \mathsf{FIP}$.

*Proof (Hirschfeldt, personal communication).* Epstein [94] (see also [1], p. 27) constructed an initial segment of the $\Delta^0_2$ Turing degrees of order type $\omega$. That is, there are sets $\varnothing = A_0 <_T A_1 <_T \cdots <_T \varnothing'$ such that if $Y \leqslant_T A_i$ for some $i$ then $Y = A_j$ for some $j$ (necessarily $j \leqslant i$). Thus, if we let $\mathcal{S}$ be the $\omega$-model $\{Y \in 2^\omega : (\exists i)[Y \leqslant_T A_i]\}$ then $\mathcal{S} = \{A_i : i \in \omega\}$. Now, for each $i$ we have $A_i <_T A_{i+1} <_T \varnothing' \leqslant_T A'_{i+1}$. Relativizing Theorem 2.8.20 to $A_i$, we see that $A_{i+1}$ is hyperimmune relative to $A_i$. By Theorem 9.10.12 this implies that $\mathcal{S} \vDash \mathsf{OPT}$. However, no $A_i$ can be 1-generic. Indeed, if we write $A_i$ as $B_0 \oplus B_1$ then $B_0$ and $B_1$ are both computable from $A_i$, hence each is equal to $A_j$ for some $j$. It follows that either $B_0 \leqslant_T B_1$ or $B_1 \leqslant_T B_0$. But by Exercise 7.8.1, the odd and even halves of a 1-generic are Turing incomparable. By Theorem 9.10.23, $\mathcal{S} \nvDash \mathsf{FIP}$. □

The proof of Theorem 9.10.23, while self-contained in terms of methods, is fairly involved and would take us a bit too far afield. One interesting observation made in [27] is that the proof seems to have an essential nonuniformity. This raises the following question.

*Question 9.10.25.* Is it the case that $\mathsf{1\text{-}GEN} \leqslant_W \mathsf{FIP}$?

Another question about the strength of intersection principles, left over from the initial investigation in [88], is whether $\mathsf{FIP}$ and $\mathsf{2IP}$ are equivalent over $\mathsf{RCA}_0$. Theorem 9.10.23 allows us to recast this in terms of genericity.

*Question 9.10.26.* Is it the case that $\mathsf{RCA}_0 \vdash \mathsf{2IP} \to \mathsf{1\text{-}GEN}$?

## 9.11 Weak weak König's lemma

The final combinatorial principle we consider in this chapter is actually a bit of an outlier among those we have been looking at so far. It is not related in any way to RT, but rather to WKL. This is the so-called *weak weak König's lemma.*

**Definition 9.11.1 (Weak weak König's lemma).**

1. WWKL is the following statement: for every $T \subseteq 2^{<\omega}$ such that
$$\lim_n \frac{|\{\sigma \in 2^n : \sigma \in T\}|}{2^n} > 0, \tag{9.7}$$
there exists an $f \in [T]$.
2. $\mathsf{WWKL}_0$ is the $\mathcal{L}_2$ theory consisting of $\mathsf{RCA}_0$ plus WWKL.

As a subsystem, $\mathsf{WWKL}_0$ was first introduced by Yu and Simpson [330]. It was further developed by Brown; Giusto, and Simpson [23], and by Giusto and Simpson [124].

Note that the ratios $|\{\sigma \in 2^n : \sigma \in T\}|/2^n$ are nonincreasing in $n$ and bounded below by 0, so (classically) the limit above always exists. The starting observation here is the following. Let $\mu$ be the "fair coin" measure on Cantor space defined in Section 9.4. Because
$$[T] = \bigcap_n [[\{\sigma \in 2^n : \sigma \in T\}]],$$
and because $[[\{\sigma \in 2^n : \sigma \in T\}]] \supseteq [[\{\sigma \in 2^{n+1} : \sigma \notin T\}]]$ for all $n$, we have
$$\mu([T]) = \lim_n \mu([[\{\sigma \in 2^n : \sigma \in T\}]]) = \lim_n 2^{-n} \cdot |\{\sigma \in 2^n : \sigma \in T\}|.$$

Hence $T$ satisfies (9.7) if and only if $\mu([T]) > 0$. For this reason, at least when working over $\omega$-models, we can think of the instances of WWKL as interchangeable with $\Pi^0_1$ classes of positive measure. (For this reason, some authors also refer to trees satisfying (9.7) as "trees of positive measure".)

It turns out that $\Pi^0_1$ classes of positive measure have interesting connections to algorithmic randomness. Recall the definition of a 1-random set from Definition 9.4.5.

**Theorem 9.11.2 (Kučera [192]).** *Let $C$ be a $\Pi^0_1$ class and $X$ a 1-random. If $\mu(C) > 0$ then $C$ has an $X$-computable member. In fact, $\tau X \in C$ for some $\tau \in 2^{<\omega}$.*

*Proof.* Fix a computable tree $T$ with $[T] = C$. We may assume $T \neq 2^{<\omega}$, as otherwise the result is obvious. Set
$$U_0 = \{\sigma \in 2^{<\omega} : \sigma \notin T \wedge (\forall \tau < \sigma)[\tau \in T]\},$$
and for each $n > 0$, let
$$U_n = \{\sigma \in 2^{<\omega} : (\exists \tau \in U_{n-1})(\exists \rho \in U_0)[\sigma = \tau\rho]\}.$$

Then the sequence $\langle U_n : n \in \omega \rangle$ is uniformly c.e. Moreover, it follows by induction that each $U_n$ is a prefix free set of strings. So,

$$\mu([[U_n]]) = \sum_{\sigma \in U_n} 2^{-|\sigma|} = \sum_{\tau \in U_{n-1}} 2^{-|\tau|} \sum_{\rho \in U_0} 2^{-|\rho|} = \mu([[U_{n-1}]])\mu([[U_0]]).$$

Inductively, $\mu([[U_n]]) = \mu([[U_0]])^{n+1}$ for all $n$. Since $0 < \mu([[U_0]]) < 1$ by assumption, we can uniformly computably extract a sequence $n_0 < n_1 < \cdots$ such that $\mu([[U_{n_k}]]) \leqslant 2^{-k}$ for all $k$. Now if $X$ is 1-random, there must consequently exist a $k$ such that $X \notin [[U_{n_k}]]$. Pick the least such $k$. If $k = 0$, then $X \in [T]$ by definition. Suppose $k > 0$. Then $X \in [[U_{k-1}]]$, so necessarily $\tau \prec X$ for some $\tau \in U_{k-1}$. Define $f \in 2^\omega$ by $f(x) = X(x + |\tau|)$. Clearly, $f \leqslant_T X$, and we claim that $f \in [T]$. If not, then some $\rho \prec f$ belongs to $U_0$. But then $\tau\rho \prec X$, witnessing that $X \in [[U_k]]$, a contradiction. □

**Corollary 9.11.3.** *Let $\mathcal{S}$ be an $\omega$-model. Then $\mathcal{S} \models$ WWKL if and only if for every $A \in \mathcal{S}$ there exists an $X \in \mathcal{S}$ such that $X$ is 1-random relative to $A$. In particular, there is an $\omega$-model of* $\mathsf{RCA}_0 + \neg\mathsf{WWKL}_0$.

*Proof.* First, suppose $\mathcal{S} \models$ WWKL. Fix any $A \in \mathcal{S}$. Relativizing Theorem 9.4.6, let $\langle \mathcal{U}_n : n \in \omega \rangle$ be a universal Martin-Löf test relative to $A$. Let $C_1 = 2^\omega \setminus \mathcal{U}_1$. Then $C_1$ is a $\Pi^{0,A}_1$ class since $f \in C_1$ if and only if $(\forall k)[f \upharpoonright k \notin U_1]$ for the $A$-c.e. set $U_1 \subseteq 2^{<\omega}$ for which $\mathcal{U}_1 = [[U_1]]$. So there is some $A$-computable tree $T \subseteq 2^{<\omega}$ with $C_1 = [T]$. In particular, $T \in \mathcal{S}$. Furthermore, $\mu(C_1) = 1 - \mu(\mathcal{U}_1) \geqslant 1 - 2^{-1} = \frac{1}{2}$, so $T$ is an instance of WWKL. Therefore, there exists $X \in \mathcal{S} \cap [T]$. Since $X \notin \bigcap_n \mathcal{U}_n$, $X$ is 1-random relative to $A$.

Now suppose that for every $A \in \mathcal{S}$ there exists an $X \in \mathcal{S}$ such that $X$ is 1-random relative to $A$. Let $T \in \mathcal{S}$ be any instance of WWKL. Then $C = [T]$ is a $\Pi^{0,T}_1$ class of positive measure. Let $X$ be any set 1-random relative to $T$ such that $X \in \mathcal{S}$. By Theorem 9.11.2, relativized to $T$, there is an $X$-computable $f \in C$. As $\mathcal{S}$ is closed under $\leqslant_T$, we have $f \in \mathcal{S}$. Since $T$ was arbitrary, we conclude that $\mathcal{S} \models$ WWKL, as desired.

For the second part, since no 1-random can be computable, REC $\nvDash$ $\mathsf{WWKL}_0$. □

**Corollary 9.11.4.** *If $X$ is 1-random, then there is an $\omega$-model of* WWKL *consisting entirely of $X$-computable sets.*

*Proof.* By Theorem 9.4.18, let $\mathcal{S}$ be an $\omega$-model consisting entirely of $X$-computable sets, and such that for every $Y \in \mathcal{S}$ there is a $Y^* \in \mathcal{S}$ which is 1-random relative to $Y$. By Corollary 9.11.3, $\mathcal{S} \models$ WWKL. □

How does WWKL compare to WKL (or $\mathsf{WKL}_0$ to $\mathsf{WWKL}_0$)? Obviously, every $\Pi^0_1$ class of positive measure is, *a fortiori*, nonempty. Hence, WWKL is a subproblem of WKL. But in fact, the extra "W" is deserved, and WWKL is indeed a weaker principle. To see this, we will use the following result of Stephan. We omit the proof, which can be readily found in other texts (e.g., Downey and Hirschfeldt [83, Theorem 8.8.4]).

**Theorem 9.11.5 (Stephan [303]).** *If $X$ is 1-random and of PA degree then $X \geqslant_T \emptyset'$.*

**Corollary 9.11.6 (Simpson and Yu [330]).** WKL $\nleqslant_\omega$ WWKL. *Hence,* $\mathsf{WWKL}_0 \nvDash \mathsf{WKL}_0$.

*Proof.* By Corollary 9.4.7, there is a nonempty $\Pi^0_1$ class consisting entirely of 1-random sets. Using the cone avoidance basis theorem (Theorem 2.8.23), we can therefore find a 1-random set $X \not\geqslant_T \varnothing'$. Apply Corollary 9.11.4 to find an $\omega$-model $\mathcal{S}$ of WWKL consisting entirely of $X$-computable sets. Now, no $Y \in \mathcal{S}$ can have PA degree. Otherwise, any $Y^* \in \mathcal{S}$ which is 1-random relative to $Y$ would be both 1-random and of PA degree, and so would compute $\varnothing'$ by Theorem 9.11.5. But $\varnothing'$ is not computable from $X$, and therefore not in $\mathcal{S}$. By Proposition 4.6.3, we conclude that $\mathcal{S} \nvDash$ WKL. □

Another striking contrast between WKL and WWKL is demonstrated with strong omniscient computable reducibility, $\leqslant_{\text{soc}}$, which was introduced in Definition 8.8.9.

**Theorem 9.11.7 (Monin and Patey [217]).**

*1.* KL $\leqslant_{\text{soc}}$ WWKL.
*2.* KL $\nleqslant_{\text{soc}}$ WKL.

*Proof.* For part (1), suppose we are given an instance $T \subseteq \omega^{<\omega}$ of KL. Thus $T$ is an infinite, finitely-branching tree, so we may pick some $f \in [T]$. Let $S = \{\chi_f \upharpoonright k : k \in \omega\}$, where $\chi_f$ is the characteristic function of $f$. Then $S$ is an instance of WKL whose only solution is $\chi_f$, which computes $f$.

For part (2), let $T \subseteq \omega^{<\omega}$ be any instance of KL with a single noncomputable path, $f$ (e.g., as in the proof of Proposition 3.6.10.) By a famous result of Sacks (see [74, Corollary 8.12.2]), the collection of sets that compute $f$ has measure 0. Hence, every instance of WWKL must have a solution that does not compute $f$. So, $T$ witnesses that KL $\nleqslant_{\text{soc}}$ WWKL. □

A similar argument to part (2) above yields a perhaps even more surprising result.

**Theorem 9.11.8 (Hirschfeldt and Jockusch [148]).** $\mathsf{RT}^1_2 \nleqslant_{\text{soc}}$ WWKL.

*Proof.* Let $S$ be a bi-hyperimmune set (That is, $S$ and $\overline{S}$ are both hyperimmune. The construction of such a set is a standard exercise.) Let $c : \omega \to 2$ be the characteristic function of $S$, regarded as an instance of $\mathsf{RT}^1_2$. Every $\mathsf{RT}^1_2$-solution to $c$ is therefore an infinite subset of either $S$ or $\overline{S}$. By Theorem 9.4.21, the collection of all sets that can compute an infinite subset of a given hyperimmune set has measure 0. Hence, the collection of sets that can compute a solution to $c$ has measure 0, too. It follows that every instance of WWKL has a solution that does not compute a solution to $c$. □

WWKL is not entirely without strength, however. The following result is an immediate consequence of Kučera's theorem (Theorem 9.4.14). We include the proof in order to show how it is formalized.

**Corollary 9.11.9 (Giusto and Simpson [124]).** $\mathsf{WWKL}_0 \vdash$ DNR.

*Proof.* We argue in $\mathsf{RCA}_0$. Let $A$ be any set. Define the following tree:

$$T = \{\sigma \in 2^{<\mathbb{N}} : (\forall e \leqslant |\sigma|)[e > 1 \wedge \Phi^A_e(e) \downarrow \in 2^{<\mathbb{N}} \to \Phi^A_e(e) \neq \sigma \upharpoonright e]\}.$$

Then that for all $n > 1$, if $\sigma \notin T$ but $\sigma \upharpoonright n - 1 \in T$ it must be that $\sigma = \Phi_n(n)$. In particular, $\sigma$ is unique with this property. Since $T$ contains both strings of length 1, it thus follows by induction that for all $n > 1$,

$$\begin{aligned}\frac{|\{\sigma \in 2^n : \sigma \in T\}|}{2^n} &\geqslant \frac{2 \cdot |\{\sigma \in 2^{n-1} : \sigma \in T\}| - 1}{2^n} \\ &= \frac{|\{\sigma \in 2^{n-1} : \sigma \in T\}|}{2^{n-1}} - \frac{1}{2^n} \\ &\geqslant 1 - \sum_{2 \leqslant i \leqslant n} 2^{-i} \\ &\geqslant 2^{-1}.\end{aligned}$$

It follows that $T$ is an instance of WWKL. Apply WWKL to obtain some $X \in [T]$. Then by definition, $X \upharpoonright e \neq \Phi_e(e)$ for all $e > 1$. Define a function $f$ so that $f(0) \neq \Phi_0(0)$, $f(1) \neq \Phi_1(1)$, and $f(e) = X \upharpoonright e$ for all $e > 1$. Then $f$ is DNC relative to $A$. Since $A$ was arbitrary, the proof is complete. □

Using a fairly sophisticated combinatorial argument, Ambos-Spies, Kjos-Hanssen, Lempp, and Slaman [2] showed that the above is actually a strict implication. We omit the proof. Thus, $\mathsf{WWKL}_0$ is intermediate between $\mathsf{WKL}_0$ and DNR.

**Theorem 9.11.10 (Ambos-Spies, Kjos-Hanssen, Lempp, and Slaman [2]).** *There is an $\omega$-model of* $\mathsf{DNR} + \neg\mathsf{WWKL}_0$. *In particular,* $\mathsf{WWKL}_0 \nvdash \mathsf{DNR}$.

An interesting question is whether the strength of WWKL is in any way changed if we look not just at $\Pi^0_1$ classes of (arbitrary) positive measure, but some prescribed amount of measure instead. To this, let us define the following class of restrictions of WWKL.

**Definition 9.11.11 (Dorais, Dzhafarov, Hirst, Mileti, and Shafer [72]).** Fix $q \in \mathbb{Q}$ with $0 < q < 1$. $q$-WWKL is the following statement: for every $T \subseteq 2^{<\omega}$ such that

$$\lim_n \frac{|\{\sigma \in 2^n : \sigma \in T\}|}{2^n} \geqslant q,$$

there exists an $f \in [T]$.

So if $0 < p < q < 1$ then $q$-WWKL is a subproblem of $p$-WWKL, which in turn is a subproblem of WWKL. It turns out that over $\mathsf{RCA}_0$, all of these principles are equivalent.

**Theorem 9.11.12 (Dorais, Dzhafarov, Hirst, Mileti, and Shafer [72]).** *Fix $p, q \in \mathbb{Q}$ with $0 < p < q < 1$.*

*1.* $p$-WWKL $\equiv_{\mathrm{sc}}$ $q$-WWKL.
*2.* $\mathsf{RCA}_0 \vdash q\text{-}\mathsf{WWKL} \leftrightarrow \mathsf{WWKL}$.

*Proof.* For part (1), we need only show that $p$-WWKL $\leqslant_{\mathrm{sc}}$ $q$-WWKL. Fix any tree $T \subseteq 2^{<\omega}$ with $\mu([T]) \geqslant p$. Let $\langle \mathcal{U}_n : n \in \omega \rangle$ be a universal Martin-Löf test relative to $T$, and let $i$ be least so that $2^{-i} \leqslant 1 - q$. Then $2^\omega \smallsetminus \mathcal{U}_i$ is a $\Pi^{0,T}_1$ class, hence there is a $T$-computable tree $S \subseteq 2^{<\omega}$ with $[S] = 2^\omega \smallsetminus \mathcal{U}_i$. By choice of $i$, $\mu([S]) \geqslant q$. Since every $X \in [S]$ is 1-random relative to $T$, it follows by Theorem 9.11.2 that $\tau X \in [T]$ for some $\tau \in 2^{<\omega}$. In particular, $X$ computes an element of $[T]$.

For part (2), we note that the above argument can be readily formalized in $\mathsf{RCA}_0$. We then argue as follows. Given an instance $T$ of WWKL, fix a positive $p \in \mathbb{Q}$ so that

$$\lim_n \frac{|\{\sigma \in 2^n : \sigma \in T\}|}{2^n} \geqslant p.$$

Then, apply the above argument. This shows that $q$-WWKL implies WWKL, hence $q$-WWKL $\leftrightarrow$ WWKL. □

But there is a detectable difference under finer reducibilities.

**Proposition 9.11.13 (Dorais, Dzhafarov, Hirst, Mileti, and Shafer [72]).** *Fix $p, q \in \mathbb{Q}$ with $0 < p < q < 1$. Then $p$-WWKL $\nleqslant_{\mathrm{W}}$ $q$-WWKL.*

*Proof.* Fix Turing functionals $\Phi$ and $\Psi$. Uniformly computably in indices for $\Phi$ and $\Psi$ we build a computable tree $T \subseteq 2^{<\omega}$ witnessing that it is not the case that $p$-WWKL $\leqslant_{\mathrm{sW}}$ $q$-WWKL via $\Phi$ and $\Psi$. From here, it follows by Proposition 4.4.5 that $p$-WWKL $\nleqslant_{\mathrm{W}}$ $q$-WWKL.

The construction of $T$ is in stages. At stage $s$, we define a finite tree $T_s \subseteq 2^{\leqslant s}$ containing at least $2^s p$ many strings of length $s$. We assume without loss of generality that $\Phi(T_s)$ is always a finite subtree of $2^{<\omega}$, and we let $n_s$ be least such that $\Phi(T_s) \subseteq 2^{\leqslant n_s}$. By our usual use conventions, we may also assume $n_s \leqslant s$ for convenience.

Let $k \in \omega$ be least so that $2^{-k} < q$, and let $m \in \omega$ be large enough so that $1 - 2^{-m}k \geqslant p$.

***Construction.*** At stage 0, let $T_0 = \{\langle\rangle\}$. Now suppose we are at stage $s+1$, and that $T_s$ has already been defined. Let $x_0 < \cdots < x_{m-1}$ be the least numbers that we have not yet *acted for* in the construction, as defined below. We consider two cases.

*Case 1: The following hold.*

1. $m \leqslant s$.
2. $x_{m-1} < s$.
3. $n_s > 0$.
4. $\Psi^\sigma(x_i) \downarrow < 2$ *for all* $\sigma \in \Phi(T_s)$ *of length* $n_s$ *and all* $i < m$.
5. $|\{\sigma \in 2^{n_s} : \sigma \in \Phi(T_s)\}| \geqslant 2^{n_s} q$.

Define $\rho \in 2^m$ as follows. If at least half of all $\sigma \in \Phi(T_s)$ of length $n_s$ satisfy $\Psi^\sigma(x_i) = 0$, let $\rho(i) = 0$. Otherwise, let $\rho(i) = 1$. (In this case, of course, at least half of all $\sigma \in \Phi(T_s)$ of length $n_s$ satisfy $\Psi^\sigma(x_i) = 1$.) We now let $T_{s+1}$ be $T_s$ together with $\sigma 0$ and $\sigma 1$ for every $\sigma \in T_s$ of length $s$ for which $(\exists i < m)[\sigma(x_i) \neq \rho(i)]$. In this case, say we have *acted* for $x_0, \ldots, x_{m-1}$.

*Case 2: Otherwise.* In this case, we let $T_{s+1}$ be $T_s$ together with $\sigma 0$ and $\sigma 1$ for every $\sigma \in T_s \cap 2^s$.

***Verification.*** Let $T = \bigcup_s T_s$. Clearly, $T$ is a tree, uniformly computable in $\Phi$ and $\Psi$. We claim that $\mu([T]) \geqslant p$, and that if $\Phi(T)$ is an infinite tree, and if $\Psi(f) \in [T]$ for every $f \in [\Phi(T)]$, then $\mu([\Phi(T)]) < q$.

In the construction, the measure of $[T]$ is only ever reduced if we enter Case 1. At each such stage, at most $2^{-m}|\{\sigma \in T_s : |\sigma| = s\}|$ many strings are not extended into $T_{s+1}$, so the eventual measure of $[T]$ decreases by at most $2^{-m}$. At the same time, the eventual measure of $[\Phi(T)]$ is halved.

Suppose $\mu([T]) < p$. Fix the least $s$ such that $T_{s+1}$ contains fewer than $2^{s+1}p$ many strings of length $s$. In particular, the construction must have entered Case 1 at stage $s$. Say there are a total of $j \geqslant 1$ many stages, up to and including $s+1$, at which the construction enters Case 1. Then we have

$$p > \frac{|\{\sigma \in 2^{s+1} : \sigma \in T_{s+1}\}|}{2^{s+1}} \geqslant 1 - 2^{-m} j.$$

By choice of $m$, we must have $j > k$. But then by choice of $k$, we have

$$\frac{|\{\sigma \in 2^{n_s} : \sigma \in \Phi(T_{n_s})\}|}{2^{n_s}} \leqslant 2^{-(j-1)} \leqslant 2^{-k} < q.$$

Thus, property (5) fails at stage $s+1$, so the construction cannot enter Case 1 at this stage after all. This is a contradiction. We conclude that $\mu([T]) \geqslant p$.

Now suppose $\Phi(T)$ is infinite and that $\Psi(f) \in [T]$ for all $f \in [\Phi(T)]$. For fixed choice of $x_0 < \cdots < x_{m-1}$, it follows by assumption on $\Phi$ and $\Psi$ that properties (1)–(4) of Case 1 hold for all sufficiently large $s$. Thus, the construction must enter Case 1 at some stage: if it did not then property (5) would always hold, so Case 1 would apply at all sufficiently large stages. At the same time, since $\mu([T]) \geqslant p$, the construction can enter Case 1 at only finitely many stages $s$. Therefore, we can let $s_0$ be the largest stage at which the construction enters Case 1. Now fix the least $x_0 < \cdots < x_{m-1}$ not yet acted for at, or prior to, stage $s_0$. Let $s \geqslant s_0$ be large enough so that properties (1)–(4) of Case 1 hold. Then since the construction does not enter Case 1 at stage $s+1$, it must be that property (5) fails. This forces $\mu([\Phi(T)]) < q$, as claimed. □

## 9.12 The reverse mathematics zoo

The collection of inequivalent principles below (or around) $\mathsf{ACA}_0$ has come to be called the *reverse mathematics zoo*. The proliferation of these principles has made understanding the increasingly complex web of relationships between them much more difficult. Diagrams help. We have already seen several, in Figures 8.2, 9.1, 9.2 and 9.4. This style of diagram, with arrows indicating implications/reductions, and double arrows indicating those that are strict, was introduced by Hirschfeldt and Shore [152], and is now used widely in the reverse mathematics literature.

Typically, such diagrams are included in a paper only to showcase the principles of interest there. Figure 9.5 displays a much larger selection of principles, assembled from across the reverse mathematics literature. One gets a sense from this of just how massive the reverse mathematics zoo has gotten over the years—the figure includes 204 principles and combinations of principles. In addition, it becomes clear that diagrams of this sort have limited utility: when too many principles are included, they are really quite useless. More generally, keeping track of what is known and what is not in the subject is quite unwieldy with such a large body of work.

Of course, this is not a problem unique to reverse mathematics. "Zoos" exist in many areas, from complexity theory to algorithmic randomness. Sometime prior to 2010, Kjos-Hanssen assembled a large diagram of downward-closed classes of Turing degrees, ordered by inclusion. Subsequently, Miller wrote an interactive command-line tool to make this information easier to work with. Starting with a database of facts, the program was able to derive new facts from old by transitive closure and display open questions. This was later adapted for reverse mathematics, and more specifically for weak combinatorial principles, by Dzhafarov. Here, one complication is that in addition to implications one has to deal with joins of principles, which makes closing off under what is known slower. The current version of this tool has been rewritten and optimized by Astor, and now also supports a variety of finer reducibilities, including $\leqslant_{\mathrm{c}}$, $\leqslant_{\mathrm{W}}$, and $\leqslant_{\omega}$. (It was used to generate Figure 9.5.)

The reverse mathematics zoo tool is a simple Python script and can be downloaded at `rmzoo.uconn.edu`. More recently, Khan, Miller, and Patey have written a visualizer for the tool, which also supports databases from areas other than reverse mathematics. The visualizer runs in a web browser, and can be accessed at `computability.org/zoo-viewer`.

## 9.13 Exercises

**Exercise 9.13.1.** Show that $\mathsf{RCA}_0$ proves the following: if $(P, \leqslant_P)$ is a partial order, there exists a linear ordering $\leqslant_L$ of $P$ that extends $\leqslant_P$. That is, $x \leqslant_P y \rightarrow x \leqslant_L y$, for all $x, y \in P$.

**Exercise 9.13.2.**

1. Let $(P, \leqslant_P)$ be a computable stable partial order, say with every element either small or isolated. Show that if $S \subseteq P$ is infinite and every element in it is small (respectively, isolated), then there is an $S$-computable infinite set $S^* \subseteq S$ which is an chain (respectively, antichain) for $\leqslant_P$.
2. Let $(L, \leqslant_L)$ be a computable stable linear order. Show that if $S \subseteq P$ is infinite and every element in it is small (respectively, large), then there is an $S$-computable infinite set $S^* \subseteq S$ which is an ascending sequence (respectively, descending sequence) for $\leqslant_P$.

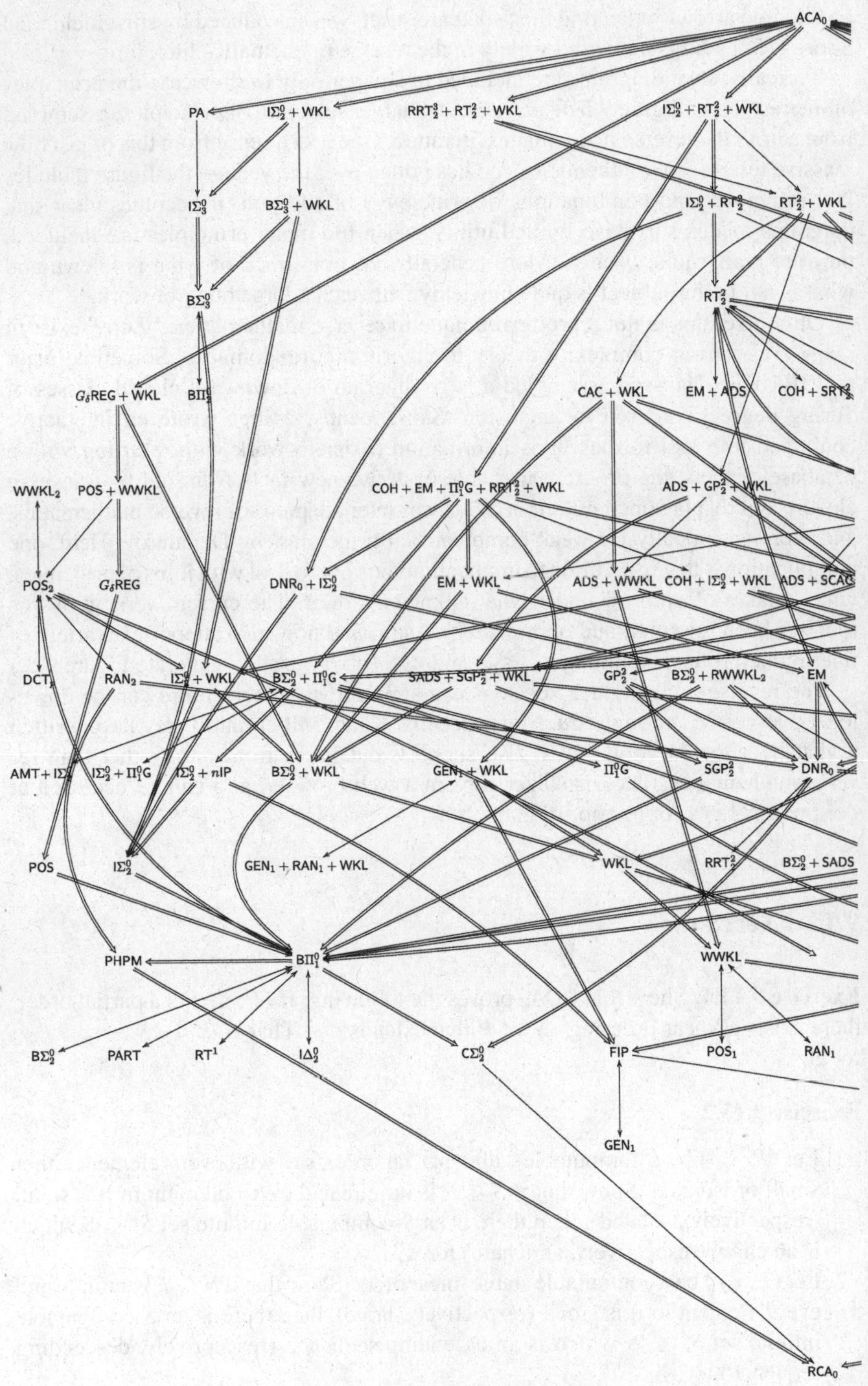
$\mathsf{ACA}_0$
$\mathsf{PA}$
$\mathsf{I}\Sigma^0_2 + \mathsf{WKL}$
$\mathsf{RRT}^3_2 + \mathsf{RT}^2_2 + \mathsf{WKL}$
$\mathsf{I}\Sigma^0_2 + \mathsf{RT}^2_2 + \mathsf{WKL}$
$\mathsf{I}\Sigma^0_3$
$\mathsf{B}\Sigma^0_3 + \mathsf{WKL}$
$\mathsf{I}\Sigma^0_2 + \mathsf{RT}^2_2$
$\mathsf{RT}^2_2 + \mathsf{WKL}$
$\mathsf{B}\Sigma^0_3$
$\mathsf{RT}^2_2$
$G_\delta\mathsf{REG} + \mathsf{WKL}$
$\mathsf{B}\Pi^0_2$
$\mathsf{CAC} + \mathsf{WKL}$
$\mathsf{EM} + \mathsf{ADS}$
$\mathsf{COH} + \mathsf{SRT}^2_2$
$\mathsf{WWKL}_2$
$\mathsf{POS} + \mathsf{WWKL}$
$\mathsf{COH} + \mathsf{EM} + \Pi^0_1\mathsf{G} + \mathsf{RRT}^2_2 + \mathsf{WKL}$
$\mathsf{ADS} + \mathsf{GP}^2_2 + \mathsf{WKL}$
$\mathsf{POS}_2$
$G_\delta\mathsf{REG}$
$\mathsf{DNR}_0 + \mathsf{I}\Sigma^0_2$
$\mathsf{EM} + \mathsf{WKL}$
$\mathsf{ADS} + \mathsf{WWKL}$
$\mathsf{COH} + \mathsf{I}\Sigma^0_2 + \mathsf{WKL}$
$\mathsf{ADS} + \mathsf{SCAC}$
$\mathsf{DCT}_2$
$\mathsf{RAN}_2$
$\mathsf{I}\Sigma^0_2 + \mathsf{WKL}$
$\mathsf{B}\Sigma^0_2 + \Pi^0_1\mathsf{G}$
$\mathsf{SADS} + \mathsf{SGP}^2_2 + \mathsf{WKL}$
$\mathsf{GP}^2_2$
$\mathsf{B}\Sigma^0_2 + \mathsf{RWWKL}_2$
$\mathsf{EM}$
$\mathsf{AMT} + \mathsf{I}\Sigma^0_2$
$\mathsf{I}\Sigma^0_2 + \Pi^0_1\mathsf{G}$
$\mathsf{I}\Sigma^0_2 + \pi\mathsf{IP}$
$\mathsf{B}\Sigma^0_2 + \mathsf{WKL}$
$\mathsf{GEN}_1 + \mathsf{WKL}$
$\Pi^0_1\mathsf{G}$
$\mathsf{SGP}^2_2$
$\mathsf{DNR}_0$
$\mathsf{POS}$
$\mathsf{I}\Sigma^0_2$
$\mathsf{GEN}_1 + \mathsf{RAN}_1 + \mathsf{WKL}$
$\mathsf{WKL}$
$\mathsf{RRT}^2_2$
$\mathsf{B}\Sigma^0_2 + \mathsf{SADS}$
$\mathsf{PHPM}$
$\mathsf{B}\Pi^0_1$
$\mathsf{WWKL}$
$\mathsf{B}\Sigma^0_2$
$\mathsf{PART}$
$\mathsf{RT}^1$
$\mathsf{I}\Delta^0_2$
$\mathsf{C}\Sigma^0_2$
$\mathsf{FIP}$
$\mathsf{POS}_1$
$\mathsf{RAN}_1$
$\mathsf{GEN}_1$
$\mathsf{RCA}_0$

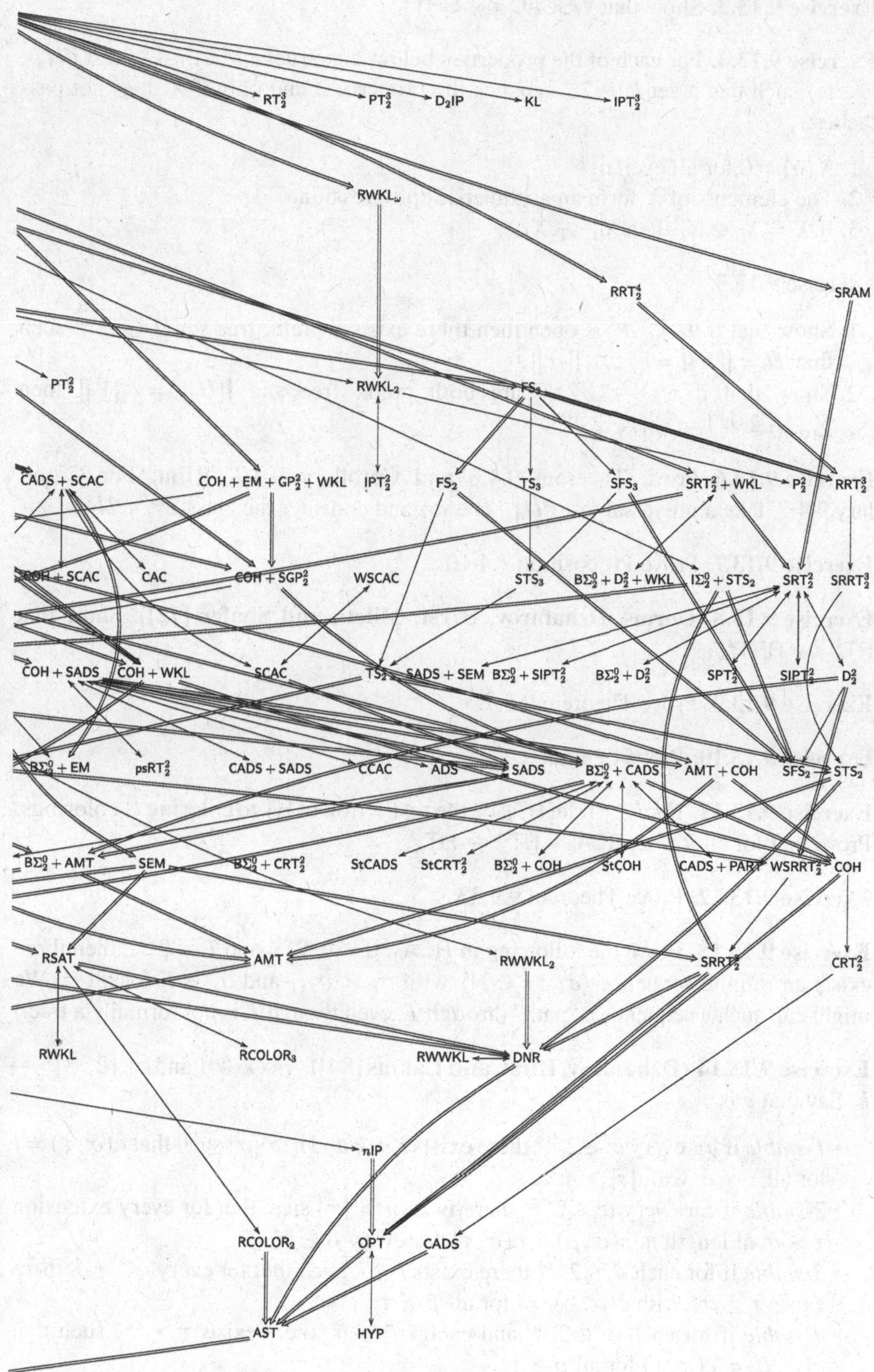


**Figure 9.5.** A snapshot of the reverse mathematics zoo, indicating the myriad principles below $ACA_0$. Arrows indicate implications over $RCA_0$; double arrows are implications that cannot be reversed.

**Exercise 9.13.3.** Show that $\mathsf{WSCAC} \leqslant_{\mathrm{W}} \mathsf{SRT}^2_2$.

**Exercise 9.13.4.** For each of the properties below, construct a Martin-Löf test $\langle \mathcal{U}_n : n \in \omega \rangle$ such that a set $X \in 2^\omega$ satisfies the property if and only if $X$ does not pass the test.

1. $X(n) = 0$ for all even $n$.
2. The elements of $X$ form an arithmetical progression.
3. If $X = X_0 \oplus X_1$ then $X_0 \leqslant_{\mathrm{T}} X_1$.

**Exercise 9.13.5.**

1. Show that if $\mathcal{U} \subseteq 2^\omega$ is open then there exists a prefix free set $U \subseteq 2^{<\omega}$ such that $\mathcal{U} = [[U]] = \bigcup_{\sigma \in U} [[\sigma]]$.
2. Show that if $U, V \subseteq 2^{<\omega}$ are both prefix free and $[[U]] = [[V]]$ then $\sum_{\sigma \in U} 2^{-|\sigma|} = \sum_{\sigma \in V} 2^{-|\sigma|}$.

**Exercise 9.13.6.** Prove Theorem 9.4.6 and Corollary 9.4.7. (Hint: For Corollary 9.4.7, take a universal test $\langle \mathcal{U}_n : n \in \omega \rangle$ and consider the class $2^\omega \setminus \mathcal{U}_1$.)

**Exercise 9.13.7.** Prove Proposition 9.4.10.

**Exercise 9.13.8 (Dorais, Dzhafarov, Hirst, Mileti, and Shafer [72]).** Show that $\mathsf{RT}^1_2 \nleqslant_{\mathrm{W}} \mathsf{RRT}^2_2$.

**Exercise 9.13.9.** Prove Theorem 9.5.7.

**Exercise 9.13.10.** Prove Lemma 9.9.4.

**Exercise 9.13.11.** For $k \geqslant 1$, let $\mathsf{HT}_k$ be the restriction of $\mathsf{HT}$ to coloring $k$-colorings. Prove that for all $k \geqslant 2$, $\mathsf{RCA}_0 \vdash \mathsf{HT}_2 \leftrightarrow \mathsf{HT}_k$.

**Exercise 9.13.12.** Prove Theorem 9.9.33.

**Exercise 9.13.13.** Prove the following in $\mathsf{RCA}_0$. If $T \subseteq 2^{<\mathbb{N}}$ and $T \cong 2^{<\mathbb{N}}$, then there exists an infinite sequence $\langle \sigma_i : i \in \mathbb{N} \rangle$ with $\sigma_i \prec \sigma_{i+1}$ and $\sigma_i \in T$ for all $i$. (We might call such a sequence a "path" through $T$, even though $T$ is not formally a tree.)

**Exercise 9.13.14 (Dzhafarov, Hirst, and Lakins [84]).** Fix $k \geqslant 1$ and $c \colon [2^{<\omega}]^2 \to k$. Say that $c$ is

- *1-stable* if for every $\sigma \in 2^{<\omega}$ there exists $i < k$ and $n \geqslant |\sigma|$ such that $c(\sigma, \tau) = i$ for all $\tau \succ \sigma$ with $|\tau| \geqslant n$.
- *2-stable* if for every $\sigma \in 2^{<\omega}$ there is an $n \geqslant |\sigma|$ such that for every extension $\tau \succ \sigma$ of length $n$, $c(\sigma, \rho) = c(\sigma, \tau)$ for every $\rho \succeq \tau$.
- *3-stable* if for each $\sigma \in 2^{<\omega}$ there exists $i < k$ such that for every $\sigma^* \succeq \sigma$ there exists $\tau \succeq \sigma^*$ with $c(\sigma, \rho) = i$ for all $\rho \succeq \tau$.
- *4-stable* if for each $\sigma \in 2^{<\omega}$ and each $\sigma^* \succeq \sigma$, there exists $\tau \succ \sigma^*$ such that $c(\sigma, \rho) = c(\sigma, \tau)$ for all $\rho \succeq \tau$.

- *5-stable* if for every $\sigma \in 2^{<\omega}$ there is a $\sigma^* > \sigma$ such that $c(\sigma,\tau) = c(\sigma,\sigma^*)$ for all $\tau \succeq \sigma^*$.
- *6-stable* if for every $\sigma \in 2^{<\omega}$ we can find a $\sigma^* > \sigma$ and an $i < k$ such that for all subtrees $T$ extending $\sigma^*$ which are isomorphic to $2^{<\omega}$, there is a $\tau \in T$ such that $c(\sigma,\tau) = i$.

1. Prove in $\mathsf{RCA}_0$ that

$$c \text{ is 1-stable} \to c \text{ is 2-stable} \to c \text{ is 4-stable} \to c \text{ is 5-stable}$$

and

$$c \text{ is 1-stable} \to c \text{ is 3-stable} \to c \text{ is 4-stable} \to c \text{ is 5-stable.}$$

2. Prove in $\mathsf{RCA}_0$ that $c$ is 5-stable if and only if it is 6-stable.
3. Prove in $\mathsf{RCA}_0$ that $c$ is 1-stable if and only if it is both 2-stable and 3-stable.
4. Fix $i \in \{1,\dots,6\}$ and let $\mathsf{S}^i\mathsf{TT}^2_k$ be the restriction of $\mathsf{TT}^2_k$ to $i$-stable colorings. Prove that $\mathsf{SRT}^2_k \leqslant_{\mathrm{W}} \mathsf{S}^i\mathsf{TT}^2_k$.

**Exercise 9.13.15.** Prove that there exists a computable map $f\colon \omega \times 2^{<\omega} \to 2$ with the following property: for each $n \in \omega$ and $e < n$, if $\Phi_e$ is a computable set $T \cong 2^{<\omega}$ then there exist incomparable nodes $\sigma_0, \sigma_1 \in T$ such that for each $i < 2$, $f(n,\tau) = i$ for all $\tau \in T$ extending $\sigma_i$.

**Exercise 9.13.16.** Complete the proof of Theorem 9.8.3.

**Exercise 9.13.17.** Fix $n, k \geqslant 1$.

1. Show that $\mathsf{TT}^n_k$ is identity reducible to $\mathsf{MTT}^n_{2^k}$.
2. Find an $m \geqslant 1$ such that $\mathsf{TT}^n_k \leqslant_{\mathrm{W}} \mathsf{MTT}^n_2 * \cdots * \mathsf{MTT}^n_2$, where the compositional product on the right is taken over $m$ many terms.
3. Show that $\mathsf{RCA}_0 \vdash \mathsf{MTT}^n_k \to \mathsf{TT}^n_k$.

**Exercise 9.13.18.** Formulate a notion of stability for instances of $\mathsf{PMTT}^2$ with the following properties. Fix a set $A$ and numbers $d, k \geqslant 1$. Let $T_0,\dots,T_{d-1} \subseteq \omega^{<\omega}$ be rooted, meet-closed, finitely branching, with height $\omega$ and no leaves.

1. If $c\colon \mathcal{S}_2(T_0,\dots,T_{d-1}) \to k$ for some $k \geqslant 1$ is a stable $A$-computable coloring then there is an $A'$-computable coloring $d\colon \mathcal{S}_1(T_0,\dots,T_{d-1}) \to k$ such that for all $\langle S_0,\dots,S_{d-1}\rangle \in \mathcal{S}_\omega(T_0,\dots,T_{d-1})$, $d$ is constant on $\mathcal{S}_1(S_0,\dots,S_{d-1})$ if and only of $c$ is constant on $\mathcal{S}_2(S_0,\dots,S_{d-1})$.
2. If $c\colon \mathcal{S}_1(T_0,\dots,T_{d-1}) \to k$ for some $k \geqslant 1$ is $A'$-computable then there is an $A$-computable coloring $d\colon \mathcal{S}_2(T_0,\dots,T_{d-1}) \to k$ such that for all $\langle S_0,\dots,S_{d-1}\rangle \in \mathcal{S}_\omega(T_0,\dots,T_{d-1})$, $d$ is constant on $\mathcal{S}_1(S_0,\dots,S_{d-1})$ if and only of $c$ is constant on $\mathcal{S}_2(S_0,\dots,S_{d-1})$.

**Exercise 9.13.19.** Use Theorems 9.10.12 and 9.10.15 to show that $\mathsf{OPT} \nleqslant_\omega \mathsf{AST}$.

**Exercise 9.13.20 (Downey, Diamondstone, Greenberg and Turetsky [65]).** Prove each of the following.

1. FIP admits a universal instance.
2. For each $n \geqslant 2$, $n$IP has a universal instance.

**Exercise 9.13.21.** Prove Proposition 9.10.19.

**Exercise 9.13.22.** Show that $\mathsf{WKL} \leqslant_{\mathrm{sW}} \widehat{\mathsf{WWKL}}$.

# Part IV
# Other areas

# Chapter 10
# Analysis and topology

Analysis and continuous mathematics have been a topic in logic and effective mathematics throughout the history of the field. While the foundations of arithmetic have sometimes been of less interest to "working mathematicians", the foundations of real analysis have traditionally been viewed as more central to mathematical practice.

Work on formalizing and arithmetizing the real line dates back to Dedekind and Cantor, and the real line has been studied by virtually every foundational program since. Notable examples include Brouwer's intuitionistic continuum and Weyl's work on predicativism. Feferman's later work with higher order logic is a direct influence on reverse mathematics. Unsurprisingly, analysis received heavy attention in reverse mathematics, especially during the 1980s and 1990s.

At the same time, continuous mathematics presents a particular challenge for reverse mathematics because of the coding required. With countable combinatorics and countable algebra, the coding required to express theorems in second order arithmetic is often straightforward. Continuous mathematics unavoidably involves uncountability, however, even when we focus on fundamental spaces like the real line.

In countable combinatorics and algebra, we can code the individual points or objects as natural numbers. With uncountable spaces, we typically code each object (for example, each real number) with a set of natural numbers. This allows us to code spaces whose set of objects has the cardinality $2^{\aleph_0}$, the size of complete separable metric spaces. This allows us to formalize results about spaces of this size, including many theorems of real and complex analysis. The cost is that additional work is required simply to verify properties of the coding systems employed. Moreover, the entire space is not represented as a single object in second order arithmetic. Thus we can quantify over points of a given space, using a set quantifier, but our ability to quantify over all spaces is more limited.

We begin this chapter with a collection of results to illustrate the key methods and phenomena, especially related to subsets of $\mathbb{R}$ and functions on the real line. (We will omit proofs that are straightforward or standard.) We then turn to some results that are less well documented in the secondary literature, including theorems



D. D. Dzhafarov, C. Mummert, *Reverse Mathematics*, Theory and Applications of Computability, https://doi.org/10.1007/978-3-031-11367-3_10

from ergodic theory and topological dynamics. We finish with a survey of work on formalizing non-metric topological spaces.

The results we study in this chapter can be particularly hard to attribute to a specific individual or group. As we have mentioned, the basic properties of the real line and complete separable metric spaces have been studied by many foundational programs, including constructive and computable analysis, since the 19th century. Some of these theorems were among the initial examples of reverse mathematics presented by Friedman [108]. Other results are incremental strengthenings of previously known results (for example, generalizing from $[0, 1]$ to compact complete separable metric spaces). As such, it can be challenging to completely describe the provenance of these theorems, especially reversals based on computable counterexamples that may date back to the mid 20th century. We have attempted to provide references with more detailed treatments of many results, but these should not always be taken as definitive.

## 10.1 Formalizations of the real line

There are several well-known ways to "arithmetize" the real line. The traditional representation of real numbers in reverse mathematics uses quickly converging Cauchy sequences. In this section, we explore some aspects of this formalization that make it especially suitable for our purposes.

We will represent a real number as the limit of a Cauchy sequence of rationals. As usual, two Cauchy sequences $\langle q_n \rangle$ and $\langle r_n \rangle$ *have the same limit* if for every $\varepsilon \in \mathbb{Q}^+$ there is an $N$ such that $|q_m - r_n| < \varepsilon$ whenever $m, n > N$. However, two Cauchy sequences of rationals with the same limit may not provide the same effective information about that limit. In order to extract information about the limit from the terms of the Cauchy sequence, we need to have some bound on how far the terms can be from the limit, in terms of their indices. The following definition is made in $\mathsf{RCA}_0$.

**Definition 10.1.1 (Quickly converging Cauchy sequence).** A sequence $\langle q_n \rangle$ of rational numbers is *quickly converging* if $|q_n - q_m| < 2^{-m}$ whenever $n > m$. A *real number* is defined to be a quickly converging sequence of rationals.

The choice of $2^{-m}$ in the previous definition is arbitrary (see Exercise 10.9.1). The spirit of the definition is that, in order to approximate the limit to within $\varepsilon = 2^{-n}$, we only need to look at term $n + 1$ of the quickly converging Cauchy sequence. Of course, there are still many quickly converging Cauchy sequences for each real number, so we define equality of reals via an equivalence relation.

**Definition 10.1.2 (Equality of Cauchy sequences).** Two real numbers $x = \langle q_n \rangle$ and $y = \langle r_n \rangle$ are *equal* if their Cauchy sequences have the same limit.

The real numbers are thus treated as a *setoid*, with a distinction between intensional equality of Cauchy sequences and extensional equality of the real numbers they represent.

There are many other representations of real numbers, including Dedekind cuts, "unmodulated" Cauchy sequences without a known modulus of convergence, and signed binary expansions (see [325]). There are several reasons for choosing quickly converging Cauchy sequences. One key reason relates to effectiveness, as shown in the next two lemmas. A second reason is that the definition via quickly converging Cauchy sequences generalizes directly to arbitrary complete separable metric spaces, unlike Dedekind cuts or binary expansions. This allows the real line to be handled as a special case in theorems about complete separable metric spaces.

**Lemma 10.1.3.** *When the real numbers are represented by quickly converging Cauchy sequences, the following hold.*

1. *The relation $x < y$ on real numbers can be represented with a $\Sigma^0_1$ formula.*
2. *The relations $x \leqslant y$ and $x = y$ on real numbers can be represented with $\Pi^0_1$ formulas.*

**Lemma 10.1.4.** *When the reals are represented by quickly converging Cauchy sequences, the operations $x + y$, $x - y$, $x \times y$, and $x/y$ when $y \neq 0$ are uniformly computable by Turing functionals with oracles for $x$ and $y$.*

Neither of these lemmas holds for the representation with Dedekind cuts, or the representation with unmodulated Cauchy sequences. Hirst [158] gives a detailed study of representations and the effectiveness of conversions between them.

Beyond the real line, we can study complete separable metric spaces with essentially the same framework. We begin with a countable dense set equipped with a metric, and take the quickly converging Cauchy sequences with an equivalence relation. The following definition is straightforward to formalize in $\mathsf{RCA}_0$.

**Definition 10.1.5 (Coding for complete separable metric spaces).** A complete separable metric space $\widehat{A}$ is coded with a set $A \subseteq \omega$ and a function $d\colon A \times A \to \mathbb{R}^+$ that is a metric on $A$.

A point in $\widehat{A}$ is a quickly converging Cauchy sequence $\langle a_i : i \in \omega \rangle$ of elements of $A$. Per our conventions, this means that $d(a_m, a_n) < 2^{-m}$ whenever $n > m$.

Each $a \in A$ is identified with a point $\widehat{a}$ in $\widehat{A}$ via a constant Cauchy sequence. We often write $a$ for $\widehat{a}$ when there is little chance of confusion.

In particular, $\mathsf{RCA}_0$ is strong enough to prove that $\mathbb{R}$, $\mathbb{R}^n$ for $n \in \mathbb{N}$, $2^{\mathbb{N}}$, and $\mathbb{N}^{\mathbb{N}}$ are complete separable metric spaces. The next lemma is a key fact about the complexity of the metric on a complete separable metric space.

**Lemma 10.1.6.** *Let $\widehat{A}$ be a complete separable metric space. For $x, y \in \widehat{A}$ and $r \in \mathbb{Q}^+$, the relation $R(x, y, r) \equiv [d(x, y) < r]$ can be expressed with a $\Sigma^0_1$ formula. The relation $E(x, y) \equiv [x = y]$ can be expressed with a $\Pi^0_1$ formula.*

## 10.2 Sequences and convergence

In this section, we look at some of the most basic theorems in real analysis, each relating to the convergence of sequences.

The definition of a Cauchy sequence in a complete separable metric space $\widehat{A} = (A, d)$ can be directly translated into $\mathsf{RCA}_0$. The terminology in the next definition follows the convention in which a point "exists" when it is an element of the model being considered.

**Definition 10.2.1 (Convergence of Cauchy sequences).** A Cauchy sequence $\langle a_n \rangle$ in a complete separable metric space $\widehat{A}$ *converges* if there is a point $l$ in $\widehat{A}$ (that is, a quickly converging Cauchy sequence of points in $A$) with $\lim a_n = l$.

$\mathsf{RCA}_0$ proves that every quickly converging Cauchy sequence (of arbitrary points) in a complete separable metric space converges (see Exercise 10.9.2). If we do not require the sequence to be quickly converging, however, we will see that $\mathsf{ACA}_0$ is required to find the limit.

**Lemma 10.2.2.** $\mathsf{ACA}_0$ *proves that every Cauchy sequence in a complete separable metric space converges.*

*Proof.* Let $\langle x_i \rangle$ be a Cauchy sequence in a complete separable metric space $\widehat{A} = (A, d)$. Use arithmetical comprehension to form the set $S$ of all pairs $\langle a, r \rangle \in A \times \mathbb{Q}^+$ for which the sequence $\langle x_i \rangle$ is eventually in the ball $B(a, r)$, that is,

$$\langle a, r \rangle \in S \leftrightarrow (\exists N)(\forall i > N)(\forall k)[d(x_i, a) < r].$$

Because $\langle x_i \rangle$ is a Cauchy sequence, for each $r \in \mathbb{Q}^+$ there is at least one $a$ with $\langle a, r \rangle \in S$.

The remainder of the proof is effective relative to the oracle $S$. Working from a fixed enumeration of $A$, we can effectively produce a sequence $a_i$ such that $(a_i, 2^{-(i+1)}) \in S$ for all $i$. We claim that $\langle a_i \rangle$ is a rapidly converging Cauchy sequence. Given $j > i$, we can choose $k$ large enough that $d(x_k, a_i) < 2^{-(i+1)}$ and $d(x_k, a_j) < 2^{-(j+1)} < 2^{-(i+1)}$. Hence $d(a_i, a_j) < 2^{-i}$. □

**Theorem 10.2.3.** *The following are equivalent over* $\mathsf{RCA}_0$.

1. $\mathsf{ACA}_0$.
2. *Every Cauchy sequence in a complete separable metric space converges.*
3. *Every Cauchy sequence of real numbers converges.*
4. *Every bounded sequence of real numbers has a convergent subsequence.*
5. *The restriction of (3) or (4) to sequences in* $[0, 1]$.

*Proof.* Lemma 10.2.2 shows that $\mathsf{ACA}_0$ implies (2), which trivially implies (3). $\mathsf{RCA}_0$ proves every Cauchy sequence is bounded, so (3) implies (4). Conversely, $\mathsf{RCA}_0$ proves that if a Cauchy sequence $\langle a_n \rangle$ has a convergent subsequence then $\langle a_n \rangle$ converges to the limit of that subsequence, so (4) implies (3). Because we can scale functions in $\mathsf{RCA}_0$, both (3) and (4) are equivalent to their restrictions to $[0,1]$. Therefore, it is sufficient to assume that every Cauchy sequence of real numbers converges, and prove $\mathsf{ACA}_0$.

Let $f\colon \mathbb{N} \to \mathbb{N}$ be given. We will show that the range of $f$ exists, which is sufficient by Theorem 5.7.2. For each $m$, let $A_m$ be the set of $n < m$ which are in the range of $f \upharpoonright m$. Let $q_m = \sum_{i \in A_m} 2^{-(2i+1)}$. Intuitively, thinking of the binary expansion of $q_m$, we use the odd-numbered bits to track values that have entered the range of $f$ by stage $m$. These odd bits cannot interact because they are spaced apart with the even-numbered bits.

We first prove in $\mathsf{RCA}_0$ that $\langle q_m \rangle$ is a Cauchy sequence. Given $\varepsilon > 0$, choose $k$ with $2^{-k} < \varepsilon$. By $\Sigma^0_1$ bounding there is an $r$ such that for all $n < k$, if $n$ is in the range of $f$ then $n$ is in the range of $f \upharpoonright k$. This means that the bits corresponding to $0, \ldots, k$ will not change their value after stage $r$. In particular, for $i > r$, we have

$$|q_i - q_r| \leqslant \sum_{j=k+1}^{\infty} 2^{-(2j+1)} \leqslant \sum_{j=k+1}^{\infty} 2^{-j} = 2^{-k} < \varepsilon.$$

Now, suppose that we have a quickly converging Cauchy sequence $\langle a_i \rangle$ that converges to the same point $x$ as $\langle q_m \rangle$. Given $n$, to determine if $n \in \operatorname{range}(f)$, we can simply consult $a_{2n+2}$. It follows from construction that the first $2n+1$ bits of $a_{2n+2}$ must agree with the first $2n+1$ bits of $x$. Otherwise, there would be some $k > 2n+2$ with $d(a_k, a_{2n+2}) \geqslant 2^{-2n+1}$, which contradicts the choice of $(a_i)$. Therefore, we have that $n \in \operatorname{range}(f)$ if and only if bit $2n+1$ of $a_{2n+2}$ is 1. □

A *Specker sequence* is a monotone increasing, bounded sequence of rationals whose limit is not a computable real number. These sequences have a long history in logic, and were first studied by Specker [299] in 1949. In the previous proof, if $f$ is a computable function with a noncomputable range, the sequence that is constructed is a Specker sequence.

Another elementary result in real analysis shows that every bounded sequence in $\mathbb{R}$ has a supremum.

**Theorem 10.2.4.** *The following are equivalent over* $\mathsf{RCA}_0$.

1. $\mathsf{ACA}_0$.
2. *Every bounded sequence of real numbers has a least upper bound.*
3. *Every bounded, monotone increasing sequence of real numbers converges.*
4. *Every bounded, monotone increasing sequence of rational numbers converges.*
5. *The restriction of (2), (3), or (4) to sequences in* $[0,1]$.

*Proof.* The implication from $\mathsf{ACA}_0$ to (2) is similar to Lemma 10.2.2, and is left to Exercise 10.9.10. The implication from (2) to (3) is straightforward: given a least upper bound $x$ for a bounded, monotone increasing sequence $\langle a_i \rangle$, $\mathsf{RCA}_0$ proves that $\langle a_i \rangle$ converges to $x$. The implication from (3) to (4) is trivial.

$\mathsf{RCA}_0$ also proves that every bounded, monotone sequence is Cauchy, which shows that (4) implies $\mathsf{ACA}_0$ via Theorem 10.2.3. Additionally, the reversal in that proof constructs a monotone sequence of rationals, providing a direct reversal. □

## 10.3 Sets and continuous functions

Individual real numbers (or points, more generally) are rarely the focus of analysis. Sets of real numbers, and continuous functions, are of more interest. As we have mentioned, this leads to a key challenge in studying continuous mathematics within second order arithmetic.

We cannot hope to code arbitrary sets of points, or arbitrary functions, using only second order objects. Each real number is represented as a Cauchy sequence of rationals, and by unwinding the codings a real number is ultimately represented as a set of natural numbers. A simple cardinality argument shows we cannot code every set of real numbers as a single set of naturals. We can code the most important sets, however. There are straightforward codings for open and closed sets, and even Borel and some projective sets can be coded by more complicated methods that we will discuss in Section 12.2. Similarly, continuous functions between complete separable metric spaces can be coded as sets of naturals, even though arbitrary functions cannot.

### 10.3.1 Sets of points

To code open sets, we rely on the fact that the open balls of rational radius, centered on points from a countable dense set, form a basis for any complete separable metric space. We use the standard notation for open and closed balls. If $\widehat{A}$ is a complete separable metric space, $x \in \widehat{A}$ and $r \in \mathbb{Q}^+$, we will use the notation $B_r(x)$ to refer to the open ball $\{w \in \widehat{A} : d(w,x) < r\}$. The closed ball $\overline{B}_r(x)$ is $\{w \in \widehat{A} : d(w,x) \leqslant r\}$. Of course, neither of these sets can be directly represented in second order arithmetic, but the relations $w \in B_r(x)$ and $w \in \overline{B}_r(x)$ can be represented by $\Sigma^0_1$ and $\Pi^0_1$ formulas, respectively.

**Definition 10.3.1 (Open and closed sets).** Let $\widehat{A}$ be a complete separable metric space.

- An *open set* is coded by a sequence $\langle a(i), r(i)\rangle$ in $A \times \mathbb{Q}^+$. A sequence of that sort codes the set $\bigcup_i B_{r(i)}(a(i))$.
- A *closed set* of real numbers is represented with a code for the complementary open set.

This definition allows us to perform basic set algebra on open and closed sets, effectively on their codes. The definition for closed sets may seem unexpected. One advantage is that it is (trivially) straightforward to convert the code for an open set to a code for its complement, and *vice versa*. We have already seen that closed sets in $2^\omega$ and $\omega^\omega$ can be represented as the set of paths through a tree, which is often a more useful coding. We will see a third method for coding closed sets in Section 10.5.1. We will see the specific coding we have available for a closed set can be a key tension in a reversal.

**Lemma 10.3.2.** *Let $\widehat{A}$ be a complete separable metric space. The following hold.*

1. *The union and intersection of two open sets $U$ and $V$ are uniformly computable from the codes for $U$ and $V$. The union of a sequence of open sets is uniformly computable from the sequence.*
2. *The union and intersection of two closed sets $U$ and $V$ are uniformly computable from the codes for $U$ and $V$. The intersection of a sequence of closed sets is uniformly computable from the sequence.*
3. *The relation that a real is inside an open set is $\Sigma^0_1$, and the relation that a real is in a closed set is $\Pi^0_1$.*

This lemma leads to a corollary that, in $\mathsf{RCA}_0$, we may form a code for the union or intersection of two coded open sets, or for the union of a sequence of coded open sets. Similarly, we can form a code for the union or intersection of two coded closed sets, or the intersection of a sequence of coded closed sets.

### 10.3.2 Continuous functions

The last basic coding needed for real analysis is a way to represent continuous functions from one complete separable metric space to another. We have seen a simple definition already, in Definition 3.5.5, that is suitable for some questions about specific continuous functions, but has limitations. For example, there is no immediate way to compose two functions coded in the style of that definition. Thus an alternate definition is used more often in the reverse mathematics literature. This definition is somewhat technical and benefits from motivation.

Although the "$\varepsilon$–$\delta$" definition of continuity is typically expressed in terms of strict inequalities (i.e., open balls), it can be stated equivalently with non-strict inequalities (i.e., closed balls). Thus a function $f\colon \mathbb{R} \to \mathbb{R}$ is continuous if, for all $x \in \mathbb{R}$, for every closed ball $\overline{B}_\varepsilon(f(x))$ there is a closed ball $\overline{B}_\delta(x)$ with $f(\overline{B}_\delta(x)) \subseteq \overline{B}_\varepsilon(f(x))$. We will be interested in closed balls because, when we approximate a point $x$ as a limit of a sequence of points $\langle a_i \rangle$, and each $a_i$ is in a particular open ball $B$, the most we know is that $x$ is in the closure of $B$.

To represent the function $f$ in second order arithmetic, we want to capture information about which closed balls map into which other closed balls without referring directly to points in the space. A code for a $f$ will be a sequence $F$ that enumerates certain facts of the form $f(B_r(a)) \subseteq B_s(b)$, for $a \in A$, $b \in B$, and $r, s \in \mathbb{Q}^+$. The inclusion of a four-tuple $\langle a, r, b, s\rangle$ in $F$ shows that this set inclusion holds.

The code $F$ does not need to include every such inclusion, however. It only needs to include enough information about $f$ for us to recover a value for $f(x)$ for each $x \in \widehat{A}$. The domain of $F$ will be the set of points of $\widehat{A}$ for which the code does have enough information to recover the value of $f$. Moreover, we only need $F$ to *enumerate* a sufficient set of facts about $f$. Thus $F$ will really be a set of five-tuples $\langle i, a_i, r_i, b_i, s_i\rangle$, but we frequently simplify notation by not writing $i$. When there is no chance of confusion, we will identify the code $F$ with the function it codes, writing $F(x)$ and $f(x)$ interchangeably.

**Definition 10.3.3 (Continuous function codes).** Let $\widehat{A}$ and $\widehat{B}$ be complete separable metric spaces. A *code for a continuous function* from $\widehat{A}$ to $\widehat{B}$ is a sequence $F$ of four-tuples $\langle a, r, b, s\rangle$ with $a \in A$, $b \in B$, and $r, s \in \mathbb{Q}^+$ satisfying the following properties.

1. If $\langle a, r, b, s\rangle$and $\langle a, r, b', s'\rangle$ are enumerated in $F$, then $d_B(b, b') \leqslant s + s'$. Intuitively, this means that if $f(\overline{B}_r(a)) \subseteq \overline{B}_s(b) \cap \overline{B}_{s'}(b')$ then $\overline{B}_s(b)$ and $B_{s'}(b')$ must overlap.
2. If $\langle a, r, b, s\rangle \in F$ and $d(a', a) + r' < r$ then $\langle a', r', b, s\rangle \in F$. Intuitively, if $\overline{B}_{r'}(a')$ is a subset of $\overline{B}_r(a)$ in the strong sense that $d(a, a') + r' < r$, and $f((\overline{B}_r(a))) \subseteq \overline{B}_s(b)$ then $f((\overline{B}_{r'}(a'))) \subseteq \overline{B}_s(b)$ as well.
3. If $\langle a, r, b, s\rangle \in F$ and $d(b, b') + s < s'$ then $\langle a, r, b', s'\rangle \in F$. This condition is analogous to (2) but for closed balls that contain $B_s(b)$ in a strong sense.

A point $x \in \widehat{A}$ is *in the domain* of $F$ if for every $t \in \mathbb{Q}^+$ there is a tuple $(a, r, s, b)$ enumerated in $F$ with $s < t$ and $d(x, a) < r$. The set $F$ is a *continuous function code* from $\widehat{A}$ to $\widehat{B}$ if the domain of $F$ is $\widehat{A}$. As mentioned, we will often identify $F$ with the function it codes.

The next result shows that $\mathsf{RCA}_0$ is strong enough to evaluate a coded continuous function at each point of its domain.

**Proposition 10.3.4.** *The following is provable in* $\mathsf{RCA}_0$. *If $F$ is a coded continuous function code from $\widehat{A}$ to $\widehat{B}$, and $x$ is in the domain of $F$, there is a point $y = f(x) \in \widehat{B}$ with $d(y, b) \leqslant s$ whenever $(a, r, b, s)$ is enumerated in $F$ and $d(a, x) < s$. Moreover, the point $y$ is unique up to equality of points in $\widehat{B}$.*

$\mathsf{RCA}_0$ is strong enough to produce codes for the functions we commonly encounter.

*Example 10.3.5.* The following are provable in $\mathsf{RCA}_0$.

1. There are codes for the addition, subtraction, multiplication, and division of functions on $\mathbb{R}$ (where they are defined).
2. There is a code for each polynomial function on $\mathbb{R}$.
3. For each $m$ there are codes for the functions $f(\bar{x}) = \sum_{i=1}^m x_i$ and $g(\bar{x}) = \prod_{i=1}^m x_i$.

*Example 10.3.6.* The following are provable in $\mathsf{RCA}_0$. Let $\widehat{A}$, $\widehat{B}$, and $\widehat{C}$ be complete separable metric spaces.

1. The identity function $f(x) = x$ has a code.
2. The metric function $d \colon \widehat{A} \times \widehat{A} \to \mathbb{R}$ has a code.
3. For each $y \in B$, there is a code for the constant function $f \colon \widehat{A} \to \widehat{B}$ given by $f(x) = y$.
4. If $f \colon \widehat{A} \to \widehat{B}$ and $g \colon \widehat{B} \to \widehat{C}$ are coded continuous functions then there is a code for the composition $g \circ f$.

The final lemma of this section shows that the definition of a continuous function code is effective in a certain sense. Recall that, for $y, z \in \mathbb{R}$, the relation $z > y$ can be expressed as a $\Sigma^0_1$ formula. If $F$ is a continuous function code from a complete separable metric space $\widehat{A}$ to $\mathbb{R}$, this gives a naively $\Delta^1_1$ method to express the relation $F(x) > y$: we can write $(\exists z)[z = F(x) \wedge z > y]$ or $(\forall z)[z = F(x) \rightarrow z > y]$. The next lemma shows we can leverage the coding to express $F(x) > y$ with no set quantifiers. The proof is Exercise 10.9.9.

**Lemma 10.3.7.** *Suppose $F$ is a continuous function code from a complete separable metric space $\widehat{A}$ to $\mathbb{R}$, and $y \in \mathbb{R}$. The relation $P(x, y) \equiv F(x) > y$ is given by a $\Sigma^0_1$ formula with a parameter for $F$.*

The preceding lemma and examples form part of the justification for representing continuous functions with the particular codes we have defined. Another justification is experience with many results in reverse mathematics, where this method has been able to facilitate meaningful results. Nevertheless, the optimality of this coding method is a question of genuine interest. In Section 12.4, we will see results in higher order reverse mathematics that help illuminate aspects of the question.

## 10.4 The intermediate value theorem

In this section, we study the reverse mathematics of the Intermediate Value Theorem from the perspectives of second order arithmetic and computable reducibility, extending the results from Chapter 3. The intermediate value theorem is a particularly interesting result because it is computably true, but not uniformly computably true. It is thus a useful example of how considering several reducibility notions can help us understand a theorem more deeply.

The first result we consider is a formalized version of Proposition 3.7.3. We include a detailed proof to demonstrate the additional verification needed to ensure the proof carries through in $\mathsf{RCA}_0$.

**Theorem 10.4.1.** *The following is provable in $\mathsf{RCA}_0$. If $F$ is a coded continuous function from an interval $[a, b]$ to $\mathbb{R}$, and $y$ is between $f(a)$ and $f(b)$, there is an $x \in [a, b]$ with $F(x) = y$.*

*Proof.* We work in $\mathsf{RCA}_0$. We assume without loss of generality that $y = 0$ and $f(a) < 0$. If there is a $q \in \mathbb{Q} \cap [a, b]$ with $F(q) = 0$ then we may simply let $x = q$. Suppose not, so $F(q) > 0$ or $F(q) < 0$ for all $q \in \mathbb{Q}^+ \cap [a, b]$. Because the relation $P(q) \equiv F(q) > 0$ and $N(q) \equiv F(q) < 0$ are each $\Sigma^0_1$, we can form the set $S = \{q \in \mathbb{Q} \cap [a, b] : F(q) < 0\}$ with $\Delta^0_1$ comprehension.

We now mimic the usual proof via subdivision. By induction, using $S$ as a parameter, there is a sequence $\langle a_i, b_i \rangle$ such at $a_0 = a$, $b_0 = b$, and

$$\langle a_{n+1}, b_{n+1} \rangle = \begin{cases} \langle (a_n + b_n)/2, b_n \rangle & \text{when } F((a_n + b_n)/2) < 0, \\ \langle a_n, (a_n + b_n)/2 \rangle & \text{when } F((a_n + b_n)/2) > 0. \end{cases}$$

It follows by induction that $F(a_n) < 0$ and $F(b_n) > 0$ for all $n$. Because $|b_n - a_n| = 2^{-n}$ for all $n$, it is straightforward to verify that $x = (a_n)$ is a quickly converging Cauchy sequence. We claim that $F(x) = 0$.

Suppose not. If $F(x) = z > 0$ then, because $x$ is in the domain of $F$, there must be some $(p, r, q, s)$ enumerated in $F$ with $s < z/2$ and $d(x, p) < r$. We may thus choose $n$ large enough that $a_n, b_n \in [p - r, p + r]$. This means that $F(a_n) \geqslant z/2$, contradicting the fact that $F(a_n) < 0$. If $F(x) < 0$, a similar argument shows that $F(b_n) < 0$ for some $n$, which is again a contradiction. □

As we discussed in Chapter 3, the proof begins with a division into cases, depending on whether there is a $q \in \mathbb{Q} \cap [a, b]$ with $F(q) = 0$. Because $\mathsf{RCA}_0$ includes the law of the excluded middle, there is no obstacle to making such ineffective choices during a proof. But the choice means that this proof does not yield an algorithm to compute the desired point $x$. In fact, there is no such algorithm, as we saw in Proposition 3.7.4, which stated that the problem $\mathsf{IVT}$ does not uniformly admit computable solutions.

Theorem 10.4.1 and Proposition 3.7.4 show that an analysis in second order arithmetic is too coarse to fully explore the intermediate value theorem. We have seen this phenomenon already with combinatorial results. $\mathsf{IVT}$ shows the phenomenon is not limited to that area. As with combinatorics, moving to finer reducibility notions allows us to understand $\mathsf{IVT}$ better.

A Weihrauch-style analysis of the intermediate value theorem was performed by Brattka and Gherardi [19]. They show that the following problem $\mathsf{B}$ is equivalent to $\mathsf{C}_I$ (the principle of choice for closed intervals), and provide much additional information on this and other choice principles for closed sets.

**Definition 10.4.2.** The Weihrauch principle $\mathsf{B}$ has as its instances all sequences of pairs of rational numbers $\langle a_n, b_n : n \in \omega \rangle$ in $[0, 1]$ such that $a_n \leqslant a_{n+1} \leqslant b_{n+1} \leqslant b_n$ for all $n$. A solution of an instance $\langle a_n, b_n \rangle$ is a real $x$ with $a_n \leqslant x \leqslant b_n$ for all $n$.

The principle $\mathsf{B}$ is implicit in some informal proofs of $\mathsf{IVT}$. The next theorem makes this implicit appearance into a formal result on computable reducibility.

**Theorem 10.4.3.** $\mathsf{IVT} \equiv_{\mathrm{sW}} \mathsf{B}$.

*Proof.* We first show that $\mathsf{IVT} \leqslant_{\mathrm{sW}} \mathsf{B}$. Let $f$ be an instance of $\mathsf{IVT}$. We build an instance $\langle a_n, b_n \rangle$ of $\mathsf{B}$ inductively. Let $a_0 = 0$ and $b_0 = 1$. At stage $k + 1$, we perform an effective search, for $k$ time steps, looking for the pair of rationals $p, q$ with $a_k \leqslant p \leqslant q \leqslant b_k$ and $f(p) \cdot f(q) < 0$ that minimizes the value of $|p - q|$. This is a finite search because we limit the number of steps of computation that can be performed. If we find such a pair, we let $a_{k+1} = p$ and $b_{k+1} = q$. Otherwise, we let $a_{k+1} = a_k$ and $b_{k+1} = b_k$. It is immediate that $\langle a_n, b_n \rangle$ is an instance of $\mathsf{IVT}$. Moreover, we must have $\lim |b_k - a_k| = 0$. Otherwise, given that the number of time steps available in the construction becomes arbitrarily large, we would eventually find a pair of rationals $p, q$ that bracket a root of $f$ with $|q - p| < \lim |b_k - a_k|$, a contradiction to the construction. The (unique) solution $x$ to this instance of $\mathsf{IVT}$ is then a root of the function $f$.

Next, we show that $\mathsf{B} \leqslant_{\mathrm{sW}} \mathsf{IVT}$. For each rational $q$, define the functions

$$P_q = \begin{cases} 1 - \frac{x}{q} & \text{if } x \leqslant q, \\ 0 & \text{if } x \geqslant q, \end{cases}$$

and

$$N_q = \begin{cases} 0 & \text{if } x \leqslant q, \\ 1 - \frac{1-x}{1-q} & \text{if } x \geqslant q. \end{cases}$$

Thus $P(q)$ and $N(q)$ are computable functions that are strictly positive on $[0, q)$ and $(q, 0]$, respectively. Given an instance $\langle a_n, b_n \rangle$ of $\mathsf{B}$, define a function $f(x)$ with $f(0) = 1$, $f(1) = -1$, and otherwise

$$f(x) = \left( \sum_{\{n:a(n)<x\}} 2^{-n} P_{a(n)}(x) \right) - \left( \sum_{\{n:b(n)>\}} 2^{-n} N_{b(n)}(x) \right).$$

Then $f(x)$ is a computable function and $f^{-1}(0)$ is exactly the set of solutions to the instance of $\mathsf{B}$. □

We have now seen two results characterizing the strength of the intermediate value theorem. Overall, the intermediate value theorem has these properties:

1. Every computable instance has a computable solution.
2. $\mathsf{RCA}_0$ proves that every instance of $\mathsf{IVT}$ has a solution. This shows that not only can the theorem be proved without noncomputable set existence principles, it can also be proved with weak induction axioms.
3. $\mathsf{IVT}$ is equivalent to $\mathsf{B}$.
4. $\mathsf{IVT}$ does not admit uniformly computable solutions.
5. Uniformly computable solutions are admitted for the special case of $\mathsf{IVT}$ in which the function has only one root, or in which the set of roots in nowhere dense (Exercise 10.9.14).

This combination of properties is a remarkable illustration of how the combination of different reducibilities can lead to a detailed understanding of a theorem. The intermediate value theorem has also been studied in constructive reverse mathematics; see Berger, Ishihara, Kihara, and Nemoto [14] for another perspective on the theorem.

## 10.5 Closed sets and compactness

In this section, we consider results about closed sets in complete separable metric spaces. The real line, unit interval, Cantor space $2^\omega$, and Baire Space $\omega^\omega$ are key examples. The theorem we study are among the most basic in analysis, including results on separability and compactness. Yet many of these results are not computably true, and some even rise to the strength of $\Pi^1_1$ comprehension.

There are many ways to characterize compactness of complete separable metric spaces, including total boundedness, the open cover property, and every sequence having a convergent subsequence. In order to study the latter two properties within $\mathsf{RCA}_0$, we use a version of total boundedness to define compactness. The following definition can be readily formalized in $\mathsf{RCA}_0$.

**Definition 10.5.1 (Compact metric spaces).** A complete separable metric space $\widehat{A}$ is *compact* if there is an infinite sequence $\langle A_i : i \in \mathbb{N}\rangle$ of finite sequences of points, $A_i = \langle a_{i,0}, \ldots, a_{i,n_i}\rangle$, so that for each $z \in \widehat{A}$ and each $i$ there is a $j \leqslant n_i$ with $d(z, a_{i,j}) < 2^{-i}$.

$\mathsf{RCA}_0$ proves that many common spaces are compact, including $[0, 1]$, the Cantor space $2^\omega$, finite products $[0, 1]^n$, and the Hilbert space $[0, 1]^\omega$. We will see that stronger systems are necessary to prove that spaces like these are compact in other senses. This difference in the strength of different definitions of compactness is the motivation the for our choice of a definition in $\mathsf{RCA}_0$.

A standard result shows that every sequence in a compact space has a convergent subsequence. The next theorem shows this result is equivalent to $\mathsf{ACA}_0$ over $\mathsf{RCA}_0$. The underlying computable counterexample is again a Specker sequence.

**Theorem 10.5.2.** *The following are equivalent over* $\mathsf{RCA}_0$.

1. $\mathsf{ACA}_0$.
2. *Each sequence in a compact metric space has a convergent subsequence.*
3. *Each sequence in* $[0, 1]$ *has a convergent subsequence.*

*Proof.* We first prove (2) in $\mathsf{ACA}_0$. Let $\widehat{A}$ be a compact metric space, witnessed by sequences $\langle A_i\rangle$ and $\langle n_i\rangle$. Let $\langle x_k\rangle$ be a sequence of points in $\widehat{A}$.

Using arithmetical comprehension, we may form the set $S$ of all sequences $\langle j_0, \ldots, j_n\rangle$ such that there are infinitely many $k$ with

$$x_k \in B(a_{0,j(0)}, 1) \cap \cdots \cap B(a_{n,j(n)}, 2^{-n}).$$

Using the pigeonhole principle in $\mathsf{ACA}_0$, there is at least one sequence of length 1 in $S$, so $S$ is nonempty. Moreover, if $\langle j_0, \ldots, j_n\rangle \in S$ then, by the pigeonhole principle again, there is at least one $j_{n+1}$ with $\langle j_0, \ldots, j_n, j_{n+1}\rangle \in S$. Hence $S$ is a nonempty tree with no dead ends.

Let $f$ be an infinite path in $S$. Define a sequence $k(m)$ inductively as follows: $k(0)$ is the least $k$ with $x_k \in B(a_{f(0)}, 1)$, and $k(m + 1)$ is the least $k > k(m)$ with

$$x_k \in B(a_{0,j(0)}, 1) \cap \cdots \cap B(a_{m+1,f(m+1)}, 2^{-(m+1)}).$$

Then $\langle x_{k(m)} : m \in \mathbb{N}\rangle$ is a convergent subsequence of $(x_k)$, as desired. The special case (3) follows immediately.

For the reversal, assume every sequence in $[0, 1]$ has a convergent subsequence. $\mathsf{RCA}_0$ proves that $[0, 1]$ is a compact space. Thus, in particular, each Cauchy sequence in $[0, 1]$ must have a convergent subsequence, which $\mathsf{RCA}_0$ can verify is the limit of the Cauchy sequence. This yields $\mathsf{ACA}_0$ by Theorem 10.2.3. □

We have seen several examples where the strength of results about $[0, 1]$ and $2^{\mathbb{N}}$ are the same. Interestingly, the sequential compactness of $2^{\mathbb{N}}$ is far weaker than sequential compactness of the unit interval. By formalizing Proposition 4.1.7, we see that sequential compactness of $2^{\mathbb{N}}$ is equivalent to $\mathsf{COH}$. The next lemma simultaneously generalizes and formalizes Proposition 3.8.1.

**Lemma 10.5.3.** $\mathsf{WKL}_0$ *proves that each open cover of a compact metric space has a finite subcover. That is, if* $\langle U_i \rangle$ *is a sequence of open sets in a compact metric space* $\widehat{A}$ *that covers the space, then there is an* $N$ *such that* $\widehat{A} = \bigcup_{i \leqslant N} U_i$.

*Proof.* Let $\widehat{A}$ be a compact metric space, witnessed by sequences $\langle A_i \rangle$ and $\langle n_i \rangle$ as in Definition 10.5.1. Let $\langle U_i \rangle$ be a sequence of open sets that covers $\widehat{A}$. Without loss of generality, we may assume each $U_i$ is an open ball $B(x_i, r_i)$.

We construct a $\Pi^0_1$ tree $T \subseteq 2^{<\mathbb{N}}$. Intuitively, we build the tree so that any infinite path would be a Cauchy sequence, consisting of points of the form $a_{i,j}$, to a hypothetical point not covered by $\langle U_i \rangle$. Under the assumption that $\langle U_i \rangle$ does cover $\widehat{A}$, there is no such point, so the tree will be finite, which will allow us to extract a finite subcovering.

For each sequence $\tau \in T$, each element $\tau(i)$ will refer to a point $t_i = a_{i,\tau(i)} \in \widehat{A}$. We put a sequence $\tau$ into $T$ if the following conditions are met.

1. $\tau(i) \leqslant n_i$ for $i \leqslant |\tau|$.
2. For all $i, j < |\tau|$, $d(t_i, t_j) \leqslant 2^{-i} + 2^{-j}$.
3. For all $i, j < |\tau|$, $d(t_i, a_j) + 2^{-i} \geqslant r_j$.

Item (1) ensures that the notation $t_i$ is well defined. Item (2) ensures that, if $f$ is an infinite path in $T$, the corresponding sequence $\langle t_i : i \in \mathbb{N} \rangle$ provides a quickly converging Cauchy sequence with some limit $x_f \in \widehat{A}$. Item (3) ensures that, if $f$ is an infinite path in $T$, then for all $j$ we have $x_f \notin B(a_j, r_j)$ (taking the limit as $i \to \infty$ in (3) yields $d(x_f, a_j) \geqslant r_j$).

Item (1) is $\Delta^0_1$, and the remaining two items are each $\Pi^0_1$, by Lemma 10.1.3. Thus the tree $T$ is a $\Pi^0_1$ tree and, by formalizing Exercise 2.9.13, $\mathsf{RCA}_0$ proves there is a tree $T' \supseteq T$ with the same infinite paths as $T$. Since $T$ and thus $T'$ have no infinite paths, by bounded König's lemma the tree $T'$ must be finite. Hence there is an $N$ such that $T'$, and hence $T$, has no sequence of length $N$.

We claim that $\langle B(x_i, r_i) : i \leqslant N \rangle$ covers $\widehat{A}_i$. If not, let $x$ be a point that is not covered. By $\Sigma^0_1$ induction we can construct a sequence $\tau$ of length $N + 1$ so that $d(t_i, x) \leqslant 2^{-i}$ for $i \leqslant |\tau|$. Then $\tau \in T$, which is a contradiction. □

Examining the forward implication in the proof yields the following corollary.

**Corollary 10.5.4.** *The following is provable in* $\mathsf{WKL}_0$. *If* $\langle \widehat{A}_n \rangle$ *is a sequence of compact metric spaces and* $\langle U_n \rangle$ *is a sequence such that each* $\langle U_{i,n} : n \in \mathbb{N} \rangle$ *is an open cover of* $\widehat{A}_i$, *then there is a sequence* $\langle N_i \rangle$ *such that* $\langle U_{i,1}, \ldots, U_{i,N_i} \rangle$ *is a cover of* $\widehat{A}_i$ *for each* $i$.

The next theorem, a version of the Heine–Borel theorem, was also among the original examples provided by Friedman.

**Theorem 10.5.5.** *The following are equivalent over* $\mathsf{RCA}_0$.

1. $\mathsf{WKL}_0$.
2. *For every compact metric space* $\widehat{A}$ *and sequence* $\langle U_i \rangle$ *of open sets that covers* $\widehat{A}$, *there is an* $n$ *such that* $\langle U_i : i \leqslant n \rangle$ *covers* $\widehat{A}$.
3. *For every sequence* $\langle V_i \rangle$ *of open sets that covers* $[0, 1]$, *there is an* $n$ *such that* $\langle V_i : i \leqslant n \rangle$ *is also a covering of* $[0, 1]$.

*Proof.* The proof that $\mathsf{WKL}_0$ implies (2) is Lemma 10.5.3, and the implication from (2) to (3) is immediate. Therefore, we work in $\mathsf{RCA}_0$ and prove that (3) implies $\mathsf{WKL}_0$.

Let $T \subseteq 2^{<\mathbb{N}}$ be a tree with no infinite paths. Our goal is to prove that $T$ is finite. The construction in Lemma 3.8.2 can be formalized directly in $\mathsf{RCA}_0$, Let $U$ be enumeration of open intervals corresponding to the tree $T$, along with the auxiliary sequences $\langle I_\tau \rangle$ and $\langle J_\tau \rangle$.

By the assumption that $T$ has no infinite path, $U$ covers the unit interval. Therefore, by (3), there is a finite initial segment $U'$ of the enumeration that covers the interval. By construction, $C$ is disjoint from $\bigcup_\tau I_\tau$, so $C$ must be covered by the finite collection of intervals of the form $J_\tau$ in $U'$. Thus the height of $T$ is bounded by the length of the largest $\tau$ for which $J_\tau$ is in $U'$, so $T$ is finite. □

### 10.5.1 Separably closed sets

An open set $U$ in a complete separable metric space is coded as a sequence of open balls. A closed set is coded with the sequence of open balls in its complement. Closed sets are thus coded with "negative information": information on which points are not in the closed set. We may want to convert this to "positive information" about which points are in the set.

Because most closed sets are not the countable union of open balls or the countable union of closed balls, we require a different way to represent this positive information. One possibility is to enumerate a dense subset $S$ of the closed set. Thus a point $y$ is in the closed set if and if there are points in $S$ arbitrarily close to $y$, which is an arithmetical property of $S$ and $y$.

**Definition 10.5.6 (Separably closed sets).** A set $F \subseteq X$ is *separably closed* if there is a sequence $\langle x_n \rangle$ of points in $C$ such that for every $y \in C$ and every $r \in \mathbb{Q}^+$ there is an $i$ with $d(x_i, y) < r$. We treat the sequence $\langle x_n \rangle$ as a code for the set $F$.

With two representations of closed sets, a natural question is the difficulty of converting between the representations. The next theorem shows that converting between positive and negative representations of closed sets is nontrivial even in the particular case of $[0, 1]$.

**Theorem 10.5.7 (Brown [22]; see [157, 11]).** *The following are equivalent over* $\mathsf{RCA}_0$.

1. $\mathsf{ACA}_0$.
2. *In a compact metric space, every nonempty closed set is separably closed.*
3. *In* $[0,1]$, *every nonempty closed set is separably closed.*
4. *In a complete separable metric space, every separably closed set is closed.*
5. *In* $[0,1]$, *every separably closed set is closed.*

*Proof.* Exercises 10.9.11 and 10.9.12 ask for proofs of (2) and (4) in $\mathsf{ACA}_0$. Part (3) follows immediately from (2) in $\mathsf{RCA}_0$, and (5) follows from (4). We will prove that each of (3) and (5) implies $\mathsf{ACA}_0$ over $\mathsf{RCA}_0$.

First, we work in $\mathsf{RCA}_0$ and assume every nonempty closed set in $[0,1]$ is separably closed. In light of Theorem 10.2.4, it is enough to prove that every bounded, increasing sequence of rationals in $[0,1]$ converges. Let $\langle a_i : i \in \mathbb{M}\rangle$ be a monotone increasing sequence of rationals in $[0,1]$. We may assume that there is no rational number that is the limit of $(a_i)$.

For each $i$, let $U_i$ be the interval $[0, a_i)$, which is open in $[0,1]$. Let $C$ be the closed set with complement $\bigcup_i U_i$. We have $1 \in C$, so $C$ is nonempty. Under our assumption, there is a sequence $S = \langle s_i\rangle$ so that $\overline{S} = C$.

Now, for each rational $q \in [0,1]$, either $q \leqslant a_i$ for some $i$, or $q \geqslant b_i$ for some $i$. Otherwise, because $q \geqslant a_i$ for all $i$, then $q \in C$ by construction, but $q < s_i$ for all $i$, so $q = \min C$, so $\lim a_n = q$, contradicting the assumption that $a_n$ does not converge to a rational number. This means that, for a rational number $q$, the properties $q \in U$ and $q \in C$ are $\Sigma^0_1$, and every rational in $[0,1]$ is in $U \cup C$. Working in $\mathsf{RCA}_0$, we may form the set $L$ of all rationals $q$ in $[0,1]$ such that $q \leqslant a_i$ for some $i$, that is, the set of rationals in $U$.

We now apply an interval-halving technique that is also seen in the proof of the intermediate value theorem in $\mathsf{RCA}_0$ (Theorem 10.4.1). We will produce sequences $\langle c_n\rangle$ and $\langle d_n\rangle$ of rationals with $c_n \leqslant c_{n+1} \leqslant d_{n+1} \leqslant d_n$ and $|c_n - d_n| \leqslant 2^{-n}$, so that each $c_n$ is in $U$ and each $d_n$ is in $C$. At stage 0, let $c_0 = 0$ and $d_0 = 1$. At stage $k+1$, assume we have constructed rationals $c_k$ and $d_k$. Let $z = (c_k + d_k)/2$, which will be rational. If $z \in L$ we let $c_{k+1} = z$ and $d_{k+1} = d_k$. Otherwise we let $c_{k+1} = c_k$ and $d_{k+1} = z$. In either case, $c_{k+1}$ and $d_{k+1}$ have the desired properties. By $\Sigma^0_1$ induction, the sequences $(c_n)$ and $(d_n)$ exist.

By construction, $\langle c_n\rangle$ and $\langle d_n\rangle$ are both quickly converging Cauchy sequences with the same limit $x$. It is impossible for $x \in U$, because then $x < a_i$ for some $i$, and hence $\lim d_n \neq x$. It is also impossible to have $b_i < x$ for any $i$, because then $\lim c_n \neq x$. Hence $x = \min C$, and $\lim a_i = x$ as desired.

For the second reversal, we work in $\mathsf{RCA}_0$ and assume that every separably closed set is closed. To establish $\mathsf{ACA}_0$, it is enough to show that the range of an arbitrary injection $f : \mathbb{N} \to \mathbb{N}$ exists. Let $S$ enumerate $\{2^{-f(m)} : m \in \mathbb{N}\} \cup \{0\}$. The set of points enumerated in $S$ is closed, but in any case we may view $S$ as a dense subset of a closed set $C$. By assumption, there is a code for the open complement $U$ of $C$. Now we have that an arbitrary $y$ is in the range of $f$ if and only $2^{-y} \in S$, and also if and only if there is no ball $B(a,r)$ in $U$ with $2^{-y} \in B(a,r)$. We may thus form the range of $f$ via $\Delta^0_1$ comprehension. □

In Theorem 10.5.7 (2), there is an additional assumption that the space is compact. For non-compact spaces, the next theorem shows that $\Pi^1_1$-$\mathsf{CA}_0$ is needed to find a dense subset of an arbitrary closed set. The fundamental idea in the proof is that for each $\Sigma^1_1$ formula $\varphi(n)$ there is a sequence of trees $T_n$ in $\omega^\omega$ such that, for each $n$, $\varphi(n)$ holds if and only if there is a path in $T_n$. At the same time, the set of paths through a tree in $\omega^\omega$ is a closed set in that space. Hence locating points in closed sets of $\omega^\omega$ can allow us to answer $\Sigma^1_1$ questions.

**Theorem 10.5.8.** *The following are equivalent over* $\mathsf{RCA}_0$.

*1.* $\Pi^1_1$-$\mathsf{CA}_0$.
*2. In a complete separable metric space, each nonempty closed set is separably closed.*
*3. In* $\mathbb{N}^\mathbb{N}$, *every nonempty closed set is separably closed.*

*Proof.* We first work in $\Pi^1_1$-$\mathsf{CA}_0$ and let $\widehat{A}$ be a complete separable metric space. Let $C$ be a nonempty closed set in $\widehat{A}$. We may use $\Pi^1_1$-$\mathsf{CA}_0$ to form the set $S$ of pairs $\langle a,r\rangle$ such that $C \cap B(a,r)$ is nonempty. We will show in Corollary 12.1.14 that $\mathsf{ATR}_0$, and thus $\Pi^1_1$-$\mathsf{CA}_0$, proves the $\Sigma^1_1$ *axiom of choice*, which is the following scheme over $\Sigma^1_1$ formulas $\psi$ that may have parameters:

$$(\forall k)(\exists X)\psi(n,X) \to (\exists (Y_n))(\forall n)\psi(n,Y_n).$$

We have shown the following formula holds, in which the matrix is arithmetical:

$$(\forall (a,r))(\exists x)[(a,r) \in S \to x \in \widehat{A} \land x \notin B(a,r)$$

Thus, using $\Sigma^1_1$ choice, we may form a sequence $\langle y_n\rangle$ of points such that, whenever $n = \langle a,r\rangle$ and $C \cap B(a,r) \neq \emptyset$, we have $y_n \in C \cap B(a,r)$. It is immediate that we can form a dense sequence in $C$ from the sequence $\langle y_n\rangle$.

The implication from (2) to (3) is immediate. To prove the reversal that (3) implies $\Pi^1_1$ comprehension, we first show in $\mathsf{RCA}_0$ that (3) implies $\mathsf{ACA}_0$ (see Exercise 10.9.15). Thus, working in $\mathsf{ACA}_0$, it is enough to show that (3) implies $\Sigma^1_1$ comprehension.

Let $\psi(n)$ be a $\Sigma^1_1$ formula with parameters. We want to form the set $\{n : \psi(n)\}$. We first apply Kleene's normal form theorem (Theorem 5.8.2) to find a formula $\theta$ such that

$$(\forall n)[\varphi(n) \leftrightarrow (\exists f \in \mathbb{N}^\mathbb{N})(\forall m)\theta(n,m,f[m])].$$

We may form a sequence of trees $T_n$ such that the set of paths through $T_n$ is exactly the set of $f$ such that $\theta(n,m,f[m])$ holds for all $m$. We may then combine the trees $\langle T_n\rangle$ to form a single tree $T$ so that a sequence $\sigma$ is in $T$ if and only if $\sigma$ is of the form $\langle i\rangle ^\frown \tau$ where $\tau \in T_i$.

The set of paths through $T$ is a closed set. As usual, the corresponding open set is coded by the set of basic open balls corresponding to all elements of $\mathbb{N}^{<\mathbb{N}}$ not in $T$. Apply (3) to find a sequence $\langle g_k \rangle$ dense in $T$. Then, for each $n$, there is some $f \in [T_n]$ if and only if there is an $g_k$ with $g_k[0] = n$, and this happens if and only if $\varphi(n)$. This allows us to form the set $\{n : \varphi(n)\}$ with arithmetical comprehension relative to the sequence $\langle g_k \rangle$. □

### 10.5.2 Uniform continuity and boundedness

In some foundational programs, particularly in constructive analysis, an assumption is made that each continuous function on the real line is accompanied by a modulus of continuity on each closed interval. The lack of that assumption in reverse mathematics is a key source of separation between the programs. For example, there are theorems that are viewed a constructive in Bishop's program, when expressed in the style of that program, but which are not provable when expressed in $\mathsf{RCA}_0$ in the style of reverse mathematics. In this section, we survey results on the reverse mathematics of uniform continuity.

**Definition 10.5.9 (Modulus of uniform continuity).**
Let $f$ be a function from a complete separable metric space $\widehat{A}$ to a complete separable metric space $\widehat{B}$. A *modulus of uniform continuity* for $f$ is a function $h\colon \mathbb{Q}^+ \to \mathbb{Q}^+$ such that, for all $x, y \in \widehat{A}$ and all $r \in \mathbb{Q}^+$, if $d(x, y) < h(r)$ then $d(f(x), f(y)) < r$.

If we let $D = \{2^{-n} : n \in \mathbb{N}\}$, there is no loss of generality in assuming $h$ is a function from $D$ to $D$ instead of $\mathbb{Q}^+$ to $\mathbb{Q}^+$.

While the moduli of continuity for polynomial functions or other common functions can be directly constructed, $\mathsf{WKL}_0$ is needed in general to obtain a modulus of uniform continuity for a continuous function on a compact space. Exercise 10.9.17 asks for a proof of the following.

**Lemma 10.5.10 (Brown; see Simpson [288, IV.2.2]).** *The following is provable in* $\mathsf{WKL}_0$. *Suppose $\widehat{A}$ is a compact metric space, $C \subseteq \widehat{A}$ is closed, and $f$ is a continuous function from $C$ to a complete separable metric space $\widehat{B}$. Then $f$ has a modulus of uniform continuity on $C$.*

The proof of the next theorem demonstrates a somewhat peculiar method: when $\mathsf{WKL}_0$ fails, there is an infinite tree with no infinite path. Trees with this property are difficult to visualize, but can be useful to construct counterexamples.

**Theorem 10.5.11 (Simpson [288, IV.2.3]).** *The following are equivalent over* $\mathsf{RCA}_0$.

1. $\mathsf{WKL}_0$.
2. *If $F$ is a coded continuous function from an interval $[a, b]$ to $\mathbb{R}$ then $F$ achieves a maximum value at some point in the interval.*
3. *If $F$ is a coded continuous function from an interval $[a, b]$ to $\mathbb{R}$ then $F$ is bounded.*

*Proof.* The implication from (1) follows in $\mathsf{WKL}_0$ from Lemma 10.5.10, because a modulus of continuity immediately provides a bound on the function. The implication from (1) to (2) is straightforward. To show (2) implies $\mathsf{WKL}_0$, we work in $\mathsf{RCA}_0$ and assume $\mathsf{WKL}_0$ fails, which means there is an infinite tree $T$ with no path. We want to show that (3) fails.

Apply the construction from Lemma 3.8.2 to $T$, letting $C$ be the Cantor set and letting $\langle J_\tau \rangle$ be the corresponding sequence of intervals. Define a set $\widetilde{T} \subseteq 2^{<\mathbb{N}}$ by putting $\tau \in \widetilde{T}$ if and only if $\tau \notin T$ but every prefix of $\tau$ is in $T$. Thus the intervals in $\{J_\tau : \tau \in \widetilde{T}\}$ are pairwise disjoint and cover $C$. Note that, in the lexicographic order, each element $\tau \in \widetilde{T}$ has an immediate successor and immediate predecessor within $\widetilde{T}$ which can be found effectively from $\tau$.

For each $x \in [0, 1]$, one of two cases holds. Case 1: $x$ is in the closure of an interval $J_\sigma$ for a unique $\sigma \in \widetilde{T}$. Case 2: $x$ is strictly between two intervals $J_\sigma$ and $J_\tau$, with $\sigma, \tau \in \overline{T}$ so that $\tau$ is the immediate successor of $\sigma$ within $\overline{T}$.

We may thus define a continuous function $f\colon [0, 1] \to \mathbb{R}^+$ as follows. In case 1, for $x \in J_\sigma$, let $f(x) = |\sigma|$. For case 2, for $x$ between $J_\sigma$ and $J_\tau$, interpolate linearly between the values on $J_\sigma$ and $J_\tau$. Because $\overline{T}$ is infinite, this function is unbounded. □

We leave the proof of the following to Exercise 10.9.17.

**Theorem 10.5.12 (Simpson [288, IV.2.3]).** *The following are equivalent over* $\mathsf{RCA}_0$.

1. $\mathsf{WKL}_0$.
2. *Every continuous function on a compact complete separable metric space has a modulus of uniform continuity.*
3. *Every continuous function on the unit interval has a modulus of uniform continuity.*

## 10.6 Topological dynamics and ergodic theory

In this section, we consider several results from topological dynamics. These results move us beyond the elementary analysis of the previous sections. They are also of interest for demonstrating a principle that is strictly between $\mathsf{WKL}_0$ and $\mathsf{ACA}_0$, and highlighting the open question of the strength of the iterated version of Hindman's Theorem.

We then move to measure theory and ergodic theory. Some basic aspects of measure theory are closely related to $\mathsf{WWKL}_0$ (Section 9.11) and to algorithmic randomness. We also state a result of Avigad and Simic on the mean ergodic theorem.

In topological dynamics, a compact dynamical system consists of a compact metric space $X$ and a continuous function $T\colon \widehat{X} \to \widehat{X}$. The *orbit* of a point $z$ is the set $\{T^k(z) : k \geqslant 0\}$. In very simple systems, the orbit of a point $z$ may be periodic, with $T^n(z) = z$ for some $n$. More complex systems will not have periodic orbits, but they will have orbits with other recurrence properties, as in the following definition.

**Definition 10.6.1 (Almost periodic and proximal points).**

1. A point $x$ of a compact dynamical system is *recurrent* if there is a sequence $\langle n_i \rangle$ such that $\lim T^{(n_i)}x = x$.
2. A point $x$ is *almost periodic* (also *uniformly recurrent*) if, for every $r \in \mathbb{Q}^+$ there is an $N_r$ such that, for every $m$, there is a $k < N_r$ with $T^{m+k}(x) \in B_r(x)$.
3. Two points $x$ and $y$ are *proximal* if, for every $r \in \mathbb{Q}^+$, there are infinitely many $n$ with $d(T^n x, T^n y) < r$.

Birkhoff's recurrence theorem shows that every compact dynamic system has a recurrent point. A stronger theorem shows that every such system has an almost periodic point. We begin this section with an examination of these two results, based on work of Day.

We then study the Auslander–Ellis theorem, which states the every point in a compact dynamical system is proximal to an almost periodic point. This theorem is closely related to an iterated version of Hindman's theorem and to several other combinatorial results. In the second subsection, we discuss the known equivalences and state several key open problems.

### 10.6.1 Birkhoff's recurrence theorem

In this section, we focuses on dynamical systems over closed subsets of Cantor space. For such spaces, we can use a particularly concrete representation of the continuous transformation of the dynamical system. Exercise 10.9.18 shows that, in $\mathsf{WKL}_0$, every coded continuous function on Cantors space is encoded as in the next definition.

**Definition 10.6.2 (Coded transformations on $2^{\mathbb{N}}$, Day [63]).** A function $f\colon 2^{<\mathbb{N}} \to 2^{<\mathbb{N}}$ *encodes a transformation* of $2^{\mathbb{N}}$ if $f$ is total and order preserving, and for every $l$ there is an $m$ such that $|f(\sigma)| \geqslant l$ for all $\sigma \in \{0,1\}^m$.

**Definition 10.6.3 (Compact dynamical systems on Cantor space, Day [63]).** A *compact dynamical system over the Cantor space* consists of:

1. a tree $C \subseteq 2^{<\mathbb{N}}$ coding a nonempty closed subset $[C]$ of $2^{\mathbb{N}}$, and
2. a transformation $F\colon 2^{\mathbb{N}} \to 2^{\mathbb{N}}$ encoded by a map $f\colon 2^{<\mathbb{N}} \to 2^{<\mathbb{N}}$.

We require that $f(\sigma) \in C$ for all $\sigma \in C$. This guarantees $F(C) \subseteq C$.

The next results relate to a form of Birkhoff's recurrence theorem.

**Lemma 10.6.4 (Day [63]).** $\mathsf{WKL}_0$ *proves that every compact dynamical system over the Cantor space has a recurrent point.*

*Proof.* Working in $\mathsf{WKL}_0$, let $(C, f)$ be a compact dynamical system over the Cantor space. It follows from definitions that a point $x$ in the system is recurrent if, for every $i$, there are $n, l > i$ such that $x \upharpoonright i \subseteq f^n(x \upharpoonright l)$. Thus the set of recurrent points $\mathcal{R}$ is $\Pi^0_2$ relative to $C$ and $f$. Our goal is to construct an infinite tree $T$ so that $[T] \subseteq \mathcal{R}$.

To do so, we will inductively construct a sequence $\langle U_i \rangle$ of finite sets of strings, viewing each $U_i$ as a closed set $[U_i] \subseteq 2^{\mathbb{N}}$. The construction will ensure that if $x \in [U_i] \cap C$ then $x \upharpoonright i \subseteq f^n(x \upharpoonright l)$ for some $n, l > i$. Hence every $x \in \bigcap_i [U_i] \cap [C]$ will be a recurrent point.

We also need to ensure that $[U_i] \cap [C]$ is nonempty for each $i$. To do so, the construction will make sure that for each $i$ there is an $s = s_i$ such that $[C] \subseteq \bigcup_{n<s} f^{-n}([U_i])$. Assuming this condition, if $[U_i] \cap [C] = \emptyset$ then there would be some $x \in [C]$ and $n < s$ with $f^n(x) \in [U_i]$, a contradiction to the assumption that $f([C]) \subseteq [C]$.

The construction will form a sequence $\langle U_i \rangle$ of finite sets of strings, a sequence $\langle s_i \rangle$ of numbers, and an auxiliary sequence $\langle V_i \rangle$ of finite sets of strings. We will ensure the following properties hold for each $i$.

1. For every $\tau \in U_i$, there is an $n$ with $i \leqslant n \leqslant s_i$ such that $\tau \upharpoonright i \preceq f^n(\tau)$.
2. For every $\sigma \in V_i$ there are $n \leqslant s_i$ and $\tau \in U_i$ such that $\tau \preceq f^n(\sigma)$.
3. $V_i$ codes a finite open cover of $[C]$. This is a $\Sigma^0_1$ property: there is a level $l$ such that every $\tau \in C$ of length $l$ extends some element of $V_i$.

The three conditions together can be written as a $\Sigma^0_1$ formula, allowing us to use only $\Sigma^0_1$ induction in the construction.

At stage 0, let $U_0 = V_0 = \langle\rangle$ and $s_0 = 0$. The three conditions are straightforward to verify. At stage $i + 1$, we define infinite sequences

$$U_{i+1}(s) = \{\sigma \in 2^{<\mathbb{N}} : (\exists \tau \in U_i)[\sigma \preceq \tau] \wedge (\exists n)[(i \leqslant n \leqslant s) \wedge (\sigma \upharpoonright i \preceq f^n(\sigma))]\},$$
$$V_{i+1}[s] = \{\sigma \in 2^{<\mathbb{M}} : (\exists m \leqslant s)(\exists \tau \in U_{i+1}(s))[f^n(\sigma) \succeq \tau]\}.$$

We immediately have $[U_{i+1}(s)] \subseteq [U_i]$, $[U_{i+1}(s)] \subseteq [U_{i+1}(s+1)]$, and $[V_{i+1}(s)] \subseteq [V_{i+1}(s+1)]$. It can be shown in $\mathsf{WKL}_0$ that $[C] \subseteq \bigcup_s [V_{i+1}(s)]$ (Exercise 10.9.19). Hence, by Theorem 10.5.5, there is some $s_{i+1}$ such that every sequence of length $s_{i+1}$ in $C$ extends some sequence in $V_{i+1}(s_{i+1})$. We define $U_{i+1} = U_{i+1}(s_{i+1})$ and $V_{i+1} = V_{i+1}(s_{i+1})$. We may verify the three properties hold.

It follow from the construction that $[U_i] \cap [C]$ is nonempty for each $i$. Hence $([U_i] \cap [C] : i \in \mathbb{N})$ is a nested sequence of nonempty closed sets of $2^{\mathbb{N}}$. Hence, by Exercise 10.9.6, there is a point in $\bigcap_i [U_i] \cap [C]$, which is the desired recurrent point of $(C, f)$. □

**Theorem 10.6.5 (Day [63]).** *The following are equivalent over* $\mathsf{RCA}_0$.

*1.* $\mathsf{WKL}_0$.
*2. Every compact dynamical system over the Cantor space has a recurrent point.*

*Proof.* In light of the previous lemma, we need to prove that (2) implies $\mathsf{WKL}_0$. Working in $\mathsf{RCA}_0$, let $T$ be an infinite subtree of $2^{<\mathbb{N}}$. We view $T$ as a subset of $3^{<\mathbb{N}}$ and construct a compact dynamical system over $3^{\mathbb{N}}$ (modifying the definitions in the obvious way). We will define a dynamical system $D$ on $3^{\mathbb{N}}$ so that every recurrent point is a path through $T$. Because there is an effective bijection $2^{<\mathbb{N}}$ to $3^{<\mathbb{N}}$ that gives an effective homeomorphism from $2^{\mathbb{N}}$ to $3^{\mathbb{N}}$, we can then apply (2) to find the desired recurrent point in $D$.

As we construct the map $f\colon 3^{\mathbb{N}} \to 3^{\mathbb{N}}$, our goal is that if $x \in [T]$ then $f(x) = x$, and otherwise $x$ is not recurrent. The additional branching in $3^{<\mathbb{N}}$ gives us space to move each point $x \in 3^{<\mathbb{N}}$ that is not a path in $T$. The general method is that the orbit of $x$ moves in increasing lexicographical order, trying to find a path in $[T]$, and wraps around when it extends $\langle 2\rangle$.

We let $F(\langle\rangle) = \langle\rangle$. Given $\sigma \neq \langle\rangle$, let $n = |\sigma| - 1$, so $\sigma(n)$ is the final value of $\sigma$. If $\sigma \in T$ then let $f(\sigma) = \sigma \upharpoonright n$. Otherwise, if $\sigma \notin T$, let $\pi$ be the shortest initial segment of $\sigma$ such that $\pi \notin T$. Hence, if $\pi$ contains a 2 then this 2 is the final value of $\pi$. Using that fact, we can define

$$f(\sigma) = \begin{cases} (\rho^\frown\langle 1\rangle^\frown 0^{\mathbb{N}}) \upharpoonright n & \text{if } \pi = \rho^\frown\langle 0\rangle \text{ or } \pi = \rho^\frown\langle 0,2\rangle, \\ (\rho^\frown\langle 2\rangle^\frown 0^{\mathbb{N}}) \upharpoonright n & \text{if } \pi = \rho^\frown\langle 1\rangle \text{ or } \pi = \rho^\frown\langle 1,2\rangle, \\ 0^{\mathbb{N}} & \text{if } \pi = \langle 2\rangle. \end{cases}$$

Then $D = (3^{\mathbb{N}}, f)$ is a compact dynamical system. For the rest of this proof, let $\leqslant_{\text{lex}}$ be the lexicographical order on $3^{<\mathbb{N}}$. Each of the following claims is straightforward to verify from the construction.

1. Let $n > 0$, $\sigma_0, \ldots, \sigma_n$ be a sequence such that $\sigma_n \preceq \sigma_0$, $f(\sigma_i) = \sigma_{i+1}$ for all $i < n$, and $\sigma_n \notin T$. Then $\langle 2\rangle \preceq \sigma_k$ for some $k$.
2. If $\tau \in T$, $|\sigma| < |\tau|$ and $\sigma \leqslant_{\text{lex}} \tau$ then $f(\sigma) \leqslant_{\text{lex}} \tau$.

Let $x$ be a recurrent point for $D$. Working towards a contradiction, assume $x \notin [T]$, and choose $\sigma \preceq x$ with $\sigma \notin T$. Because $x$ is recurrent, we may choose a sequence $\sigma_0, \ldots, \sigma_n$ with $\sigma_n = \sigma \preceq \sigma_0 \preceq x$ and $f(\sigma_i) \preceq \sigma_{i+1}$. By the first claim, we have $\langle 2\rangle \preceq \sigma_k$ for some $k < n$.

Now choose $\tau \in T$ such that $|\tau| > |\sigma_i|$ for all $i \leqslant n$. Because $\sigma_{k+1}$ is a string of all 0s, and $|\sigma_{k+1}| < |\tau|$, we have $\sigma_{k+1} \leqslant_{\text{lex}} \tau$. Then, by induction on the second claim, we have $\sigma_n \leqslant_{\text{lex}} \tau$. Now $\sigma_n \npreceq \tau$ because $\sigma_n \notin T$, and $\sigma_n \preceq \sigma$, so $\sigma_0 \leqslant_{\text{lex}} \tau$.

By another induction over the second claim, we have $\sigma_i \leqslant_{\text{lex}} \tau$ for all $i$, and hence $\langle 2\rangle \npreceq \sigma_i$ for all $i$, contradicting $\langle 2\rangle \preceq \sigma_k$. Thus $x \in [T]$, as desired. □

We now turn to the principle AP which states that every compact topological system over Cantor space has an almost periodic point. We state the following results without proof. Here, a system $(C, F)$ is *minimal* if, for every system $(D, f)$ with $[D] \subseteq [C]$, we have $[D] = [C]$.

**Lemma 10.6.6 (Day [63]).** $\mathsf{WKL}_0$ *proves that every point in a minimal system is almost periodic.*

**Lemma 10.6.7 (Day [63]).** *Over* $\mathsf{WKL}_0$, $\mathsf{ACA}_0$ *is equivalent to the proposition that every compact dynamical system on Cantor space contains a minimal subsystem.*

The two lemmas combine to give the following bound on the strength of AP.

**Corollary 10.6.8 (Day [63]).** *The principle* AP *is provable in* $\mathsf{ACA}_0$.

Building on this, Day showed that AP is strictly between $\mathsf{WKL}_0$ and $\mathsf{ACA}_0$. This result requires two theorems, one to show AP is stronger than $\mathsf{WKL}_0$ and one to show $\mathsf{ACA}_0$ is stronger than AP. Each of the proofs constructs an $\omega$-model to give the separation. We prove one of the two separations here. In the following results, we use a fixed effective enumeration of all $\Pi^0_1$ classes. We state the next lemma without proof.

**Lemma 10.6.9 (Day [63]).** *Let $f$ be the left shift on Cantor space and let $P \subseteq 2^\omega$ be a $\Pi^0_1$ class. There is a $\Pi^0_1$ class $C$, whose index is uniformly computable from an index for $P$, such that $(C, f)$ is a compact dynamical system on Cantor space and one of the following holds.*

1. *$C \cap P = \varnothing$.*
2. *There is a nonempty $\Pi^0_1$ class $\widehat{P} \subseteq P$ with the property that no element of $\widehat{P}$ is an almost period point of $(C, f)$.*

**Theorem 10.6.10 (Day [63]).** *There is an $\omega$-model of* $\mathsf{WKL}_0 + \neg\mathsf{AP}$.

*Proof.* Let $\langle Q_i \rangle$ be an effective enumeration of all $\Pi^0_1$ classes on Cantor space and let $f$ be the left shift map. Let $\pi_e$ be the projection from a point in Cantor space onto its $e$th coordinate. Using Lemma 10.6.9, and applying pullbacks as needed, we can build a single system

$$(C, g) = \prod_{e \in \omega} (C_e, f)$$

so that for each $e$, either $\pi_e(Q_e) \cap C_e$ is empty or there is a nonempty $\Pi^0_1$ class $\widehat{Q}_e$ so that no element of $\pi_e(\widehat{Q})$ is almost periodic.

We next form a nested sequence $\langle P_i \rangle$ of $\Pi^0_1$ classes. Let $\Phi_s$ be an effective enumeration of all Turing functionals. At state 0, let $P_0$ be a nonempty $\Pi^0_1$ class all of whose members have PA degree. Now, at stage $s+1$, consider $\Phi_s$. If there is an $n$ such that $\{x \in P_s : \Phi^X_s(n)\uparrow\}$, let $P_{s+1}$ be this set for the least such $n$.

If $\Phi_s$ is total on all elements of $P_s$, consider the $\Pi^0_1$ class $Q = \Phi_s(P_s)$. Let $e$ be an index for $Q$. There are two cases. In the first case, if $\pi_e(Q) \cap C = \varnothing$, let $P_{s+1} = P_s$. In this case, it is clear that no element of $P_{s+1}$ computes an element of $C$. In the second case, there is a nonempty $\Pi^0_1$ set $\widehat{Q} \subseteq Q$ such that no element of $\pi_e(\widehat{Q})$ is almost periodic in $C_e$, and thus no element of $\widehat{Q}$ is almost periodic in $C$. In this case, let $P_{s+1} = \{x \in P_s : \Phi^x_s \in \widehat{Q}\}$. Again, $P_{s+1}$ will be a nonempty $\Pi^0_1$ class and no element of $P_{s+1}$ computes an almost periodic point in $(C, g)$ under $\Phi_s$.

Because $\langle P_s \rangle$ is a nested sequence of nonempty closed sets in Cantor space, by compactness there is a point $x \in \bigcap_s P_s$. In particular $x \in P_0$, so $x$ has PA degree, and by construction $x$ cannot compute an almost periodic point of $(C, g)$. The $\omega$-model of sets computable from $x$ will thus satisfy $\mathsf{WKL}_0$ and not AP. □

We omit the proofs of the following results, which separate AP from $\mathsf{ACA}_0$ and also provide a conservation result related to AP.

**Theorem 10.6.11 (Day [63]).** *There is an $\omega$-model of* AP *consisting entirely of low sets. In particular, this model satisfies* $\mathsf{AP} + \neg\mathsf{ACA}_0$.

**Corollary 10.6.12 (Day [63]).** *Over* $\mathsf{RCA}_0$, *the principle* $\mathsf{AP}$ *is strictly between* $\mathsf{ACA}_0$ *and* $\mathsf{WKL}_0$.

**Theorem 10.6.13 (Day [63]).** $\mathsf{AP}$ *is conservative over* $\mathsf{WKL}_0$ *for* $\Pi^1_1$ *sentences.*

### 10.6.2 The Auslander–Ellis theorem and iterated Hindman's theorem

We turn now to the Auslander–Ellis theorem. Because this theorem is known to imply $\mathsf{ACA}_0$, we can often work over $\mathsf{ACA}_0$ when proving equivalences, making the coding systems much easier to manage.

We begin by summarizing the equivalences between two iterated versions of Hindman's theorem, a principle on ultrafilters, and a version of the Milliken–Taylor theorem. Full definitions, and proofs yielding the following theorem, are given by Hirst [155], who studied the possibilities of formalizing proofs of Hindman's theorem via ultrafilters into second order arithmetic.

**Theorem 10.6.14.** *The following are equivalent over* $\mathsf{RCA}_0$.

1. $\mathsf{IHT}$ $(= \mathsf{IHT}_2)$: *For every sequence of 2-colorings* $\langle C_i : i \in \mathbb{N}\rangle$ *there is an increasing sequence* $\langle x_i : i \in \mathbb{N}\rangle$ *of numbers such that for every* $j \in \mathbb{N}$ *the set* $\{x_i : i > j\}$ *satisfies Hindman's Theorem for* $C_j$ *(see [155, Theorem 3]).*
2. $\mathsf{IHT}_{<\omega}$: *For every sequence of finite colorings* $\langle C_i : i \in \mathbb{N}\rangle$ *there is an increasing sequence* $\langle x_i \in \mathbb{N} : i \in \mathbb{N}\rangle$ *such that for every* $j \in \mathbb{N}$ *the set* $\{x_i : i > j\}$ *satisfies Hindman's Theorem for* $C_i$ *(see [155, Lemma 4]).*
3. $\mathsf{AUF}$: *Every countable downward translation algebra has an almost downward translation invariant ultrafilter (see [155, p. 2]).*
4. *For any* $n \geqslant 3$, *the statement* $\mathsf{MT}_n$: *If* $f : [\mathbb{N}]^n \to k$ *then there is an increasing sequence* $X = \langle x_i \in \mathbb{N} : i \in \mathbb{N}\rangle$ *such that* $f$ *is constant on* $\mathrm{FS}_n(X)$ *(see [155, Lemma 5]).*

Blass, Hirst, and Simpson [17, Theorem 4.13] show that $\mathsf{IHT}$ is provable in $\mathsf{ACA}_0^+$, and hence so are the equivalent principles above.

The *Auslander–Ellis theorem* is the following statement: Let $X$ be a compact metric space and $T : X \to X$ a continuous function. For every $x \in X$ there is a $y \in X$ such that $y$ is almost periodic under $T$ and $x$ is proximal to $y$ under $T$ (See [17, pp. 147–148] and [120, Theorem 8.7]).

This theorem is known to be closely related to $\mathsf{IHT}$. Blass, Hirst, and Simpson [17, Theorem 5.11] use $\mathsf{IHT}$ as a lemma to prove the Auslander–Ellis theorem in $\mathsf{ACA}_0^+$. Kreuzer [188, Theorem 11] notes that $\mathsf{IHT}$ follows from $\mathsf{ACA}_0$ and the Auslander–Ellis Theorem using ultrafilter techniques.

We will sketch a slightly different proof in the next theorem, using well-known results described by Furstenberg [120], and assuming the reader has access to that book. In particular, see [120, Theorems 8.8, 8.10, and 8.11; Remark, p. 163; and p. 127].

**Theorem 10.6.15.** *The following are equivalent over* $\mathsf{ACA}_0$.

*1.* IHT.
*2. The Auslander–Ellis theorem.*

*Proof.* The implication from (1) to (2) in $\mathsf{ACA}_0$ is given by Blass, Hirst, and Simpson [17, Theorem 5.11]. We work in $\mathsf{ACA}_0$ and prove the converse.

Recall that $\mathsf{ACA}_0$ (actually $\mathsf{WKL}_0$) proves that every continuous function on a compact metric space has a modulus of uniform continuity. Thus for every compact metric space $X$, continuous map $T: X \to X$, $\varepsilon > 0$, and $p \in \mathbb{N}$ we can find (uniformly in $\varepsilon$ and $p$) a $\delta > 0$ such that $d(x, y) < \delta$ implies $d(T^p x, T^p y) < \varepsilon$ for all $x, y \in X$.

Let $\langle C_i : i \in \mathbb{N}\rangle$ be a sequence of 2-colorings of $\mathbb{N}$. We write the color sets of $C_i$ as $C_{i,0} \cup C_{i,1} = \mathbb{N}$. Let $Z = 2^{\mathbb{N}\times\mathbb{N}}$, with each element $z \in Z$ regarded as a doubly indexed sequence $\langle z_i(j) \in \{0, 1\} : i, j \in \mathbb{N}\rangle$. Give $Z$ the product topology, so $Z$ is a compact metric space.

Let $T$ be the function on $X$ such that $Tz_i(n) = z_i(n + 1)$ for all $i$ and $n$. This is a kind of modified left shift map. $\mathsf{ACA}_0$ proves that $T$ is continuous. Form a point $x \in X$ by letting $x_i(n)$ be, for each $i$ and $n$, the unique color assigned to $n$ by $C_i$.

Apply the Auslander–Ellis Theorem to obtain an almost periodic point $y \in Z$ such that $x$ and $y$ are proximal. Let $Y$ be the closure of $\{T^i y : i \in \mathbb{N}\}$ in $Z$. By proximality, $x \in Y$. $\mathsf{ACA}_0$ proves that $Y$ is a compact metric space and is able to form a code for the restriction of $T$ to $Y$.

The remainder of the proof is parallel to the proof of [120, Proposition 8.10] (see also [120, Remark, p. 163]). We will construct a strictly increasing sequence $\langle p_i : i \in \mathbb{N}\rangle$ such that for each $j \in \mathbb{N}$ the set $\{p_i : i \geqslant j\}$ satisfies Hindman's Theorem for $C_j$. We require the following fact.

*Claim. For any open ball* $U = B_t(y)$ *around* $y$ *in* $Y$*, there is a* $p \in \mathbb{N}$ *such that* $T^p x \in U$ *and* $T^p y \in U$*. Moreover,* $p$ *may be obtained uniformly arithmetically from* $t$.

To prove the claim, we first define a value $r = t/2$. Because $y$ is almost periodic, we can choose $N_r$ such that for every $m$ there is some $k < N_r$ with $T^{m+k} y \in B_r(y)$. Moreover, $N_r$ is arithmetically definable relative to $r$ and parameters. Next, using a modulus of uniform continuity, choose $s \in \mathbb{Q}^+$ such that $d(z, w) < s$ implies $d(T^i z, T^i w) < t/2$ for all $i \leqslant N_r$.

Because $x$ and $y$ are proximal, we may choose $m$ such that $d(T^m x, T^m y) < s$, and by construction we may choose $k \leqslant N_r$ such that $d(T^{m+k} y, y) < t/2$. Then $d(T^{m+k} y, y) < t$ and $d(T^{m+k} x, y) < t$, so the claim holds with $p = m + k$. The construction of $p$ can be carried out uniformly arithmetically. This completes the proof of the claim.

We now return to the proof of the theorem. Fix a sequence of sets $\langle V_i : i \in \mathbb{N}\rangle$ of open balls such that $y \in V_i$ for all $i$ and, moreover, $w_i(0) = z_i(0)$ for all $w, z \in V_i$. We construct a sequence $U_1 \supseteq U_2 \supseteq \cdots$ of open balls around $y$, and simultaneously build an increasing sequence $\langle p_i \in \mathbb{N} : i \in \mathbb{N}\rangle$, by induction. Let $U_1 = V_1$. By induction, assume $U_i$ has been constructed such that $y \in U_i$. Use the claim to choose $p_i \in \mathbb{N}$ such that $T^{p_i} x \in U_i$ and $T^{p_i} y \in U_i$. Let $U_{i+1}$ be an open ball around $y$ that is contained in $U_i \cap T^{-p_i} U_i \cap V_{i+1}$. This completes the construction of $\langle U_i \rangle$ and $\langle p_i \rangle$. Note that $U_i \subseteq U_j$ whenever $i \leqslant j$, by induction on $j$.

Now fix $k \in \mathbb{N}$ and let $k \leqslant i(0) < i(1) < \cdots < i(n)$. We want to show that

$$T^{p_{i(0)}+p_{i(1)}+\cdots+p_{i(n)}} x \in V_k.$$

The proof is by induction on $n$. By construction,

$$\begin{aligned} T^{p_{i(0)}+p_{i(1)}+\cdots+p_{i(n)}} x &\in T^{p_{i(0)}+p_{i(1)}+\cdots+p_{i(n-1)}} U_{p_{i(n)}} \\ &\subseteq T^{p_{i(0)}+p_{i(1)}+\cdots+p_{i(n-2)}} U_{p_{i(n-1)}} \\ &\subseteq \cdots \subseteq U_{i(0)} \subseteq V_k. \end{aligned}$$

Because $V_k$ was chosen to fix $z_k(0)$ for all $z \in Y$, we see that $\{p_i : i > k\}$ satisfies Hindman's Theorem for the coloring $C_k$. □

**Corollary 10.6.16.** *The following are equivalent over* $\mathsf{ACA}_0$.

1. IHT.
2. AUF.
3. $\mathsf{MT}_n$ *for each* $n \geqslant 3$.
4. *The Auslander–Ellis Theorem.*

The precise strength of the principles from this corollary is unknown, as is their precise relationship to Hindman's theorem.

*Question 10.6.17.* The principles in Corollary 10.6.16 are known to imply $\mathsf{ACA}_0$ and are provable in $\mathsf{ACA}_0^+$. Determine their precise strength, or show the strength is strictly between those two systems.

*Question 10.6.18.* Does HT imply IHT over $\mathsf{RCA}_0$, or even over $\mathsf{ACA}_0$?

Kreuzer obtained additional characterizations of the Auslander–Ellis theorem in terms of the existence of certain ultrafilters. This adds additional principles to the cluster with IHT.

**Theorem 10.6.19 (Kreuzer [188]).** *The following are equivalent over* $\mathsf{ACA}_0$.

1. *The Auslander–Ellis Theorem (and hence all the additional equivalent results from Corollary 10.6.16).*
2. *Every countable downward translation algebra has a partial minimal idempotent ultrafilter.*
3. *Every countable downward translation algebra has a partial idempotent ultrafilter.*

### 10.6.3 Measure theory and the mean ergodic theorem

Ergodic theory is another key area of dynamical systems. It relies on measure theory and measure preserving transformations. A significant amount of work has been done on effective measure theory, especially as it relates to algorithmic randomness.

We saw in Corollary 9.11.3 that $\mathsf{WWKL}_0$ is equivalent to the principle that, for each $X$, there is a $Y$ that is 1-random relative to $X$, which can be restated informally as the principle that the complement of a measure zero set is nonempty. $\mathsf{WWKL}_0$ is equivalent to many other statements from measure theory. Avigad and Simic [11] provide a thorough summary.

**Theorem 10.6.20 (Yu and Simpson [331]; see Brown, Giusto, and Simpson [23]).** *The following are equivalent over* $\mathsf{RCA}_0$.

1. $\mathsf{WWKL}_0$.
2. *For any covering of the closed unit interval with a sequence of rational intervals* $(a_i, b_i)$, *we have* $\sum_{i=0}^{\infty} |b - i - a_i| \geqslant 1$.

Compared to compact dynamical systems, less work has been done on the reverse mathematics of ergodic theory. Avigad and Simic [11] studied the strength of a version of the mean ergodic theorem, and the existence of certain projections.

**Definition 10.6.21 (Partial averages and fixed points).** If $T$ is an isometry of a Hilbert space $H$, and $x \in H$, the sequence of partial averages is the sequence

$$S_n(x) = \frac{1}{n}(x + Tx + T^2x + \cdots + T^{n-1}x).$$

The set of fixed points of $T$ is $\mathrm{Fix}(T) = \{x \in H : Tx = x\}$.

**Theorem 10.6.22 (Avigad and Simic [11]).** *The following are equivalent over* $\mathsf{RCA}_0$.

1. $\mathsf{ACA}_0$.
2. *For every Hilbert space $H$, isometry $T$ on $H$, and point $x \in H$, the sequence of partial averages $S_n(x)$ converges.*
3. *For every Hilbert space $H$, isometry $T$ on $H$, and point $x \in H$, the projection of $x$ onto* $\mathrm{Fix}(T)$ *exists.*

*Moreover, the above equivalence holds if we consider nonexpansive linear operators in place of isometries.*

## 10.7 Additional results in real analysis

In this section we survey without proof the reverse mathematics of a few theorems that have been of particular interest in computable analysis, including fixed point theorems, existence theorems for ordinary differential equations, and the Hahn–Banach theorem.

The first result we mention shows that a form of the Stone–Weierstrass theorem is provable in $\mathsf{RCA}_0$. For definitions, see Brown [21].

**Theorem 10.7.1 (Brown [21, Theorem 3.27]).** *The following is provable in* $\mathsf{RCA}_0$. *Let* $\widehat{A}$ *be a complete separable metric space and let* $S$ *be an algebra on* $C(\widehat{A})$ *that separates points in* $\widehat{A}$ *and contains* 1. *Let* $f\colon \widehat{A} \to \mathbb{R}$ *be continuous and have a modulus of uniform continuity. Then for any* $\varepsilon > 0$ *there is a function* $g \in \overline{S}$ *such that* $|f(x) - g(x)| < \varepsilon$ *for all* $x \in \widehat{A}$.

The next result shows that forms of Brouwer's fixed point theorem and Schauder's fixed point theorem are equivalent to $\mathsf{WKL}_0$.

**Theorem 10.7.2 (Shioji and Tanaka [278]; see Simpson [288, Theorems IV.7.7 and IV.7.9]).** *The following are equivalent over* $\mathsf{RCA}_0$.

1. $\mathsf{WKL}_0$.
2. *Every continuous function from a nonempty closed convex set in* $[-1,1]^n$ *to itself has a fixed point.*
3. *Let* $C$ *be the convex hull of a nonempty set of finite points in* $\mathbb{R}^n$ *for some* $n \in \mathbb{N}$. *Then every continuous function* $C \to C$ *has a fixed point.*
4. *Every continuous function from the unit square to itself has a fixed point.*

The formalization of Shauder's fixed point theorem leads to the following characterization of Peano's existence theorem of solutions of ODEs.

**Theorem 10.7.3 (Simpson [288, Theorem IV.8.2]).** *The following are equivalent over* $\mathsf{RCA}_0$.

1. $\mathsf{WKL}_0$.
2. *Suppose* $f(x,y)$ *is a continuous real-valued function on the rectangle* $-a \leqslant x \leqslant a, -b \leqslant y \leqslant b$, *where* $a, b > 0$. *Then the initial value problem* $dy/dx = f(x,y)$, $y(0) = 0$ *has a continuously differentiable solution on the interval* $-\alpha \leqslant x \leqslant \alpha$, *where* $\alpha = \min(a.b/M)$ *where*
$$M = \max\{|f(x,y)| : -a \leqslant x \leqslant a, -b \leqslant y \leqslant b\}.$$
3. *If* $f(x,y)$ *is continuous and has a modulus of uniform continuity in some neighborhood of* $(0,0)$ *then the initial value problem* $dy/dx = f(x,y)$, $y(0) = 0$ *has a continuously differentiable solution in some interval containing* $x = 0$.

We end this subsection with a result on a formal version of the Hahn–Banach theorem. For definitions, see [288].

**Theorem 10.7.4 (Simpson [288, Theorem IV.9.3]).** *The following are equivalent over* $\mathsf{RCA}_0$.

1. $\mathsf{WKL}_0$.
2. *Let* $\widehat{A}$ *be a separable Banach space and let* $\widehat{S}$ *be a subspace of* $\widehat{A}$. *Let* $f\colon \widehat{S} \to \mathbb{R}$ *be a bounded linear functional with* $\|f\| < \alpha$ *for some* $\alpha > 0$. *Then there exists a bounded linear functional* $\widetilde{f}\colon \widehat{A} \to \mathbb{R}$ *with* $\|\widetilde{f}\| < \alpha$.

## 10.8 Topology, MF spaces, CSC spaces

While metric spaces are the heart of analysis, non-metric topological spaces also appear. There is a natural question of how much non-metric topology can be formalized in second order arithmetic. Of course, we cannot expect to represent every topological space: there are $2^{2^{\aleph_0}}$ pairwise non-homeomorphic topologies on $\omega$ [193], too many to code each one with an element of $2^\omega$. Therefore, we have to find a representation of some special class of topological spaces and then study that class. We will survey two of these special classes: countable, second countable (CSC) spaces introduced by Dorais, and maximal filter (MF) spaces introduced by Mummert.

Hunter [164], Normann and Sanders [235], and Sanders [269, 270] have also studied topology using higher order arithmetic, which has advantages in this setting compared to second order arithmetic. For example, higher types make it possible to treat a topology on a space directly as a collection of subsets of the space. As with much higher order reverse mathematics, however, the link to classical computability theory is more tenuous.

The programs of locale theory and domain theory provide another approach towards effective topology. The usual approach in domain theory is more category theoretic than proof theoretic. However, Mummert and Stephan [227] show the class of second countable MF spaces is the same as the class of second countable domain representable spaces, giving a link between these programs. Sanders [268] has also studied the reverse mathematics of domain theory.

### 10.8.1 Countable, second countable spaces

Dorais [70, 71] initiated the study of countable, second countable spaces in reverse mathematics. These are spaces that have a countable set of points and also a countable basis. The countability of the points allows for the spaces to be formalized in second order arithmetic. To use systems weaker than $\mathsf{ACA}_0$, it will be convenient to use enumerated sets rather than ordinary (decidable) sets to represent open sets of points.

*Convention 10.8.1 (Enumerated sets).* For the remainder of this section, an enumerated set $V$ is coded by a function $E_V : \omega \to \omega$ so that $x \in V \leftrightarrow x + 1 \in \text{range}(E_V)$. This definition allows the empty set to be enumerated, e.g. by $\lambda x.0$, and in general the function $E_V$ need not be injective. As usual, we use ordinary notation for set operations on enumerated sets: we write $V \cap W$, $V \cup W$, etc., to represent the corresponding enumerated sets.

**Definition 10.8.2 (Strong and weak bases).** The following definitions are made in $\mathsf{RCA}_0$. A *strong basis* for a topology on a countable set $X \subseteq \mathbb{N}$ is a sequence $U = \langle U_i \rangle$ of subsets of $X$ and a function $k\colon X \times \mathbb{N} \times \mathbb{N} \to \mathbb{N}$ such that:

1. For every $x \in X$ there is some $i$ with $x \in U_i$.
2. For $x \in X$ and all $i, j \in \mathbb{N}$, if $x \in U_i \cap U_j$ then $x \in U_{k(x,i,j)} \subseteq U_i \cap U_j$.

A *weak basis* for a topology on a set $X$ is a sequence $V = \langle V_i \rangle$ of enumerated subsets of $X$ and a function $k$ satisfying properties (1) and (2).

**Definition 10.8.3 (Strong and weak CSC spaces).**

- A *strong CSC space* is a triple $(X, U, k)$ where $X \subseteq \mathbb{N}$ and $(U, k)$ is a strong basis for a topology on $X$.
- A *weak CSC space* is a triple $(X, U, k)$ where $X \subseteq \mathbb{N}$ and $(U, k)$ is a weak basis for a topology on $X$.

When the distinction is not important, we simply write *CSC spaces*. A theorem referring to CSC spaces is really one theorem for strong CSC spaces and another for weak CSC spaces.

One motivation for considering weak CSC spaces is that $\mathsf{RCA}_0$ is able to form a weak basis for a space given by a countable set $X$ and a metric $d\colon X \to \mathbb{R}^+$.

A set $W \subseteq X$ is *open*, relative to a (weak or strong) basis $\langle U_i \rangle$, if for each $x \in W$ there is an $i$ with $x \in U_i \subseteq W$. It is natural to ask whether we can obtain a suitable $i$ from $x$ and $U$. The following proposition characterizes when this is possible; the proof is Exercise 10.9.21.

**Proposition 10.8.4 (Dorais [71]).** *The following is provable in* $\mathsf{RCA}_0$. *Let $X$ be a (weak or strong) CSC space with basis $\langle U_i \rangle$. If $U \subseteq X$ is a nonempty set or enumerated set, the following are equivalent.*

1. *$U$ is an effective union of basic open sets: there is an enumerated set $J \subseteq \omega$ with $U = \bigcup_{j \in J} U_j$.*
2. *$U$ is uniformly open: there is a partial function $n\colon X \to \mathbb{N}$, which we will call a* neighborhood function, *such that if $x \in U$ then $n(x) \downarrow$ and $x \in U_{n(x)} \subseteq U$.*

A set $U$ with those equivalent properties is *effectively open*. We can perform many manipulations with effectively open sets in $\mathsf{RCA}_0$.

**Proposition 10.8.5.** *The following are provable in* $\mathsf{RCA}_0$. *Let $X$ be a CSC space.*

1. *If $U, V \subseteq X$ are effectively open, then so are $U \cap V$ and $U \cup V$.*
2. *If $\langle U_i \rangle$ is a sequence of effectively open sets and $n(k,x)$ is a partial function such that $\lambda x.n(k,x)$ is a neighborhood function for $U_k$ for each $k$, then $\bigcup U_i$ is effectively open.*

A key aspect of CSC spaces is that we can represent a function from a CSC space $X$ to a CSC space $Y$ directly in second order arithmetic. We can thus define continuous functions using the usual topological definition. The next definition and proposition describe an effective version of continuity, analogous to the distinction between open and effectively open sets.

**Definition 10.8.6 (Effectively continuous maps).** Let $(X, \langle U_i \rangle, k)$ and $(Y, \langle V_i \rangle, l)$ be a CSC space. A function $f\colon X \to Y$ is *effectively continuous* if there is a partial function $\varphi\colon X \times \mathbb{N} \to \mathbb{N}$ such that, if $f(x) \in V_j$ then $\varphi(x, j) \downarrow$ and $x \in U_{\varphi(x,j)} \subseteq f^{-1}(V_j)$.

**Proposition 10.8.7.** *The following is provable in* $\mathsf{RCA}_0$. *If $X$ and $Y$ are CSC spaces, $f: X \to Y$ is effectively continuous, and $V$ is an effectively open subset of $Y$ then $f^{-1}(V)$ is an effectively open subset of $X$.*

The preceding definitions allow us to study many theorems of general topology, within $\mathsf{RCA}_0$, in the restricted setting of CSC spaces. For example, Dorais [71] has obtained results on discrete spaces and Hausdorff spaces. We state the following definition and theorem which examine the compactness of CSC spaces.

**Definition 10.8.8 (Basically compact space).** The following definitions are made in $\mathsf{RCA}_0$.

- A CSC space $X$ is *basically compact* if, for every enumerated set $I$ with $X = \bigcup_{i \in I} U_i$, there is a finite $F \subseteq I$ with $X = \bigcup_{i \in F} U_i$.
- A CSC space $X$ is *sequentially compact* if every sequence of points in $X$ has an accumulation point (in the usual topological sense).

**Theorem 10.8.9 (Dorais [71]).** $\mathsf{RCA}_0$ *proves that every sequentially compact strong CSC space is basically compact. Moreover, the following are equivalent over* $\mathsf{RCA}_0 + \mathsf{B}\Sigma^0_2$.

*1.* $\mathsf{ACA}_0$.
*2. Every sequentially compact weak CSC space is basically compact.*

Many interesting questions about the reverse mathematics of CSC spaces are open, including questions on compactness and metrizability of these spaces.

The collection of ordered spaces has been of particular interest in topology. It has also received attention in reverse mathematics. For example, given a linear $(L, \leqslant_L)$, $\mathsf{RCA}_0$ can form a strong CSC space for the order topology on $L$.

Shafer [277] studied aspects of compactness for ordered spaces. He uses the term *compact with respect to honest open covers* to refer to the spaces we call basically compact. The "honesty" is that the cover comes with an explicit enumeration of which basic open sets it uses.

**Theorem 10.8.10 (Shafer [277]).** *The following are equivalent over* $\mathsf{RCA}_0$.

*1.* $\mathsf{WKL}_0$.
*2. For every linear order $L$ on $\mathbb{N}$, if $L$ is complete then the order topology on $L$ is basically compact as a CSC space.*

A weaker kind of cover would be a sequence $\langle U_i \rangle$ of open sets that covers the space, with no additional information on which basic open sets are contained in the sets $U_i$. Shafer [277] also obtains equivalences between $\mathsf{ACA}_0$ and forms of compactness using these weaker covers.

### 10.8.2 MF spaces

Our second representation for non-metric topological spaces uses a variation of Stone duality to create a topology on the set of maximal filters of an arbitrary partial order. This class of MF spaces includes all complete separable metric spaces and additional spaces. Several theorems about MF spaces have high reverse mathematics strength. In particular, there is a metrization theorem for these spaces that is equivalent to $\Pi^1_2$ comprehension over $\Pi^1_1$-$\mathsf{CA}_0$. This is one of very few theorems known to have a strength higher than $\Pi^1_1$-$\mathsf{CA}_0$.

Recall the definition of a filter in the context of forcing (Definition 7.2.7). This is a general definition of partial orders, which we will use here in a different way. We recall the terminology, for convenience.

**Definition 10.8.11 (Filters and maximal filters).** Let $(P, \leqslant_P)$ be a partial ordering.

1. A set $F \subseteq P$ is a *filter* if it is closed upward ($q \in F$ and $q \leqslant_P p$ implies $p \in F$) and consistent (if $p, q \in F$ there exists $r \in F$ with $r \leqslant_P p$ and $r \leqslant_P q$).
2. A filter is *maximal* if it is not a proper subset of any other filter.

**Definition 10.8.12 (MF spaces).** Let $P$ be a partial ordering. The space $\mathrm{MF}(P)$ has as its points the set of all maximal filters on $P$. For each $q \in P$ there is a basic open set $N_q = \{f \in \mathrm{MF}(P) : q \in f\}$, and the topology on $\mathrm{MF}(P)$ is the one generated by this basis.

A space of the form $\mathrm{MF}(P)$ is an *MF space*. If $P$ is countable, the space is said to be *countably based*.

MF spaces are always $T_1$, but not always Hausdorff. Even Hausdorff MF spaces need not be metrizable. However, as the next example shows, every complete metric space is an MF space.

*Example 10.8.13.* Suppose that $\widehat{A}$ is a complete metric space. We can construct an MF space homeomorphic to $\widehat{A}$ as follows. Let $P$ consist of all rational open balls $(a, r)$ where $a \in A$ and $r \in \mathbb{Q}^+$. Set $(a, r) \leqslant_P (b, s)$ if and only if $d(a, b) + r < s$; this relation is sometimes known as *formal inclusion*. Then $\mathrm{MF}(P)$ is homeomorphic to $\widehat{A}$.

*Example 10.8.14.* The *Gandy–Harrington topology* is the topology on $\omega^\omega$ with the topology generated by the collection of all lightface $\Sigma^1_1$ sets. This space has been studied for its applications to descriptive set theory (see, e.g., Kechris [177]). The set $\omega^\omega$ with the Gandy–Harrington topology is homeomorphic to a countably based MF space. This topology is not regular, however, and therefore not metrizable.

**Theorem 10.8.15 (Mummert and Stephan [227]).** *The class of MF spaces has the following topological properties.*

1. *An MF space $X$ is homeomorphic to a countably based MF space if and only if $X$ is second countable.*
2. *The class of MF spaces is closed under arbitrary topological products.*

3. *The class of MF spaces is closed under taking $G_\delta$ subspaces.*
4. *Every MF space has the property of Baire.*
5. *Every countably based Hausdorff MF space has either countably many points or contains a perfect closed set.*

Mummert and Stephan obtained a characterization of the second countable MF spaces using a particular topological game.

**Definition 10.8.16 (Strong Choquet game, Choquet [44]; see Kechris [177]).** The *strong Choquet game* is a two-player game played with a fixed topological space $X$.

1. On the first move, Player I chooses an open set $U_0$ and a point $x_0 \in U$.
2. Player II then chooses an open set $V_0$ with $x \in V_0 \subseteq U_0$.
3. On a subsequent move, say move $k+1$, Player I chooses an open set $U_{k+1}$ and a point $x \in U_{k+1}$ with $x_{k+1} \subseteq U_{k+1} \subseteq V_k$.
4. Player II then chooses an open set $V_{k+1}$ with $x_{k+1} \in V_{k+1} \subseteq U_{k+1}$.

Play continues for $\omega$ rounds. At the end, Player I wins if $\bigcap_k U_k$ is empty (this is equivalent to $\bigcap_k V_k$ being empty). Player II wins otherwise, if $\bigcap_k U_k$ is nonempty.

A topological space has the *strong Choquet property* if Player II has a winning strategy for the game on that space.

Strong Choquet games are a particular kind of *game of perfect information content*, discussed further in Section 12.3. Choquet introduced his game to characterize complete metrizability of metric spaces.

**Theorem 10.8.17 (Choquet [44]).** *A separable metric space has a topologically equivalent, complete metric if and only if the space has the strong Choquet property.*

The strong Choquet property cannot be directly formulated in second order arithmetic, and therefore we cannot study its reverse mathematics directly. We can use the property to understand the class of MF spaces, however. Every MF space has the strong Choquet property, and beyond a small amount of separation the strong Choquet property is sufficient to characterize countably based MF spaces.

**Theorem 10.8.18 (Mummert and Stephan [226]).** *A topological space is homeomorphic to a countably based MF space if and only if it is second countable, $T_1$ and has the strong Choquet property.*

Mummert and Stephan [227] also showed that a second countable space domain representable (via a dcpo) if and only if the space is $T_1$ and has the strong Choquet property. We omit the definitions here. Representability via a dcpo is a key topic in *domain theory*, a field in topology and computer science that has been applied to and motivated by the semantics of programming languages. Domain theory does not lend itself directly to reverse mathematics analysis in second order arithmetic due to the complexity of the basic definitions. The tie between domain theory and MF spaces suggests that more analysis may be possible.

The characterization results above make essential use of second countability. It is known there are Hausdorff spaces with the strong Choquet property that are not homeomorphic to MF spaces.

### 10.8.3 Reverse mathematics of MF spaces

It is possible to formalize MF spaces in second order arithmetic. To do so, we need to represent both the spaces and the continuous functions between them.

The representation of an MF space within $\mathsf{RCA}_0$ is immediate. We begin with a countable partial ordering $P = (\mathbb{N}, \leqslant_P)$. Each point in $\mathrm{MF}(P)$ is represented as a maximal filter, which is now a subset of $\mathbb{N}$, similar to the way that a point in a complete separable metric space is coded as a subset of $\mathbb{N}$. One key difference is that the definition of a quickly converging Cauchy sequence is $\Pi^0_1$, while the definition of a maximal filter is $\Pi^1_1$.

The representation of continuous functions between MF spaces is inspired by a basic property of continuous functions: if $f\colon X \to Y$ is continuous and $z \in X$, then $f(z)$ is in an open set $V \subseteq Y$ if and only if there is an open set $U \subseteq X$ with $z \in U$ and $f(U) \subseteq V$. The following definition is made in $\mathsf{RCA}_0$.

**Definition 10.8.19 (Coded continuous function).** A *code for a continuous function* between $\mathrm{MF}(P)$ and $\mathrm{MF}(Q)$ is a subset $F$ of $\mathbb{N} \times P \times Q$. Each code induces at least a partial map $f$ from $\mathrm{MF}(P)$ to $\mathrm{MF}(Q)$ in which

$$f(x) = \{q \in Q : (\exists n)(\exists p \in P)[(n, p, q) \in F]\}$$

when this set is a point in $\mathrm{MF}(Q)$, and $f(x)$ is undefined otherwise. A *coded continuous function* is a function that has a total code.

Spaces $\mathrm{MF}(P)$ and $\mathrm{MF}(Q)$ are *homeomorphic* if there is a coded continuous bijection from $\mathrm{MF}(P)$ to $\mathrm{MF}(Q)$ whose inverse is also a coded continuous function.

A space $\mathrm{MF}(P$ is *homeomorphic to a complete separable metric space* if there is a complete metric space $\widehat{A}$ such that there is a homeomorphism between $\mathrm{MF}(P)$ and the MF space constructed from $\widehat{A}$ as in Example 10.8.13.

It can be shown in ZFC that every continuous function from a countably based MF space to a countably based MF space has a code. The role of $\mathbb{N}$ in the definition is to allow $\mathsf{RCA}_0$ to construct the composition of two coded continuous functions.

Mummert and Simpson [226] obtained the following characterization analogous to Urysohn's metrization theorem. We will provide only a broad sketch of the argument here. See Mummert and Simpson [226] for a longer sketch, and see Mummert [225] for full proofs and additional results.

**Theorem 10.8.20.** *The following are equivalent over* $\Pi^1_1$-$\mathsf{CA}_0$.

1. $\Pi^1_2$ *comprehension.*
2. *Every regular, countably based MF space is homeomorphic to a complete separable metric space.*

*Proof (sketch).* The forward implication is a formalization of the proof of Urysohn's metrization theorem and Choquet's characterization of complete metrizability. In this part of the proof, given a regular, countably based space $\mathrm{MF}(P)$, we use $\Pi^1_2$ comprehension to form the set $S$ of pairs $\langle p, q\rangle \in P \times P$ such that the closure of

$N_q$ is contained in $N_p$. For a complete separable metric space, the property that the closure of one open ball is contained in a second open ball is $\Pi^1_1$ complete in general; for MF spaces the analogous relation is formally $\Pi^1_2$.

With the set $S$ in hand, it is possible to produce a metric for MF($P$), using methods inspired by work of Schröder [273] in effective topology. This metric may not be complete, however. A second phase of the proof interpolates ideas from Choquet's characterization theorem to show that every countably based MF space that is metrizable is completely metrizable. This two-step proof provides one direction of the theorem.

For the reversal, we work in $\Pi^1_1$-$\mathsf{CA}_0$ and begin with a $\Sigma^1_2$ formula $\varphi(n)$. We want to form the set $S = \{n \in \mathbb{N} : \varphi(n)\}$. The reversal constructs a particular partial order $P$ so that MF($P$) is regular, and moreover there is a closed set $C$ in MF($P$) whose points are in effective correspondence (relative to $\mathsf{ACA}_0$) to the elements of $S$. Moreover, the construction ensures that every point of $C$ is an isolated point relative to $C$.

Intuitively, the construction ensures that, in order to construct $S$, it is sufficient to enumerate a dense subset of $C$. This set will be $C$ itself, and by the correspondence between $C$ and $S$ we are able to produce $S$. The construction depends on Kondo's theorem ($\Pi^1_1$ uniformization), which is provable in $\Pi^1_1$-$\mathsf{CA}_0$.

Having constructed $P$ and verified it is regular, by assumption we obtain a complete metric space $\widehat{A}$ and a homeomorphism between MF($P$) and $\widehat{A}$. This allows us to find a closed set $C' \subseteq \widehat{A}$ that is homeomorphic to $C$. Because we are working in $\Pi^1_1$-$\mathsf{CA}_0$, we can apply Theorem 10.5.8 to show that $C'$ is separably closed. This gives an enumeration of the points of $C'$, which allows us to enumerate $C$ and thus $S$, completing the reversal. □

It is also possible to study the descriptive set theory of MF spaces, although even basic properties quickly lead to set theoretic results. We will sketch a few additional results on the reverse mathematics of descriptive set theory in the final chapter.

It is also possible to study the descriptive set theory of MF spaces, although even basic properties quickly lead to set theoretic results. We will sketch a few additional results on the reverse mathematics of descriptive set theory in the final chapter. For the definition of $L$, see Section 12.3.3.

**Theorem 10.8.21 (Mummert [225]).** *The proposition that every closed subset of a countably based MF space is either countable or has a perfect subset is equivalent over* $\Pi^1_1$-$\mathsf{CA}_0$ *to the principle that* $\aleph_1^{L(A)}$ *is countable for every* $A \subseteq \mathbb{N}$. *In particular, the principle is independent of* ZFC *and false if* $V = L$.

The principle that every countably based MF space is either countable or has a perfect subset is provable in ZFC, but the reverse mathematics strength is not known.

## 10.9 Exercises

**Exercise 10.9.1.** We will call a function $f: \mathbb{N} \to \mathbb{Q}^+$ a *modulus of convergence* if $f$ is strictly decreasing and $\lim_{n\to\infty} f(n) = 0$. If $f$ is a modulus of convergence, a sequence $\langle q_n \rangle$ of rationals is *$f$-quickly converging* if $|q_n - q_m| < f(m)$ when $n > m$. Thus a quickly converging Cauchy sequence is $h$-quickly converging for $h(n) = 2^{-n}$.

1. $\mathsf{RCA}_0$ proves that if $f$ and $g$ are moduli of convergence, for every $f$-quickly converging Cauchy sequence there is a $g$-quickly converging Cauchy sequence with the same limit.
2. If $f$ and $g$ are moduli of convergence there is a uniform procedure, computable relative to $f$ and $g$, the converts each $f$-quickly converging sequence to a $g$-quickly converging sequence.

**Exercise 10.9.2.** As usual, a sequence $\langle x_n \rangle$ in a complete separable metric space $\widehat{A}$ is quickly converging if $d_A(x_n, x_m) < 2^{-m}$ when $n > m$. Prove in $\mathsf{RCA}_0$ that every quickly converging sequence $\langle x_n \rangle$ in $\widehat{A}$ converges. That is, there is a quickly converging sequence $\langle r_n \rangle$ of points in $A$ with the same limit as $\langle x_n \rangle$.

**Exercise 10.9.3.** An *(open) Dedekind cut* is a set $X$ of rational numbers so that:

1. $X$ is nonempty and $X \neq \mathbb{Q}$.
2. If $a, b \in \mathbb{Q}$, $a \in X$ and $b < a$ then $b \in X$.
3. $X$ has no maximum element.

As usual, an Dedekind cut $X$ represents the real number $\sup(X)$, and every real is represented by some Dedekind cut.

Show that the operation $+_D$ of addition on Dedekind cuts is not uniformly computable. In particular, the Turing jump operation is Weihrauch reducible to the parallelization $\widehat{+_D}$. (Similar results hold if we do not require the Dedekind cuts to be open.)

**Exercise 10.9.4 (Simpson [288, II.4.2]).** The following *nested completeness theorem* is provable in $\mathsf{RCA}_0$. Assume that $\langle a_i : i \in \mathbb{N} \rangle$ and $\langle b_i : i \in \mathbb{N} \rangle$ are sequences of real numbers such that, for all $n$, $a_n \leqslant a_{n+1} \leqslant b_{n+1} \leqslant b_n$, and $\lim |b_n - a_n| = 0$. Then there is a real number $x$ such that $\lim a_n = x = \lim b_n$.

**Exercise 10.9.5.** $\mathsf{RCA}_0$ proves that $\mathbb{R}$ is uncountable: there is no sequence $\langle r_i \rangle$ of real numbers that includes a code for every real number.

**Exercise 10.9.6.** Let $\langle C_i : i \in \mathbb{N} \rangle$ be a sequence of nonempty closed sets in $2^{\mathbb{N}}$ with $C_{i+1} \subseteq C_i$ for each $i$. Prove in $\mathsf{WKL}_0$ that $\bigcap_i C_i$ is nonempty.

**Exercise 10.9.7.** Prove Example 10.3.5.

**Exercise 10.9.8.** Prove Example 10.3.6.

**Exercise 10.9.9.** Prove Lemma 10.3.7.

**Exercise 10.9.10.** Prove in $\mathsf{ACA}_0$ that every bounded sequence of real numbers has a least upper bound.

**Exercise 10.9.11.** Prove in $\mathsf{ACA}_0$ that, in a compact metric space, every closed set is separably closed.

**Exercise 10.9.12.** Prove in $\mathsf{ACA}_0$ that, in a complete separable metric space, every separably closed set is closed.

**Exercise 10.9.13 (see Avigad and Simic [11]).** A set $C$ in a complete separable metric space $\widehat{A}$ is *located* if the distance function $d(x, C) = \inf\{d(x, y) : y \in C\}$ exists. Prove the following in $\mathsf{RCA}_0$.

1. If $\widehat{A}$ is a complete separable metric space then a set $C \subseteq \widehat{A}$ is closed and located if and only if the set $\{\langle a, r\rangle \in A \times \mathbb{Q}^+ : B_r(a) \cap C = \emptyset\}$ exists.
2. $\mathsf{ACA}_0$ is equivalent to the principle that every closed set in a compact metric space is located (Giusto and Simpson [124]).
3. $\mathsf{ACA}_0$ is equivalent to the principle that every separably closed set in a complete separable metric space is located (Giusto and Marcone [123, Theorem 7.3]).

**Exercise 10.9.14.** The following results demonstrate several aspects of the nonuniformity of the principle $\mathsf{IVT}$.

1. The following principle is equivalent to $\mathsf{WKL}_0$ over $\mathsf{RCA}_0$: Given a sequence of coded continuous functions $f_i \colon [0, 1] \to \mathbb{R}$ with $f_i(0) < 0$ and $f_i(1) > 0$ for all $i$, there is a sequence $\langle x_i\rangle$ of points in $[0, 1]$ with $f_i(x_i) = 0$ for all $i$.
2. Uniformly computable solutions are admitted for the special case of $\mathsf{IVT}$ where the function $f$ has only one root. (Hint: Use a trisection argument. For example, because $f(1/3)$ and $f(2/3)$ cannot both be zero, by approximating both with enough accuracy we will eventually determine that one or the other is nonzero.)
3. Uniformly computable solutions are admitted for the special case of $\mathsf{IVT}$ where the set $\{x : f(x) = 0\}$ is nowhere dense.

**Exercise 10.9.15.** Prove in $\mathsf{RCA}_0$ that the following statement implies $\mathsf{ACA}_0$: In $\mathbb{N}^\mathbb{N}$, every nonempty closed set is separably closed.

**Exercise 10.9.16 (Simpson [288, IV.1.7, IV.1.8]).** The following are provable in $\mathsf{WKL}_0$. Let $X$ be a compact metric space.

1. The property that a closed set $C$ is nonempty is expressible by a $\Pi^0_1$ formula with a parameter for the code for $C$.
2. If $\langle C_i : i \in \mathbb{N}\rangle$ is a sequence of nonempty closed sets in $X$, there is a sequence $\langle x_i : i \in \mathbb{N}\rangle$ of points of $X$ with $x_i \in C_i$.

**Exercise 10.9.17.** Prove the following:

1. Lemma 10.5.10 (Use Corollary 10.5.4).
2. Theorem 10.5.12 (Use a method similar to Theorem 10.5.11).

3. Prove that $\mathsf{WKL}_0$ is equivalent over $\mathsf{RCA}_0$ to the principle that every bounded function $f\colon [0,1] \to \mathbb{R}^+$ achieves a maximum. (Combine the method of Theorem 10.5.11 with a Specker sequence.)

**Exercise 10.9.18.** Prove in $\mathsf{WKL}_0$ that, given a continuous function code $F$ for a total function from $2^{\mathbb{N}}$ to itself, there is a function $f\colon 2^{<\mathbb{N}} \to 2^{<\mathbb{N}}$ that encodes $f$ as in Definition 10.6.2.

**Exercise 10.9.19.** Working in $\mathsf{WKL}_0$, prove the claim in the proof of Lemma 10.6.4.

**Exercise 10.9.20.** Prove Lemma 10.6.6,

**Exercise 10.9.21.** Prove Proposition 10.8.4.

**Exercise 10.9.22.** Characterize the strength of the principles "every open set in a strong CSC space is effectively open" and "every open set in a weak CSC space is effectively open".

# Chapter 11
# Algebra

Computable algebra forms a core area of research in modern computability theory. It has a long history, stretching back to the first half of the 20th century, with pioneering early papers by Fröhlich and Shepherdson [119], Rabin [251], Mal'cev [200], Seidenberg [276], and Ershov [101]. Naturally, there is also a significant intersection with reverse mathematics, which we explore in this chapter. An interesting caveat is that, perhaps more so that in any other subject, many applications of computability to algebra are not readily accessible to reverse mathematics. This is particularly the case for computable structure theory. Questions about degree spectra, categoricity, etc., do not lend themselves naturally to investigation within our framework. But, of course, they often find at least partial reflections in questions that do.

## 11.1 Groups, rings, and other structures

Throughout this chapter, all structures we will consider will be countable, and we will omit repeating this explicitly. Under this assumption, we can represent common algebraic structures for consideration in computability theory or reverse mathematics. Notably, we have the following.

**Definition 11.1.1 (Groups and rings).**

1. A *group* is a tuple $(G, *_G, e_G)$ such that:
   - $G$ is a subset of $\omega$,
   - $*_G$ is a function $G^2 \to G$,
   - $e_G \in G$,

   and such that the group axioms are satisfied.
2. A *ring* is a tuple $(R, +_R, \cdot_R, 0_R, 1_R)$ such that:
   - $R$ is a subset of $\omega$,
   - $+_R, \cdot_R$ are functions $R^2 \to R$,

 
D. D. Dzhafarov, C. Mummert, *Reverse Mathematics*, Theory and Applications of Computability, https://doi.org/10.1007/978-3-031-11367-3_11

- $0_R, 1_R \in G$,

and such that the ring axioms are satisfied.

Fields are rings, so they do not require separate representation, though we will typically write a field as $(K, +_K, \cdot_K, 0_K, 1_K)$ in keeping with more customary notation. We define *subgroups* and *ideals* in the obvious way, with ideals always meaning proper ideals (not containing the unity). In Section 11.5, we will consider a subtly different way of defining subgroup in a specific example, but otherwise we will use the definitions above.

We will follow all usual conventions. For example, we may refer to a group $(G, *_G, e_G)$ or ring $(R, +_R, \cdot_R, 0_R, 1_R)$ simply as $G$ or $R$, for short. If $G$ is abelian, we will write it instead as $(G, +_G, 0_G)$, and use $-a$ for the $+_G$-inverse of $a \in G$. Similarly in $R$, we will write $-r$ and $r^{-1}$ for the inverses of $r \in R$ under $+_R$ and $\cdot_R$, respectively. In this case we will also write, e.g., $a -_R b$ for $a +_R (-b)$, etc.

Various other common algebraic objects and constructions can be accommodated in computability theory and, by extension, $\mathsf{RCA}_0$. For example, if $I$ is an ideal of a ring $R$, the *quotient ring* $R/I$ is defined as the set of all $r \in R$ such that there is no $r^* < r$ (in the standard order of the natural numbers) with $r -_R r^* \in I$. Addition and multiplication in $R/I$ can be defined in a straightforward manner so as to give $R/I$ an effective ring structure: e.g., if $r_0, r_1 \in R$ then $r_0 \cdot_{R/I} r_1$ is the least $r \in R$ such that $r - (r_0 \cdot_R r_1) \in I$, etc. Note that $R/I$ exists by $\Delta^0_1$ comprehension, and $\mathsf{RCA}_0$ can prove that it is a ring. Similar constructions work for other kinds of quotient structures.

In $\mathsf{RCA}_0$, we can also consider finite products of algebraic structures. For example, if $G$ is a group and $\alpha \in G^{<\mathbb{N}}$ we define $\prod_{a \in \alpha} a$ inductively on the length of $\alpha$: if $|\alpha| = 0$ then $\prod_{a \in \alpha} a = 1_G$; given $n \in \mathbb{N}$, having defined $\prod_{a \in \alpha^*} a$ for all $\alpha^*$ of length $n$, and given $\alpha$ of length $n+1$, we define $\prod_{a \in \alpha} a = \prod_{a \in \alpha \upharpoonright n} a \ *_G \ \alpha(n)$. $\mathsf{RCA}_0$ suffices to prove the usual properties of this operation, such as that if $G$ is abelian and $\alpha, \beta \in G^{<\mathbb{N}}$ have the same range then $\prod_{a \in \alpha} a = \prod_{a \in \beta} a$ (see Exercise 11.7.2). In this case, we may unambiguously write this instead as $\sum_{a \in F} a$, for $F$ a finite subset of $G$, as is customary.

Also important are formal sums. As we will see, polynomial rings play a big role in facilitating reversals in the reverse mathematics of algebra.

**Definition 11.1.2 (Polynomials).** Let $K$ be a field and fix $n \in \omega$.

1. A *monomial (in n variables)* is an element $m$ of $\omega^n$.
2. A *polynomial (over K in n variables)* is an element $p$ of $\omega^{<\omega}$ as follows:
    - For all $i < |p|$, $p(i) = \langle k, m \rangle$, where $k \in K$ and $m$ is a monomial in $n$ variables.
    - For all $i < j < |p|$, if $p(i) = \langle k_0, m_0 \rangle$ and $p(j) = \langle k_1, m_1 \rangle$ then $m_0$ lexicographically precedes $m_1$.

We are representing $K[x_0, \ldots, x_{n-1}]$ here. The idea is that a monomial $m$ represents $x_0^{m(0)} \cdots x_{n-1}^{m(n-1)}$. A polynomial $p$ with $p(i) = \langle k_i, m_i \rangle$ for all $i < |p|$ represents the formal sum $\sum_{i<|p|} k_i m_i$. We will use this more familiar notation, with suitable accompanying terminology. For example, for each $i < |p|$ we refer to $p(i)$ as a *term* of $p$. If $m$ is a monomial with $m(i) = j \in \mathbb{N}$ for some $i < |m|$ we say $x_i^j$ is a *factor* of $m$. We identify each monomial $m$ as the single term polynomial $1_K m$, and for each $k \in K$ identify the single term polynomial $kx_0^0 \cdots x_{n-1}^0$ with $k$. From here, we can define scalar multiplication, addition, and multiplication for polynomials as usual, and it is easy to see that these operations are computable. That is, the map taking (codes for) polynomials $p$ and $q$ to (a code for) their product is a computable map $\omega \to \omega$, etc.

**Definition 11.1.3 (Polynomial rings).** Let $K$ be a field.

1. For $n \in \omega$, $K[x_0, \ldots, x_{n-1}]$ or $K[x_i : i < n]$ is the structure with domain the set of all (codes for) polynomials over $K$ in $n$ variables, with the usual ring operations and identities.
2. $K[x_0, x_1, \ldots]$ or $K[x_i : i \in \omega]$ is the structure with domain the set of all (codes for) polynomials over $K$ in $n$ variables for some $n > 1$, with the usual ring operations and identities.

These definitions directly formalize in $\mathsf{RCA}_0$, which can prove that $K[x_i : i < n]$ and $K[x_i : i \in \mathbb{N}]$ are indeed rings.

A thorough survey of coding techniques for representing algebraic objects can be found in the pioneering works of Rabin [251] and Fröhlich and Shepherdson [119]. The main takeaway is that, as in our earlier examples, these codings are natural and benign in the sense that once we have made them we can largely forget about them. Worth noting, too, is that many small variations on such codings are possible. Typically, the specific choice of such variation makes no difference.

## 11.2 Vector spaces and bases

Before venturing off to abstract algebra for the remainder of this chapter, we take a brief excursion to linear algebra. Our aim is to prove the following result, in large part to give a first impression of some of the coding techniques commonly employed in the reverse mathematics investigation of algebraic theorems.

**Theorem 11.2.1 (Friedman, Simpson, and Smith [117], after Dekker (unpublished) and Metakides and Nerode [209]).** *The following are equivalent under* $\leqslant_c$ *and over* $\mathsf{RCA}_0$.

1. $\mathsf{TJ}$.
2. *Every vector space over a field K has a basis.*
3. *Every vector space over* $\mathbb{Q}$ *has a basis.*

*Hence, (2) and (3) is equivalent over* $\mathsf{RCA}_0$ *to* $\mathsf{ACA}_0$.

Here, we use $\mathbb{Q}$ to refer to the field $\langle \mathbb{Q}, +, \cdot, 0, 1\rangle$, as usual. We begin with an elaboration on Definition 11.1.1.

**Definition 11.2.2 (Vector spaces).** Let $K$ be a field. A *vector space over* $K$ is an abelian group $(V, +_V, 0_V)$ together with a function $\cdot_V : K \times V \to V$ satisfying the axioms of scalar multiplication.

We can add some standard terminology. Given a finite set $F \subseteq V$ and $v \in V$, we say $v$ is a *linear combination* of (the vectors in) $F$ if for each $b \in F$ there exists $k_b \in K$ with $v = \sum_{b \in F} k_b b$; the linear combination is *nontrivial* if $k_b \neq 0$ for some $b \in F$. A finite set $F \subseteq V$ is *linearly dependent* if $0_V$ is a nontrivial linear combination of $F$; otherwise, $F$ is *linearly independent*. Finally, a *basis* for $V$ is a set $B \subseteq V$ satisfying the following usual properties.

- Every finite subset of $B$ is linearly independent.
- Every $v \in W$ is a linear combination of some finite subset of $B$.

$\mathsf{RCA}_0$ can verify the standard fact that for every nonzero $v \in V$ there is a unique finite $F \subseteq B$ and coefficients $k_b$ such that $v = \sum_{b \in F} k_b b$ and $k_b \neq 0_K$ for all $b \in F$. We will call $F$ the *canonical representation* of $v$ in terms of the basis $B$. (See Exercise 11.7.1.)

*Proof (of Theorem 11.2.1).* We prove the equivalence over $\mathsf{RCA}_0$.

(1) $\to$ (2): We argue in $\mathsf{ACA}_0$. Let $V$ be given. If there is a finite set $B \subseteq V$ that forms a basis for $V$, we are done. So assume not. Using arithmetical comprehension we inductively define a sequence $\langle b_i : i \in \mathbb{N} \rangle$, as follows. Let $b_0 \in V$ be arbitrary. Then fix $n > 0$, and assume that we have defined $b_i$ for all $i < n$ and that $F = \langle b_i : i < n \rangle$ is linearly independent. By assumption, there is a $v \in V$ that is not a linear combination of $F$. Choose the least such $v$ and let $b_n = v$. Now $F \cup \{b_n\}$ is easily seen to be linearly independent, so the inductive assumption is maintained, and $B = \{b_i : i \in \mathbb{N}\}$ is a basis for $V$.

(2) $\to$ (3): Obvious.

(3) $\to$ (1): We now argue in $\mathsf{RCA}_0$. Fix an injective function $f : \mathbb{N} \to \mathbb{N}$. Our task is to show that range$(f)$ exists. We will work with the set of all formal sums $\sum_{i<n} q_i x_i$ over $\mathbb{Q}$, which naturally forms a vector space $V$ over $\mathbb{Q}$. Our coding of sequences from Chapter 5 ensures that for each $i$, the code of the vector $x_i$ is minimal among the codes of vectors $\sum_{i<n} q_i x_i$ with $q_i \neq 0$.

For each $j \in \mathbb{N}$, let $x'_j = x_{2f(j)} + (j+1)x_{2f(j)+1}$. Let $U_0$ be the subspace of $V$ generated by the linear span of $\{x'_j : j \in \mathbb{N}\}$. Then $U_0$ exists, because a vector $\sum_{i<n} q_i x_i$ belongs to $U_0$ if and only if the following conditions hold for all $i$:

- If $2i < n$ and $q_{2i} \neq 0$ then $2i + 1 < n$ and $q_{2i+1} = (j+1)q_{2i}$ for some $j$ with $f(j) = i$.
- If $2i + 1 < n$ and $q_{2i} = 0$ then $q_{2i+1} = 0$.

Furthermore, $B_0 = \{x'_j : j \in \mathbb{N}\}$ forms a basis for $U_0$.

Let $V_1$ be the quotient space $V/U_0$ (using minimal representatives, as discussed in the previous section). Applying (3), we may fix a basis $B_1$ for $V_1$. Then $B = B_0 \cup B_1$ is a basis for $V$. To see this, we verify the clauses in the definition of a basis.

- A finite set $F \subseteq B$ may be written as $F_0 \cup F_1$, where $F_0 \subseteq B_0$ and $F_1 \subseteq B_1$. Each of $F_0$ and $F_1$ are linearly independent. Hence, if $F$ were linearly dependent, then some nontrivial linear combination of $F_1$ would be an element of $U_0$. But this means that, in $V_1$, this linear combination is equal to 0, and so $F_1$ is linearly dependent after all.
- Fix any $v \in V$. We may then fix the unique $w$ such that $w \in V_1$ and $v - w \in U_0$. (If $v \in U_0$ then $w = 0$.) Fix $F_0 \subseteq B_0$ and $F_1 \subseteq B_1$ such that $v - w$ is a linear combination of $F_0$ and $w$ is a linear combination of $F_1$. Then $v$ is a linear combination of $F_0 \cup F_1$.

To complete the proof, let $R$ be the set of all $i \in \mathbb{N}$ such that either the canonical representation of $x_{2i}$ in terms of $B$ or the canonical representation of $x_{2i+1}$ in terms of $B$ contains $x'_j \in B_0$ for some $j$ with $f(j) = i$. Clearly, $R$ exists by $\Delta^0_1$ comprehension, and we claim that $R = \text{range}(f)$. The containment $R \subseteq \text{range}(f)$ is obvious. Conversely, suppose $i \in \text{range}(f)$, say with $f(j) = i$. Thus, $x'_j = x_{2f(j)} + (j + 1)x_{2f(j)+1} = x_{2i} + (j + 1)x_{2i+1}$ belongs to $U_0$, and in fact to $B_0$. In particular, the canonical representation of $x'_j$ in terms of $B$ is $\{x'_j\}$. We now consider two cases.

*Case 1:* $x_{2i} \in V_1$. Let $F$ be the canonical representation of $x_{2i}$ in terms of $B$. By assumption, $F \subseteq B_1$. Since $x'_j \notin B_1$ and canonical representations are unique, it follows that $F \cup \{x'_j\}$ is the canonical representation in terms of $B$ of $(j + 1)x_{2i+1} = x'_j - x_{2i}$. But then $F \cup \{x'_j\}$ is also the canonical representation of $x_{2i+1}$, since this is just a nonzero scalar multiple of $(j + 1)x_{2i+1}$.

*Case 2:* $x_{2i} \notin V_1$. Since $V_1$ is a quotient space, there exist vectors $v \in V_1$ and $u \in U_0$ such that $v < x_{2i}$ and $x_{2i} - v = u$. By our assumption on the order of codings, $v$ cannot contain $x_{2i}$. By definition of $U_0$, this means that $u = w + x'_j$ for some $w$ that does not contain $x_{2i}$ of $x_{2i+1}$. Let $F_v$ be the canonical representation of $v$ in terms of $B$, and $F_w$ the canonical representation of $w$ in terms of $B$ . Thus, $F_v \subseteq B_1$ and $F_w \subseteq B_0 \smallsetminus \{x'_j\}$. Since $B_0$ and $B_1$ are disjoint, it follows that $F_v \cup F_w \cup \{x'_j\}$ is the canonical representation of $x_{2i} = v + w + x'_j$.

In either case, we see that $i \in R$, as was to be shown. □

The implication from (1) to (2) featured the initial case analysis about whether or not $V$ is "finite dimensional", meaning "has a finite basis". If so, then of course the theorem is provable in $\mathsf{RCA}_0$ (trivially). But we could also define "finite dimensional" to mean "there exists a $n$ such that every finite set $F \subseteq V$ of size $n$ is linearly dependent". The assertion that every such space has a basis is computably true (since every solution is, in particular, a finite set). However, this is no longer as straightforward to prove.

**Theorem 11.2.3 (Hirst and Mummert [160]).** *The following are equivalent over* $\mathsf{RCA}_0$.

1. $\mathsf{I}\Sigma^0_2$.
2. *Let $V$ be a vector space such that, for some $n \in \mathbb{N}$, every finite set $F \subseteq V$ of size $n$ is linearly dependent. Then $V$ has a basis.*

Finite dimensional vector spaces also turn out to be interesting when viewed under some of our stronger reducibilities. Consider the following family of problems.

**Definition 11.2.4 (Hirst and Mummert [160]).** Fix $n \geqslant 2$. $\mathsf{VSB}_n$ is the problem whose instances are all $n$-dimensional vector spaces over $\mathbb{Q}$, with the solutions to any such vector space being all its bases.

As an $\forall\exists$ theorem, $\mathsf{VSB}_n$ is vacuous. But not so as a problem under $\leqslant_{\mathrm{W}}$. Recall the choice problem $\mathbb{C}_{\mathbb{N}}$ from Exercise 4.8.10.

**Theorem 11.2.5 (Hirst and Mummert [160]).** *For all $n \geqslant 2$, $\mathsf{VSB}_n \equiv_{\mathrm{W}} \mathsf{C}_{\mathbb{N}}$.*

This is a somewhat surprising result, perhaps more so by virtue of the fact that it holds for all $n \geqslant 2$. That is, the specific dimension of a finite dimensional vector space does not affect the uniform computational complexity of finding a basis. This stands in contrast to some results we have seen earlier involving parameterized principles, where the specific value of the parameter was more significant. For example, recall Proposition 4.3.7, stating that $\mathsf{RT}^1_k \nleqslant_{\mathrm{W}} \mathsf{RT}^1_j$ whenever $k > j$.

The proof of Theorem 11.2.5 is an elaboration on that of Theorem 11.2.1. By taking a suitable quotient of the vector space $V_1$ constructed above, one obtains a 2-dimensional vector space every basis of which can be used to select an element not enumerated by a given instance of $\mathbb{C}_{\mathbb{N}}$.

## 11.3 The complexity of ideals

We begin by considering two classical theorems of ring theory whose analysis in reverse mathematics is by now classical in its own right, and which is often extolled as a hallmark of the subject's capacity to reveal hidden differences between mathematical results. These theorems concern the existence of prime and maximal ideals in commutative rings.

Recall that an ideal $I$ of a ring $R$ is

- *prime* if for all $g, h \in G$, if $g \cdot_R h \in I$ then $g \in I$ or $h \in I$,
- *maximal* if there is no ideal $I^*$ of $R$ such that $I \subseteq I^*$ and $I \neq I^*$.

Every commutative ring has both a prime and maximal ideal. The typical way to prove this is as follows. First, observe that every maximal ideal is prime, so it suffices to show the existence of the former. To this end, note that any union of ideals is still an ideal, and so a maximal ideal exists by an application of Zorn's lemma.

The above is a completely nonconstructive proof, and of course it completely obscures any potential distinction between building a maximal ideal and building a prime ideal. As it turns out, a distinction does exist, and it can be measured exactly using subsystems of $\mathsf{Z}_2$.

We begin by calibrating the strength of the principle asserting the existence of prime ideals.

**Theorem 11.3.1 (Friedman, Simpson, and Smith [117, 118]).** *The following are equivalent under $\leqslant_c$ and over* $\mathsf{RCA}_0$.

*1.* WKL.
*2. Every commutative ring has a prime ideal.*

*Proof.* We focus on the equivalence over $\mathsf{RCA}_0$. The equivalence under computable reducibility is similar, but actually quite a bit easier, as we comment on below.

(1) → (2): We argue in $\mathsf{WKL}_0$. Let $R$ be an infinite commutative ring, say with domain $\{r_0, r_1, \ldots\}$, where $r_0 = 0_R$ and $r_1 = 1_R$. Define a tree $T \subseteq 2^{<\mathbb{N}}$ by letting $\sigma \in T$ if and only if

- $\sigma(0) = 1$,
- $\sigma(1) = 0$,
- for all $x, y, z < |\sigma|$ with $r_z = r_x +_R r_y$, if $\sigma(x) = \sigma(y) = 1$ then $\sigma(z) = 1$,
- for all $x, y, z < |\sigma|$ with $r_z = r_x \cdot_R r_y$, if $\sigma(x) = 1$ then $\sigma(z) = 1$,
- for all $x, y, z < |\sigma|$ with $r_z = r_x \cdot_R r_y$, if $\sigma(x) = \sigma(y) = 0$ then $\sigma(z) = 0$.

We will show that $T$ is infinite. Once we do this, we can apply WKL to fix a path $f$ through $T$. Then $I = \{r_x : f(x) = 1\}$ is $\Delta^0_1$ definable from $f$, and it is easy to verify, using the properties above, that it is a prime ideal. For instance, if $r_z = r_x \cdot_R r_y \in I$ then $f(z) = 1$, so by definition either $f(x) = 1$ or $f(y) = 1$, meaning $r_x \in I$ or $r_y \in I$.

To show that $T$ is infinite, we begin by defining a function $p$ from $2^{<\mathbb{N}}$ to (codes for) finite subsets of $\mathbb{N}$ by induction on $\sigma \in 2^{<\mathbb{N}}$. Let $p(\langle\rangle)$ be (a code for) the finite set $\{0_R\}$, and assume inductively that $p(\sigma)$ is defined and equal to (a code for) a finite set for some $\sigma \in 2^{<\mathbb{N}}$. Now fix $x, y, z, k \in \mathbb{N}$.

- If $|\sigma| = 4\langle x, y, z, k\rangle$ with $r_z = r_x +_R r_y$ and $r_x, r_y \in p(\sigma)$, then let $p(\sigma 0) = p(\sigma) \cup \{r_z\}$ and $p(\sigma 1) = \varnothing$.
- If $|\sigma| = 4\langle x, y, z, k\rangle + 1$ with $r_z = r_x \cdot_R r_y$ and $r_x \in p(\sigma)$, then let $p(\sigma 0) = p(\sigma) \cup \{r_z\}$ and $p(\sigma 1) = \varnothing$.
- If $|\sigma| = 4\langle x, y, z, k\rangle + 2$ with $r_z = r_x \cdot_R r_y$ and $r_z \in p(\sigma)$, then let $p(\sigma 0) = p(\sigma) \cup \{r_x\}$ and $p(\sigma 1) = p(\sigma) \cup \{r_y\}$.
- If $|\sigma| = 4\langle x, y, z, k\rangle + 3$ with $1_R \in p(\sigma)$ then let $p(\sigma 0) = p(\sigma 1) = \varnothing$.

Otherwise, let $p(\sigma 0) = p(\sigma 1) = p(\sigma)$. This completes the definition of $p$, which exists by $\Delta^0_1$ comprehension.

Let $U$ be the set of all $\sigma \in 2^{<\mathbb{N}}$ such that $p(\sigma) \neq \varnothing$. By $\Pi^0_1$ induction on $n$, if $\sigma \in 2^{<\mathbb{N}}$ has length $n$ then $x \leqslant n$ for all $x \in p(\sigma)$. So $U$ exists, and it is readily seen that $U$ is a tree. We claim that it is infinite. To this end, we prove the stronger fact that for every $n$, there is a $\sigma \in 2^{<\mathbb{N}}$ of length $n$ such that $\varnothing \neq p(\sigma)$ and $\langle p(\sigma)\rangle$, the subring of $R$ generated by the elements of $p(\sigma)$, does not contain $1_R$. Note that, as there are only finitely many strings of length $n$, the condition to verify is actually $\Pi^0_1$ in $n$. Hence, we can again proceed using $\Pi^0_1$ induction. The result is clear for $n = 0$, so fix $n$ and assume the result is true for this $n$. Fix a witnessing string $\sigma$. If $|\sigma| = 4\langle x, y, z, k\rangle$ or $|\sigma| = 4\langle x, y, z, k\rangle + 1$, then $p(\sigma 0)$ adds at most the sum or product of two elements already in $p(\sigma)$, and hence $\langle p(\sigma 0)\rangle = \langle p(\sigma)\rangle$. In this case, $\sigma 0$ witnesses the result for $n+1$. If $|\sigma| = 4\langle x, y, z, k\rangle + 2$ and $r_z = r_x \cdot_R r_y \in p(\sigma)$, then either $\langle p(\sigma) \cup \{r_x\}\rangle = \langle p(\sigma 0)\rangle$ or $\langle p(\sigma) \cup \{r_y\}\rangle = \langle p(\sigma 1)\rangle$ does not contain $1_R$. So, either $\sigma 0$ or $\sigma 1$ witnesses the result for $n + 1$. And if $|\sigma| = 4\langle x, y, z, k\rangle + 3$ then by definition, both $\sigma 0$ and $\sigma 1$ witness the result for $n + 1$. This proves the claim.

Apply WKL to obtain a path $h$ through $U$. By construction, $\varnothing \neq p(h \upharpoonright n)$ for all $n$. Moreover, we must have $1_R \notin \langle p(h \upharpoonright n)\rangle$ for all $n$. If not, fix the largest $n$ such that $1_R \notin \langle p(h \upharpoonright n)\rangle$. Then necessarily $n = 4(x, y, z, k) + 2$ for some $x$, $y$, $z$, and $k$, and either $h(n) = 0$ and $r_x^{-1} \in \langle p(h \upharpoonright n)\rangle$, or $h(n) = 1$ and $r_y^{-1} \in \langle p(h \upharpoonright n)\rangle$. Say it is the former; the latter case is symmetric. Fix $w$ such that $r_x^{-1} = r_w$, and choose the least $k$ so that $m = 4(x, w, 1, k) + 2 > n$. Then $p(h \upharpoonright m)$ contains both $r_x$ and $r_w$, and since $r_1 = 1_R = r_x \cdot_R r_w$ by assumption, it follows that $p(h \upharpoonright (m + 1))$, being nonempty, equals $p(h \upharpoonright m) \cup \{1_R\}$. But then by construction, $p(h \upharpoonright (m + 2)) = \varnothing$, a contradiction.

Now fix any $\ell$; we exhibit an element of $T$ of length $\ell$. Let $b = \max\{r_x : x < \ell\}$. Form the set $F = \{r_x < b : x < \ell \wedge (\exists n)[r_x \in p(h \upharpoonright n)]\}$, which exists by bounded $\Sigma^0_1$ comprehension. Define $\sigma \in 2^{<\mathbb{N}}$ of length $\ell$ by $\sigma(x) = 1$ if and only if $r_x \in F$, for all $x < \ell$. Then it is not difficult to check that $\sigma$ satisfies each of the defining conditions in the definition $T$, hence $\sigma \in T$.

(2) $\rightarrow$ (1): We argue in $\mathsf{RCA}_0$. Fix functions $f : \mathbb{N} \rightarrow \mathbb{N}$ and $g : \mathbb{N} \rightarrow \mathbb{N}$ with disjoint ranges. We use (2) to obtain a set $Z$ such that for all $x$, if $x$ is in the range of $f$ then $x \in Z$ and if $x$ is in the range of $g$ then $x \notin Z$. By Exercise 5.13.18, this establishes WKL.

Consider the polynomial ring $\mathbb{Q}[x_i : i \in \mathbb{N}]$, and consider the ideal

$$J = \langle \{x_i^{j+1} : f(j) = i\} \cup \{x_i^{j+1} - 1 : g(j) = i\}\rangle.$$

A polynomial $p \in \mathbb{Q}[x_i : i \in \mathbb{N}]$ belongs to $J$ if and only contains a monomial with a factor of the form $x_i^s$, where $f(j) = i$ or $g(j) = i$ for some $j < s$. Hence, $J$ exists by $\Delta^0_1$ comprehension. Consequently, so does the quotient ring $R = \mathbb{Q}[x_i : i \in \mathbb{N}]/J$. Write $[p]$ for the representative of $p$ in $R$, and notice that the function $p \mapsto [p]$ exists.

Apply (2) to find a prime ideal $I$ of $R$. Let $Z = \{i : [x_i] \in I\}$, which exists because the function $p \mapsto [p]$ exists. If $i \in \mathrm{range}(f)$, say with $f(j) = i$, then $x_i^{j+1} \in J$, hence $[x_i]^{j+1} = [x_i^{j+1}] = 0_R \in I$. By primeness, it follows that $[x_i] \in I$. Thus, $i \in Z$. On the other hand, if $i \in \mathrm{range}(g)$, say with $g(j) = i$, then $[x_i]$ cannot belong to $I$ else so would $[x_i]^{j+1} = [x_i^{j+1}] = 1_R$. Thus, $i \notin Z$. This completes the proof. □

That (1) and (2) are equivalent under $\leqslant_c$ follows by an analogous argument. But the reduction of (2) to (1) is an interesting example of how a computability theoretic construction may be delicate to formalize in $\mathsf{RCA}_0$ for reasons *other than* induction. Indeed, consider again the tree $T \subseteq 2^{<\omega}$ constructed in the proof of the (1) $\rightarrow$ (2) implication. This tree is computable in the given ring $R$, and as noted, every path through $T$ defines a prime ideal of the ring $R$. But it is also easy to see that if $I$ is any prime ideal of $R$ then the characteristic function of $I$ is a path through $T$. Hence, since we know the classical result that $R$ has a prime ideal, we immediately know that $T$ is infinite. But the classical result is exactly what we are proving in $\mathsf{RCA}_0$, so we cannot invoke it in the course of our proof, and that is why $T$ being infinite requires separate justification. (We saw the opposite phenomenon in the proof of Proposition 4.1.3.) The ability to use "external facts" like this can make working with finer reducibilities easier than working in $\mathsf{RCA}_0$.

**Theorem 11.3.2 (Friedman, Simpson, and Smith [117]).** *The following are equivalent under $\leqslant_c$ and over $\mathsf{RCA}_0$.*

*1.* $\mathsf{TJ}$.
*2. Every commutative ring has a maximal ideal.*

*Hence, (2) is equivalent over $\mathsf{RCA}_0$ to $\mathsf{ACA}_0$.*

*Proof.* Here, let us prove the equivalence under computable reducibility. The arguments can be readily formalized (albeit the proof becomes longer).

(2) $\leqslant_c$ (1): Let $R$ be an infinite commutative ring, say with domain $\{r_0 < r_1 < \cdots\}$ where $r_0 = 0_R$. Define a function $f$ from $\omega$ to (codes for) finite sets, as follows: $f(0) = \{0_R\}$, and for all $x > 0$,

$$f(x) = \begin{cases} f(x-1) \cup \{r_x\} & \text{if } 1_R \notin \langle f(x-1) \cup \{r_x\} \rangle, \\ f(x-1) & \text{otherwise.} \end{cases}$$

Clearly, $f \leqslant_T R'$. Let $I = \text{range}(f)$. Since $f$ is nondecreasing, we have $I \leqslant_T R \oplus I \leqslant_T R'$. Clearly, $0_R \in I$. By induction, $1_R \notin f(x)$ for all $x$, hence $1_R \notin I$. From here it is easily verified that $I$ is an ideal of $R$. We claim that it is maximal. To this end, fix any $r_z \notin I$. Then in particular $r_z \notin f(z)$, so by definition it must be that $1_R \in \langle f(z-1) \cup \{r_z\} \rangle$. Since $f(z-1) \subseteq I$, it follows that no ideal contains all of $I$ as well as $r_z$. This proves the claim.

(1) $\leqslant_c$ (2): We show that (2) codes the jump. Fix $A \in 2^\omega$. Let $K$ be the field of fractions of the polynomial ring $\mathbb{Q}[x_i : i \in \omega]$. Formally, for all $p, q \in \mathbb{Q}[x_i : i \in \omega]$ with $q \neq 0$, let $p/q$ denote the least pair $\langle r, s \rangle$ such that $r, s \in \mathbb{Q}[x_i : i \in \omega]$, $s \neq 0$, and $rq = ps$. Then $K$ is the set of all such $p/q$, with the obvious operations. Clearly, $K$ is computable.

Let $S$ be the set of all polynomials $q \in \mathbb{Q}[x_i : i \in \omega]$ that contain at least one nontrivial monomial in which only $x_i$ with $i \in A'$ appear. It is easy to see that $S$ is a multiplicative ring, and so

$$R = \{\frac{p}{q} \in K : q \in S\}$$

is a ring. (It is isomorphic to the localization of $\mathbb{Q}[x_i : i \in \omega]$ by $S$.) Of course, $R$ need not be $A$-computable. But it is $A$-c.e. Therefore, we may fix an $A$-computable bijection $f : \omega \to R$. (In the argument over $\mathsf{RCA}_0$, the existence of $f$ follows by Theorem 6.1.6.)

We now pull back the structure in $R$ via $b$ to get an $A$-computable ring $R^*$ with domain $\omega$, as follows: $0_{R^*} = f^{-1}(0_R)$; $1_{R^*} = f^{-1}(1_R)$; and for all $x, y \in \omega$, $x +_{R^*} y = f^{-1}(f(x) +_R f(y))$ and $x \cdot_{R^*} y = f^{-1}(f(x) \cdot_R f(y))$. Since $+_R$ and $\cdot_R$ are computable operations (they are the same operations as in $K$), it follows that $R^*$ is computable from $f$ and hence from $A$, as wanted. Moreover, $R^*$ is isomorphic to $R$ via $f$.

Suppose $I$ is any maximal ideal of $R^*$. We claim that $i \in A'$ if and only if $f^{-1}(x_i) \notin I$, which implies that $A' \leqslant_T A \oplus I$, completing the proof. We prove the equivalent fact that $i \in A'$ if and only if $x_i \notin f(I)$, which is notationally lighter. Notice that as $f$ is an isomorphism, $f(I)$ is a maximal ideal of $R$.

First, suppose $i \in A'$. Then $x_i \in S$ and hence $1/x_i \in R$. If $x_i$ belonged to $b(I)$ then as $b(I)$ is an ideal, so would $1/x_i \cdot_R x_i = 1_R$, which is impossible. Conversely, suppose $x_i \notin b(I)$. By maximality of $b(I)$, there must exist $p/q \in R$ and $r/s \in b(I)$ such that

$$\frac{p}{q}x_i + \frac{r}{s} = 1_R,$$

or equivalently, $psx_i = qs - qr$. Since $q$ and $s$ belong to $S$, so does their product; thus, $qs$ contains a monomial $m$ in which only $x_j$ with $j \in A'$ appear. By contrast, $r \notin S$, else $r/s \in b(I)$ would be invertible in $R$. So every monomial in $r$ has a factor of the form $x_j$ for some $j \notin A'$, and the same must consequently be true of $qr$. It follows that $m$ cannot cancel with any term in $qr$, and so must also be a monomial of $psx_i$. But this means that $x_i$ is a factor of $m$, and therefore $i \in A'$, as was to be shown. □

Further results concerning the complexity of ideals have been obtained by Downey, Lempp, and Mileti [77]. For example, consider the fact that a ring has no nonzero proper ideal if and only if it is a field. The "if" direction of this is easily proved in $\mathsf{RCA}_0$, and the "only if" direction is easily proved in $\mathsf{ACA}_0$. (As pointed out in [77], if $R$ is not a field then it has an element $r$ that is not a unit, and then $\langle r \rangle$ is a nontrivial ideal. Since $\langle r \rangle$ is arithmetically definable, $\mathsf{ACA}_0$ suffices to prove that it exists.) We omit the proof of the following result showing that, with more effort, $\mathsf{WKL}_0$ suffices.

**Theorem 11.3.3 (Downey, Lempp, and Mileti [77]).** *The following are equivalent under $\leqslant_c$ and over* $\mathsf{RCA}_0$.

1. $\mathsf{WKL}$.
2. *A ring R has no nonzero proper ideal if and only if it is a field.*

## 11.4 Orderability

In this section, we turn to theorems concerning the existence of orderings of groups, rings, and fields. Here, an *ordering* is a linear ordering of the domain that is compatible with the algebraic structure. The definitions are as follows.

**Definition 11.4.1.**

1. A group $(G, *_G, e_G)$ is *orderable* if there exists a linear ordering $\leqslant_G$ of $G$ such that for all $a, b \in G$, if $a \leqslant_G b$ then $a *_G c \leqslant_G b *_G c$ and $c *_G a \leqslant_G c *_G b$ for all $c \in G$.
2. A ring $(R, +_R, \cdot_R, 0_R, 1_R)$ is *orderable* if there exists a linear ordering $\leqslant_R$ of $R$ such that for all $r, s \in R$:
   - if $r \leqslant_R s$ then $r +_R t \leqslant_R s +_R t$ for all $t \in R$,
   - if $r \leqslant_R s$ then $r \cdot_R t \leqslant_R s \cdot_R t$ and $t \cdot_R r \leqslant_R t \cdot_R s$ for all $t \in R$ with $t \geqslant_R 0_R$.
3. A field is *orderable* if it is orderable as a ring.

A number of classical results in group theory and field theory from the early part of the 20th century (referred to by Metakides and Nerode [209] as the "Steinitz–Artin period" in algebra) concern conditions under which different algebraic structures admit orderings.

One prominent such result, due to Artin and Schreier [3], is that every formally real field is orderable. Recall that a field $(K, +_K, \cdot_K, 0_K, 1_K)$ is *formally real* if there is no $k \in K$ such that $k \cdot_K k +_K 1_K = 0_K$. (Note that if $K$ is orderable them it is necessarily formally real, so the Artin–Schreier result is actually a characterization. But the converse direction is trivial.) Ershov [101], and independently Metakides and Nerode [209], proved the existence of a computable formally real field having no computable ordering. Hence, as a $\forall\exists$ theorem, the Artin–Schreier result is not computably true. Metakides and Nerode [209] actually showed more. Namely, they proved that the class of orderings of a field forms a $\Pi^0_1$ class in $2^\omega$, and conversely, every nonempty $\Pi^0_1$ class is computably homeomorphic to the class of orderings of some computable field. Building on this, Friedman, Simpson, and Smith [117] obtained the following equivalence.

**Theorem 11.4.2 (Friedman, Simpson, and Smith [117]).** *The following are equivalent under $\leqslant_c$ and over* $\mathsf{RCA}_0$.

1. $\mathsf{WKL}$.
2. *Every formally real field is orderable.*

We refer to Simpson [288, Chapter IV.4] for a proof, as well as for much more content about the reverse mathematics of formally real fields.

Next we shift to groups. Recall that an abelian group $(G, +_G, 0_G)$ is *torsion free* if for every nonzero $a \in G$ and every natural number $n > 0$, $na \neq 0_G$. (Here, $na$ is defined inductively: $0a = a$, and for $n > 0$, $na = (n-1)a +_G a$.) A classical result of Levi [197] states that an abelian group is orderable if and only if it is torsion free. The effective content of this result was first considered in computable structure theory, by Downey and Kurtz [73]. They gave an explicit construction of a computable torsion free abelian group with no computable ordering. Thus, Levi's theorem is not computably true. In turn, Downey and Kurtz asked whether an analogue of the Metakides and Nerode result above holds also in this setting. A close connection between $\Pi^0_1$ classes and orderings of torsion free abelian groups was provided by Hatziriakou and Simpson [140], as follows.

**Theorem 11.4.3 (Hatziriakou and Simpson [140]).** *The following are equivalent under $\leqslant_c$ and over* $\mathsf{RCA}_0$.

1. WKL.
2. *Every torsion free abelian group is orderable.*
3. *An abelian group is torsion free if and only if it is orderable.*

*Proof.* That every orderable abelian group is torsion free can be proved by induction in $\mathsf{RCA}_0$ (see Exercise 11.7.8). Thus, it suffices to prove the equivalence of (1) with (2).

To prove this, will make use of the following auxiliary notion. Suppose $(G, +_G, 0_G)$ is an abelian group. A *positive cone* for $G$ is a set $P \subseteq G$ such that:

- $G$ is closed under $+_G$.
- For all $a \in G$, either $a$ or $-a$ belongs to $G$.
- For all $a \in G$, $a$ and $-a$ belong to $G$ if and only if $a = 0_G$.

Then $G$ is orderable if and only if it has a positive cone, and this fact is provable in $\mathsf{RCA}_0$. (See Exercise 11.7.4.) In fact, if $\leqslant_G$ is an ordering of $G$ then there is a positive cone $P$ for $G$ computes in $G \oplus \leqslant_G$; conversely, given a positive cone $P$ for $G$, there is an ordering $\leqslant_G$ of $G$ computable in $G \oplus P$.

We now proceed to the argument. We prove the equivalence over $\mathsf{RCA}_0$; as usual, the $\leqslant_c$ reduction is similar.

(1) $\to$ (2): We argue in WKL. Let $(G, +_G, 0_G)$ be an abelian group, say with domain $\{a_0 < a_1 < \cdots\}$ where $r_0 = 0_G$. Let $T \subseteq 2^{<\mathbb{N}}$ be the tree of all $\sigma$ with the following properties:

- If $|\sigma| > 0$ then $\sigma(0) = 0$.
- For all nonempty $F < |\sigma|$ and all $z < |\sigma|$ with $a_z = \sum_{x \in F} a_x$, if $\sigma(x) = 1$ for all $x \in F$ then $\sigma(z) = 1$.
- For all $x, y < |\sigma|$ with $a_x +_G a_y = 0_G$, $\sigma(x) = 1 - \sigma(y)$.

We claim that $T$ is infinite, and hence is an instance of WKL. In fact, we proceed by induction to show the following stronger fact: for every $s$, there is a $\sigma \in T$ of length $s$ such that $0_G \notin \langle\{a_x : \sigma(x) = 1\}\rangle$. This is trivial for $s = 0$. So fix $s > 0$ and assume the claim holds for $s - 1$, as witnessed by $\sigma \in T$. If $\sigma 0 \in T$ then we are done because for all $x$, $\sigma(x) = 1$ if and only if $\sigma 0(x) = 1$. Otherwise, from the definition there are two possibilities:

- *Case 1:* There is a nonempty $F < s - 1$ such that $\sum_{x \in F} a_x = a_{s-1}$ and $\sigma(x) = 1$ for all $x \in F$.
- *Case 2:* There is an $x^* < s - 1$ such that $a_{x^*} + a_{s-1} = 0_G$ and $\sigma(x^*) = 0$.

In the first case we have $a_{s-1} \in \langle\{a_x : \sigma(x) = 1\}\rangle$, so in particular $0_G \notin \langle\{a_x : \sigma(x) = 1\}\rangle$. Thus, $a_{s-1}$ cannot equal $0_G$ or $-a_w$ for any $w < s-1$ such that $\sigma(w) = 1$. It follows that $\sigma 1 \in T$. In the second case, we have only to check that $-a_{s-1}$ is not equal to $\sum_{x \in F} a_x$ for some nonempty $F < s - 1$ with $\sigma(x) = 1$ for all $x \in F$. But if this were the case then we would have $\sigma(x^*) = 1$, since $\sigma \in T$. Hence, again $\sigma 1 \in T$, and the claim is proved.

The induction in the above argument is on a $\Pi^0_1$ formula of $s$, and hence its verification goes through in $\mathsf{RCA}_0$.

Let $f \in [T]$ be arbitrary and let $P = \{0\} \cup \{a_x : f(x) = 1\}$. Then by definition of $T$, $P$ is a positive cone for $G$.

(2) → (1): We now argue in $\mathsf{RCA}_0$. Fix computable injections $f, g\colon \mathbb{N} \to \mathbb{N}$ with disjoint ranges. We construct a separating set for the ranges of $f$ and $g$, thereby obtaining WKL. Let $\{p_x : x \in \mathbb{N}\}$ be an enumeration of the primes in increasing order. Let $G$ be the abelian group generated by elements $y, x_0, x_1, \ldots$, subject to the following relations for all $i$:

$$p_{2j}x_{f(j)} -_G y = 0_G,$$
$$p_{2j+1}x_{g(j)} +_G y = 0_G.$$

A typical element of $G$ can be put in the form $cy + \sum_{i\in F} d_i x_i$, where $c \in \mathbb{N}$, the $d_i$ are positive integers, $F$ is a (possibly empty) finite subset of $\mathbb{N}$, and for all $i \in F$ we have

$$(\forall j)[p_{2j} \leqslant d_i \to f(j) \neq i \wedge p_{2j+1} \leqslant d_i \to g(j) \neq i]. \tag{11.1}$$

Since $n + 1 < p_n$ for all $n$ (easily verified in $\mathsf{I}\Sigma^0_1$), (11.1) is equivalent to

$$(\forall j < d_i)[p_{2j} \leqslant d_i \to f(j) \neq i \wedge p_{2j+1} \leqslant d_i \to g(j) \neq i].$$

Thus, by considering representatives in the above form and defining addition appropriately, we can regard $G$ as a group in the sense of Definition 11.1.1. Clearly, $G$ is abelian.

We claim that $G$ is torsion free. To this end, fix $a = cy + \sum_{i\in F} d_i x_i \in G$ and suppose $na = 0_G$ for some $n > 0$. For the $y$ term to cancel with the $x_i$ terms, for every $i \in F$ there must be a $j_i$ such that either $i = f(j_i)$ and $p_{2j_i} \mid nd_i$ or $i = g(j_i)$ and $p_{2j_i+1} \mid nd_i$. For each $i \in F$, let $q_i$ be $p_{2j_i}$ if $i = f(j_i)$ and $-p_{2j_i+1}$ if $i = g(j_i)$. Thus, $q_i x_i = y$ for all $i \in F$. Also, by (11.1), we have $d_i < |q_i|$, hence $q_i \nmid d_i$; since $|q_i|$ is prime, this means $q_i \mid n$. Putting these facts together, we obtain

$$nd_i x_i = nd_i q_i^{-1}(q_i x_i) = nd_i q_i^{-1} y.$$

That $q_i \mid n$ is used to conclude that the rightmost term above is defined. The above yields that

$$0_G = na = ncy + \sum_{i\in F} nd_i x_i = (nc + \sum_{i\in F} nd_i q_i^{-1})y,$$

and so $c + \sum_{i\in F} d_i q_i^{-1}$ must be equal to 0. Let $r = \prod_{i\in F} q_i$. Then for each $i \in F$ we have

$$cr + \sum_{j\in F\setminus\{i\}} d_j q_j^{-1} r = d_i q_i^{-1} r. \tag{11.2}$$

Since $f$ and $g$ are injective, the map $i \mapsto |q_i|$ is injective on $F$. It follows that $q_i \mid q_j^{-1} r$ for all $j \neq i$ and therefore that $q_i$ divides the left hand side of (11.2). But by the same token, $q_i \nmid q_i^{-1}r$. Hence, $q_i$ must divide $d_i$, which we already noted above is not the case. This proves the claim.

Apply (2) to obtain a positive cone $P$ for $G$. Note that if $x_i \in P$ then by induction, $nx_i \in P$ for all $n$. Conversely, if $nx_i \in P$ for some $n > 0$ then $x_i \in P$. This is because if $x_i \notin P$ then $-x_i \in P$ since $P$ is a positive cone, and so also $(n-1)(-x_i)$ and $nx_i +_G (n-1)(-x_i)$ belong to $P$. But the latter is equal to $x_i$ (again, by induction). So in particular, for all $j$ we have $x_{f(j)} \in P$ if and only if $p_{2j}x_{f(j)} \in P$, and $x_{g(j)} \in P$ if and only if $p_{2j+1}x_{f(j)} \in P$. Also, for all $j$, $p_{2j}x_{f(j)} = y$ and $p_{2j+1}x_{g(j)} = -y$, and exactly one of $y$ and $-y$ belongs to $P$. Consequently, either $x_{f(j)} \in P$ and $x_{g(j)} \notin P$ for all $j$, or $x_{g(j)} \in P$ and $x_{f(j)} \notin P$ for all $j$. Thus, if we let $Z = \{i : x_i \in P\}$, then we have either that range$(f) \subseteq Z$ and $Z \cap$ range$(g) = \emptyset$, or range$(g) \subseteq Z$ and $Z \cap$ range$(f) = \emptyset$, meaning that $Z$ is the desired separating set. □

Further results about orderability of algebraic structures have been obtained by Solomon [296, 297] (see also Solomon [298]). For a group $G$, let $Z(G)$ denote the center of $G$, i.e., $\{a \in G : (\forall x \in G)[a *_G x = x *_G a]\}$. For each $N \triangleleft G$, let $\pi_N : G \to G/N$ be the natural homeomorphism sending elements to cosets. Then in $\mathsf{RCA}_0$, we say a group $G$ is *nilpotent* if there exists an $n > 0$ and a sequence $\langle N_i : i < n\rangle$ of normal subgroups of $G$ such that

- $N_0 = \{e_G\}$,
- for all $i < n$, $N_i = \pi^{-1}_{N_{i-1}} Z(G/N_i)$,
- $N_{n-1} = G$.

Using this definition, the following result can be obtained.

**Theorem 11.4.4 (Solomon [297]).** *The following are equivalent under $\leqslant_c$ and over* $\mathsf{RCA}_0$.

1. $\mathsf{WKL}$.
2. *Every torsion free nilpotent group is orderable.*

In general, $\mathsf{RCA}_0$ cannot prove for every group $G$ that $Z(G)$ exists (see Exercise 11.7.9). But if $Z(G)$ does exist then $\mathsf{RCA}_0$ can verify its basic properties, e.g., that $Z(G)$ is a normal subgroup of $G$ (see [297]).

We can find orderability results also at the level of $\mathsf{RCA}_0$ and $\mathsf{ACA}_0$. By replacing "linear order" with "partial order" in Definition 11.4.1, we obtain the notion of a *partially orderable* group. This, too, can be characterized in terms of positive cones, only with the condition that a cone contains every group element or its inverse removed. If $\leqslant_G$ is a partial ordering of $G$, a subgroup $N \triangleleft G$ is *convex* (with respect to this ordering) if $N$ contains every $x \in G$ such that $a \leqslant_G x \leqslant_G b$ for some $a, b \in N$. If $N$ is convex, then $\leqslant_G$ induces an ordering of the quotient group $G/N$. Namely, let $P$ be the positive cone on $G$ associated to $\leqslant_G$; then $\{a \in G/N : (\exists b \in N)[a *_G b \in P]\}$ is a positive cone on $G/H$. It can be checked that the associated ordering is linear if $\leqslant_G$ is.

**Theorem 11.4.5 (Solomon [297]).** *The principle "if G is a (linearly) orderable group and N is a convex normal subgroup of G, then the induced ordering on G/N exists" admits computable solutions and is provable in* $\mathsf{RCA}_0$.

On the other hand, for partial orderings in general, the existence of the induced ordering on convex subgroups codes the jump.

**Theorem 11.4.6 (Solomon [297]).** *The following are equivalent under $\leqslant_c$ and over* $\mathsf{RCA}_0$.

1. TJ.
2. *If $G$ is a partially orderable group and $N$ is a convex normal subgroup of $G$, then the induced ordering on $G/N$ exists.*

*Hence, (2) is equivalent over* $\mathsf{RCA}_0$ *to* $\mathsf{ACA}_0$.

## 11.5 The Nielsen–Schreier theorem

One theorem of algebra with especially interesting reverse mathematics behavior is the *Nielsen–Schreier theorem*, which asserts that every subgroup of a free group is free. It turns out that the strength of this theorem is affected by how, precisely, we choose to think of subgroups. As we will see, this consideration yields two versions of the Nielsen–Schreier theorem, one provable in $\mathsf{RCA}_0$ and the other equivalent over $\mathsf{RCA}_0$ to $\mathsf{ACA}_0$. We will state both of these results carefully in this section, and give a proof of the latter.

**Definition 11.5.1.** Fix a set $X \in 2^\omega$.

1. $\mathrm{Word}_X$ denotes the set $(X \times \{-1, 1\})^{<\omega}$; its elements are *words* (over $X$).
2. A word $w$ is *reduced* if for all $i < |w| - 1$, if $w(i)(0) = w(i+1)(0)$ then $w(i)(1) = w(i+1)(1)$. The set of reduced words over $X$ is denoted $\mathrm{Red}_X$.
3. Two words $w_0$ and $w_1$ are 1*-step equivalent* (over $X$) if there is a $b < 2$ such that $|w_b| = |w_{1-b}| + 2$, and there is an $i < |w_b| - 1$ as follows:
   - $w_{1-b}(j) = w_b(j)$ for all $j < i$,
   - $w_{1-b}(j) = w_b(j+2)$ for all $j$ with $i \leqslant j < |w_{1-v}|$,
   - $w_b(i)(0) = w_b(i+1)(0)$ and $w_b(i)(1) = -w_b(i+1)(1)$.
4. Two words $w_0$ and $w_1$ are *freely equivalent* (over $X$) if there is an $n > 2$ and a sequence of words $v_0, \ldots, v_{n-1}$ such that $w_0 = v_0$, $w_1 = v_{n-1}$, and for all $i < n-1$, $v_i$ is equal or 1-step equivalent to $v_{i+1}$.

Note that we are using $w, v, u, \ldots$ for words, instead of Greek letters as we ordinarily do for sequences; thus, e.g., $wv$ indicates the concatenation of $w$ by $v$, etc. Likewise, following customary (and more legible) notation, we may write $a_0^{\varepsilon_0} \cdots a_{k-1}^{\varepsilon_{k-1}}$ for the word $w$ with $|w| = k$ and $w(i) = \langle a_i, \varepsilon_i \rangle$ for all $i < k$. Thus, $w$ is reduced if and only if it does not have the form $\cdots a^1 a^{-1} \cdots$ or $\cdots a^{-1} a^1 \cdots$ for some $a \in A$. The act of "deleting" some such occurrence of $a^1 a^{-1}$ or $a^{-1} a^1$ corresponds to a 1-step equivalence, and doing this iteratively gives free equivalence.

The above definition readily formalizes in $\mathsf{RCA}_0$, and $\mathsf{RCA}_0$ suffices to prove various basic facts concerning it, most notably that every word $w$ is freely equivalent to a unique reduced word, which we denote by $\rho_X(w)$. (See Exercise 11.7.6.) This facilitates the following definition, central to this section.

**Definition 11.5.2.** Fix $X \in 2^\omega$. The *free group on* $X$, denoted $F_X$, is the structure $(\mathrm{Red}_X, \cdot_X, 1_X)$ in the language of groups, where $1_X$ is the empty string (as a sequence in $\mathrm{Word}_X$) and for $w, v \in \mathrm{Word}_X$, $w \cdot_X v = \rho_X(wv)$ (with $wv = w^\frown v$ meaning the concatenation of $w$ by $v$, as strings).

In $\mathsf{RCA}_0$, we can verify that $F_X$ is indeed a group (Exercise 11.7.7).

One remark, which will be important shortly, is that the map $\rho_X$ naturally extends to a map $\widehat{\rho}_X : \mathrm{Word}_{\mathrm{Word}_X} \to \mathrm{Red}_X$.

**Definition 11.5.3.**

1. For $w \in \mathrm{Word}_X$, let $\neg w \in \mathrm{Word}_X$ to be the sequence of length $|w|$ such that for all $i < |w|$,
$$\neg w(i)(0) = w(|w| - i - 1)(0),$$
$$\neg w(i)(1) = -w(|w| - i - 1)(1).$$
That is, if $w = a_0^{\varepsilon_0} \cdots a_{k-1}^{\varepsilon_{k-1}}$ then $\neg w = a_{k-1}^{-\varepsilon_{k-1}} \cdots a_0^{-\varepsilon_0}$.
2. For $w \in \mathrm{Word}_X$ and $\varepsilon \in \{-1, 1\}$, let
$$\mathrm{sgn}(w, \varepsilon) = \begin{cases} w & \text{if } \varepsilon = 1, \\ \neg w & \text{if } \varepsilon = -1. \end{cases}$$
3. For $(w_0, \varepsilon_0) \cdots (w_{n-1}, \varepsilon_{n-1}) \in \mathrm{Word}_X$, let
$$\widehat{\rho}_X((w_0, \varepsilon_0) \cdots (w_{n-1}, \varepsilon_{n-1})) = \rho_X(\mathrm{sgn}(w_0, \varepsilon_0) \cdots \mathrm{sgn}(w_{n-1}, \varepsilon_{n-1})).$$

So for example, if we take $X = \{a, b\}$ and apply $\widehat{\rho}_X$ to the word
$$(a^1 b^{-1} a^1 a^1)^{-1} (a^1 b^{-1} 1 b^{-1})^1$$
over $\mathrm{Word}_X$, we obtain
$$\rho_X(a^{-1} a^{-1} b^1 a^{-1} a^1 b^{-1} b^{-1}) = a^{-1} a^{-1} b^{-1}.$$

Using the notation and definitions above, we now give the definition of what it means (in our setting) for a subgroup of $F_X$ to be free.

**Definition 11.5.4.** Fix $X \in 2^\omega$ and let $H$ be a subgroup of $F_X$. Then $H$ is *free* if there is a set $B \subseteq H$ as follows:

- For every $h \in H$ there is a $w \in \mathrm{Red}_B \subseteq \mathrm{Word}_{\mathrm{Word}_X}$ such that $h = \widehat{\rho}_X(w)$.
- For all $w_0, w_1 \in \mathrm{Red}_B$, if $w_0 \neq w_1$ then $\widehat{\rho}_X(w_0) \neq \widehat{\rho}_X(w_1)$.

Thus, a subgroup $H$ of $F_X$ is free if it is generated by a subset of the elements of $H$ among which there are no nontrivial relations (in $H$). This, of course, is the standard definition in algebra.

The Nielsen–Schreier theorem may at first glance seem trivial. But it is not, precisely because finding the "right" basis $B$ in the above definition is nontrivial. (Indeed, a subgroup $H$ of a free group $F_X$ may also be generated by a set of elements that *do* satisfy nontrivial relations.) There is a quick topological proof of the theorem, which uses the fact that a group is free if and only if is the fundamental group of a graph, and that every subgroup of such a group is the fundamental group of a covering of this graph. This argument is difficult to formalize in second order arithmetic. But there is another, using so-called Schreier transversals, which is more direct and more constructive. (See Igusa [165] for both proofs in detail.) Building on the method, Downey, Hirschfeldt, Lempp, and Solomon [76] obtained the following.

**Theorem 11.5.5 (Downey, Hirschfeldt, Lempp, and Solomon [76]).** $\mathsf{RCA}_0$ *proves that for every* $X \in 2^{\mathbb{N}}$, *every subgroup of* $F_X$ *is free.*

We do not include the proof here, which would take us a somewhat afield. In broad outline, however, it follows the classical version.

There is another way to talk about a subgroup which is often more convenient in algebra, and using a presentation rather than an explicit subset. The next definition shows how this can be accommodated in our setting.

**Definition 11.5.6.** Fix $X \in 2^{\omega}$.

1. For $S \subseteq F_X$, the *subgroup presented by* $S$ is the structure $\langle S\rangle$ with domain
$$\{h \in F_X : (\exists w \in \mathrm{Words}_S)[\widehat{\rho}_X(w) = h]\},$$
with multiplication inherited from $F_X$.
2. $\langle S\rangle$ is *free* if there is a set $B \subseteq F_X$ as follows:
   - For every $w \in B$ there is a $w \in \mathrm{Words}_S$ such that $w = \widehat{\rho}_X(v)$.
   - For every $h \in \langle S\rangle$ there is a $w \in \mathrm{Red}_B$ such that $h = \widehat{\rho}_X(w)$.
   - For all $w_0, w_1 \in \mathrm{Red}_B$, if $w_0 \neq w_1$ then $\widehat{\rho}_X(w_0) \neq \widehat{\rho}_X(w_1)$.

In what follows, when we say that $\langle S\rangle$ is free we do not mean necessarily that $\langle S\rangle$ exists. Indeed, $\mathsf{RCA}_0$ cannot in general prove this existence, as we are about to see. But if $\langle S\rangle$ does exist then $\mathsf{RCA}_0$ can prove it is a subgroup of $F_X$ (Exercise 11.7.10). And it is not difficult to see that if $\langle S\rangle$ exists then it is free in the sense of Definition 11.5.4 if and only if it is free in the sense of Definition 11.5.6. Now, classically, moving between $S$ and $\langle S\rangle$ is unproblematic and natural. But it has significance for the strength of the Nielsen–Schreier theorem.

**Theorem 11.5.7 (Downey, Hirschfeldt, Lempp, and Solomon [76]).** *The following are equivalent under* $\leqslant_{\mathrm{c}}$ *and over* $\mathsf{RCA}_0$.

1. $\mathsf{TJ}$.
2. *For every* $X \in 2^{\omega}$ *and every* $S \subseteq F_X$, $\langle S\rangle$ *exists.*
3. *For every* $X \in 2^{\omega}$ *and every* $S \subseteq F_X$, $\langle S\rangle$ *is free.*

*Hence, (2) and (3) are equivalent over* $\mathsf{RCA}_0$ *to* $\mathsf{ACA}_0$.

*Proof.* We give the equivalence over $\mathsf{RCA}_0$.

(1) → (2): Arguing in $\mathsf{ACA}_0$, fix $X$ and $S \subseteq F_X$. As $\langle S \rangle$ is $\Sigma^0_1$-definable in $S$, it exists by arithmetical comprehension.

(2) → (3): Arguing in $\mathsf{RCA}_0$, fix $X$ and $S \subseteq F_X$. By (2), $\langle S \rangle$ exists and this is a subgroup of $F_X$. By Theorem 11.5.5, $\langle S \rangle$ is free in the sense of Definition 11.5.4 and hence in the sense of Definition 11.5.6.

(3) → (1): We argue in $\mathsf{RCA}_0$, assuming (3). Fix an injective function $f : \mathbb{N} \to \mathbb{N}$; we prove that range$(f)$ exists. Let $X = \{a_n : n \in \mathbb{N}\}$ be a set of infinitely many generators, and define

$$S = \{a_n^2 : y \in \mathbb{N}\} \cup \{a_n^{2x+1} : f(x) = n\}.$$

Here, for $z \in \mathbb{Z}$, $a_i^n$ is the obvious shorthand, defined inductively for $z > 1$ by $a_i^z = a_i^{z-1} a_i^1$ and for $z < 1$ by $a_i^z = a_i^{z+1} a_i^1$. Apply (2) to find a set $B \subseteq F_X$ witnessing that $\langle S \rangle$ is free. The proof is completed by the following two claims.

*Claim 1: For all $n \in \mathbb{N}$, $n \in$ range$(f)$ if and only if $a_n \in \langle S \rangle$.* For every $n$ we have $a_n^2 \in S$ and hence $a_n^{-2x} \in \langle S \rangle$ for every $x \in \mathbb{N}$. Now if $n \in$ range$(f)$ then $f(x) = n$ for some $x$, hence $a_i^{2x+1}$ belongs to $S$ and consequently $a_n = \widehat{\rho}_X(a_n^{2x+1} a_n^{-2x})$ belongs to $\langle S \rangle$.

In the other direction, suppose $n \notin$ range$(f)$. We will show there is no $w \in \text{Words}_S$ with $a_n = \widehat{\rho}_X(w)$, implying that $a_n \notin \langle S \rangle$, as needed.

For $w, v \in F_X$, say $v$ is an *$a_n$ block in $w$* if there exists $i_0 < |w|$ as follows:

- $v(i)(0) = a_n$ for all $i < |v|$,
- $i_0 + |v| \leqslant |w|$ and $v(i) = w(i_0 + i)$ for all $i < |v|$,
- if $i_0 > 0$ then $w(i_0 - 1)(0) \neq a_n$,
- if $i_0 + |v| < |w|$ then $w(i_0 + |v|)(0) \neq a_n$.

Note that since $v \in F_X = \text{Red}_X$, the first clause implies that $v(i)(1)$ is the same for all $i < |v|$. Thus, $v$ is an $a_n$ block in $w$ just if it has the form $a_n^1 \cdots a_n^1$ or $a_n^{-1} \cdots a_n^{-1}$, and $w$ has the form $\cdots a_{m_0} v a_{m_1} \cdots$ for some $m_0, m_1 \neq n$.

We claim that for all $w \in \text{Words}_S$ and all $v \in F_X$, if $v$ is an $a_n$ block in $\widehat{\rho}_X(w)$ then $|v|$ is even. In particular, this means $\widehat{\rho}_X(w) \neq a_n$, as is to be shown. The proof is by induction on $|w|$. (Note that the statement to be proved is $\Pi^0_1$.) If $|w| = 1$ then $w = (a_m^2)^{\pm 1}$ for some $m$, so any $a_n$ block in $\widehat{\rho}_X(w) = a_m^{\pm 2}$ has length 0 or 2. Assume then that $|w| > 1$, and that we have already proved the result for all words over $S$ of length $|w| - 1$. Then $w = w^*(a_m^k)^{\pm 1}$ for some $w^* \in \text{Words}_S$ with $|w^*| < |w|$ and some $m, k \in \mathbb{N}$. By inductive hypothesis, any $a_n$ block in $\widehat{\rho}_X(w^*)$ has even length. If $m \neq n$, then $\widehat{\rho}_X(w^*)$ and $\widehat{\rho}_X(w)$ have the same $a_n$ blocks, so the claim holds. So suppose $m = n$. In this case, we necessarily have $k = 2$ since $n \notin$ range$(f)$ by assumption, and $\widehat{\rho}_X(w) = \rho(\widehat{\rho}(w^*) a_n^{\pm 2})$. If $\widehat{\rho}(w^*)$, as a sequence, ends in an $a_n$ block $v$, then $|v|$ is even so $\rho(v a_n^{\pm 2}))$ is another $a_n$ block in $\widehat{\rho}_X(w)$ of even length. Otherwise, $\rho(\widehat{\rho}(w^*) a_n^{\pm 2}) = \widehat{\rho}(w^*) a_n^{\pm 2}$, and the $a_n$ blocks in this word are those of $\widehat{\rho}(w^*)$ as well as $a_n^{\pm 2}$, which has length two.

*Claim 2:* $\{n : a_n \in \langle S\rangle\}$, *and hence* range$(f)$, *exists.* For every $n$, $a_n^2 \in S \subseteq \langle S\rangle$, hence by definition of $B$ there is a $v_n \in \mathrm{Red}_B$ with $\widehat{\rho}_X(v_n) = a_n^2$. Moreover, this $v$ is unique, so the map $n \mapsto v_n$ exists by $\Sigma^0_0$ comprehension. (See also Exercise 11.7.5.)

We claim that for each $n$, $a_n \in \langle S\rangle$ if and only if there is a $w \in \mathrm{Red}_B$ satisfying the following:

- $\widehat{\rho}_X(w) = a_n$,
- $|w| < |v_n|$,
- $\{w(i)(0) : i < |w|\} \subseteq \{v_n(j)(0) : j < |v|\}$.

Clearly, once this is proved then it follows by $\Sigma^0_0$ comprehension that $\{n : a_n \in \langle S\rangle\}$ exists. Note that the last clause simply says that every word occurring in $w$ as an element of $\mathrm{Words}_S$ also occurs in $v_n$.

Fix $n$. By definition, $a_n \in \langle S\rangle$ if and only if there exists a $w \in \mathrm{Red}_B$ satisfying the first clause above. Thus, to complete the proof it suffices to show that if $\widehat{\rho}_X(w) = a_n$ for some $w$ then $w$ necessarily satisfies the second and third clause above as well.

Since $v_n \in \mathrm{Red}_B$, we have that $\rho_B(w^2) = v_n$. First, suppose $w$ has odd length as an element of $\mathrm{Red}_B$, so that we can write it as $b_0^{\varepsilon_0} \cdots b_{2k}^{\varepsilon_{2k}}$ for some $k$. Then

$$w^2 = (b_0^{\varepsilon_0} \cdots b_k^{\varepsilon_k} b_{k+1}^{\varepsilon_{k+1}} \cdots b_{2k}^{\varepsilon_{2k}})(b_0^{\varepsilon_0} \cdots b_{k-1}^{\varepsilon_{k-1}} b_k^{\varepsilon_k} \cdots b_{2k}^{\varepsilon_{2k}}),$$

which has length $2|w|$. Since $b_k^{\varepsilon_k}$ cannot cancel with $b_k^{\varepsilon_k}$ in the free reduction of $w^2$ to $v_n$, the most that can happen is for $b_{k+1}^{\varepsilon_{k+1}} \cdots b_{2k}^{\varepsilon_{2k}}$ to cancel with $b_0^{\varepsilon_0} \cdots b_{k-1}^{\varepsilon_{k-1}}$, leaving the reduced word

$$b_0^{\varepsilon_0} \cdots b_k^{\varepsilon_k} b_k^{\varepsilon_k} \cdots b_{2k}^{\varepsilon_{2k}}.$$

Similarly, if $w$ has even length as an element of $\mathrm{Red}_B$ then we can write it as $b_0^{\varepsilon_0} \cdots b_{2k-1}^{\varepsilon_{2k-1}}$ for some $k$, in which case the most that can happen in the reduction of

$$w^2 = (b_0^{\varepsilon_0} \cdots b_k^{\varepsilon_k} b_{k+1}^{\varepsilon_{k+1}} \cdots b_{2k-1}^{\varepsilon_{2k-1}})(b_0^{\varepsilon_0} \cdots b_{k-2}^{\varepsilon_{k-2}} b_{k-1}^{\varepsilon_{k-1}} \cdots b_{2k-1}^{\varepsilon_{2k-1}}),$$

to $v_n$ is for $b_{k+1}^{\varepsilon_{k+1}} \cdots b_{2k-1}^{\varepsilon_{2k-1}}$ to cancel with $b_0^{\varepsilon_0} \cdots b_{k-2}^{\varepsilon_{k-2}}$. This is because $b_k^{\varepsilon_k}$ cannot cancel with $b_{k-1}^{\varepsilon_{k-1}}$, as these also appear next to each other in $w$, which is already reduced over $B$. In this case, we are thus left with the reduced word

$$b_0^{\varepsilon_0} \cdots b_k^{\varepsilon_k} b_{k-1}^{\varepsilon_{k-1}} b_k^{\varepsilon_k} \cdots b_{2k-1}^{\varepsilon_{2k-1}}.$$

Either way, we have that $|v_n| = |\rho_B(w^2)| > |w|$, and every word occurring in $w$ occurs also in $v_n$. This completes the proof. □

We conclude this section with a comment on representation. To algebraists, being a free group is a property up to isomorphism. Thus, just as being a cyclic group means to have a generator, and not necessarily to *be* $\mathbb{Z}/n\mathbb{Z}$ for some $n$, so too being a free group means more than just being $F_X$ for some $X$. It is important to notice that the above treatment accommodates this view, even if the terminology does not. Namely, Definition 11.5.2 does not define a group to be free if it is *isomorphic* to some $F_X$. But the proofs of Theorems 11.5.5 and 11.5.7 would go through just the same for such groups. This is because in a model of $\mathsf{RCA}_0$, if a group $G$ is asserted to be isomorphic to $F_X$ then the isomorphism must exist (in the model). So for instance, if we have a subgroup $H$ of $G$ that we wish to show is free, we first apply the isomorphism to $H$ to get a subgroup $\widehat{H}$ of $F_X$ (which exists, since the isomorphism is surjective), and then apply the relevant theorem to conclude $F_X$ is free, and hence that $H$ is free on account of being isomorphic to $\widehat{H}$.

## 11.6 Other topics

Many other corners of algebra have been explored in reverse mathematics. In this section, we take a look at a smattering of these results. This survey is incomplete but illustrates several interesting theorems. We will omit the proofs, some of which are quite involved. Many of the reverse mathematics tools they use are implicit in the proofs we have seen already, for example, coding with polynomials over $\mathbb{Q}$. In addition to these methods, a deep understanding of the underlying algebraic results is required.

Algebra principles turn up at all levels of the "big five" hierarchy. We have seen principles at the level of $\mathsf{WKL}_0$ and $\mathsf{ACA}_0$ above. An interesting principle at the level of $\mathsf{RCA}_0$ is the following version of Hilbert's *Nullstellensatz*.

**Definition 11.6.1 (Sakamoto and Tanaka [266]).** Fix $n, m \geqslant 1$. $\mathsf{NSS}_{n,m}$ is the following statement: for all $p_0, \ldots, p_{m-1} \in \mathbb{C}[x_0, \ldots, x_{n-1}]$ having no common zeroes there exists $q_0, \ldots, q_{m-1} \in \mathbb{C}[x_0, \ldots, x_{n-1}]$ such that $\sum_{i<m} p_i q_i = 1$.

**Theorem 11.6.2 (Sakamoto and Tanaka [266]).** $\mathsf{RCA}_0$ *proves* $(\forall m)(\forall n)\mathsf{NSS}_{n,m}$.

The main difficulty here is that $\mathsf{NSS}_{n,m}$ speaks about real (and complex) polynomials, and $\mathsf{RCA}_0$ only understands the reals as quickly converging Cauchy sequences (hence, as sets). It can thus be quite complex to write down even basic statements about the reals in the language $\mathcal{L}_2$.

What is needed, therefore, is a satisfaction predicate for sentences about the reals that better lends itself to manipulating the reals in $\mathsf{RCA}_0$, and in particular, suffices to prove that the real numbers satisfy the usual axioms of real closed fields. The latter includes the axiom that every polynomial of odd degree has a root. For polynomials of standard length, a suitable predicate was developed by Simpson and Tanaka [290] and Tanaka and Yamazaki [307]. As pointed out by Sakamoto and Tanaka [266], this is enough to prove $(\forall m)\mathsf{NSS}_{n,m}$ in $\mathsf{RCA}_0$ for every standard $n$. The key innovation in [266] is the development of a stronger satisfaction predicate, and an accompanying soundness theorem for $\mathsf{RCA}_0$, that pushes the argument through also for $(\forall n)(\forall m)\mathsf{NSS}_{n,m}$.

Several results related to Ulm's theorem characterizing countable abelian $p$-groups lie at the level of $\mathsf{ATR}_0$. For a group $G$, let $\widehat{G}$ denote $\bigoplus_{i\in\omega} G$, i.e., the direct sum of countably many copies of $G$.

**Theorem 11.6.3 (Friedman [111]).** *The following are equivalent over* $\mathsf{RCA}_0$.

1. $\mathsf{ATR}_0$.
2. *For all p-groups* $G_0$ *and* $G_1$, *either* $G_0$ *is embeddable into* $\widehat{G}_1$ *or* $G_1$ *is embeddable into* $\widehat{G}_0$.
3. *For all p-groups G and H, there is a direct summand K of G and H such that every direct summand of G and H is embeddable into K.*
4. *For all p-groups G and H, there is a direct summand J of* $\widehat{G}$ *and* $\widehat{H}$ *such that every direct summand of* $\widehat{G}$ *and* $\widehat{H}$ *is a direct summand of J.*
5. *For every sequence* $\langle G_0, G_1, \ldots \rangle$ *of p-groups, there exist* $i < j$ *such that* $G_i$ *is embeddable in* $G_j$.
6. *For every sequence* $\langle G_0, G_1, \ldots \rangle$ *of p-groups with* $G_0 \geqslant G_1 \geqslant \cdots$, *there exist* $i < j$ *such that* $G_i$ *is embeddable in* $G_j$.

In Chapter 12, we will see another equivalence to $\mathsf{ATR}_0$, similar in shape to (2): given two countable well orders, one necessarily embeds into the other. As observed by Greenberg and Montalbán [129], both of these results can be obtained by iterating a certain kind of derivative operator along a countable well order. In fact, the authors present a general framework for results of this kind, including several further examples.

Recall that an abelian group $G$ is *divisible* if for every $a \in G$ and every $n > 0$ there is a $b \in G$ such that $nb = a$, and that $G$ is *reduced* if its only divisible subgroup is the trivial group, $\{0_G\}$. A well-known result of algebra states that every abelian group is a direct sum of a divisible and a reduced group. This provides an example of a theorem at the level of $\Pi^1_1$-$\mathsf{CA}_0$.

**Theorem 11.6.4 (Friedman, Simpson, and Smith [117], after Feferman [103]).** *The following are equivalent over* $\mathsf{RCA}_0$.

1. $\Pi^1_1$-$\mathsf{CA}_0$.
2. *For every abelian group G there exists a divisible group D and a reduced group R such that* $G = D \oplus R$.

Our discussion thus far has exposed a conspicuous lack of examples of natural theorems of algebra that would fall outside the "big five". As discussed in Section 1.4, experience has shown that the primary objects of study in algebra are of sufficiently rich structure to facilitate coding, pushing most theorems not provable in $\mathsf{RCA}_0$ into one of the other four main subsystems. But this is an empirical observation only, and Conidis [52] (see also Montalbán [220]) has initiated a program to find a counterexample: that is, a theorem known to algebraists (the miniaturization of) which is neither provable in $\mathsf{RCA}_0$ nor equivalent to any of $\mathsf{WKL}_0$, $\mathsf{ACA}_0$, $\mathsf{ATR}_0$, or $\Pi^1_1$-$\mathsf{CA}$.

An early contender for such a theorem was the classical result that every Artinian ring is Noetherian. Recall that a ring $R$ is *Artinian* if it has no infinite strictly descending sequence of ideals; $R$ is *Noetherian* if it has no infinite strictly ascending sequence of left ideals. Conidis [52] showed that the principle that every Artinian ring is Noetherian follows from $\mathsf{ACA}_0$ and implies $\mathsf{WKL}_0$ over $\mathsf{RCA}_0$, leaving open the tantalizing possibility of it lying strictly in between. In subsequent work, this was shown not to be the case.

**Theorem 11.6.5 (Conidis [53]).** *The following are equivalent over* $\mathsf{RCA}_0$.

1. $\mathsf{WKL}_0$.
2. *Every Artinian ring is Noetherian.*
3. *Every local Artinian ring is Noetherian.*
4. *Every Artinian ring is a finite direct product of local Artinian rings.*
5. *The Jacobson radical of an Artinian ring exists and is nilpotent.*

Here we recall that a ring is *local* if it has a unique maximal ideal, and the *Jacobson ideal* of a (not necessarily commutative) ring is the intersection of all its maximal ideals.

The implication from (1) to (2) in the theorem is especially noteworthy. Indeed, the standard proof of (2) only goes through in $\mathsf{ACA}_0$, so Conidis gives, in particular, a new proof of this well-known algebraic theorem. This is reminiscent of the proof in $\mathsf{WKL}_0$ of the fact that every commutative ring has a prime ideal (Theorem 11.3.1), which likewise proceeded in a very different way from the standard one. Of course, we have also seen aspects of this in Chapter 8 with Ramsey's theorem. The takeaway is that sometimes the system that can naturally formalize a classical proof coincides with the weakest system in which the theorem can be proved, but not always, and in those cases a separate, more refined argument is needed in the weaker system. In the case of (2) above, this necessitated the use of some fairly novel algebraic techniques.

Remarkably, the intuition that the theory of Noetherian rings might yield an example of an algebraic result provably distinct from each of the "big five" turns out to be correct. In recent work, Conidis [54] has identified such a principle:

**Definition 11.6.6 (Conidis [54]).** $\mathsf{NFP}$ is the following statement: every Noetherian ring has only finitely many prime ideals.

This is in turn closely related to a purely combinatorial principle.

**Definition 11.6.7 (Conidis [54]).**

1. A tree $T \subseteq 2^{<\omega}$ is *completely branching* if for all $\sigma \in T$, if $\sigma i \in T$ for some $i < 2$ then also $\sigma^\frown(1-i) \in T$.
2. $\mathsf{TAC}$ is the following statement: every $\Sigma^0_1$-definable infinite completely branching tree has an infinite antichain.

Conidis (see [54], Remark 2.4) gives a simple proof of $\mathsf{TAC}$ in $\mathsf{RCA}_0 + \mathsf{CAC}$. Fix a $\Sigma^0_1$-definable infinite completely branching $T \subseteq 2^{<\mathbb{N}}$. By Theorem 6.1.6, there is an injective function $f \colon \mathbb{N} \to 2^{<\mathbb{N}}$ with range$(f) = T$. (As usual, we write this merely as shorthand; we do not mean to assert that $T$ exists.) We can use $f$ to define a partial order $\leqslant_P$ on $\mathbb{N}$, namely $x \leqslant_P y$ if and only if $f(x) \preceq f(y)$. By $\mathsf{CAC}$, $(\mathbb{N}, \leqslant_P)$ admits an infinite chain or antichain, $S$. If $S$ is an antichain, then we are done; so suppose $S$ is a chain. Say $S = \{x_0 < x_1 < \cdots\}$, so that $f(x_0) \prec f(x_1) \prec \cdots$. For each $i$, let $b_i = f(x_{i+1})(|f(x_i)|)$, so that $f(x_i) \prec f(x_i)b_i \preceq f(x_{i+1})$. But since $T$ is completely branching, we must also have $f(x_i)^\frown(1-b_i) \in \text{range}(f)$. Now the set $\{f(x_i)^\frown(1-b_i) : i \in \mathbb{N}\}$ exists and is an infinite antichain in $T$.

Conidis [54] also established the following far less straightforward results.

**Theorem 11.6.8 (Conidis [54]).**

*1.* $\mathsf{RCA}_0 \vdash \mathsf{WKL} + \mathsf{CAC} \to \mathsf{NFP} \to \mathsf{TAC}$.
*2.* $\mathsf{RCA}_0 \vdash \text{2-}\mathsf{RAN} \to \mathsf{TAC} \to \mathsf{OPT}$.

Since $\mathsf{WKL}_0$ has an $\omega$-model consisting entirely of sets of hyperimmune free degree (Exercise 4.8.13), we immediately obtain the following:

**Corollary 11.6.9.** $\mathsf{NFP} \not\leq_\omega \mathsf{WKL}$.

On the other hand, by the low basis theorem and Corollary 8.5.9 there is also an $\omega$-model of $\mathsf{WKL} + \mathsf{CAC}$ (indeed, of $\mathsf{WKL} + \mathsf{RT}^2_2$) consisting entirely of $\text{low}_2$ sets. In particular, such a model does not contain $\emptyset'$.

**Corollary 11.6.10.** $\mathsf{TJ} \not\leq_\omega \mathsf{NFP}$.

We conclude that $\mathsf{WKL}_0 \nvdash \mathsf{NFP}$ and $\mathsf{RCA}_0 \nvdash \mathsf{NFP} \to \mathsf{ACA}_0$. In particular, NFP is not provable in $\mathsf{RCA}_0$, nor equivalent over $\mathsf{RCA}_0$ to any of the other four main subsystems of $\mathsf{Z}_2$.

Conidis (personal communication) has conjectured that NFP implies $\mathsf{WKL}_0$ over $\mathsf{RCA}_0$. As of this writing, this remains open.

*Question 11.6.11 (Conidis).* Does $\mathsf{RCA}_0 \vdash \mathsf{NFP} \to \mathsf{WKL}$?

A positive answer would give another example (along with the principle AP of the previous chapter) of a natural theorem to lie strictly between $\mathsf{ACA}_0$ and $\mathsf{WKL}_0$.

The principle TAC is of independent interest, especially alongside the various combinatorial principles discussed in Chapter 9. The bounds of Theorem 11.6.8 (2) match those established for $\mathsf{RRT}^2_2$ in Theorems 9.4.8 and 9.4.15. However, we know $\mathsf{RRT}^2_2$ implies DNR (Theorem 9.4.13), whereas CAC does not (Theorem 9.2.21). Since CAC implies TAC, as we saw above, it follows that TAC cannot imply $\mathsf{RRT}^2_2$. However, the reverse implication remains open:

*Question 11.6.12 (Conidis [54]).* Does $\mathsf{RCA}_0 \vdash \mathsf{RRT}^2_2 \to \mathsf{TAC}$?

## 11.7 Exercises

**Exercise 11.7.1.** Prove the following in $\mathsf{RCA}_0$. Let $V$ be a vector space over a field $K$ and let $B$ be a basis for $V$. Prove that for every vector $v \in V$ there is a unique finite set $F \subseteq B$ such that for each $b \in B$ there is a scalar $k_b \neq 0_K$ satisfying $v = \sum_{b \in F} k_b b$.

**Exercise 11.7.2.** Prove the following in $\mathsf{RCA}_0$. If $G$ is an abelian group and $\alpha, \beta \in G^{<\mathbb{N}}$ have the same range then $\prod_{a \in \alpha} a = \prod_{a \in \beta} a$.

**Exercise 11.7.3 (Friedman, Simpson, and Smith [117]).** Prove the following in $\mathsf{RCA}_0$. Let $K$ be a field.

1. For all $p, q \in K[x]$ there exist $d, r \in K[x]$ such that $q = pd + r$ and $\deg(r) < \deg(p)$.
2. For all $p, q \in K[x]$ with $p \neq q$ and $q \neq 0$ there exist $p^*, q^* \in K[x]$ such that $pq^* = p^*q$ and for all $r, s, t \in K[x]$, if $rs = p$ and $rt = q$ then $r = 1$.

**Exercise 11.7.4.** Show that an abelian group is orderable if and only if it has a positive cone.

**Exercise 11.7.5.** Working in $\mathsf{RCA}_0$, fix $X \in 2^{\mathbb{N}}$, and show that $\mathrm{Red}_X$ exists, and so does the set of all pairs $(w, v) \in \mathrm{Word}_X \times \mathrm{Word}_X$ such that $w$ is freely equivalent to $v$.

**Exercise 11.7.6 (Downey, Hirschfeldt, Lempp, and Solomon [76]).** Working in $\mathsf{RCA}_0$, fix $X \in 2^{\mathbb{N}}$ and define a function $\rho_X$ on $\mathrm{Word}_X$ recursively as follows:

- $\rho_X(1_X) = 1_X$,
- $\rho_X(a^\varepsilon) = a^\varepsilon$ for every $a \in A$ and $\varepsilon \in \{-1, 1\}$,
- given $w \in \mathrm{Word}_X$ with $\rho_X(w) = a_0^{\varepsilon_0} \cdots a_{k-1}^{\varepsilon_{k-1}} \in \mathrm{Word}_X$, and given $a \in A$ and $\varepsilon \in \{-1, 1\}$,

$$\rho_X(wa^\varepsilon) = \begin{cases} 1_X & \text{if } k = 1 \wedge a_{k-1} = a \wedge \varepsilon_{k-1} = -\varepsilon, \\ a_0^{\varepsilon_0} \cdots a_{k-2}^{\varepsilon_{k-2}} & \text{if } k > 1 \wedge a_{k-1} = a \wedge \varepsilon_{k-1} = -\varepsilon, \\ a_0^{\varepsilon_0} \cdots a_{k-1}^{\varepsilon_{k-1}} a^\varepsilon & \text{otherwise.} \end{cases}$$

1. Prove that $\rho_X(w) \in \mathrm{Red}_X$ for all $w \in \mathrm{Word}_X$.
2. Prove that if $w \in \mathrm{Red}_X$ then $\rho_X(w) = w$.
3. Prove that $w$ is freely equivalent to $\rho_X(w)$ for all $w \in \mathrm{Word}_X$.
4. Prove that if $w, v \in \mathrm{Word}_X$ are freely equivalent then $\rho_X(w) = \rho_X(v)$.
5. Conclude that every word $w$ over $A$ is freely equivalent to a unique reduced word, and this is $\rho_X(w)$.

**Exercise 11.7.7 (Downey, Hirschfeldt, Lempp, and Solomon [76]).** Prove in $\mathsf{RCA}_0$ that for every $X \subseteq \mathbb{N}$, the free group on $X$ is a group.

**Exercise 11.7.8 (Hatziriakou and Simpson [140]).** Prove in $\mathsf{RCA}_0$ that every orderable abelian group is torsion free.

**Exercise 11.7.9 (Solomon [297]).** Working in $\mathsf{RCA}_0$, fix an injective function $f\colon \mathbb{N} \to \mathbb{N}$. Let $G$ be the free group generated by $\{x_i : i \in \mathbb{N}\} \cup \{y_i : i \in \mathbb{N}\}$ subject to the following relations for all $i, j$:

- $x_i x_j = x_j x_i$,
- $y_i y_j = y_j y_i$,
- $y_i x_j = x_j y_i \leftrightarrow (\forall k \leqslant j)[f(k) \neq i]$.

1. Prove by induction on the length of words that every element of $G$ is freely equivalent to a unique word of the form

$$x_{i_0}^{\varepsilon_0} \cdots x_{i_{n-1}}^{\varepsilon_{n-1}} y_{j_0}^{\delta_0} \cdots y_{j_{m-1}}^{\delta_{m-1}} z,$$

where $i_0 < \cdots < i_{n-1}$, $j_0 < \cdots < j_{m-1}$, $\varepsilon_i \neq 0$ and $\delta_j \neq 0$ for all $i < n$ and $j < m$, and $z$ is a product of nontrivial commutators of elements of $G$. (Recall that the commutator of $w_1, w_2 \in G$ is $[w_1, w_2] = w_1^{-1} w_2^{-1} w_1 w_2$.)
2. Conclude that $G$ exists as a group in the sense of Definition 11.1.1.
3. Show that range$(f)$ is $\Sigma^0_0$ definable from $Z(G)$.
4. Conclude that the statement that for every group $G$, the center $Z(G)$ exists, is equivalent to $\mathsf{ACA}_0$.

**Exercise 11.7.10.** Working in $\mathsf{RCA}_0$, fix $X \in 2^\omega$ and $S \subseteq F_X$. Verify that if $\langle S \rangle$ exists then it is a subgroup of $F_X$.

**Exercise 11.7.11 (Conidis [54]).** Let $T \subseteq 2^{<\omega}$ be an infinite completely branching tree. A sequence $\langle T_s : s \in \omega \rangle$ is a *enumeration* of $T$ if the following properties hold.

1. $T_0 = \{\langle\rangle\}$.
2. $T_s$ is a finite subset of $T$, for all $s$.
3. For each $s$ there is a unique $\sigma \in T_s$ such that $\sigma 0, \sigma 1 \in T_{s+1} \setminus T_s$.
4. $T = \bigcup_s T_s$.

Prove in $\mathsf{RCA}_0$ that there is an enumeration for every $\Sigma^0_1$-definable infinite completely branching tree.

# Chapter 12
# Set theory and beyond

Simpson [288, p. 176] famously remarked that "$\mathsf{ATR}_0$ is the weakest set of axioms which permits the development of a decent theory of countable ordinals". We cannot easily talk about *ordinals* (equivalence classes of well orderings) as such in $\mathsf{Z}_2$, but many properties of the ordinals can be formulated in terms of specific well orderings instead. We have already seen that $\mathsf{ATR}_0$ can express many such properties quite naturally. In this chapter, we investigate additional properties of countable ordinals. We will also look at several other topics from set theory, including basic results concerning Borel sets and determinacy.

Our account here presupposes some familiarity with these topics and the basic background surrounding them. The reader who has not seen this material before may wish to consult a set theory text first, e.g., Jech [166] or Kunen [191].

## 12.1 Well orderings and ordinals

It will be convenient to lay out some terminology and notation that we will use throughout this section and the next.

**Definition 12.1.1.** Let $X$ and $Y$ be linear orderings.

1. A function $f\colon X \to Y$ is an *embedding* if it is injective and $x <_X y \leftrightarrow f(x) <_Y f(y)$ for all $x, y \in X$.
2. An embedding $f\colon X \to Y$ is a *strong embedding* if range$(Y)$ is an initial segment of $Y$ under $\leqslant_Y$.
3. *X embeds into Y*, written $X \leftrightarrows Y$, if there is an embedding $f\colon X \to Y$. If range$(f) \subseteq Y^{[<_Y y]}$ for some $y \in Y$ then we write $X \hookrightarrow Y$.
4. *X strongly embeds into Y*, written $X \leftrightarrows_s Y$, if there is a strong embedding $f\colon X \to Y$.

The assertion that $f$ is an embedding (or strong embedding) of $X$ into $Y$ is arithmetical in $X$ and $Y$. $\mathsf{RCA}_0$ can easily verify that a composition of embeddings is an

 
D. D. Dzhafarov, C. Mummert, *Reverse Mathematics*, Theory and Applications of Computability, https://doi.org/10.1007/978-3-031-11367-3_12

embedding, that a composition of strong embeddings is a strong embedding, and other basic properties.

**Definition 12.1.2.** If $S$ is a nonempty subset of $\mathbb{N}$ and $\langle X_n : n \in S \rangle$ is a sequence of linear orderings, then $\sum_{n \in S} X_n$ denotes the partial ordering $U$ with domain $\bigcup_{n \in S} X_n \times \{n\}$ and linear ordering $\leqslant_U$ defined by $\langle a, n \rangle <_U \langle b, m \rangle$ if and only if $n < m$ or $n = m$ and $a <_{X_n} b$.

If $X$ and $Y$ are linear orderings we write $X + Y$ for $\sum_{n \in \{0,1\}} X_n$, where $X_0 = X$ and $X_1 = Y$. We also write $X + 1$ for the linear $\sum_{n \in \{0,1\}} X_n$ where $X_0 = X$ and $X_1$ is the linear ordering with domain $\{0\}$ and $\leqslant_{X_1} = \{\langle 0, 0 \rangle\}$.

Clearly, $\bigcup_{n \in S} X_n \times \{n\}$ is always linear ordering. We also have the following, which is a straightforward consequence of the definitions.

**Lemma 12.1.3.** *The following is provable in* $\mathsf{RCA}_0$.

1. *If $S$ is a nonempty subset of $\mathbb{N}$, and $\langle X_n : n \in S \rangle$ is a sequence of well orderings, then $\sum_{n \in S} X_n$ is a well ordering.*
2. *If $X_0$, $X_1$, and $Z$ are well orderings with $X_1 \neq \emptyset$, and if $Y = X_0 + X_1$ and $Y \hookrightarrow\!\!\!\!\!\!\!\!\!\;=\; Z$, then $X_0 \hookrightarrow Z$.*

## 12.1.1 The $\Sigma^1_1$ separation principle

Our starting point is to prove Theorem 5.8.20, which states that the following statements are equivalent over $\mathsf{RCA}_0$.

1. $\mathsf{ATR}_0$.
2. The $\Sigma^1_1$ separation principle (Definition 5.8.19).

Recall that, if $T \subseteq \omega^{<\omega}$ is an infinite tree, then the *rank* of an element $\alpha \in T$ is defined to be

$$\mathrm{rk}_T(\alpha) = \sup\{\mathrm{rk}_T(\beta) + 1 : \beta \in T \wedge \beta > \alpha\},$$

and the *rank of $T$* is then defined to be $\mathrm{rk}(T) = \mathrm{rk}_T(\langle\rangle)$. This definition is difficult to formalize in $\mathsf{RCA}_0$ (or even $\mathsf{ACA}_0$). Since the rank of a well founded tree can be an infinite ordinal, that definition seems to require at least $\mathsf{ATR}_0$ to state and work with. So, instead, we make do with the following.

**Definition 12.1.4 (Rank of a tree).** The following definition is made in $\mathsf{RCA}_0$. Fix a well founded tree $T \subseteq \mathbb{N}^{<\mathbb{N}}$ and a well ordering $X$. We write $\mathrm{rk}(T) \preceq X$ if there is a function $f : T \to X$ such that $(\forall \alpha, \beta \in T)[\alpha \prec \tau \to f(\alpha) >_X f(\tau)]$.

The following says that $\mathsf{ATR}_0$ basically understands this to be the "correct" definition of rank.

**Lemma 12.1.5.** *The following is provable in* $\mathsf{ATR}_0$. *Fix a well founded tree $T \subseteq \mathbb{N}^{<\mathbb{N}}$ and a well ordering $X$ such that $\mathrm{rk}(T) \preceq X$. Then there is a function $f : T \to X$ such that for all $\alpha \in T$,*

$$f(\alpha) = \sup\{S_X(f(\beta)) : \beta \in T \wedge \beta > \alpha\},$$

*where $S_X(x)$ denotes the successor $x \in X$ under $\leqslant_X$.*

The proof is left to Exercise 12.5.1. Another important property of Definition 12.1.4 is the following.

**Lemma 12.1.6.** *The following is provable in* $\mathsf{RCA}_0$. *Fix a well founded tree $T \subseteq \mathbb{N}^{<\mathbb{N}}$. Then* $\mathrm{rk}(T) \preceq \mathrm{KB}(T)$.

*Proof.* Recall from Proposition 5.8.8 that $\mathrm{WF}(T)$ implies $\mathrm{WO}(\mathrm{KB}(T))$. By the definition of the Kleene–Brouwer ordering, the identity map $T \to \mathrm{KB}(T)$ witnesses that $\mathrm{rk}(T) \preceq \mathrm{KB}(T)$. □

The next lemma we will need is the following, showing essentially that $\mathsf{ATR}_0$ suffices to prove the existence of initial segments of Kleene's $\mathcal{O}$.

**Lemma 12.1.7.** *The following is provable in* $\mathsf{ATR}_0$. *Fix $X$ such that* $\mathrm{WO}(X)$. *For every set $S$, the set*

$$\mathcal{O}^S_X = \{e \in \mathbb{N} : \Phi^S_e \text{ codes a linear ordering } Y_e \rightrightarrows X\}$$

*exists.*

*Proof.* In this proof, given a linear order $Y$ and $y \in Y$, write $Y \upharpoonright y$ in place of $Y^{[<_Y y]}$ for ease of notation. $\mathsf{ATR}_0$ can prove the existence of the set $Z$ of all $e \in \mathbb{N}$ such that $\Phi^S_e$ codes a linear ordering $Y_e$ and

$$(\forall y \in Y_e)(\exists x \in X)[Y_e \upharpoonright y \rightrightarrows X \upharpoonright x].$$

(See Exercise 12.5.2.) We claim that $Z = \mathcal{O}^S_X$.

If $e \in \mathcal{O}^S_X$, we may fix an embedding $f \colon Y_e \to X$. Then for each $y \in Y_e$, $Y_e \upharpoonright y \rightrightarrows X \upharpoonright f(y)$. Thus, $e \in Z$. (This is provable even in $\mathsf{RCA}_0$.)

Conversely, fix $e \in Z$. Then $Y_e$ must be a well ordering. For suppose we had $y_0 >_{Y_e} y_1 >_{Y_e} \cdots$ for some $y_0, y_1, \ldots \in Y_e$. Yet $f \colon Y_e \upharpoonright y_0 \to X$ be an embedding. Then $f(y_1) >_X f(y_2) >_X \cdots$ would be an infinite descending sequence in $X$, contradicting that $X$ is a well ordering. So, we may apply arithmetic transfinite recursion along $Y_e$. In particular, for each $y \in Y_e$, we can define $f_y \colon Y_e \upharpoonright y \to X$ as follows: if $y = 0_{Y_e}$ then $f_y = \varnothing$; if $y$ is a limit under $\leqslant_{Y_e}$ then $f_y = \bigcup_{z <_{Y_e} y} f_z$; if $y$ is the successor under $\leqslant_{Y_e}$ of $w \in Y_e$ then $f_y = f_w \cup \{\langle w, m\rangle\}$ for the $\leqslant_X$-least $m \in X \smallsetminus \mathrm{range}\, f_w$, or $\varnothing$ if no such $m$ exists.

We claim that $f_y \neq \varnothing$ for all $y >_{Y_e} 0_Y$. If not, then by arithmetical transfinite induction, we may fix the $\leqslant_{Y_e}$-least such $y$. By construction, $y$ must be a successor under $\leqslant_{Y_e}$, say of $w \in Y_e$. Our supposition thus implies that range $f_w = X$, so $f_w$ is an isomorphism. Fix any embedding $h \colon Y_e \upharpoonright y \to X$. Then $g = f_z^{-1} \circ h$ is an embedding $Y_e \upharpoonright y \to Y_e \upharpoonright z$, which is impossible (see Exercise 12.5.3). This proves the claim.

By induction along $Y_e$, we find that each $f_y$ is a strong embedding. Letting $f = \bigcup_{y \in Y_e} f_y$ we thus obtain an embedding of $Y_e$ into $X$. We conclude that $e \in \mathcal{O}^S_X$, as was to be shown. □

The final ingredient we will need to prove Theorem 5.8.20 is the following construction of a so-called *double descent tree*.

**Definition 12.1.8.** If $X, Y$ are linear orderings, then $X * Y$ is the set containing $\langle\rangle$ and all finite strings of the form

$$\langle\langle x_0, y_0\rangle, \ldots, \langle x_{n-1}, y_{n-1}\rangle\rangle \in \omega^{<\omega},$$

where $n > 0$, $x_0 <_X \cdots <_X x_{n-1}$, and $y_0 <_Y \cdots <_Y y_{n-1}$.

The basic properties of this definition, listed below, are easy to verify. They are also straightforward to formalize in $\mathsf{RCA}_0$.

**Lemma 12.1.9.** *Let X and Y be linear orderings..*

1. *$X * Y$ is a tree.*
2. *If $\neg\,\mathrm{WF}(X * Y)$ then $\neg\,\mathrm{WO}(X)$ and $\neg\,\mathrm{WO}(Y)$.*
3. *If $\neg\,\mathrm{WO}(X)$ then $Y \hookrightarrow X * Y$.*
4. *If $\mathrm{WO}(X)$ and $\mathrm{WO}(Y)$ and $X \hookrightarrow Y$, then $X \hookrightarrow X * Y$.*

Since $X * Y$ and $Y * X$ are clearly isomorphic, all occurrences of $X * Y$ above could be replaced by $Y * X$. With reference to (4), note that $X$ need *not* embed into $X * Y$ if it does not embed into $Y$.

We are ready to assemble the pieces to prove the theorem.

*Proof (of Theorem 5.8.20).* (1) $\to$ (2): We argue in $\mathsf{ATR}_0$. Let $\varphi(x)$ and $\psi(x)$ be $\Sigma^1_1$ formulas such that $(\forall x)[\varphi(x) \to \neg\psi(x)]$. We must prove that there is a separating set for $\varphi$ and $\psi$, meaning a set $Z$ such that $(\forall x)[\varphi(x) \to x \in Z]$ and $(\forall x)[\psi(x) \to x \notin X]$.

By Corollary 5.8.4, there exist sequences of trees $\langle T_n : n \in \mathbb{N}\rangle$ and $\langle S_n : n \in \mathbb{N}\rangle$ such that for all $n \in \mathbb{N}$,

$$\varphi(n) \leftrightarrow \neg\,\mathrm{WF}(T_n)$$

and

$$\psi(n) \leftrightarrow \neg\,\mathrm{WF}(S_n).$$

Thus, our assumption that $\varphi$ and $\psi$ are disjoint means that for all $n$,

$$\mathrm{WF}(T_n) \vee \mathrm{WF}(S_n). \tag{12.1}$$

By (12.1) and Lemma 12.1.9 (2), it follows that $\mathrm{KB}(T_n) * \mathrm{KB}(S_n)$ is well founded for all $n$. And by Lemma 12.1.6 and Lemma 12.1.9 (3), if $\varphi(n)$ holds then necessarily $\mathrm{rk}(S_n) \preceq \mathrm{KB}(T_n) * \mathrm{KB}(S_n)$.

Let $U = \sum_n \mathrm{KB}(T_n) * \mathrm{KB}(S_n)$. By Lemma 12.1.3, $\mathrm{WO}(U)$. Clearly, $\mathrm{KB}(T_n) * \mathrm{KB}(S_n) \hookrightarrow U$ for every $n$. Hence, by the discussion above, if $\varphi(n)$ holds then $\mathrm{rk}(S_n) \preceq U$. On the other hand, if $\psi(n)$ holds then $\neg\,\mathrm{WF}(S_n)$, so we cannot have $\mathrm{rk}(S_n) \preceq U$ (or $\mathrm{rk}(S_n) \preceq X$, for any well ordering $X$). Thus, $Z = \{n : S_n \preceq Z\}$ is a separating set for $\varphi$ and $\psi$. And $Z$ exists by applying Lemma 12.1.7 to $S = \langle S_n : n \in \mathbb{N}\rangle$ and noting that $X = O^S_X$.

(2) → (1): First, notice that $\mathsf{RCA}_0$ together with the $\Sigma^1_1$ separation principle implies $\Sigma^0_1$ comprehension. Indeed, given any $\Sigma^0_1$ formula $\varphi(x)$ both it and $\psi(x) \equiv \neg\varphi(x)$ are $\Sigma^1_1$, and a separating set for $\varphi$ and $\psi$ is precisely $\{x \in \mathbb{N} : \varphi(x)\}$. Hence, in the rest of the proof we may argue over $\mathsf{ACA}_0$.

Fix $X$ and $L$ and assume $\mathrm{WO}(L)$. We must show that $(\exists Z)[Z = X^{(L)}]$. For a set $Y$ and $n \in L$, let $Y^{[n]} = \{y \in \mathbb{N} : \langle n, y\rangle \in Y\}$ and $Y^{[<_L n]} = \bigcup_{m <_L n} Y^{[m]}$. Notice that for all sets $Y$ and $V$, if

$$Y^{[0_L]} = X \wedge (\forall n <_L x)[n >_L 0_L \to Y^{[n]} = Y^{[<_L n]'}]$$

and

$$V^{[0_L]} = X \wedge (\forall n <_L x)[n >_L 0_L \to V^{[n]} = V^{[<_L n]'}],$$

then $Y^{[<_L x]} = V^{[<_L x]}$. This is proved easily by arithmetical transfinite induction along $L$. (Recall that, by Exercise 5.13.19, $\mathsf{ACA}_0$ proves arithmetical transfinite induction.)

Define $\varphi(x, y)$ and $\psi(x, y)$ to be the $\Sigma^1_1$ formulas

$$(\exists Y)[Y^{[0_L]} = X \wedge (\forall n <_L x)[n >_L 0_L \to Y^{[n]} = Y^{[<_L n]'}] \wedge y \in Y^{[<_L x]'}]$$

and

$$(\exists Y)[Y^{[0_L]} = X \wedge (\forall n <_L x)[n >_L 0_L \to Y^{[n]} = Y^{[<_L n]'}] \wedge y \notin Y^{[<_L x]'}],$$

respectively. Then for all $x, y$, we must have $\neg\varphi(x, y) \vee \neg\psi(x, y)$. For suppose $\varphi(x, y)$ and $\psi(x, y)$ were both true, as witnessed by $Y$ and $V$, respectively. Then by the observation above, we would have $Y^{[<_L x]} = V^{[<_L x]}$ and hence also $Y^{[<_L x]'} = V^{[<_L x]'}$. But then this set would both contain and not contain $y$, a contradiction.

We may consequently apply the $\Sigma^1_1$ separation principle to obtain a set $Z$ such that for all $x, y$, if $\varphi(x, y)$ then $\langle x, y\rangle \in Z$, and if $\psi(x, y)$ then $\langle x, y\rangle \notin Z$. Another arithmetical transfinite induction along $L$ now establishes that for all $n \in L$, $Z^{[n]} = Z^{[<_L n]'}$. Hence, $Z = X^{(L)}$, which is what we wanted. □

### 12.1.2 Comparability of well orderings

We now turn to what is arguably one of the most fundamental properties of ordinals: any two ordinals are *comparable*, meaning one is an initial segment of the other. Using Definition 12.1.1, we can formalize this in two ways.

**Definition 12.1.10.**

1. *The weak comparability of well orderings* (WCWO) is the statement that for all $X, Y$, if $\mathrm{WO}(X)$ and $\mathrm{WO}(Y)$, then either $X \hookrightarrow Y$ or $Y \hookrightarrow X$.
2. *The strong comparability of well orderings* (SCWO) is the statement that for all $X, Y$, if $\mathrm{WO}(X)$ and $\mathrm{WO}(Y)$, then either $X \hookrightarrow_{\mathrm{s}} Y$ or $Y \hookrightarrow_{\mathrm{s}} X$.

In this section, we will show that these principles are equivalent to each other and to $\mathsf{ATR}_0$. We will need the following two preliminary results.

**Proposition 12.1.11 (Friedman; see Hirst and Friedman [114]).** $\mathsf{RCA}_0 + \mathsf{WCWO} \vdash \mathsf{ACA}_0$.

*Proof.* We argue in $\mathsf{RCA}_0$. Fix an injective function $f\colon \mathbb{N} \to \mathbb{N}$. Let $X$ be the set of all triples $\langle y, x, z\rangle$ such that either $x = z = 0$, or $f(x) = y$ and $z \leqslant x$. Let $\leqslant_X$ be the lexicographic ordering of $X$. Using $\mathsf{I}\Sigma^0_1$, we can verify that $\mathsf{WO}(X)$. Let $Y = \mathbb{N} + 1$. For ease of notation, we may identify the largest element of $Y$ under $\leqslant_Y$ with 0, and identify $\{i \in Y : i <_Y 0\}$ under $\leqslant_Y$ with the positive integers under their usual ordering. We clearly have $\mathsf{WO}(Y)$.

By $\mathsf{WCWO}$, either $X \hookrightarrow Y$ or $Y \hookrightarrow X$. We claim the latter is impossible. For suppose otherwise, and fix an embedding $g\colon Y \to X$. For each $i \in \mathbb{N}$, write $g(i) = \langle y_i, x_i, z_i\rangle$. Since $g(1) <_X g(2) <_X \cdots <_X g(0)$ and $\leqslant_X$ is the lexicographic order, we must have $y_1 \leqslant l_2 \leqslant \cdots \leqslant y_0$. By bounded $\Sigma^0_1$ comprehension, the set $F = \{y_i : i > 0\}$ exists and so is finite. Choose an $i > 0$ so that $y_i = \max F$. Then $y_j = y_i$ for all $j \geqslant i$. If $x_j = 0$ for all $j \geqslant i$ then we must also have $z_j = 0$ by definition of $X$, which contradicts the fact that $g(i) <_X g(i+1) <_X \cdots$.

So, without loss of generality, suppose $x_i > 0$. Then $x_j \geqslant x_i > 0$ for all $j \geqslant i$, hence $f(x_j) = y_j = y_i$ by definition of $X$, so actually $x_j = x_i$ since $f$ is injective. This means that $z_j \leqslant x_j = x_i$ for all $j \geqslant 1$. But then we cannot have $z_i < z_{i+1} < \cdots$ by Proposition 6.2.7, so again we obtain a contradiction to the fact that $g(i) <_X g(i+1) <_X \cdots$.

We conclude that $X \hookrightarrow Y$, say via $h\colon X \to Y$. Now if $f(x) = y$ then $\langle y, x, z\rangle <_X \langle y+1, 0, 0\rangle$ for all $z \leqslant x$. Thus

$$h(\langle y, x, 0\rangle) < \cdots < h(\langle y, x, x\rangle) < h(\langle y+1, 0, 0\rangle).$$

In particular, $x < h(\langle y+1, 0, 0\rangle)$. So we see that

$$y \in \operatorname{range}(f) \leftrightarrow (\exists x < h(\langle y+1, 0, 0\rangle))[f(x) = y],$$

and therefore $\operatorname{range}(f)$ exists. □

We leave the proof of the second lemma to the next subsection. Recall the $\Sigma^1_1$ axiom of choice introduced in Chapter 10, which is the scheme containing

$$(\forall n)(\exists X)\psi(n, X) \to (\exists Y)(\forall n)\psi(n, Y^{[n]})$$

for all $\Sigma^1_1$ formulas $\psi$.

**Proposition 12.1.12 (Friedman; see Hirst and Friedman [114]).** $\mathsf{RCA}_0 + \mathsf{WCWO}$ *proves the $\Sigma^1_1$ axiom of choice.*

The proof is combinatorially involved, but does not feature any elements we have not already encountered. We will see that the $\Sigma^1_1$ axiom of choice is strictly weaker than $\mathsf{ATR}_0$ in Corollary 12.1.14 and Proposition 12.1.15 at the end of this section.

We can now prove the comparability theorem.

**Theorem 12.1.13 (Friedman; see Hirst and Friedman [114]).** *The following statements are equivalent over* $\mathsf{RCA}_0$.

1. $\mathsf{ATR}_0$.
2. SCWO.
3. WCWO.

*Proof.* (1) → (2): We argue in $\mathsf{ATR}_0$. Fix $X$ and $Y$ such that WO($X$) and WO($Y$). The proof is similar to that of Lemma 12.1.7, only this time we do not know which of $X$ or $Y$ should embed into the other. We write $X \upharpoonright x$ and $Y \upharpoonright y$ for initial segments under $\leqslant_X$ and $\leqslant_Y$, as in that proof. By arithmetical transfinite recursion along $Y$, define $f_y : Y \upharpoonright y \to X$ as follows: if $y = 0_Y$ then $f_y = \emptyset$; if $y$ is a limit under $\leqslant_Y$ then $f_y = \bigcup_{z <_X y} f_z$; and if $y$ is the successor under $\leqslant_X$ of some $w \in Y$ then $f_y = f_w \cup \{\langle w, m \rangle\}$ for the $\leqslant_Y$-least $m \in X \setminus \text{range } f_w$, or $\emptyset$ if no such $m$ exists. We consider two cases.

*Case 1:* $f_y \neq \emptyset$ *for all* $y >_Y 0_Y$. By induction along $Y$, each $f_y$ is a strong embedding. Letting $f = \bigcup_{y \in Y} f_y$ yields an embedding of $Y$ into $X$.

*Case 2:* $f_y = \emptyset$ *for some* $y >_Y 0_Y$. By arithmetical transfinite induction, fix the least such $y$. By construction, $y$ must be a successor under $\leqslant_Y$, say of $w \in Y$. By hypothesis, range $f_w = X$, so $f_w$ is an isomorphism. Then $f_w^{-1}$ is an embedding of $X$ into $Y$.

(2) → (3): Clear.

(3) → (1): Assume WCWO. By Proposition 12.1.11, we may argue over $\mathsf{ACA}_0$. By Theorem 5.8.20, it suffices to prove the $\Sigma^1_1$ separation principle. Let $\varphi(x)$ and $\psi(x)$ be $\Sigma^1_1$ formulas such that for all $x$, $\neg\varphi(x) \vee \neg\psi(x)$. By Corollary 5.8.4, there exist sequences of trees $\langle T_n : n \in \mathbb{N}\rangle$ and $\langle S_n : n \in \mathbb{N}\rangle$ such that for all $n \in \mathbb{N}$,

$$\varphi(n) \leftrightarrow \neg\, \mathrm{WF}(T_n)$$

and

$$\psi(n) \leftrightarrow \neg\, \mathrm{WF}(S_n).$$

We claim that for all $n$, either KB($T_n$) or KB($S_n$) embeds into KB($T_n$) $*$ KB($S_n$). If $\varphi(n)$ or $\psi(n)$ holds, then this follows by Lemma 12.1.9 (3) just as in the proof of Theorem 5.8.20 above. If neither $\varphi(n)$ nor $\psi(n)$ holds, then KB($T_n$) and KB($T_n$) are both well orderings. Thus one embeds into the other by WCWO, and hence at least one embeds into KB($T_n$) $*$ KB($S_n$) by Lemma 12.1.9 (4).

Let $\theta(n, f)$ be the formula asserting that $f$ is an embedding from KB($T_n$) or KB($S_n$) into KB($T_n$)$*$KB($S_n$). Note that $\theta$ is arithmetical, hence $\Sigma^1_1$. By the argument just given we have $(\forall n)(\exists f)\theta(n, f)$. By Proposition 12.1.12 we may apply the $\Sigma^1_1$ axiom of choice. This yields a sequence $\langle f_n : n \in \mathbb{N}\rangle$ such that $\theta(n, f_n)$ holds for every $n$. Let $Z$ be the set of all $n$ such that $f_n$ is an embedding from KB($S_n$) into KB($T_n$) $*$ KB($S_n$). Then $Z$ exists by arithmetical comprehension, and we claim it is a separating set for $\varphi$ and $\psi$. Indeed, if $\varphi(n)$ holds then $\neg$ WF($T_n$), so there is no embedding from KB($T_n$) into (the well founded tree) KB($T_n$) $*$ KB($S_n$). Thus,

$n \in Z$. Similarly, if $\psi(n)$ holds then there is no embedding from $\mathrm{KB}(S_n)$ into $\mathrm{KB}(T_n) * \mathrm{KB}(S_n)$, and hence $n \notin Z$. □

One consequence is the following, which follows immediately from Proposition 12.1.12 and Theorem 12.1.13.

**Corollary 12.1.14.** $\mathsf{ATR}_0$ *proves the* $\Sigma^1_1$ *axiom of choice.*

Of course, we have yet to prove Proposition 12.1.12, and we do this in the next section. For completeness, we point out that the above corollary cannot be improved to an equivalence. Recall the $\omega$-model HYP from Chapter 5, and the fact that $\mathsf{ATR}_0$ fails in this model (Proposition 5.8.21).

**Proposition 12.1.15 (Kreisel [186]).** *The* $\omega$*-model* HYP *satisfies the* $\Sigma^1_1$ *axiom of choice. Hence, the* $\Sigma^1_1$ *axiom of choice does not imply* $\mathsf{ATR}_0$ *over* $\mathsf{RCA}_0$.

*Proof.* Kreisel [186] showed that if $\varphi(x, y)$ is a $\Pi^1_1$ formula then

$$(\forall x)(\exists y)\varphi(x, y) \rightarrow (\exists h \in \mathrm{HYP})(\forall x)\varphi(x, h(x)). \tag{12.2}$$

See Sacks [264, Lemma II.2.6] for a proof. Another standard fact of hyperarithmetical theory is that for every $b$ in Kleene's $O$, the predicates $x \in H_b$ and $x \notin H_b$ are both uniformly $\Pi^1_1$ in $b$ (see [264, Lemma II.1.3]).

Now consider any $\Sigma^1_1$ formula $\varphi(n, X)$. By Kleene's normal form theorem (Theorem 5.8.2), we can write this as $(\exists f \in \mathbb{N}^{\mathbb{N}})(\forall k)\theta(X \upharpoonright k, f \upharpoonright k, n)$ for some arithmetical formula $\theta$. Suppose

$$\mathrm{HYP} \vDash (\forall n)(\exists X)\varphi(n, X). \tag{12.3}$$

Then, for each $n$, there is a hyperarithmetical set $X$ and hyperarithmetical function $f$ such that $(\forall k)\theta(X \upharpoonright k, f \upharpoonright k, n)$. Then (12.3) can be rewritten as

$$(\forall n)(\exists b, e_0, e_1)\varphi(b, e_0, e_1, n),$$

where $\varphi$ is the formula

$$b \in O \wedge \Phi^{H_b}_{e_0} \in 2^\omega \wedge \Phi^{H_b}_{e_1} \in \omega^\omega \wedge (\forall k)\theta(\Phi^{H_b}_{e_0} \upharpoonright k, \Phi^{H_b}_{e_1} \upharpoonright k, n).$$

By the remarks above, $\varphi$ is $\Pi^1_1$. So, by Kreisel's theorem, we may fix $h$ satisfying $(\forall n)[h(n) = \langle b, e_0, e_1\rangle \wedge \varphi(b, e_0, e_1, n)]$. Since $h \in \mathrm{HYP}$, also $\langle X_n : n \in \omega\rangle \in \mathrm{HYP}$, where $X_n = \Phi^{H_b}_{e_0}$ for the $b$ and $e_0$ so that $h(n) = \langle b, e_0, e_1\rangle$. Now, by construction we have $\varphi(n, X_n)$ for all $n$. Since $\varphi$ was arbitrary, we conclude that HYP satisfies the $\Sigma^1_1$ axiom of choice. □

### 12.1.3 Proof of Proposition 12.1.12

Our aim now is to fill in the gap left in our proof of Theorem 12.1.13 by showing how to derive the $\Sigma^1_1$ axiom of choice from WCWO. There are three main definitions we will need: *indecomposable well orderings*, the *explosion* of a tree, and the *wedge product*. We will also need a series of technical lemmas concerning these notions. Some of these are straightforward and left to the exercises. Others require a somewhat delicate uses of WCWO.

**Definition 12.1.16.** Let $X$ be a well ordering.

1. A *coinitial segment* of $X$ is the restriction of $\leqslant_X$ to $\{x \in X : x >_X y\}$ for some $y \in X$. We call this the coinitial segment *above* $x$.
2. $X$ is *indecomposable* if $X \leftrightarrows Y$ for every coinitial segment $Y$ of $X$.

**Lemma 12.1.17.** *The following is provable in* $\mathsf{RCA}_0$*. Suppose $X$, $Y$, and $Z$ are well orderings, $X$ is indecomposable, and $X \hookrightarrow Y + Z$. Then either $X \leftrightarrows Y$ or $X \hookrightarrow Z$.*

The proof is left to Exercise 12.5.5.

**Lemma 12.1.18.** *The following is provable in* $\mathsf{RCA}_0$+WCWO*. Suppose $X_0, \ldots, X_{m-1}$ are well orderings such that $X_{m-1} \hookrightarrow \cdots \hookrightarrow X_1 \hookrightarrow X_0$. Then $X_{m-1} \hookrightarrow X_0$.*

*Proof.* We argue in $\mathsf{RCA}_0$. Assume WCWO. For each $n < m-1$, let $Y_n = X_{n+1} + X_n$, and let $Y_{m-1} = X_{m-1}$. Then, let $W_0 = \sum_{n<m} Y_n$, and let $W_1 = (\sum_{n<m} X_n) + 1$. An element of $W_0$ is thus formally a pair $\langle y, n\rangle$, where $n < m$ and $y \in Y_n$. For ease of notation, we will refer to this element simply by $y$, and in this way identify each $Y_n \times \{n\} \subseteq W_0$ with $Y_n$. Similarly, we identify each $(X_n \times \{n\}) \times \{0\} \subseteq W_1$ with $X_n$, each $X_{n+1} \times \{0\} \subset Y_n$ with $X_{n+1}$, etc. The context will make the usage clear.

By Lemma 12.1.3, each of $W_0$ and $W_1$ is a well ordering. Hence, by WCWO, we have one of the following.

*Case 1:* $W_0 \leftrightarrows W_1$. Fix a witnessing embedding $f\colon W_0 \to W_1$. We claim that for all $n < m$, $f(0_{Y_n}) \geqslant_{W_1} 0_{X_n}$ and $f(Y_n) \subseteq X_n$. Note that these are both $\Pi^0_1$ statements in $n$ (using $f$, $W_0$, and $W_1$ as parameters), so we can proceed by induction (or really, $\mathsf{L}\Pi^0_1$).

We begin with the first claim. This is obvious for $n = 0$, since $0_{X_0}$ is also the least element of $W_1$. If the claim fails, we can find the least positive $n < m$ such that $f(0_{Y_n}) <_{W_1} 0_{X_n}$. Since $f$ is order preserving, $y \leqslant_{W_0} 0_{Y_n}$ for every $y \in Y_{n-1}$, and $f(0_{Y_{n-1}}) \geqslant_{W_1} 0_{X_{n-1}}$ by assumption, it follows that $Y_{n-1} \cup \{0_{Y_n}\} \leftrightarrows X_{n-1}$ via $f$. But $Y_{n-1} = X_n + X_{n-1}$, so this implies that $X_{n-1} \hookrightarrow X_{n-1}$,which is impossible (Exercise 12.5.3). So the first claim holds.

Now for the second claim. In light of the first, we have this for $n = m-1$ since $Y_{m-1} = X_{m-1}$. So if the claim fails, we can fix the least positive $n < m$ such that $f(Y_n) \subseteq X_n$. In particular, $f(0_{Y_n}) \in X_n$, and by the first claim we have $f(0_{Y_{n-1}}) \geqslant_{W_1} 0_{X_{n-1}}$. Since $f$ is order preserving and $0_{Y_{n-1}} \leqslant_{W_0} y <_{W_0} 0_{Y_n}$ for all $y \in Y_{n-1}$, the fact that $f(0_{Y_{n-1}}) \nsubseteq X_{n-1}$ means that there is a coinitial segment $Z$ of

$Y_{n-1}$ such that $f(Z) \subseteq X_n$. Thus $f$ witnesses that $Z \trianglelefteq X_n$. Since $Y_{n-1} = X_n + X_{n-1}$, we may take $Z$ to be a coinitial segment of $X_{n-1}$. But $X_{n-1}$ is indecomposable, so then $X_{n-1} \trianglelefteq Z \trianglelefteq X_n \hookrightarrow X_{n-1}$, which is impossible. So the second claim holds, too.

By the second claim, $f$ witnesses that $Y_n \trianglelefteq X_n$ for all $n < m$. For $n < m - 1$, we have $Y_n = X_{n+1} + X_n$, hence $f$ also witnesses that $X_{n+1} \hookrightarrow X_n$. For each $n < m-1$, let $f_n$ denote the restriction of $f$ to $X_{n+1}$ (inside $Y_n$). Then the composition $h_0 \circ \cdots \circ h_{m-2}$ witnesses that $X_{m-1} \hookrightarrow X_0$.

*Case 2:* $W_1 \trianglelefteq W_0$. We will show this case cannot hold. Fix an embedding $g\colon W_1 \to W_0$. We claim that for each $n < m$, $g(0_{X_n}) \geqslant_{W_0} 0_{Y_n}$. We proceed by induction. The claim is obvious for $n = 0$, since $0_{Y_0}$ is also the $\leqslant_{W_0}$-least element of $W_0$. If the claim fails, we may consequently fix the least positive $n < m$ such that $g(0_{X_n}) <_{W_0} 0_{Y_n}$. Then also $g(x) <_{W_0} 0_{Y_n}$ for every $x \in X_{n-1}$. Since $g(x) \geqslant_{W_0} 0_{Y_{n-1}}$ for every $x \in X_{n-1}$ by assumption, $g$ witnesses that $X_{n-1} \cup \{0_{X_n}\}$ embeds into $Y_{n-1}$. Thus $X_{n-1} \hookrightarrow Y_{n-1} = X_n + X_{n-1}$. By Lemma 12.1.17, and the fact that $X_n \hookrightarrow X_{n-1}$, we conclude that $X_{n-1} \hookrightarrow X_{n-1}$, which is impossible. The claim is proved.

Now, in particular, we have $g(0_{X_{m-1}}) \geqslant_{W_0} 0_{Y_{m-1}}$. Thus, $g$ witnesses that $X_{m-1} \cup \{1_{W_1}\}$ embeds into $Y_{m-1} = X_{m-1}$, where $1_{W_1}$ is the $\leqslant_{W_1}$-largest element of $W_1$. But this means that $X_{m-1} \hookrightarrow X_{m-1}$, which is again impossible. This completes the proof. □

**Lemma 12.1.19.** *The following is provable in* $\mathsf{RCA}_0 + \mathsf{WCWO}$. *There is no sequence* $\langle X_n : n \in \mathbb{N}\rangle$ *of indecomposable well orderings such that for each n,* $X_{n+1} \hookrightarrow X_n$.

*Proof.* Assume WCWO and suppose, towards a contradiction, that we had such a sequence, $\langle X_n : n \in \mathbb{N}\rangle$. As in the previous lemma, let $Y_n = X_{n+1} + X_n$ for all $n$, and let $W_0 = \sum_{n\in\mathbb{N}} Y_n$. Let $W_1 = (\sum_{n\in\mathbb{N}} X_n) + 1$. By Lemma 12.1.3, each of $W_0$ and $W_1$ is a well ordering, so we have the following case analysis.

*Case 1:* $W_0 \trianglelefteq W_1$. Fix an embedding $f\colon W_0 \to W_1$. As in Case 1 of the previous lemma, we can argue that $f(0_{Y_n}) \geqslant_{W_1} 0_{X_n}$ for all $n$. We claim that also $f(Y_n) \subseteq X_n$. Suppose not, as witnessed by $n$. Choose $n^*$ so that $f(0_{Y_{n+1}}) \in X_{n^*}$. Then by $\mathsf{L}\Sigma^0_1$ we may choose the largest $m \leqslant n^*$ so that $f(Y_n) \cap X_m \neq \emptyset$. Then $f(Z) \subseteq X_m$ for some coinitial segment of $Y_n$, and since $Y_n = X_{n+1} + X_n$, we may take $Z$ to be a coinitial segment of $X_n$. Since $X_n$ is indecomposable, it follows that $X_n \trianglelefteq X_m$. But $f(Y_n) \nsubseteq X_n$, thus necessarily $m > n$. But then $X_m \hookrightarrow X_n$ and so $X_n \hookrightarrow X_n$, which is impossible.

*Case 2:* $W_1 \trianglelefteq W_0$. Let $g\colon W_1 \to W_0$ be an embedding. As in Case 2 of the previous lemma, we can argue that $g(0_{X_n}) \geqslant_{W_0} 0_{Y_n}$ for all $n$. Fix $n$ so that $g(1_{W_1}) \in Y_n$, where $1_{W_1}$ is the $\leqslant_{W_1}$-largest element of $W_1$. Then

$$0_{Y_{n+1}} \leqslant_{W_0} g(0_{X_{n+1}}) \leqslant_{W_0} g(1_{W_1}) <_{W_0} 0_{Y_{n+1}},$$

which is a contradiction. □

We now add the following definition of the so-called *explosion* of a tree $T \subseteq \mathbb{N}^{<\mathbb{N}}$, which is similar to the double descent tree $T * \mathbb{N}^{<\mathbb{N}}$ (Definition 12.1.8).

**Definition 12.1.20.** Let $T$ be a subtree of $\mathbb{N}^{<\mathbb{N}}$. Then $\mathrm{E}(T)$ is the set containing $\langle\rangle$ and all finite strings of the form

$$\langle\langle x_0, k_0\rangle, \cdots \langle x_{n-1}, k_{n-1}\rangle\rangle \in \mathbb{N}^{<\mathbb{N}},$$

where $\langle x_0, \ldots, x_{n-1}\rangle \in T$ and $k_0, \ldots, k_{n-1} \in \mathbb{N}$ are arbitrary.

Clearly, $\mathrm{E}(T)$ is always a tree.

**Lemma 12.1.21.** *The following is provable in* $\mathsf{ACA}_0$. *If $T$ is a well founded subtree of $\mathbb{N}^{<\mathbb{N}}$ then $\mathrm{KB}(\mathrm{E}(T))$ is an indecomposable well ordering.*

*Proof.* We argue in $\mathsf{ACA}_0$. Fix $T$. Any path through $\mathrm{E}(T)$ has the form $f \oplus g$, where $f, g \in \mathbb{N}^{\mathbb{N}}$ and $f \in [T]$. So by assumption, $\mathrm{E}(T)$ must be well founded. Hence, $\mathrm{KB}(\mathrm{E}(T))$ is a well ordering by Proposition 5.8.8. It thus remains to show that $\mathrm{KB}(\mathrm{E}(T))$ is indecomposable. Consider any $\alpha \in \mathrm{E}(T)$ and consider the coinitial segment of $\mathrm{KB}(\mathrm{E}(T))$ above $\alpha$. Say $\alpha(0) = \langle x, k\rangle$ and fix $m$ so that $\langle 0, m\rangle > \langle x, k\rangle$ as codes (i.e., as numbers). By the definition of the pairing function, it follows that $\langle y, j\rangle > \langle x, k\rangle$ for all $y$ and all $j \geqslant m$. Let $S$ be the set of all $\beta \in \mathrm{E}(T)$ such that if $\beta(0) = \langle y, j\rangle$ then $j \geqslant m$. Then $S \subseteq Z$, and we define $f\colon \mathrm{E}(T) \to S$ as follows: for $\beta \in \mathrm{E}(T)$ with $\beta(0) = \langle y, j\rangle$, let $f(\beta) = \langle y, j+m\rangle ^\frown \beta(1)\cdots\beta(|\beta|-1)$. Now it is not difficult to verify that $f$ is an embedding, so $\mathrm{E}(T) \hookrightarrow Z$. Since $Z$ was arbitrary, we are done. □

The last technical definition we need is the following.

**Definition 12.1.22.** The following definition is made in $\mathsf{RCA}_0$. Let $\langle T_n : n \in \mathbb{N}\rangle$ be a sequences of subtrees of $\mathbb{N}^{<\mathbb{N}}$.

1. A *wedge* of $\langle T_n : n \in \mathbb{N}\rangle$ is a sequence $w = \langle \alpha_0, \ldots, \alpha_{m-1}\rangle$ for some $m \in \mathbb{N}$ where $\alpha_n \in T_n$ and $|\alpha_n| = m - n$. We say $w$ has *size m*.
2. A wedge $w^*$ *extends* a wedge $w$ if $w^* = \langle \alpha^*_0, \ldots, \alpha^*_{m^*-1}\rangle$ and $w = \langle \alpha_0, \ldots, \alpha_{m-1}\rangle$ for some $m^* \geqslant m$ and $\alpha^*_n \succeq \alpha_n$ for all $n < m^*$.
3. The *wedge product* of $\langle T_n : n \in \mathbb{N}\rangle$, denoted $\prod_{n\in\mathbb{N}} T_n$, is the set of all (codes for) sequences of the form $\langle w_0, \ldots, w_{m-1}\rangle$, where each $w_n$ is a wedge of $\langle T_n : n \in \mathbb{N}\rangle$ of size $n+1$ and $w_{n+1}$ extends $w_n$.

Clearly, $\prod_{n\in\mathbb{N}} T_n$ is itself a subtree of $\mathbb{N}^{<\mathbb{N}}$. Our final two lemmas establish basic properties of this tree.

**Lemma 12.1.23 (Friedman; see Hirst and Friedman [114]).** *The following is provable in* $\mathsf{RCA}_0$. *For each sequence $\langle T_n : n \in \mathbb{N}\rangle$ of subtrees of $\mathbb{N}^{<\mathbb{N}}$ such that $\neg\,\mathrm{WF}(\prod_{n\in\mathbb{N}} T_n)$, there exists a sequence $\langle f_n : n \in \mathbb{N}\rangle$ such that $f_n \in [T_n]$ for each n.*

The proof is left to Exercise 12.5.4.

**Lemma 12.1.24 (Friedman; see Hirst and Friedman [114]).** *The following is provable in* $\mathsf{RCA}_0 + \mathsf{WCWO}$. *If* $\langle T_n : n \in \mathbb{N}\rangle$ *is a sequence of subtrees of* $\mathbb{N}^{<\mathbb{N}}$ *such that* $\neg\,\mathrm{WF}(T_n)$ *for all n, then* $\neg\,\mathrm{WF}(\prod_{n\in\mathbb{N}} T_n)$.

*Proof.* Assume $\mathsf{WCWO}$. By Proposition 12.1.11, we can argue over $\mathsf{ACA}_0$. Let $\langle T_n : n \in \mathbb{N}\rangle$ be given such that $\neg\,\mathrm{WF}(T_n)$ for all $n$. Seeking a contradiction, assume $\mathrm{WF}(\prod_{n\in\mathbb{N}} T_n)$. Fix $p\colon \mathbb{N} \to \mathbb{N}$ so that for each $i \in \mathbb{N}$ there are infinitely many $n$ with $p(n) = i$. Then for each $s$, any path through $\prod_{n\geqslant s} T_{p(n)}$ arithmetically defines a path through $\prod_{n\in\mathbb{N}} T_n$ (using $p$ and $\langle T_n : n \in \mathbb{N}\rangle$ as parameters). Thus, we must have $\mathrm{WF}(\prod_{n\geqslant s} T_{p(n)})$. By Lemma 12.1.21, $\mathrm{KB}(\mathrm{E}(\prod_{n\geqslant s} T_{p(n)}))$ is an indecomposable well ordering. We will show that for each $s$,

$$\mathrm{KB}(\mathrm{E}(\prod_{n\geqslant s+1} T_{p(n)})) \hookrightarrow \mathrm{KB}(\mathrm{E}(\prod_{n\geqslant s} T_{p(n)})), \tag{12.4}$$

thereby contradicting Lemma 12.1.19.

So fix $s$, along with any $f \in [T_s]$. We will use $f$ to help us define an embedding $h\colon \mathrm{KB}(\mathrm{E}(\prod_{n\geqslant s+1} T_{p(n)})) \to \mathrm{KB}(\mathrm{E}(\prod_{n\geqslant s} T_{p(n)}))$. First, we define a function $g$ on the set of all pairs $\langle w, k\rangle$ where $w$ is a wedge of $\langle T_n : n \geqslant s+1\rangle$ and $k \in \mathbb{N}$. If $w$ has size $m$, then let

$$g(\langle w, k\rangle) = \langle\langle f \upharpoonright m+1, w(0), \ldots, w(m-1)\rangle, \max\{g(x) : x < \langle w, k\rangle\} + 1\rangle.$$

Note that $\langle f \upharpoonright m+1, w(0), \ldots, w(m-1)\rangle$ is a wedge of $\langle T_n : n \geqslant s\rangle$ of size $m+1$. (That is, $w(0)$ is an element of $T_{s+1}$ of length $m$, so we prepend $f \upharpoonright m+1$, an element of $T_s$ of length $m+1$.) Now define $h$ as follows. Given an element $z = \langle\langle w_0, k_0\rangle, \ldots, \langle w_{m-1}, k_{m-1}\rangle\rangle \in \mathrm{KB}(\mathrm{E}(\prod_{n\geqslant s+1} T_{p(n)}))$, let

$$h(z) = \langle\langle f(0), 0\rangle, g(\langle w_0, k_0\rangle)\rangle, \ldots, g(\langle w_{m-1}, k_{m-1}\rangle)\rangle.$$

Then $h(z) \in \mathrm{KB}(\mathrm{E}(\prod_{n\geqslant s} T_{p(n)}))$, and it is not difficult to check that $h$ defines an embedding. Moreover, by the usual properties of the pairing function, $h(z) <_{\mathrm{KB}} \langle f(0), 1\rangle$ for all $z$, so $h$ actually witnesses (12.4). □

At last, we are ready to prove Proposition 12.1.12.

*Proof (of Proposition 12.1.12).* Assume $\mathsf{WCWO}$. By Proposition 12.1.11, we can argue over $\mathsf{ACA}_0$. Let $\psi(n, X)$ be a $\Sigma^1_1$ formula such that $(\forall n)(\exists X)\psi(n, X)$. By Corollary 5.8.4, there is a sequence of trees $\langle T_n : n \in \mathbb{N}\rangle$ such that for every $n$ we have $(\forall X)[\psi(n, X) \leftrightarrow (\exists g \in \mathbb{N}^{\mathbb{N}})[X \oplus g \in [T_n]]]$. Since $\neg\,\mathrm{WF}(T_n)$ for all $n$ by assumption, so also $\neg\,\mathrm{WF}(\prod_{n\in\mathbb{N}} T_n)$ by Lemma 12.1.24. Applying Lemma 12.1.23 we find a sequence $\langle f_n : n \in \mathbb{N}\rangle$ such that $f_n \in [T_n]$ for each $n$. Writing each $f_n$ as $X_n \oplus g_n$, we have that $\psi(n, X_n)$ holds. Then $\langle X_n : n \in \mathbb{N}\rangle$ satisfies the conclusion of the $\Sigma^1_1$ axiom of choice, and the proof is complete. □

## 12.2 Descriptive set theory

We have already seen some ways we can "speak about" collections of sets and functions in $\mathcal{L}_2$, even though the language itself only has variables for numbers and sets of numbers. In this section, we survey representations for Borel and analytic subsets of Baire space and Cantor space. We frame our discussion in terms of the former, but an analogous development works for the latter.

The concept of a code for a Borel subset of $\omega^\omega$ is familiar from effective descriptive set theory. (See, e.g., Mansfield and Weitkamp [201].) Computable codes are used to define the lightface Borel hierarchy. We will use the following formulation. In $\mathsf{RCA}_0$, fix an enumeration $\langle \alpha_0, \alpha_1, \ldots \rangle$ of all elements of $\mathbb{N}^{<\mathbb{N}}$.

**Definition 12.2.1 (Borel codes).** The following definition is made in $\mathsf{RCA}_0$. A *Borel code* is a well founded tree $B \subseteq \omega^\omega$ with the property that there is a unique $m_B \in \omega$ such that $\langle m_B \rangle \in B$.

The precise nature of the coding is better explained with the help of the following definition.

**Definition 12.2.2 (Evaluation maps).** The following definition is made in $\mathsf{RCA}_0$. Let $B \subseteq \mathbb{N}^{<\mathbb{N}}$ be a Borel code, and fix any $X \in \omega^\omega$. Then $f : B \to 2$ is a *evaluation map for $B$ at $X$* if the following conditions hold for all $\alpha \in B$.

1. If $\alpha$ is a leaf of $B$, then:
   - if $\alpha(|\alpha| - 1) < 2$ then $f(\alpha) = \alpha(|\alpha| - 1)$,
   - if $\alpha(|\alpha| - 1) = 2i + 2$ for some $i \in \omega$ then $f(\alpha) = 1 \leftrightarrow \alpha_i \preceq X$,
   - if $\alpha(|\alpha| - 1) = 2i + 3$ for some $i \in \omega$ then $f(\alpha) = 1 \leftrightarrow \alpha_i \npreceq X$.
2. If $\alpha \neq \langle\rangle$ and $\alpha$ is not a leaf of $B$, then:
   - if $\alpha(|\alpha| - 1)$ is even, then $f(\alpha) = 1 \leftrightarrow (\exists x)[\alpha x \in B \wedge f(\alpha x) = 1]$,
   - if $\alpha(|\alpha| - 1)$ is odd, then $f(\alpha) = 1 \leftrightarrow (\forall x)[\alpha x \in B \wedge f(\alpha x) = 1]$.
3. $f(\langle\rangle) = f(\langle m_B \rangle)$.

*$X$ belongs to $B$*, written $X \in B$, if there is an evaluation map $f$ for $B$ at $X$ such that $f(\langle\rangle) = 1$. *$X$ does not belong to $B$*, written $X \notin B$, if there is an evaluation map $f$ for $B$ at $X$ such that $f(\langle\rangle) = 0$. Note that $\in$ here is being used as an abbreviation, not as the symbol of $\mathcal{L}_2$.

The idea is that among the leaves of $B$, those with last bit 0 code the empty set. Those with last bit 1 code $\mathbb{N}^\mathbb{N}$. And those with last bit $2i + 2 + j$ for $j$ equal to 0 or 1 code $[[\alpha_i]]$ and $\mathbb{N}^\mathbb{N} \setminus [[\alpha_i]]$, respectively. All other nonempty strings in $B$ code unions or intersections, depending as their last bit is even or odd. For example, the open set $\bigcup_{\alpha \in U} [[\alpha]]$ can be coded by $B \subseteq \mathbb{N}^{<\mathbb{N}}$ containing the strings $\langle\rangle$ and $\langle 0 \rangle$, and then for each $i$ such that $\alpha_i \in U$, the string $0^\frown(2i + 2)$. Clearly, Borel codes are in general not unique.

Given $B$ and $X$ as above, an evaluation map for $B$ at $X$ can be constructed by transfinite recursion along the rank of $B$. The following theorem formalizes this fact.

**Theorem 12.2.3 (Simpson [288]).** $\mathsf{ATR}_0$ *proves that for every Borel code* $B \subseteq \mathbb{N}^{\mathbb{N}}$ *and* $X \in \mathbb{N}^{\mathbb{N}}$, *there exists a unique evaluation map for* $B$ *at* $X$.

*Proof.* Uniqueness is actually provable in $\mathsf{ACA}_0$, and this is left to Exercise 12.5.6. To prove existence, we argue in $\mathsf{ATR}_0$. Fix $B$ and $X$. Since $\mathrm{WF}(B)$, we have by Lemma 12.1.6 that $\mathrm{rk}(B) \leq \mathrm{KB}(B)$. Let $g\colon B \to \mathrm{KB}(X)$ be the function given by Lemma 12.1.5. So

$$g(\alpha) = \sup\{S_{\mathrm{KB}(B)}(g(\beta)) : \beta \in B \wedge \beta \succ \alpha\}$$

for all $\alpha \in B$. Let $Z = \mathrm{range}(g) \smallsetminus \{g(\langle\rangle)\}$, regarded as a suborder of $\mathrm{KB}(B)$. Clearly, $\mathrm{WO}(Z)$. We define a sequence of sets $\langle F_x : x \in Z\rangle$ by arithmetical transfinite recursion along $Z$. Let $F_{0_Z}$ be the set of all pairs $\langle \alpha, v\rangle$ such that $g(\alpha) = 0_Z$ (so $\alpha$ is a leaf of $B$) and $v < 2$, with $v = 1$ if and only if $\alpha(|\alpha| - 1) = 1$, or $\alpha(|\alpha| - 1) = 2i + 2$ and $\alpha_i \preceq X$, or $\alpha(|\alpha| - 1) = 2i + 3$ and $\alpha_i \npreceq X$. For $x >_Z 0_Z$, we let $F_x$ consist of all pairs $\langle \alpha, v\rangle$ such that $g(\alpha) = x$ and $v < 2$, with $v = 1$ if and only if either $\alpha(|\alpha| - 1)$ is even and $\langle \beta, 1\rangle \in F_{g(\beta)}$ for some $\beta \succ \alpha$ of length $|\alpha| + 1$, or $\alpha(|\alpha| - 1)$ is even and $\langle \beta, 1\rangle \in F_{g(\beta)}$ for all $\beta \succ \alpha$ of length $|\alpha| + 1$. Let $v_B$ be such that $\langle\langle m_B\rangle, v_B\rangle \in F_{g(\langle m_B\rangle)}$. Then $f = \bigcup_{x \in Z} F_x \cup \{\langle\langle\rangle, v_B\rangle\}$ is an evaluation map for $B$ at $X$. □

Classically, given a Borel code $B$, the collection of $X \in \omega^\omega$ such that the evaluation map $f$ for $B$ at $X$ satisfies $f(\langle\rangle) = 1$ is a Borel set. Conversely, given any Borel set $\mathcal{B}$ there is a code $B \subseteq \omega^{<\omega}$ satisfying Definition 12.2.1 such that the elements of $\mathcal{B}$ are precisely those $X \in \omega^\omega$ such that the evaluation map for $B$ at $X$ satisfies $f(\langle\rangle) = 1$. In this sense, Borel codes do a good job of representing Borel sets.

There is one subtlety, though, which is that the proof of Theorem 12.2.3 does not go through in the absence of $\mathsf{ATR}_0$. In fact, it does not even go through for *trivial* Borel codes (so called in [78]), which are the codes such that $\alpha(|\alpha| - 1) < 2$ for all leaves $\alpha$. The only Borel sets such codes can represent are $\emptyset$ or $\omega^\omega$, so defining an evaluation map may seem trivial. But, per Definition 12.2.2, evaluation maps must consistently assign values to all the elements of a given code, which can be quite complicated even if the "eventual" Borel set is trivial. This is an insurmountable problem.

**Theorem 12.2.4 (Dzhafarov, Flood, Solomon, and Westrick [78]).** *The following are equivalent over* $\mathsf{RCA}_0$.

1. $\mathsf{ATR}_0$.
2. *For every Borel code* $B \subseteq \mathbb{N}^{\mathbb{N}}$ *and* $X \in \mathbb{N}^{\mathbb{N}}$, *there exists an evaluation map for* $B$ *at* $X$.
3. *For every trivial Borel code* $B \subseteq \mathbb{N}^{\mathbb{N}}$ *and* $X \in \mathbb{N}^{\mathbb{N}}$, *there exists an evaluation map for* $B$ *at* $X$.

This is a key issue because, without an evaluation map, the relations $X \in B$ and $X \notin B$ for a Borel code $B$ hold no meaning. In particular, the preceding theorem has the following somewhat surprising consequence.

**Corollary 12.2.5.** *The following are equivalent over* $\mathsf{RCA}_0$.

1. $\mathsf{ATR}_0$.
2. *For every Borel code* $B \subseteq \mathbb{N}^\mathbb{N}$ *and* $X \in \mathbb{N}^\mathbb{N}$, *either* $X \in B$ *or* $X \notin B$.
3. *For every Borel code* $B \subseteq \mathbb{N}^\mathbb{N}$, *there exists* $X \in \mathbb{N}^\mathbb{N}$ *such that* $X \in B$ *or* $X \notin B$.

The consequence of this phenomenon is that even very basic results about Borel sets tend to immediately jump up at least to the level of $\mathsf{ATR}_0$ in complexity. (In the parlance of Section 1.3, Borel codes impose a significant coding overhead.) Recently, Westrick (see [78]) has proposed the concept of a *determined Borel code*, which consists of a Borel code $B$ together with the assertion that for each $X$ an evaluation for $B$ at $X$ exists. Formulating a theorem about Borel sets in terms of determined Borel codes escapes the problems suggested above and, in practice, the reverse mathematics analysis seems more faithful to the "true" strength of the theorem. (Examples may be found in [78], as well as in newer papers by Astor, Dzhafarov, Montalbán, Solomon, and Westrick [5] and by Towsner, Weisshaar, and Westrick [312]).

Another important collection of subsets of Baire (or Cantor) space is the collection of analytic sets. These are the subsets definable by $\Sigma^1_1$ formulas of $\mathcal{L}_2$. (We may regard a $\Sigma^1_1$ formula $\varphi(X)$ as being defined on $X \in \mathbb{N}^\mathbb{N}$ by identifying $X$ with the characteristic function of its graph as a function.) Analytic sets, too, admit a more combinatorial representation.

**Definition 12.2.6 (Analytic codes).** The following definition is made in $\mathsf{RCA}_0$. Given a tree $A \subseteq \mathbb{N}^{<\mathbb{N}}$, an element $X \in \mathbb{N}^\mathbb{N}$ is a *point* of $A$ if $(\exists f \in \mathbb{N}^\mathbb{N})[f \oplus X \in [A]]$.

It is customary to call the tree $A$ an *analytic code* in this context. Clearly, the collection of all points of a given analytic code is definable by a $\Sigma^1_1$ formula. The next result establishes the converse.

**Theorem 12.2.7 (Simpson [288]).** *If* $\varphi(X)$ *is a* $\Sigma^1_1$ *formula, then* $\mathsf{ACA}_0$ *proves there is an analytic code* $A \subseteq \mathbb{N}^{<\mathbb{N}}$ *such that for all* $X$, $\varphi(X) \leftrightarrow X$ *is a point of* $A$.

*Proof.* By Kleene's normal form theorem (Theorem 5.8.2), there is an arithmetical formula $\theta$ such that

$$\mathsf{ACA}_0 \vdash (\forall X)[\varphi(X) \leftrightarrow (\exists f \in \mathbb{N}^\mathbb{N})(\forall k)[\theta(X \upharpoonright k, f \upharpoonright k)]].$$

Let $A$ be the set of all sequences $\langle\langle\alpha_0,\beta_0\rangle,\dots,\langle\alpha_{n-1},\beta_{n-1}\rangle\rangle \in \mathbb{N}^{<\mathbb{N}}$ such that:

- $\alpha_0 \preceq \cdots \preceq \alpha_{n-1}$,
- $\beta_0 \preceq \cdots \preceq \beta_{n-1}$,
- $|\alpha_i| = |\beta_i| = i$ for all $i < n$,
- $\theta(\alpha_{n-1},\beta_{n-1})$ holds.

Then $A$ is a tree and $\varphi(X) \leftrightarrow X$ is a point of $A$. □

The theorem sets the stage for deriving most of the standard facts about the relationship between analytic and Borel sets. For starters, note that by Definition 12.2.2, belonging and not belonging to a given Borel code are each definable by a $\Sigma^1_1$ formula. Thus, Theorem 12.2.7 yields the following corollary, formalizing the fact that every Borel set is analytic.

**Corollary 12.2.8.** *The following is provable in* $\mathsf{ATR}_0$. *If* $B \subseteq \mathbb{N}^{<\mathbb{N}}$ *is a Borel code, there exist trees* $A_0, A_1 \subseteq \mathbb{N}^{<\mathbb{N}}$ *such that for all* $X$, $X \in B \leftrightarrow X$ *is a point of* $A_0$, *and* $X \notin B \leftrightarrow X$ *is a point of* $A_1$.

Note that, by the uniqueness of evaluation maps in $\mathsf{ATR}_0$, no $X$ can both belong and not belong to a given Borel code, so the trees $A_0$ and $A_1$ in the corollary are (codes for) each other's complements.

$\mathsf{ATR}_0$ can also prove the converse: any analytic set whose complement is also analytic must be Borel. This follows from the following theorem on the strength of *Lusin's separation theorem* from descriptive set theory, which is part (2) below.

**Theorem 12.2.9 (Simpson [288]).** *The following are equivalent over* $\mathsf{RCA}_0$.

1. $\mathsf{ATR}_0$.
2. *If* $A_0, A_1 \subseteq \mathbb{N}^{<\mathbb{N}}$ *are analytic codes such that no* $X$ *is a point of both* $A_0$ *and* $A_1$, *then there is a Borel code* $B$ *such that for all* $X$, *if* $X$ *is a point of* $A_0$ *then* $X \in B$, *and if* $X$ *is a point of* $A_1$ *then* $X \notin B$.

We omit the proof for brevity. The reversal follows by observing that Lusin's separation theorem easily implies the $\Sigma^1_1$ separation principle (Theorem 5.8.20), whose equivalence to $\mathsf{ATR}_0$ we saw in Section 12.1.1. The following corollary is now immediate.

**Corollary 12.2.10.** *The following is provable in* $\mathsf{ATR}_0$. *If* $A_0, A_1 \subseteq \mathbb{N}^{<\mathbb{N}}$ *are analytic codes such that every* $X \in \mathbb{N}^{\mathbb{N}}$ *is a point of exactly one of* $A_0$ *and* $A_1$, *then there is a Borel code* $B \subseteq \mathbb{N}^{<\mathbb{N}}$ *such that for all* $X$, $X$ *is a point of* $A_0 \leftrightarrow X \in B$.

Corollaries 12.2.8 and 12.2.10 give a proof in $\mathsf{ATR}_0$ of the classical *Souslin's theorem*, which states that a set is Borel if and only if it and its complement are analytic. This theorem is not equivalent to $\mathsf{ATR}_0$, however (see Simpson [288, Remark V.3.12]).

We conclude this section by mentioning one further consequence of Theorem 12.2.7, which will be of independent interest to us in Section 12.3.2. The proof features a satisfying use of self reference.

**Corollary 12.2.11.** *If* $\varphi(X)$ *is a* $\Sigma^1_1$ *formula then* $\mathsf{ACA}_0$ *proves there is an* $X$ *such that* $\neg(\varphi(X) \leftrightarrow \mathsf{WO}(X))$.

*Proof.* We argue in $\mathsf{ACA}_0$. Given a tree $A \subseteq \mathbb{N}^{<\mathbb{N}}$ and $X \in \mathbb{N}^{\mathbb{N}}$, define

$$T_A(X) = \{\alpha \in \mathbb{N}^{<\mathbb{N}} : (\forall k \leqslant |\alpha|)[\langle\langle \alpha \upharpoonright 0, X \upharpoonright 0\rangle, \ldots, \langle \alpha \upharpoonright k, X \upharpoonright k\rangle\rangle \in A]\}.$$

Then $T_A(X)$ is a tree, and $X$ is a point of $A$ if and only if $\neg\,\mathsf{WF}(T_A(X))$. Now fix $\varphi$, and let $\psi(X)$ be the formula $\varphi(\mathsf{KB}(T_X(X)))$, which is again $\Sigma^1_1$. By Theorem 12.2.7, there is an $A$ such that for all $X$, $\psi(X) \leftrightarrow \neg\,\mathsf{WF}(T_A(X))$. Then we have

$$\varphi(\mathrm{KB}(T_A(A))) \leftrightarrow \psi(A) \leftrightarrow \neg\,\mathrm{WF}(T_A(A)) \leftrightarrow \neg\,\mathrm{WO}(\mathrm{KB}(T_A(A))),$$

by Proposition 5.8.8. Taking $X = \mathrm{KB}(T_A(A))$ yields the result. □

## 12.3 Determinacy

In this section, we consider so-called *Gale–Stewart games*, also known as *games of perfect information*. These consist of two players who alternate playing natural numbers, with the first player winning if the resulting sequence of numbers produced during the play belongs to a predetermined class of elements of $\omega^\omega$, called a *winning class*, and the second player winning otherwise. A game is said to be *determined* if one of the players has a winning strategy. In set theory, the *axiom of determinacy* states that every game is determined. This has many interesting consequences including, perhaps most famously, that every set of reals is Lebesgue measurable (Mycielski and Świerczkowski [229]). Of course, this means the axiom of determinacy is provably false in ZFC because, as is well known, the axiom of choice implies the existence of nonmeasurable sets.

The tension between choice and determinacy inspired a great deal of work into just "how much" determinacy ZFC *does* admit. This is usually calibrated by the topological complexity of the winning class. One of the earliest results here is due to Gale and Stewart [121], who showed that open determinacy (determinacy for games in which the winning class is open) is provable in ZFC. Wolfe [328] extended this to $F_\sigma$ determinacy, and Davis [62] to $F_{\sigma\delta}$ determinacy. Then, in a sweeping generalization, Martin [203] showed that ZFC actually proves Borel determinacy. And this, it turns out, is the limit: even analytical determinacy already requires stronger set theoretic axioms.

Our interest, of course, is in the reverse mathematics content of the above results. As we will see in Section 12.3.4, the amount of determinacy provable in $\mathsf{Z}_2$, as opposed to ZFC, stops well short of Borel. And even low levels of determinacy require comparatively strong axioms to prove (Section 12.3.2). We begin in the next section with some basic definitions. A more thorough account of the classical theory of determinacy can be found in many texts on set theory or descriptive set theory (see, e.g., Kechris [177]).

### 12.3.1 Gale–Stewart games

The main object of interest for determinacy is the following.

**Definition 12.3.1.** Fix $\mathcal{S} \subseteq \omega^\omega$. The *(Gale–Stewart) game* $G(\mathcal{S})$ is a two-player game in which Players I and II alternate playing elements of $\omega$, with Player I going first and playing $n_0, n_2, \ldots$ and Player II playing $n_1, n_3, \ldots$. If the sequence

$n_0 n_1 n_2 n_3 \cdots \in \omega^\omega$ belongs to $\mathcal{S}$ then Player I wins; otherwise, Player II wins. We call $\mathcal{S}$ the *winning class* for $G(\mathcal{S})$.

A play of $G(\mathcal{S})$ is typically denoted by $r \otimes s$, where $r, s \in \omega^\omega$, with $r(i) = n_{2i}$ indicating the moves of Player I, and $s(i) = n_{2i+1}$ indicating the moves of Player II. (Technically, $r \otimes s$ is just $r \oplus s$, but we will use the traditional notation where appropriate.)

**Definition 12.3.2.** Fix $\mathcal{S} \subseteq \omega^\omega$.

1. A *strategy* is a function $f\colon \omega^{<\omega} \to \omega$.
2. A strategy $f\colon \omega^{<\omega} \to \omega$ is *winning for Player I in* $G(\mathcal{S})$ if for all $s \in \omega^\omega$, if $r \in \omega^\omega$ is defined by $r(i) = f(s \upharpoonright i)$ for all $i$, then $r \otimes s \in \mathcal{S}$.
3. A strategy $f\colon \omega^{<\omega} \to \omega$ is *winning for Player II in* $G(\mathcal{S})$if for all $r \in \omega^\omega$, if $s \in \omega^\omega$ is defined by $s(i) = f(r \upharpoonright i+1)$ for all $i$, then $r \otimes s \notin \mathcal{S}$.
4. $G(\mathcal{S})$ is *determined* if there is a winning strategy for Player I or Player II.
5. If $\Gamma$ is a collection of classes of elements of $\omega^\omega$, $\Gamma$ *determinacy* is the assertion that $G(\mathcal{S})$ is determined for every $\mathcal{S} \in \Gamma$.

We consider collections $\Gamma$ corresponding to the usual topological classes, e.g., open, $F_\sigma$, Borel, etc. These are usually described in terms of the boldface Borel hierarchy, e.g., $\mathbf{\Sigma}^0_1, \mathbf{\Sigma}^0_2, \mathbf{\Delta}^1_1$, etc. It is known that $\mathbf{\Sigma}^0_n$ determinacy is equivalent to $\mathbf{\Pi}^0_n$ determinacy, for all $n$ (see, e.g., Hachtman and Palumbo [131, Section 4] for a concise proof).

In what Simpson [288, p. 210] calls "one of the earliest results of reverse mathematics", Steel [302] showed that clopen (i.e., $\mathbf{\Delta}^0_1$) and open (i.e., $\mathbf{\Sigma}^0_1$) determinacy are both equivalent to $\mathsf{ATR}_0$, which we will see proved in the next section. On the other hand, Friedman [107] showed that $\mathbf{\Sigma}^0_5$ determinacy is not provable even in full $\mathsf{Z}_2$. A proof of Friedman's theorem, in fact of a strengthening to $\mathbf{\Sigma}^0_4$ due to Martin (unpublished; see [204]), appears in Section 12.3.4.

Understanding the levels between $\mathbf{\Sigma}^0_1$ and $\mathbf{\Sigma}^0_4$ has been the focus of considerable work. Tanaka [305] showed that determinacy for intersections of open and closed classes is equivalent to $\Pi^1_1$-$\mathsf{CA}_0$. Tanaka [306] also characterized $\mathbf{\Sigma}^0_2$ determinacy, albeit not in terms of any of the usual subsystems of $\mathsf{Z}_2$. MedSalem and Tanaka [208] showed that $\mathbf{\Delta}^0_3$ determinacy follows from $\Delta^1_3$-$\mathsf{CA}_0$ plus a strong transfinite recursion principle; and Welch [326] showed that $\mathbf{\Sigma}^0_3$ determinacy is provable in $\Pi^1_3$-$\mathsf{CA}_0$. More recently, Montalbán and Shore [222] showed that the dividing line between classes for which determinacy is and is not provable in $\mathsf{Z}_2$ falls strictly between $\mathbf{\Sigma}^0_3$ and $\mathbf{\Sigma}^0_4$ (in fact, $\mathbf{\Delta}^0_4$). For $n \geqslant 1$, say a set $\mathcal{S} \subseteq \omega^{<\omega}$ is $n$-$\mathbf{\Pi}^0_3$ if there exists $\mathbf{\Pi}^0_3$ classes $\mathcal{S}_0, \mathcal{S}_1, \ldots, \mathcal{S}_n$ such that $\mathcal{S}_n = \emptyset$ and $\mathcal{S} = (\mathcal{S}_0 \setminus \mathcal{S}_1) \cup (\mathcal{S}_2 \setminus \mathcal{S}_3) \cup \cdots$. Thus, the 1-$\mathbf{\Pi}^0_3$ classes coincide with $\mathbf{\Pi}^0_3$ classes, the 2-$\mathbf{\Pi}^0_3$ classes are differences of $\mathbf{\Pi}^0_3$ classes, etc. Let $n$-$\mathbf{\Pi}^0_3$ also denote the collection $\Gamma$ of all $n$-$\mathbf{\Pi}^0_3$ classes.

**Theorem 12.3.3 (Montalbán and Shore [222]).** *Fix $n \geqslant 1$.*

*1.* $\Pi^1_{n+2}$-$\mathsf{CA}_0$ *proves* $n$-$\mathbf{\Pi}^0_3$ *determinacy.*
2. $\Delta^1_{n+2}$-$\mathsf{CA}_0$, *even together with full induction, does not prove* $n$-$\mathbf{\Pi}^0_3$ *determinacy.*

Since $\mathsf{Z}_2$ can be written as $\bigcup_{n\geqslant 1}\Delta^1_n$-$\mathsf{CA}_0$, it follows by compactness (of first order logic) that $\mathsf{Z}_2 \nvdash (\forall n)[n\text{-}\mathbf{\Pi}^0_3$ determinacy$]$. So $\mathsf{Z}_2$ cannot prove, e.g., $\mathbf{\Delta}^0_4$ determinacy, which is a further extension of Friedman's theorem. The proof lies outside the scope of this book, but it uses some of the same set theoretic methods that we survey in Section 12.3.3.

For a more complete overview of the above development, including many additional related results leading up to, and stemming from, Theorem 12.3.3, see the discussion of Montalbán and Shore [222, Section 1].

## 12.3.2 Clopen and open determinacy

In this section, we analyze the reverse mathematics strength of clopen and open determinacy. These statements turn out to be equivalent, and require precisely arithmetical transfinite recursion to prove.

By Proposition 2.8.2, if $\mathcal{S}$ is open there is a tree $T \subseteq \omega^\omega$ with $\mathcal{S} = \omega^\omega \setminus [T]$. Thus, Player I wins a run of $G(\mathcal{S})$ if and only if the constructed sequence is not an element of $[T]$.

If $\mathcal{S}$ is clopen, we may fix two trees, $T_1, T_2 \subseteq \omega^\omega$, with $[T_0] = \mathcal{S}$ and $[T_1] = \omega^\omega \setminus \mathcal{S}$. Here, it is convenient to think of a slightly different game. Let $T = T_0 \cap T_1$. Players I and II play $n_0, n_1, n_2, \ldots$ as before, but now subject to the condition that for all $i \in \omega$ and $j < 2$, $n_{2i+j}$ can only be played if $\langle n_0 n_1 \cdots n_{2i+j}\rangle \in \mathrm{Ext}(T_j)$. Since $T$ is well founded, it follows that every run of this game is finite. We declare whichever player reaches a leaf of $T$ first to be the winner. It is then easy to check that a player has a winning strategy in this game if and only if that player has a winning strategy in $G(\mathcal{S})$.

**Theorem 12.3.4.** *The following statements are equivalent over* $\mathsf{RCA}_0$.

1. $\mathsf{ATR}_0$.
2. *Open determinacy.*
3. *Clopen determinacy.*

*Proof.* (1) $\to$ (2): Fix a tree $T \subseteq \mathbb{N}^{<\mathbb{N}}$. Given a well ordering $X$, we define $O_x \subseteq \mathbb{N}^{<\mathbb{N}}$, $x \in X$, by transfinite recursion. Let $O_{0_X} = \{\alpha : \alpha \notin T\}$, and for $x >_X 0_X$, let

$$O_x = \{\alpha : |\alpha| \text{ even} \wedge (\exists n)[\alpha n \in \bigcup_{y<_X x} O_y] \text{ or } |\alpha| \text{ odd} \wedge (\forall n)[\alpha n \in \bigcup_{y<_X x} O_y]\}.$$

Let $O_X = \bigcup_{x\in X} O_x$. It is not difficult to see that if there is an $X$ such that $\mathrm{WO}(X)$ and $\langle\rangle \in O_X$, then Player I has a winning strategy. So suppose instead that there is no such $X$. We claim that, in this case, Player II has a winning strategy.

Let $\psi(X)$ be the formula that says that $X$ is a linear ordering with least element $0_X$ and there exists $P \subseteq X \times \mathbb{N}^{<\mathbb{N}}$ such that $P^{[0_X]} = \{\alpha : \alpha \notin T\}$, and for $x >_X 0_X$,

$$P^{[x]} = \{\alpha : |\alpha| \text{ even} \wedge (\exists n)[\alpha n \in P^{[y<_X x]}] \text{ or } |\alpha| \text{ odd} \wedge (\forall n)[\alpha n \in P^{[y<_X x]}]\},$$

and $\langle\rangle \notin P_Y = \bigcup_{x \in Y} P_x$.

By hypothesis, $\mathrm{WO}(X) \to \psi(X)$ for all $X$, because if $X$ is a well ordering we can form $\langle O_x : x \in X\rangle$ and let this be $P$ (i.e., $P = \bigcup_{x \in X}\{x\} \times O_x$). Now since $\psi$ is $\Sigma^1_1$, it follows by Corollary 12.2.11 that we cannot have $\psi(X) \to \mathrm{WO}(X)$ for all $X$. So there is some set $X$ such that $\psi(X)$ holds and $\neg\,\mathrm{WO}(X)$. Let $x_0 >_Y x_1 >_Y \cdots$ be an infinite descending sequence in $X$. We want to show that Player II can always stay outside of $P_{0_X}$ (and thus inside of $T$).

We prove that on the $i$th move, Player II can make sure its move is outside of $P_{x_i}$. The strategy then is: if $|\alpha| = i$, play an $n$ such that $\alpha n \notin P_{x_i}$, and hence such that $\alpha n \in T$ (since $T$ is closed downwards under $\preceq$). We can prove that this is always possible. Indeed, using the definition of $P_x$, and the fact that $\langle\rangle \notin P_Y$, it follows by induction that

- if $|\alpha|$ is even and $\alpha \notin P_{x_i}$ then $(\forall n)[\alpha n \notin P_{x_{i+1}}]$,
- if $|\alpha|$ is odd and $\alpha \notin P_{x_i}$ then $(\exists n)[\alpha n \notin P_{x_{i+1}}]$.

This proves the claim.

(2) $\to$ (3): Clear.

(3) $\to$ (1): We use the alternate formulation of clopen games described above. We argue in $\mathsf{RCA}_0$, and first claim that (3) implies $\mathsf{ACA}_0$. To better match the rest of the argument, we prove that $Z'$ exists for ever set $Z$, which is equivalent (Corollary 5.6.3). We take $Z = \varnothing$. The general case is analogous. We define a game, a play of which proceeds as follows.

- Player I plays $n \in \mathbb{N}$.
- Player II plays 1 or 0, which we think of as "yes" or "no".
- Player I plays $s \in \mathbb{N}$.
- Player II plays $t \in \mathbb{N}$.

Player II wins if either it has played "yes" and then $\Phi_n(n)[t]\downarrow$, or it has played "no" and $\Phi_n(n)[s]\uparrow$. Intuitively, Player I proposes a candidate $n$ for membership in $\varnothing'$, and Player II says whether or not it thinks $n$ is in $\varnothing'$. If Player II thinks $n \notin \varnothing'$, the burden of proof is on Player I to find an $s$ so that $\Phi_n(n)[s]\downarrow$. If Player II thinks $n \in \varnothing'$, the burden of proof is on it to find a $t$ such that $\Phi_n(n)[t]\downarrow$.

This is a clopen game, if we think of the players as always having one more turn if they have not yet lost (so for example, if Player II says "no" and Player II plays an $s$ such that $\Phi_n(n)[s]\uparrow$, then Player II can play once more and the game stops). The longest run of a play is therefore 5, which is why the tree of plays is well founded.

Now Player I cannot have a winning strategy for this game, since for each $n \in \mathbb{N}$, Player II can always respond with the correct answer about whether or not $n \in \varnothing'$. Thus, by clopen determinacy, Player II has a winning strategy $f$, and we have

$$(\forall n)[n \in \varnothing' \leftrightarrow f(\langle n\rangle) = \text{"yes"}].$$

So, we can argue in $\mathsf{ACA}$ from now on.

Now fix $X$ such that $\mathrm{WO}(X)$. We extend the above argument to show that $\varnothing^{(X)}$ exists. By Theorem 5.8.18 (and relativization), this implies $\mathsf{ATR}_0$. Without loss of generality, we may assume $X$ has no $\leqslant_X$-largest element. We build trees $T_{n,x}$ for $x \in X$ such that Player I wins in $T_{n,x}$ if and only if $n \in \varnothing^{(X \restriction x)}$. To begin, let $T_{n,0_X}$ be the tree corresponding to the game in which Player I plays an $s \in \mathbb{N}$, and then Player II plays some $t \in \mathbb{N}$ if and only if $\Phi_n(x)[s] \uparrow$. Then Player I wins this game if and only if $n \in \varnothing'$.

Now suppose $x \in X$ is the successor of $y \in X$ under $\leqslant_X$, and suppose $T_{m,y}$ has been defined for all $m$. Let $T_{n,x}$ be the tree corresponding to the following game.

- Player I plays a (code for a) $\sigma \in 2^{<\mathbb{N}}$ such that $\Phi_n^\sigma(n) \downarrow$.
- Player II chooses $m < |\sigma|$.
- If $\sigma(m) = 1$, Player II plays $m$, and on subsequent moves we start playing $T_{m,y}$, with Player I starting.
- If $\sigma(m) = 0$, we start playing $T_{m,y}$, with Player II starting.

Intuitively, the play of $\sigma$ above represents an initial segment of $\varnothing^{(X \restriction y)}$ long enough so that $\Phi_n^\sigma(n) \downarrow$. Since we have convergence, we only need to check that $\sigma$ is indeed an initial segment of $\varnothing^{(X \restriction y)}$, and for this Player II queries the bits of $\sigma$ to see if they are in $\varnothing^{(X \restriction y)}$. If $m < |\sigma|$ is claimed (by $\sigma$) to be in $\varnothing^{(X \restriction y)}$, the burden of proof is on Player I to win $T_{m,y}$, whereas if $m$ is claimed not to be in $\varnothing^{(X \restriction y)}$, the burden of proof is on Player II to win $T_{m,y}$, acting as the first Player In that game (which means Player I does not win that game).

Now suppose $x$ is a limit under $\leqslant_X$ and that $T_{m,y}$ has been defined for all $m$ and all $y <_X x$. Say $n = \langle m, y\rangle$. Then let $T_{n,x} = \{\langle\rangle\}$ if $y \geqslant_X x$ or $y \notin X$, and otherwise let $T_{n,x} = T_{m,y}$. Then Player I wins $T_{n,x}$ if and only if $n = \langle m, y\rangle$ for some $y <_X x$ and Player I wins $T_{m,y}$.

In the definition above, the sequence $\langle T_{n,x} : n \in \mathbb{N}\rangle$ is uniformly $\Delta_1^0$ definable from $\langle T_{m,y} : m \in \mathbb{N}, y <_X x\rangle$, and the distinction between whether $x$ ix $0_X$, a successor, or a limit, is arithmetical in $X$. Hence, using arithmetical comprehension and arithmetical induction in $\mathsf{ACA}_0$, we can form $\langle T_{n,x} : x \in X, n \in \mathbb{N}\rangle$ (Exercise 5.13.20). Thus, we can define the following game.

1. Player I plays $\langle n, x\rangle$ for some $x \in X$.
2. Player II chooses "yes" or "no".
3. If "no", Player II plays "no" and on subsequent moves we start playing $T_{n,x}$, with Player I starting.
4. If "yes", we start playing $T_{n,x}$, with Player II starting.

This is a clopen game since all of these trees $T_{n,x}$ are well founded. By clopen determinacy, one of the two players has a winning strategy for it, but as above, it cannot be Player I. Thus, Player II has a winning strategy $f$, and we have

$$(\forall n)[n \in \varnothing^{(X)} \leftrightarrow (\exists x \in X)[f(\langle n, x\rangle) = \text{“yes”}].$$

This completes the proof. □

### 12.3.3 Gödel's constructible universe

In this section, we quickly collect several definitions and lemmas concerning *Gödel's constructible universe*, $L$, which we will need to prove Friedman's theorem. Let $\mathsf{ZFC}^-$ denote ZFC without the power set axiom. The following gives an important connection between models of $\mathsf{ZFC}^-$ and models of $\mathsf{Z}_2$. The proof is left to Exercise 12.5.7.

**Proposition 12.3.5.** *If* $\mathcal{M} \vDash \mathsf{ZFC}^-$*, then* $(\omega, \mathcal{M} \cap \mathcal{P}(\omega)) \vDash \mathsf{Z}_2$.

Thus, one way to show that $\mathsf{Z}_2$ does not prove a certain result (for example, some amount of determinacy) is to build a model of $\mathsf{ZFC}^-$ and appeal to the above theorem. This is what we do below. Let us now describe the model of interest.

We begin with the definition of $L$. Let Ord denote the class of all ordinals. We use lowercase letters to range over sets here, as is customary in set theory, but stick to using capitals when wanting to emphasize that a set is a subset of $\omega$. (The sets $L_\alpha$ and $L$ below are an exception.)

**Definition 12.3.6.** We define $L$ recursively along Ord.

1. $L_0 = \emptyset$.
2. $L_{\alpha+1} = \{\{x \in L_\alpha : L_\alpha \vDash \varphi(x, \vec{a})\} : \vec{a} \in L_\alpha, \varphi \text{ is a first order formula}\}$.
3. $L_\gamma = \bigcup_{\alpha<\gamma} L_\alpha$.

Then $L$ is the proper class $L = \bigcup_{\alpha \in \mathrm{Ord}} L_\alpha$.

$L$ was introduced by Gödel [126], in his famous work showing the consistency with ZF of the continuum hypothesis and the axiom of choice. We identify $L_\alpha$ and $L$ with the structures $(L_\alpha, \in)$ and $(L, \in)$, respectively. $L$ satisfies ZFC + "there is a well ordering of the universe". This well ordering, called $<_L$, is defined by putting a canonical well ordering on each $L_\alpha$, and ordering all elements in $L_{\alpha+1} \smallsetminus L_\alpha$ above all elements of $L_\alpha$.

We will make use of the following two lemmas.

**Lemma 12.3.7.** *There are unboundedly many* $\alpha < \omega_1$ *such that* $L_\alpha$ *is an elementary substructure of* $L_{\omega_1}$.

*Proof.* Fix any $\alpha < \omega_1$. Let $\beta_0 = \alpha$, and given $\beta_n$, let $\beta_{n+1}$ be the least $\beta$ with $\beta_n < \gamma < \omega_1$ such that for all $\varphi$ and $\vec{a} \in L_{\beta_n}$, if $L_{\omega_1} \vDash (\exists x)\varphi(x, \vec{a})$ then there is some $b$ in $L_\gamma$ witnessing this. (That such a $\gamma$ exists for a single choice of $\varphi$ and $\vec{a}$ follows from the fact that $\omega_1$ is a limit. That we can choose a single $\gamma$ that works for all $\varphi$ and $\vec{a}$ simultaneously follows from the fact that there are only countably many such formulas and tuples, and $\omega_1$ has cofinality $\omega_1$.) Let $\beta = \sup_n \alpha_n$. By the Tarski–Vaught test, $L_\beta$ is an elementary substructure of $L_{\omega_1}$. □

**Lemma 12.3.8.** $L_{\omega_1} \vDash \mathsf{ZFC}^-$.

*Proof.* We check each axiom, one by one. Comprehension for subsets of a set $z$ follows from taking some $\alpha < \omega_1$ such that $z \in L_\alpha$ and such that $L_\alpha$ is an elementary substructure of $L_{\omega_1}$. Other axioms are similar, or are satisfied simply because $\in$ is represented by actual set membership. □

We can now define the model that will be of interest to us.

**Definition 12.3.9.** $\beta_0$ is the least $\beta \in \mathrm{Ord}$ such that $L_\beta \models \mathsf{ZFC}^-$.

By the previous two lemmas, $\beta_0 < \omega_1$. In particular, $L_{\beta_0}$ is countable. The following theorem lists one of its crucial properties: $\Sigma^1_1$ formulas are absolute to $L_{\beta_0}$.

**Theorem 12.3.10.** *If $\varphi(X)$ is a $\Sigma^1_1$ formula of $\mathcal{L}_2$, then $L_{\beta_0} \models (\exists X)\varphi(X)$ if and only if $V \models (\exists X)\varphi(X)$.*

*Proof.* First, note that if $X \in L_{\beta_0}$ with $X \subseteq \omega^2$, then $\mathrm{WO}(X)$ is absolute to $L_{\beta_0}$. Indeed, if $L_{\beta_0} \models \mathrm{WO}(X)$ then there is an isomorphism $f : X \to (\alpha, \in)$ in $L_{\beta_0}$ for some $\alpha \in \mathrm{Ord}^{L_{\beta_0}}$. Since $\mathrm{Ord}^{L_{\beta_0}} \subseteq \mathrm{Ord}$, it follows that $f$ witnesses $\mathrm{WO}(X)$ in the universe. Conversely, if $L_{\beta_0} \models \neg\,\mathrm{WO}(X)$ then there is some infinite descending sequence $f : \omega \to X$ in $L_{\beta_0}$, and this also witnesses $\neg\,\mathrm{WO}(X)$ in the universe.

Now suppose $\varphi(X)$ is $\Sigma^1_1$. Then by Corollary 5.8.4, there is a tree $T \subseteq \omega^\omega$ such that $\mathsf{ACA}_0 \vdash (\exists X)\varphi(X) \leftrightarrow \mathrm{WF}(T) \leftrightarrow \mathrm{WO}(\mathrm{KB}(T))$. By Proposition 12.3.5, this equivalence holds in $L_{\beta_0}$. Hence, by the remarks above, we have $L_{\beta_0} \models (\exists X)\varphi(X) \leftrightarrow V \models (\exists X)\varphi(X)$, as was to be shown. □

More generally, $\omega$-models with the property that a $\Sigma^1_1$ formula of $\mathcal{L}_2$ is satisfied if and only if it is true are called *$\beta$-models*. By the preceding theorem, the $\omega$-model $(\omega, L_{\beta_0} \cap \mathcal{P}(\omega))$ is actually a $\beta$-model. There is a rich theory of $\beta$-models and their applications in reverse mathematics. Simpson [288, Chapter VII] provides a detailed exposition.

We close with one final definition and lemma, which will be needed for technical reasons. After that, we will be ready to prove Friedman's theorem.

**Definition 12.3.11.** We say a model $\mathcal{M}$ is *well founded* if for every $\alpha \in \mathcal{M}$, if $\mathcal{M} \models \alpha \in \mathrm{Ord}$, then $\{\beta \in \mathcal{M} : \beta \in^{\mathcal{M}} \alpha\}$ is well ordered by $\in^{\mathcal{M}}$.

**Lemma 12.3.12.** *Every well founded model $\mathcal{M}$ of $\mathsf{ZFC}^- + V = L$ is isomorphic to $L_\alpha$ for some $\alpha$.*

*Proof.* Let $\alpha$ be the least ordinal not in $\mathcal{M}$. Then there is a bijection $f : \alpha \to \mathrm{Ord}^{\mathcal{M}}$. By transfinite recursion, build, for $\beta < \alpha$, an isomorphism $g_\beta : L_\beta \to (L_\beta)^{\mathcal{M}}$. Let $g = \bigcup_\beta g_\beta$. Then $g$ is the desired isomorphism. □

### 12.3.4 Friedman's theorem

Our goal in this section is to prove Friedman's theorem that $\mathbf{\Sigma}^0_5$ determinacy is not provable in $\mathsf{Z}_2$. We actually prove a strengthening, due to Martin (unpublished), that the same is true for $\mathbf{\Sigma}^0_4$ determinacy. Martin [204, p. 39] says of this proof that it is "essentially the same" as Friedman's.

At the level of $\mathbf{\Sigma}^0_4$, we may restrict ourselves to games $G(\mathcal{S})$ where $\mathcal{S} \subseteq 2^\omega$ and where Players I and II alternate playing 0s and 1s (rather than arbitrary numbers). We will call such games *binary*, to emphasize the difference. It is not difficult to see that $\mathbf{\Sigma}^0_4$ determinacy for games in (in the full sense of Definition 12.3.1) is equivalent to $\mathbf{\Sigma}^0_4$ determinacy for binary games. In fact, the same is true even for $\mathbf{\Sigma}^0_3$ determinacy. However, for lower levels in the arithmetical hierarchy there is a difference. A nice discussion is given by Montalbán and Shore [222, Section 2].

In what follows, we keep in mind a fixed enumeration of all sentences in the language of set theory with parameters from $L_{\beta_0}$, which is a countable collection. We can then regard infinite sequences of 0s and 1s (such as plays of binary games) as specifying a subset of this collection via its characteristic function.

**Lemma 12.3.13.** *Suppose $G$ is a binary game in $L_{\beta_0}$ such that:*

- *If Player I plays $\mathrm{Th}_{L_{\beta_0}}$ then Player I wins.*
- *If not, and if Player II plays $\mathrm{Th}_{L_{\beta_0}}$ then Player II wins.*

*Then $L_{\beta_0} \nvDash$ "$G$ is determined".*

*Proof.* First, we prove that $L_{\beta_0} \vDash$ "Player II has no winning strategy". Suppose there is an $f \in L_{\beta_0}$ such that $L_{\beta_0} \vDash$ "$f$ is a winning strategy for Player II". Then since $s$ is winning, for every $r \in 2^\omega$, $r \otimes s \notin G$, so being a winning strategy is a $\Pi^1_1$ statement. By Theorem 12.3.10, $s$ must be a winning strategy also in $V$. But no strategy can be winning for Player II in $V$, since Player I can always play $r = \mathrm{Th}_{L_{\beta_0}}$.

Next, we prove that $L_{\beta_0} \vDash$ "Player I has no winning strategy". Suppose otherwise, so that there is some winning strategy $s \in L_{\beta_0}$ for Player I. Then again $s$ is a winning strategy in $V$. Now suppose Player II plays the "copy cat" strategy, which means that whenever Player I plays $i \in \{0, 1\}$, Player II plays $i$ on the next move. Then the outcome of this play must be $\mathrm{Th}_{L_{\beta_0}} \otimes \mathrm{Th}_{L_{\beta_0}} = \mathrm{Th}_{L_{\beta_0}} \oplus \mathrm{Th}_{L_{\beta_0}}$. This is because, if Player I ever played an $i$ which was "wrong" in terms of playing $\mathrm{Th}_{L_{\beta_0}}$ (i.e., either $i = 0$ and $\varphi_i \in \mathrm{Th}_{L_{\beta_0}}$, or $i = 1$ and $\varphi_i \notin \mathrm{Th}_{L_{\beta_0}}$), then at the first place that this happens, Player II could play $1 - i$ on the next move, and thereafter continue to play $\mathrm{Th}_{L_{\beta_0}}$. Since Player I could never correct its mistake, it would end up not playing $\mathrm{Th}_{L_{\beta_0}}$, while Player II would, so Player II would win, contrary to the fact that Player I is playing a winning strategy.

Therefore, we have that Player I plays $s(\langle\rangle)$, then Player II plays $s(\langle\rangle)$, then Player I plays $s(s(\langle\rangle))$, and so on, meaning that

$$s(\langle\rangle)s(s(\langle\rangle))s(s(s(\langle\rangle)))\cdots = \mathrm{Th}_{L_{\beta_0}},$$

and therefore $\mathrm{Th}_{L_{\beta_0}} \leqslant_T s$. But then $\mathrm{Th}_{L_{\beta_0}}$ is definable by a single formula $\varphi$, so for every formula $\psi$ we have $\psi \in \mathrm{Th}_{L_{\beta_0}} \leftrightarrow \mathrm{Th}_{L_{\beta_0}} \vdash \varphi(\ulcorner\psi\urcorner)$. This cannot be, by Tarski's theorem on the undefinability of truth. □

**Theorem 12.3.14 (Friedman [107]; Martin, unpublished).** $\mathsf{Z}_2$ *does not prove* $\mathbf{\Sigma}^0_4$ *determinacy.*

Our aim is to build a binary $\mathbf{\Sigma}^0_4$ game $G$ which satisfies the conditions of Theorem 12.3.13. This will imply the theorem by Proposition 12.3.5. In $G$, Players I and II will play strings of 0s and 1s, as before. Let us call the collections of formulas (coded by the sequences) they play $T_{\mathrm{I}}$ and $T_{\mathrm{II}}$, respectively.

Let $P$ be I or II, indicating a player. Let $\mathcal{M}_P$ be the term model of $T_P$ defined as follows.

- For each $\varphi(x)$ such that $T_P \vdash (\exists! x)\varphi(x)$ add a term $t_\varphi$.
- Let $t_\varphi \sim t_\psi$ if and only if $T_P \vdash (\forall x)[\varphi(x) \leftrightarrow \psi(x)]$.
- Let $\mathcal{M}_P = \{t_\varphi\}/\sim$.

Then we can prove $\mathcal{M}_P \models T_P$, where $\mathcal{M}_P \models t_\varphi \in t_\psi$ if and only if $T \models (\forall x, y)[\varphi(x) \wedge \psi(y) \rightarrow x \in y]$. For this we use the fact that if $T_P \models (\exists x)\psi(x)$ then $T_P \models \psi(t_\varphi)$, where $\varphi(x) \equiv \psi(x) \wedge (\forall y <_L x)[\neg\psi(y)]$. So $\mathcal{M}_P$ contains witnesses.

*Proof (of Theorem 12.3.14; see [204]).* The game $G$ is defined as follows.

- Players have to play consistent complete extensions of $\mathsf{ZFC}^- +$ "$V = L$" $+$ "$(\forall \alpha \in \mathrm{Ord})[L_\alpha \not\models \mathsf{ZFC}^-]$". (Note that this is a $\Pi^0_1$ condition.)
- Players have to play $\omega$-models. So if $x \in \mathcal{M}_I$ and $\mathcal{M}_I \models x \in \omega$ then for some $n \in \omega$, $\mathcal{M}_I \models x = n$. Similarly for Player II. (Note that this is a $\Pi^0_2$ clause.)
- Player I wins if one of the following conditions hold.
  - $\mathrm{Ord}^{\mathcal{M}_{\mathrm{I}}}$ is isomorphic to an initial segment of $\mathrm{Ord}^{\mathcal{M}_{\mathrm{II}}}$.
  - There is an $\alpha \in \mathrm{Ord}^{\mathcal{M}_{\mathrm{I}}}$ such that $\alpha$ is isomorphic to a proper initial segment of $\mathrm{Ord}^{\mathcal{M}_{\mathrm{II}}}$ but $\alpha + 1$ is not.

  Otherwise, Player II wins.

*Claim 1: If Player I plays* $\mathrm{Th}_{L_{\beta_0}}$ *then* $\mathcal{M}_{\mathrm{I}} \cong L_{\beta_0}$ *and Player I wins.* The first part follows because everything in $\mathrm{Th}_{L_{\beta_0}}$ is definable from a term. To prove the second part, there are three possibilities for what can happen.

*Case 1:* $\mathrm{Ord}^{\mathcal{M}_{\mathrm{I}}}$ *is an initial segment of* $\mathrm{Ord}^{\mathcal{M}_{\mathrm{II}}}$. Then Player I wins by definition.

*Case 2:* $\mathrm{Ord}^{\mathcal{M}_{\mathrm{II}}}$ *is a proper initial segment of* $\mathrm{Ord}^{\mathcal{M}_{\mathrm{I}}}$. Then if $\mathcal{M}_{\mathrm{II}}$ models $V = L$, $\mathcal{M}_{\mathrm{II}} \cong L_\alpha$ for some $\alpha < \beta_0$ by Lemma 12.3.12. But then $\mathcal{M}_{\mathrm{II}}$ does not satisfy $\mathsf{ZFC}^-$ by definition of $\beta_0$, so this cannot happen.

*Case 3:* $\mathrm{Ord}^{\mathcal{M}_{\mathrm{I}}}$ *and* $\mathrm{Ord}^{\mathcal{M}_{\mathrm{II}}}$ *are incomparable.* Then it cannot be that each $\alpha \in \mathrm{Ord}^{\mathcal{M}_{\mathrm{I}}}$ is isomorphic to a proper initial segment of $\mathrm{Ord}^{\mathcal{M}_{\mathrm{II}}}$, and it is not difficult to see that the least $\alpha$ for which this is not the case cannot be a limit. That is, there is an $\alpha \in \mathrm{Ord}^{\mathcal{M}_{\mathrm{I}}}$ such that $\alpha$ is isomorphic to a proper initial segment of $\mathrm{Ord}^{\mathcal{M}_{\mathrm{II}}}$ but $\alpha + 1$ is not. So Player I wins.

This proves Claim 1.

*Claim 2: If Player I does not play* $\mathrm{Th}_{L_{\beta_0}}$ *but Player II does, then* $\mathcal{M}_{\mathrm{I}} \not\cong L_{\beta_0}$ *and* $\mathcal{M}_{\mathrm{II}} \cong L_{\beta_0}$ *and Player II wins.* The first part is clear. For the second, there are the same three possibilities as in the previous claim.

*Case 1:* $\mathrm{Ord}^{\mathcal{M}_{\mathrm{I}}}$ *is an initial segment of* $\mathrm{Ord}^{\mathcal{M}_{\mathrm{II}}}$. Player II wins by the same argument used to prove that Player I wins in case 2 of the preceding claim, but with $\mathcal{M}_{\mathrm{I}}$ and $\mathcal{M}_{\mathrm{II}}$ reversed.

*Case 2:* $\mathrm{Ord}^{\mathcal{M}_{\mathrm{II}}}$ *is a proper initial segment of* $\mathrm{Ord}^{\mathcal{M}_{\mathrm{I}}}$. Player II wins because neither of the two conditions for Player I to win are met.

*Case 3:* $\mathrm{Ord}^{\mathcal{M}_{\mathrm{I}}}$ *and* $\mathrm{Ord}^{\mathcal{M}_{\mathrm{II}}}$ *are incomparable.* Suppose $\alpha \in \mathrm{Ord}^{\mathcal{M}_{\mathrm{I}}}$ is isomorphic to a proper initial segment of $\mathrm{Ord}^{\mathcal{M}_{\mathrm{II}}}$. Since $\mathrm{Ord}^{\mathcal{M}_{\mathrm{II}}}$ is a subclass of the real ordinals, there must be some $\beta \in \mathrm{Ord}^{\mathcal{M}_{\mathrm{II}}}$ that is isomorphic to this initial segment. But then $\alpha + 1$ is isomorphic to $\beta + 1$, and $\beta + 1$ is isomorphic to a proper initial segment of $\mathrm{Ord}^{\mathcal{M}_{\mathrm{II}}}$, namely the initial segment that $\beta$ is isomorphic to followed by the least element of $\mathrm{Ord}^{\mathcal{M}_{\mathrm{II}}}$ not in this segment. Thus, Player I does not win, so Player II wins.

This proves Claim 2.

It remains to show that $G$ is $\mathbf{\Sigma}^0_4$, i.e., that the winning class for Player I has a $\mathbf{\Sigma}^0_4$ definition. Note that $\mathrm{Ord}^{\mathcal{M}_{\mathrm{I}}}$ is an initial segment of $\mathrm{Ord}^{\mathcal{M}_{\mathrm{II}}}$ if and only if for all $X \in \mathcal{M}_{\mathrm{I}}$, if $\mathcal{M}_{\mathrm{I}} \models X \subseteq \omega^2 \wedge \mathrm{WO}(x)$ then there is a $Y \in \mathcal{M}_{\mathrm{II}}$ such that $\mathcal{M}_{\mathrm{II}} \models Y \subseteq \omega^2 \wedge \mathrm{WO}(y)$ and $X$ and $Y$ represent the same subset of $\omega$, which means that for every $n \in \omega$, $\mathcal{M}_{\mathrm{I}} \models n \in x \leftrightarrow \mathcal{M}_{\mathrm{II}} \models n \in y$. This is a $\mathbf{\Pi}^0_3$ definition. Similarly, saying there is an $\alpha \in \mathrm{Ord}^{\mathcal{M}_{\mathrm{I}}}$ such that $\alpha$ is isomorphic to an initial segment of $\mathrm{Ord}^{\mathcal{M}_{\mathrm{II}}}$ but $\alpha + 1$ is not is $\mathbf{\Sigma}^0_4$. □

## 12.4 Higher order reverse mathematics

To this point, we have considered reverse mathematics in the setting of second order arithmetic. Using higher order arithmetic instead gives a different perspective on many theorems. Higher order arithmetic features higher types, such as functions from $\omega^\omega$ to $\omega^\omega$, which allow for more direct coding methods. At the same time, the higher order computability theory implicit in higher order arithmetic is more complicated than the classical computability theory implicit in second order arithmetic. Thus higher order reverse mathematics is not an extension of second order reverse mathematics: it is a genuinely alternative approach with its own techniques, motivations, and interpretations. In this section we will sketch the fundamental definitions and several results to show how higher order reverse mathematics can help fill out our understanding of particular theorems.

The approach most commonly used for higher order reverse mathematics was proposed by Kohlenbach [185]. This approach leverages systems of higher order arithmetic that are well known in proof theory (see Feferman [104], Kohlenbach [183], Avigad and Feferman [10] and Troelstra [313]). In particular, rather than simply extending to third order arithmetic or fourth order arithmetic, higher order reverse mathematics uses arithmetic in all finite types. In practice, this makes the definitions more straightforward, although the higher types (e.g. above fourth order) have little role.

Rather than using a set-based language like $\mathsf{RCA}_0$, this approach uses a function-based language. The first order part is the same, with operations for numerical addition, multiplication, and an equality relation on numbers. At higher levels, instead of a set membership operator $\in$, we have an operation for function composition. However, the functions we compose will have various higher types.

**Definition 12.4.1.** The collection of *finite types* is defined inductively as follows.

- Type 0 consists of the natural numbers.
- If $\rho$ and $\tau$ are types, then type $\rho \to \tau$ consists of all functions from type $\rho$ to type $\tau$. This type is also written $\tau(\rho)$ in the literature. When there is no confusion, we write $\sigma \to \tau \to \rho$ in place of $\sigma \to (\tau \to \rho)$; this means that $\to$ is treated as "right associative".

The *pure* types are denoted by natural numbers, and also defined inductively.

- 0 is a pure type.
- If $n$ is a pure type, $n+1$ is the type $n \to 0$. In particular, type $1 = 0 \to 0$ is the type of functions from $\omega$ to $\omega$, and $2 = (0 \to 0) \to 0$ is the type of functions from $\omega^\omega$ to $\omega$.

Every type is equivalent to some pure type, in the sense that the function spaces are effectively isomorphic. These equivalences are often easiest to demonstrate by also considering product types $\rho \times \tau$ and using the equivalence known as *Currying*:

$$\rho \to (\tau \to \sigma) \equiv (\rho \times \tau) \to \sigma \equiv \tau \to (\rho \to \sigma).$$

For example:

- Type $0 \to 1 = 0 \to (0 \to 0)$ is equivalent to $(0 \times 0) \to 0$, which is equivalent to type $1 = 0 \to 0$ through the use of a pairing function.
- Type $1 \to 1 = (0 \to 0) \to (0 \to 0)$ is equivalent to $((0 \to 0) \times 0) \to 0$. Using an effective pairing method again to convert $(0 \to 0) \times 0$ to $0 \to 0$, we see that $1 \to 1$ is equivalent to type $2 = (0 \to 0) \to 0$.

The *degree* of a type $\rho$ is the number $n$ so that $\rho$ is equivalent to pure type $n$.

We use $\lambda$ notation to informally describe a function of a given type. If $T^\sigma(x^\tau)$ is a term of type $\sigma$ with a parameter for a variable $x$ of type $\tau$, the notation $\lambda x^\tau.T^\sigma(x)$ refers to the function $f\colon \tau \to \sigma$ sending each $z$ of type $\tau$ to $T(z)$.

The precise collection of allowable terms $T^\sigma$ will vary with the particular formal system we are using, but at the very least we will have an infinite collection of variables of each type. The notation $x^\tau$ indicates that the variable $x$ is limited to values of type $\tau$; we can omit the superscript when the type is clear. Moreover, for each term $T^{\sigma\to\rho}$ and each term $s^\sigma$ we have a term $T(s)$ of type $\rho$. We will see several examples below that clarify the way $\lambda$ notation is used.

In function-based systems, rather than using set existence axioms, we use axioms that assert the existence of various higher order functions, which in this context are sometimes called *combinators*. For example:

- For each pair of types $\sigma, \tau$ there is a combinator $K_{\sigma,\tau}$ (also written $\Sigma_{\sigma,\tau}$) of type $\sigma \to \tau \to \sigma$ that allows us to form constant functions:

$$K_{\sigma,\tau}(x^\sigma) = \lambda y^\tau.[x].$$

- For each triple of types $\rho, \sigma, \tau$ there is a combinator $S_{\rho,\sigma,\tau}$ of type

$$(\rho \to \sigma \to \tau) \to (\rho \to \sigma) \to (\rho \to \tau).$$

To understand this combinator, use Currying to view a function $f^{\sigma\to\tau}(x^\rho)$ as if were a function $f^\tau(x^\rho, y^\sigma)$. Then, given a function $g\colon \rho \to \sigma$, we can form

$$S_{\rho,\sigma,\tau}(f,g) = \lambda x^\rho.[f(x, g(x))].$$

- For each type $\tau$, we have a combinator $R_\tau$ of type

$$\sigma \to (0 \to \sigma \to \sigma) \to 0 \to \sigma.$$

We view $R_\sigma$ as a higher order functional of the form $R(f^\sigma, g^{0\to\sigma\to\sigma}, n^0)$. The axioms for this combinator allow us to define functions by recursion:

$$R(f^\sigma, g, 0) = f;$$
$$R(f^\sigma, g, n+1) = g(n, R(f, g, n)).$$

**Definition 12.4.2.** The system E-PRA$^\omega$ is a formal system in many-sorted classical logic in the language of arithmetic in all finite types, along with:

- Terms and defining axioms for$K_{\sigma,\tau}$ and $S_{\rho,\sigma,\tau}$, for all types $\rho, \sigma, \tau$;
- A term and defining axioms for the combinator $R_0$, that is, the $R$ combinator in type 0 only;
- Terms for the number 0, and terms and defining axioms for the successor, addition, and multiplication operations, and the order relation. These axioms state that $\omega$ is an ordered semiring;
- For all types $\rho$ and $\tau$, an axiom of extensionality

$$(\forall x^\rho, y^\rho, z^{\rho\to\tau})[x =_\rho y \to z(x) =_\tau z(y)];$$

- The axiom scheme for quantifier-free induction. For each quantifier-free formula $A_0$ this scheme includes

$$(A_0(0) \wedge (\forall x)[A_0(x) \to A_0(x+1)]) \to (\forall x)A_0(x).$$

**Definition 12.4.3.** In the language of E-PRA$^\omega$, the principle of *quantifier free choice*, QF-AC$^{\rho,\tau}$, is the axiom scheme containing each formula of the form

$$(\forall x^\rho)(\exists y^\tau)A_0(x,y) \to (\exists Y^{\rho\to\tau})(\forall x^\rho)A_0(x, Y(x))$$

where $A_0$ is quantifier-free and may have parameters.

The quantifier free choice principle QF-AC$^{0,0}$ can be used to construct many useful functions of type 1. For example, given $m^0$, for each $x^0$ there is a $y^0$ such that $y = 0$ and $x = m$, or $y = 1$ and $x \neq m$. Hence there is a function $\mathrm{Eq}_m(x)$ so that $\mathrm{Eq}_m(x) = 0$ if $x = m$ and $\mathrm{Eq}_m(x) = 1$ otherwise.

We now have enough definitions to define the standard base systems for higher order reverse mathematics

**Definition 12.4.4.** The system $\mathsf{RCA}_0^\omega$ is defined to be $\mathsf{E\text{-}PRA}^\omega$ + $\mathsf{QF\text{-}AC}^{1,0}$. The system $\mathsf{RCA}_0^2$ is defined to consist of $\mathsf{E\text{-}PRA}^2$ (the second order part of $\mathsf{E\text{-}PRA}^\omega$) along with $\mathsf{QF\text{-}AC}^{0,0}$.

Although the definition only includes quantifier-free induction, $\mathsf{RCA}_0^\omega$ proves $\Sigma_1^0$ induction, using $\mathsf{QF\text{-}AC}^{0,0}$ (Exercise 12.5.9). The following result is Exercise 12.5.10.

**Proposition 12.4.5 (see Kohlenbach [185]).** *The systems* $\mathsf{RCA}_0$ *and* $\mathsf{RCA}_0^2$ *are bi-interpretable.*

**Corollary 12.4.6.** $\mathsf{RCA}_0^2$ *has an* $\omega$*-model in which the set of objects of type* 1 *is exactly the set of computable functions.*

As with second order reverse mathematics, higher order reverse mathematics results will show that a particular statement is provable from, or equivalent to, a function existence principle in higher order arithmetic. One key function principle is $(\exists^2)$:

$$(\exists^2) \colon (\exists E^2)(\forall f^1)(E(f) = 0 \leftrightarrow (\exists x^0)[f(x) = 0]).$$

The strength of principles such as $\exists^2$ comes from the ability to combine them with other functions and combinators.

*Example 12.4.7.* $\mathsf{RCA}_0^\omega + (\exists^2)$ proves that every function $g^{0\to 0}$ has a range. In this setting, a set is represented by its characteristic function. Given a function $g^1$, in $\mathsf{RCA}_0^\omega$ we can form the function

$$h(m^0, x^0) = \begin{cases} 0 & \text{if } g(x) = m; \\ 1 & \text{otherwise.} \end{cases}$$

Thus $h(m) = \lambda x^0.h(m, x)$ is type 1 for each $m$. Then $r(m) = 1 \dot{-} E^2(h(m))$ is the characteristic function of the range of $g$.

The example suggests that $\mathsf{RCA}_0^\omega + (\exists^2)$ can serve as a higher order version of $\mathsf{ACA}_0$; sometimes this system is denoted $\mathsf{ACA}_0^\omega$. The statements shown equivalent to $(\exists^2)$ in the following result are genuinely third-order statements, and could not be stated in second order arithmetic.

**Proposition 12.4.8 (Kohlenbach [185]).** *The following are equivalent to* $(\exists^2)$ *over* $\mathsf{RCA}_0^\omega$*:*

1. *There exists a function* $f \colon \mathbb{R} \to \mathbb{R}$ *that is not everywhere sequentially continuous.*
2. *There is a function* $P \colon \mathbb{R} \to \mathbb{R}$ *such that*

$$P(x) = \begin{cases} 0 & \text{if } x \leqslant 0; \\ 1 & \text{if } x > 0 \end{cases}$$

Moving to a higher order setting also allows us to state principles without the countability restrictions inherent in second order arithmetic. For example, we saw a second order version of the Heine–Borel theorem in Theorem 10.5.5. Normann and Sanders [234] and Sanders [270] have studied the following higher order version of Heine–Borel theorem:

$$(\mathsf{HBU})\colon (\forall \Phi\colon \mathbb{R} \to \mathbb{R}^+)(\exists y_0, \ldots, y_k \in [0,1])(\forall x \in [0,1])[x \in \bigcup_{i \leqslant k} B(y_i, \Phi(y_i))].$$

Intuitively, this principle is considering an open cover of $[0,1]$ in which each point $x$ is covered by the interval $B(x, \Phi(x))$; the conclusion is that $[0,1]$ is contained in a finite number of these intervals.

**Proposition 12.4.9 (Normann and Sanders [234]).** *The principle* HBU *is provable from the system* $\mathsf{RCA}_0^\omega + (\exists^3)$*, where*

$$(\exists^3)\colon (\exists E^3)(\forall x^2)(E(x) = 0 \leftrightarrow (\exists w^2)[x(w) = 0]).$$

*Moreover,* HBU *is not provable in* $\mathsf{RCA}_0^\omega$ *plus* $\Pi^1_k$ *comprehension, for any* $k$.

Sanders and Normann obtain similar results for a number of additional theorems of analysis, including versions of Cousin's lemma, Lindelöf's lemma, and theorems about the gauge integral. Sanders [272] has also used higher order arithmetic to study the coding inherent in second order reverse mathematics.

**Definition 12.4.10.** The following definitions are made in $\mathsf{RCA}_0^\omega$.

- A *set of reals* is coded by its characteristic function. If $A, B$ are sets of reals, we can state $A \subseteq B$ using the characteristic functions.
- A set $A$ of reals is *open* if, for each $x \in A$, there is some $r \in \mathbb{Q}^+$ with $B(x, r) \subseteq A$.
- A set $A \subseteq \mathbb{R}$ is *countable* if there is a function $F\colon \mathbb{R} \to \mathbb{N}$ such that $F \upharpoonright A$ is injective.
- A set $A \subseteq \mathbb{R}$ is *second countable* if there is a countable sequence of open balls such that every open covering of $A$ can be written as a union of balls from the sequence.

**Proposition 12.4.11 (Sanders [271]).**

1. *Over* $\mathsf{RCA}_0^\omega$, $(\exists^2)$ *is equivalent to the principle that every sequence* $\langle x_n \rangle$ *of reals is a countable set.*
2. *Over* $\mathsf{RCA}_0$*, the principle that the unit interval is second-countable implies* $(\exists^2)$.

Higher order systems can also be used to examine the strength of our coding system for continuous functions, as in the following theorem.

**Proposition 12.4.12 (Kohlenbach [184]).** $\mathsf{E\text{-}PRA}^\omega + \mathsf{QF\text{-}AC}^{1,0} + \mathsf{QF\text{-}AC}^{0,1}$ *does not prove that every continuous functional* $\Phi^2$ *(that is,* $\mathbb{N}^\mathbb{N} \to \mathbb{N}$*) can be represented though a continuous function code as in Definition 10.3.3.*

Constructive versions of higher order arithmetic can also be studied. These use a similar framework to $\mathsf{E\text{-}PRA}^\omega$, using constructive logic (without the law of the excluded middle) rather than classical logic. Constructive reverse mathematics is typically carried out in informal systems, like ordinary constructive mathematics, but can be formalized into constructive systems of higher order arithmetic.

Using methods from proof theory, constructive higher order systems can also be used to study the principle of the excluded middle [159] and formalized Weihrauch reducibility [161]. Uftring [315] studies formalized Weihrauch reducibility with a version of higher order arithmetic incorporating features from linear logic.

## 12.5 Exercises

**Exercise 12.5.1.** Prove Lemma 12.1.5.

**Exercise 12.5.2.** Give a careful proof in $\mathsf{ATR}_0$ of the existence of the set $Z$ in Lemma 12.1.7.

**Exercise 12.5.3.** Prove the following in $\mathsf{RCA}_0$.

1. For all well orderings $X$, it is not the case that $X \hookrightarrow X$.
2. If $x <_X y$, then there is no embedding of $X^{[<_X y]}$ into $X^{[<_X x]}$.

**Exercise 12.5.4.** Prove Lemma 12.1.23.

**Exercise 12.5.5.** Prove Lemma 12.1.17.

**Exercise 12.5.6 (Simpson [288]).** Show that $\mathsf{ACA}_0$ proves that if $B \subseteq \mathbb{N}^\mathbb{N}$ is a Borel code and $X \in \mathbb{N}^\mathbb{N}$ then the evaluation map for $B$ at $X$ (if it exists) is unique.

**Exercise 12.5.7.** Prove Proposition 12.3.5.

**Exercise 12.5.8.** Prove that every finite type is equivalent to a pure type. Moreover, $\deg(\rho \to \tau)$ is $\max\{\deg(\rho) + 1, \deg(\tau)\}$.

**Exercise 12.5.9.** Show that the $\Sigma^0_1$ induction scheme is provable in $\mathsf{RCA}^\omega_0$.

**Exercise 12.5.10.** Prove Proposition 12.4.5.

**Exercise 12.5.11.** Let $U \subseteq \omega^\omega$ be an undetermined set. Use $U$ to construct a set $A \subseteq \omega^\omega$ so that $A$ is determined but its complement is not.

# References


1. Klaus Ambos-Spies and Peter A. Fejer, *Degrees of unsolvability*, Computational logic, Handb. Hist. Log., vol. 9, Elsevier/North-Holland, Amsterdam, 2014, pp. 443–494. MR 3362163
2. Klaus Ambos-Spies, Bjørn Kjos-Hanssen, Steffen Lempp, and Theodore A. Slaman, *Comparing DNR and WWKL*, J. Symbolic Logic **69** (2004), no. 4, 1089–1104. MR 2135656
3. Emil Artin and Otto Schreier, *Algebraische Konstruktion reeller Körper*, Abh. Math. Sem. Univ. Hamburg **5** (1927), no. 1, 85–99. MR 3069467
4. C. J. Ash and J. Knight, *Computable structures and the hyperarithmetical hierarchy*, Studies in Logic and the Foundations of Mathematics, vol. 144, North-Holland Publishing Co., Amsterdam, 2000. MR 1767842
5. Eric P. Astor, Damir D. Dzhafarov, Antonio Montalbán, Reed Solomon, and Linda Brown Westrick, *The determined property of Baire in reverse math*, J. Symb. Log. **85** (2020), no. 1, 166–198. MR 4085059
6. Eric P. Astor, Damir D. Dzhafarov, Reed Solomon, and Jacob Suggs, *The uniform content of partial and linear orders*, Ann. Pure Appl. Logic **168** (2017), no. 6, 1153–1171. MR 3628269
7. Jeremy Avigad, *Notes on $\Pi^1_1$-conservativity, $\omega$-submodels, and the collection schema*, Tech. Report CMU-PHIL-125, Carnegie Mellon University, 2002.
8. ———, *Mathematical logic and computation*, Cambridge University Press, 2022.
9. Jeremy Avigad, Edward T. Dean, and Jason Rute, *Algorithmic randomness, reverse mathematics, and the dominated convergence theorem*, Ann. Pure Appl. Logic **163** (2012), no. 12, 1854–1864. MR 2964874
10. Jeremy Avigad and Solomon Feferman, *Gödel's functional ("Dialectica") interpretation*, Handbook of proof theory, Stud. Logic Found. Math., vol. 137, North-Holland, Amsterdam, 1998, pp. 337–405. MR 1640329
11. Jeremy Avigad and Ksenija Simic, *Fundamental notions of analysis in subsystems of second-order arithmetic*, Ann. Pure Appl. Logic **139** (2006), no. 1-3, 138–184. MR 2206254
12. James E. Baumgartner, *A short proof of Hindman's theorem*, J. Combinatorial Theory Ser. A **17** (1974), 384–386. MR 354394
13. David R. Belanger, *Conservation theorems for the cohesiveness principle*, preprint, 2015.
14. Josef Berger, Hajime Ishihara, Takayuki Kihara, and Takako Nemoto, *The binary expansion and the intermediate value theorem in constructive reverse mathematics*, Arch. Math. Logic **58** (2019), no. 1-2, 203–217. MR 3902812
15. Josef Berger, Hajime Ishihara, and Peter Schuster, *The weak Kőnig lemma, Brouwer's Fan theorem, de Morgan's law, and dependent choice*, Rep. Math. Logic (2012), no. 47, 63–86. MR 3185436
16. Andreas Blass, *Combinatorial cardinal characteristics of the continuum*, Handbook of set theory. Vols. 1, 2, 3, Springer, Dordrecht, 2010, pp. 395–489. MR 2768685
17. Andreas R. Blass, Jeffry L. Hirst, and Stephen G. Simpson, *Logical analysis of some theorems of combinatorics and topological dynamics*, Logic and combinatorics (Arcata, Calif., 1985), Contemp. Math., vol. 65, Amer. Math. Soc., Providence, RI, 1987, pp. 125–156. MR 891245


 
D. D. Dzhafarov, C. Mummert, *Reverse Mathematics*, Theory and Applications of Computability, https://doi.org/10.1007/978-3-031-11367-3

18. Robert Bonnet, *Stratifcations et extension des genres de chaînes dénombrables*, C. R. Acad. Sci. Paris Sér. A-B **269** (1969), A880–A882. MR 252282
19. Vasco Brattka and Guido Gherardi, *Effective choice and boundedness principles in computable analysis*, Bull. Symbolic Logic **17** (2011), no. 1, 73–117. MR 2760117 (2012c:03108)
20. Vasco Brattka and Tahina Rakotoniaina, *On the uniform computational content of Ramsey's theorem*, The Journal of Symbolic Logic **82** (2017), no. 4, 1278–1316. MR 3743611
21. Douglas K. Brown, *Functional analysis in weak subsystems of second order arithmetic*, Ph.D. thesis, Pennsylvania State University, 1987. MR 2635633
22. ———, *Notions of closed subsets of a complete separable metric space in weak subsystems of second-order arithmetic*, Logic and computation (Pittsburgh, PA, 1987), Contemp. Math., vol. 106, Amer. Math. Soc., Providence, RI, 1990, pp. 39–50. MR 1057814
23. Douglas K. Brown, Mariagnese Giusto, and Stephen G. Simpson, *Vitali's theorem and WWKL*, Arch. Math. Logic **41** (2002), no. 2, 191–206. MR 1890192
24. Lorenzo Carlucci, *A weak variant of Hindman's Theorem stronger than Hilbert's Theorem*, Arch. Math. Logic **57** (2018), no. 3-4, 381–389. MR 3778965
25. ———, *"Weak yet strong" restrictions of Hindman's finite sums theorem*, Proc. Amer. Math. Soc. **146** (2018), no. 2, 819–829. MR 3731714
26. Lorenzo Carlucci, Leszek Aleksander Koł odziejczyk, Francesco Lepore, and Konrad Zdanowski, *New bounds on the strength of some restrictions of Hindman's theorem*, Unveiling dynamics and complexity, Lecture Notes in Comput. Sci., vol. 10307, Springer, Cham, 2017, pp. 210–220. MR 3678752
27. Peter Cholak, Rodney G. Downey, and Greg Igusa, *Any FIP real computes a 1-generic*, Trans. Amer. Math. Soc. **369** (2017), no. 8, 5855–5869. MR 3646781
28. Peter Cholak, Damir D. Dzhafarov, Denis R. Hirschfeldt, and Ludovic Patey, *Some results concerning the* $\mathsf{SRT}^2_2$ *vs.* $\mathsf{COH}$ *problem*, Computability **9** (2020), no. 3–4, 193–217. MR 4133713
29. Peter Cholak, Sergey Goncharov, Bakhadyr Khoussainov, and Richard A. Shore, *Computably categorical structures and expansions by constants*, J. Symbolic Logic **64** (1999), no. 1, 13–37. MR 1683891
30. Peter Cholak and Ludovic Patey, *Thin set theorems and cone avoidance*, Trans. Amer. Math. Soc. **373** (2020), no. 4, 2743–2773. MR 4069232
31. Peter A. Cholak, Damir D. Dzhafarov, Denis R. Hirschfeldt, Antonio Montalbán, and Linda Brown Westrick, *Abstract of "FRG: Collaborative Research: Computability-Theoretic Aspects of Combinatorics", United States National Science Foundation*, 2019.
32. Peter A. Cholak, Mariagnese Giusto, Jeffry L. Hirst, and Carl G. Jockusch, Jr., *Free sets and reverse mathematics*, Reverse mathematics 2001, Lect. Notes Log., vol. 21, Assoc. Symbol. Logic, La Jolla, CA, 2005, pp. 104–119. MR 2185429 (2006g:03101)
33. Peter A. Cholak, Carl G. Jockusch, and Theodore A. Slaman, *On the strength of Ramsey's theorem for pairs*, J. Symbolic Logic **66** (2001), no. 1, 1–55. MR 1825173 (2002c:03094)
34. C. T. Chong, Steffen Lempp, and Yue Yang, *On the role of the collection principle for* $\Sigma^0_2$*-formulas in second-order reverse mathematics*, Proc. Amer. Math. Soc. **138** (2010), no. 3, 1093–1100. MR 2566574 (2011b:03013)
35. C. T. Chong, Wei Li, Wei Wang, and Yue Yang, *On the strength of Ramsey's theorem for trees*, Adv. Math. **369** (2020), 107180, 39. MR 4093609
36. C. T. Chong, Wei Li, and Yue Yang, *Nonstandard models in recursion theory and reverse mathematics*, Bull. Symb. Log. **20** (2014), no. 2, 170–200. MR 3230838
37. C. T. Chong and K. J. Mourad, $\Sigma_n$ *definable sets without* $\Sigma_n$ *induction*, Trans. Amer. Math. Soc. **334** (1992), no. 1, 349–363. MR 1117216
38. C. T. Chong, Lei Qian, Theodore A. Slaman, and Yue Yang, $\Sigma_2$ *induction and infinite injury priority arguments. III. Prompt sets, minimal pairs and Shoenfield's conjecture*, Israel J. Math. **121** (2001), 1–28. MR 1818378
39. C. T. Chong, Theodore A. Slaman, and Yue Yang, *The metamathematics of stable Ramsey's theorem for pairs*, J. Amer. Math. Soc. **27** (2014), no. 3, 863–892. MR 3194495
40. ———, *The inductive strength of Ramsey's Theorem for Pairs*, Adv. Math. **308** (2017), 121–141. MR 3600057

41. C. T. Chong and Yue Yang, *$\Sigma_2$ induction and infinite injury priority argument. I. Maximal sets and the jump operator*, J. Symbolic Logic **63** (1998), no. 3, 797–814. MR 1649062
42. Chi Tat Chong, Wei Li, Lu Liu, and Yue Yang, *The strength of Ramsey's theorem for pairs over trees: I. Weak König's lemma*, Trans. Amer. Math. Soc. **374** (2021), no. 8, 5545–5581. MR 4293780
43. Chitat Chong, Wei Wang, and Yue Yang, *Conservation strength of the infinite pigeonhole principle for trees*, preprint, arXiv:2110.06026, 2021.
44. Gustave Choquet, *Lectures on analysis. Vol. I: Integration and topological vector spaces*, Edited by J. Marsden, T. Lance and S. Gelbart, W. A. Benjamin, Inc., New York-Amsterdam, 1969. MR 0250011 (40 #3252)
45. Jennifer Chubb, Jeffry L. Hirst, and Timothy H. McNicholl, *Reverse mathematics and partitions of trees*, unpublished, 2005.
46. V. Chvátal, *On finite polarized partition relations*, Canad. Math. Bull. **12** (1969), 321–326. MR 260606
47. Paul Cohen, *The independence of the continuum hypothesis*, Proc. Nat. Acad. Sci. U.S.A. **50** (1963), 1143–1148. MR 157890
48. Paul J. Cohen, *The independence of the continuum hypothesis. II*, Proc. Nat. Acad. Sci. U.S.A. **51** (1964), 105–110. MR 159745
49. ______, *Independence results in set theory*, Theory of Models (Proc. 1963 Internat. Sympos. Berkeley), North-Holland, Amsterdam, 1965, pp. 39–54. MR 0195710
50. W. W. Comfort, *Ultrafilters: some old and some new results*, Bull. Amer. Math. Soc. **83** (1977), no. 4, 417–455. MR 454893
51. Chris J. Conidis, *Classifying model-theoretic properties*, J. Symbolic Logic **73** (2008), no. 3, 885–905. MR 2444274
52. ______, *Chain conditions in computable rings*, Trans. Amer. Math. Soc. **362** (2010), no. 12, 6523–6550. MR 2678985
53. ______, *The computability, definability, and proof theory of Artinian rings*, Adv. Math. **341** (2019), 1–39. MR 3872843
54. ______, *Computability theoretic aspects of an antichain theorem for infinite extendible trees of nontrivial rank*, preprint, 2020.
55. Chris J. Conidis and Theodore A. Slaman, *Random reals, the rainbow Ramsey theorem, and arithmetic conservation*, J. Symbolic Logic **78** (2013), no. 1, 195–206. MR 3087070
56. Jared Corduan, Marcia J. Groszek, and Joseph R. Mileti, *Reverse mathematics and Ramsey's property for trees*, J. Symbolic Logic **75** (2010), no. 3, 945–954. MR 2723776
57. Barbara F. Csima, Damir D. Dzhafarov, Denis R. Hirschfeldt, Carl G. Jockusch, Jr., Reed Solomon, and Linda Brown Westrick, *The reverse mathematics of Hindman's theorem for sums of exactly two elements*, Computability **8** (2020), no. 3–4, 253–263. MR 4016726
58. Barbara F. Csima, Denis R. Hirschfeldt, Julia F. Knight, and Robert I. Soare, *Bounding prime models*, J. Symbolic Logic **69** (2004), no. 4, 1117–1142. MR 2135658
59. Barbara F. Csima and Joseph R. Mileti, *The strength of the rainbow Ramsey theorem*, J. Symbolic Logic **74** (2009), no. 4, 1310–1324. MR 2583822 (2011b:03086)
60. Paul-Elliot Anglès d'Auriac, *Infinite computations in algorithmic randomness and reverse mathematics*, Ph.D. thesis, Université Paris-Est, 2019.
61. Paul-Elliot Anglès d'Auriac, Peter A. Cholak, Damir D. Dzhafarov, Benoit Monin, and Ludovic Patey, *Milliken's tree theorem and its applications: a computability-theoretic perspective*, preprint, arXiv:2007.09739, 2021.
62. Morton Davis, *Infinite games of perfect information*, Advances in Game Theory, Princeton Univ. Press, Princeton, N.J., 1964, pp. 85–101. MR 0170727
63. Adam R. Day, *On the strength of two recurrence theorems*, J. Symb. Log. **81** (2016), no. 4, 1357–1374. MR 3580476
64. Walter Dean and Sean Walsh, *The prehistory of the subsystems of second-order arithmetic*, Rev. Symb. Log. **10** (2017), no. 2, 357–396. MR 3650982
65. David Diamondstone, Rodney G. Downey, Noam Greenberg, and Dan Turetsky, *The finite intersection principle and genericity*, Math. Proc. Cambridge Philos. Soc. **160** (2016), no. 2, 279–297. MR 3458954

66. David E. Diamondstone, Damir D. Dzhafarov, and Robert I. Soare, $\Pi^0_1$ *classes, Peano arithmetic, randomness, and computable domination*, Notre Dame J. Form. Log. **51** (2010), no. 1, 127–159. MR 2666574 (2011g:03101)
67. Hannes Diener and Hajime Ishihara, *Bishop-style constructive reverse mathematics*, Handbook of computability and complexity in analysis, Theory Appl. Comput., Springer, Cham, 2021, pp. 347–365. MR 4300760
68. Natasha Dobrinen, *A list of problems on the reverse mathematics of Ramsey theory on the Rado graph and on infinite, finitely branching trees*, unpublished, arXiv:1808.10227, 2018.
69. Natasha Dobrinen, Claude Laflamme, and Norbert Sauer, *Rainbow Ramsey simple structures*, Discrete Math. **339** (2016), no. 11, 2848–2855. MR 3518438
70. François G. Dorais, *Reverse mathematics of compact countable second-countable spaces*, preprint, arXiv:1110.6555, 2011.
71. ———, *Countable topological spaces in subsystems of second-order arithmetic*, unpublished, `github.com/fgdorais/CSC`, 2018.
72. François G. Dorais, Damir D. Dzhafarov, Jeffry L. Hirst, Joseph R. Mileti, and Paul Shafer, *On uniform relationships between combinatorial problems*, Trans. Amer. Math. Soc. **368** (2016), no. 2, 1321–1359. MR 3430365
73. R. G. Downey and Stuart A. Kurtz, *Recursion theory and ordered groups*, Ann. Pure Appl. Logic **32** (1986), no. 2, 137–151. MR 863331
74. Rodney G. Downey and Denis R. Hirschfeldt, *Algorithmic randomness and complexity*, Theory and Applications of Computability, Springer, New York, 2010. MR 2732288 (2012g:03001)
75. Rodney G. Downey, Denis R. Hirschfeldt, Steffen Lempp, and Reed Solomon, *A* $\Delta^0_2$ *set with no infinite low subset in either it or its complement*, J. Symbolic Logic **66** (2001), no. 3, 1371–1381. MR 1856748 (2002i:03046)
76. Rodney G. Downey, Denis R. Hirschfeldt, Steffen Lempp, and Reed Solomon, *Reverse mathematics of the Nielsen–Schreier theorem*, Proc. International Conf. on Math. Logic (2002), 59–71.
77. Rodney G. Downey, Steffen Lempp, and Joseph R. Mileti, *Ideals in computable rings*, J. Algebra **314** (2007), no. 2, 872–887. MR 2344588
78. Damir Dzhafarov, Stephen Flood, Reed Solomon, and Linda Westrick, *Effectiveness for the Dual Ramsey Theorem*, Notre Dame Journal of Formal Logic **62** (2021), no. 3, 455 – 490. MR 4323042
79. Damir D. Dzhafarov, *Cohesive avoidance and strong reductions*, Proc. Amer. Math. Soc. **143** (2015), no. 2, 869–876. MR 3283673
80. ———, *Strong reductions between combinatorial principles*, J. Symbolic Logic **81** (2016), no. 4, 1405–1431. MR 3579116
81. Damir D. Dzhafarov, Jun Le Goh, Denis R. Hirschfeldt, Ludovic Patey, and Arno Pauly, *Ramsey's theorem and products in the Weihrauch degrees*, Computability **9** (2020), no. 2, 85–110. MR 4100139
82. Damir D. Dzhafarov, Denis R. Hirschfeldt, and Sarah Reitzes, *Reduction games, provability, and compactness*, preprint, arXiv:2008.00907, 2021.
83. Damir D. Dzhafarov and Jeffry L. Hirst, *The polarized Ramsey's theorem*, Arch. Math. Logic **48** (2009), no. 2, 141–157. MR 2487221
84. Damir D. Dzhafarov, Jeffry L. Hirst, and Tamara J. Lakins, *Ramsey's theorem for trees: the polarized tree theorem and notions of stability*, Arch. Math. Logic **49** (2010), no. 3, 399–415. MR 2609990
85. Damir D. Dzhafarov, Carl G. Jockusch, Reed Solomon, and Linda Brown Westrick, *Effectiveness of Hindman's theorem for bounded sums*, Computability and Complexity: Essays Dedicated to Rodney G. Downey on the Occasion of His 60th Birthday (Adam Day, Michael Fellows, Noam Greenberg, Bakhadyr Khoussainov, Alexander Melnikov, and Frances Rosamond, eds.), Springer, 2017, pp. 134–142. MR 3629719
86. Damir D. Dzhafarov and Carl G. Jockusch, Jr., *Ramsey's theorem and cone avoidance*, J. Symbolic Logic **74** (2009), no. 2, 557–578. MR 2518811

87. Damir D. Dzhafarov and Carl Mummert, *Reverse mathematics and properties of finite character*, Ann. Pure Appl. Logic **163** (2012), no. 9, 1243–1251. MR 2926282
88. ———, *On the strength of the finite intersection principle*, Israel J. Math. **196** (2013), no. 1, 345–361. MR 3096595
89. Damir D. Dzhafarov and Ludovic Patey, *Coloring trees in reverse mathematics*, Adv. Math. **318** (2017), 497–514. MR 3689748
90. Damir D. Dzhafarov, Ludovic Patey, Reed Solomon, and Linda Brown Westrick, *Ramsey's theorem for singletons and strong computable reducibility*, Proc. Amer. Math. Soc. **145** (2017), no. 3, 1343–1355. MR 3589330
91. Benedict Eastaugh, *Set existence principles and closure conditions: unravelling the standard view of reverse mathematics*, Philos. Math. (3) **27** (2019), no. 2, 153–176. MR 4011555
92. Benedict Eastaugh and Sam Sanders, *Reverse mathematics and coding overhead*, preprint, 2021.
93. Herbert B. Enderton, *A mathematical introduction to logic*, second ed., Harcourt/Academic Press, Burlington, MA, 2001. MR 1801397
94. Richard L. Epstein, *Degrees of unsolvability: structure and theory*, Lecture Notes in Mathematics, vol. 759, Springer, Berlin, 1979. MR 551620
95. P. Erdős and A. Hajnal, *Some results and problems on certain polarized partitions*, Acta Math. Acad. Sci. Hungar. **21** (1970), 369–392. MR 281642
96. P. Erdős, A. Hajnal, and E. C. Milner, *Set mappings and polarized partition relations*, Combinatorial theory and its applications, I (Proc. Colloq., Balatonfüred, 1969), 1970, pp. 327–363. MR 0299537
97. ———, *Polarized partition relations for ordinal numbers*, Studies in Pure Mathematics (Presented to Richard Rado), Academic Press, London, 1971, pp. 63–87. MR 0277390
98. P. Erdős and L. Lovász, *Problems and results on 3-chromatic hypergraphs and some related questions*, Infinite and finite sets, Vol. II, North-Holland, 1975, pp. 609–627. Colloq. Math. Soc. János Bolyai, Vol. 10. MR 0382050
99. P. Erdős and L. Moser, *On the representation of directed graphs as unions of orderings*, Magyar Tud. Akad. Mat. Kutató Int. Közl. **9** (1964), 125–132. MR 168494
100. ———, *A problem on tournaments*, Canad. Math. Bull. **7** (1964), 351–356. MR 166773
101. Ju. L. Eršov, *Theorie der Numerierungen. III*, Z. Math. Logik Grundlagen Math. **23** (1977), no. 4, 289–371. MR 439603
102. Solomon Feferman, *Some applications of the notions of forcing and generic sets: Summary*, Theory of Models (Proc. 1963 Internat. Sympos. Berkeley), North-Holland, Amsterdam, 1965, pp. 89–95. MR 0202577
103. ———, *Impredicativity of the existence of the largest divisible subgroup of an abelian p-group*, Model theory and algebra (A memorial tribute to Abraham Robinson), Springer-Verlag, 1975, pp. 117–130. Lecture Notes in Math., Vol. 498. MR 0401446
104. ———, *Theories of finite type related to mathematical practice*, Handbook of mathematical logic, Stud. Logic Found. Math., vol. 90, North-Holland, Amsterdam, 1977, pp. 913–971. MR 3727428
105. L. Feiner, *The strong homogeneity conjecture*, J. Symbolic Logic **35** (1970), 375–377. MR 286655
106. Johanna N. Y. Franklin and Christopher P. Porter (eds.), *Algorithmic randomness—progress and prospects*, Lecture Notes in Logic, vol. 50, Cambridge University Press, Cambridge; Association for Symbolic Logic, Ithaca, NY, 2020. MR 4382437
107. Harvey M. Friedman, *Higher set theory and mathematical practice*, Ann. Math. Logic **2** (1970/71), no. 3, 325–357. MR 284327
108. ———, *Some systems of second order arithmetic and their use*, Proceedings of the International Congress of Mathematicians (Vancouver, B. C., 1974), Vol. 1, Canad. Math. Congress, Montreal, Que., 1975, pp. 235–242. MR 0429508 (55 #2521)
109. ———, *Systems of second-order arithmetic with restricted induction, I, II*, J. Symbolic Logic **41** (1976), no. 2, 551–560.
110. ———, *53:free sets/reverse math*, 1999, FOM email list.

111. ______, *Metamathematics of Ulm Theory*, preprint, 32 pages, 2001.
112. ______, *The inevitability of logical strength: strict reverse mathematics*, Logic Colloquium 2006, Lect. Notes Log., vol. 32, Assoc. Symbol. Logic, Chicago, IL, 2009, pp. 135–183. MR 2562551
113. ______, *The emergence of (strict) reverse mathematics*, preprint, 110 pages, 2021.
114. Harvey M. Friedman and Jeffry L. Hirst, *Weak comparability of well orderings and reverse mathematics*, Ann. Pure Appl. Logic **47** (1990), no. 1, 11–29. MR 1050559
115. Harvey M. Friedman, Kenneth McAloon, and Stephen G. Simpson, *A finite combinatorial principle which is equivalent to the* 1*-consistency of predicative analysis*, Patras Logic Symposion (Patras, 1980), Studies in Logic and the Foundations of Mathematics, vol. 109, North-Holland, Amsterdam-New York, 1982, pp. 197–230. MR 694261
116. Harvey M. Friedman and Stephen G. Simpson, *Issues and problems in reverse mathematics*, Computability theory and its applications (Boulder, CO, 1999), Contemp. Math., vol. 257, Amer. Math. Soc., Providence, RI, 2000, pp. 127–144. MR 1770738
117. Harvey M. Friedman, Stephen G. Simpson, and Rick L. Smith, *Countable algebra and set existence axioms*, Ann. Pure Appl. Logic **25** (1983), no. 2, 141–181. MR 725732
118. ______, *Addendum to: "Countable algebra and set existence axioms" [Ann. Pure Appl. Logic* **25** *(1983), no. 2, 141–181; MR0725732 (85i:03157)]*, Ann. Pure Appl. Logic **28** (1985), no. 3, 319–320. MR 790391
119. A. Fröhlich and J. C. Shepherdson, *On the factorisation of polynomials in a finite number of steps*, Math. Z. **62** (1955), 331–334. MR 71385
120. H. Furstenberg, *Recurrence in ergodic theory and combinatorial number theory*, Princeton University Press, 1981, M. B. Porter Lectures. MR 603625
121. David Gale and F. M. Stewart, *Infinite games with perfect information*, Contributions to the theory of games, vol. 2, Annals of Mathematics Studies, no. 28, Princeton University Press, Princeton, N.J., 1953, pp. 245–266. MR 0054922
122. Guido Gherardi and Alberto Marcone, *How incomputable is the separable Hahn-Banach theorem?*, Proceedings of the Fifth International Conference on Computability and Complexity in Analysis (CCA 2008), Electron. Notes Theor. Comput. Sci., vol. 221, Elsevier Sci. B. V., Amsterdam, 2008, pp. 85–102. MR 2873349
123. Mariagnese Giusto and Alberto Marcone, *Lebesgue numbers and Atsuji spaces in subsystems of second-order arithmetic*, Arch. Math. Logic **37** (1998), no. 5-6, 343–362, Logic Colloquium '95 (Haifa). MR 1634278
124. Mariagnese Giusto and Stephen G. Simpson, *Located sets and reverse mathematics*, J. Symbolic Logic **65** (2000), no. 3, 1451–1480. MR 1791384 (2003b:03085)
125. Kurt Gödel, *Über formal unentscheidbare sätze der principia mathematica und verwandter systeme, i*, Monatshefte für Mathematik und Physik **38** (1931), 173–198.
126. Kurt Gödel, *The consistency of the axiom of choice and of the generalized continuum-hypothesis*, Proceedings of the National Academy of Sciences **24** (1938), no. 12, 556–557.
127. S. S. Gončarov and A. T. Nurtazin, *Constructive models of complete decidable theories*, Algebra i Logika **12** (1973), 125–142, 243. MR 0398816
128. Ronald L. Graham, Bruce L. Rothschild, and Joel H. Spencer, *Ramsey theory*, second ed., John Wiley & Sons, Inc., New York, 1990. MR 1044995
129. Noam Greenberg and Antonio Montalbán, *Ranked structures and arithmetic transfinite recursion*, Trans. Amer. Math. Soc. **360** (2008), no. 3, 1265–1307. MR 2357696
130. Marcia J. Groszek and Theodore A. Slaman, *Foundations of the priority method I: finite and infinite injury*, preprint, 2021.
131. Sherwood Hachtman and Justin Palumbo, *Notes on determinacy*, unpublished lecture notes, UCLA Logic Summer School: Determinacy, `homepages.math.uic.edu/~shac/determinacy/determinacy2015.pdf`, 2015.
132. Petr Hájek, *Interpretability and fragments of arithmetic*, Arithmetic, proof theory, and computational complexity (Prague, 1991), Oxford Logic Guides, vol. 23, Oxford Univ. Press, New York, 1993, pp. 185–196. MR 1236462
133. Petr Hájek and Antonín Kučera, *On recursion theory in* $I\Sigma_1$, J. Symbolic Logic **54** (1989), no. 2, 576–589. MR 997890

134. Petr Hájek and Pavel Pudlák, *Metamathematics of first-order arithmetic*, Perspectives in Mathematical Logic, Springer-Verlag, Berlin, 1993. MR 1219738 (94d:03001)
135. J. D. Halpern and H. Läuchli, *A partition theorem*, Trans. Amer. Math. Soc. **124** (1966), 360–367. MR 200172
136. Valentina S. Harizanov, *Pure computable model theory*, Handbook of recursive mathematics, Vol. 1, Stud. Logic Found. Math., vol. 138, North-Holland, Amsterdam, 1998, pp. 3–114. MR 1673621
137. Leo Harrington, *Recursively presentable prime models*, J. Symbolic Logic **39** (1974), 305–309. MR 351804
138. Kenneth Harris, *The complexity of classical theorems on saturated models*, Ph.D. thesis, University of Chicago, 2007, 114 pages. MR 2710193
139. Matthew Harrison-Trainor, *There is no classification of the decidably presentable structures*, J. Math. Log. **18** (2018), no. 2, 1850010, 41. MR 3878472
140. Kostas Hatzikiriakou and Stephen G. Simpson, $\mathsf{WKL}_0$ *and orderings of countable abelian groups*, Logic and computation (Pittsburgh, PA, 1987), Contemp. Math., vol. 106, Amer. Math. Soc., Providence, RI, 1990, pp. 177–180. MR 1057821
141. Leon Henkin, *A generalization of the concept of $\omega$-consistency*, J. Symbolic Logic **19** (1954), 183–196. MR 0063324 (16,103d)
142. E. Herrmann, *Infinite chains and antichains in computable partial orderings*, J. Symbolic Logic **66** (2001), no. 2, 923–934. MR 1833487 (2003e:03082)
143. Neil Hindman, *Finite sums from sequences within cells of a partition of $N$*, J. Combinatorial Theory Ser. A **17** (1974), 1–11. MR 349574
144. Neil Hindman, Imre Leader, and Dona Strauss, *Open problems in partition regularity*, Combin. Probab. Comput. **12** (2003), no. 5-6, 571–583. MR 2037071
145. Neil Hindman and Dona Strauss, *Algebra in the Stone-Čech compactification*, De Gruyter Textbook, Walter de Gruyter & Co., Berlin, 2012, Second edition. MR 2893605
146. Denis R. Hirschfeldt, *Computable trees, prime models, and relative decidability*, Proc. Amer. Math. Soc. **134** (2006), no. 5, 1495–1498. MR 2199197
147. ———, *Slicing the truth: On the computable and reverse mathematics of combinatorial principles*, Lecture notes series / Institute for Mathematical Sciences, National University of Singapore, World Scientific Publishing Company Incorporated, 2014. MR 3244278
148. Denis R. Hirschfeldt and Carl G. Jockusch, Jr., *On notions of computability-theoretic reduction between $\Pi^1_2$ principles*, J. Math. Log. **16** (2016), no. 1, 1650002, 59. MR 3518779
149. Denis R. Hirschfeldt, Carl G. Jockusch, Jr., Bjørn Kjos-Hanssen, Steffen Lempp, and Theodore A. Slaman, *The strength of some combinatorial principles related to Ramsey's theorem for pairs*, Computational prospects of infinity. Part II. Presented talks, Lect. Notes Ser. Inst. Math. Sci. Natl. Univ. Singap., vol. 15, World Sci. Publ., Hackensack, NJ, 2008, pp. 143–161. MR 2449463 (2009i:03038)
150. Denis R. Hirschfeldt, Karen Lange, and Richard A. Shore, *Induction, bounding, weak combinatorial principles, and the homogeneous model theorem*, Mem. Amer. Math. Soc. **249** (2017), no. 1187, iii+101. MR 3709722
151. Denis R. Hirschfeldt and Sarah Reitzes, *Thin set versions of Hindman's theorem*, preprint, arXiv:2203.08658, 2022.
152. Denis R. Hirschfeldt and Richard A. Shore, *Combinatorial principles weaker than Ramsey's theorem for pairs*, J. Symbolic Logic **72** (2007), no. 1, 171–206. MR 2298478
153. Denis R. Hirschfeldt, Richard A. Shore, and Theodore A. Slaman, *The atomic model theorem and type omitting*, Trans. Amer. Math. Soc. **361** (2009), no. 11, 5805–5837. MR 2529915
154. Jeffry L. Hirst, *Combinatorics in subsystems of second order arithmetic*, Ph.D. thesis, The Pennsylvania State University, 1987. MR 2635978
155. Jeffry L. Hirst, *Hindman's theorem, ultrafilters, and reverse mathematics*, J. Symbolic Logic **69** (2004), no. 1, 65–72. MR 2039345
156. ———, *Minima of initial segments of infinite sequences of reals*, MLQ Math. Log. Q. **50** (2004), no. 1, 47–50. MR 2029605
157. ———, *Reverse mathematics of separably closed sets*, Arch. Math. Logic **45** (2006), no. 1, 1–2. MR 2209734

158. ———, *Representations of reals in reverse mathematics*, Bull. Pol. Acad. Sci. Math. **55** (2007), no. 4, 303–316. MR 2369116
159. Jeffry L. Hirst and Carl Mummert, *Reverse mathematics and uniformity in proofs without excluded middle*, Notre Dame J. Form. Log. **52** (2011), no. 2, 149–162. MR 2794648
160. ———, *Reverse mathematics of matroids*, Computability and complexity, Lecture Notes in Comput. Sci., vol. 10010, Springer, Cham, 2017, pp. 143–159. MR 3629720
161. ———, *Using Ramsey's theorem once*, Arch. Math. Logic **58** (2019), no. 7-8, 857–866. MR 4003638
162. Paul Howard and Jean E. Rubin, *The axiom of choice for well-ordered families and for families of well-orderable sets*, J. Symbolic Logic **60** (1995), no. 4, 1115–1117. MR 1367198 (96k:03113)
163. Noah A. Hughes, *Applications of computability theory to infinitary combinatorics*, Ph.D. thesis, University of Connecticut, 2021.
164. James Hunter, *Higher-order reverse topology*, Ph.D. thesis, University of Wisconsin - Madison, 2008, p. 97. MR 2711768
165. Kiyoshi Igusa, *Notes on the Nielsen–Schreier theorem*, 2019, lecture notes, `people.brandeis.edu/~igusa/Math131b/NS.pdf`.
166. Thomas Jech, *Set theory*, Springer Monographs in Mathematics, Springer-Verlag, Berlin, 2003, The third millennium edition, revised and expanded. MR 1940513 (2004g:03071)
167. Carl Jockusch and Frank Stephan, *A cohesive set which is not high*, Math. Logic Quart. **39** (1993), no. 4, 515–530. MR 1270396 (95d:03078)
168. Carl G. Jockusch, Jr., *Ramsey's theorem and recursion theory*, J. Symbolic Logic **37** (1972), 268–280. MR 0376319 (51 #12495)
169. ———, *Degrees of generic sets*, Recursion theory: its generalisation and applications (Proc. Logic Colloq., Univ. Leeds, Leeds, 1979), London Math. Soc. Lecture Note Ser., vol. 45, Cambridge Univ. Press, Cambridge, 1980, pp. 110–139. MR 598304 (83i:03070)
170. ———, *Degrees of functions with no fixed points*, Logic, methodology and philosophy of science, VIII (Moscow, 1987), Stud. Logic Found. Math., vol. 126, North-Holland, Amsterdam, 1989, pp. 191–201. MR 1034562 (91c:03036)
171. Carl G. Jockusch, Jr., Bart Kastermans, Steffen Lempp, Manuel Lerman, and Reed Solomon, *Stability and posets*, J. Symbolic Logic **74** (2009), no. 2, 693–711. MR 2518820 (2010g:03064)
172. Carl G. Jockusch, Jr. and Robert I. Soare, *Degrees of members of* $\Pi^0_1$ *classes*, Pacific J. Math. **40** (1972), 605–616. MR 0309722
173. ———, $\Pi^0_1$ *classes and degrees of theories*, Trans. Amer. Math. Soc. **173** (1972), 33–56. MR 0316227 (47 #4775)
174. Asher M. Kach, Manuel Lerman, and Reed Solomon, *Cappable CEA sets and Ramsey's theorem*, Proceedings of the 11th Asian Logic Conference, World Sci. Publ., Hackensack, NJ, 2012, pp. 114–127. MR 2868509
175. Xiaojun Kang, *Combinatorial principles between* $\mathrm{RRT}^2_2$ *and* $\mathrm{RT}^2_2$, Front. Math. China **9** (2014), no. 6, 1309–1323. MR 3261000
176. Richard Kaye, *Models of Peano arithmetic*, Oxford Logic Guides, vol. 15, The Clarendon Press Oxford University Press, New York, 1991. MR 1098499 (92k:03034)
177. Alexander S. Kechris, *Classical descriptive set theory*, Graduate Texts in Mathematics, vol. 156, Springer-Verlag, New York, 1995. MR 1321597
178. Mushfeq Khan and Joseph S. Miller, *Forcing with bushy trees*, Bull. Symb. Log. **23** (2017), no. 2, 160–180. MR 3664721
179. Makoto Kikuchi and Kazuyuki Tanaka, *On formalization of model-theoretic proofs of Gödel's theorems*, Notre Dame J. Formal Logic **35** (1994), no. 3, 403–412. MR 1326122
180. S. C. Kleene, *On notation for ordinal numbers*, J. Symbolic Logic **3** (1938), no. 4, 150–155.
181. ———, *Recursive predicates and quantifiers*, Trans. Amer. Math. Soc. **53** (1943), 41–73. MR 7371
182. Julia F. Knight, *Algebraic independence*, J. Symbolic Logic **46** (1981), no. 2, 377–384. MR 613290

183. U. Kohlenbach, *Applied proof theory: proof interpretations and their use in mathematics*, Springer Monographs in Mathematics, Springer-Verlag, Berlin, 2008. MR 2445721 (2009k:03003)
184. Ulrich Kohlenbach, *Foundational and mathematical uses of higher types*, Reflections on the foundations of mathematics (Stanford, CA, 1998), Lect. Notes Log., vol. 15, Assoc. Symbol. Logic, Urbana, IL, 2002, pp. 92–116. MR 1943304
185. ———, *Higher order reverse mathematics*, Reverse mathematics 2001, Lect. Notes Log., vol. 21, Assoc. Symbol. Logic, La Jolla, CA, 2005, pp. 281–295. MR 2185441
186. G. Kreisel, *The axiom of choice and the class of hyperarithmetic functions*, Nederl. Akad. Wetensch. Proc. Ser. A 65 = Indag. Math. **24** (1962), 307–319. MR 0140418
187. Alexander P. Kreuzer, *Primitive recursion and the chain antichain principle*, Notre Dame J. Form. Log. **53** (2012), no. 2, 245–265. MR 2925280
188. ———, *Minimal idempotent ultrafilters and the Auslander–Ellis theorem*, 2015, unpublished, arXiv:1305.6530v2.
189. Masahiro Kumabe, *A fixed point free minimal degree*, unpublished, 1996.
190. Masahiro Kumabe and Andrew E. M. Lewis, *A fixed-point-free minimal degree*, J. Lond. Math. Soc. (2) **80** (2009), no. 3, 785–797. MR 2559129
191. Kenneth Kunen, *Set theory*, Studies in Logic and the Foundations of Mathematics, vol. 102, North-Holland Publishing Co., Amsterdam, 1980. MR 597342 (82f:03001)
192. Antonín Kučera, *Measure, $\Pi^0_1$-classes and complete extensions of* PA, Recursion theory week (Oberwolfach, 1984), Lecture Notes in Math., vol. 1141, Springer, Berlin, 1985, pp. 245–259. MR 820784
193. Roland E. Larson and Susan J. Andima, *The lattice of topologies: a survey*, Rocky Mountain J. Math. **5** (1975), 177–198. MR 388306
194. Manuel Lerman, *Degrees of unsolvability*, Perspectives in Mathematical Logic, Springer-Verlag, Berlin, 1983, Local and global theory. MR 708718 (85h:03044)
195. Manuel Lerman, Reed Solomon, and Henry Towsner, *Separating principles below Ramsey's theorem for pairs*, J. Math. Log. **13** (2013), no. 2, 1350007, 44. MR 3125903
196. H. Lessan, *Models of arithmetic*, New studies in weak arithmetics, CSLI Lecture Notes, vol. 211, CSLI Publ., Stanford, CA, 2013, pp. 389–448. MR 3220622
197. Friedrich Wilhelm Levi, *Arithmetische gesetze im gebiete diskreter gruppen,*, Rend. Circ. Mat. Palermo **35** (1913), 225–236.
198. Lu Liu, Benoit Monin, and Ludovic Patey, *A computable analysis of variable words theorems*, Proc. Amer. Math. Soc. **147** (2019), no. 2, 823–834. MR 3894920
199. Lu Liu and Ludovic Patey, *The reverse mathematics of the thin set and Erdős–Moser theorems*, J. Symb. Log. (2022), no. 4, 2743–2773. MR 4069232
200. A. I. Mal'cev, *Constructive algebras. I*, Uspehi Mat. Nauk **16** (1961), no. 3 (99), 3–60. MR 0151377
201. Richard Mansfield and Galen Weitkamp, *Recursive aspects of descriptive set theory*, Oxford Logic Guides, vol. 11, The Clarendon Press, Oxford University Press, New York, 1985, With a chapter by Stephen Simpson. MR 786122
202. David Marker, *Model theory*, Graduate Texts in Mathematics, vol. 217, Springer-Verlag, New York, 2002. MR 1924282
203. Donald A. Martin, *Borel determinacy*, Ann. of Math. (2) **102** (1975), no. 2, 363–371. MR 403976
204. ———, *Determinacy of infinitely long games*, unpublished manuscript, 2018.
205. Per Martin-Löf, *The definition of random sequences*, Information and Control **9** (1966), 602–619. MR 223179
206. A. R. D. Mathias, *Happy families*, Ann. Math. Logic **12** (1977), no. 1, 59–111. MR 0491197 (58 #10462)
207. Kenneth McAloon, *Paris-Harrington incompleteness and progressions of theories*, Recursion theory (Ithaca, N.Y., 1982), Proc. Sympos. Pure Math., vol. 42, Amer. Math. Soc., Providence, RI, 1985, pp. 447–460. MR 791070

208. MedYahya Ould MedSalem and Kazuyuki Tanaka, *Weak determinacy and iterations of inductive definitions*, Computational prospects of infinity. Part II. Presented talks, Lect. Notes Ser. Inst. Math. Sci. Natl. Univ. Singap., vol. 15, World Sci. Publ., Hackensack, NJ, 2008, pp. 333–353. MR 2449473
209. G. Metakides and A. Nerode, *Effective content of field theory*, Ann. Math. Logic **17** (1979), no. 3, 289–320. MR 556895
210. Joseph Mileti, *Modern mathematical logic*, Cambridge Mathematical Textbooks, Cambridge University Press, 2022.
211. Joseph R. Mileti, *Partition theorems and computability theory*, Ph.D. thesis, University of Illinois at Urbana-Champaign, 2004.
212. Terrence S. Millar, *Foundations of recursive model theory*, Ann. Math. Logic **13** (1978), no. 1, 45–72. MR 482430
213. Webb Miller and Donald A. Martin, *The degrees of hyperimmune sets*, Z. Math. Logik Grundlagen Math. **14** (1968), 159–166. MR 0228341
214. Keith R. Milliken, *A Ramsey theorem for trees*, J. Combin. Theory Ser. A **26** (1979), no. 3, 215–237. MR 535155
215. ———, *A partition theorem for the infinite subtrees of a tree*, Trans. Amer. Math. Soc. **263** (1981), no. 1, 137–148. MR 590416
216. Benoit Monin, *Higher randomness and forcing with closed sets*, Theory Comput. Syst. **60** (2017), no. 3, 421–437. MR 3627418
217. Benoit Monin and Ludovic Patey, *$\Pi^0_1$-encodability and omniscient reductions*, Notre Dame J. Form. Log. **60** (2019), no. 1, 1–12. MR 3911103
218. ———, $\mathsf{SRT}^2_2$ *does not imply* $\mathsf{RT}^2_2$ *in $\omega$-models*, Adv. Math. **389** (2021), Paper No. 107903, 32. MR 4288219
219. ———, *Calculabilité*, Calvage & Mounet, 2022.
220. Antonio Montalbán, *Open questions in reverse mathematics*, Bull. Symbolic Logic **17** (2011), no. 3, 431–454. MR 2856080 (2012h:03044)
221. Antonio Montalbán, *Computable structure theory: Within the arithmetic*, Perspectives in Mathematical Logic, Cambridge University Press, 2021. MR 4274028
222. Antonio Montalbán and Richard A. Shore, *The limits of determinacy in second-order arithmetic*, Proc. Lond. Math. Soc. (3) **104** (2012), no. 2, 223–252. MR 2880240
223. Daniel Mourad, *Computability theory: Constructive applications of the lefthanded local lemma, characterizations of some classes of cohesive powers, and intuitionistic reduction games*, Ph.D. thesis, University of Connecticut, 2023.
224. Karim Joseph Mourad, *Fragments of arithmetic and the foundations of the priority method*, Ph.D. thesis, University of Chicago, 1988. MR 2611881
225. Carl Mummert, *Reverse mathematics of MF spaces*, J. Math. Log. **6** (2006), no. 2, 203–232. MR 2317427
226. Carl Mummert and Stephen G. Simpson, *Reverse mathematics and $\Pi^1_2$ comprehension*, Bull. Symbolic Logic **11** (2005), no. 4, 526–533. MR 2198712
227. Carl Mummert and Frank Stephan, *Topological aspects of poset spaces*, Michigan Math. J. **59** (2010), no. 1, 3–24. MR 2654139
228. Roman Murawski, *Pointwise definable substructures of models of Peano arithmetic*, Notre Dame J. Formal Logic **29** (1988), no. 3, 295–308. MR 953701
229. Jan Mycielski and S. Świerczkowski, *On the Lebesgue measurability and the axiom of determinateness*, Fund. Math. **54** (1964), 67–71. MR 161788
230. Michael Mytilinaios, *Finite injury and $\Sigma_1$-induction*, J. Symbolic Logic **54** (1989), no. 1, 38–49. MR 0987320
231. Jaroslav Nešetřil and Vojtěch Rödl, *Ramsey classes of set systems*, J. Combin. Theory Ser. A **34** (1983), no. 2, 183–201. MR 692827
232. David Nichols, *Effective techniques in reverse mathematics*, Ph.D. thesis, University of Connecticut, 2019.
233. André Nies, *Computability and randomness*, Oxford Logic Guides, vol. 51, Oxford University Press, Oxford, 2009. MR 2548883 (2011i:03003)

234. Dag Normann and Sam Sanders, *On the mathematical and foundational significance of the uncountable*, J. Math. Log. **19** (2019), no. 1, 1950001, 40. MR 3960896
235. ———, *Open sets in computability theory and reverse mathematics*, J. Logic Comput. **30** (2020), no. 8, 1639–1679. MR 4182786
236. Steven Orey, *On ω-consistency and related properties*, J. Symbolic Logic **21** (1956), 246–252. MR 82936
237. J. B. Paris and L. A. S. Kirby, *$\Sigma_n$-collection schemas in arithmetic*, Logic Colloquium '77 (Proc. Conf., Wrocław, 1977), Stud. Logic Foundations Math., vol. 96, North-Holland, Amsterdam-New York, 1978, pp. 199–209. MR 519815
238. Jeff Paris and Leo Harrington, *A mathematical incompleteness in Peano arithmetic*, Handbook of mathematical logic, Stud. Logic Found. Math., vol. 90, North-Holland, Amsterdam, 1977, pp. 1133–1142. MR 3727432
239. Charles Parsons, *On a number theoretic choice schema and its relation to induction*, Intuitionism and Proof Theory (Proc. Conf., Buffalo, N.Y., 1968), North-Holland, Amsterdam, 1970, pp. 459–473. MR 0280330
240. Ludovic Patey, *Somewhere over the rainbow Ramsey theorem for pairs*, preprint, arXiv: 1501.07424, 2015.
241. ———, *Partial orders and immunity in reverse mathematics*, Pursuit of the universal, Lecture Notes in Comput. Sci., vol. 9709, Springer, [Cham], 2016, pp. 353–363. MR 3535177
242. ———, *The strength of the tree theorem for pairs in reverse mathematics*, J. Symb. Log. **81** (2016), no. 4, 1481–1499. MR 3579119
243. ———, *The reverse mathematics of Ramsey-type theorems*, Theses, Université Paris Diderot (Paris 7) Sorbonne Paris Cité, February 2016, PhD thesis.
244. ———, *Dominating the Erdős-Moser theorem in reverse mathematics*, Ann. Pure Appl. Logic **168** (2017), no. 6, 1172–1209. MR 3628270
245. ———, *Iterative forcing and hyperimmunity in reverse mathematics*, Computability **6** (2017), no. 3, 209–221. MR 3689068
246. Ludovic Patey and Keita Yokoyama, *The proof-theoretic strength of Ramsey's theorem for pairs and two colors*, Adv. Math. **330** (2018), 1034–1070. MR 3787563
247. Arno Pauly, *On the (semi)lattices induced by continuous reducibilities*, MLQ Math. Log. Q. **56** (2010), no. 5, 488–502. MR 2742884
248. David Pincus, *On the independence of the Kinna-Wagner principle*, Z. Math. Logik Grundlagen Math. **20** (1974), 503–516. MR 369066
249. W. Pohlers, *Proof theory and ordinal analysis*, Arch. Math. Logic **30** (1991), no. 5-6, 311–376. MR 1087371 (92k:03025)
250. Wolfram Pohlers, *Subsystems of set theory and second order number theory*, Handbook of proof theory, Stud. Logic Found. Math., vol. 137, North-Holland, Amsterdam, 1998, pp. 209–335. MR 1640328
251. Michael O. Rabin, *Computable algebra, general theory and theory of computable fields*, Trans. Amer. Math. Soc. **95** (1960), 341–360. MR 113807
252. Richard Rado, *Studien zur Kombinatorik*, Math. Z. **36** (1933), no. 1, 424–470. MR 1545354
253. F. P. Ramsey, *On a Problem of Formal Logic*, Proc. London Math. Soc. (2) **30** (1929), no. 4, 264–286. MR 1576401
254. Helena Rasiowa and Roman Sikorski, *The mathematics of metamathematics*, Monografie Matematyczne, Tom 41, Państwowe Wydawnictwo Naukowe, Warsaw, 1963. MR 0163850
255. Michael Rathjen, *The realm of ordinal analysis*, Sets and proofs (Leeds, 1997), London Math. Soc. Lecture Note Ser., vol. 258, Cambridge Univ. Press, Cambridge, 1999, pp. 219–279. MR 1720577
256. Sarah Reitzes, *Computability theory and reverse mathematics: Making use of the overlaps*, Ph.D. thesis, University of Chicago, 2022.
257. Brian Rice, *The Thin Set Theorem for Pairs and Substructures of the Muchnik Lattice*, Ph.D. thesis, University of Wisconsin - Madison, 2014. MR 3251073
258. H. G. Rice, *Classes of recursively enumerable sets and their decision problems*, Trans. Amer. Math. Soc. **74** (1953), 358–366. MR 53041

259. Hartley Rogers, Jr., *Theory of recursive functions and effective computability*, McGraw-Hill Book Co., New York-Toronto, Ont.-London, 1967. MR 0224462
260. Joseph G. Rosenstein, *Recursive linear orderings*, Orders: description and roles (L'Arbresle, 1982), North-Holland Math. Stud., vol. 99, North-Holland, Amsterdam, 1984, pp. 465–475. MR 779865
261. Herman Rubin and Jean E. Rubin, *Equivalents of the axiom of choice*, Studies in Logic and the Foundations of Mathematics, North-Holland Publishing Co., Amsterdam-London, 1970. MR 0434812
262. ______, *Equivalents of the axiom of choice. II*, Studies in Logic and the Foundations of Mathematics, vol. 116, North-Holland Publishing Co., Amsterdam, 1985. MR 798475 (87c:04004)
263. Andrei Rumyantsev and Alexander Shen, *Probabilistic constructions of computable objects and a computable version of Lovász local lemma*, Fund. Inform. **132** (2014), no. 1, 1–14. MR 3214660
264. Gerald E. Sacks, *Higher recursion theory*, Perspectives in Mathematical Logic, Springer-Verlag, Berlin, 1990. MR 1080970 (92a:03062)
265. ______, *Saturated model theory*, second ed., World Scientific Publishing Co. Pte. Ltd., Hackensack, NJ, 2010. MR 2568247
266. Nobuyuki Sakamoto and Kazuyuki Tanaka, *The strong soundness theorem for real closed fields and Hilbert's Nullstellensatz in second order arithmetic*, Arch. Math. Logic **43** (2004), no. 3, 337–349. MR 2052887
267. Sam Sanders, *Reverse-engineering reverse mathematics*, Ann. Pure Appl. Logic **164** (2013), no. 5, 528–541. MR 3022748
268. ______, *Reverse mathematics and computability theory of domain theory*, Logic, language, information, and computation, Lecture Notes in Comput. Sci., vol. 11541, Springer, Berlin, 2019, pp. 550–568. MR 3976098
269. ______, *Reverse mathematics of topology: dimension, paracompactness, and splittings*, Notre Dame J. Form. Log. **61** (2020), no. 4, 537–559. MR 4200349
270. ______, *Nets and reverse mathematics: a pilot study*, Computability **10** (2021), no. 1, 31–62. MR 4212352
271. ______, *Countable sets versus sets that are countable in reverse mathematics*, Computability **11** (2022), no. 1, 9–39. MR 4371416
272. ______, *Representations and the foundations of mathematics*, Notre Dame J. Form. Log. **63** (2022), no. 1, 1–28. MR 4405527
273. Matthias Schröder, *Effective metrization of regular spaces*, Computability and Complexity in Analysis (K. Ko, A. Nerode, K. Pour-El, and K. Weihrauch, eds.), Informatik-Berichte, vol. 235, FernUniversität Hagen, 1998, pp. 63–80.
274. Dana Scott, *Algebras of sets binumerable in complete extensions of arithmetic*, Proc. Sympos. Pure Math., Vol. V, American Mathematical Society, Providence, R.I., 1962, pp. 117–121. MR 0141595
275. David Seetapun and Theodore A. Slaman, *On the strength of Ramsey's theorem*, Notre Dame J. Formal Logic **36** (1995), no. 4, 570–582. MR 1368468 (96k:03136)
276. A. Seidenberg, *Constructions in algebra*, Trans. Amer. Math. Soc. **197** (1974), 273–313. MR 349648
277. Paul Shafer, *The strength of compactness for countable complete linear orders*, Computability **9** (2020), no. 1, 25–36. MR 4072502
278. Naoki Shioji and Kazuyuki Tanaka, *Fixed point theory in weak second-order arithmetic*, Ann. Pure Appl. Logic **47** (1990), no. 2, 167–188. MR 1055926
279. Richard A. Shore, *Splitting an $\alpha$-recursively enumerable set*, Trans. Amer. Math. Soc. **204** (1975), 65–77. MR 379154
280. ______, *The homogeneity conjecture*, Proc. Nat. Acad. Sci. U.S.A. **76** (1979), no. 9, 4218–4219. MR 543312
281. ______, *On homogeneity and definability in the first-order theory of the Turing degrees*, J. Symbolic Logic **47** (1982), no. 1, 8–16. MR 644748
282. ______, *Reverse mathematics: the playground of logic*, Bull. Symbolic Logic **16** (2010), no. 3, 378–402. MR 2731250

283. ———, *The Turing degrees: an introduction*, Forcing, iterated ultrapowers, and Turing degrees, Lect. Notes Ser. Inst. Math. Sci. Natl. Univ. Singap., vol. 29, World Sci. Publ., Hackensack, NJ, 2016, pp. 39–121. MR 3411034
284. Stephen G. Simpson, *Degrees of unsolvability: a survey of results*, Handbook of mathematical logic, Stud. Logic Found. Math., vol. 90, North-Holland, Amsterdam, 1977, pp. 631–652. MR 3727420
285. ———, *Recursion theoretic aspects of the dual Ramsey theorem*, Recursion theory week (Oberwolfach, 1984), Lecture Notes in Math., vol. 1141, Springer, Berlin, 1985, pp. 357–371. MR 820790
286. ———, *Partial realizations of Hilbert's Program*, J. Symbolic Logic **53** (1988), no. 2, 349–363. MR 947843
287. ———, *53:free sets/reverse math*, 1999, FOM email list.
288. ———, *Subsystems of second order arithmetic*, second ed., Perspectives in Logic, Cambridge University Press, Cambridge, 2009. MR 2517689 (2010e:03073)
289. ———, *The Gödel hierarchy and reverse mathematics*, Kurt Gödel: essays for his centennial, Lect. Notes Log., vol. 33, Assoc. Symbol. Logic, La Jolla, CA, 2010, pp. 109–127. MR 2668193
290. Stephen G. Simpson and Kazuyuki Tanaka, *On the strong soundness of the theory of real closed fields, extended abstract*, Proceedings of the Fourth Asian Logic Conference, Tokyo (1990), 7–10.
291. Theodore A. Slaman, *$\Sigma_n$-bounding and $\Delta_n$-induction*, Proc. Amer. Math. Soc. **132** (2004), no. 8, 2449–2456. MR 2052424
292. Theodore A. Slaman and Keita Yokoyama, *The strength of Ramsey's theorem for pairs and arbitrarily many colors,* J. Symb. Log. **83** (2018), no. 4, 1610–1617. MR 3893291
293. Robert I. Soare, *Recursively enumerable sets and degrees*, Perspectives in Mathematical Logic, Springer-Verlag, Berlin, 1987. MR 882921 (88m:03003)
294. ———, *Turing–Post relativized computability and interactive computing*, Computability : Turing, Gödel, Church, and beyond (B. Jack Copeland, Carl J. Posy, and Oron Shagrir, eds.), The MIT Press, Cambridge, Massachusetts, 2013. MR 3156390
295. ———, *Turing computability*, Theory and Applications of Computability, Springer-Verlag, Berlin, 2016. MR 3496974
296. David Reed Solomon, *Reverse mathematics and ordered groups*, Ph.D. thesis, Cornell University, 1998. MR 2697413
297. Reed Solomon, *Reverse mathematics and fully ordered groups*, Notre Dame J. Formal Logic **39** (1998), no. 2, 157–189. MR 1714964
298. ———, *Ordered groups: a case study in reverse mathematics*, Bull. Symbolic Logic **5** (1999), no. 1, 45–58. MR 1681895
299. Ernst Specker, *Nicht konstruktiv beweisbare Sätze der Analysis*, J. Symbolic Logic **14** (1949), 145–158. MR 31447
300. ———, *Ramsey's theorem does not hold in recursive set theory*, Logic Colloquium '69 (Proc. Summer School and Colloq., Manchester, 1969), North-Holland, Amsterdam, 1971, pp. 439–442. MR 0278941
301. Joel Spencer, *Asymptotic lower bounds for Ramsey functions*, Discrete Math. **20** (1977/78), no. 1, 69–76. MR 491337
302. J.R. Steel, *Determinateness and subsystems of analysis*, Ph.D. thesis, University of California, Berkeley, 1977. MR 2627251
303. Frank Stephan, *Martin-Löf random and PA-complete sets*, Logic Colloquium '02, Lect. Notes Log., vol. 27, Assoc. Symbol. Logic, La Jolla, CA, 2006, pp. 342–348. MR 2258714
304. John Stillwell, *Reverse mathematics*, Princeton University Press, Princeton, NJ, 2018. MR 3729321
305. Kazuyuki Tanaka, *Weak axioms of determinacy and subsystems of analysis. I. $\Delta^0_2$ games*, Z. Math. Logik Grundlag. Math. **36** (1990), no. 6, 481–491. MR 1114101
306. ———, *Weak axioms of determinacy and subsystems of analysis. II. $\Sigma^0_2$ games*, Ann. Pure Appl. Logic **52** (1991), no. 1-2, 181–193, International Symposium on Mathematical Logic and its Applications (Nagoya, 1988). MR 1104060

307. Kazuyuki Tanaka and Takeshi Yamazaki, *Manipulating the reals in* $\mathrm{RCA}_0$, Reverse mathematics 2001, Lect. Notes Log., vol. 21, Assoc. Symbol. Logic, La Jolla, CA, 2005, pp. 379–393. MR 2185447
308. Stanley Tennenbaum, *Non-archimedean models for arithmetic*, Notices of the American Mathematical Society **6** (1959), 270, abstract.
309. Neil Thapen, *A note on* $\Delta_1$ *induction and* $\Sigma_1$ *collection*, Fund. Math. **186** (2005), no. 1, 79–84. MR 2163104
310. Stevo Todorcevic, *Introduction to Ramsey spaces*, Annals of Mathematics Studies, vol. 174, Princeton University Press, Princeton, NJ, 2010. MR 2603812
311. Henry Towsner, *A simple proof and some difficult examples for Hindman's theorem*, Notre Dame J. Form. Log. **53** (2012), no. 1, 53–65. MR 2925268
312. Henry Towsner, Rose Weisshaar, and Linda Brown Westrick, *Borel combinatorics fail in* HYP, preprint, arXiv:2106.13330, 2022.
313. A. S. Troelstra (ed.), *Metamathematical investigation of intuitionistic arithmetic and analysis*, Lecture Notes in Mathematics, Vol. 344, Springer-Verlag, Berlin-New York, 1973. MR 0325352
314. A. M. Turing, *On Computable Numbers, with an Application to the Entscheidungsproblem*, Proceedings of the London Mathematical Society **s2-42** (1937), no. 1, 230–265.
315. Patrick Uftring, *The characterization of Weihrauch reducibility in systems containing* $E - PA^{\omega} + QF - AC^{0,0}$, J. Symb. Log. **86** (2021), no. 1, 224–261. MR 4282707
316. Jean van Heijenoort, *From Frege to Gödel. A source book in mathematical logic, 1879–1931*, Harvard University Press, Cambridge, Mass., 1967. MR 0209111
317. Michael van Lambalgen, *Random sequences*, Ph.D. thesis, University of Amsterdam, 1987.
318. Peter Vojtáš, *Generalized Galois-Tukey-connections between explicit relations on classical objects of real analysis*, Set theory of the reals (Ramat Gan, 1991), Israel Math. Conf. Proc., vol. 6, Bar-Ilan Univ., Ramat Gan, 1993, pp. 619–643. MR 1234291
319. Vítězslav Švejdar, *The limit lemma in fragments of arithmetic*, Comment. Math. Univ. Carolin. **44** (2003), no. 3, 565–568. MR 2025821
320. Hao Wang, *Popular lectures on mathematical logic*, second ed., Van Nostrand Reinhold Co., New York; Science Press Beijing, Beijing, 1981. MR 737563
321. Wei Wang, *Some logically weak Ramseyan theorems*, Adv. Math. **261** (2014), 1–25. MR 3213294
322. ______, *The definability strength of combinatorial principles*, J. Symbolic Logic **81** (2016), no. 4, 1531–1554. MR 3579121
323. Klaus Weihrauch, *Computability*, EATCS Monographs on Theoretical Computer Science, vol. 9, Springer-Verlag, Berlin, 1987. MR 892102
324. ______, *The degrees of discontinuity of some translators between representations of the real numbers*, Technical report TR-92-050, International Computer Science Institute, Berkeley, 1992.
325. ______, *Computable analysis*, Texts in Theoretical Computer Science. An EATCS Series, Springer-Verlag, Berlin, 2000. MR 1795407
326. P. D. Welch, *Weak systems of determinacy and arithmetical quasi-inductive definitions*, J. Symbolic Logic **76** (2011), no. 2, 418–436. MR 2830409
327. Wikipedia, *Lovász local lemma*, `en.wikipedia.org/wiki/Lovász_local_lemma`, 2021, accessed 2021-11-28.
328. Philip Wolfe, *The strict determinateness of certain infinite games*, Pacific J. Math. **5** (1955), 841–847. MR 73909
329. Liang Yu, *Lowness for genericity*, Arch. Math. Logic **45** (2006), no. 2, 233–238. MR 2209745 (2006j:03060)
330. Xiaokang Yu and Stephen G. Simpson, *Measure theory and weak König's lemma*, Arch. Math. Logic **30** (1990), no. 3, 171–180. MR 1080236
331. ______, *Measure theory and weak König's lemma*, Arch. Math. Logic **30** (1990), no. 3, 171–180. MR 1080236
332. Liao Yuke, *A recursive coloring without* $\Delta_3$ *solutions for Hindman's Theorem*, slides from talk at National University of Singapore, 2022.

# Index


$\dot{-}$, 19
$=^*$, 10
$\subseteq^*$, 10
$\cong$, 306
$\equiv$, 12
$\Vdash$, 186
$\Vdash^{\mathcal{M}}$, 195
$\equiv_c$, 81
$\leqslant_c$, 81
$\leqslant_{KB}$, 134
$\leqslant_{oc}$, *see* reducibility, strong omniscient computable
$\equiv_\omega$, 96
$\leqslant_\omega$, 96
$\equiv_{sc}$, 89
$\leqslant_{sc}$, 89
$\leqslant_{soc}$, *see* reducibility, strong omniscient computable
$\equiv_{sW}$, 89
$\leqslant_{sW}$, 89
$\equiv_T$, 25
$\leqslant_T$, 25
$\equiv_W$, 84
$\leqslant_W$, 84
$\hookrightarrow$, 427
$\hookrightarrow$, 427
$\hookrightarrow_s$, 427
$[T]^n$, 305
1-random, *see* Martin-Löf, random

$\mathsf{ACA}_0$, 124
  and $\mathsf{ACA}_0'$, 125
  and $\mathsf{ACA}_0^+$, 126
  and $\mathsf{KL}$, 131
  and $\Sigma^0_1$ comprehension, 124
  and $\mathsf{WKL}_0$, 128
  and existence of ranges, 129
  and existence of vector space bases, 403, 404
  and free subgroups, 417
  and maximal ideals, 409
  and orderability, 415
  and the jump, 125
  minimum $\omega$-model of, 125
  models of, 125
$\mathsf{ACA}_0'$, 125
  and compactness, 125
$\mathsf{ACA}_0^+$, 126
$\mathsf{ADC}$, 288, 289
admitting
  preservation, *see* preservation, of a class
  solutions in a class, *see* solutions, admitting
  strong solutions, *see* strong, solutions
$\mathsf{ADS}$, 272, 286, 289
  and $\mathsf{COH}$, 278
  cohesive, *see* $\mathsf{CADS}$
  stable, *see* $\mathsf{SADS}$
$\forall\exists$ theorem, *see* theorem, $\forall\exists$
almost
  contained, 10
  equal, 10
almost periodic point, 380
$\alpha$-recursion theory, 167
Ambos-Spies, K., 352
American Association of University of Professors, xi
$\mathsf{AMT}$, 341
analytic codes, 441
  and Borel codes, 442
  point of, 441
analytical hierarchy
  of formulas, 113
Anglè s d'Auriac, P.-E., xii, 311, 324, 329
antichain, 272
apartness property, 320




D. D. Dzhafarov, C. Mummert, *Reverse Mathematics*, Theory and Applications of Computability, https://doi.org/10.1007/978-3-031-11367-3

arithmetic
  first order, *see* first order arithmetic
  Peano, *see* Peano arithmetic
  second order, *see* second order arithmetic
arithmetical comprehension axiom, *see* $\mathsf{ACA}_0$
arithmetical hierarchy, 33
  notation, 33
  of formulas, *see* formula, arithmetical
  of relations, 112
arithmetical transfinite induction, *see* induction, arithmetical transfinite
arithmetical transfinite recursion, *see* $\mathsf{ATR}_0$
arithmetization, 3, 364
Artin, E., 411
Artinian ring, *see* ring, Artinian
ascending chain, 287
ascending sequence, 272
ascending/descending sequence principle, *see* ADS
Ash, C. J., 338
AST, 344
Astor, E. P., xii, 287, 288, 355, 441
atom, 340
atomic model, *see* model, atomic
atomic model theorem, *see* AMT
  subenumerable types, *see* AST
atomic theory, *see* theory, atomic
$\mathsf{ATR}_0$, 136, 138
  and $\Delta^1_1$-$\mathsf{CA}_0$, 138, 139
  and HYP, 138
  and $\Pi^1_1$-$\mathsf{CA}_0$, 140
  and $\Sigma^1_1$ axiom of choice, 434
  and $\Sigma^1_1$ separation principle, 139
  and existence of transfinite jumps, 139
  and Ulm's theorem, 420
AUF, 385, 387
Auslander–Ellis theorem, 385, 387
Avigad, J., viii, xii, 199, 302, 376, 387, 388, 398, 452
avoidance
  cone, 66, 225
  PA, 65
  strong cone, 67, 226
  strong PA, 67
axiom
  of choice, 116, 345, 346, 448
  of choice, $\Sigma^1_1$, *see* $\Sigma^1_1$ axiom of choice
  of determinacy, *see* determinacy, axiom of
  power set, 448
axiomatic strength, 1

B, 372
Baire space, 11
Banff International Research Station, xi
bar, 58
basic functions, 17
basis for a CSC space
  strong, 390
  weak, 390
basis for a vector space, *see* vector space, basis
basis theorem
  cone avoidance, 45
  Gandy, 140
  hyperimmune free, 45
  Kleene, 43
  low, 44
Baumgartner, J. E., 317
Belanger, D. R., xii, 249
Benham, H., xii
Berger, J., 58, 373
$\beta$ function, 23
$\beta$-model, *see* model, $\beta$-
$\mathsf{B}\Gamma$, 154
bi-hyperimmune, *see* hyperimmune, bi-
Bienvenu, L., xii
"big five", 4, 124
  phenomenon, 8–10
Birkhoff's recurrence theorem, 381
Black Forest, xi
Blass, A. R., 51, 52, 77, 317, 320, 323, 385
blob, *see* Seetapun, blob
Bolzano–Weierstrass theorem, 62
Bonnet, R., 52
Borel codes, 439
  and analytic codes, 442
  determined, 441
  evaluation map, 439
  membership, 439
  trivial, 440
Borel determinacy, *see* determinacy, Borel
bounded $\Sigma^0_1$ comprehension, *see* induction, bounded $\Sigma^0_n$ comprehension
bounded quantifiers, 21, 108
bounded sequence, 61
bounding scheme, 154
  $\Sigma^0_2$, 168–173
  $\Sigma^0_3$, 239
Brattka, V., xii, 86, 103, 265, 372
Brodhead, K., xii
Brouwer's fixed point theorem, 388
Brouwer, L. E. J., 8, 363
Brown, D. K., 349, 376, 388
$\mathsf{B}\Sigma^0_2$, *see* bounding scheme, $\Sigma^0_2$
$\mathsf{B}\Sigma^0_3$, *see* bounding scheme, $\Sigma^0_3$
BW, 62

c.e.
  in a model, 166

set, 31
$\mathsf{C}_{\mathbb{N}}$, 75
$\mathsf{C}_n$, *see* choice problem, on *n*
CAC, 272, 286, 289
cohesive, 278
stable, *see* SCAC
CADS, 278, 286
Cai, M., xii
Cantor middle-thirds set, 71
Cantor space, 11
measure, 296, 349
Cantor, G., 363
cardinality, 10
Carlucci, L., xii, 331, 333, 334
Carones, M., xii
Casa Matemática Oaxaca, xi
Catalan number, 317
Cauchy sequence
convergence, 366
equality, 364
quickly converging, 364
unmodulated, 364
Cenzer, D., xii
$\mathsf{C}\Gamma$, 303
chain, 272
chain/antichain principle, *see* CAC
challenger and responder, 51
Charles University, xii
choice problem
on $\mathbb{N}$, 75, 103
on *n*, 75, 103
Cholak, P. A., xi, xii, 193, 194, 196, 199, 216, 223, 225, 226, 228, 232, 239, 249, 252, 255, 259, 311–314, 317, 340, 348
Cholak–Jockusch–Slaman decomposition, 223
Chong, C. T., xii, 167, 168, 172, 220, 249, 253, 257, 294, 308
Chong–Slaman–Yang theorem, 253, 257
Choquet, G., 394
Chubb, J., xii, 306
Chubb–Hirst–McNicholl, *see* tree theorem, Chubb–Hirst–McNicholl
Church's thesis, 97
Church, A., 203
Church–Turing thesis, 15
Chvátal, V., 291
class
$\Pi^0_1$, 39
closed set, 368
coding, 6, 59
directness, 6
finite sequences, 22
fullness, 7
Jockusch's, 211
jump into solutions, 66
negative and positive information, 376
of analytic sets, *see* analytic codes
of Borel sets, *see* Borel codes
of continuous functions, 62
of finite sequences, 127
of finite sets, 120
of functions on $\mathbb{N}$, 127
of one problem into another, 81
of partial orderings, 133
of real numbers, 61
overhead, 6, 441
PA into solutions, 65
utility, 6
COH, *see also* cohesive, 286
and cone avoidance, 225
and PA avoidance, 232
characterization of solutions, 221
Cohen forcing, *see* forcing, Cohen
Cohen P., 175
cohesive, 81
$\vec{R}$-, 81
*p*-, 81
*r*-, 81
coloring, 55
*k*-bounded, 295
induced, 217
level, 306
normal, 297
semi-transitive, 273, 274, 277
stable, 217
stable, of trees, 308, 358, 359
transitive, 273, 274, 277
combinator, 453
combinatorial core, 9
compact dynamical system, 381
compactness
of second order arithmetic, 109
comparability of well orderings, *see* WCWO, SCWO
complete separable metric space, 365
closed set, 368
compact, 374
open set, 368
complete type, *see* type, complete
completeness, 109
compositional product, 94
comprehension
$\Delta^0_1$, *see also* $\mathsf{RCA}_0$
arithmetical, *see also* $\mathsf{ACA}_0$
axiom, 116
computability
applied, 2
informal characterization, 15

theory, vii
Computability in Europe, xi
computable
equivalence, *see* equivalence, computable
reducibility, *see* reducibility, computable
function, *see* computable function
mathematics, viii
model theory, *see* computable, structure theory
structure theory, 338, 401, 411
computable function, 25
index, 26
low, 32
monotonicity, 28
partial, 16, 24
primitive recursive, *see* primitive recursive
relative, 24, 36
total, 24
Turing functional, 27
universal, 27
use principle, 28
computable mathematics, 2
computable solutions, *see* solutions, computable
computably enumerable
relation, 31
set, 31
computably true problem, 97
cone, 66
avoidance, *see* avoidance, cone
Conidis, C. J., xii, 302, 303, 346, 421–423, 425
Conrad, K., xii
conservation result, *see* conservativity
conservativity
$\Pi^1_1$, 251
and ∀∃ theorems, 142
and $\omega$-submodels, 141, 142
and $\Sigma^0_1$ induction, 142, 143
and model extensions, 142
for structures, 141
for theories, 141
of $\mathsf{ACA}_0$ over $\mathsf{PA}$, 144
of $\mathsf{RCA}_0$ over $\mathsf{PA}^- + \mathsf{I}\Sigma^0_1$, 144
restricted $\Pi^1_2$, 250
consistent, *see* theory, consistent
constant symbol, 339
constructible universe, see Gödel's constructible universe, 448
constructive mathematics, 2
continuous function
coding, 370
domain, 370
continuous mathematics, 363
continuum hypothesis, 448
convergent
computation, 23
sequence of real numbers, 62
Cooper, S. B., xi, xii
Corduan, J., 174, 307
CSC space, 390
Csima, B. F., xii, 280, 295, 297, 298, 300–302, 336, 338, 341, 342
cut, 158
bi-tame, 172
proper, 158
Czech–U.S. Fulbright Commission, xii

$\mathsf{D}^2_k$, 219
Davis, M. D., xii, 443
Day, A. R., xii, 381–384
Dean, E. T., 302
Dean, W., 4
Dedekind cut, 397
Dedekind, R., 363
deduction, 339
definability
$\Sigma^0_1$, and ranges of functions, 156
degree structure, 25
$\Delta^0_1$ definability, 137
$\Delta^0_1$-$\mathsf{CA}_0$, *see* $\mathsf{RCA}_0$
$\Delta^0_2$ $k$-partition subset principle, *see also* $\mathsf{D}^2_k$, 219
$\Delta^1_1$ definability, 137
$\Delta^1_1$-$\mathsf{CA}_0$, 137
and $\mathsf{ACA}_0$, 137
minimum $\omega$-model of, 137
models of, 137
Denisov, A. S., 275, 280
dense set, *see* forcing, dense set
dependently hyperimmune, 283
descending chain, 287
descending sequence, 52, 133, 272
determinacy, 443, 444
$F_\sigma$, 443
$F_{\sigma\delta}$, 443
$\boldsymbol{\Sigma}^0_4$, 444
$\boldsymbol{\Sigma}^0_5$, 444
$\boldsymbol{\Delta}^0_3$, 444
$\boldsymbol{\Sigma}^0_2$, 444
$n$-$\boldsymbol{\Pi}^0_3$, 444
and $\mathsf{Z}_2$, 443, 444
axiom of, 443
Borel, 443
clopen, 444
open, 443, 444
determined Borel code, *see* Borel codes, determined

diagonally noncomputable
  problem, *see* DNR
diagonally noncomputable function, 42
Diamondstone, D., xii, 43, 347, 359
Disj-HBT$_{[0,1]}$, 57
Disj-WKL, 57
divergent computation, 24
DNC, *see* diagonally noncomputable function, *see also* DNR
DNR, *see also* diagonally noncomputable, function, 87, 252, 281, 286, 301, 315, 352
  and COH, 252
  and $RT^2_2$, 257
  and $WWKL_0$, 351, 352
$DNR_k$, 87
DNR(∅′), 294, 298, 305
Dobrinen, N. L., xii, 310
domain theory, 394
dominating function, *see* function, dominates
Dorais, F. G., xii, 84, 86, 94, 265, 315, 316, 352, 353, 358
double descent tree, *see* tree, double descent
Downey, R. G., viii, xii, 15, 166, 253, 297, 301, 347, 348, 350, 359, 410, 411, 417, 424
Dzhafarov, D. D., 43, 84, 86, 94, 170, 174, 225, 226, 228, 254, 255, 260, 262, 265, 270, 287, 288, 291–293, 304, 308, 311, 315, 316, 333, 336, 338, 345–347, 352, 353, 355, 358, 440, 441

$E(T)$, 437
Eastaugh, B., xii, 6, 9
effective forcing, *see* forcing, effective
effectively calculable, 15, 16
elementary substructure, 448
EM, 303
embedding, *see* ordering, embedding
Enderton, H. B., viii
*Entscheidungsproblem*, 203
Epstein, R. L., 348
equivalence
  $\omega$-model, 96
  computable, 81
  strong computable, 89
  strong Weihrauch, 89
  Weihrauch, 84
Erdős, P., 291, 303, 336
Erdős–Moser theorem, 303
Ershov, Y. L., 401, 411
essential formula, 282
evaluation map (for a Borel code), *see* Borel codes, evaluation map
existence, 115
  set vs. class, 115
Exp, 157
extension
  finite, of Cohen conditions, 175
  finite, of Mathias conditions, *see* forcing, Mathias forcing

f.i.p., *see* finite intersection, property
factor (of a polynomial), 403
"fair coin" measure, 296, 349
family
  maximal subfamily, 346
  nontrivial, 346
  subfamily, 346
FAN, 58
fan theorem, 58
Feferman, S., 3, 175, 421, 452
Feiner, L., 37
field
  algebraic, 402
  formally real, 411
  orderable, *see* ordering, of an algebraic structure
filter, 393
  $n$-generic, 195
  generic, 182
  in forcing, 180
  maximal, 393
  valuation property, 185
finitary function, *see* function, finitary
finitary Ramsey's theorem, *see* Ramsey's theorem, finitary
finitary relation, *see* relation, finitary
finite coloring, *see* coloring
finite extension method, *see* extension, finite, of Cohen conditions
finite injury, 166
  computably bounded type, 166
  unbounded type, 166
finite intersection
  principle, 345
  property, 345, 346
finite set, *see* set, finite
finite unions principle, *see* FUF
finite unions theorem, *see* FUT
finitistic reductionism, 8
FIP, 346
first order arithmetic, 113
first order part
  of a structure, 143
  of a theory, 143
fixed point, 388
fixed point free function, 50
Flood, S., xii, 440

Focused Research Grant, xi
Fokina, E., xii
forcing, 91, 175
  *n-B*-generic, 187
  *n*-generic, 187
  avoiding, 182
  Cohen, 178
    compatible (with Mathias condition), 260
  condition, 175, 178
  dense set, 182
  effective, 189
  extension, 176, 178
  filter, *see* filter
  generic, 176, 182, 183
  ground, 282
  in models, 195
  iterated, 282
  Jockusch–Soare, 43, 178
    with computable subtrees, 181
  language, 186
  Mathias, 179
    *k*-fold, 213
    large, 243
    pseudo-forcing, 245, 246
    small, 243
    with computable reservoirs, 181
    with cone avoiding reservoirs, 181
  meeting, 182
  notion, 177
  object, 178
  object determined by, 180
  pre-condition, 232
  precondition, 280
  product, 181
  relation, 175, 186
  sequence constructed via, 178
  stronger/weaker condition, 178
  sufficiently generic, 183
  valuation, 178
  with $\Pi^0_1$ classes, 181
  with bushy trees, 308
formalization, 117
formula, 339
  arithmetical, 111
FPF, *see* fixed point free function
Franklin, J. N. Y., xii
free set, 312
  theorem, 312
FRG, xi
Friedberg, R. L., 231, 232
Friedberg–Muchnik theorem, 166, 167
Friedman's theorem, 443, 449
Friedman, H., xii, 1, 4, 5, 116, 126, 312–314, 364, 376, 403, 407, 409, 411, 420, 421, 423, 432, 433, 437, 438, 443, 444, 450
Frittaion, E., xii
Fröhlich, A., 401, 403
FRT, *see* Ramsey's theorem, finitary, 58, 203
FS, 317
$\mathsf{FS}^{\leqslant n}$, 329
$\mathsf{FS}^{\leqslant n}$, 329
$\mathsf{FS}^n_\omega$, 312
FUF, 56, 63
full
  semantics, 146
  structure, 146
function
  characteristic, 11
  dominates, 10, 44, 45
  finitary, 11
  multivalued, 52
  principal, 10
  symbol, 339
Furstenberg, H., 323, 385
FUT, 322, 324, 334
FVP, *see* filter, valuation property

Gale, D., 443
Gale–Stewart game, *see* game, Gale–Stewart
Galois–Tukey connection, 77
Galvin, D., xii
Galvin, F., 317
game
  binary, 450
  determined, 444
  Gale–Stewart, 443
  Hirschfeldt–Jockusch, 100
  of perfection information, *see* game, Gale–Stewart
  strong Choquet, 394
Γ cardinality scheme, 303
Γ conservative, *see* conservativity
Γ-CA, 116
Γ-large, *see* forcing, Mathias, large
Γ-small, *see* forcing, Mathias, small
Gandy, R. O., 140, 157
General-IPHP, 57, 83
General-$\mathsf{RT}^n_k$, 212
generalized composition, 17
generic (forcing), *see* forcing, generic
generic (topological space), 185
Gherardi, G., 84, 103, 372
Giusto, M., 312–314, 349, 351, 388, 398
Glazer, S., 317
Gödel, K., 8, 9, 23, 119, 142, 339, 340, 448
Gödel hierarchy, 9

Gödel's constructible universe, 448
Gödel's incompleteness theorem, 8, 142, 339
Goh, J. L., xii, 225
Goncharov, S. S., 275, 340, 341, 344
Graham, R. L., 59, 336
Greenberg, N., xii, 347, 359, 421
Groszek, M. J., xii, 167, 174, 307
group, 401
  divisible, 421
  free, *see also* subgroup, free
  nilpotent, 414
  orderable, *see* ordering, of an algebraic structure
  partially orderable, 414
  reduced, 421
  subgroup, *see* subgroup
  torsion free, 411
Gura, K., xii

Hachtman, S., 444
Hájek, P., 119, 157, 162, 167, 199, 204
Hajnál, A., 291
Halpern, J. D., 311
halting problem, 32
Hamkins, J. D., xii, 72
Harizanov, V. S., xii, 338
Harrington's theorem, 196
Harrington, L. A., 196, 258, 341
Harris, K. A., 345
Harrison-Trainor, M., xii, 340
Hatziriakou, K., 411, 412, 424
$\mathsf{HBT}_{[0,1]}$, 57, 63
Heine–Borel theorem, 376
  disjunctive form, 57
  for $[0, 1]$, 57
Henkin semantics, 147
Henkin, L., 341
Herrmann, E., 275, 290
hierarchy
  analytic, *see* analytic hierarchy
  arithmetical, *see* arithmetical hierarchy
  Kirby–Paris, *see* Kirby–Paris hierarchy
  low/low$_n$, 32
higher order reverse mathematics, *see* reverse mathematics, higher order
Hilbert space, 388
Hilbert's program, 8
Hilbert, D., 8
Hindman's theorem, 317
  and finite unions theorem, *see* FUT
  apartness, 320, 322, 330
  for bounded sums, 330
  for exact sums, 329
  full-matching, 324
  gap, 321
  half-matching, 324
  short gap, 321
  simple proof, 323
  very short gap, 321
Hindman, N., 317, 323, 330
Hirschfeldt, D. R., viii, xi, xii, 10, 15, 86, 100, 101, 166, 170–174, 225, 250–253, 255, 257, 259, 272–275, 277–282, 297, 301, 336, 338, 341–346, 348, 350, 351, 355, 417, 424
Hirschfeldt–Jockusch game, *see* game, Hirschfeldt–Jockusch
Hirst's theorem, 169
Hirst, J. L., xi, xii, 7, 75, 84, 86, 94, 169, 265, 291–293, 306, 308, 312–317, 320, 323, 352, 353, 358, 376, 385, 405, 406, 432, 433, 437, 438, 457
HLLP, *see* Halpern–Laüchli theorem
Hölzl, R., xii
homogeneous, 55
  $p$-, 291
  for colorings of trees, 305
  increasing $p$-, 291
  limit, 170, 217
  pre-, 204
homogeneous model, *see* model, homogeneous
how to read this book, ix
Howard, Paul, 345
$\mathsf{HT}$, 318
$\mathsf{HT}^{=n}_k$, 329
$\mathsf{HT}^{\leqslant n}_k$, 330
$\mathsf{HT}^{=n}$, 330
Hughes, N. A., xii, 288–290
Hunter, J., 390
HYP, 137
hyperarithmetical reducibility, *see* reducibility, hyperarithmetical
hyperdegrees, 137
hyperimmune, 266
  bi-, 50
  free, 45, 191
hyperimmune free basis theorem, *see* basis theorem, hyperimmune free
hyperjump, 134

ideal, 402
  Jacobson, 422
  jump, 96
  maximal, 406
  prime, 406
  Turing, *see* ideal, Turing
$\mathsf{I}\Delta^0_2$, *see* induction, $\Delta^0_1$
$\mathsf{I}\Delta^0_n$, *see* induction, $\Delta^0_n$

identity reducibility, *see* reducibility, identity
$\mathsf{I}\Gamma$, 153
Igusa, G., xii, 348, 417
IHT, 385, 387
$\mathsf{IHT}_2$, 385
$\mathsf{IHT}_\omega$, 385
increasing $p$-homogeneous, *see* homogeneous, increasing $p$-
index
 $\Delta^0_1$, 27
 $\Pi^0_1$ class, 40
 computable function, 26
 Turing functional, 27
induction, 153
 and bounding scheme, *see* bounding scheme
 and least number principle, *see* least number principle
 and Peano arithmetic, 114
 arithmetical transfinite, 139
 bounded $\Sigma^0_n$ comprehension, 122, 161
 $\Delta^0_1$, 157
 $\Delta^0_n$, 157
 overspill, *see* overspill
 set, 116
 $\Sigma^0_1$, 116, 117, 172–174
 $\Sigma^0_2$, 239, 405
 strong, 174
 underspill, *see* underspill
infinitary pigeonhole principle, *see* IPHP
 over general sets, *see* General-IPHP
Ingall, C., xii
instance–solution problem, *see* problem, instance–solution
instances
 of a problem, 51
intermediate value theorem, 63, 371–373
intuitionism, 8, 61, 363
IP set, 318
IPHP, 54, 80, 83
 and $\mathsf{RT}^1$, 56, 90
 over general sets, *see* General-IPHP
 parallelization of, 80
$\mathsf{IPHP}_+$, 86
IPT, 291
$\mathsf{IPT}^n$, 291
$\mathsf{IPT}^n_k$, 291
Ishihara, H., 58, 373
$\mathsf{I}\Sigma^0_1$, *see* induction, $\Sigma^0_1$
$\mathsf{I}\Sigma^0_2$, *see* induction, $\Sigma^0_2$
isolated type, *see* type, principal
isometry, 388
isomorphic to $2^{<\omega}$, 306
iterating and dovetailing, 99
iteration of a functional, 136
 and the Turing jump, 136, 137
 formalized, 138
IVT, 63, 398

Jacobson ideal, *see* ideal, Jacobson
Jech, T., 427
Jockusch's coding theorem, 211
Jockusch, C. G., xii, 45, 86, 87, 100, 101, 175, 178, 193, 194, 196, 199, 203, 204, 209, 211, 216, 221, 223–226, 228, 232, 239, 249, 251–253, 257, 259, 276, 312–314, 333, 336, 338, 351
Jockusch–Soare forcing, *see* forcing, Jockusch–Soare
join
 Turing, *see* Turing, join
jump
 coding into solutions, *see* coding, jump into solutions
jump ideal, *see* ideal, jump

$k$-coloring, *see* coloring
König's lemma, 53, 54, 127
Kach, A. M., 253, 303, 304
Kastermans, B., 276
Kaye, R., 114, 162, 258
KB, 134
Kechris, A. S., 393, 394, 443
Khan, M., 308, 355
Khoussainov, B., 340
Kihara, T., 373
Kikuchi, M., 142
Kirby, L. A., 154, 162, 164
Kirby–Paris hierarchy, 162, 164
Kjos-Hanssen, B., xii, 251, 253, 257, 352, 355
KL, 54, 127
Kleene's fixed point theorem, *see* recursion theorem
Kleene's normal form, *see* normal form theorem
Kleene, S. C., 21, 30, 43, 132, 134, 137, 175, 434
Kleene–Brouwer ordering, *see* ordering, Kleene–Brouwer
Knight, J. F., xii, 313, 338, 341, 342
Kohlenbach, U., xii, 111, 452, 455, 456
Kołodziejczyk, L. A., xii, 331, 333, 334
König's lemma, 38
Kossak, R., xii
Kreisel, G., 3, 434
Kreuzer, A. P., 305, 385, 387
Kučera, A., xii, 167, 299, 349, 351
Kumabe, M., 308

Kunen, K., 427
Kurtz, S. A., 411

$L$, 448
$\mathcal{L}_1$, 113
$\mathcal{L}_2$, 107
 equality relation, 108, 110
 parameters, 109
 signature, 107
 structure, 108
 theory, 108
$\mathcal{L}_2(\mathcal{B})$, 109
Lachlan's disjunction, 213, 233
Laflamme, C., 310
Lakins, T. J., 308, 358
$L_\alpha$, 448
Lange, K. M., xii, 345
language, 339
Laüchli, H., 311
Laver, R., 311
$\mathsf{L}\Delta^0_n$, *see* least number principle, $\Delta^0_n$
Leader, I., 330
least number principle, 156
 $\Delta^0_n$, 157
Lempp, S., xii, 172, 220, 224, 251, 253, 257, 276, 352, 410, 417, 424
Lepore, F., 331, 333, 334
Lerman, M., xii, 166, 276, 282, 303, 304
Lessan, H., 164
level coloring, *see* coloring, level
Levi, F. W., 411
$\mathsf{L}\Gamma$, *see* least number principle
Li, W., xii, 168, 308
limit color, 217
limit lemma, 35
linear combination
 see vector space, linear combination, 404
linear extension, 52
linear ordering, *see* ordering, linear
linearly independent, *see* vector space, linear independence
Liu's theorem, 231, 252, 257
Liu, L., 231, 232, 252, 254, 257, 308, 317, 338
located set, 398
Löwe, B., xii
Löwenheim–Skolem theorem, 109, 148
Lovász local lemma, 336, 337
Lovász, L., 336
low
 function, 32
 in a model, 253
low basis theorem, *see* basis theorem, low
low solutions, *see* solutions, low
Lubarsky, R. S., xii

Lusin's separation theorem, *see* separation principle, Lusin's

$\mathcal{M}$-bounded set, *see* set, bounded
$\mathcal{M}$-coded set, *see* coding, of finite sets
$\mathcal{M}$-finite, *see* set, $\mathcal{M}$-finite
Mal'cev, A. I., 401
Mansfield, R., 439
Marcone, A., xii, 84, 398
Marfori, Marianna Antonutti, xii
Marker, D., 340
Marks, A. S., xii
Marshall University, xi
Martin, D. A., 443, 444, 449, 450
Martin-Löf
 random, 296
 test, 296
 universal test, 297
Martin-Löf, P., 296, 297
Mathematisches Forschungsinstitut Oberwolfach, xi
Mathias forcing, *see* forcing, Mathias
Mathias, A. R. D., xii, 179
maxim
 relativization, 37
maximal set, 168
McAloon, K., 207
McNicholl, T. H., 306
MedSalem, M., 444
Metakides, G., 403, 411
metatheorem, 109
metatheory vs. object theory, 130
MF space, 393
 countably based, 393
Mileti, J. R., viii, xii, 84, 86, 94, 174, 224, 265, 280, 295, 297, 298, 300–302, 307, 315, 316, 352, 353, 358, 410
Millar, T. S., 341, 343
Miller, J. S., xii, 298, 299, 308, 343, 355
Miller, R., xii
Milliken's tree theorem, *see* tree theorem, Milliken's
Milliken, K. R., 309
Milliken–Taylor theorem, 385
Milner, E. C., 291
miniaturization, 6, 345
minimal
 dynamical system, 383
 pair, 168, 199
model
 $\beta$-, 449
 $\omega$-, 96, 110
 $\omega$-submodel, 110
 atomic, 340

cut, *see* cut
homogeneous, 345
proper cut, *see* cut, proper
saturated, 345
submodel, 110
topped, 119, 345
modulus
convergence, 397
uniform continuity, 379
Monin, B., viii, xii, 254, 257, 300, 311, 329, 338, 351
Monin–Patey theorem, 254, 257
monomial (over a field), 402
Montalbán, A., xii, 9, 338, 421, 441, 444, 445, 450
morphism, 77
Moser, L., 303
Mourad, D., 338
Mourad, K. J., 167
$\mathsf{MT}_n$, 385, 387
MTT, 310
$\mathsf{MTT}^n$, 310
$\mathsf{MTT}^n_k$, 310
$\mu$ operator, 24
Muchnik, A. A., 232
multifunction, *see* function, multivalued
Mummert, C., 345–347, 393–396, 405, 406, 457
Murawski, R., 174
Mycielski, J., 443
Mytilinaios, M., 167

$n$ intersection principle, 346
$n$-generic, *see* forcing, $n$-generic
$n$-random, *see* Martin-Löf, random
National Science Foundation, xi
Nemoto, T., 373
Nerode, A., 403, 411
Nešetřil, J., 309
nested completeness theorem, 397
NFP, 422
Ng, K. M., xii
Nichols, D., xii, 293
Nicholson, M., xii
Nielsen–Schreier theorem, 415
Nies, A., 297
$n$IP, 346
Noetherian ring, *see* ring, Noetherian
$\mathsf{NonSplit}_\omega$, 72
normal form
for $\Sigma^1_1$ formulas, 132, 434
for partial computable functions, 26
normal form theorem
for partial computable functions, 29
Normann, D., 9, 390, 456
$n$-$\Pi^0_3$, 444
Nugent, R., xii
Nurtazin, A. T., 275, 341, 344

observer effect, 6
$\omega$ jump, 126
$\omega$-extension, *see* model, $\omega$-submodel
$\omega$-model, *see* model, $\omega$-
absoluteness of arithmetical formulas, 195
equivalence, *see* equivalence, $\omega$-model
reducibility, *see* reducibility, $\omega$-model
$\omega$-ordered, *see* ordering, $\omega$-ordered
$\omega$-submodel, *see* model, $\omega$-submodel
omitting
solutions in a class, *see* solutions, omitting
type, *see* type, omitting
omitting partial types principle, 343
omitting types theorem, 341
omniscient computable reducibility, *see* reducibility, omniscient computable
open determinacy, *see* determinacy, open
open set, 368
OPT, 343
oracle, 18, 24, 36
orbit, 380
order type
$\omega$, 272
$\omega + \omega^*$, 172, 275
$\omega^*$, 272
strongly $\omega + \omega^*$, 172
ordering
$\omega$-ordered, 288
embedding, 427
finite, 271
infinite, 271
isolated element, 276
Kleene–Brouwer, 134, 135, 429
large element, 276
linear, 271
of an algebraic structure, 410, 411
partial, 271, 393
positive cone, 412
preorder, 25
quasiorder, 25
small element, 276
stable, linear, *see also* SADS, 276
stable, partial, 276
stable, partial vs. linear, 276
strong comparability, 431
strong embedding, 427
weak comparability, 431
weakly stable, *see also* WSCAC, 276
ordinal number, 427, 431

ordinary mathematics, 1
Orey, S., 341

$p$-homogeneous, *see* homogeneous, $p$-
PA degree, 46
$\mathsf{PA}^2$, 147
padding, 27
pairing function, 12, 115
Palumbo, J., 444
$\mathsf{PA}^-$, 114
parallel product, 53
parallelization, 53
parameterization theorem, *see* $S^m_n$ theorem
parameters, 109
Paris, J. B., 154, 162, 164, 258
Paris–Harrington theorem, 257
Parsons, C., 154, 164
partial computable function, *see* computable function, partial
partial ordering, *see* ordering, partial
partial type, *see* type, partial, *see* type, principal
Patey, L., viii, xii, 67, 225, 249, 254, 255, 257, 260, 265–267, 282, 283, 288, 293, 294, 297, 300, 304, 308, 311, 314, 315, 317, 329, 338, 343, 351, 355
path
 dense set of, 341
 isolated, 341
 listing of, 341
Pauly, A., xii, 103, 225
Peano arithmetic, 114
 avoidance in solutions, *see* avoidance, PA
 axioms, 114
 coding into solutions, *see* coding, PA into solutions
 induction scheme, 114
Peano axioms, second order, 147
$\mathsf{P}^{\text{fe}}$, 221
PH, 257
$\Pi^0_1$ class, 39
$\Pi^0_n$, 111
$\Pi^1_1$-$\mathsf{CA}_0$, 132
 and divisible and reduced groups, 421
 and existence of hyperjumps, 134
 models of, 135
$\Pi^1_n$, 113
pigeonhole principle
 IPHP, 54
 finitary, 161
 vs. Ramsey's theorem, 55
Pincus, D., 311
playground of logic, x
PMTT, 311
$\mathsf{PMTT}^n$, 310
$\mathsf{PMTT}^n_k$, 310
Pohlers, W., 173
Poincaré, H., 8
polarized Ramsey's theorem, *see* Ramsey's theorem, polarized
polynomial (over a field), 402
Porter, C. P., xii
Post, E., 34, 175
pre-homogeneous, *see* homogeneous, pre
predicative mathematics, 8, 9, 363
predicative reductionism, 9
prefix free set of strings, 296
preorder, *see* ordering, preorder
preservation, 99
 of $m$ among $k$ among hyperimmunities, strong, 267
 of $m$ among $k$ hyperimmunities, 266
 of a class, 67
 of definitions, 304
 of dependent hyperimmunity, 283
 of hyperimmunity, 267
prime
 bounding, 342
 model, 342
primitive recursive, 19
 function, 17–23
  definition, 18
 functional, 18
 relative, 18
principal function, *see* function, principal
priority method, 166
probabilistic method, 336
problem, 51
 instance, *see* instances, of a problem
 instance–solution, 51
 multiple forms, 56
 solution, *see* solutions, to an instance of a problem
 vs. $\forall\exists$ theorem, 53, 56
 with finite errors, 221
product forcing, *see* forcing, product
proof, 107
proof theoretic ordinal, 173
proximal points, 380
pseudowellordering, 134, 138
PT, 291
$\mathsf{PT}^n$, 291
$\mathsf{PT}^n_k$, 291
Pudlák, P., xii, 119, 157, 162, 204

Qiang, L., 168
quasiorder, *see* ordering, quasiorder
Quine, W., 31
quotient ring, *see* ring, quotient

r-$\Pi^1_2$, *see* theorem, restricted $\Pi^1_2$
Rabin, M. O., 401, 403
Rado, R., 318
rainbow, 295
rainbow Ramsey's theorem, 295
Rakotoniaina, T., 86, 265
Ramsey number, 58
  bounding, problem of, 58
  finding, problem of, 58
Ramsey's theorem, viii, *see also* RT, $\mathsf{RT}^n$, $\mathsf{RT}^n_k$, $\mathsf{RT}^1$, $\mathsf{RT}^2_k$, 55, 203
  and Erdős–Moser theorem, 303
  arithmetical bounds, 204, 209
  coding the jump, 211
  finitary, 58
  infinitary, 55
  polarized, 291
  rainbow, *see* rainbow Ramsey's theorem
  stable, *see also* $\mathsf{SRT}^2_k$, 219
  vs. infinitary pigeonhole principle, 55
Ramsey, F. P., 203
random, *see* Martin-Löf, random
range of a function, 129, 130
rank, *see* tree, rank
Rasiowa, H., 183
Rasiowa–Sikorski theorem, 183
Rathjen, M., xii
$\mathsf{RCA}_0$, 118
  and $\mathsf{WKL}_0$, 128
  and $\mathsf{WWKL}_0$, 350
  and bounded $\Sigma^0_1$ comprehension, *see* induction, bounded $\Sigma^0_n$ comprehension
  and finite sets, 119–121
  and formalized computability, 122
  models of, 118, 119
$\mathsf{RCA}^*_0$, 173
REC, 97
recurrent point, 380
recursion theorem, 30
recursive
  synonym for computable, 118
recursive comprehension, *see* $\mathsf{RCA}_0$
reducibility, 77, 78
  $\omega$-model, 96
  comparison of various, 144
  computable, viii, 81
  computable, to *m* applications, 92
  hyperarithmetical, 137
  identity, 78
  notion, 144
  omniscient computable, 254
  strong computable, 89
  strong omniscient computable, 255
  strong Weihrauch, 89
  subproblem, 78
  $\not\leq_{\mathrm{sW}}$ to $\not\leq_{\mathrm{W}}$ trick, 90
  Turing, 25
  uniform identity, 78
  Weihrauch, viii, 84
  Weihrauch, to *m* applications, 93
reducibility notion, 25
reduction, *see* reducibility
Reimann, J., xii
Reitzes, S., xii, 170, 174, 338
relation
  finitary, 11
  on $\omega^\omega$ and $2^\omega$, 36
relativization, 18, 36
representation, 6, *see also* coding, 59
  canonical, in a vector space, *see* vector space, canonical representation
  for a finite set, *see* coding, of finite sets
  nonuniqueness of, 60
  of groups and rings, 401
  of polynomials, 402
represented space, *see* representations, represented space
reservoirs, *see* forcing, Mathias
reversal, 1, 130
reverse mathematics, vii, viii
  higher order, 5, 6
  history, 3–5, 444
  strict, 5, 6
  zoo, 9, 257, 354
reverse recursion theory, 123, 166
Rice, B., 315
Rice, H. G., 49
ring, 401
  Artinian, 421
  ideal, *see* ideal
  local, 422
  Noetherian, 421
  orderable, *see* ordering, of an algebraic structure
  quotient, 402
Ripley, D., xii
Rödl, V., 309
Rogers, H., 112
Rosenstein, J. G., 275
Rossberg, M., xii
Rossegger, D., xii
Rothschild, B. L., 59, 336
$\mathsf{RRT}^n_k$, 295
RT, 55
RT, 203
$\mathsf{RT}^1$, 55
  and $\mathsf{B}\Sigma^0_2$, 169
  and IPHP, 56, 90

and strong cone avoidance, 226
and strong PA avoidance, 232
$\mathsf{RT}_2^2$
and cone avoidance, 212
and PA avoidance, 231
vs. $\mathsf{SRT}_2^2$, 251, 252
$\mathsf{RT}_k^2$, 212
$\mathsf{RT}^n$, 55
$\mathsf{RT}_2^n$
and $\mathsf{RCA}_0$, 210
$\mathsf{RT}_k^n$, 55
Rubin, H., 345
Rubin, J. E., 345
Rumyantsev, A., 336, 338
Rumyantsev–Shen theorem, 336, 338
Rute, J., 302

$S_n(T)$, 309
$S_\omega(T)$, 309
Sacks splitting theorem, 166
Sacks, G. E., 137, 140, 340, 351, 434
SADC, 288, 289
SADS, 276, 286, 289
Sakamoto, N., 420
Sanders, S., xii, 6, 9, 390, 456
saturated model, *see* model, saturated
Sauer, C., 310
SCAC, 276, 286, 289
Schauder's fixed point theorem, 388
Schloss Dagstuhl–Leibniz-Zentrum für Informatik, xi
Schmerl, J. S., xii
Schröder number, 316
Schreier, O., 411
Schuster, P., 58
Schweber, N. D., xii
Scott set, 96
Scott, D., 46
SCWO, 431, 433
WCWO, 433, 435–438
second order arithmetic, *see* $\mathsf{Z}_2$, *see also* $\mathcal{L}_2$
Seetapun
blob, 215
Seetapun's theorem, 212, 225, 232, 259
sequence, 215
tree, 215
Seetapun, D., 212, 216, 232, 259, 302, 303
Seidenberg, A., 401
semiring axioms, 114
separably closed set, 376
relation to closed sets, 376, 378
separating set, 139
separation principle, 139
$\Sigma_1^0$, 151
$\Sigma_1^1$, 139
Lusin's, 442
$\mathsf{SeqCompact}_{2\omega}$, 72
sequence of sets, 12
set
bounded, 120
coded, *see* coding, of finite sets
finite, 121
$\mathcal{M}$-finite, 121
set induction, *see* induction, set
setoid, 364
Shafer, P., xii, 84, 86, 94, 265, 315, 316, 352, 353, 358, 392
Shen, A., 336, 338
Shepherdson, J. C., 401, 403
Shioji, N., 389
Shoenfield, J. R., 35
Shore blocking, 167, 342
Shore, R. A., x–xii, 37, 167, 171–173, 177, 250, 257, 272–275, 277–282, 340–346, 355, 444, 445, 450
$\Sigma_1^0$ completeness, 123
$\Sigma_n^0$, 111
universal $\Sigma_n^0$ formula, 150
$\Sigma_1^1$ absoluteness, 449
$\Sigma_1^1$ axiom of choice, 378, 432
and $\mathsf{ATR}_0$, 434
and weak comparability of well orderings, 432
models of, 434
$\Sigma_1^1$ separation principle, *see* separation principle, $\Sigma_1^1$
$\Sigma_n^1$, 113
Sikorski, R., 183
Simic, K., 376, 387, 388, 398
Simons Foundation, xi
Simpson, S. G., vii, ix, xi, xii, 1, 8, 9, 26, 48, 115, 119, 128, 139, 311, 312, 317, 320, 323, 349–351, 379, 380, 385, 388, 389, 395, 397, 398, 403, 407, 409, 411, 412, 420, 421, 423, 424, 427, 440–442, 444, 449, 457
Slaman, T. A., xii, 9, 155, 157, 167, 168, 193, 194, 196, 199, 216, 223, 225, 226, 228, 232, 239, 249, 251–253, 257, 259, 294, 302, 303, 341–346, 352
Smith, R. L., 403, 407, 409, 411, 421, 423
$S_n^m$ theorem, 29, 30
Soare, R. I., viii, xii, 12, 15, 26, 43, 45, 46, 166, 178, 341, 342
Solomon, D. R., xii, 253–255, 257, 260, 265, 276, 282, 287, 288, 303, 304, 333, 336, 338, 414, 415, 417, 424, 440, 441

Solovay, R., 46
solutions
 $\Delta^0_2$, 64
 admitting, 64
 arithmetical, 64
 computable, 63
 low, 64
 $\text{low}_2$, 64
 omitting, 64
 PA, 65
 to an instance of a problem, 51
 uniformly computable, 68
Soskova, M. I., xii
soundness, 109
Souslin's theorem, 442
Sovine, S., xii
Specker sequence, 367, 374
Specker, E., 203, 367
Spencer, J. H., 59, 336
sports, 303
$\mathsf{SRT}^2_2$
 and low solutions, 253
$\mathsf{SRT}^2_2$ vs. COH problem, 251
$\mathsf{SRT}^2_k$, 219, 286
 and $\mathsf{B}\Sigma^0_2$, 220
 and cone avoidance, 225
stable coloring, *see* coloring, stable
stable Ramsey's theorem, *see* Ramsey's theorem, stable
Stahl, R., xii
standard semantics, 147
Steel, J. R., 4, 444
Stephan, F., xii, 221, 350, 393, 394
Stewart, F. M., 443
Stillwell, J., viii
Stone–Čech compactification, 323
Stone–Weierstrass theorem, 388
strategy
 computable, 101
 in Gale–Stewart games, 444
 in Hirschfeldt–Jockusch games, 101
 winning, 101, 444
Strauss, D., 323, 330
strict reverse mathematics, *see* reverse mathematics, strict
strong
 computable equivalence, *see* equivalence, strong computably
 computable reducibility, *see* reducibility, strong computable
 Weihrauch equivalence, *see* equivalence, strong Weihrauch
 Weihrauch reducibility, *see* reducibility, strong Weihrauch
 cone avoidance, *see* avoidance, strong cone
 PA avoidance, *see* avoidance, strong PA
 solutions, 67
strong comparability of well orderings, *see* SCWO
strong omniscient computable reducibility, *see* reducibility, strong omniscient computable
strong subtree, *see* tree, strong subtree
structural Ramsey theory, 309
structure
 computable, 340
 decidable, 339
subgroup, 402
 convex, 414
 free, *see also* group, free, 417
 presented, 417
submodel, *see* model, submodel
subproblem, *see* reducibility, subproblem
subsystem, 116
 robust, 9
 stronger, 117
 subscript 0, 117
sufficiently generic, *see* forcing, sufficiently generic
Suggs, J., xii, 287, 288
Švejdar, V., 168

TAC, 422
Tanaka, K., 142, 389, 420, 444
Tarski–Kuratowski algorithm, 112
Tarski–Vaught test, 448
Tedder, A., xii
Tennenbaum, S., 275, 280, 340
term, 108, 339, 453
 of a polynomial, 403
Thapen, N., xii, 158
theorem
 $\forall\exists$, 2, 53–56
 restricted $\Pi^1_2$, 250
theory, 339
 atomic, 340
 consistent, 109
 decidable, 339
 two sorted, 107
thin set, 312
 theorem, 312
Thomas, M., xii
Todorčević, S., 309
topped model, *see* model, topped
tournament principle, *see* Erdős–Moser theorem
Towsner, H., xii, 282, 288, 289, 304, 317, 323, 441

tree, 37
  $k$-branching, 309
  and $\Sigma^1_1$ formulas, 134
  and Kleene–Brouwer ordering, 135
  completely branching, 422
  double descent, 430
  enumeration, 425
  explosion, 435, 437
  finitely branching, 309
  height, 309
  leaf, 309
  level of a node, 309
  meet-closed, 309
  of approximations, 37
  rank, 428
  rooted, 309
  strong subtree, 309
  wedge, 437
  wedge product, 435, 437
  well founded, *see* well founded, tree
tree labeling method, 254, 262
  labeled subtree, 262
  labels of nodes, 262
tree theorem
  Chubb–Hirst–McNicholl, 305, 306
  Milliken's, 309, 330
  Milliken's, product form, 310, 330
trivial Borel code, *see* Borel codes, trivial
Troelstra, A. S., 111, 452
$\mathsf{TS}^n_\omega$, 312
$\mathsf{TT}$, 306
$\mathsf{TT}^n$, 306
$\mathsf{TT}^n_k$, 306
Turetsky, D., xii, 347, 359
Turing
  degree, 25, 46
  functional, 27
  ideal, 96
  join, 12
  jump, 32
  machine, 15, 16
  reducibility, *see* reducibility, Turing
Turing, A., 15, 32, 203
two sorted theory, *see* theory, two sorted
type
  complete, 340
  equivalence of, 344
  omitting, 340
  partial, 340
  principal, 340
  realizing, 340
  subenumeration, 344

Uftring, P., 457
ultrafilter
  almost downward translation invariant, 385
  idempotent, 387
  minimal, 387
unbounded search operator, *see* $\mu$ operator
uniform identity reducibility, *see* reducibility, uniform identity
uniformly $\Sigma^0_1$ subset of $2^\omega$, 296
uniformly computable solutions, *see* solutions, uniformly computable
uniformly recurrent point, 380
universal $\Sigma^0_n$ formula, *see* $\Sigma^0_n$, universal $\Sigma^0_n$ formula
universal instances, *see* instances, universal
University of Chicago, xii
University of Connecticut, xi

Valenti, M., xii
van Lambalgen's theorem, 301
van Lambalgen, M., 301
vector space
  basis, 404
  canonical representation, 404
  linear combination, 404
  over a field, 404
Villano, J. D., xii
Vojtáš, P., xii, 52, 77

Walsh, S., xii, 4
Wang, H., 125, 126
Wang, W., xii, 67, 279, 296, 304, 305, 308, 314, 316
Wansner, L., xii
Wcisło, B., xii
$\mathsf{WCWO}$, 431
weak comparability of well orderings, *see* $\mathsf{WCWO}$
weak König's lemma, *see* $\mathsf{WKL}$, *see also* $\mathsf{WKL}_0$
weak weak König's lemma, *see* $\mathsf{WWKL}$
Weber, R., 303
Weiermann, A., xii
Weihrauch
  equivalence, *see* equivalence, Weihrauch
  reducibility, *see* reducibility, Weihrauch
Weihrauch, K., 52, 84, 365
Weisshaar, R., xii, 441
Weitkamp, G., 439
Welch, P. D., 444
well founded
  ordering, *see also* well ordering, 133
  relations, and $\Sigma^1_1$ formulas, 133
  tree, 133
well ordering, 136
  coinitial segment, 435

comparability, *see* WCWO, SCWO
indecomposable, 435
of the universe, 448
Westrick, L. B., xii, 254, 255, 260, 265, 333, 336, 338, 440, 441
Weyl, H., 3, 8
WF, 133
Wheeler, W., xii
winning class, 444
WKL, 54, 127
disjunctive form, 57
$WKL_0$, 127
and $ACA_0$, 128
and $RCA_0$, 128
and $WWKL_0$, 350
and Artinian and Noetherian rings, 422
and orderability, 411, 412
and prime ideals, 407
models of, 128
nonexistence of minimal $\omega$-models, 128
vs. WKL, 128
WO, 133
Wolfe, P., 443
word
1-step equivalence, 415
free equivalence, 415
over an alphabet, 415
reduced, 415
$Word_X$, 415
WSCAC, 276, 286, 289
WWKL, 349
$WWKL_0$, 349, 387
and DNR, 351, 352
and $RCA_0$, 350
and $WKL_0$, 350

Yamazaki, T., 420
Yang, Y., xii, 168, 172, 220, 249, 253, 257, 294, 308
Yokoyama, K., xii, 249
Yu, L., 199
Yu, X., 349, 350, 388
Yuke, L., 323

$Z_2$, 107, 116
full, 116
language of, *see* $\mathcal{L}_2$
models of, 448
subsystems/fragments, 116, 117
Zdanowski, K., 331, 333, 334
Zermelo–Fraenkel set theory, *see* ZF, ZFC
ZF, 345, 346, 448
ZFC, 116, 443
$ZFC^-$, 448

Printed in the United States
by Baker & Taylor Publisher Services

Printed in the United States
by Baker & Taylor Publisher Services